Terrane Accretion and Orogenic Belts

Geodynamics Series

Geodynamics Series

1. Dynamics of Plate Interiors
   A. W. Bally, P. L. Bender, T. R. McGetchin, and R. I. Walcott (Editors)

2. Paleoreconstruction of the Continents
   M. W. McElhinny and D. A. Valencio (Editors)

3. Zagros, Hindu Kush, Himalaya, Geodynamic Evolution
   H. K. Gupta and F. M. Delany (Editors)

4. Anelasticity in the Earth
   F. D. Stacey, M. S. Patterson, and A. Nicholas (Editors)

5. Evolution of the Earth
   R. J. O'Connell and W. S. Fyfe (Editors)

6. Dynamics of Passive Margins
   R. A. Scrutton (Editor)

7. Alpine-Mediterranean Geodynamics
   H. Berckhemer and K. Hsü (Editors)

8. Continental and Oceanic Rifts
   G. Pálmason, P. Mohr, K. Burke, R. W. Girdler, R. J. Bridwell, and G. E. Sigvaldason (Editors)

9. Geodynamics of the Eastern Pacific Region, Caribbean and Scotia Arcs
   Ramón Cabré, S. J. (Editor)

10. Profiles of Orogenic Belts
    N. Rast and F. M. Delany (Editors)

11. Geodynamics of the Western Pacific-Indonesian Region
    Thomas W. C. Hilde and Seiya Uyeda (Editors)

12. Plate Reconstruction From Paleozoic Paleomagnetism
    R. Van der Voo, C. R. Scotese, and N. Bonhommet (Editors)

13. Reflection Seismology: A Global Perspective
    Muawia Barazangi and Larry Brown (Editors)

14. Reflection Seismology: The Continental Crust
    Muawia Barazangi and Larry Brown (Editors)

15. Mesozoic and Cenozoic Oceans
    Kenneth J. Hsü (Editor)

16. Composition, Structure and Dynamics of the Lithosphere-Asthenosphere System
    K. Fuchs and C. Froidevaux (Editors)

17. Proterozoic Lithospheric Evolution
    A. Kröner (Editor)

18. Circum-Pacific Orogenic Belts and Evolution of the Pacific Ocean Basin
    J. W. H. Monger and J. Francheteau (Editors)

Terrane Accretion and Orogenic Belts

Edited by Evan C. Leitch
Erwin Scheibner

Geodynamics Series
Volume 19

American Geophysical Union
Washington, D.C.
Geological Society of America
Boulder, Colorado
1987

Published under the aegis of AGU Geophysical Monograph Board.

Library of Congress Cataloging-in-Publication Data

Terrane accretion and orogenic belts.

(Geodynamic series; v. 19)
1. Geology, Structural. 2. Orogeny. I. Leitch, Evan C. II. Scheibner, Erwin III. Series.
QE601.T47 1987 551.8 87-1772
ISBN 0-87590-516-1
ISSN 0277-6669

Printed in the United States of America.

CONTENTS

PREFACE

The unifying theme of this collection of papers is the realisation that, in the analysis of orogenic belts, the present spatial juxtaposition of elements does not necessarily reflect their relative position throughout the history of the belt. Orogenic belts are assemblages of tectonostratigraphic terranes. The constituent rocks of individual terranes are clearly linked, but separated from those of adjacent terranes by dislocations, mainly faults, across which original continuity cannot be established. Thus many papers in this volume deal with the recognition of terranes, the nature of terrane boundaries, and the relationship of deformation and metamorphism to the amalgamation of terranes and their docking at cratonic margins. Movements on a scale of thousands of kilometres are involved across some terrane boundaries but these are only presently demonstrable in some orogens from detailed paleomagnetic and biogeographic studies, subjects also addressed in this volume.

Nearly all of the papers were presented at the Third Circum-Pacific Terrane Conference held in Sydney, Australia in August 1985. Extended abstracts of the 72 conference papers were published by the Geological Society of Australia at the time of the meeting (Abstracts Volume number 14) and over one third of these, substantially amplified and modified in the light of ensuing discussions, are published here. Contributions span the broad field of geology, from geophysics to paleontology, and show both the unifying elements of terrane analysis and the necessity in terrane studies of considering a great range of geological information. As suggested by the conference title most of the papers are regionally based. The First Circum-Pacific Terrane Conference, held in Japan, resulted in the volume 'Accretion Tectonics in the Circum-Pacific Regions' in which most of the papers addressed aspects of the terranes of northeast Asia, the Second Conference, in California, led to 'Tectonostratigraphic Terranes of the Circum-Pacific Region' in which a majority of papers dealt with the northeast quadrant of the Circum-Pacific. In the present work the regional bias is towards Australia and the southwest Pacific. The change in geographic emphasis has been paralleled by a change in the antiquity of the terrane assemblages discussed. In the proceedings of the first two meetings most of these assemblages were ones that had emerged in the Mesozoic and Tertiary and in many the complimentary processes of accretion and dispersion are continuing. Many of the regions discussed at the third meeting are in the hinterland of the active Pacific rim. However they once faced a pre-cursor ocean of comparable scale and complexity to the present Pacific, and an understanding of the evolution of the modern Pacific realm is critical to understanding terrane accretion and the nature of the older orogenic belts.

Terranes are the building blocks of orogenic belts. Their modern analogues include microcontinents, magmatic arcs, intraplate volcanic chains, accretionary subduction complexes and oceanic plateaux, which are brought together by plate motions. Rarely are these motions simple unidirectional conveyor belt ones of the type envisaged in the Wilson Cycle, but rather they involve complex movement paths, not haphazard, but dictated by the changing vectors associated with lithosphere motion oblique to plate boundaries and migrating triple junctions. The potential for complexity can be seen in the present day plate tectonics of the Pacific, and the products of past movements are manifest in the encircling fold belts. The Circum-Pacific domain is unique in revealing all stages in evolution of orogenic belts.

The meeting from which this publication arises received financial and logistical support from professional organisations, resource companies and government agencies. The editors thank the authors for their contributions and the reviewers for their constructive comments. We thank AGU for agreeing to publish the volume and Dr. R.A. Price who was instrumental in arranging publication. The help and patience of Mr. Steven Mansberg, production coordinator at AGU, is gratefully acknowledged. This volume would not have appeared without the support of the Earth Resources Foundation of the University of Sydney and the Geological Survey of New South Wales of the Department of Mineral Resources.

E. C. Leitch
Department of Geology and Geophysics
University of Sydney
New South Wales, Australia

E. Scheibner
Geological Survey of New South Wales
Department of Mineral Resources
Sydney, New South Wales, Australia

STRATOTECTONIC TERRANES OF THE EASTERN AUSTRALIAN TASMANIDES

E. C. Leitch

Department of Geology and Geophysics, The University of Sydney, Sydney, 2006, Australia

E. Scheibner

Geological Survey of New South Wales, GPO Box 5288, Sydney, 2001, Australia

Abstract. Rigorous analysis of regional relationships within the Paleozoic Tasmanides of eastern Australia reveals many structural discontinuities across which the older rocks of the orogenic belt cannot be unequivocally linked. These discontinuities, most of which are faults or assumed faults buried beneath younger cover, form the boundaries to some thirty-six tectonostratigraphic terranes. The development of the Tasmanides involved the amalgamation of these terranes into larger units and their accretion at the eastern cratonic edge of Gondwanaland. Major episodes of amalgamation/accretion are assumed to have coincided with widespread deformational episodes, although deformation also accompanied the dispersal of the terranes and the formation of rift and pull-apart basins in which the earliest overlap sequences accumulated. Although many of the terranes in the western and central parts of the Tasmanides show evidence suggestive of their development close to a cratonic mass the common assumption that these were the adjacent parts of the Precambrian craton along the western edge of the orogenic belt has yet to be demonstrated. It is possible that present relationships are the result of large scale strike-slip or thrust movements and the terranes must be treated as suspect. There is no data requiring the presence of large exotic terranes, although there is evidence that some elements of terranes interpreted as accretionary subduction complexes may be far travelled, and terrane amalgamation to form the New England Fold Belt might have occurred distant from the cratonic margin. Despite long lived subduction along the east Gondwanaland margin in the Paleozoic and the Mesozoic there is no evidence in the Tasmanides, nor in the New Zealand Rangitata Orogen, for major continental collision.

Introduction

Terrane analysis in orogenic belts focuses attention on structural discontinuities, mainly faults, and uses them as boundaries to crustal elements which are considered to have evolved independently of adjacent elements until the opposite can be demonstrated [Coney et al., 1980; Jones et al., 1983; Howell et al., 1985]. Thus many of the boundaries on our terrane map of the Tasmanides of eastern Australia are familiar (Fig. 1). They coincide with the margins of masses variously referred to as highs, troughs, blocks, platforms, geanticlines, zones etc. The important feature of interpreting tectonic units as terranes is that we admit the possibility that they are collectively a collage, rather than a pattern that has maintained an essential coherency during the development of the major orogenic belt. The methodology of terrane analysis requires that, where major rock masses now juxtaposed show contrasting geological development over much of their histories, the possibility they evolved in widely separated regions be considered [Jones et al., 1983]. Conversely the possibility that masses now far apart, but which show marked similarities in their geological histories and faunal elements, evolved contiguously and were only displaced towards the end of their development must be admitted.

In defining terranes an attempt is made to identify large crustal blocks that show a common geological history. By definition their boundaries are either tectonic, or obscured by later rocks which provide linkages constraining the age of terrane amalgamation.

Disruption of amalgamated terranes is common. This can occur along old sutures or, in part or entirely, transverse to these structures. Movements on a scale of kilometres to perhaps a few hundred kilometres may not destroy the overall pattern of terranes within an orogenic belt, but the accretion of an exotic terrane formed in part of cratonic rocks within one belt, necessarily requires the prior dispersion of elements from an earlier belt.

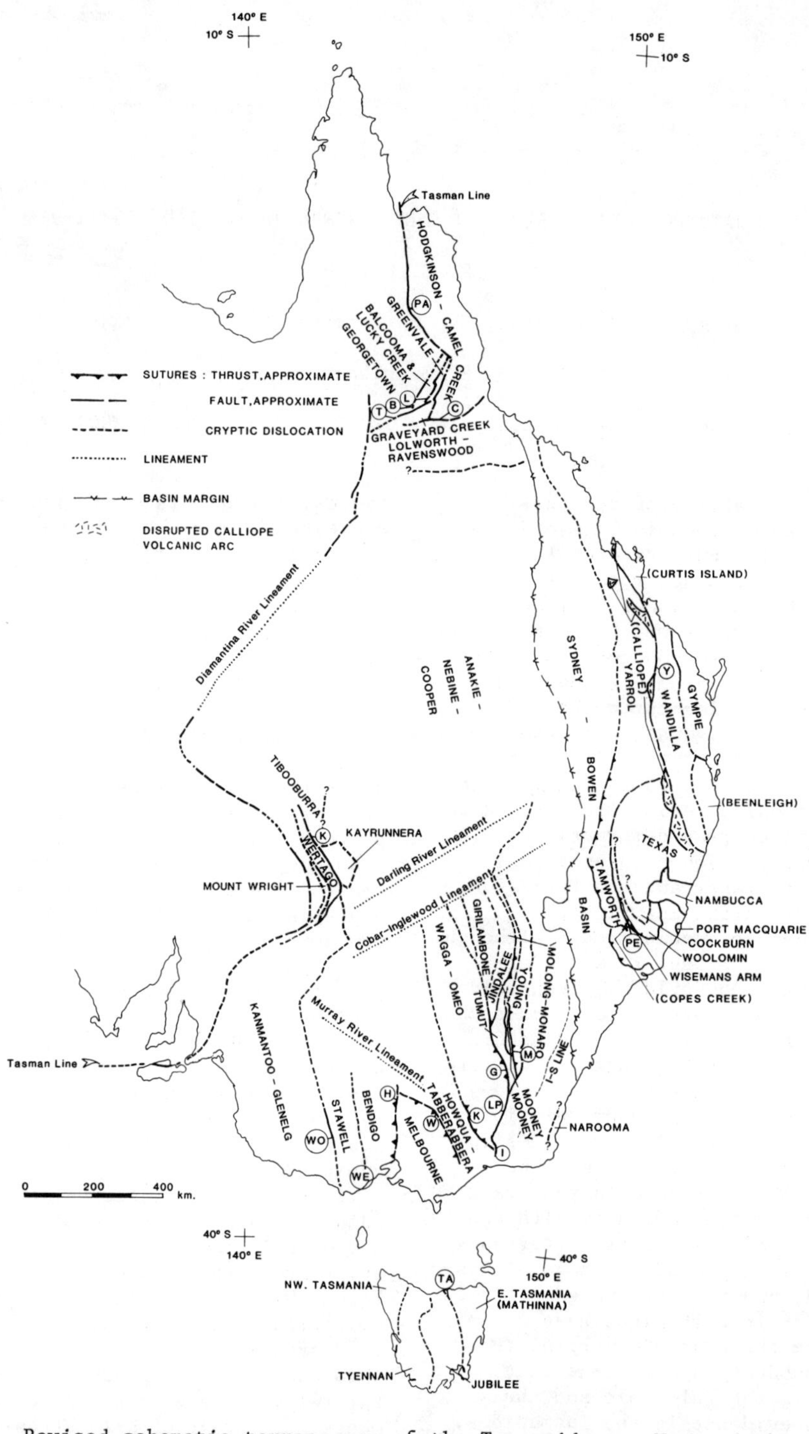

Fig. 1. Revised schematic terrane map of the Tasmanides. Names in brackets are possible exotic elements in defined terranes. B - Balcooma mylonite zone; C - Clarke River fault; G - Gilmore suture; H - faults bounding the Heathcote greenstone belt; I - Indi fault; K- Kiewa thrust; L - Lucky Creek mylonite zone; LP - Long Plain fault; M - Mooney Mooney thrusts; P - Parkes thrust; PA - Palmerville fault; PE - Peel fault system; T - Teddy Mount fault; TA - Tamar fracture system; W - Faults bounding the Mt Wellington greenstone belt; WE - Wedderburn line; WO - Woorndoo fault; Y - Yarrol fault.

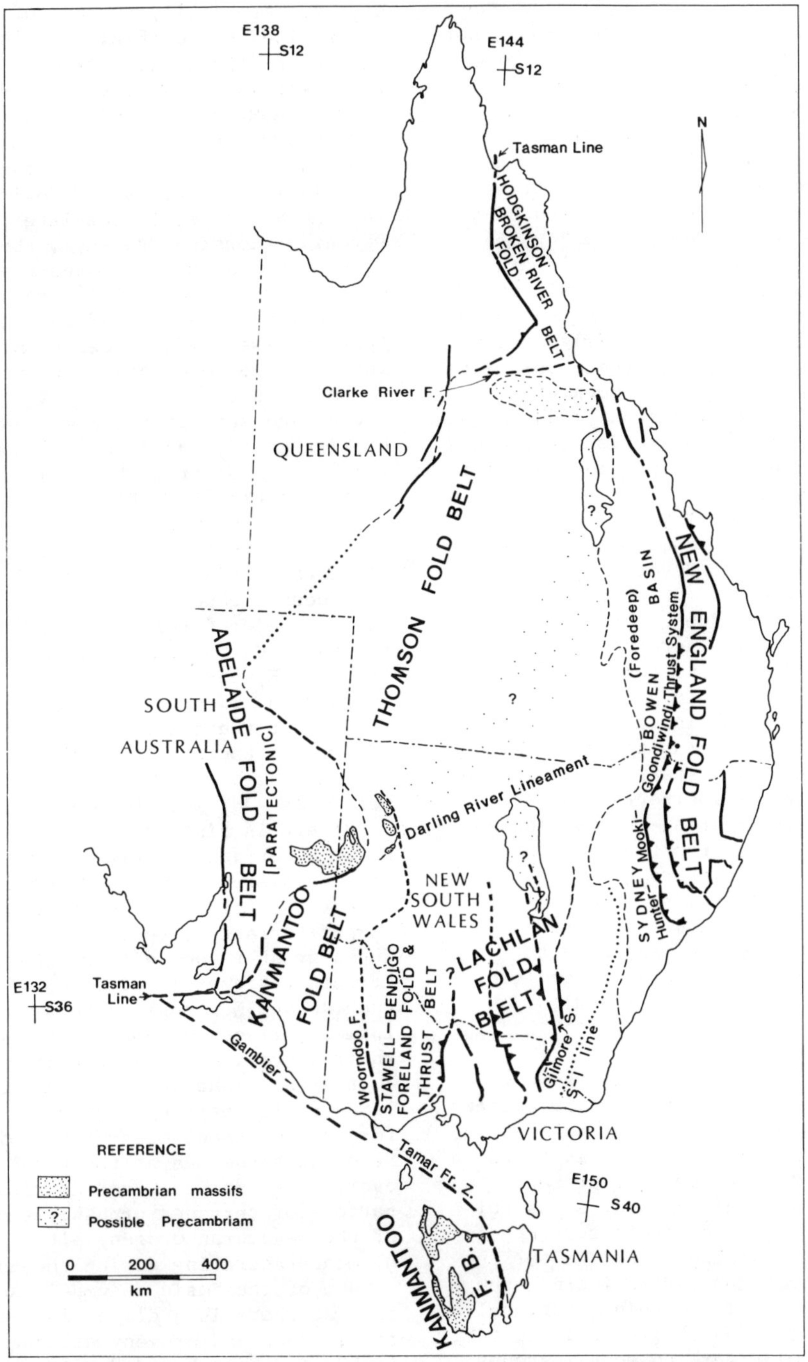

Fig. 2. Schematic structural map of the Tasmanides (or Tasman Fold Belt System) [Scheibner, 1987].

Tasmanide Terrane Analysis

Systematic terrane analysis of the Tasmanides following the methodology laid down by Jones et al. [1983] was attempted by Scheibner [1985] for the whole of the composite orogenic belt and Cawood and Leitch [1985] for the New England Fold Belt (Fig. 2). The existence of an active plate margin associated with the Tasmanides throughout the Paleozoic had long previously been recognised [e.g. Oversby, 1971; Solomon and Griffiths, 1972; Scheibner, 1972; Packham, 1973], and many writers on the accretionary subduction complexes of the region have tacitly recognised that some components were long travelled (ocean-floor rocks and remnants of oceanic islands), but the possibility that more substantial elements might be allochthonous has been little discussed. Leitch [1974] suggested that two major components, the New England and Lachlan Fold Belts, which have dissimilar Middle Paleozoic histories, might have been juxtaposed by transcurrent faulting. Crook [1980] postulated an allochthonous origin for a number of major elements, notably volcanic belts which he interpreted as oceanic arcs, found in the eastern half of the fold belt, and he used criteria similar to those of Jones et al. [1983] to trace an amalgamation history. Leitch [1980] discussed the possibility that the Hastings block might have been of allochthonous origin, and Harrington [1983] argued that 'The possibility that the Gympie terrane was tectonically transported to its present position must therefore be admitted'.

The 'Third Circum Pacific Terrane Conference' in 1985 in Sydney catalysed Tasmanide terrane studies, with some 23 papers devoted to this orogen appearing in extended abstract [Leitch, 1985]; many of these have been amplified for the present volume. This rapid increase in the publication of ideas, and the opportunity afforded by the Conference both for interaction amongst Tasmanide workers as well as input from a highly-knowledgeable international contingent, has led to a need to reassess earlier terrane schemes and also to attempt to rationalise terrane nomenclature. It became clear during the meeting and associated field excursions that just as the geology of the Tasmanides differs from that of the North American Cordillera, so also the approaches taken by Tasmanide workers are not always the same as those pioneered in North America. While we accept the methodology laid down by, for example Jones et al. [1983], differing experiences lead to different balancing of various lines of evidence.

The distribution of the various terranes we recognised as outlined in Figure 1, and parameters important to their definition and interpretation, are summarised in Table 1: many are discussed in detail elsewhere in this volume. At present biogeographic and paleomagnetic data, so useful elsewhere in recognising large-scale terrane displacements, are just beginning to emerge for the Tasmanides, and accompanying papers include the first to discuss these topics within a terrane framework (see also extended abstracts in Leitch [1985]). More systematic work on fossil groups and a greater array of well-dated pole positions should considerably help in tracing the development of this large composite orogenic belt.

Terranes and Major Structural Divisions of the Tasmanides

The Tasmanides (or Tasman Fold Belt System) comprise a number of major structural divisions distinguished particularly by their deformational history, especially the age of orogeny, and their generalised trend (Fig. 2). The most clearly defined of these divisions are (i) the Kanmantoo Fold Belt in the southwest, characterised by major pre-Middle Ordovician deformation (Delamarian Orogeny), (ii) the Lachlan Fold Belt of the southeastern Tasmanides, that underwent important deformation in the Late Ordovician - Early Silurian (Benambran Orogeny), the latest Silurian - early Devonian (Bowning Orogeny), the Middle Devonian (Tabberabberan Orogeny) and the terminal Early Carboniferous (Kanimblan Orogeny), and (iii) the New England Fold Belt of the eastern Tasmanides that underwent widespread Permian deformation (Hunter-Bowen Orogeny). Largely concealed rocks of Cambrian to Carboniferous age in central Queensland west of the Bowen Basin were grouped by Kirkegaard (1974) in the Thomson Fold Belt, which he considered showed an evolution similar to the Lachlan Fold Belt but had a dominant northeast-southwest trend. More recently [Murray and Kirkegaard, 1978] the northern elements originally included in this division have been grouped in the Hodgkinson Fold Belt, wherein major deformation probably took place in the Early Carboniferous, and the affinity of the Lolworth-Ravenswood block has been questioned [Henderson, 1986].

Each of these structural divisions is made up of several tectonostratigraphic terranes but no terrane has been recognised in more than one fold belt. In addition the amalgamation history of each fold belt is distinctive, and the age of linkages between terranes within individual folt belts predates linkages between adjacent belts. To some extent the fold belts might be looked on as 'super-terranes', their individual terranes showing a greater community of character than exists between terranes in different fold belts. The Lachlan and New England Fold Belts probably formed collages of amalgamated terranes prior to their accretion to the continental edge of Australia, although there was later addition or large-scale rearrangement of elements in the former. We treat most of the Silurian and Devonian sequences of the Lachlan Fold Belt as overlap assemblages, and consider that the basic pattern of terranes probably emerged at the end of the Benambran Orogeny although with some elements accreting during the Bowning Orogeny.

One of the basic assumptions of this account is that there is a close link between the joining of terranes and orogeny although, as Packham [1987] and Korsch and Harrington [1987] make clear, terrane dispersion may also be accompanied by deformation, metamorphism and granite emplacement.

Terranes of the Kanmantoo Fold Belt in Tasmania and Western Victoria

The Tamar fracture system separates redeposited quartzose sandstone and siltstone of Early Ordovician and Late Silurian to Early Devonian age in eastern Tasmania from a complex of Early Paleozoic graben ('troughs') and intervening Precambrian horsts ('geanticlines') farther west. Three belts of Precambrian rocks, those of the Rocky Cape block, the Tyennan (Prince of Wales Range and Cradle Mountain) and Forth blocks, and the Badger Road and Jubilee blocks [Collins and Williams, 1986] differ in lithologic makeup, metamorphic character, structural style and age of deformation, and form the basis for three terranes. At least the Northwest Tasmania terrane might be composite.

The status of the intervening grabens is less clear. The Eocambrian-Cambrian sequences of the Dundas trough lie unconformably on the Northwest Tasmania terrane, and the Cambrian strata interfinger eastward with rocks of the Mt Read Volcanic Belt that in turn rest unconformably on the Tyennan terrane. Thus at least part of the Dundas trough succession provides an overlap sequence and a minimum age for the amalgamation of the two terranes. It appears that terrane accretion occurred prior to Eocambrian transgression, a departure from earlier ideas [Scheibner, 1985]. The mafic and ultramafic intrusions associated with mafic volcanics within the Dundas trough are indicative of rifting [Brown, 1986] and do not define a continuous suture. Some of these bodies are Alaskan-type intrusions [Brown, Page and Love, 1987].

Cambrian rocks of the Adamsfield trough appear to overlap the Tyennan and Jubilee terranes. Thus the western two-thirds of Tasmania probably comprised a single composite terrane by the Middle Cambrian and possibly from as early as Eocambrian times. During Eocambrian and Cambrian rifting some dispersal of the three terranes probably occurred, but it is difficult to assess its extent. Deformation of the terranes, and the overlap sequences, occurred during Devonian (Tabberabberan) accretion of the East Tasmania terrane, but was preceded by Middle-Late Cambrian deformation which caused inversion of troughs and tightening of the Proterozoic terranes along old sutures. The latter deformation also affected rocks in the Kanmantoo - Glenelg terrane in western Victoria. These comprise (?) Late Proterozoic - Early Cambrian quartz and lithic greywacke, slate, uncommon dolomitic lithologies, and mafic igneous rocks of the Glenelg zone, and possibly the discontinuous outcrops of mafic and ultramafic intrusive rocks, basalt, andesite and dacite that make up the Mount Stavely greenstone belt (see below). Deformation of the Kanmantoo-Glenelg terrane probably involved inversion (collapse) of a passive margin sequence possibly overlain by arc rocks [for further discussion see Scheibner, 1987].

Terranes of the Kanmantoo Fold Belt in Far Western New South Wales

Leitch et al. [1987] indicated that Proterozoic and Early Paleozoic rocks found east of the Bancannia trough in far western New South Wales should be considered as constituting four terranes rather than the single composite body suggested by Scheibner [1985]. Of these, three show clear evidence of major Middle Cambrian (Delamerian) deformation and are appropriately considered to be part of the Kanmantoo Fold Belt [Scheibner, 1972], whereas the fourth, based on isolated inliers in the northwest corner of the state, appears to be more closely allied with terranes of the Lachlan Fold Belt and is described in the following section.

The Mount Wright terrane comprises a thick sequence of Early Cambrian mafic volcanics with intercalated siltstone and limestone overlain by silicic tuff and sandstone and early Middle Cambrian sandstone, siltstone and limestone. It is juxtaposed against the composite Wertago terrane along the Mt Wright fault. The latter terrane, which extends east to the Koonenberry fault, includes at least three units of metamorphic rocks, the youngest of which may be Early Paleozoic. East of the Koonenberry fault pre-Late Cambrian green sandstone and slate make up the Kayrunnera terrane that contrasts greatly in aeromagnetic signature with that of the Wertago terrane. Although some lithological correlations of early rocks across the Koonenberry fault have been suggested they require confirmation.

The latest movement on both the Mt Wright and Koonenberry faults was post-Late Devonian but fluvial to shallow marine mostly clastic strata of Late Cambrian - Early Ordovician age that occur unconformably above the Middle Cambrian and older rocks of the Mount Wright and Kayrunnera terranes, and are infaulted into the basement rocks of the Wertago terrane and may also possibly overlie these rocks [Brunker, Offenberg and Rose, 1967], are considered to constitute an overlap sequence and provide a minimum age for terrane amalgamation.

Terranes of the Lachlan Fold Belt in Western and Central Victoria and Eastern Tasmania

The terrane scheme adopted for the Lachlan Fold Belt in Victoria west of the Indi fault is a modification of that originally put forward by Scheibner [1985]. East of the Glenelg zone variably metamorphosed twice-deformed unfossiliferous quartz greywacke, slate and basalt, probably of Late Cambrian age, make up the Stawell terrane. Scheibner [1985] located the boundary between this and the Kanmantoo - Glenelg terrane along the Woorndoo fault, which places the Mount Stavely belt in the eastern part of the latter. Alternatively the boundary may lie further west, masked by the non-marine quartzose sediments and silicic volcanics of the Grampian Group [Ramsay and VandenBerg, 1986] in which case the Mount Stavely rocks are a component of the Stawell terrane and a possible correlative of amphibolites found a little further east [Buckland and Ramsay, 1982]. The Grampian Group rests

TABLE 1. Nature of Tectonostratigraphic

Terrane Name	Bounding elements (west, east)	Age of Pre-accretionary complexes	Inferred Basement
N.W. Tasmania	Cryptic, along continental margin. Cryptic, fault(s) beneath Dundas Trough	Proterozoic to Eocambrian	Continental
Tyennan	Cryptic, fault(s) beneath Dundas Trough. Cryptic fault(s) beneath Adamsfield Trough.	Proterozoic to Eocambrian	Continental to (?)partly ensimatic
Jubilee	Cryptic, fault(s) beneath Adamsfield Trough. Tamar Fracture System.	Proterozoic to Eocambrian	Continental to (?)partly ensimatic
Kanmantoo-Glenelg	Cryptic, Tasman Line. Woorndoo Fault or to west beneath Grampian Group.	Late Proterozoic - Cambrian	(?)Thinned ensimatic
Mt. Wright	Cryptic, beneath Bancannia Synclinorial Zone. Mt. Wright Fault.	Early-Middle Cambrian	(?)Thinned continental
Wertago	Mt. Wright Fault. Koonenberry Fault Zone.	Proterozoic to (?) Middle Cambrian	?
Kayrunnera	Koonenberry Fault zone. Cryptic, possibly along major gravity lineament	Early to Middle Cambrian	?
Stawell	Woorndoo Fault or beneath Grampian Group. Wedderburn Line	(?) Cambrian	(?)Thinned continental
Bendigo	Wedderburn Line, Djerriwarh, McIvor, Mt. Ida and Mt. William Faults.	Cambrian - Ordovician	Thinned continental, partly ensimatic
Melbourne	Djerriwarh and associated faults. Mt. Wellington Fault Zone	Cambrian - Late Ordovician (? - Middle Devonian)	Continental, ensimatic in east
Howqua - Tabberabbera	Mt. Wellington Fault Zone. Kiewa Fault	Ordovician	Thinned continental
Wagga - Omeo	Kiewa Fault. Gilmore-Long Plain-Indi Fault (Gilmore Suture)	Ordovician	Thinned continental to ensimatic

Terranes of the Tasmanides

Terrane type	Time of Accretion/ Amalgamation	Overlap sequences and/or Stitching plutons	Inferred Tectonic Character
Metamorphic (basement) and stratigraphic	Cambrian	Late Cambrian sequences	Passive margin
Metamorphic (basement) and stratigraphic	Cambrian	Late Cambrian sequences	Passive margin or microcontinent
Metamorphic (basement) and stratigraphic	Cambrian	Late Cambrian sequences (west) Middle Devonian pluton (west)	Passive margin or microcontinent
(?)Composite	(?)Middle to Late Cambrian	Siluro-Devonian sequence	Continental margin to back arc basin plus overlying volcanic arc
Stratigraphic	Middle to Late Cambrian	Latest Cambrian sequence	Continental margin possibly active
Metamorphic and stratigraphic	Middle Cambrian	Late Cambrian sequence	Continental margin
Stratigraphic	Middle Cambrian	Late Cambrian sequence	(?)Continental margin
Stratigraphic	Silurian - Late Ordovician(?)	(?)Late Silurian-Early Devonian sequence	(?)Ocean floor and trench (accretionary prism) or passive margin
Stratigraphic	Silurian - Late Ordovician (or Middle Devonian)	Early Devonian plutons	Ocean floor and trench (accretionary prism), or passive margin plus arc
Stratigraphic	Early Silurian (or Middle Devonian)	Middle-Late Devonian pluton Late Devonian sequence	Marginal plateau or continental margin plus volcanic arc
Stratigraphic	Early Silurian	Early Devonian sequence	Marginal sea
Metamorphic	Early Silurian	Middle Silurian sequence	Marginal sea

TABLE 1.

Terrane Name	Bounding elements (west, east)	Age of Pre-accretionary complexes	Inferred Basement
East Tasmania	Tamar Fracture System. Cryptic along continental margin	Early Ordovician to Early Devonian	?
Tibooburra	Northern extension of Koonenberry Fault. Cryptic, possibly along major gravity lineament	(?) Ordovician	?
Girilambone	Gilmore Suture. Cryptic, below Tullamore Syncline	(?) Cambrian or older Ordovician	(?)Ensimatic
Tumut	Gilmore Suture - Parkes Thrust. Gundagai - Killimicat - Cootamundra Faults	Ordovician, (?) Silurian	Thinned continental to ensimatic
Jindalee	Cootamundra - Gundagai - Killimicat Faults. Mooney Mooney Fault System	Early Paleozoic (?) Silurian	(?)Ensimatic
Mooney Mooney	Mooney Mooney Fault System	Early Silurian	Ensimatic
Young	Mooney Mooney Fault System. Long Plain Fault and cryptic extension	Silurian	Continental
Molong - Monaro	Long Plain Fault. Cryptic below Tullamore Syncline. Cryptic	Ordovician	Composite (in part continental)
Narooma	Cryptic. Continental margin	Ordovician	Ensimatic
Anakie - Nebine - Cooper	Cryptic, along the Tasman Line. Cryptic suture beneath Bowen Basin	(?)Proterozoic - Early Paleozoic	Continental and (?)ensimatic
Lolworth - Ravenswood	Clarke River Fault. Cryptic suture to south	Proterozoic - Early Ordovician	Continental
Balcooma	Balcooma Mylonite Zone. Suture intruded by stitching pluton	(?)Early Paleozoic	?
Lucky Creek	Suture intruded by stitching pluton. Lucky Creek Mylonite Zone.	(?)Early Paleozoic	?

(continued)

Terrane type	Time of Accretion/ Amalgamation	Overlap sequences and/or Stitching plutons	Inferred Tectonic Character
Stratigraphic	Middle Devonian	Middle Devonian pluton	(?) Continental margin perhaps in back-arc basin
Stratigraphic	Probably pre-Devonian	-	(?) Continental margin
Metamorphic	Early Silurian	Middle Silurian sequence	Marginal sea (?accretionary complex)
Stratigraphic	Earliest Devonian	Early Devonian pluton and sequence	Rift or pull-apart basin
Composite	Earliest Devonian	Early Devonian pluton and sequence	Rift or pull-apart basin (basement ? accretionary prism)
Igneous (ophiolitic)	Earliest Devonian	Early Devonian sequence	Rift or pull-apart basin
Igneous	Earliest Devonian	Early Devonian sequence	Volcanic arc
Composite	Early Silurian	Silurian-Devonian sequences	Volcanic arc - fore arc basin
Disrupted	(?) Early Silurian	Early Devonian pluton	Ocean floor and trench (accretionary prism)
Composite	?	Early Permian (or earlier) sequence	(?)Ocean floor and trench (accretionary prism) plus continental margin
Stratigraphic	Middle Ordovician	Middle Ordovician pluton	Passive margin plus overlying volcanic arc
Metamorphic	pre-Middle Silurian	Middle-Late Silurian pluton	Volcanic arc
Metamorphic	pre-Middle Silurian	Middle-Late Silurian pluton	Volcanic arc

TABLE 1.

Terrane Name	Bounding elements (west, east)	Age of Pre-accretionary complexes	Inferred Basement
Greenvale	Lucky Creek Mylonite Zone. Burdekin Fault System	Late Proterozoic	Continental and (?)ensimatic
Graveyard Creek	Teddy Mount Fault. Gray Creek Fault. Clarke River Fault.	Ordovician - Devonian	Thinned continental
Hodgkinson - Camel Creek	Palmerville - Herbert River - Gray Creek Fault. Cryptic along coast. Clarke River Fault	(?)Ordovician - Devonian	?
Everetts Creek - Carriers Well	Gray Creek Fault	Ordovician	?
Tamworth - Yarrol	Cryptic suture below Sydney-Bowen Basin. Peel and Yarrol Fault Systems	Cambrian - Ordovician	Thinned continental, locally ensimatic
Wisemans Arm	Peel Fault System. Mt. Abundance, and Goat Mt. Faults	(?)Devonian	Accretionary prism
Woolomin	Peel Fault System (south), Mt. Abundance (north). Spring Creek Fault.	(?)Early Paleozoic	Ensimatic
Cockburn	Spring Creek Fault. Bundarra granites	(?)Devonian	Ensimatic
Texas	Bundarra granites. Cryptic	Carboniferous	Ensimatic
Wandilla	Yarrol Fault System. (?)Widgee Serpentinite Belt	Carboniferous (and possibly older)	Ensimatic
Nambucca	Mihi Fault. Bellingen Fault System, Parrabel Fault	Early Permian	Accretionary complex
Gympie	Fault beneath Esk Rift or Widgee Serpentine Belt. Cryptic	(?)Carboniferous - Early Triassic	(?)Thinned continental

(continued)

Terrane type	Time of Accretion/ Amalgamation	Overlap sequences and/or Stitching plutons	Inferred Tectonic Character
Metamorphic	(?)Late Ordovician	Early Silurian sequence	?
Stratigraphic	(?)Middle Devonian	Late Devonian sequence	Fore-arc basin
Disrupted	(?)Middle Devonian	Late Devonian or Early Carboniferous sequence	Ocean floor and trench (accretionary prism)
Stratigraphic	(?)Middle Devonian	?	Volcanic arc
Stratigraphic	Visean	Early Permian sequence	Arc - fore-arc basin
Stratigraphic	?	Late Carboniferous - Early Permian pluton	Outer fore-arc basin
Disrupted	(?)pre-Early Devonian	Early Permian sequence	Ocean floor (accretionary prism)
Disrupted	Late Devonian	Late Carboniferous - Early Permian pluton	Ocean floor and trench (accretionary prism)
Disrupted	Late Carboniferous	Late Carboniferous - Early Permian pluton	Ocean floor and trench (accretionary prism)
Disrupted	Late Carboniferous	Late Carboniferous pluton	Ocean floor and trench (accretionary prism)
Metamorphic	Middle Permian	Early Triassic plutons	Rift or pull-apart basin
Stratigraphic	Middle Triassic	Triassic pluton	Volcanic arc plus fore-arc

unconformably on rocks of the Kanmantoo - Glenelg terrane and forms an overlap sequence providing a minimum age for amalgamation of this and the Stawell if a western position for the terrane boundary is adopted. If the eastern location proves more applicable then they comprise part of the Kanmantoo - Glenelg terrane. In either case there is little justification for a Grampians - Ararat terrane as suggested by Fergusson et al. [1986].

The eastern limit of the Stawell terrane is placed along the Wedderburn line, a poorly understood boundary, possibly a major structural break, that separates unfossiliferous rocks of the Stawell terrane from lithologically similar but fossiliferous Early Ordovician rocks further east. Fergusson et al. [1986] noted a change in structural grain 'indicating the presence of a major fault' along the eastern boundary of their "Grampians - Ararat terrane" which seems to approximate the Wedderburn line, and Ramsay and VandenBerg [1986] recorded a difference in structural history across the line.

Early Paleozoic rocks lying between the Wedderburn line and the fracture system defined by the Mt William, Mt Ida, McIvor and Djerriwarh faults are grouped in the Bendigo (equals Ballarat of Fergusson et al. [1986]) terrane. This comprises Cambrian mafic volcanic and volcaniclastic rocks, confined to close to the eastern boundary (Heathcote greenstone belt - Crawford et al. [1984]), and the overlying thick sequence of Ordovician quartz greywacke and slate.

The Melbourne terrane, extending east from the faulted edge of the Bendigo terrane as far as the Mt Wellington fault zone [Fergusson et al., 1986], is notable for its thick conformable sequence of mostly redeposited quartz-rich clastics that range in age from Ordovician to early Middle Devonian. Cambrian ultramafic rocks, mafic and intermediate volcanics, and associated volcaniclastic materials, locally overlain by sandstone, siltstone and bioclastic limestone, that comprise the Mt Wellington belt, were included in this terrane by Fergusson et al. [1986]. However most contacts are faults and stratigraphic continuity with Ordovician strata has not been demonstrated [Ramsay and VandenBerg, 1986, Table 1].

The region east of the Mt Wellington fault zone was included by Fergusson et al. [1986] and Fergusson [1987] in an all-embracing Benambra terrane. This region consists of fault-juxtaposed units of similar geological history according to Fergusson, but correlation of units sufficiently sound to allow estimates of fault displacements have not been achieved and we consider it prudent to continue to recognise several terranes.

Thus we show a distinct Howqua - Tabberabbera terrane (Fig. 1) composed of redeposited quartz-rich sedimentary rocks, perhaps deposited on rocks now represented in the Mt Wellington greenstone belt [Scheibner, 1985], situated between the Mt Wellington fault and the Kiewa fault. East of the latter, and extending to the Indi fault in Victoria, are high T/low P regional metamorphic rocks of the Wagga - Omeo terrane. These were possibly derived from a protolith similar to the Howqua - Tabberabbera terrane and are associated with Silurian S-type granites. A mylonite zone up to 2 km wide is at least locally associated with the Kiewa fault [Fagan in Fergusson, 1987] but Fergusson indicated that early structures and metamorphic grade are similar on either side of the fault in the Tambo River district.

Many accounts of the geological evolution of Victoria have treated the rocks of the Wagga - Omeo terrane and those further west as closely related. Each terrane shows a relatively thick sequence of Early Paleozoic (or older) redeposited craton-derived clastics of similar composition and, in some cases, the same lithostratigraphic names have been applied on either side of terrane boundaries (for example, the Early Ordovician Howqua Shale and the Middle-Late Ordovician Mount Easton Shale have been mapped in both the Melbourne and Howqua - Tabberabbera terranes). However recent studies stress the importance of thrusting on faults here regarded as terrane boundaries, with suggested movements of up to several hundred kilometres [Crawford et al., 1984]. Fergusson et al. [1986] showed the faults on either side of the Melbourne terrane as flattening westward at depth and suggested that the Bendigo terrane has been thrust east over the western edge of the Melbourne terrane, which has in turn overthrust the Howqua - Tabberabbera terrane on the Mt Wellington fault zone. They envisaged that these movements were associated with the folding of the Bendigo, Melbourne and Howqua - Tabberabbera terranes in the Middle Devonian and postulated that the rocks east of the Mt Wellington fault zone were juxtaposed against those of the Melbourne terrane by earlier (Silurian - (?)Middle Devonian) dextral transcurrent movements. This could account for the absence of evidence for Early Silurian (Benambran) movements in the Melbourne terrane rocks despite their presence in those immediately east, and for some anomalous paleogeographic patterns. Other explanations for these features have been suggested [Packham, 1973; Scheibner, 1985] and much of the thrusting may have been associated with movements grouped in the Late Ordovician - Early Silurian Benambran Orogeny [Crawford et al., 1984; Scheibner, 1985].

Paleozoic strata found east of the Tamar fracture system in Tasmania are redeposited quartzose sandstone and siltstone that contain Early Ordovician fossils in the west and Early Devonian fossils in the east. The Ordovician rocks contrast markedly in facies with those of similar age further west. Baillie [1985] has suggested that the Tamar fracture system is part of a fault-displaced major suture that continues as the Woorndoo fault in Victoria, and the Tasmanian rocks, grouped as the East Tasmania terrane, are probably an extension of the Bendigo, Melbourne or Howqua - Tabberabbera terrane.

Lachlan Fold Belt Terranes in New South Wales and Eastern Victoria

The Tibooburra terrane in the northwestern corner of New South Wales is made up of a thick sequence of alternating beds of graded sandstone and slate with scattered intercalations of mafic igneous rocks [Leitch et al., 1987]. An upper age limit is provided by a 410 Ma post-kinematic granite body. Although previously the rocks were allied with those of the Kanmantoo Fold Belt to the south, and accorded a Precambrian age, their facies and the presence in them of simple trace fossils suggest that they may be Ordovician and considered part of the Lachlan Fold Belt. Rocks of the Tibooburra terrane form inliers surrounded by Cretaceous strata and their relationship to other terranes is obscure.

The northern continuation of most of the terranes of western and central Victoria is masked by the Tertiary sediments of the Murray Basin and they have not been recognised in New South Wales. However, the Wagga - Omeo terrane is readily identified in the latter state. Its eastern edge is marked by the first major suture to be identified in the Lachlan Fold Belt, the Gilmore suture [Scheibner, 1985]. The suture is made up of sections of a number of faults, principally the Indi, Long Plains and Gilmore. In the south the Gilmore suture has a strong gravity expression, with a relatively steep westerly gradient south of West Wyalong indicating a west dip, in agreement with the steep west dip on the Gilmore fault in the Snowy Mountains region. It is also an impressive aeromagnetic boundary [Wyatt et al., 1980; Tennison-Woods and Webster, 1985; Suppel et al., 1986]. To the east are magnetic Ordovician volcanic rocks, and magnetic intrusive rocks have been emplaced along and close to the suture, some during Silurian rifting. North of the Lachlan River the suture is less well defined and it may divide into two strands, one running towards Nymagee and east of Cobar, and the other towards Fifield and Tottenham along a line defined by several Alaskan-type intrusives possibly emplaced along the suture [Scheibner, 1985; 1987].

Rocks abutting the Gilmore suture on the east belong to several major terranes. In the south the large Molong - Monaro terrane [Scheibner, 1985] is in contact with the Indi - Long Plain section. This terrane, which we consider was accreted at the time of the Benambran Orogeny (latest Ordovician - Early Silurian), is characterised by the presence of abundant intercalated limestone, and closely associated volcaniclastic and more abundant quartz-rich greywacke and slate. It was the site of extensive basin formation, of rift and/or pull-apart origin, in the Silurian (e.g. Cowra trough, Hill End trough). Various restorations of the original form of the Molong - Monaro terrane have been postulated [Scheibner, 1974; Cas et al., 1980]. Most recently Packham [1987] has suggested that the Ordovician volcanic rocks formed an originally continuous volcanic arc (the Molong volcanic arc) that has been displaced along sinistral transform faults into three individual segments, a western Parkes segment, a central Molong segment, and an eastern New England segment, the last causing a pronounced gravity ridge beneath the Sydney - Bowen Basin.

The Ordovician volcanics around Parkes have strong aeromagnetic expression [Tennison-Woods and Webster, 1985] and according to Agostini [1984, 1987] can be followed north to the Cobar-Inglewood lineament and, with lesser conviction, farther north [cf. Packham, 1987]. They extend south to the Victorian border. To the east of the volcanics occur coeval mainly quartz-rich turbiditic sediments with some chert and black slate. Locally the volcanics have prograded eastward over the sediments (Rockley Volcanics and Triangle Group). This geometry presents some difficulties for Packham's [1987[suggestion that the central Molong segment of the Molong volcanic arc has been rotated in such a way that originally the sediments were west of the arc and contiguous with those of the Wagga - Omeo terrane.

An interesting feature in the eastern part of the composite Molong - Monaro terrane is the I-S line of Chappell and White [1974], which separates contrasting granitoid provinces. Isotopic data suggest that the source rocks of the granitoids in both provinces were of Proterozoic age (up to 1400 Ma old). The S-type rocks to the west of the line might have been derived from a Precambrian basement [Chappell and White, 1977], perhaps that of the Molong microcontinent [Scheibner, 1983, 1985] but it is more difficult to explain the mainly Devonian I-type granitoids in the east. The inferred age of their source rocks is too old to be related to Paleozoic subduction. If they were crustal melts they could have been derived either from material underplated by igneous activity or from tectonically underplated igneous rocks. Glen and VendenBerg [1985] concluded that the upper crustal rocks on either side of the I-S line had similar histories and thus the line is not a terrane boundary in exposed rocks but appears to mark a deep crustal boundary. Thicker rigid crust in the west and thinner crust in the east are indicated by structural differences across the line documented by Glen and VandenBerg. Their conclusions are compatible with an earlier model [Scheibner, 1983] which suggested tectonic underplating by a Precambrian volcanic arc rafted in during Ordovician subduction beneath the Molong volcanic arc.

In the north the Gilmore suture is inferred to separate Ordovician rocks of the Wagga - Omeo terrane from metasedimentary rocks with scattered mafic volcanic lenses that form the basement of Scheibner's [1985] Girrilambone terrane. As discussed by Scheibner, some workers group these rocks and those of the Wagga - Omeo terrane together, and this possibly gains support from the discovery of late Early Ordovician fossils in both terranes [Sherwin, 1983; Stewart and Glen,

1986]. However Stewart and Glen noted that the Girrilambone fossils come from strata of low metamorphic grade and simple structural history, which contrast with widespread higher grade more complexly deformed rocks. It is thus possible that much of the Girrilambone terrane comprises older rocks than those of the Wagga - Omeo terrane and for this reason we continue to recognise two distinct terranes. The Girrilambone terrane is characterised by an overall positive Bouger gravity pattern in contrast to the negative pattern over adjacent parts of the Wagga - Omeo terrane.

The relationship between the Girrilambone terrane and the Molong - Monaro terrane is unclear because of the absence of exposed contacts between pre-Silurian units in the region where the two are juxtaposed. Scheibner [1985] included Ordovician arc volcanics, limestone and clastic rocks between Grenfell and Narromine in his Girrilambone terrane. However the great stratigraphic and faunal similarities between at least the volcanic-limestone sequence here [Webby et al., 1981; Webby, 1987] and those rocks further east suggest these rocks are part of the Molong - Monaro terrane. There are strong though indirect arguments for considering the Ordovician volcanic sequence in the western part of the Molong - Monaro terrane to have accumulated on older-continental crust [Scheibner, 1985; Packham, 1987], and Scheibner [1985] suggested that the Girrilambone rocks might extend into this region. If this is demonstrated then there will be little case for recognising a separate Girrilambone terrane. At present we place the Girrilambone - Molong - Monaro boundary along a hypothetical cryptic suture extending north-northeast from near West Wyalong beneath alluvium and the Devonian strata of the Tullamore syncline, even though gravity data suggest that the Girrilambone rocks extend at depth up to the Parkes thrust. The Devonian rocks provide a minimum age for amalgamation of the two terranes.

Basden et al. [1987] have identified several terranes east of the Gilmore suture and south of the Girrilambone terrane. Theirs is the first attempt to subdivide the Molong - Monaro terrane, recognised as a composite mass at the time of its formal definition [Scheibner, 1985]. The Tumut and Jindalee terranes are most clearly distinguished in terms of their basement rocks according to Basden et al. [1987]. Overlying strata are thick Silurian volcaniclastic and volcanic sequences (Tumut trough fill). The Silurian strata are broken by an apparently discontinuous fault system taken as separating the two terranes [Basden et al., 1987]. Until the Silurian geology is better known it will remain unclear as to whether these rocks should be treated as parts of the terranes or a complex overlap sequence.

The northern continuation of the Tumut and Jindalee terranes has not been traced in detail but the Ordovician rocks of the Tumut terrane are possibly continuous below younger cover with those exposed between Grenfell and Narromine. A little-known discontinuous strip of serpentinite, mafic volcanics and manganiferous chert lying further east is tentatively included in the Jindalee terrane (Fig. 1).

Fault-bounded ophiolitic rocks along the eastern side of the Jindalee terrane make up the highly disrupted Mooney - Mooney terrane which is faulted against the large Silurian Young Granodiorite, a major component of the Young terrane [Basden et al., 1987]. The northern and eastern extent of the latter are still to be defined but we show it extending north through a zone of Ordovician and Silurian-Early Devonian rocks characterised by the absence of Ordovician volcanics.

The Tumut, Jindalee, Mooney - Mooney and Young terranes are linked by Early Devonian stitching plutons and overlap sequences. They were probably accreted in the latest Silurian-earliest Devonian, coincident with widespread deformation (Bowning Orogeny) including possible major strike-slip faulting and thrusting [Basden et al., 1987].

Highly deformed but little metamorphosed greywacke, slate, chert and mafic volcanics exposed in coastal areas south of the Sydney Basin are grouped in the Narooma terrane [Powell, 1983; Scheibner, 1985]. The rocks, at least in part of Ordovician age, have a poorly defined boundary with those of the Molong - Monaro terrane, much of the contact being obscured by Devonian stitching plutons and overlap sequences.

Terrane Recognition in Queensland West of the Bowen Basin

Early Paleozoic and possibly older rocks are known only from the Anakie Inlier and scattered drill holes in the vast area of Queensland west of the Bowen Basin that is considered to be underlain by the Tasmanides. Attempts at subdivision into terranes [e.g. Scheibner, 1985] are frustrated by a lack of data and, following C.G. Murray (pers. comm. 1986) we now group all of the rocks in a single almost certainly composite Anakie - Nebine - Cooper terrane. Middle Devonian-Carboniferous rocks of the Adavale and Drummond Basins probably post-date accretion of the terrane to the Australian craton.

Terranes of Northeastern Queensland

Proterozoic rocks at the eastern edge of the Australian craton are juxtaposed against Paleozoic material along a series of intersecting faults in northeastern Queensland. Although contrasts between the regionally metamorphosed craton and the mostly less recrystallised younger rocks are marked, the work of Bain and Withnall [1985] and Withnall [1985] indicates that some units previously considered part of the craton may better be regarded as Tasmanide terranes. They distinguish the Balcooma, Lucky Creek and

Greenvale terranes, citing lithological evidence that suggests the first two might be of Early Paleozoic age. The Balcooma and Lucky Creek terranes separate the Proterozoic Greenvale terrane from the more extensive Precambrian rocks of the Georgetown block to the west. According to Withnall [1985] the Balcooma and Lucky Creek might be exposures of a single terrane thrust beneath the eastern part of the craton (including the Greenvale terrane) and exposed in a later major antiform. Alternatively he suggested the three terranes may have amalgamated elsewhere and been accreted as a composite mass. A minimum age of Silurian is placed on the age of accretion both by overlap sequences and stitching plutons.

Isolated fragments of the Greenvale terrane extend south along the Grey Creek fault zone which comprises an important boundary in the Paleozoic rocks. To the west lie rocks of the Graveyard Creek terrane, a coherent sequence of (?) Ordovician-Carboniferous rocks, mainly craton-derived clastic sediments and limestone but with volcanics near the base and scattered volcaniclastic-rich horizons. The Camel Creek subprovince of Arnold and Henderson [1976] lies east of the Grey Creek fault zone. This comprises a highly disrupted complex of redeposited clastic sediments, chert, and mafic volcanic rocks of inferred Ordovician to Devonian age. Similar rocks characterise the Hodgkinson province with which rocks of the Camel Creek subprovince appear to merge [e.g. Henderson, 1985] and we include them together in the Hogkinson - Camel Creek terrane.

A diverse assemblage of rocks, including massive and oolitic limestone, mafic volcaniclastic sediment, redeposited quartzose sandstone and siltstone, radiolarian chert and mafic-intermediate volcanic rocks mapped as Everetts Creek Volcanics and Carriers Well Limestone, occurs within the Grey Creek fault zone. The limey rocks have yielded Ordovician fossils but the relationship of these units to rocks of supposedly similar age to the east and west is unclear and they are best regarded as part of a separate terrane (C.G. Murray, pers. comm., 1986).

The Lolworth-Ravenswood terrane is characterised by the presence of a thick sequence of calc-alkaline volcanics and sedimentary rocks of Late Cambrian and Early Ordovician age [Henderson, 1986] that has been disrupted by the emplacement of the large Late Ordovician Ravenswood Granodiorite Complex and younger (Early Devonian-Early Permian) granites. Scattered occurrences of regional metamorphic rocks possibly include Precambrian basement (e.g. Cape River Metamorphics), for the oldest parts of the Paleozoic sequence are of cratonic provenance.

Terranes of the New England Fold Belt

Despite the masking effects of a lobe of Mesozoic sediment in southern Queensland there is a large degree of agreement concerning the continuity of major Paleozoic structures in the Tasmanides east of the Sydney - Bowen Basin. Paleozoic volcaniclastic rocks with scattered limestone lenses and intercalated volcanics that form a coherent sequence immediately east of the Basin constitute the Tamworth - Yarrol terrane [Scheibner, 1985; Cawood and Leitch, 1985]. Similar rocks of Early Devonian age near Warwick (Silverwood Group) were included in this terrane but Flood and Fergusson [1982] have suggested these comprise a separate possibly exotic terrane. Isolated exposures of comparable strata occur further north and have been grouped in the same general tectonic element (Calliope island arc, of Day et al. [1978]) but although these are mostly fault-bounded, they are unconformably overlain by rocks of the Tamworth - Yarrol terrane west of Rockhampton and hence cannot collectively be considered a separate terrane. However some elements included in this unit might be exotic.

We now include the Hastings and Copes Creek terranes of Scheibner [1985] within the Tamworth-Yarrol terrane. Recent mapping in the Devonian of the Hastings has revealed great similarities to the Devonian succession in the southern part of the Tamworth - Yarrol, and unpublished mapping by P.A. Cawood has shown that the Early Paleozoic rocks which form the basis for the Copes Creek terrane are unconformably overlain by rocks of the Tamworth - Yarrol terrane (a problem similar to that of the Calliope volcanic arc rocks discussed earlier).

East of the Tamworth - Yarrol terrane, and separated from it by a major fault system, are much more deformed rocks ranging in age from (?) Early Paleozoic to Carboniferous. Rock types include basalt, chert, siliceous siltstone and coarser redeposited volcaniclastic sediments. Although some coherent sequences are present, especially in younger parts of this complex, much of the region is disrupted by inferred early thrust faults [Cawood, 1982] or is melange [Cross et al., 1987]. In the south most workers have grouped these rocks in a single composite terrane (Scheibner [1985] - Texas - Woolomin terrane; Fergusson et al. [1958] - Gwydir terrane; Korsch and Harrington [1987]), although Cawood and Leitch [1985] suggested some four terranes could be distinguished as indicated in Table 1. Further north only a single Paleozoic terrane, the Wandilla terrane was identified east of the Tamworth - Yarrol terrane by Scheibner [1985], and it is uncertain whether this includes the same range of strata found in the south, or only equivalents of the younger (Carboniferous) Texas terrane.

The Woolomin, Cockburn, Texas and Wandilla terranes have been widely interpreted as accretionary subduction complexes. Each is composite, being broken by faults and/or melange zones most of which are inferred to mark where the upper levels of the crust on the subducting plate were stripped from the downgoing lithospheric slab. A strict reading of the definition of tectonostratigraphic terranes requires that each slice be considered a terrane, for the history of each differs from that of adjacent slices. However such division seems, at least at the moment, to be of little value. Instead divisions have been suggested based on inferred age, internal structure and the relative proportions of basaltic ocean floor, pelagic sediment, and trench turbidite sandstones, in the slices. Differences in the latter two properties reflect important changes in both accumulative and tectonic history and appear sensibly related to the inferred ages of off-scraping and the structural character of the terranes [Cross et al., 1987].

The boundaries of the terranes, and important divisions within the Texas terrane [Cross et al., 1987], provide valuable structural markers revealing the effects of rifting, strike-slip faulting and oroclinal bending that disrupted the region, mainly in the Permian [e.g. Flood and Fergusson, 1982; Korsch and Harrington, 1987].

Korsch and Harrington [1987] have emphasised differences in the detrital mineralogy of sandstones in two parts of what we show as Wandilla terrane in introducing their Curtis Island and Beenleigh terranes. The validity of these divisions can only be assessed once a much fuller account is available of the character of the Wandilla terrane; the Beenleigh sandstones certainly appear distinctive but whether those of the Curtis Island differ significantly from other Wandilla sandstones is yet to be demonstrated.

The status of Early Permian rocks grouped as the Nambucca terrane is debated. Scheibner [1985] considered they were sufficiently distinctive to justify recognition as a separate terrane, whereas Cawood and Leitch [1985] interpreted them as a deformed and metamorphosed part of the Early Permian overlap sequence.

Although provenance studies suggest a close association existed between these various terranes from at least the Devonian [Cawood and Leitch, 1985] no stitching plutons or overlap sequences older than Early Permian have been identified.

Harrington [1983] first suggested that rocks of the Gympie terrane in easternmost Queensland constitute a distinctive stratotectonic unit of possible exotic character. The inferred Middle Triassic age of accretion is the youngest such event recognised in the Tasmanides, and Harrington and Korsch [1985] have speculated that it was associated with very widespread deformation of epicontinental basin sequences in eastern Australia.

Concluding Remarks

A major aim in defining terranes in any complex orogenic belt is the identification of tectonic elements the relative positions of which must be defined through time before meaningful paleogeographic and paleotectonic reconstructions can be formulated. Nearly all such reconstructions of the Tasmanides assume that there has been little or no relative movement among the elements we have here designated terranes. This may be the case but the experience gained in more intensely studied orogenic belts, and those more amenable to paleomagnetic investigations, renders this an unwarranted assumption and one no more intrinsically probable than that which holds that all the terranes are mutually allochthonous and that the situation is one of tectonic anarchy. As we have already indicated there is evidence that suggests some of the terranes of particular Tasmanide fold belts were associated throughout their development. Biogeographic studies [e.g. Roberts, 1987; Webby, 1987], investigations of sediment provenance [e.g. Cawood, 1983], and the arrangement of inferred morphotectonic elements [e.g. Scheibner, 1985; Cawood and Leitch, 1985] all provide suggestive albeit not totally conclusive evidence. At least a tenuous linkage of the Lachlan Fold Belt terranes to a cratonic mass is indicated by the widespread quartzose sandstones that characterise many Ordovician sequences. Although this linkage is generally assumed to be with the shield areas of eastern Gondwanaland this has never been conclusively established, and studies have been insufficiently detailed to show whether one or several distinct sources were involved. It has yet to be demonstrated that the whole of the New England Fold Belt is not allochthonous with respect to those parts of the Tasmanides farther west. Unfortunately preliminary paleomagnetic results that suggested that this might indeed be the case [Klootwijk, 1985] have not been confirmed (C. Klootwijk, pers. comm. 1986). Ordovician faunas suggest some links with the Lachlan Fold Belt [Webby, 1987] but the distinctive Carboniferous fauna from near Murgon in the Wandilla terrane [Roberts, 1987] indicates an exotic origin for some accretionary prism components.

Suggestions of overthrusting involving displacements of perhaps several hundred kilometers on some major faults are gaining credence from detailed structural studies and indirect evidence derived from petrological and geochemical studies of granitic rocks, despite the rarity of observed low angle thrusts, widespread flat lying foliation or major recumbent folds. Transcurrent faulting involving up to several hundred kilometers of displacement has been suggested in the last decade to account for many displacements in the Tasmanides. Movements of an order of magnitude greater parallel to the trend of the orogenic belt could be present, even in the Kanmantoo Fold Belt where they could account for the apparent differences between the 'Proterozoic' geology of adjacent terranes in western Tasmania.

Terrane movements are a direct result of plate motions and once it proves possible to trace out the movement paths of the terranes, plate-tectonic interpretations of the Tasmanides will become increasingly more sophisticated and soundly based. Most plate tectonic schemes so far promulgated have been based on the identification of plate boundary-specific stratotectonic elements and only a few interpretations [e.g. Korsch and Harrington, 1985] have attempted the more difficult kinematic reconstructions, and then often from disputed assumptions.

Although much remains to be discovered about the Tasmanides we consider it is of considerable significance that, despite evidence of subduction of the oceanic lithosphere of the Ur-Pacific for a total period of in excess of 200 Ma during their development, no major continental collision has been recognised. In terms of east Gondwanaland tectonics this period can be extended by perhaps a further 130 Ma in considering the history of New Zealand, still without evidence of a major collision. The main exotic elements carried to the convergent eastern margin of Gondwanaland throughout the Paleozoic and Mesozoic were magmatic arcs and microcontinental blocks. In the case of the latter it is possible, although unestablished, that they are masses originally rifted from the margin.

Acknowledgments. We would like to acknowledge the contributions by C.G. Murray, A.H.M. VandenBerg and E. Williams who reviewed the first draft of Table 1 and made valuable comments and amendments. Discussions with many colleagues, including the authors of Tasmanide papers in this volume, have been most helpful but the authors alone are responsible for the opinions expressed herein. The final manuscript was typed by Miss S.M. Binns.

E.S. publishes with the permission of the Secretary, New South Wales Department of Mineral Resources.

References

Agostini, A., Nyngan 1:250 000 sheet - a preliminary geological interpretation from regional aeromagnetic and gravity data, N.S.W. Geol. Surv. Quart. Notes, 54, 13-23, 1984.

Agostini, A., Walgett 1:250 000 sheet - a geological interpretation from regional aeromagnetic and gravity data, N.S.W. Geol. Surv. Quart. Notes, 66, 21-31, 1987.

Arnold, G.O., and R.A. Henderson, Lower Palaeozoic history of the southwestern Broken River Province, north Queensland, J. Geol. Soc. Aust., 23, 73-93, 1975.

Baillie, P.A., A Paleozoic suture in eastern Gondwanaland, Tectonics, 4, 653-660, 1985.

Bain, J.H.C., and I.W. Withnall, The significance of suspect terranes in the study of Proterozoic and Early Paleozoic metallogeny in northeast Queensland, Geol. Soc. Aust., Abstr., 14, 9-12, 1985.

Basden, H., B.J. Franklin, B. Marshall, and A.E. Waltho, Terranes of the Tumut district, southeastern New South Wales, Australia, in Terrane Accretion and Orogenic Belts, edited by E.C. Leitch and E. Scheibner, Amer. Geophys. Union, Geodynamics Series, (this volume), 1987.

Brown, A.V., Geology of the Dundas - Mt Lindsay - Mt Youngback region, Geol. Surv. Tasmania Bull., 62, 221pp., 1986.

Brown, A., N.J. Page, and A.H. Love, Geology and platinum-group element (PGE) geochemistry of the Serpentine Hill Complex, Dundas Trough, western Tasmania, Canad. Mineral., (in press), 1987.

Buckland, G.L., and W.R.H. Ramsay, The geological setting and mineral potential of the Mount Stavely volcanic complex, western Victoria, Aust. Inst. Min. Metal. Conf. Proc. Melbourne, 419-432, 1982.

Cas, R.A.F., C.McA. Powell, and K.A.W. Crook, Ordovician palaeogeography of the Lachlan Fold Belt: a modern analogue and tectonic constraints, J. Geol. Soc. Aust., 27, 19-32, 1980.

Cawood, P.A., Structural relations in the subduction complex of the Paleozoic New England Fold Belt, eastern Australia, J. Geol., 90, 381-392, 1982.

Cawood, P.A., Accretionary tectonics and terrane dispersal within the New England Fold Belt, eastern Australia, Proc. Circum-Pacif. Terrane Conf., Stanford Univ., Publ. Geol. Ser., XVIII, 50-52, 1983.

Cawood, P.A., and E.C. Leitch, Accretion and dispersal tectonics of the southern New England Fold Belt, eastern Australia, in Tectonostratigraphic Terranes in the Circum-Pacific Region, edited by D.G. Howell, Circum-Pacific Council Energy Min. Res., Earth Sci. Ser., 1, 481-492, 1985.

Chappell, B.W., and A.J.R. White, Two contrasting granite types, Pacific Geology, 8, 173-174, 1974.

Chappell, B.W., and A.J.R. White, Ultrametamorphism and granitoid genesis, Tectonophysics, 43, 7-22, 1977.

Collins, P.L.F., and E. Williams, Metallogeny and tectonic development of the Tasman Fold Belt System in Tasmania, Ore Geol. Rev., 1, 153-201, 1986.

Coney, P.J., D.L. Jones, and J.W.H. Monger, Cordilleran suspect terranes, Nature, 288, 329-333, 1980.

Crawford, A.J., W.E. Cameron and R.R. Keays, The association of boninite low-Ti andesite - tholeiite in the Heathcote Greenstone Belt, Victoria; ensimatic setting for the early Lachlan Fold Belt, Aust. J. Earth Sci., 31, 161-175.

Crook, K.A.W., Forearc evolution in the Tasman Geosyncline: the origin of southeast Australian continental crust, J. Geol. Soc. Aust., 27, 215-232, 1980.

Cross, K.C., C.L. Fergusson, and P.G. Flood, Contrasting structural styles in the Paleozoic subduction complex of the southern New England Orogen, eastern Australia, in Terrane Accretion and Orogenic Belts, edited by E.C. Leitch and E. Scheibner, Amer. Geophys. Union, Geodynamics Series, (this volume), 1987.

Day, R.W., C.G. Murray, and W.G. Whitaker, The eastern part of the Tasman Orogenic Zone, Tectonophysics, 48, 327-364, 1978.
Fergusson, C.L., Early Paleozoic back-arc deformation in the Lachlan Fold Belt, southeastern Australia: implications for terrane translations in eastern Gondwanaland, in Terrane Accretion and Orogenic Belts, edited by E.C. Leitch and E. Scheibner, Amer. Geophys. Union, Geodynamics Series, (this volume), 1987.
Fergusson, C.L., P.G. Flood, and K.C. Cross, The Gwydir terrane: a Palaeozoic subduction complex in the southern New England Fold Belt of eastern Australia, Geol. Soc. Aust., Abstr., 14, 66-70, 1985.
Fergusson, C.L., D.R. Gray, and R.A.F. Cas, Overthrust terranes in the Lachlan Fold Belt, southeastern Australia, Geology 14, 519-522, 1986.
Flood, P.G., and C.L. Fergusson, Tectonostratigraphic units and structure of the Texas - Coffs Harbour region, in New England Geology, edited by P.G. Flood and B. Runnegar, 71-78, Univ. of New England and AHV Club, Armidale, N.S.W., 1982.
Glen, R.A., and A.H.M. VandenBerg, Evaluation of the I-S line in the Delegate area, southeastern Australia, as a possible terrane boundary, Geol. Soc. Aust. Abstr., 14, 91-95, 1985.
Harrington, H.J., Correlation of the Permian and Triassic Gympie Terrane of Queensland with the Brook Street and Maitai Terranes of New Zealand, in Permian Geology of Queensland, 431-436, Geol. Soc. Aust., Qld. Div., Brisbane, Qld, 1983.
Harrington, H.J., and R.J. Korsch, Deformation associated with the accretion of the Gympie Terrane in eastern Australia, Geol. Soc. Aust. Abstr., 14, 104-108, 1985a.
Harrington, H.J., and R.J. Korsch, Tectonic model for the Devonian to middle Permian of the New England Orogen, Aust. J. Earth Sci., 32, 163-179, 1985b.
Henderson, R.A., Evolution of the Broken River Province - a Mid-Palaeozoic active margin terrane of the northern Tasman Orogenic Zone, Geol. Soc. Aust. Abstr., 14, 109-110, 1985.
Henderson, R.A., Geology of the Mt Windsor subprovince - a Lower Palaeozoic volcano - sedimentary terrane in the northern Tasman Orogenic Zone, Aust. J. Earth Sci., 33, 343-364, 1986.
Howell, D.G., D.L. Jones, and E.R. Schermer, Tectonostratigraphic terranes of the Circum-Pacific Region, in Tectonostratigraphic Terranes of the Circum-Pacific Region, edited by D.G. Howell, Circum-Pacif. Council Energy Min. Res., Earth Sci. Ser., 1, 3-30, 1985.
Jones, D.L., D.G. Howell, P.J. Coney and J.W.H. Monger, Recognition, character, and analysis of tectonostratigraphic terranes in western North America, in Accretion Tectonics in the Circum-Pacific Regions, edited by M. Hashimoto and S. Uyeda, 21-35, Terra, Tokyo, 1983.
Kirkegaard, A.G., Structural elements of the northern part of the Tasman Geosyncline, in The Tasman Geosyncline - a symposium, edited by A.K. Denmead, G.W. Tweedale, and A.F. Wilson, 47-64, Geol. Soc. Aust., Qld. Div., Brisbane, Qld., 1974.
Klootwijk, C., Palaeomagnetism of the Tasman Fold Belt: indication for mid-Carboniferous large-scale southward displacement of the New England region, Geol. Soc. Aust., Abstr. 14, 124-127, 1985.
Korsch, R.J., and H.J. Harrington, Oroclinal bending, fragmentation and deformation of terranes in the New England Orogen, eastern Australia, in Terrane Accretion and Orogenic Belts, edited by E.C. Leitch and E. Scheibner, Amer. Geophys. Union, Geodynamics Series, (this volume), 1987.
Leitch, E.C., The geological development of the New England fold belt, J. Geol. Soc. Aust., 21, 133-156, 1974.
Leitch, E.C., The Great Serpentine Belt of New South Wales: diverse mafic-ultramafic complexes set in a Palaeozoic arc, in Ophiolites, edited by A. Panayiotou, 637-648, Geological Survey Department, Cyprus, 1980.
Leitch, E.C., (editor), Third Circum-Pacific Terrane Conference, Extended Abstracts, Geol. Soc. Aust., Abstr., 14, 254 pp., 1985.
Leitch, E.C., B.D. Webby, K.J. Mills, and P. Kolbe, Terranes of the Wonominta Block, far western New South Wales, Australia, in Terrane Accretion and Orogenic Belts, edited by E.C. Leitch and E. Scheibner, Amer. Geophys. Union, Geodynamics Series, (this volume), 1987.
Murray, C.G., and A.G. Kirkegaard, The Thomson Orogen of the Tasman Orogenic Zone, Tectonophysics, 48, 299-325, 1978.
Oversby, G., Palaeozoic plate tectonics in the southern Tasman Geosyncline, Nature Phys. Sci., 234, 45-48, 1971.
Packham, G.H., A speculative Phanerozoic history of the south-west Pacific, in The Western Pacific: Island Arcs, Marginal Seas, Geochemistry, edited by P.J. Coleman, 369-388, Univ. of Western Australia, Perth, 1973.
Packham, G.H., The eastern Lachlan Fold Belt of southeast Australia: a possible Late Ordovician to Early Devonian sinistral strike slip regime, in Terrane Accretion and Orogenic Belts, edited by E.C. Leitch and E. Scheibner, Amer. Geophys. Union, Geodynamics Series (this volume), 1987.
Powell, C.McA., Geology of N.S.W., South Coast, Geol. Soc. Aust., S.G.T.S.G. Field Guide No 1, 118 pp., 1983.
Ramsay, W.R.H., and A.H.M. VandenBerg, Metallogeny and tectonic development of the Tasman Fold Belt System in Victoria, Ore Geol. Rev., 1, 213-257, 1986.
Roberts, J., Carboniferous faunas: their role in the recognition of tectonostratigraphic terranes in the Tasman Belt, eastern Australia, in Terrane Accretion and Orogenic Belts, edited by E.C. Leitch and E. Scheibner, Amer. Geophys. Union, Geodynamics Series, (this volume), 1987.

Scheibner, E., The Kanmantoo pre cratonic province in New South Wales, N.S.W. Geol. Surv. Quart. Notes, 7, 1-10, 1972.

Scheibner, E., A plate tectonic model of the Palaeozoic tectonic history of New South Wales, J. Geol. Soc. Aust., 20, 405-426, 1974.

Scheibner, E., Suspect terranes in the Tasman Fold Belt System (Eastern Australia), Proc. Circum-Pacif. Terrane Conf., Stanford Univ., Publ. Geol. Ser., XVIII, 170-174, 1983.

Scheibner, E., Suspect terranes in the Tasman Fold Belt System, Eastern Australia, in Tectonostratigraphic Terranes in the Circum-Pacific Region, edited by D.G. Howell, Circum-Pacific Council Energy Min. Res., Earth Sci. Ser., 1, 493-514, 1985.

Scheibner, E., Paleozoic tectonic development of eastern Australia in relation to the Pacific region, in Circum-Pacific Orogenic Belts and the Evolution of the Pacific Ocean Basin, edited by J.W.H. Monger and J. Francheteau, Amer. Geophys. Union, Geodynamics Series, 18, 133-165, 1987.

Sherwin, L., New Occurrence of Ordovician graptolites from central N.S.W., N.S.W. Geol. Surv. Quart. Notes, 53, 1-4, 1983.

Solomon, M., and J.R. Griffiths, Tectonic evolution of the Tasman Orogenic Zone, eastern Australia, Nature Phys. Sci., 237, 3-6, 1972.

Stewart, I.R., and R.A. Glen, An Ordovician age for part of the Girilambone Group of Yanda Creek, east of Cobar, N.S.W. Geol. Surv. Quart. Notes, 64, 23-25, 1986.

Suppel, D.W., A.Y.E. Warren, J.J. Watkins, J. Chapman, K. Tennison-Woods and L. Barron, A reconnaissance study of the geology and gold deposits of the West Wyalong - Temora - Adelong district, N.S.W. Geol. Surv. Quart. Notes, 64, 1-23, 1986.

Tennison-Woods, K., and S.S. Webster, Geophysical signature of gold and porphyry copper mineral deposits in the Lachlan Fold Belt, N.S.W., Explor. Geophys., 16, 325-331, 1985.

Webby, B.D., (editor), The Ordovician system in Australia, New Zealand and Antarctica: correlation chart and explanatory notes, Int. Union Geol. Sci. Publ., 6, 1-64, 1981.

Webby, B.D., Biogeographical significance of some Ordovician faunas in relation to east Australian Tasmanide suspect terranes, in Terrane Accretion and Orogenic Belts, edited by E.C. Leitch and E. Scheibner, Amer. Geophys. Union, Geodynamics Series, (this volume), 1987.

Withnall, I.W., Suspect terranes along the Precambrian/Palaeozoic margin, Greenvale area, north Queensland, Aust. Geol. Soc. Abstr., 14, 247-250, 1985.

Wyatt, B.W., A.N. Yeates, and D.H. Tucker, A regional review of the geological sources of magnetic and gravity fields in the Lachlan Fold Belt of N.S.W., BMR J. of Aust. Geol. and Geophys., 5, 289-300, 1980.

A CRITIQUE OF PALEOMAGNETIC RESULTS FROM AUSTRALIAN PALEOZOIC FOLD BELTS AND DISPLACED TERRANES

P. W. Schmidt and B. J. J. Embleton

CSIRO Division of Mineral Physics and Mineralogy, P.O. Box 136, North Ryde, NSW 2113

Abstract. The paleomagnetic results from Paleozoic fold belts of Australia have given rise to various tectonic models for the development of southeastern Australia. These range from allochthonous (displaced terrane) models to autochthonous (in situ) models. The paleomagnetic results are not discriminatory enough to resolve between the extremes, there being much uncertainty in the ages of rocks and their magnetizations. A critical assessment of the results shows that many magnetic directions can not be confidently related to the paleohorizontal. Although most of the rock formations involved are from fold belts there is a glaring absence of fold tests [Graham, 1949 and McFadden and Jones, 1981]. If magnetizations are regarded to be syndeformational or postdeformational, a more simple picture emerges in which most rock units appear to have suffered remagnetization during the Carboniferous. The prevalence of secondary magnetizations is consistent with observations made in many fold belts in other continents, and highlights the importance of fold tests in folded terranes. Examples of two new studies of Devonian rocks, where fold tests have been applied, confirms the presence of Carboniferous postfolding magnetizations, although prefolding magnetizations have also been found. The prefolding magnetizations yield pole positions at variance with that of a recent study of Upper Devonian limestone reefs from Western Australia [Hurley and Van der Voo, 1986] and may be revealing for the first time reliable evidence of displaced terranes in southeastern Australia.

Introduction

Since Cox [1957] discovered anomalous paleomagnetic directions in the Eocene Siletz River volcanics in Oregon, USA, there has been a steady accumulation of such occurrences. Now it is widely recognized that much of the west coast of North America is composed of displaced terranes, each with its own geologic and paleomagnetic signature [see Beck ,1976, Irving, 1979 and McWilliams, 1983]. These terranes, which are associated with converging plate margins, have docked during the late Mesozoic and Tertiary. There is now evidence that supports the existence of much older displaced terranes, such as parts of the east coast of the USA [Van der Voo, 1981]. Paleomagnetically, the detection and demonstration of Paleozoic displaced terranes is problematical, not only because such rocks have suffered a more complex geological history than their Mesozoic and Tertiary counterparts, but also because the reference apparent polar wander path (APWP) for a particular continent may not be as firmly established for the Paleozoic as it is for younger times. This is indeed the problem faced by paleomagnetic workers in Australia and other Gondwanaland continents.

There are four separate APWPs that have been proposed for Gondwana during the Paleozoic, based largely on data from Australian rocks. Embleton et al [1974] and McElhinny and Embleton [1974] argued for an allochthonous, or exotic origin for the Lachlan Fold Belt (LFB) of south eastern Australia to explain the apparent difference between the APWP derived for the LFB and the APWP derived for central and western Australia. This interpretation was not incontrovertible though, as pointed out by Schmidt and Morris [1977] who used the same data set to construct a single APWP for the LFB and the rest of Australia by reassigning the age of the pole position of the Mereenie Sandstone [Embleton, 1972a] to the Early to Middle Devonian rather than the Silurian, and by invoking the 'polarity option' for early Paleozoic poles. This option admits the possibility that the paleomagnetic record is incomplete and there may be an unrecognized gap of 180 degrees of polar motion in the Paleozoic. The age of magnetization and the age of deposition of many sedimentary rocks are often different, and even when they are not, it is a difficult task to demonstrate such.

Following this alternative model, Morel and Irving [1978] reverted to the original polarity convention but constructed a more complicated path to incorporate the LFB data and the central and western Australian data in a single path, thus favoring an autochthonous, or in situ development of the LFB similarly to Schmidt and Morris [1977]. Clearly these syntheses of the paleomagnetic poles are of limited value if such loose constraints apply to the basic data. Attempts to redress this situation have served to emphasize just how complex this problem is [Goleby, 1980a and b]. Resolving multiple magnetizations of uncertain origin or age is proving to be one of the most difficult and challenging aspects of paleomagnetic studies of fold belts. Nevertheless, paleomagnetism offers the only quantitative means of establishing the motions of continental fragments during the Paleozoic (since no oceanic crust from that era has survived), and perseverance with such studies in conjunction with sound geological guidance will enable at least some of the questions surrounding displaced (suspect and cryptic) terranes to be answered.

Multicomponent Magnetizations

The magnetic records stored in rocks are rarely of high fidelity. This fact was realized very early in paleomagnetic studies and cleaning techniques were developed to enhance the signal. The three main methods used are i) alternating field (AF) demagnetization, ii) thermal demagnetization and, iii) chemical demagnetization. The details of these techniques are found in several textbooks on paleomagnetic methods e.g. see Irving, 1964, Collinson et al, 1967 McElhinny, 1973 and Collinson, 1983. Briefly the objective of cleaning the magnetization of rock samples is to eliminate secondary magnetic components that overprint the primary magnetization. Of course, the simple notion of a primary and a secondary magnetization is an ideal, and as a general rule magnetizations of Paleozoic rocks are complex. For instance, quite often it is not possible to decide whether the magnetization removed by the cleaning is older than that which remains. Also it is rare to be able to find definitive evidence for the originality of a magnetization.

The natural remanent magnetization (NRM) of a rock may be considered to consist of the vector sum of a number of individual magnetic components, each component representing an event in the history of the rock (there being no guarantee that all events are represented or were even ever recorded). Cleaning methods attempt to retrace this history by stripping off the most recently acquired magnetizations exposing the more ancient ones in a stepwise procedure. If a rock unit, such as a volcanic say, were subjected to mild burial after initial formation and magnetization then a part of the original magnetization may be reset, depending upon the temperature the buried unit attains and whether the rock possesses magnetic grains that are unstable, either magnetically or chemically, at such temperatures. If there are chemical changes there is a real possibility of the new magnetization having a greater stability than the original magnetization and this component will be that which is isolated by cleaning. For instance, if magnetite becomes oxidized to hematite, then, after partial AF or thermal demagnetization, the hematite magnetization may remain while the magnetite magnetization is eliminated because hematite has both a higher coercivity of remanence and a higher unblocking temperature (and Curie Temperature) than magnetite. On recrystallization or alteration the new mineral formed will acquire a magnetization that may partially reflect the magnetization direction of the original mineral, and also reflect the ambient magnetic field at the time of alteration [Bailey and Hale, 1981]. Chemical demagnetization will also preferentially remove the magnetite since it has a higher solubility than does hematite. It should be emphasized though, that it is often possible to determine the removed component by fitting least-squares lines and planes to the demagnetization steps using any of a number of multi- component techniques or packages that are available [Kirschvink,1980, Schmidt,1982 and Kent et al, 1983].

A further complication in making the transition from a paleomagnetic direction to a pole position can arise from the assumption of a simple dipole geomagnetic field model, which can produce errors of up to 10 to 15 degrees [Idnurm,1985]. But for the fact that the Earth's crust is far from the source of the geomagnetic dynamo, the geomagnetic field would appear to be very complex indeed.

While the above aspects of paleomagnetic studies are well known to practicing paleomagnetists they are often conveniently ignored. However, the problems touched on above should give non- specialists some feel for the technical hurdles that must be cleared before paleomagnetic data can be used to infer ancient pole positions.

A New Direction in Paleomagnetism

As mentioned above it is not always possible to find unequivocal evidence for the originality of a magnetic component. However, one method that has been successfully applied to igneous rocks is the contact test [Everitt and Clegg, 1962]. By examining the variation of the magnetization of the contact rock with distance from an igneous intrusion, it is sometimes possible to infer whether the igneous rock still retains the magnetization it acquired at the time of original cooling. This is quite difficult to apply in nature though, because it demands good outcrop and well preserved contacts. In deeply weathered regions, which are often encountered in Australia, contacts tend to weather preferentially. A more generally applicable test, easily implemented in folded terranes, is the fold test [Graham, 1949]. This test is often thought of as being less powerful than the contact test, in so far as it only dates the magnetization age with respect to the age of folding. However, it has one overwhelming advantage in that it can relate

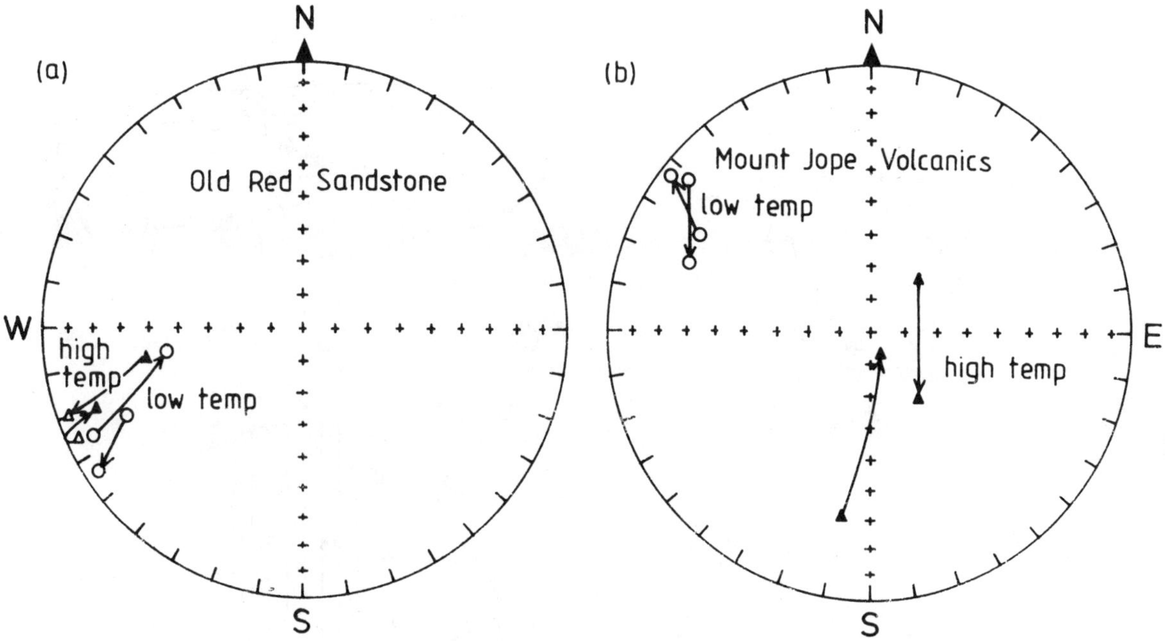

Fig. 1. Equal-area projections of magnetization directions from two fold belts, before and after unfolding, a) the Mill Haven fold after McClelland Brown [1983] and b) the Rocklea Anticline from Schmidt and Embleton [1985]. The main feature is the crossover signature indicating that the magnetizations have formed during deformation and thus there is uncertainty regarding the paleo-horizontal. These magnetizations are termed syndeformational remagnetizations (SDR) in the text. The convention is that closed(open) symbols refer to the lower(upper) hemispheres.

the direction of the magnetization to a paleohorizonal (except for the case where deformation or tilting proceeds after remagnetization). It can be appreciated that from the relationship between the directions of magnetization determined from the limbs of folded strata, before and after unfolding the limbs, the age of the magnetization with respect to the age of folding may be established. From this the paleohorizonal may be inferred. This is clearly of crucial importance if the magnetization direction is to be used to calculate a paleopole position. However, with the contact test this relationship is not always established.

A critical look at paleomagnetic results from Paleozoic rocks from Australia repeatedly raises questions of the ages of magnetizations, and whether or not directions of magnetization should be referred to the present horizontal or the paleohorizontal. For instance, the result from the Mereenie Sandstone [Embleton, 1972a] is from one fold limb only. The samples collected from another fold limb had been completely remagnetized in the late Tertiary. Thus, it was not clear if the thermally cleaned directions ought to have been referred to the present or the paleohorizontal, although on the balance of evidence Embleton [1972a] chose the paleohorizontal. The pole position calculated from the direction in its present orientation was more difficult to interpret than that after bedding correction. There are now more data available from Central Australian rock formations that are relevant and ought to be taken into account. In the past there seems to have been a tendency to fit new data to preconceived schemes, rather than to pool all the information and attempt to reinterpret the whole data set each time some new results become available. This is understandable because the paleomagnetic data set is immense and the prospect of a major rethink is daunting; however, we believe the current state of affairs requires such action. What we have attempted below is to distill the essential information from the results now available, and thus arrive at the most probable synthesis.

There is an increasing awareness that, more often than not, paleomagnetic studies of fold belts provide information that relates to events after the folding or during folding. As stated by Van der Voo and McCabe [1985], "Ancient, yet secondary, magnetizations are the rule, rather than the exception in the rocks of the Appalachians and adjacent areas." We have no doubt that this edict also applies to rocks of Australian fold belts. In fact, a recent study of the New England Orocline [Klootwijk, 1985] identified no primary magnetic component. The

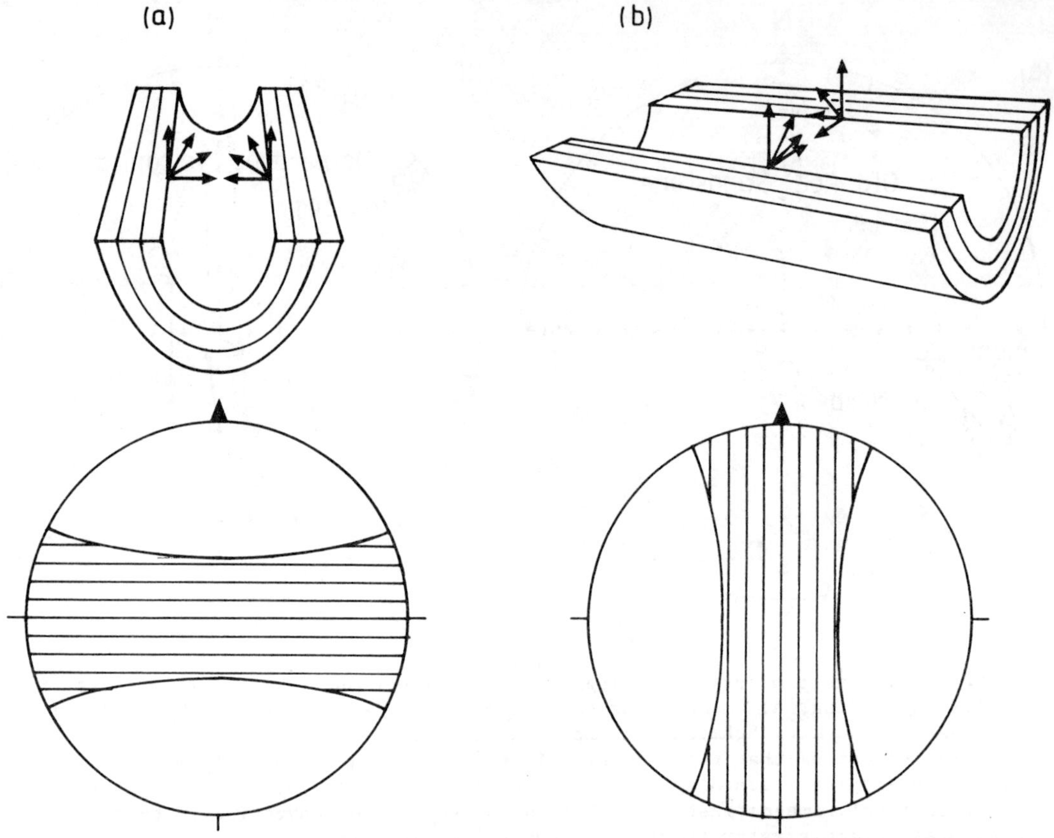

Fig. 2. A hypothetical model of two fold belts with fold axes, a) north-south, and b) east-west, which acquire magnetizations at variable times during folding in a vertical magnetic field. The former corresponds to the Lachlan Fold Belt(LFB) while the latter corresponds to the Amadeus Basin(AB).

overprinting in this region is posttectonic and complete. In general, the most likely time for rock units from fold belts to acquire secondary magnetizations is during deformation, or during times of enhanced heat flow that accompanies such events. While the nature of the mechanisms of overprinting may be difficult to establish, they may range from being purely thermal, to thermo-chemical or piezomagnetic. If the timing of remagnetization just predates actual folding, then the magnetization directions will yield a positive fold test, and vice versa for remagnetization that follows folding. More elusive of character, though, is remagnetization during deformation, yet this is possibly the most common mode of occurrence of secondary magnetizations in fold belts. However, because its signature is the most complex and difficult to decipher, it is the least well documented. Nevertheless, it is now a widely recognized phenomenon.

McClelland Brown [1983] appears to be the first worker to fully exploit syndeformational remagnetization(SDR) in an elegant study of the Pembrokeshire Old Red Sandstone (UK). What is now recognized as the fingerprint of SDR is the crossover signature (Figure 1). On correction for bedding attitude i.e.unfolding, the magnetization directions from two opposing limbs are seen to converge and then diverge. From the Mill Haven fold both the high temperature and the low temperature magnetization components in the Old Red Sandstone crossover (Figure 1a). In detail, the high temperature components converge between quarterway and halfway through folding, while the low temperature components converge at the three-quarter mark, indicating that while these secondary magnetizations are all SDR, the high temperature component appears to be older than the low temperature component. Note that these temperatures do not necessarily relate to the paleotemperatures that where attained during folding, since the mechanisms of remagnetization probably involved thermochemical effects which may produce high unblocking temperature components even at relatively low temperatures. The low temperature components may be thermoviscous in origin, acquired during the last stages of folding while the rocks cooled slowly. We

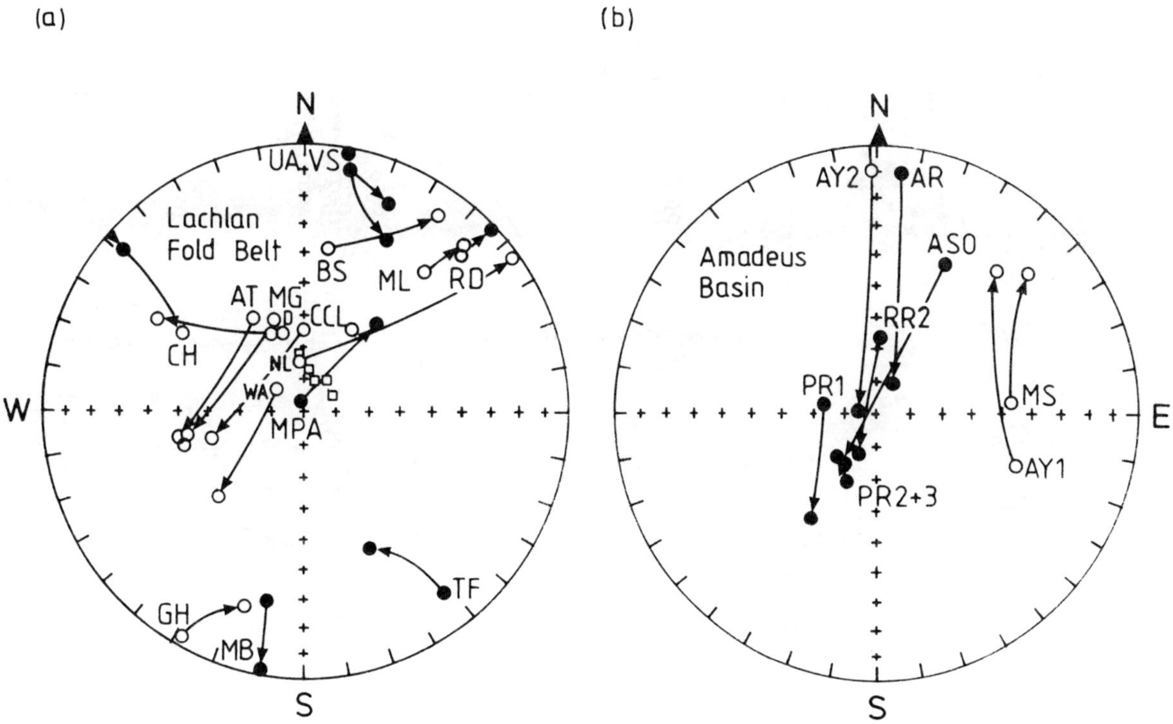

Fig. 3. Equal-area projection of magnetization directions from rock units of, a) the Lachlan Fold Belt, and b) the Amadeus Basin. Mnemonics are those used by Goleby [1980], Embleton [1981] and Schmidt et al [1986a and b], while the conventions are those adopted for Figure 1.

consider this mechanism to be responsible for SDR observed in some Precambrian volcanics from the Pilbara (Australia). In Figure 1b, the low temperature magnetizations display the crossover signature and are statistically different in either their in situ or fully unfolded orientations [Schmidt and Embleton, 1985]. The directions most closely agree in their 65 percent folded positions. The high temperature magnetizations are statistically different in their folded orientations, and are thought to be primary thermoremanent magnetizations dating from the time of formation. Also, the directions of the high temperature and the low temperature magnetizations are greatly different, reflecting a large difference in their times of acquisition or otherwise a period of rapid polar wandering.

Miller and Kent [1985] have recently observed SDR in the Catskill Formation of the Appalachians while Vick et al [1985] report SDR from the Carolina Slate Belt. Watts [1985] has found SDR that predate one deformation phase, but postdate an older deformation phase in the metamorphosed Barrovian zones of Scotland. There is little doubt of the ubiquity of these magnetizations in fold belts. This being so has some important ramifications for interpreting the paleomagnetic data from folded terranes, not the least of which is deciding upon a paleohorizontal. If the magnetizations of two fold limbs firstly converge, then diverge on progressive unfolding, the correct attitude is ambiguous because the folding may not have been symmetrical and the timing of folding and magnetization acquisition may not have been identical for both limbs. This is the major problem with interpreting the paleomagnetic data from Paleozoic fold belts of Australia. Below we examine in detail some of the more recent (post1970) results from the Lachlan Fold Belt(LFB) and the Amadeus Basin(AB) to illustrate this problem.

Before considering the actual data, it is instructive to consider some salient features of the LFB and the AB in the light of the above mode of magnetization acquisition. The LFB comprises numerous synclinoria and anticlinoria with a well developed north south structural grain. The last major orogeny in the LFB was the mid Carboniferous Kanimblan Orogeny. On the other hand the AB is an east west aligned intracratonic depression with a protracted history of sedimentation and deformation. Most of the folding in the AB occurred during the last deformational phase in the Carboniferous Alice Springs Orogeny. Throughout much of the Late Carboniferous Australia occupied high paleolatitudes, which is reflected by the steep inclinations of paleomagnetic directions of rocks magnetized at these

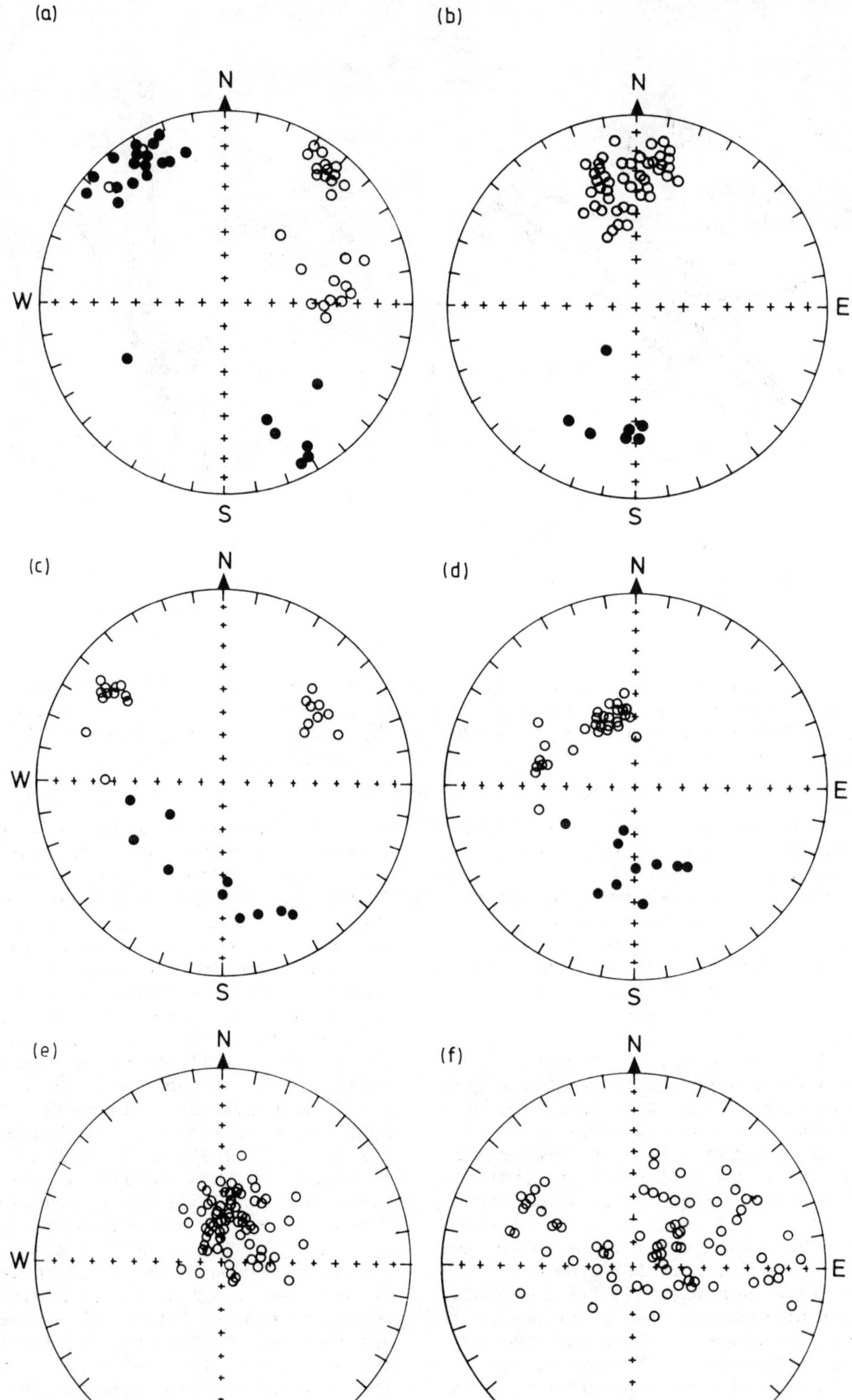
(a)
(b)
(c)
(d)
(e)
(f)
N
S
W
E

times. If we consider the folding of a sedimentary sequence about a north south axis, acquiring SDR at various stages, then the rocks will become variously magnetized in a vertical east west plane (Figure 2a). In a simplistic sense, this is our model for the LFB. Our model for the AB is similar except folding is about an east west axis resulting in rocks being remagnetized with directions spread in a vertical north south plane (Figure 2b). It is clear that reconstructing the paleomagnetic field from rocks magnetized in this fashion is challenging.

Carboniferous Remagnetization

Turning now to the data it is apparent that there is evidence of remagnetization in the Lachlan Fold Belt(LFB) and the Amadeus Basin(AB) similar to that proposed above. Plotted in Figure 3 are paleomagnetic directions, both before and after unfolding (corrections for bedding attitude assume a simple tilt). Although the models proposed for the fold belts are crude (Figure 2), the value of considering the data this way is self evident. Each circle plotted in Figure 3 represents one rock unit or statigraphic section. None of the data points internally satisfies a fold test, and all points must therefore be viewed as a whole. The statistical significance of the fold test is fully discussed by McFadden and Jones [1981], who show that the significance test devised by McElhinny [1964] is invalid. Some results that may have failed the McElhinny test, may be significant by the McFadden and Jones test. However, most of the studies of Australian fold belts have not been designed to take advantage of the fold test, and can not be tested using the McFadden and Jones test. Many of the results from the LFB scatter on unfolding (using structural information of Goleby, 1980b), while the rest form two groups to the north north east and to the south, both before and after unfolding. After unfolding the data form four loose groups that may be interpreted as original magnetizations reflecting changes in the geomagnetic field at different geologic times. Indeed this was the interpretation favored by Goleby [1980a]. However the divergence of directions D, AT, MG, CCL, NL, WA and MPA on unfolding, coupled with their general agreement with magnetization directions identified as secondary [Goleby, 1980a] before unfolding (plotted as squares in Figure 3a), suggests that these directions also might best be treated as secondary. This is particularly so considering the attention afforded secondary magnetizations from fold belts elsewhere [McClelland Brown, 1983, Schmidt and Embleton, 1985, Miller and Kent, 1985, Vick et al, 1985 and Watts, 1985]. Exactly when the magnetizations date from is difficult to say but their directions would group best at about the threequarter folded mark during the Carboniferous along with many other units of the Amadeus Basin. The Alice Springs Orogeny was widespread and intense enough to reset isotopic systems in basement rocks exposed to the north of the basin [Armstrong and Stewart, 1985]. The overriding lesson to be learnt from studies of fold belts is the absolute necessity for fold tests and whether or not data yield a positive or a negative result, at least the paleohorizontal corresponding to the age of magnetization will be established.

Some New Pole Positions

Two paleomagnetic pole positions acquired recently from some Paleozoic volcanics in the LFB satisfy the fold test as formulated by McFadden and Jones [1981]. Since folding is thought to have followed closely after their formation, the ages of the poles are relatively tightly constrained. The first of these is the late Middle to early Late Devonian Comerong Volcanics pole position [Schmidt et al, 1986a], which was also shown to predate the mid Carboniferous megakinking proposed by Powell [1984]. This pole is located at 76.9°S, 330.7°E (A_{95}=7.2°). The magnetization directions before and after unfolding (and unkinking) are depicted in Figures 4a and b. The grouping is clearly enhanced after the structural corrections are performed indicating that the directions refer to the paleohorizontal. This in itself does not prove that the directions are from primary thermoremanent magnetizations, but with the deformation following so closely after the formation of the rocks and the fact that the magnetizations have survived the orogeny supports such a conclusion. At the very least we are justified in calculating a pole position from these directions because we can confidently refer them to the paleohorizontal.

The second of these pole positions is from the latest Silurian to Early Devonian Snowy River Volcanics [Schmidt et al, 1986b] located at 68.9°S, 220.0°E (A_{95}=6.5°). These rocks were deformed during the Late Devonian Tabberabberan Orogeny. The magnetization directions from these rocks fall into two classes. The high temperature magnetization directions, like those from the Comerong Volcanics, cluster more tightly after applying the appropriate structural corrections (Figures 4c and d). However, the grouping of the low temperature magnetization

Fig. 4. Equal-area projection of magnetization directions from the high temperature directions of the, a) Comerong Volcanics before structural correction, b) Comerong Volcanics after structural correction, c) Snowy River Volcanics before structural correction, d) Snowy River Volcanics after structural correction, and low temperature overprints from, e) Snowy River Volcanics before structural correction, and f) Snowy River Volcanics after structural correction.

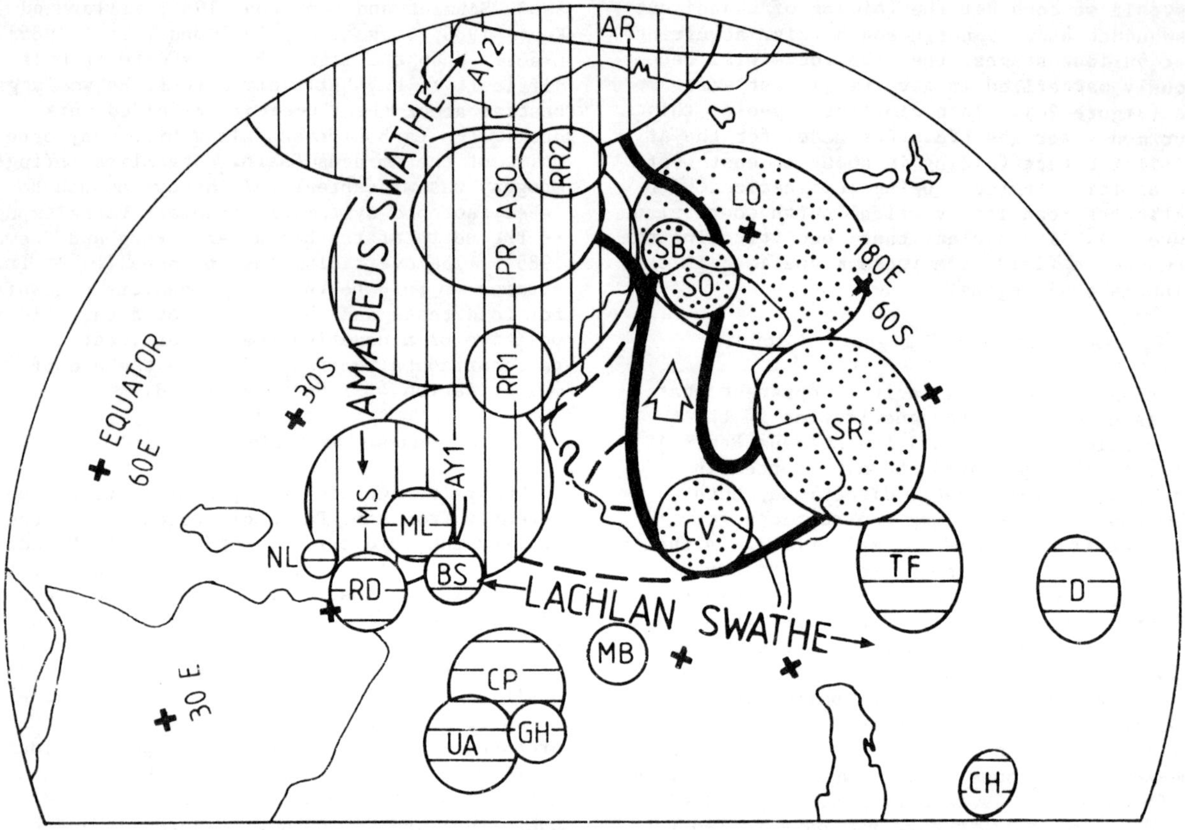

Fig. 5. Apparent pole positions calculated from various magnetization directions determined in rocks from the Lachlan Fold Belt and the Amadeus Basin. The formation mnemonics are those used by Goleby [1980a], Embleton [1981] and Schmidt et al [1986 a and b]. The horizontal hatching denotes LFB poles and the vertical hatching denotes the AB poles.

directions deteriorates after applying the structural corrections (Figures 4e and f), and is interpreted as a partial overprint having formed during the cooling phases of the Tabberabberan Orogeny i.e.latest Devonian to Early Carboniferous. The prefolding magnetizations clearly predate the midDevonian and are interpreted to date from the formation of the volcanics. We can confidently refer these magnetization directions to the paleohorizontal while the low temperature magnetization directions are referred to the present horizontal, and pole positions calculated accordingly. The immediately overlying Buchan Cave Limestone is completely overprinted [Schmidt et al, 1986b] and the two overprint pole positions from the volcanics and the limestone repectively are located at 68.7°S, 132.5°E (A_{95}=5.5°) and 64.7°S, 127.9°E (A_{95}=4.5°).

Discussion

We have emphasized the importance of the fold test [Graham, 1949 and McFadden and Jones, 1981], not so much to constrain the age of the magnetizations, but more to constrain the directions. Only if the paleohorizontal is confidently known is it justifiable to calculate pole positions. If pole positions are calculated indiscriminantly from magnetization directions that cannot be confidently referred to their correct paleohorizontal, then clearly the apparent polar wander path (APWP) may have little meaning in terms of past plate motions and may simply reflect the structural deformation of the fold belt. In Figure 5, pole positions derived from the magnetization directions discussed above are plotted with their confidence areas. The confidence areas reflect the precision of the estimates but do not account for any possible systematic errors such as incorrectly assigned paleohorizontal. It is clear that many poles from the LFB form a swathe that runs east west, while the poles from the AB form a swathe that runs north south. Therefore our model is not inconsistent with the data and provides a simpler (and apparently more reasonable) interpretation. On this basis we prefer this interpretation.

Directions considered here as probably being

secondary yield a pole position (LO) that closely resembles the postdeformation pole positions from the Snowy River Volcanics (SO) and the Buchan Caves Limestone (SB). This is further evidence that the LFB has suffered a considerable amount of remagnetization (Figure 5). Poles that may be confidently referred to a paleohorizontal form an APWP in the vicinity of presentday Antarctica. The implied approximate coincidence of SiluroDevonian paleomagnetic directions with the presentday geomagnetic field direction has possibly also served to confuse workers, and again emphasizes the necessity for fold tests.

The proposed track may be representative of Australia as a whole, although recently Hurley and Van der Voo [1986] report results from the Late Devonian Canning limestone reef complex of Western Australia, that may be providing the first definitive evidence of large scale motion between the LFB and the rest of Australia. Much of the reef complex is essentially flat lying, having been deposited on a stable Precambrian basement, and is not amenable to a fold test. However, a consistent sequence of normal and reverse polarities has been identified and is interpreted as evidence for the magnetization being original. The low to intermediate inclination of the magnetization certainly rules out a Permian or Late Carboniferous origin, but does not rule out an Early Carboniferous time of acquisition. There is evidence of six separate diagenetic events, including recrystallization and periods of fluctuating oxidation/reduction conditions [Hurley, 1986]. While it is not clear how much time these events represent, the reef complex was not exhumed and eroded until the Late Carboniferous which suggests that the possibility of Early Carboniferous magnetization must be entertained. The confirmation of a Late Devonian age for the Canning limestone pole position will require a major reevaluation of tectonics in the Late Devonian/Early Carboniferous of southeastern Australia.

To solve the problem of displaced terranes in eastern Australia and the broader problem of Gondwanaland and Laurasian tectonics, future paleomagnetic work in Australia and other Gondwanaland continents, must not only address the question of the age of magnetizations, but also that of the correct paleohorizontal. When dealing with folded strata it is essential to execute properly designed fold tests [Graham, 1949 and McFadden and Jones, 1981].

Acknowledgements. The authors are grateful for the constructive comments made on an early version of this paper by Mr David Clark, Drs Phil McFadden and Chris Powell and Mr Bruce Goleby. Drs Mart Idnurm and Erwin Scheibner are sincerely thanked for refereeing the paper.

References

Armstrong, R.L. and Stewart, A.J., 1975. Rubidium-strontium dates and extraneous argon in the Artlunga Nappe Complex, Northern Territory, J. Geol. Soc. Aust., 22, 103-115.

Bailey, M.E. and Hale, C.J., 1981. Anomalous magnetic directions recorded by laboratory induced chemical remanent magnetisation, Nature, 294, 739-741.

Beck, M.E., 1976. Discordant paleomagnetic pole positions as evidence of regional shear in the western Cordillera of North America, Am. J. Sci., 276, 694-712.

Collinson, D.W., 1983. Methods in Rock Magnetism and Palaeomagnetism, Chapman and Hall, London, 503pp.

Collinson, D.W., Creer K.M. and Runcorn, S.K. 1967. Methods in Palaeomagnetism, Elsevier 609pp.

Cox, A.V., 1957. Remanent magnetism of lower to middle Eocene basalt flows from Oregon, Nature, 179, 685-686.

Embleton, B.J.J., 1972a. The palaeomagnetism of some Palaeozoic sediments from central Australia, J. Proc. R. Soc. N.S.W., 105, 86-93.

Embleton, B.J.J., 1972b. The palaeomagnetism of some Proterozoic-Cambrian sediments from the Amadeus Basin, Central Australia, Earth Planet. Sci. Lett., 17, 217-226.

Embleton, B.J.J., 1981. A review of the paleomagnetism of Australia and Antarctica, in Paleoreconstruction of the Continents, ed. by M.W.McElhinny and D.A.Valencio, Geodyn. Ser. vol.2, pp.77-92.

Embleton, B.J.J., McElhinny, M.W., Crawford, A.R. and Luck, G.R., 1974. Palaeomagnetism and the tectonic evolution of the Tasman Orogenic Zone, J. Geol. Soc. Aust., 21, 187-194.

Everitt, C.W.F. and Clegg, J.A., 1962. A field test for palaeomagnetic stability, Geophys. J. R. astr. Soc., 6, 312- 319.

Goleby, B.R., 1980a. Early Palaeozoic Palaeomagnetism of South East Australia, Ph.D. thesis, ANU, 162pp.

Goleby, B.R., 1980b. Early Palaeozoic palaeomagnetism in south east Australia, J. Geomag. Geoelectr., 32, Suppl. III, 11-21.

Graham, J.W., 1949. The stability and significance of magnetism in sedimentary rocks, J. Geophys. Res., 54, 131-167.

Hurley, N.F. and Van der Voo, R., 1986. Paleomagnetism of Canning Basin reefal limestones-new constraints on the Gondwana apparent polar wander path, abstr. 12th Int. Sed. Congress, Canberra, Australia, p.149.

Hurley, N.F., 1986. Diagenesis of Devonian reefs in the Oscar Range, Canning Basin, Western Australia, abstr. 12th Int. Sed. Congress, Canberra, Australia, p.148.

Idnurm, M., 1985. Late Mesozoic and Cenozoic palaeomagnetism of Australia-II. Implications for geomagnetism and true polar wander, Geophys. J. R. astr. Soc., 83, 419-433.

Irving, E., 1964. Palaeomagnetism and its Application to Geological and Geophysical Problems, John Wiley, pp. 399.

Irving,E., 1979. Paleopoles and paleolatitudes of North America and speculations about displaced terranes, Can. J. Earth Sci., 16, 669-694.

Kent,J.T.,Briden,J.C. and Mardia,K.V., 1983. Linear and planar structure in ordered multivariate data, as applied to progressive demagnetization of palaeomagnetic remanence, Geophys. J. R. astr. Soc., 75, 593-621.

Kirschvink,J.L., 1978. The Precambrian-Cambrian boundary problem: primary, secondary, and transitional palaeomagnetic directions from the Amadeus Basin, Central Australia, Geol. Mag., 115, 139-150.

Kirschvink,J.L., 1980. The least squares line and plane and the analysis of palaeomagnetic data, Geophys. J. R. astr. Soc., 62, 699-718.

Klootwijk,C.T., 1980. Early Palaeozoic palaeomagnetism in Australia, III, Middle to early-Late Cambrian results from the Amadeus Basin, Northern Territory, Tectonophysics, 64, 249-332.

Klootwijk,C.T., 1985. Palaeomagnetism of the Tasman Fold Belt: Indication for Mid-Carboniferous large-scale southward displacement of the New England region, 3rd Circum-Pacific Terrane Conference, pp.124-127.

McClelland Brown,E., 1983. Palaeomagnetic studies of fold development and propagation in the Pembrokeshire Old Red Sandstone, Tectonophysics, 98, 131-149.

McElhinny,M.W., 1964. Statistical significance of the fold test in palaeomagnetism, Geophys. J. R. astr. Soc., 8, 338-340.

McElhinny,M.W., 1973. Palaeomagnetism and Plate Tectonics, Cambridge Academic Press, 358pp.

McElhinny,M.W. and Embleton,B.J.J., 1974. Australian palaeomagnetism and the Phanerozoic plate tectonics of Eastern Gondwanaland, Tectonophysics, 22, 1-29.

McFadden,P.L. and Jones,D.L., 1981. The fold test in palaeomagnetism, Geophys. J. R. astr. Soc., 67, 53-58.

McWilliams,M.O., 1977. Late Precambrian Palaeomagnetism of Australia and Africa, unpub. Ph.D. thesis, Australian National University, 138pp.

McWilliams,M.O., 1983. Paleomagnetism and the motion of large and small plates, Rev. Geophys. Space Phys., 21, 644-651.

Miller,J.D. and Kent,D.V., 1985. Another look at the Catskill, Eos Trans. AGU, 66, 257.

Morel,P. and Irving,E., 1978. Tentative palaeocontinent maps for the early Phanerozoic and Proterozoic, J. Geol., 86, 535-561.

Powell,C.McA., 1984. Terminal fold-belt deformation: Relationship of the mid-Carboniferous megakinks in the Tasman fold belt to coeval thrusts in cratonic Australia, Geology, 12, 546-549.

Powell,C.McA., Cole,J.P. and Cudahy,T.J., 1985. Megakinking in the Lachlan Fold Belt, J. Struct. Geol., 7, 281-300.

Schmidt,P.W., 1982. Linearity spectrum analysis of multicomponent magnetizations and its application to some igneous rocks from south-eastern Australia, Geophys. J. R. astr. Soc., 70, 647- 665.

Schmidt,P.W. and Embleton,B.J.J., 1985. Prefolding and overprint magnetic signatures in Precambrian (~2.9-2.7 Ga) igneous rocks from the Pilbara Craton and Hamersley Basin, NW Australia, J. Geophys. Res., 90, 2967-2984.

Schmidt,P.W.,Embleton,B.J.J.,Cudahy,T.J. and Powell,C.McA., 1986a. Prefolding and premegakinking magnetizations from the Devonian Comerong Volcanics, New South Wales, Australia, and their bearing on the Gondwana Pole Path, Tectonics, 5, 135-150.

Schmidt,P.W., Embleton,B.J.J. and Palmer, H.C., 1986b. Pre- and post-folding magnetizations from the Devonian Snowy River Volcanics and Buchan Caves Limestones, Victoria, Geophys. J. R. astr. Soc., in press.

Schmidt,P.W. and Morris,W.A., 1977. An alternative view of the Gondwana Paleozoic apparent polar wander path, Can. J. Earth Sci, 14, 2674-2678.

Van der Voo,R., 1981. Paleomagnetism of North America: A brief review, in Paleoreconstruction of the Continents, Geodyn. Ser., volume 2, edited by M.W.McElhinny and D.A.Valencio, pp.159-176, AGU, Washington, D.C.

Van der Voo,R. and McCabe,C., 1985. Remagnetizations in the Eastern U.S. and Canada, Eos Trans. AGU, 66, 257.

Vick,H., Channell,J.E.T. and Opdyke,N.D., 1985. Paleomagnetism of low grade metamorphic rocks from the Carolina Slate Belt, Eos Trans. AGU, 66, 258.

Watts, D.R., 1985. Palaeomagnetic resetting in the Barrovian zones of Scotland and its relationship to the late structural history, Earth Planet. Sci. Lett., 75, 258-264.

TERRANES OF THE WONOMINTA BLOCK, FAR WESTERN NEW SOUTH WALES

E.C. Leitch, B.D. Webby, K.J. Mills, and P. Kolbe

Department of Geology and Geophysics, University of Sydney, New South Wales 2006, Australia

Abstract. The Wonominta block in far western New South Wales comprises Precambrian and Lower Paleozoic rocks that belong to four tectonostratigraphic terranes. The Mount Wright terrane is made up of a Lower and lower Middle Cambrian sequence of volcanics and sedimentary rocks. The Wertago terrane includes three principal units, multi-deformed metamorphics and lower grade simply folded slate, sandstone, metabasalt and calcareous rocks of Proterozoic age, and (?) Cambrian sandstone and siltstone. The Kayrunnera terrane also consists of (?) Cambrian sandstone and siltstone, but aeromagnetic data suggest that it has a basement of different character from that of the Wertago terrane. These three terranes are separated by major long-active faults. Each was deformed in the Middle-Late Cambrian and the oldest linkage sequence is provided by uppermost Cambrian-Lower Ordovician paralic and shelf craton-derived clastic sediments. Isolated inliers of slate, redeposited sandstone and mafic volcanics further north that are possibly as young as Early Ordovician are grouped in the Tibooburra terrane. These rocks were folded and suffered low-grade regional metamorphism prior to emplacement of a 410 Ma granite. The rocks of the Wertago and Kayrunnera terranes are probably autochthonous for they show several features linking them to the Precambrian craton exposed further west. However the Mount Wright terrane, which possibly incorporates arc volcanics, is seemingly anomalously located and is regarded as suspect. The oldest linkage sequence tying the Tibooburra terrane to adjacent terranes are Mesozoic, but regional paleogeographic considerations suggest that it had assumed its present position at least by the Devonian.

Introduction

The Wonominta block encompasses Precambrian rocks and associated Lower Paleozoic strata that emerge from beneath a cover of Devonian and younger sediments in the Wilcannia - Tibooburra - Scopes Range region of far western New South Wales (Fig. 1). To the west the older rocks disappear beneath the thick (up to 7000m [Packham in Cooper, 1975, Fig. 7]) Devonian fill of the Bancannia trough which masks their contact with the Proterozoic rocks of the Broken Hill block, and to the east they are buried by rocks of similar age that comprise the Darling basin [Evans, 1977]. Mesozoic units surround inliers of the block around Tibooburra and completely cover it further north, whereas to the south it is obscured by Quaternary materials of the Darling River flood plain and Tertiary strata of the Murray basin. Neither Precambrian nor Lower Cambrian rocks have been conclusively identified east of the Wonominta block in New South Wales [Webby in Cooper and Grindley, 1982].

Three generalised sequences of pre-Devonian rocks have been recognised in previous, mostly reconnaissance, investigations of the block [Warris, 1967; Rose and Brunker, 1969; Cooper, 1975; Brunker et al., 1971]. Tightly folded, steeply dipping slaty and schistose metasedimentary rocks and associated mafic and silicic meta-igneous rocks have been grouped in the Wonominta Beds, mostly accorded a Precambrian age, although shown on some maps as ranging into the Cambrian [e.g. Pogson, 1972]. Lower and Middle Cambrian altered volcanic rocks, silicic tuff, limestone and mostly fine-grained clastic sedimentary rocks constitute the Gnalta Group [Rose, 1967; Kruse, 1982]. Widespread paralic to moderately deep water dominantly quartzose conglomerate, sandstone and siltstone that rest unconformably on the two preceding divisions are widely distributed. In the west they collectively make up the Mootwingee Group of latest Cambrian - Early Ordovician age [Webby, 1978; Webby et al., 1981], but further east the oldest part of the correlative sequence is early Late Cambrian (Mindyallan) in age.

Stevens [1985] discussed some aspects of aeromagnetic data collected by the Australian Bureau of Mineral Resources, and attempted to infer basement character over much of the area surrounding the Wonominta block. His account is mainly couched in terms of structural blocks defined in as yet unpublished studies by E. Scheibner of the N.S.W. Geological Survey. Scheibner [1985] grouped all of the rocks in a single terrane which he termed the Wonominta terrane but our recent investigations show that there is great variety in the Wonominta Beds and that there is no

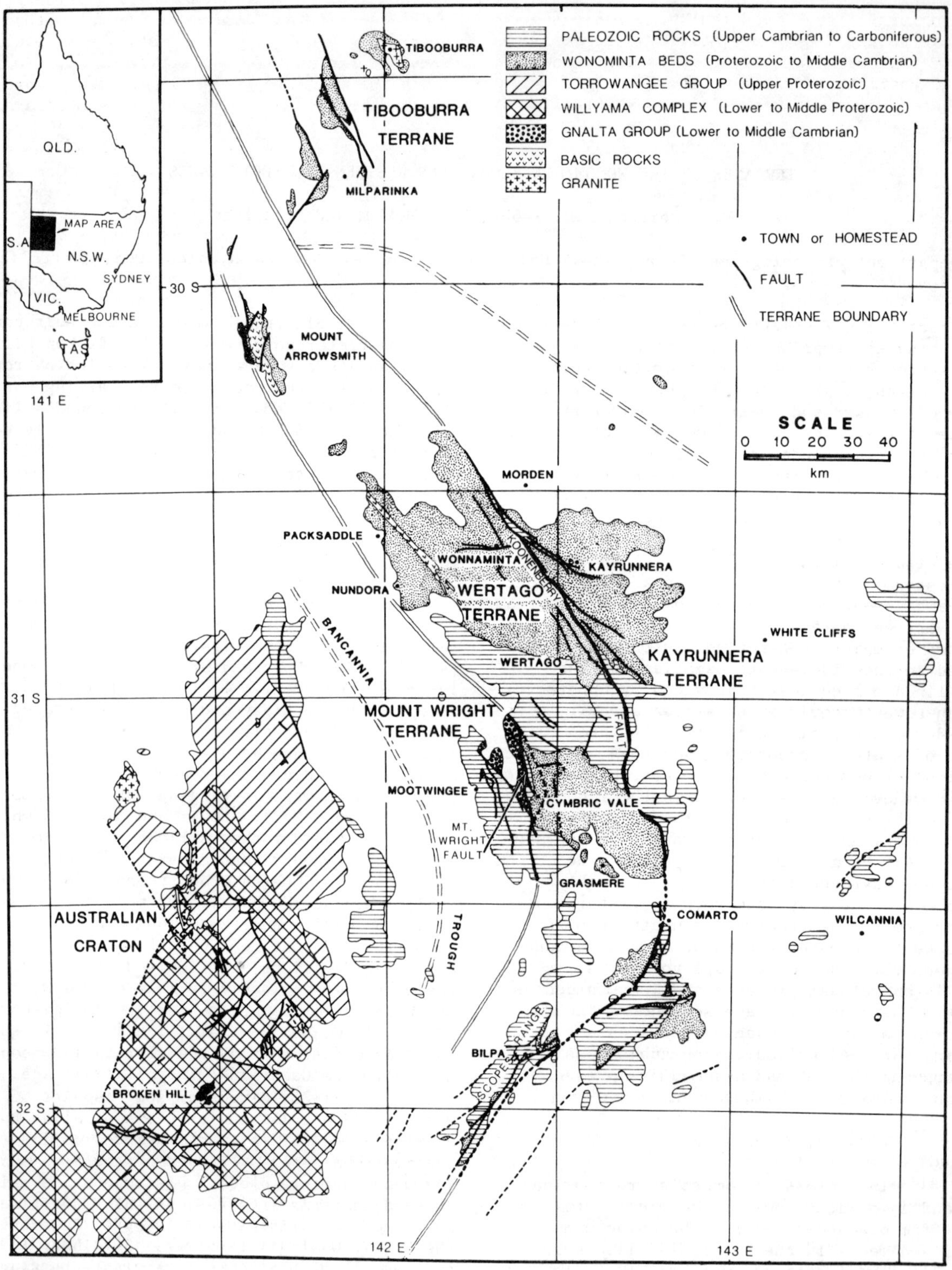

Fig. 1. Terranes and inferred terrane boundaries in far western New South Wales.

clear stratigraphic relationship between any of these rocks and those of the Gnalta Group. In addition large, long-lived faults, notably those that make up the Koonenberry fault system, separate regions of greatly contrasting aeromagnetic character suggesting they separate different basement provinces. This information is the basis for our recognition of four pre-Late Cambrian terranes in the region (Fig. 1).

Wonominta Block Terranes

In drawing our terrane map of far western New South Wales we have placed great importance on two fault systems that are well defined in the central part of the region but the northwestern and southwestern extensions of which are drawn mainly from aeromagnetic data (Fig. 1). The Mount Wright fault is readily mapped in the area west of Cymbric Vale where it brings Gnalta Group rocks against foliated metamorphics of the Wonominta Beds. The Upper Devonian - (?)Carboniferous Ravensdale Formation is broken by the fault but rocks of this unit are recognised on both sides of the fracture and the main movements were much earlier. Two serpentinite bodies occur in foliated Wonominta Beds adjacent to the Mount Wright fault. These appear to be isofacial with the surrounding rocks suggesting relatively early emplacement along an already existing line of weakness. To the north the Mount Wright fault merges with a fault cutting across the Wertago terrane just before disappearing under Quaternary sands. Its further continuation is projected parallel to the aeromagnetic strike, which positions it west of Mount Arrowsmith and collinear with fractures separating Wonominta rocks in this outlier from Lower Paleozoic strata [Wopfner, 1967; Edwards, 1980]. South of Cymbric Vale the Mount Wright fault lies within Devonian rocks and then passes below Quaternary cover. Extension parallel to the main aeromagnetic trends suggests that it swings southwest to run west of Bilpa (Fig. 1).

The Koonenberry fault is marked by a brittle crush zone in rocks as young as Late Devonian or Early Carboniferous. In the Wonnaminta area it forms a major structural boundary parallel to the strike of the Wonominta Beds and coincides with a marked change in aeromagnetic character of the basement rocks. Northnorthwest trending magnetite-bearing slate and phyllite and metamorphosed mafic tuffs and lavas intercalated in the sequence on the southwest side of the fault produce a parallel pattern of long linear high-amplitude anomalies that are locally interrupted by anomalies associated with magnetic metadolerite and metagabbro bodies. The linear trends are also broken by across strike faults, many of which show only small displacements but are in places the emplacement sites of magnetic lamprophyre dikes. Northeast of the Koonenberry fault the aeromagnetic pattern is diffuse. Most anomalies are of low intensity and they show no obvious preferred trend. The boundary separating these contrasting domains continues northnorthwest along the line of the Koonenberry fault at least as far as the Queensland border. To the south the magnetic grain swings into a north-south orientation around Comarto and then the anomalies on the west take on a more compressed form and assume a northeast strike in the Bilpa area. The Koonenberry fault is continued along a set of fractures that parallel the anomalies and on this basis can be traced to the South Australia border south of Broken Hill.

Mount Wright Terrane

Lower to Middle Cambrian rocks exposed west of the Mount Wright fault and collectively referred to the Gnalta Group comprise the Mount Wright terrane. The oldest rocks are probably unnamed limestone, dolomite and tuff exposed immediately west of the fault. These give way to the west to the Mount Wright Volcanics, a sequence of altered mafic volcanics with intercalated fine sandstone, siltstone, dolomite and limestone of total thickness at least 2500m. Conformably overlying the Mount Wright Volcanics are some 1500m of silicic ash-fall and possible distal ash-flow tuff, sandstone and archaeocyathid limestone of the Cymbric Vale Formation. According to Kruse [1982] archaeocyath faunas from the limestone lenses of the upper Mount Wright Volcanics and lower Cymbric Vale Formation suggest an Early Cambrian (late Atdabanian or possibly early Lenian) age, and a younger assemblage from the upper Cymbric Vale Formation, an Early Cambrian (early Lenian) age. This is succeeded by about 100m of limestone and shale of the Coonigan Formation of Middle Cambrian (Ordian-Templetonian) age [Öpik, 1967], and then by an unnamed silicic ash-flow and volcanic conglomerate formation about 200m thick. Published regional maps and sections [e.g. Rose, 1968] show the Gnalta Group resting on Wonominta Beds but nowhere can this be demonstrated, and we have been unable to discover any outcrops of the latter unit west of the Mount Wright fault.

The rocks of the Gnalta Group are simply folded and have suffered only burial metamorphism; they show no signs of the penetrative deformation and regional metamorphism that characterises the Wonominta Beds. Some folding occurred prior to the end of the Cambrian for some of the formations are conformably overlain by uppermost Cambrian (Payntonian) to basal Ordovician quartz-rich clastic rocks of the Nootambulla Sandstone [Webby, 1978].

Middle Cambrian strata, mainly siltstone, sandstone and conglomerate, that occur in the Mount Arrowsmith district [Wopfner, 1967; Warris in Packham, 1969] west of the projected position of the Mount Wright fault, are included in the Mount Wright terrane. The sequence, some 600m thick, is in part a time correlative of the Coonigan Formation [Wopfner, 1967] and is notable for the presence of a large amount of volcanic detritus.

Rocks of the Mount Wright terrane disappear to the west beneath uppermost Cambrian-Lower Ordovician and Devonian sequences. They extend west of the Lawrence fault of Powell et al. [1982] and are presumed to be in fault contact with Proterozoic rocks of the Australian craton beneath the Bancannia Trough (Fig. 1). The possibility that rocks of the Gnalta Group occur in the deeper levels of the trough was first suggested by Freeman [1966]. While an oil exploration hole (Bancannia South No. 1) was abandoned on entering andesite and dacite underlying Middle Devonian sandstone and shale [Cooper, 1975], correlation of the volcanic rocks with those of the Gnalta Group is highly speculative. There is no compelling evidence for deposition in the trough before the Devonian, and it does not seem to have prevented the spread of craton-derived sediment onto the Wertago and Kayrunnera terranes during the Cambrian and Early Ordovician.

Wertago Terrane

The Wertago terrane comprises all pre-Upper Cambrian rocks lying between the Mount Wright and Koonenberry faults, formerly entirely grouped within the Wonominta Beds. It is a composite terrane in which the oldest 'lower' unit appears to be everywhere in fault contact with younger rocks. This unit includes multiply deformed foliated phyllites and schists of pelitic, psammitic and mafic igneous parentage, together with uncommon foliated limestone and more massive metaquartzite, metarhyolite and amphibolitic greenstone. Metamorphic grade ranges from greenschist (chlorite zone) to lower amphibolite, with biotite zone rocks being widely present. These rocks occur in belts adjacent to the Koonenberry fault north of Wertago and to the Mount Wright fault north and south of Cymbric Vale, and also outcrop in a belt from north of Grasmere, south to near Bilpa. There is no internal evidence for the age of these rocks and their position as the oldest unit in the terrane is based on a complex structural history and metamorphic grade.

Lying between the belts of the multiply deformed rocks is a very thick fault-bounded 'middle' unit of simply but tightly to isoclinally folded and cleaved slate, slaty sandstone, dolomite, limestone, pyritic siltstone and quartzite. Two horizons of mafic volcanics have been mapped within this unit, one extending south from Mount Arrowsmith and passing a little east of Packsaddle and east of Nundora (Fig. 1), and the other exposed west and southwest of Wonnaminta. Rocks in the western belt are commonly highly vesicular and associated with limestone lenses. Pillow lavas occur in this belt at Mount Arrowsmith where the rocks are considered to be alkaline basalts [Edwards, 1979].

Folds affecting the middle unit are upright, have a large amplitude and wavelength, and are associated with a sub-vertical axial plane cleavage. A weak crenulation cleavage is only developed in a zone of secondary folding east of Nundora. Regional metamorphic grade does not exceed that of the chlorite zone.

Near Nundora redeposited quartz-rich lithic sandstone and shale conformably overlying older slates are the basis for the youngest 'upper' unit found in the Wertago terrane. Correlative strata form a wide fault-bounded block between Wertago and Wonnaminta. In both areas this unit is characterised by graded sandstone beds and well preserved associated sedimentary structures. Paleocurrent indicators suggest a southern source. In the Nundora area, the presence of gritty layers near the base of some sandstone beds, and both red and green shales, suggest a more proximal and oxidised depositional environment to the west. A recent find by K.J. Mills of siliceous sponge spicules in a grey cleaved siltstone near Wonnaminta homestead, suggests this unit is of Early Paleozoic (possibly Early-Middle Cambrian) age. A notable feature of the unit is its flat magnetic signature.

Now altered dolerite and gabbro bodies, at least some of which are tholeiitic [Edwards, 1979], have been emplaced in all three units. Near Nundora and Packsaddle syenitic intrusions are common and lamprophyre dikes occur around Wonnaminta. Altered rhyolite dikes and sills are also widespread.

Stratigraphic constraints on the ages of the three units in the Wertago terrane are only provided by unconformably overlying rocks. In the south near Bilpa there are polymictic conglomerates at the base of the Scopes Range Beds of latest Cambrian-early Ordovician age [Webby in Cooper and Grindley, 1982], and in the north at Mount Arrowsmith, the Lower Ordovician deposits of Wopfner [1967]. Lithological and structural comparisons suggest a correlation of the oldest unit with the Lower Proterozoic Willyama Supergroup of the Broken Hill district [cf. Rose and Brunker, 1969]. The middle slaty unit may be a correlative of the Upper Proterozoic-(?)Lower Cambrian Torrowangee Supergroup that unconformably overlies the Willyama Supergroup, and it has been suggested that the mafic volcanics are correlatives of the basal Willangee Volcanics of this sequence [Rose and Brunker, 1969], although they are at a relatively higher stratigraphic level in the middle Wonominta unit. The youngest unit, with its monotonous redeposited sandstone beds bears some similarity to parts of the Lower-Middle Cambrian Kanmantoo Group of South Australia [Daily and Milnes, 1972].

Kayrunnera Terrane

The Kayrunnera terrane lies east of the Koonenberry fault in an area where most of the pre-Upper

Cambrian rocks are redeposited quartz-rich lithic sandstone and slate (Copper Mine Range Beds of Pogson and Scheibner [1971]). They have yielded rare trace fossils [Webby, 1984) and siliceous sponge spicules (D.B. Collins, pers. comm. 1984) suggesting an Early or Middle Cambrian age. Lithological similarities exist between these rocks and those of the upper unit of the Wertago terrane. Despite this there is a great change in magnetic character across the fault, suggesting a contrast in basement character, and so a terrane boundary is postulated. It is conceivable that the change is a product of the presence of a very thick section of isoclinally folded weakly magnetic rocks masking a basement similar to that further west but this lacks confirmatory evidence. Little-known schists that occur east of the postulated position of the Koonenberry fault south of Comarto argue against the concept of a deeply buried basement.

Pogson and Scheibner [1971] suggest that the Copper Mine Range Beds accumulated in Early Cambrian times offshore from the setting in which the Gnalta Group of the Mount Wright terrane was contemporaneously deposited. However the sedimentary detritus in the Copper Mine Range Beds is of typical cratonic derivation and lacks a significant volcanic input rendering any sedimentary linkage with the Gnalta Group tenuous.

The rocks of the Copper Mine Range Beds were tightly folded, cleaved and subject to low grade (chlorite zone) regional metamorphism prior to the deposition of unconformably overlying fluvial and marine strata, the oldest of which are the lower Upper Cambrian (Mindyallan) deposits at Kayrunnera [Rose and Brunker, 1969].

Tibooburra Terrane

Quartzose and lithic sandstone and slaty siltstone sequences in the inliers around Tibooburra superficially resemble those of the Copper Mine Range Beds but are distinguished from them by a dark grey to black rather than a green colour, by the presence of intercalated mafic volcanics, and by rare occurrences of branching and sinuous trace fossils. The last were described by Webby [1984] who indicated they were most likely to be of Cambrian or even Early Ordovician age. There seems to be considerable lithological resemblances between these deposits and the successions of the Dullingari Group, especially the 'basinal' sequences of quartzitic sandstone, dark grey siltstone and black shales, which have been intersected in wells in the northeastern corner of South Australia [Gatehouse, 1986]. A few of the Dullingari cores have produced Ordovician graptolites (A.A. Opik and P.J. Jones, B. Daily, unpublished data, as cited by Gatehouse [1986]). The general facies of the Tibooburra rocks is also similar to that of Lower Paleozoic rocks widespread further east in New South Wales and Victoria, where thick often poorly fossiliferous sequences of turbidite sandstone and slate are common. For this reason we place these rocks in a separate terrane.

The rocks of the Tibooburra terrane are closely to tightly folded fine-grained rocks that possess a steeply inclined axial plane cleavage. The age of this deformation is poorly constrained but occurred prior to the emplacement of a massive granite pluton at Tibooburra that has yielded biotite Rb/Sr ages of 410 Ma [Shaw in Cooper and Grindley, 1982]. Although fine-grained post-tectonic intrusives of silicic composition are found further south the Tibooburra pluton is the only coarse-grained granitic body known from the Wonominta block.

The nature and position of the boundary between the Tibooburra and Kayrunnera terranes is obscured by Mesozoic and younger deposits north of Morden (Fig. 1). In the south it may lie immediately west of a northwest trending line that marks the boundary between the magnetically diffuse domain immediately east of the Koonenberry fault and a domain of more linear character further east that projects towards Milparinka. However the trace of this line is confused by a prominent isolated anomaly around which the boundary possibly swings to meet the projected trace of the Koonenberry fault north of Mount Arrowsmith (Fig. 1, see also Stevens [1985]).

Overlap Sequences

Rocks of the Mount Wright, Wertago and Kayrunnera terranes are unconformably overlain by marine and non-marine mainly quartzose clastic sediments with thin intercalations of shallow-water carbonate. The oldest unit of this sequence resting on the Mount Wright terrane is the uppermost Cambrian (Payntonian) Nootumbulla Sandstone [Webby, 1978], the lowest formation of the 1200m+ Mootwingee Group which may range up into the base of the Middle Ordovician. Conglomerates of comparable age, becoming increasingly quartzose upwards, (lower part of the Scopes Range Beds as presently defined [see Webby, 1983]), rest on the Wertago terrane in the Bilpa (Scopes Range) district, but at Mount Arrowsmith the oldest cover rocks are Lower Ordovician orthoquartzite and shale [Wopfner, 1967]. Conglomerate, sandstone and siltstone of the Cupala Creek Formation [Powell et al., 1982] rest unconformably on the Kayrunnera terrane south of Kayrunnera and probably correlative sandstone near Kayrunnera contains faunas as old as Mindyallan (early Late Cambrian) in age. They are part of a sequence which extends into the basal Ordovician [Webby et al., 1987]. The depositional system in which these rocks accumulated was such as to prevent the development of very extensive lithostratigraphic units, but their common provenance, the faunal similarities, and the consistent paleogeographic relationships indicate they are all part of the same general sequence and provide a minimum age for the amalgamation of the three terranes. The provenance of the linkage sequence indicates that by this time these terranes were adjacent to the Australian craton. They were linked in a continental shelf-like configuration at the margin of Gondwana, usually referred to as the Gnalta Shelf [Webby, 1978; 1983], a region dominated by a major delta with successive influxes of clastic material spreading over the shelf from the south and west. In the latest Cambrian to earliest Ordovician interval, shallower 'inner detrital' trilobite faunas established over extensive areas of the Mount Wright and Wertago terranes and in the southern part of the Kayrunnera terrane, whereas deeper 'outer detrital' assemblages lived in the area of the Kayrunnera terrane near Kayrunnera.

Discussion

If the lithological correlation of the two older units of the Wertago terrane with those of the Lower Proterozoic-basal Paleozoic cratonic margin in the Broken Hill district is valid, then this terrane is essentially autochthonous. The oldest unit (Willyama Supergroup correlatives) comprises old continental crust that subsided during the formation of a passive continental margin in the Late Proterozoic, and the Torrowangee Supergroup correlative strata are probably slope and rise sediments that accumulated above the foundering blocks. The inferred basement contrasts across the Koonenberry fault may have developed very early in the history of the passive margin, perhaps by juxtaposition of dissimilar elements by strike-slip movements associated with the initial rifting. The bulk of such movements had certainly been completed by the Early Cambrian if the projected correlation between the Copper Mine Range Beds and the upper unit of the Wertago terrane is accepted. The age of the rifting that formed the ocean floor towards which the younger two units in the Wertago terrane prograded is uncertain but it is most unlikely that it was as young as the Precambrian-Cambrian boundary favoured by von der Borch [1980] for breakup further southwest in the Adelaide Fold Belt [see also Degeling et al., 1986]. Although the mafic volcanics in the middle unit of the Wertago terrane have been considered to indicate an island arc [Scheibner, 1972] we believe their geochemical character [Edwards, 1979] is more akin to that of non-orogenic rocks and interpret them either as early alkaline rift volcanics or products of intra-plate volcanism.

Some workers have linked the mafic volcanics of the Wertago terrane with those of the Mount Wright terrane [Scheibner, 1972; Pogson, 1972] but reports disçounting alleged Cambrian fossils associated with the former [Scheibner, 1976] weakens this correlation. Our observations indicate that the stratigraphic setting of the two units is very different and we consider correlation most unlikely. The present position of the Mount Wright terrane is anomalous, sitting as it does with its volcanic-dominated Lower Cambrian sequence between the cratonic mass of Australia and probable Lower Cambrian rocks of the Wertago and Kayrunnera terranes which contain abundant cratonic detritus but none of volcanic provenance. Although it is possible that these sequences are of basal Cambrian (Tommotian) age and predate the Mount Wright volcanism, their thickness and the presence of Atdabanian faunas in the upper part of the Mount Wright Volcanics makes this unlikely. Thus the Mount Wright terrane may be allochthonous.

Alteration of the volcanics of the Mount Wright terrane has so far prevented determination of their magmatic affinities and hence their tectonic significance. The wide range of rock types indicated possibly argues for a convergent margin setting and they are possibly representatives of the very widely distributed Cambrian volcanics of this type that occur in terranes close to the cratonic margin from north Queensland to Tasmania. However, whereas such volcanism in Tasmania, and less certainly in western Victoria, may have extended into the Early Cambrian [Cooper and Grindley, 1982], that in north Queensland is probably Late Cambrian [Henderson, 1986].

The isolated position of the Tibooburra terrane makes it very difficult to link its development to that of the other terranes in far western New South Wales. Although linkage sequences are no older than Mesozoic regional geological considerations, especially inferred Devonian paleogeographies suggest that it was probably in its present position relative to the other terranes by the Early Devonian at the latest.

Acknowledgments. Our research in far western New South Wales has been supported by the Australian Research Grants Scheme (grant 8219257) and funds from the University of Sydney. We are indebted to many landholders for help and access and particularly thank the Anderson's of Cymbric Vale, the Clarke's of Kayrunnera, the Halfpenny's of Nundora, the Keurschner's formerly of Bilpa, the Langford's of Wertago, and the White's of Wonnaminta, for their interest, kindness and hospitality. The N.S.W. National Parks and Wildlife Service allowed entry to Mootwingee National Park.

References

Brunker, R.L., A.C. Offenberg, and G. Rose, Koonenberry Sheet, 1:500 000 Geological Series (1st edn.), N.S.W. Geol. Surv., Sydney, 1971.

Cooper, P.F., Broken Hill Sheet, 1:250 000 Geological Series, Explanatory Notes, 56pp, N.S.W. Geol. Surv., Sydney, 1975.

Cooper, R.A., and G.W. Grindley (editors), Late Proterozoic to Devonian sequences of southeastern Australia, Antarctica and New Zealand and their correlation, Geol. Soc. Aust. Spec. Publ. 9, 103pp, 1982.

Daily, B., and A.R. Milnes, Revision of the

stratigraphic nomenclature of the Cambrian Kanmantoo Group, South Australia. J. Geol. Soc. Aust., 19, 197-202, 1972.

Degeling, P.R., L.B. Gilligan, E. Scheibner and D.W. Suppel, Metallogeny and tectonic development of the Tasman Fold Belt System in New South Wales, Ore Geology Rev., 1, 259-313, 1986.

Edwards, A.C., Tectonic implications of the immobile trace-element geochemistry of mafic rocks bounding the Wonaminta block, J. Geol. Soc. Aust., 25, 459-465, 1979.

Edwards, A.C., The geology and petrology of the igneous rocks of the Mt. Arrowsmith and related areas of north western New South Wales, Ph.D. Thesis, University of N.S.W., 287pp, 1980.

Evans, P.R., Petroleum geology of western New South Wales, Aust. Petroleum Explor. Assoc. J., 17, 42-49, 1977.

Freeman, R.N., The Lake Frome embayment area, Aust. Petroleum Explor. Assoc. J., 1, 93-99, 1966.

Gatehouse, C.G., The geology of the Warburton Basin, Aust. J. Earth Sci., 33, 161-180, 1986.

Henderson, R.A., Geology of the Mt Windsor subprovince - a Lower Palaeozoic volcano-sedimentary terrane in the northern Tasman orogenic zone, Aust. J. Earth Sci., 33, 343-364, 1986.

Kruse, P.D., Archaeocyathan biostratigraphy of the Gnalta Group at Mount Wright, New South Wales, Palaeontographica, Abt. A, 177, 129-212, 1982.

Öpik, A.A., The Ordian Stage of the Cambrian and its Australian Metadoxididae, Bull. Bur. Min. Res. Geol. Geophys. Aust., 92, 133-169, 1967.

Packham, G.H., (editor), The geology of New South Wales, J. Geol. Soc. Aust., 16, 1-654, 1969.

Pogson, D.J., Geological map of New South Wales, Scale 1:1 000 000, N.S.W. Geol. Surv., Sydney, 1972.

Pogson, D.J., and E. Scheibner, Pre-Upper Cambrian sediments east of Copper Mine Range, New South Wales, N.S.W. Geol. Surv. Quart. Notes, 4, 3-8, 1971.

Powell, C.M., G. Neef, D. Crane, P. Jell, and I.G. Percival, Significance of Late Cambrian (Idamean) fossils in the Cupala Creek Formation, northwestern New South Wales, Linn. Soc. N.S.W. Proc., 106, 127-150, 1982.

Rose, G., Broken Hill Sheet, 1:250 000 Geological Series (Prov. edn), N.S.W. Geol. Surv., Sydney, 1968.

Scheibner, E., The Kanmantoo pre-cratonic province in New South Wales, N.S.W. Geol. Surv. Quart. Notes, 7, 1-10, 1972.

Scheibner, E., Explanatory notes on the tectonic map of New South Wales, N.S.W. Geol. Surv., Sydney, 283pp, 1976.

Scheibner, E., Suspect terranes in the Tasman Fold Belt System, eastern Australia, in Tectonostratigraphic terranes in the Circum-Pacific region, edited by D.B. Howell, Circum-Pacific Council Energy Min. Res., Earth Sci. Ser., 1, 493-514, 1985.

Stevens, B.P.J., Preliminary interpretation of regional basement geology in northwestern New South Wales, N.S.W. Geol. Surv. Quart. Notes, 61, 9-22, 1985.

von der Borch, C.C., Evolution of Late Proterozoic to Early Palaeozoic Adelaide Foldbelt, Australia: comparisons with post-Permian rifts and passive margins, Tectonophysics, 70, 115-134, 1980.

Warris, B.J., The Palaeozoic stratigraphy and palaeontology of north-western New South Wales, Ph.D. thesis, University of Sydney, 470pp, 1967.

Webby, B.D., History of the Ordovician continental platform and shelf margin of Australia, J. Geol. Soc. Aust., 25, 41-63, 1978.

Webby, B.D., (editor), The Ordovician System in Australia, New Zealand and Antarctica: correlation chart and explanatory notes, Internat. Union Geol. Sci. Publ., 6, 1-64, 1981.

Webby, B.D., Lower Ordovician arthropod trace fossils from western New South Wales, Proc. Linn. Soc. N.S.W., 107, 59-74, 1983.

Webby, B.D., Precambrian-Cambrian trace fossils from western New South Wales, Aust. J. Earth Sci., 31, 427-437, 1984.

Webby, B.D., Q.Z. Wang, and K.J. Mills, Upper Cambrian - basal Ordovician trilobites from western New South Wales, Alcheringa, (in press), 1987.

Wopfner, H., Cambro-Ordovician sediments from the north-eastern margin of the Frome embayment (Mt Arrowsmith, N.S.W.), J. Proc. Roy. Soc. N.S.W., 100, 163-177, 1967.

EARLY PALEOZOIC BACK-ARC DEFORMATION IN THE LACHLAN FOLD BELT, SOUTHEASTERN AUSTRALIA: IMPLICATIONS FOR TERRANE TRANSLATIONS IN EASTERN GONDWANALAND

Christopher L. Fergusson

Department of Earth Sciences, Monash University, Clayton, Vic. 3168, Australia[1]

Abstract. Much of the eastern part of the Lachlan Fold Belt in southeastern Australia consists of the Benambra terrane which in the Ordovician contained an eastward facing island arc with a back-arc basin to the west. Part of the back-arc basin is preserved in the Tabberabbera Belt of the southwestern Benambra terrane. The southern Tabberabbera Belt is dominated by east-west trending folds that formed in a Silurian back-arc deformation and is divided by a fault (the Wonnangatta Line) into two major domains. The southwestern domain is dominated by shallowly plunging steeply inclined folds with wavelengths of up to 10 km. These folds gradually tighten towards the north where an early recumbent fold is rooted in a zone of tectonic melange up to 2 km in width at the domain boundary. The horizontal enveloping surfaces and the low amplitude-to-wavelength ratios of regional folds are consistent with a thin-skinned tectonic style and fold vergence indicates tectonic transport towards the south. The northeastern domain consists of upright close to isoclinal folds with large amplitude-to-wavelength ratios reflecting thick-skinned deformation. There is no overall sense of tectonic transport and younger structures have caused significant multiple deformation.

The Silurian back-arc deformation developed due to Chilean-style oblique-slip subduction along the convergent boundary east of the Ordovician island arc. Regional dextral shear caused the east-west folding and also formed several major northwest-trending strike-slip faults that caused southwest translation of the Benambra terrane. Structural contrasts across the Tabberabbera Belt reflect southwesterly thickening pre-Ordovician crust that probably formed as the western passive continental margin of the back-arc basin. The tectonic pattern is consistent with strike-slip translation and convergence along the Early Paleozoic eastern Gondwanaland margin. There is little evidence for accretion of major allochthonous oceanic terranes as has occurred in western North America.

Introduction

The Lachlan Fold Belt of southeastern Australia is characterized by an extensive sequence of Ordovician cratonic-derived deep-marine turbidites [Cas, 1983]. Baillie [1985] recognized a Paleozoic suture between these deep-marine rocks and essentially shallow-marine rocks farther west. The suture has been traced southwards through Tasmania and into northern Victoria Land. The location of the suture in the Victorian part of the Lachlan Fold Belt is not well established. In this region Fergusson et al. [1986] recognize four tectono-stratigraphic terranes in the sense of Jones et al. [1983]. From east to west these terranes are (Figure 1): (a) the Benambra terrane, a complex mainly arc/back-arc basin assemblage; (b) the Melbourne terrane, a continental margin sequence; (c) the Ballarat terrane, another continental margin sequence; and (d) the Grampians-Ararat terrane, a complex association of sedimentary sequences and volcanics.

The Benambra terrane contains an Ordovician mafic-intermediate volcanic island arc bordered to the west by a quartz-rich turbidite sequence in a back-arc basin [Cas et al., 1980; Powell, 1983, 1984]. Basement to the arc and back-arc basin is not exposed but isotope characteristics of Siluro-Devonian granitoids appear to indicate that the Lachlan Fold Belt is partly underlain by Precambrian crust [McCulloch and Chappell, 1982]. In the southwestern Benambra terrane much of the back-arc basin was affected by a major Early to Middle Silurian deformation [Bolger, 1982].

[1] Now at Department of Geology, University of Wollongong, P.O. Box 1144, Wollongong, N.S.W. 2500, Australia.

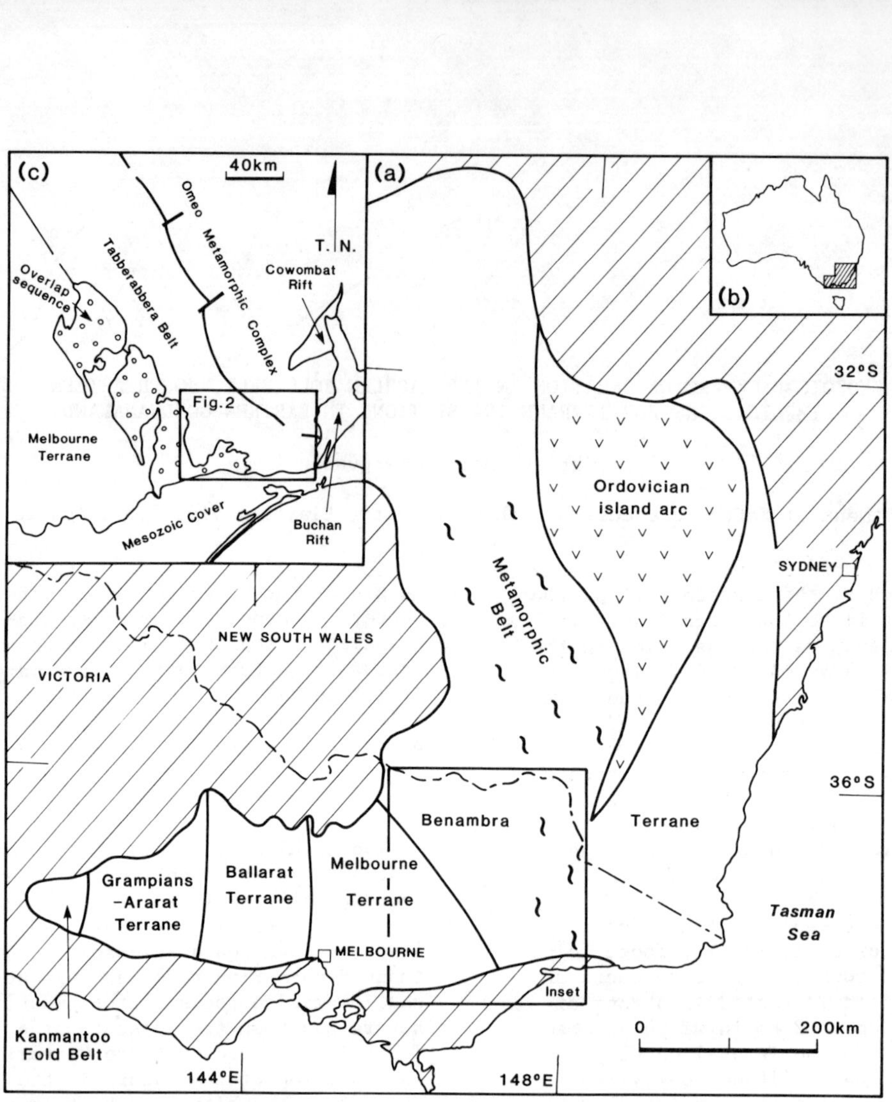

Fig. 1. (a) Map of the Lachlan Fold Belt showing the distribution of tectono-stratigraphic terranes, the Ordovician island arc and metamorphic belt [after Fergusson et al., 1986]. (b) Location of (a) in Australia. (c) Major subdivisions of the southwestern Benambra terrane and location of Figure 2.

The Late Silurian to early Middle Devonian history of the Benambra terrane is dominated by widespread silicic to intermediate volcanism with deposition in subaerial to shallow-marine environments and in deep-marine troughs [Cas, 1983]. Abundant granitoids were intruded in this phase which was terminated by regional deformation and uplift in the Middle Devonian [Fergusson et al., 1986].

The Melbourne and Ballarat terranes west of the Benambra terrane consist mainly of continental margin sequences overlying Cambrian rocks of oceanic affinities [Crawford and Keays, 1978]. These terranes were thrust over the western portion of the Benambra terrane during a major Middle Devonian deformation. Fergusson et al. [1986] suggest dextral displacement occurred between the Melbourne and Benambra terranes in the Silurian and possibly in the Early Devonian.

The aim of this paper is to describe the style of the back-arc deformation in the southwestern Benambra terrane and to examine its significance for terrane translations along the ancient continental margin of eastern Gondwanaland. Structures formed in the Early to Middle Silurian back-arc deformation are well exposed in dissected plateau country in the Tabberabbera Belt of eastern Victoria (Figures 1 and 2). Most of the Tabberabbera Belt consists of Ordovician quartz-rich turbidites that are metamorphosed to sub-greenschist facies (chlorite-white mica). The southern Tabberabbera Belt is subdivided by a major fault line, named herein the Wonnangatta Line, into two structural domains. Ordovician rocks of both domains are overlain, with a high

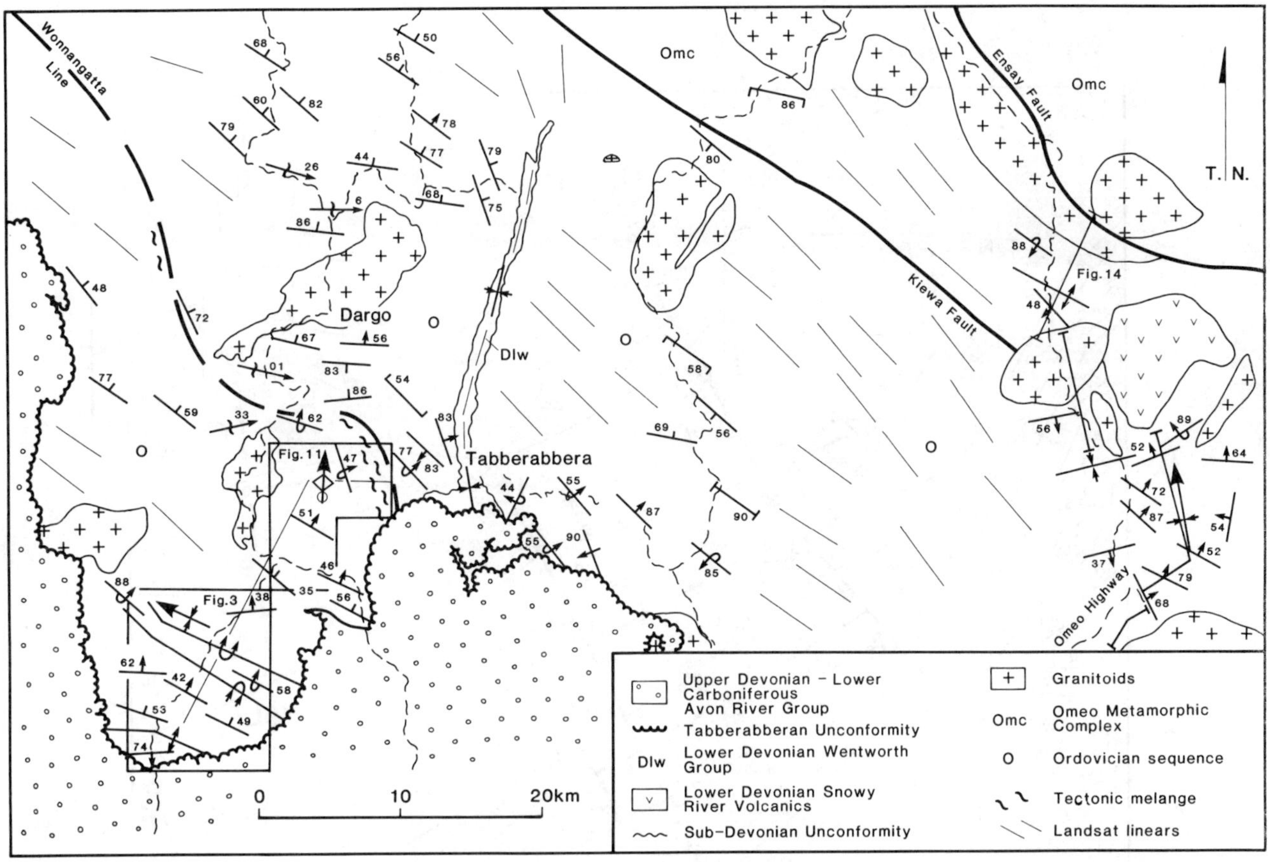

Fig. 2. Map of the southern Tabberabbera Belt. See Figure 3 for a key to structural symbols.

angular unconformity, by a sequence of Upper Devonian to Lower Carboniferous fluviatile sediments and volcanics that overlap the Melbourne and Benambra terranes. Northeast of the Tabberabbera Belt are greenschist to upper amphibolite facies metamorphics of the Omeo Metamorphic Complex which are derived from the Ordovician turbidite sequence. Upper Silurian sequences in fault contact with the deformed and metamorphosed Ordovician sequence of the metamorphic belt were, in part, derived from an uplifted source of quartz-rich sandstones and granite and do not show the deformation and metamorphism affecting the underlying rocks [Bolger, 1982].

Structure Southwest of the Wonnangatta Line

This domain is dominated by one main phase of east-southeast trending folds that are cross-cut by granitoids of Early Devonian age [Richards and Singleton, 1981]. The structure is described from two areas that have been mapped in detail.

Freestone Creek Area

In this area, the Ordovician sequence consists of three units (Figures 3 and 4) which from oldest to youngest are: (i) 2500m of sandstone-rich turbidite facies which include amalgamated sandstones, graded thick sandstone-thin mudstones and less common mudstone-thin sandstones, (ii) 550m of passage beds, and (iii) 800m of mudstone.

F_1 mesoscopic folds are widely distributed and are east-southeast trending shallowly plunging folds with steep to vertical axial surfaces (Figures 3, 5, 6, 7, 8, 9 and 10a-c). Interlimb angles range from broad to tight but are commonly close to open. Amplitude-to-wavelength ratios are typically between 0.1 and 0.6. In the F_1 folds competent sandstone layers have rounded class 1B profiles and the less competent mudstone layers have class 1C and class 3 profiles. Folds in medium- to thick-bedded sandstones, with no or very thin interbedded mudstones, are angular and are commonly disrupted by fault accommodation structures along their axial surfaces. These latter

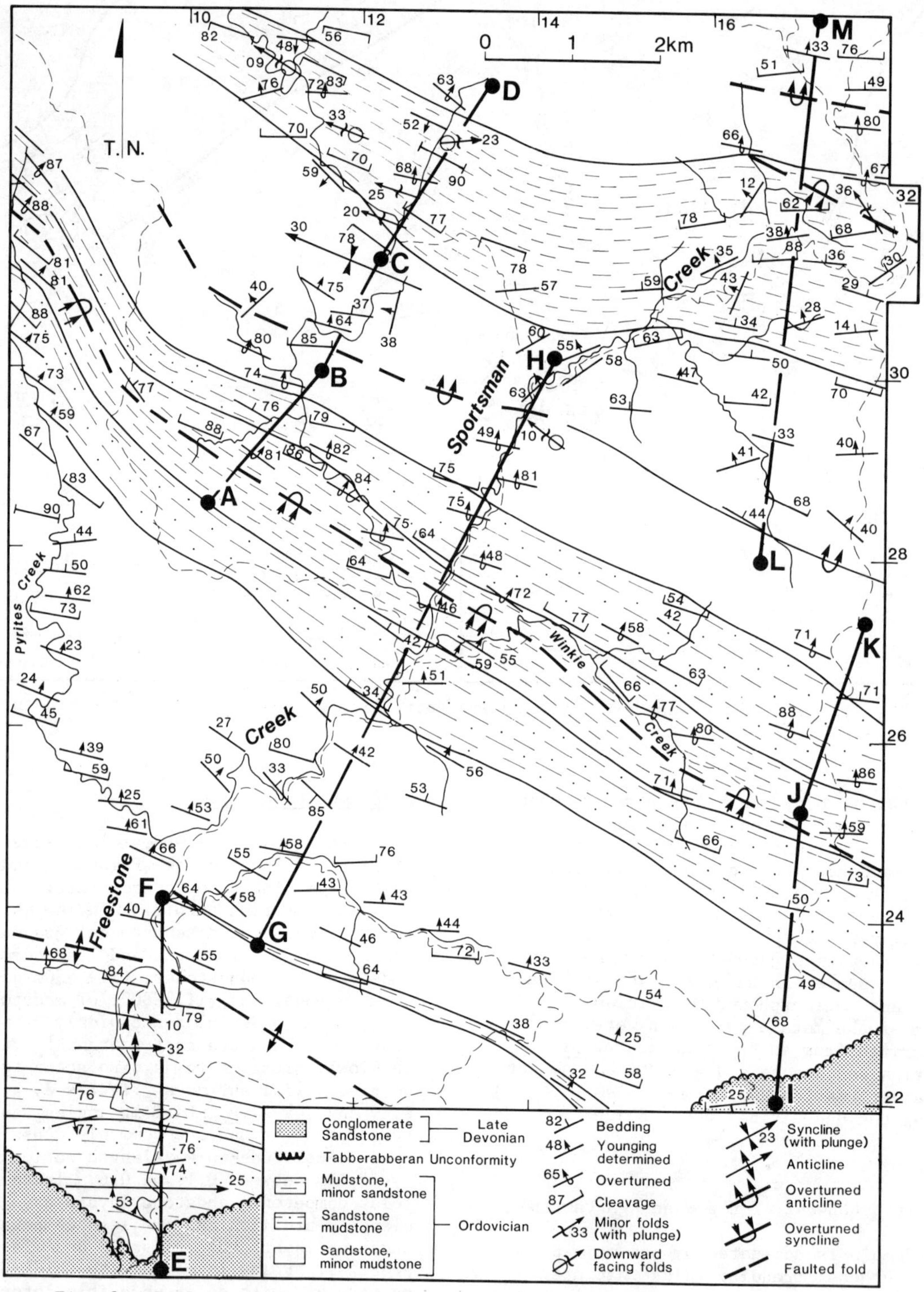

Fig. 3. Geology of the Ordovician sequence in the Freestone Creek area.

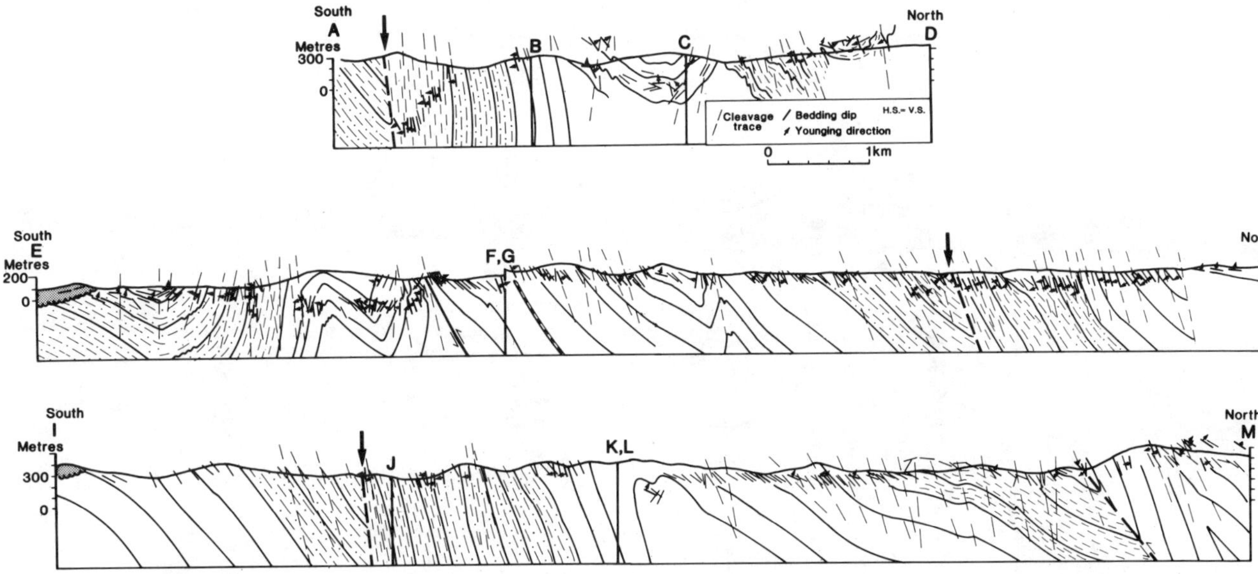

Fig. 4. Regional cross sections of the Freestone Creek area. See Figure 3 for locations and a key to symbols. The arrow above each section indicates the location of the map-scale syncline near Winkie Creek.

folds are often disharmonic and die out up and down their fold profiles.

Within domain 8 at least five F_1 mesoscopic fold pairs are downward facing (Figures 5, 8 and 9). Other F_1 folds in domain 8 are upward facing but no early fold hinges responsible for these facing changes have been found. Most of these F_1 downward facing folds occur on the overturned limb of a large asymmetric fold that has a vergence consistent with other map-scale F_1 folds (Figure 8, cross section XY). A single downward facing synform occurs in Sportsman Creek (GR 139 295) outside of domain 8 (Figure 3).

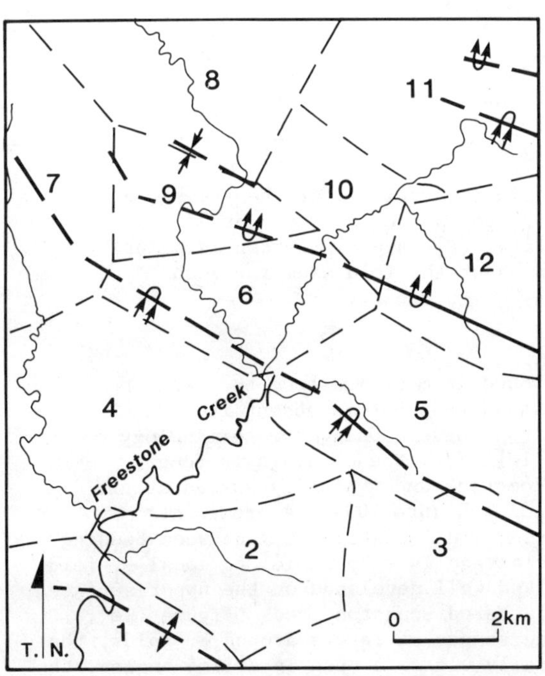

Fig. 5. Structural domains for the Ordovician sequence in the Freestone Creek area.

Map-scale structure of the Freestone Creek area is dominated by seven east-southeast trending main-phase folds. These are shallowly plunging folds spaced between 1 and 5.2 km (Figure 3) and with axial surfaces either vertical or steeply inclined to the north. Interlimb angles range from open to tight and tighten towards the north (Figure 4). Hinges are narrow and limbs are long and planar. The upper mudstone unit occupies the cores of map-scale synclines whereas main-phase anticlines contain the lower sandstone-rich unit. Thus enveloping surfaces of the map-scale folds are subhorizontal (Figures 3 and 4).

Several of the main-phase folds have major faults developed along their axial surfaces and these contain crush zones up to 2 m in width. These are reverse faults except for the faulted syncline at Winkie Creek (Figure 3) which has an apparently anomalous normal movement (Figure 4). One explanation for this anomaly is that the fault formed as a thrust and was reactivated in an episode of north-south extension and graben formation which affected the region during deposition of the Upper Devonian-Lower Carboniferous Avon River Group [e.g. Powell, 1984, Fig. 221].

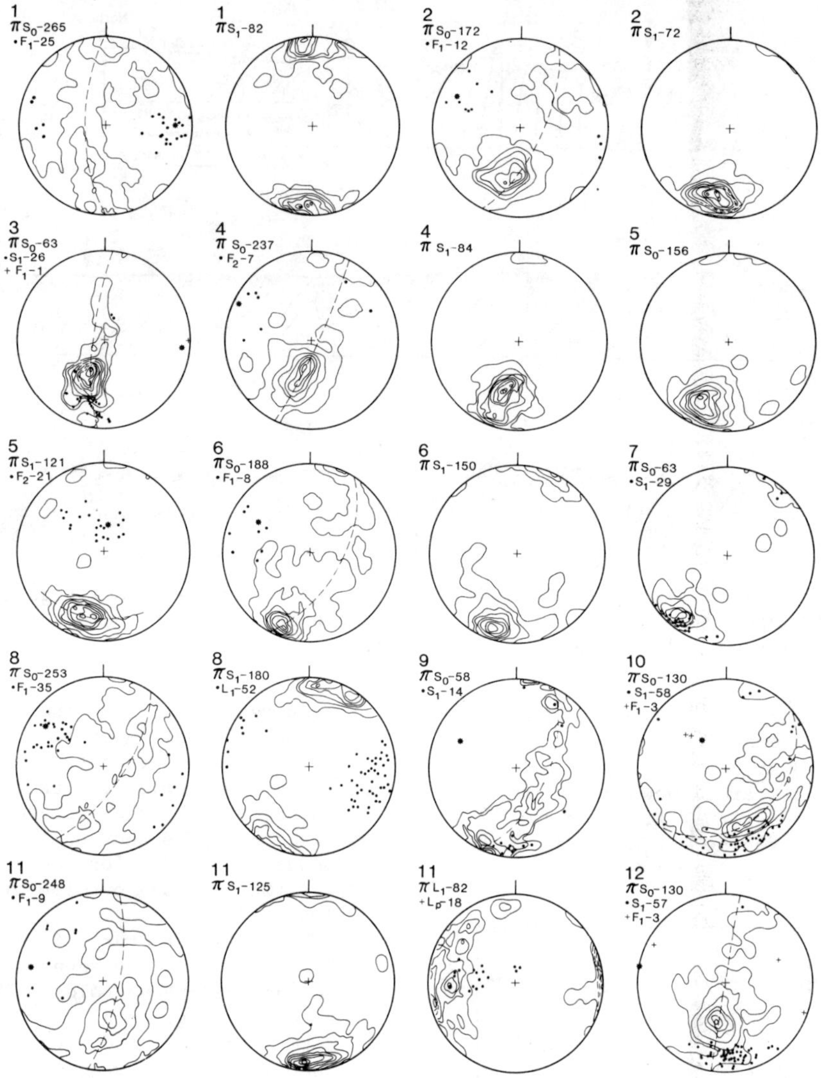

Fig. 6. Lower hemisphere equal-area stereographic projections for the Freestone Creek area. Bedding (So) poles and cleavage (S_1) poles are contoured at 3% intervals per 1% area. Cleavage (S_1) and fold axes (F_1) are also shown as poles to planes and lines respectively. Aterisks represent the fold axes for each domain.

The map-scale folds have relatively low amplitude-to-wavelength ratios which imply a shallow depth to folding (5-8 km). They have narrow, steep, south-younging limbs and shallow long north-younging limbs which indicates a vergence to the south and tectonic transport from the north (Figure 4).

A weakly developed bedding-parallel, grain alignment, compaction fabric [e.g. Maltman, 1981] is present in most mudstone beds (Figure 10c). The fabric is contained within randomly oriented mudstone clasts in intraformational breccias and is therefore of depositional origin. S_1 cleavage is developed in most mudstone beds and some thin sandstone beds but is absent from thick sandstone beds. Using the terminology of Powell [1979], the cleavage is rough, anastomosing and typically spaced at the 1-5 mm scale (Figure 10d). A prominent intersection lineation (L_1) between bedding and cleavage is subparallel to local F_1 fold axes and well developed on the upper surfaces of weathered sandstone beds (Figure 10e).

The shapes of layers around F_1 folds, the generally close to open interlimb angles, the weak cleavage fabric and the lack of appreciable cleavage fans indicates that

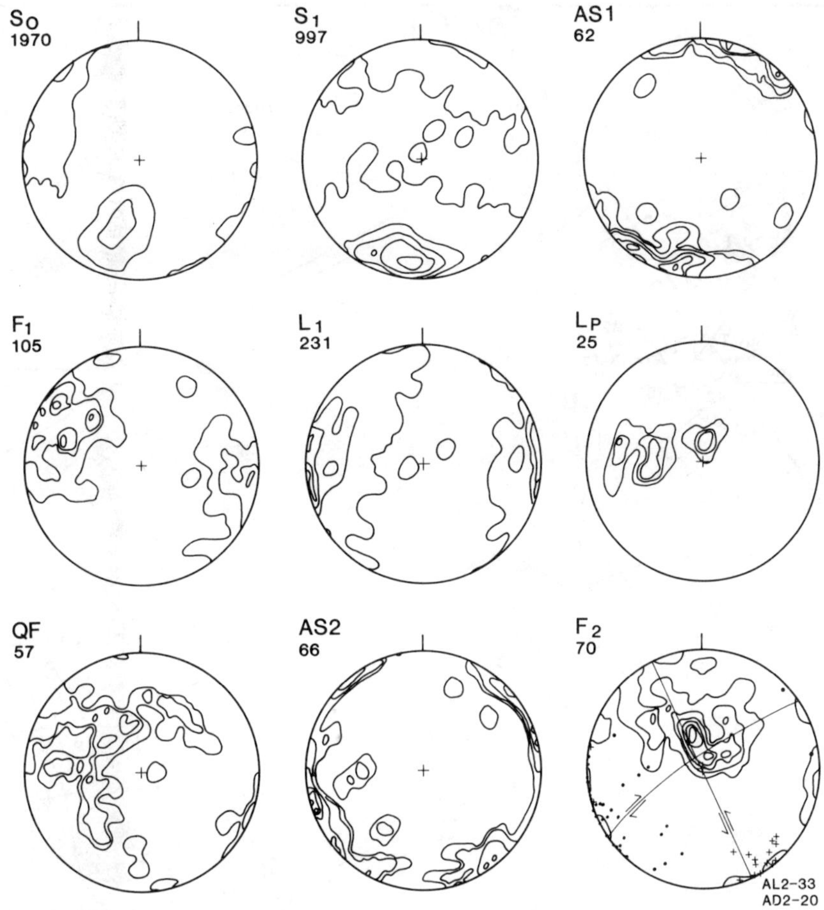

Fig. 7. Synoptic lower hemisphere equal-area stereographic projections for structural elements in the Freestone Creek area. Points are contoured at 3% intervals for 1% area. The structural elements include: bedding poles (S_o), cleavage poles (S_1), axial surface poles (AS1), F_1 fold axes, L_1 cleavage-bedding intersections, pencil fragment elongations (L_p), quartz fibre lineations (QF) on bedding planes, axial surface poles (AS2), and F_2 fold axes. Axial surfaces of F_2 folds are plotted as crosses (+) for dextral sense of vergence (AD2) and plotted as dots (.) for a sinistral sense of vergence (AL2). Orientations of the F_2 folds indicate they formed due to east-west shortening.

there has been minimal flattening associated with the development of these folds. Fold styles are consistent with folding by flexural slip and elastic buckling in competent layers. Quartz fibres are developed on bedding planes on fold limbs and many of these indicate movement directions perpendicular to fold axes (Figure 7).

Within domain 8 several of the downward facing fold couples have anomalous planar S_1 cleavage (Figures 5, 8 and 9). These are asymmetric close to tight fold couples with one near-vertical limb and two shallower limbs. Cleavage dips subparallel to or slightly steeper than bedding on the near-vertical limb (Figure 9). The lack of cleavage fans implies the non-axial planar cleavage was superimposed late in the folding episode [cf. Gray, 1981].

Cleavage has a similar non-axial planar relation with the main-phase syncline at Winkie Creek and several main-phase folds north of this structure (Figures 3 and 4). On the overturned limbs S_1 cleavage dips steeper than the overturned bedding (Figure 10f). This relation is anomalous since in a normal bedding-cleavage relation the cleavage should dip shallower than an overturned fold with a shallow plunge.

The lone F_1 downward facing synform in Sportsman Creek occurs on the northern overturned limb of the regional F_1 syncline at Winkie Creek. Because the synform is downward facing it, and the associated axial

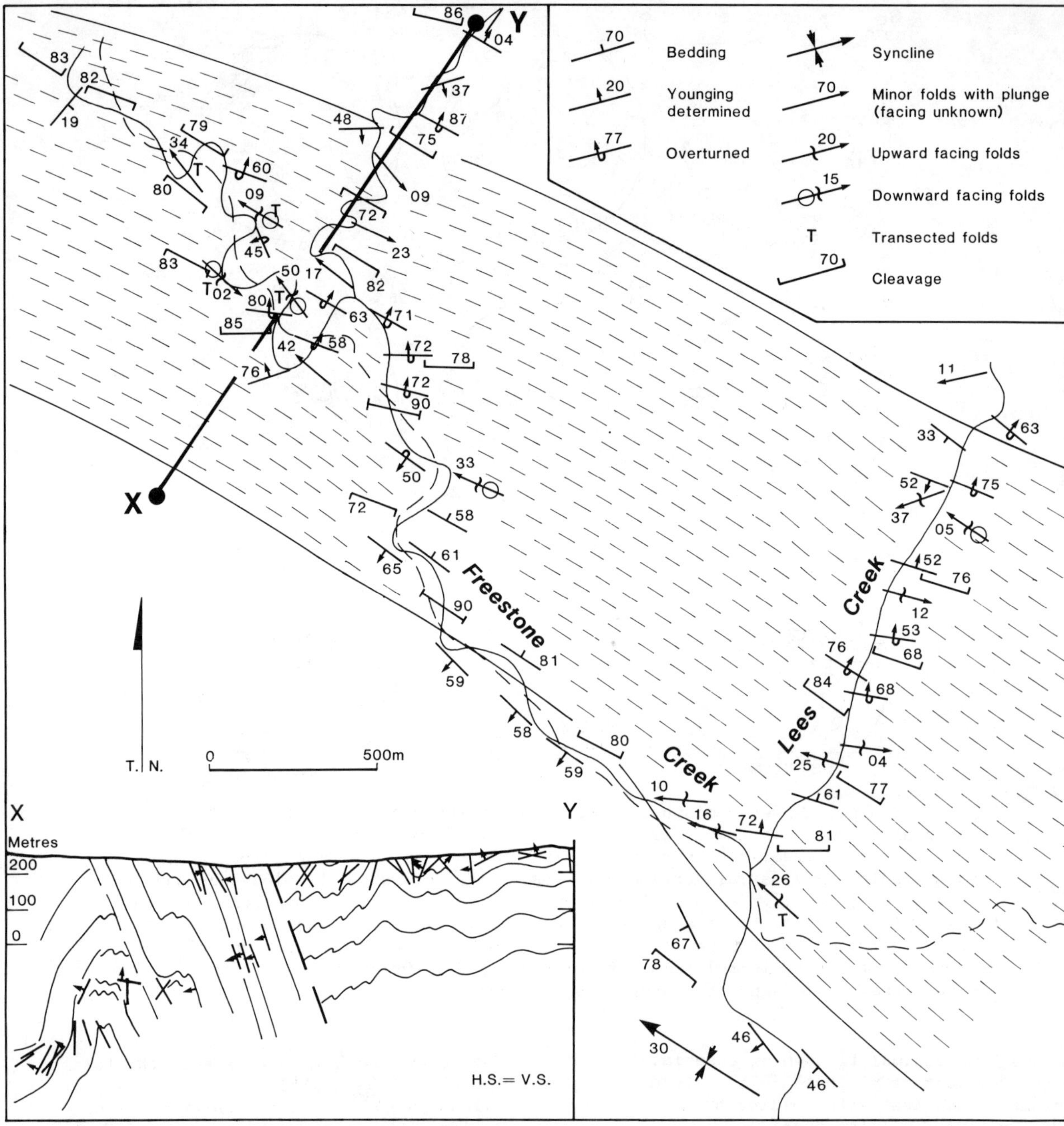

Fig. 8. Detailed map of domain 8 showing the distribution of downward facing F_1 folds and F_1 folds transected by the S_1 cleavage.

surface S_1 cleavage, must have formed late in the F_1 folding after overturning of the regional syncline limb. Cleavage on the overturned limb has non-axial planar relations with the regional syncline and is axial planar to the downward facing synform (Figure 4, cross section GH). These relations indicate the cleavage was superimposed late in the folding episode in a history of non-coaxial deformation [cf. Powell, 1974].

Bulgaback Creek Area

The Ordovician turbidite sequence in the Bulgaback Creek area (Figures 11 and 12) consists of monotonous interbedded quartz sandstone and mudstone. A mappable unit

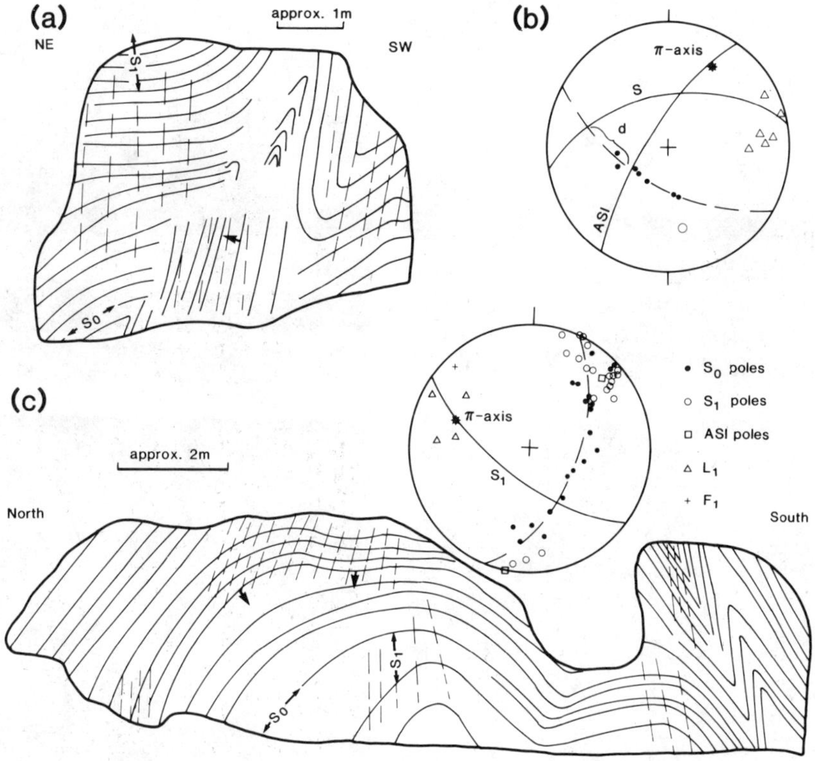

Fig. 9. (a) A sketch from a photograph of a downward facing F_1 fold couple with a transecting S_1 cleavage (GR 109 333). (b) Sketch from photographs of a downward facing F_1 antiform at GR 116 328 with lower hemisphere equal-area stereographic projection. Younging has been determined at the position of the arrows from sawn faces of sandstone layers showing truncated micro-crosslamination (Bouma C layers).

with 500 m of dominantly thick-bedded sandstone occurs in lower Bulgaback Creek (Figure 11).

The regional structure of this area is dominated by several main-phase folds. In the southwest there is a tight z-shaped fold couple which plunges shallowly to the southeast and is overturned to the southwest (Figure 12). A spaced cleavage is sporadically developed on the limbs of these folds. Towards the north structural trends gradually swing from east to southeast to south (Figure 11). Along Bulgaback Creek the map-scale structure is dominated by an open main-phase downward-facing antiform with several z-shaped parasitic folds on its western limb (Figures 11 and 12). These folds have an axial planar S_1 spaced cleavage and are downward facing as they fold an early recumbent structure, which has no associated cleavage (Figures 10g and 12). The early fold is an anticline and is downward facing to the west. Farther up Bulgaback Creek there is a downward facing s-shaped parasitic fold pair developed on the upright limb of the early structure (Figure 12, cross section QR). From the cross section ST (Figure 12) it is clear the early recumbent fold is rooted in the Wonnangatta Line (see below).

A number of younger groups of less intense structures are superimposed on the F_1 folds in the southwestern domain (Figure 7). For example, the swing in trends from northwest to west along Bulgaback Creek reflects regional refolding by northerly trending folds that are more prominent in the northeastern domain.

Structural Evolution of the Southwestern Domain

Overall, there is an increase in structural complexity from south to north in this domain. This is interpreted as the result of a single episode of progressive deformation involving tectonic transport southwards from the Wonnangatta Line. The early recumbent fold in Bulgaback Creek originated from strong shortening near the Wonnangatta Line and was refolded later in the deformation.Farther south, the local downward facing folds and non-axial planar S_1

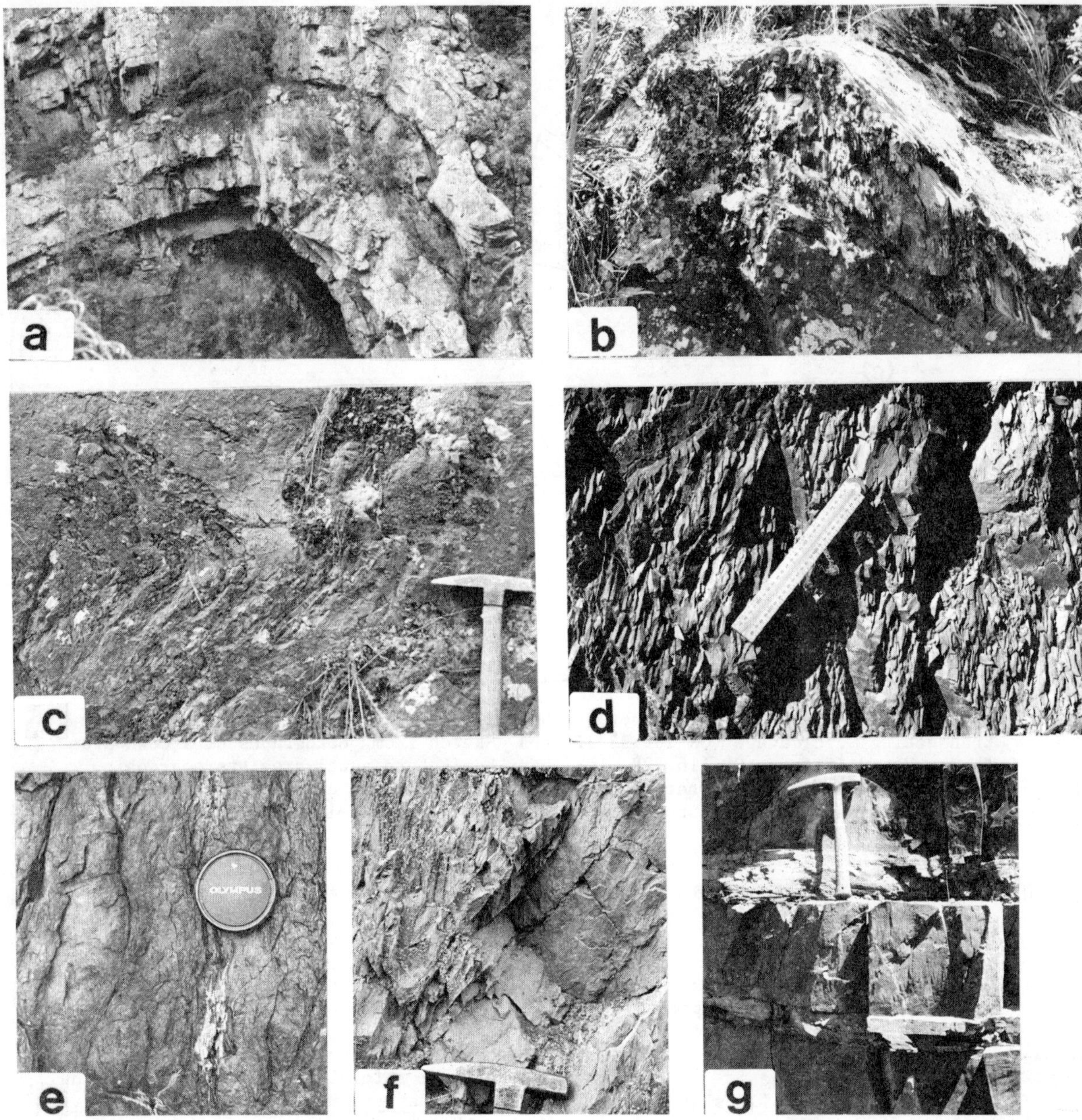

Fig. 10. (a) Open steeply inclined F_1 fold in amalgamated sandstones from domain 1, southern Freestone Creek area. Width of photograph is approximately 30 m. (b) Close upright F_1 fold from domain 8, Freestone Creek area. Cleavage appears to be axial planar in the profile plane but in the lateral view it is clearly non-axial planar. Lens cap is 5.5 cm across. (c) Overturned F_1 fold from domain 11 showing an early bedding-parallel compaction fabric in mudstones. Hammer head is 12.5 cm in width. (d) Rough S_1 cleavage oblique to bedding in mudstones. Intersection of cleavage with joints forms a pencil structure. Rule is 30 cm in length. (e) Anastomosing bedding-cleavage intersections (L_1) developed on the top of a sandstone bed. Lens cap is 5.5 cm across. (f) Anomalous bedding-cleavage relation from the northern overturned limb of the map-scale syncline near Winkie Creek (GR 132284). Cleavage (S_1) dips steeper than overturned bedding. (g) Flat-lying overturned beds from near the core of the downward facing antiform in Bulgaback Creek.

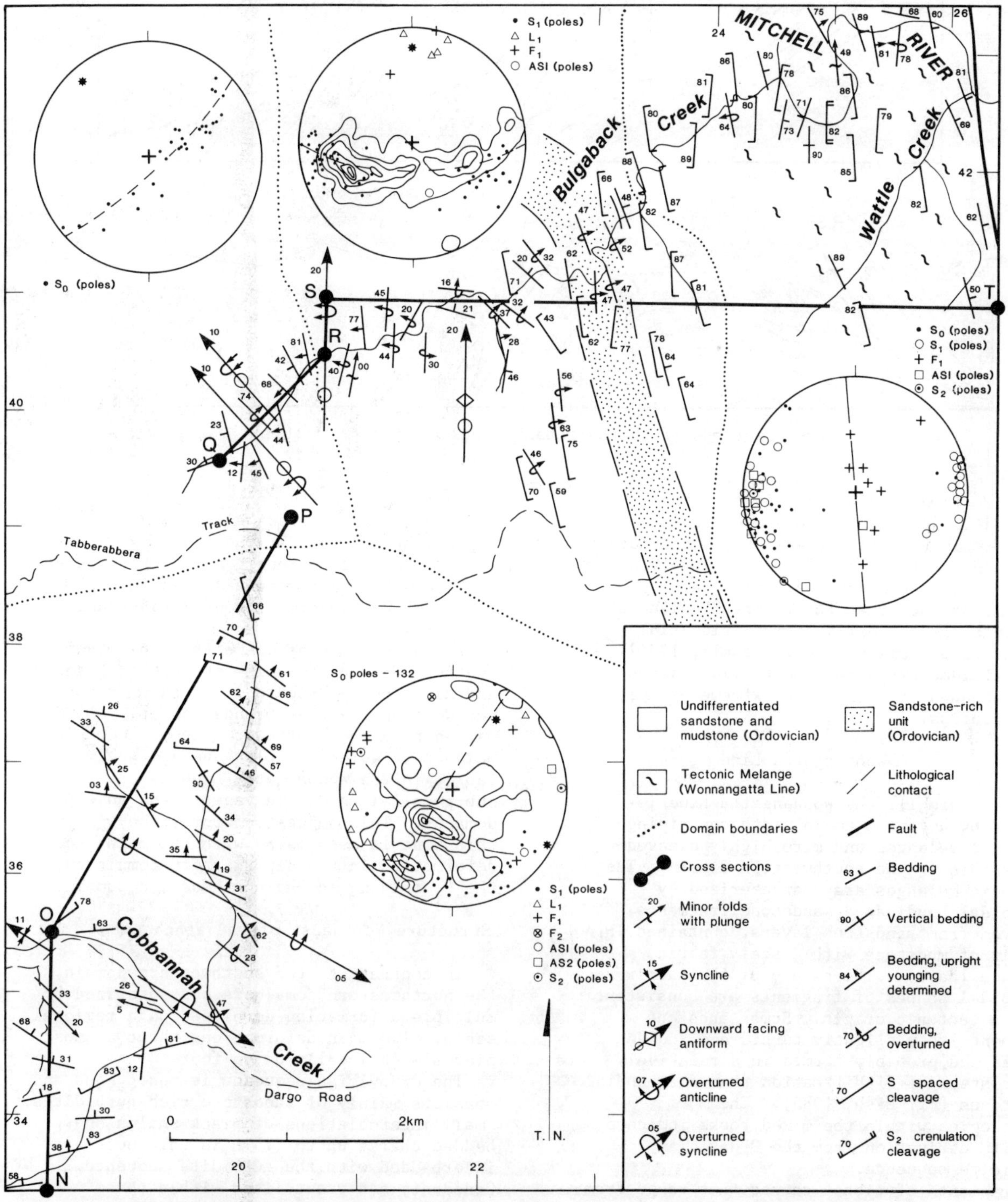

Fig. 11. Map of the Bulgaback Creek area. Southern hemisphere equal area stereographic projections are shown for four domains. Bedding (S_0) poles are contoured at 3% intervals for 1% area.

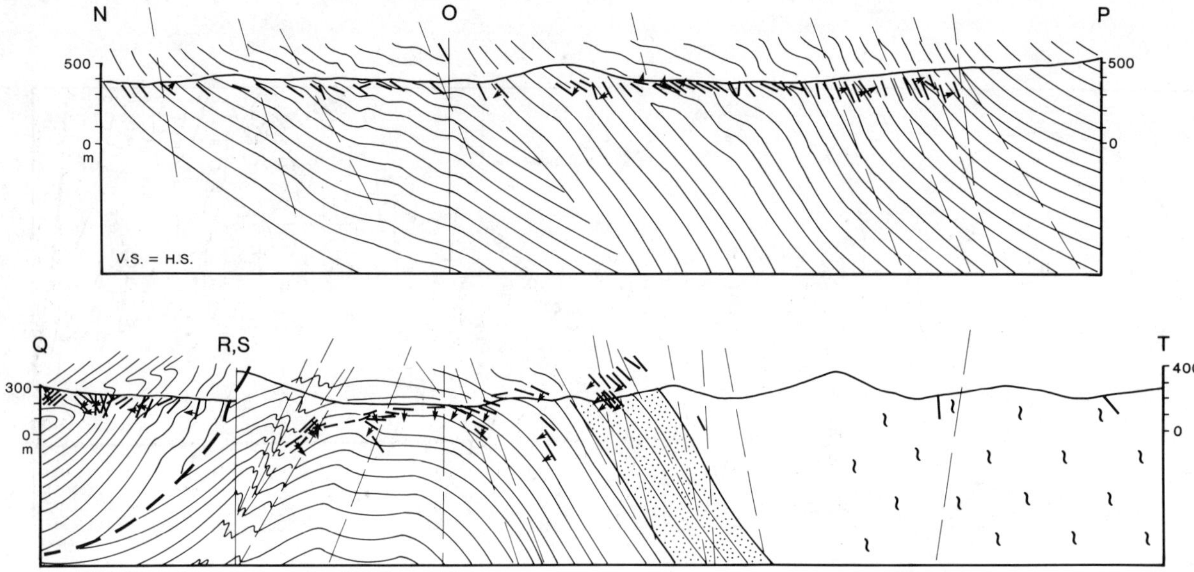

Fig. 12. Cross sections of the Bulgaback Creek area. See Figure 1 for location.

cleavage formed relatively late in the deformation as did the downward facing antiform in Bulgaback Creek. They indicate that the progressive deformation involved superimposition of non-coaxial strain increments such that the finite XY plane of strain cross-cut early formed fold axial planes [e.g. Ramsay, 1965; Powell, 1974]. The refolded early recumbent fold adjacent to the Wonnangatta Line is an extreme example of this process.

Wonnangatta Line

On Figure 11, the Wonnangatta Line is marked by a zone 2 km in width containing tectonic melanges and more highly cleaved rocks than in the southwestern domain. The tectonic melanges are characterized by phacoidal boudinaged sandstone fragments, derived from sandstone layers, contained in a matrix of mudstone with a scaly fabric (Figure 13a-d). The nature of the matrix and phacoidal shapes of fragments are consistent with a tectonic origin. Some sandstone fragments show slightly turbid and rounded shapes and probably formed in a relatively lower state of lithification than surrounding mudstones [cf. Lash, 1985]. The lithologies within the mixed rocks indicate in situ derivation from the Ordovician turbidite sequence.

Fragments of ribbon cherts that range from several centimetres across up to slabs several kilometres in length occur along the Line. Conodonts within one such slab indicate a Tremadocian age which is slightly older than the oldest recorded age from the Ordovician turbidite sequence to the northeast (I.R. Stewart and C.L. Fergusson, unpublished data). To the west the proportion of sandstone gradually increases and these have a strong anastomosing spaced cleavage mainly formed by solution transfer [cf. Gray, 1977].

Planar structural elements within the southern Wonnangatta Line strike north-south and dip steeply (Figure 11). Farther north the Wonnangatta Line trends easterly and is marked by a much narrower zone of tectonic melange (Figure 2). Cleavage and scaly fabric in the Wonnangatta Line is sub-parallel to S_1 cleavage in the adjoining domains. This suggests deformation within this zone is associated with the main-phase deformation in the southwestern domain and the F_1 folding in the northeastern domain.

Structure Northeast of the Wonnangatta Line

In contrast to the southwestern domain, the northeastern domain is characterized by multiple deformation developed on a regional scale. Two main deformations occur at most places with locally up to four.

The Ordovician sequence is monotonous and consists mainly of sandstone-rich turbidites. Small intercalations of black shales and bedded cherts up to 1.5 m in thickness are interbedded with the turbidite sequence. Conodonts and graptolites within these horizons indicate that most of the sequence is Lower Ordovician with less abundant Middle and Upper Ordovician (Stewart and Fergusson, unpublished data). In the Tabberabbera district (Figure 2), the Ordovician sequence

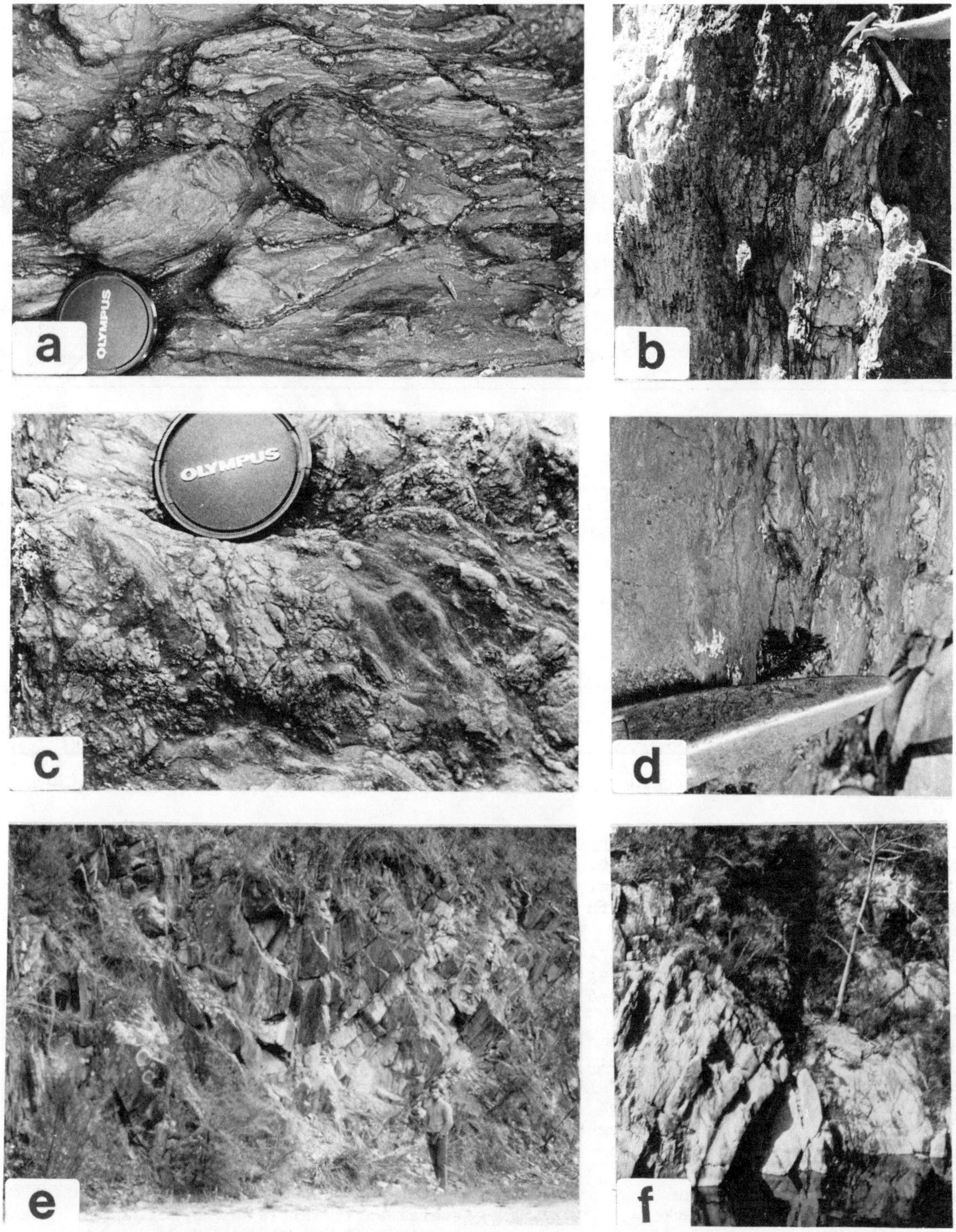

Fig. 13. (a)-(d) Examples of small-scale disruption of formerly interbedded sandstones and mudstones in matrix of tectonic melange along the Wonnangatta Line. Disruption and injection of mudstone within (c) indicate deformation of at least partly unconsolidated sediments. (e) Upright close F_1 fold hinge from the core of a major F_1 anticlinorium alongthe Tambo River in the northeastern domain. (f) Faulted F_1 anticline from the Tambo River. Width of photograph is approximately 30 m.

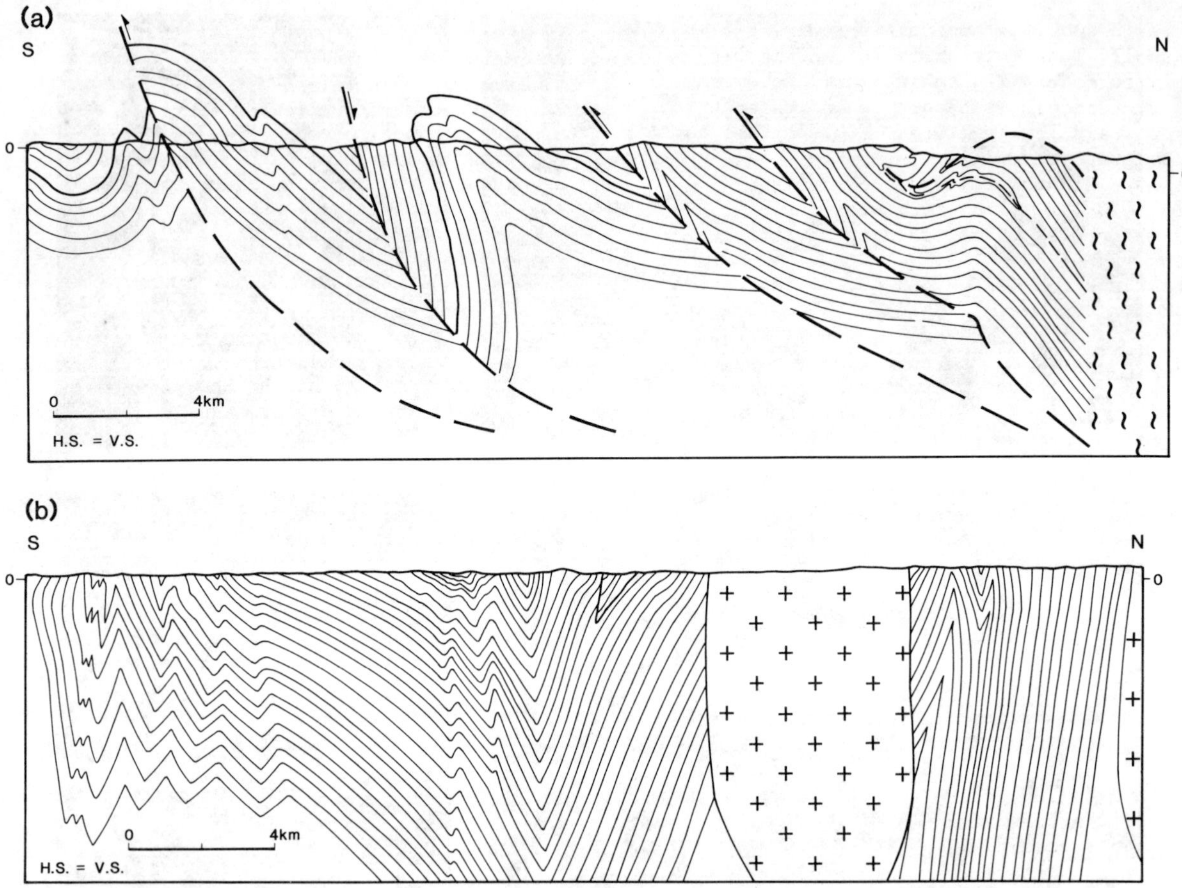

Fig. 14. Regional cross section for the Tabberabbera Belt. See Figure 2 for location.

is overlain with an angular unconformity by the Lower Devonian Wentworth Group [Talent, 1963]. This unit consists of conglomerate, sandstone, mudstone and rare limestone with the clastic components derived from the underlying unit. Farther east the Ordovician sequence is unconformably overlain by Lower Devonian Snowy River Volcanics and intruded by granitoids of a similar age.

Throughout much of the northeastern domain there is a well-developed bedding-parallel slaty cleavage. This structure is folded around F_1 fold hinges and therefore formed early in the tectonic history. The strength of the fabric indicates it is of tectonic origin and not caused by compaction. The early slaty cleavage [S*, using the terminology of Wilson and de Hedouville, 1985] is often the strongest fabric developed in the rocks.

The first major folding (F_1) in the northeastern domain consists of east-west trending shallowly plunging folds that correlate with the main-phase folds in the southwestern domain. F_1 folds are close to isoclinal with vertical to steeply dipping axial surfaces (Figure 13e-f). They have narrow hinges and long planar limbs with class 1C profiles in sandstone layers and class 3 profiles in mudstone layers. An axial surface crenulation cleavage is associated with the F_1 folding. In mudstones, the cleavage is formed by crenulation of the bedding-parallel slaty cleavage with dissolution along the steep limbs of microfolds [cf. Gray, 1979]. In some areas sandstones contain a stripy S_1 cleavage characterized by planar selvages of pressure solution residue up to 0.5 cm in width separated by zones of sandstone with quartz and chlorite overgrowths on quartz grains.

On a map scale the F_1 folds consist of several anticlinoria and synclinoria. These have long planar limbs and narrow hinge regions containing few parasitic folds (Figure 14). Northerly younging limbs are steep to overturned and contain large s-shaped fold pairs with a shallowly dipping central limb characterized by abundant F_1 parasitic folds. In the Tambo river area one of these central limbs is 4 km in width and is extensively refolded by younger structures (Figures 2 and 14). Southerly younging limbs are usually less internally folded and dip steeply to moderately southwards. No overall sense of tectonic transport is indicated by the map-scale F_1 structures. In contrast to the southwestern domain, the F_1 map-scale folds have higher amplitude-to-wavelength ratios (0.6-1.5) which indicates a thick-skinned style of folding in this domain.

The Kiewa Fault dips steeply to the west and in the north contains a mylonite zone up to 2 km in width (R.H. Fagan, unpublished data, 1979). R. Scott (unpublished data, 1985) has demonstrated from shear sense criteria that these mylonites have had an early history of dextral shear with only minor dip-slip motion. In the Tambo River area F_1 structures and low metamorphic grades are similar across the fault which is consistent with a history of limited strike-slip movement (i.e. less than 100 km). An Early Devonian pluton truncates the fault in this area (Figure 2) restricting movement to the Silurian [Richards and Singleton, 1981].

The main post-F_1 deformation formed the north-trending, tight to isoclinal, syncline containing the Lower Devonian Wentworth Group (Figure 2). This deformation is represented by north trending, close to open, folds with low amplitude-to-wavelength ratios in the Ordovician rocks. Axial surfaces are steep and an axial surface crenulation cleavage is developed throughout the domain. In the Tambo River area interference fold patterns are developed which will be described elsewhere. Because the F_1 folding affects the Wentworth Group and not the unconformably overlying Avon River Group, the timing of this deformation is constrained to the Middle Devonian.

Discussion

The formation of the Ordovician back-arc basin in the Benambra terrane has been attributed to a Precambrian slice being rifted off the Gondwanaland craton and rotated eastwards forming a marginal sea in its wake [Powell, 1983, 1984]. No unequivocal remnants of oceanic crust in the back-arc basin have been identified. The back-arc basin probably had a composite basement consisting of oceanic crust in its interior and older extended continental crust at the basin margins [Cas et al., 1980; Powell, 1983; Scheibner, 1986]. Modern analogues with extensive aprons of deep-marine turbidites are usually mainly underlain by oceanic crust. For example, the Amazon Cone rests mainly on oceanic crust and has an area of 3 x 10^5 km^2 [Damuth and Kumar, 1975] which compares with an area of 1.5 x 10^5 km^2 for the deformed back-arc basin in the Benambra terrane.

An Ordovician island arc developed on the rifted Precambrian slice above a westward dipping subduction zone [Powell, 1983]. Two different models have been proposed to account for the Early to Middle Silurian back-arc deformation. Scheibner [1985, 1986] suggested that Marianas-type subduction changed to Chilean-type due to the arrival of an oceanic plateau in the Late Ordovician. This caused flipping of the subduction zone and consumption of the oceanic crust in the back-arc basin. Subduction eventually caused intense deformation of the back-arc basin fill and formation of sutures along the Kiewa Fault and related faults farther north. The main problem with this model is that the Kiewa Fault does not appear to be a suitable candidate for a suture.

The alternative model of Powell [1983, 1984] concentrates on explaining the unusual east-west trends noted for several Early to Middle Silurian folds in the Lachlan Fold Belt. Chilean-type subduction in the Early Silurian was accompanied by dextral

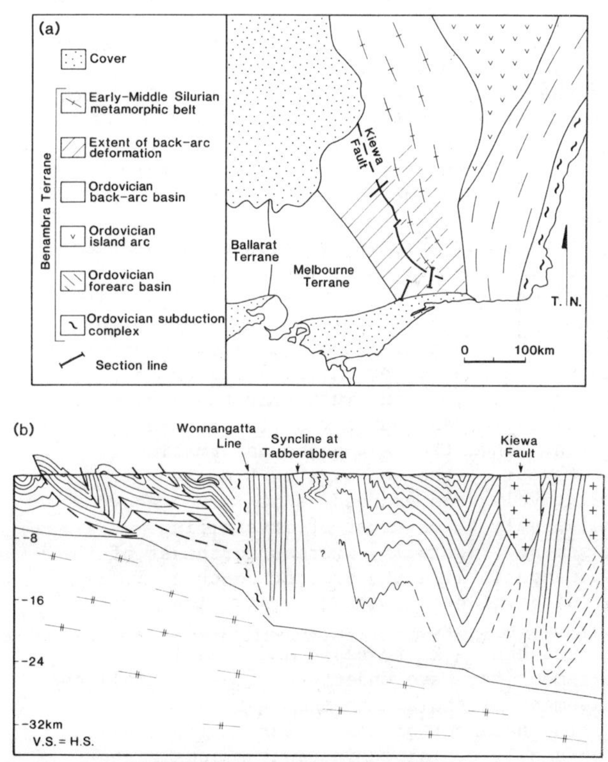

Fig. 15. (a) Tectonic elements of the Benambra terrane. (b) Crustal section showing the inferred thickening of rigid crust underneath the Tabberabbera Belt (see (a) for location).

oblique-slip resulting in a transpressional regime in parts of the arc and back-arc basin. This mechanism explains the extent of the east-west folds throughout the Tabberabbera Belt where they represent the major Early to Middle Silurian back-arc deformation (Figure 15).

The back-arc deformation is therefore attributed to transpressional tectonics with east-west folding reflecting northwest oriented dextral shear. The Kiewa, Ensay and related northwest-trending faults along the ancestral Melbourne-Benambra terrane boundary probably formed during this deformation and had predominantly strike-slip movement. Fergusson et al., [1986] suggested that dextral strike-slip movement occurred in the Silurian between the Benambra and Melbourne terranes to account for differences in their Ordovician-Silurian histories. The boundary between these terranes is a major fault zone formed by thrusting of the Benambra terrane over the Melbourne terrane in the Middle Devonian deformation [Fergusson et al., 1986].

If the Benambra terrane formed at the ancient Gondwanaland margin, and was subsequently moved to its present location by strike-slip faulting in the Silurian [Fergusson et al., 1986], then it may contain part of the pre-Ordovician attenuated passive continental margin developed during opening of the back-arc basin. Clearly this feature, if preserved, must partly underlie the craton-derived Ordovician turbidite sequence in the western part of the Benambra terrane. Differences in the structural style across the Wonnangatta Line reflect a southwesterly thickening rigid crust at depth which is interpreted as the passive continental margin (Figures 14 and 15). The Wonnangatta Line itself is a zone of intense localized shortening and may have formed above a series of extensional faults in thinned Precambrian crust near the ancient continent-ocean boundary of the back-arc basin. Thus the main-phase deformation southwest of the Wonnangatta Line is relatively thin-skinned as it formed above a relatively thick crust of crystalline basement (Figure 15) whereas northeast of the Wonnangatta Line the greater depth of F_1 folding reflects thinner crystalline crust and oceanic crust. The deformation has closed the back-arc basin and oceanic basement may have underthrust the crystalline basement of the arc in the east.

The basement of the eastern Melbourne terrane is contained in fault slices along the terrane boundary [Fergusson et al., 1986]. These rocks are Cambrian greenstones of oceanic affinities [Crawford and Keays, 1978] in contrast to the inferred continental substrate of the southwestern Benambra terrane. Dextral strike-slip movement of at least 100 km along the ancestral Melbourne-Benambra terranes boundary is required to account for this anomaly. Both terranes were formerly continuous along the eastern Gondwanaland continental margin, although the Benambra terrane has undergone a more complex history with formation of the Ordovician back-arc basin and island arc. A sudden change in plate motions at the Ordovician-Silurian boundary may have caused strong oblique convergence resulting in the back-arc deformation and strike-slip faulting.

One enigmatic aspect of the back-arc deformation is that it was accompanied by high-grade metamorphism and plutonism in the Omeo Metamorphic Complex and its continuation farther north [Powell, 1983]. Scheibner [1985, 1986] attributed this event to the deformation and associated crustal thickening whereas Powell [1983] opted for a model with a subcrustal heat source, left over from the old spreading ridge in the back-arc basin, causing the metamorphism. I consider that the metamorphics and S-type plutonism reflect a jump of the Ordovician island arc approximately 100 km to the west. The low-P high-T style of metamorphism is consistent with an Andean-style magmatic chain [cf. Vernon, 1982; England and Richardson, 1977]. In the Silurian and Early Devonian, a linear zone of plutonism has episodically migrated to the east of the metamorphic belt [Powell, 1983, 1984] and may be related to continuing underthrusting along a west-dipping subduction zone at the eastern side of the Benambra terrane.

Conclusions

1. A Silurian back-arc deformation in the southwestern Benambra terrane formed upright east-west folds and associated faults. These structures show contrasting basement-cover relations either side of a fault containing tectonic melange up to 2 km in width (the Wonnangatta Line). This fault appears to have been a zone of strong shortening and formed the root zone for a southward travelling recumbent fold. The contrast in structural style across the Wonnangatta Line reflects a southwesterly thickening basement at depth.

2. The Ordovician tectonic arrangement of the Benambra terrane consisted of a passive continental margin in the west (part of eastern Gondwanaland), a back-arc basin with some oceanic crust in the centre, and an island arc founded on older continental crust in the east [cf. Cas et al., 1980]. A change in plate motions at the Silurian-Ordovician

boundary brought about oblique convergence and strong plate coupling in Chilean-type subduction caused the back-arc deformation and strike-slip faulting. The large size and variety of tectonic elements within the Benambra terrane indicate it may have a composite character and consist of several smaller terranes [e.g. Scheibner 1985, 1986].

3. Dextral strike-slip faulting associated with the Silurian plate motions re-arranged the Ordovician eastern Gondwanaland margin. The Benambra terrane, with a history of back-arc basin formation and subsequent active tectonism, was juxtaposed with the Melbourne terrane, which developed as a continental margin sediment prism along a quiet sector of the Gondwanaland margin. Major overthrusting in the Middle Devonian has complicated this pattern [Fergusson et al., 1986].

4. Recognition of terranes in the Lachlan Fold Belt has helped to clarify the tectonic history, but many problems still remain. One special feature of the southwestern Benambra terrane is that it consists of fault-bound units which have had similar geologic histories. The absolute values of movements along the boundary faults (e.g. Wonnangatta Line, Kiewa Fault) are unknown and for the sake of simplicity in this analysis they are assumed to be relatively small (i.e. <100 km).

Acknowledgements. This work was carried out at the Department of Earth Sciences, Monash University and funded by grants from the Australian Research Grants Scheme (ARGS E83 315675; principal investigators Drs R.A.F. Cas and D.R. Gray) and Monash University. N. Murphy, S. Winter and I.R. Stewart provided valuable assistance in the field. Dr D.R. Gray kindly gave valuable comments on a draft. Draga Gelt drafted the diagrams. Jenny Jarman and Therese Carmody typed the final drafts. Fons VandenBerg and Erwin Scheibner gave valuable comments on the manuscript.

References

Baillie, P. W., A Paleozoic suture in eastern Gondwanaland, Tectonics, 4, 653-660, 1985.

Bolger, P. F., Ordovician and Silurian stratigraphy and structure in the Wombat Creek-Benambra area, northeast Victoria, Proc. R. Soc. Vict., 94, 35-47, 1982.

Cas, R. A. F., A review of the palaeogeographic and tectonic development of the Paleozoic Lachlan Fold Belt of southeastern Australia, Spec. Publ.,Geol. Soc. Aust., 10, 104 pp, 1983.

Cas, R. A. F., C. McA. Powell, and K. A. W. Crook, Ordovician palaeogeography of the Lachlan Fold Belt: a modern analogue and tectonic constraints, J. Geol. Soc. Aust., 27, 19-31, 1980.

Crawford, A. J., and R. R. Keays, Cambrian greenstone belts in Victoria: marginal sea-crust slices in the Lachlan Fold Belt of southeastern Australia, Earth Planet. Sci. Lett., 41, 197-208, 1978.

Damuth, J. E., and N. Kumar, Amazon Cone: morphology,sediments, age and growth pattern, Bull. Geol. Soc. Am., 86, 863-878, 1975.

England, P. C., and S. W. Richardson, The influence of erosion upon the mineral facies of rocks from different metamorphic environments, J. Geol. Soc. Lond., 134, 201-213, 1975.

Fergusson, C. L., Tectonic history of the Tabberabbera Belt, eastern Victoria and implications for Ordovician reconstructions of southeastern Australia, Geol. Soc. Aust. Abstr., 14, 63-65, 1985.

Fergusson, C. L., D. R. Gray, and R. A. F. Cas., Overthrust terranes in the Lachlan Fold Belt, southeastern Australia, Geology, 14, 519-522, 1986.

Gray, D. R., Cleavages in deformed psammitic rocks from southeastern Australia: Their nature and origin, Bull. Geol. Soc. Am., 89, 577-590, 1977.

Gray, D. R., Microstructure of crenulation cleavages: an indicator of cleavage origin, Am. J. Sci, 279, 97-128, 1979.

Gray, D. R., 1981, Cleavage-fold relationships and their implications for transected folds: an example from southwest Virginia, USA, J. Struct. Geol., 3, 265-277.

Jones, D. L., D. G. Howell, P. J. Coney, and H. W. H. Monger, Recognition, character and analysis of tectono-stratigraphic terranes in western North America, J. Geol. Educat., 31, 295-303, 1983.

Lash, G. G., Accretion-related deformation of an ancient (Early Paleozoic) trench-fill deposit, central Appalachian Orogen, Bull. Geol. Soc. Am., 96, 1167-1178, 1985.

Maltman, A. J., Primary bedding-parallel fabrics in structural geology, J. Geol. Soc. Lond., 139, 363-373, 1981.

McCulloch, M. T., and B. W. Chappell, Nd isotopic characteristics of S- and I-type granites, Earth Planet. Sci. Lett., 58, 51-64, 1982.

Powell, C. McA., Timing of slaty cleavage during folding of Precambrian rocks, northwest Tasmania, Bull. Geol. Soc. Am., 85, 1043-1060, 1974.

Powell, C. McA., Tectonic relationship between the Late Ordovician and Late Silurian palaeogeographies of southeastern Australia, J. Geol. Soc. Aust., 30, 353-373, 1983.

Powell, C. McA., Ordovician to earliest Silurian: marginal sea and island arc; Silurian to mid-Devonian dextral transtensional margin; Late Devonian and Early Carboniferous: continental magmatic arc along the eastern edge of the Lachlan Fold Belt, in Phanerozoic earth history of Australia, edited by J. J. Veevers, pp. 290-340, Oxford University Press, Oxford, 1984.

Ramsay, J. G., Structural investigations in the Baberton Mountain Land, eastern Transvaal, Trans. Proc. Geol. Soc. S. Afr., 66, 353-401, 1965.

Richards, J. R., and O. P. Singleton, Paleozoic Victoria, Australia: igneous rocks, ages and their interpretation, J. Geol. Soc. Aust., 28, 395-421, 1981.

Scheibner, E. C., Suspect terranes in the Tasman Fold Belt System, eastern Australia, in Tectono-stratigraphic Terranes of the Circum-Pacific Region , edited by D. G. Howell, pp. 493-514, Circum-Pacific Council for Energy and Mineral Resources, Earth Science Series, No.1, 1985.

Scheibner, E. C., Paleozoic tectonic development of eastern Australia in relation to the Pacific region, in Circum-Pacific Orogenic Belts and the Evolution of the Pacific Ocean Basin, edited by J. W. H. Monger, and J. Francheteau, Am. Geophy. Un., Geodynamic Ser., in press, 1986.

Talent, J. A., The Devonian of the Mitchell and Wentworth rivers. Geol. Surv. Vict. Mem., 24, 118 pp., 1963.

Vernon, R. H., Isobaric cooling of two regional metamorphic complexes related to igneous intrusions in southeastern Australia, Geology, 10, 76-81, 1982.

Wilson, C. J. L., and P. de Hedouville, Early cleavage development in the Late Ordovician of northeast Victoria, Australia, J. Struct. Geol., 7, 401-408, 1985.

TERRANES OF THE TUMUT DISTRICT, SOUTHEASTERN NEW SOUTH WALES, AUSTRALIA

H. Basden

New South Wales Department of Mineral Resources, Sydney, Australia

B. J. Franklin, B. Marshall, and A. E. Waltho

Department of Applied Geology, The New South Wales Institute of Technology, Sydney, Australia

Abstract. The Tumut Region of the Lachlan Fold Belt of southeastern New South Wales, has traditionally been interpreted in terms of intra-arc rifting to form the Tumut Trough, followed by rift-basin closure and inversion as part of the accretionary process. Implicit in this tectonic model is the notion that the relative position of all rock bodies, other than ophiolite slivers, was the same in the Late Silurian as it is today; that is, the rock bodies are parautochthonous. A dearth of "hard" palaeomagnetic, palaeobiogeographic and radiometric data, however, allows allochthonous interpretations and requires that terranes of the region be designated "suspect". Five suspect terranes are recognised on the basis of lithostratigraphic, structural and metamorphic evidence, plus some radiometric dating and limited fossil data. The terranes are the Wagga-Omeo, Tumut, Jindalee, Mooney Mooney, and Young terranes. Both the Jindalee and Tumut terranes are now composite and, perhaps with the Mooney Mooney terrane, may have formed a single composite terrane on which the Silurian Tumut Trough sequence was deposited as an overlapping assemblage. This would suggest terrane dispersion and amalgamation before the emplacement of Early Devonian stitching plutons and overlapping volcanics. The principal faults delimiting the terranes are the Gilmore fault zone, the Cootamundra - Gundagai - Killimicat fault system, and the Mooney Mooney fault system. Mapping suggests that the structures are steeply dipping, but the amounts and nature of displacements are generally unresolved, since they have mainly been determined from stratigraphic arguments based on a parautochthonous rift-basin model. Such displacement data are only as strong as the tectonic model. However, on the eastern branch of the Mooney Mooney fault system, kinematic indicators within mylonites suggest that the structure is an east-dipping, high angle reverse fault with a right lateral component. This is consistent with west-southwest thrusting of the Young terrane over the Mooney Mooney terrane. Major geophysical contrasts occur across the Gilmore fault zone. Geophysical data also support a high angle reverse fault interpretation of the Mooney Mooney fault system. It is unclear, however, as to whether or not the steeply dipping (at surface) faults pass down into a flat-lying thrust system, but it is anticipated that at least the Tumut, Jindalee, Mooney Mooney, and Young terranes are interpreted slices above a major plane of delamination. This interpretation is consistent with a thrust-tectonic, allochthonous model of amalgamation and accretion, and is opposed to the rift basin model of the Tumut Trough.

Introduction

The Tumut district is located in the southeastern part of the Lachlan Fold Belt, New South Wales, Australia (Fig. 1). In structural terms it incorporates the central portion of the Tumut Synclinorial Zone, which is bounded to the west by the Wagga Anticlinorial Zone, and to the east by the Young Anticlinorial Zone [Degeling et al. 1986]. In applying the terrane concept, Basden et al. [1985] equated the structural zones (from west to east), with terranes they termed the Wagga-Omeo, Girilambone, and Molong-Monaro. Their Girilambone terrane was bounded by the Gilmore fault zone on the west, and the Mooney Mooney fault system on the east.

Our recent investigations have shown that the Girilambone terrane is a composite feature which has now been sub-divided into the Tumut, Jindalee and Mooney Mooney terranes. The western portion of the similarly composite Molong-Monaro terrane is now the basis for the Young terrane (Fig. 2). The principal objectives of this paper are to outline tectonostratigraphic and structural attributes of the five terranes, to describe the characteristics of faults delimiting them, and to comment on tectonic interpretations of the terrane assemblage.

Tectonic models pertinent to the immediate region are presented by, amongst others, Scheibner [1973, 1975, 1985], Cas et al. [1980], Crook

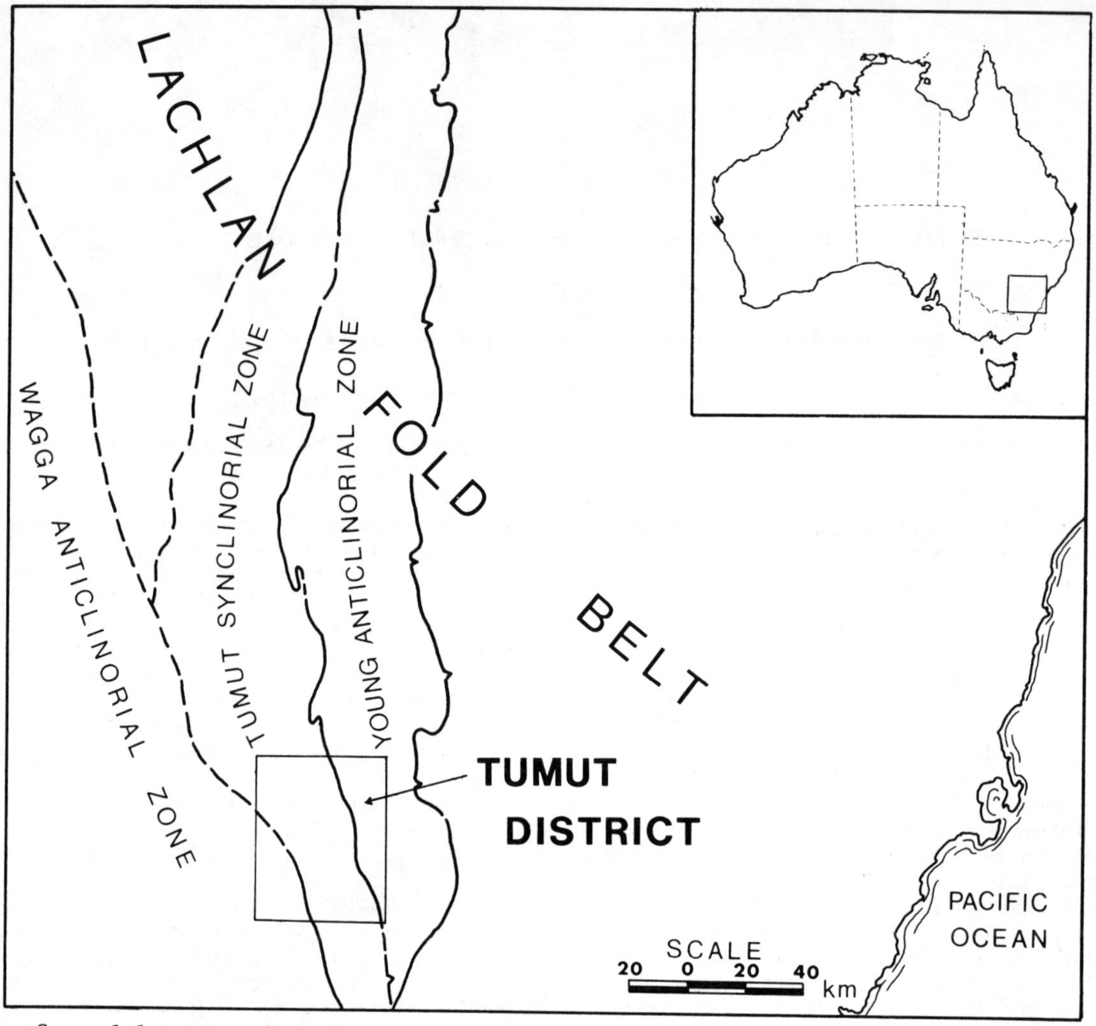

Fig. 1. General location of the Tumut region in the southern Tumut Synclinorial Zone of the Lachlan Fold Belt, based on Scheibner [1985, pers. comm.].

[1980], Cas [1983], Basden [1982], Leitch and Scheibner [1985], and Basden et al. [1985]. The essential components of most of these models are:

(a) An Ordovician volcanic arc (Molong arc), and associated convergent plate boundary to the east of the Gondwanaland craton and separated from it by the Wagga marginal sea, now represented by the rocks of the Wagga-Omeo metamorphic belt.

(b) Plate convergence giving way (in the earliest Silurian) to largely transcurrent motion and the consequent development of rift basins in the volcanic arc. One of these basins, the Tumut Trough, was controlled by the proto-Gilmore and proto-Mooney Mooney fault zones.

(c) Development of oceanic crust, now preserved as the Coolac ophiolite suite, in parts of the rift system.

(d) Emplacement, during the Late Silurian, of batholithic granodiorite bodies into the regions flanking the trough.

(e) Deformation of the basin and emplacement of the Coolac ophiolite suite along the eastern margin of the trough during the Late Silurian.

Implicit to models engrossing the above components is the notion that, with the exception of the ophiolite slivers, the relative positions of rock bodies in and flanking the Tumut Trough were essentially the same in the Late Silurian as they are today. That is, the Wagga-Omeo, Tumut, Jindalee and Young terranes are basically parautochthonous, even though rifting is a form of dispersion, and closure and inversion of the rift basins involves accretionary processes with some degree of relative displacement. Only in recent years has it been acknowledged that terranes in the Lachlan Fold Belt are "suspect" [as used by Coney et al., 1980; Scheibner, 1985], that they might be grossly allochthonous, and that potential exists to identify major exotic elements [e.g. Powell, 1983]. But despite appreciating their suspect nature, and

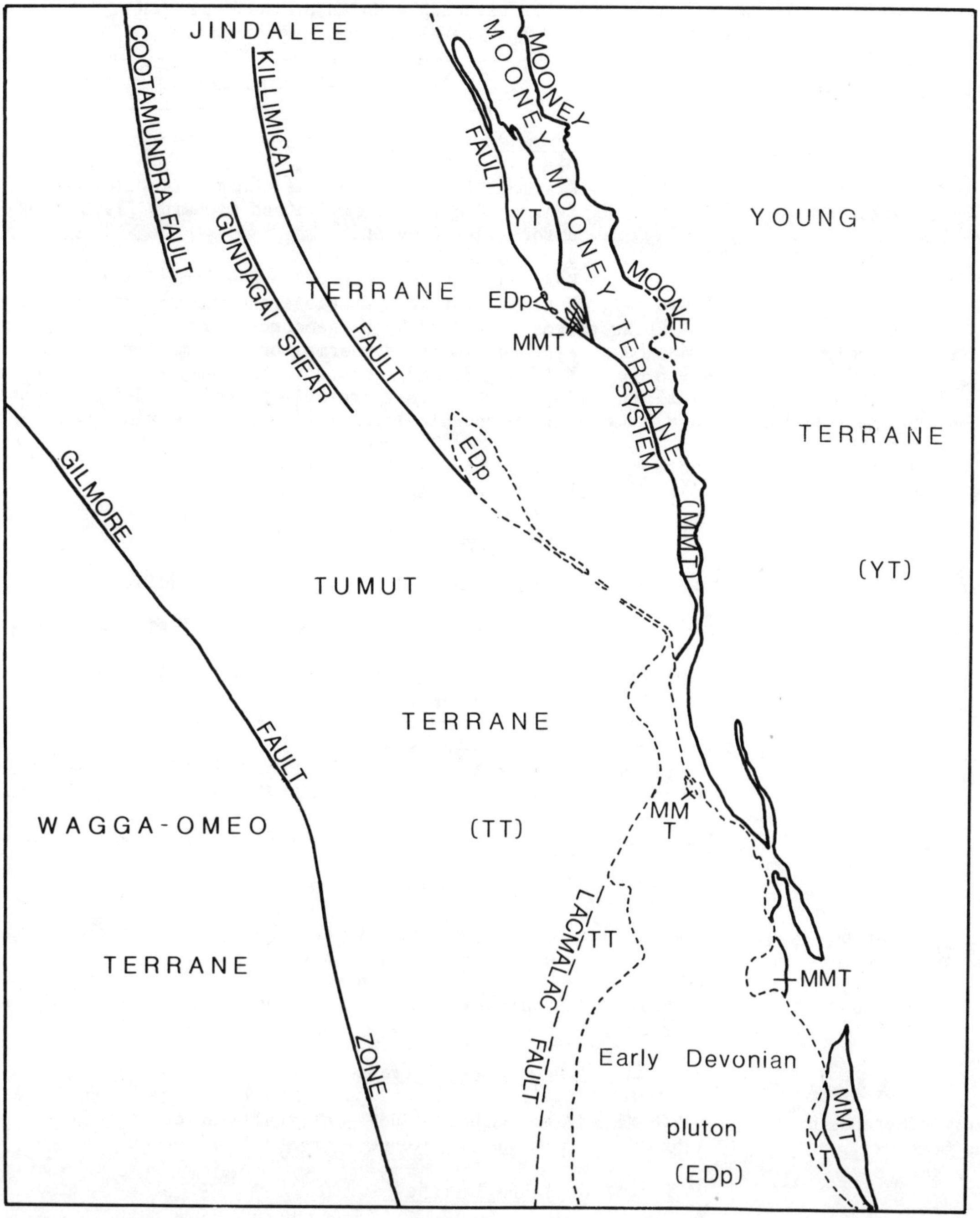

Fig. 2. Suspect tectonostratigraphic terranes of the Tumut region. The Tumut, Jindalee and Mooney Mooney Terranes are subdivisions of the Girilambone Terrane of Basden et al. [1985].

noting that this reflects a dearth of palaeomagnetic and palaeobiogeographic data, and inadequate amounts of reliable radiometric dating [Scheibner, 1985], lithostratigraphically based, parautochthonous, tectonic interpretations of the terranes of the Tumut district have still been favoured.

Tectonostratigraphy of the Terranes

Throughout this section, the evidence on which the ages are based is presented in Table 1; the distribution of the terranes is shown in Fig. 2.

The Wagga-Omeo Terrane. The Wagga Omeo terrane

TABLE 1. Fossil and Radiometric Age Dates for Lithologic Units of the Wagga-Omeo, Tumut, Jindalee, Mooney Mooney and Young Terranes in the Tumut Region

Stratigraphic Unit	Fossil/Radiometric Age Dates	
	Wagga-Omeo Terrane	
Ellerslie Granodiorite	404±6 Ma	K-Ar on biotite - probable thermal or deformational event
Wondalga Granodiorite	302 Ma	K-Ar on biotite, Evernden and Richards [1962] - thermal or deformational event
	136±10 Ma	Fission track on sphene
Avenall Basic Intrusive Complex	401±6 Ma	K-Ar on biotite - thermal or deformational event
	414±6 Ma	K-Ar on biotite - probable thermal or deformational event
	Low Rb:Sr ratios prevented Rb-Sr geochronology	
Green Hills Granodiorite	404±6 Ma	K-Ar on biotite - thermal or deformational event
	419±6 Ma	K-Ar on biotite - probable thermal or deformational event
	422±6 Ma	K-Ar on biotite - probable thermal or deformational event
Wagga-Omeo Metamorphic Belt Metasediments	Late Ordovician (Gisbornian) graptolites, L. Sherwin [pers. comm.]	
Nacka Nacka Metabasic Igneous Complex	465±6 Ma	K-Ar on hornblende - crystallisation
	467±6 Ma	K-Ar on hornblende - crystallisation
	Tumut Terrane	
Gocup Granite	409±2 Ma	K-Ar on muscovite, Richards et al. [1977] - crystallisation
	402±3 Ma	Rb-Sr whole rock, Richards et al. [1977] - thermal or deformational event
	360±6 Ma	K-Ar on biotite, Richards et al. [1977] - thermal or deformational event
	212±2 Ma	K-Ar on biotite, Richards et al. [1977] - thermal or deformational event
Bogong Granite	410±16 Ma	K-Ar on biotite, Ashley et al. [1972] - crystallisation
	390±16 Ma	K-Ar on biotite, Barkas [1976] - thermal or deformational event
Minjary Volcanics	Late Early Devonian fossils, Barkas [1976], L. Sherwin [pers. comm.]	
	Jindalee Terrane	
Blowering Formation	Lower Middle Ludlovian conodonts in allochthonous limestone blocks	
Blacks Flat Diorite	417±6 Ma	K-Ar on hornblende - crystallisation
	175±10 Ma	Fission track on sphene
Wyangle Formation	Probable Upper Llandoverian to Lower Wenlockian conodonts in limestone clasts	
	Mooney Mooney Terrane	
Honeysuckle Metabasic Igneous Complex	425±6 Ma	K-Ar on hornblende - crystallisation
	426±6 Ma	K-Ar on hornblende - crystallisation
	Young Terrane	
Young Granodiorite	417±6 Ma	K-Ar on biotite, Evernden and Richards [1962] - crystallisation
	405±16 Ma	K-Ar on biotite, Basden et al. [1978] - thermal or deformational event
Micalong Swamp Basic Igneous Complex	430±9 Ma	K-Ar on hornblende, Owen and Wyborn [1979] - crystallisation
Goobarragandra Volcanics	Middle Wenlockian fossils in limestone pod, Basden et al. [1978]	
	429±16 Ma	Rb-Sr whole rock, Owen and Wyborn [1979] - crystallisation
	417±8 Ma	Rb-Sr whole rock/biotite, Owen and Wyborn [1979] - thermal or deformational event
	429±9 Ma	K-Ar on biotite, Owen and Wyborn [1979] - crystallisation

Constants for K-Ar dates: $\lambda\varepsilon=0.581\times10^{-10}$/yr; $\eta_\beta=4.962\times10^{-10}$/yr; $^{40}K/K_{total}=1.167\times10^{-4}$(moI/moI). Constants for Rb/Sr dates: $\lambda(^{87}Rb)=1.42\times10^{-11}$/yr; atomic ratios $^{85}Rb/^{87}Rb=2.59265$; $^{86}Sr/^{88}Sr=0.1194$; $^{84}Sr/^{86}Sr=0.056584$.

[Scheibner, 1985] is a medium to high grade, regional metamorphic terrane which, in the Tumut district, comprises a Middle to Late Ordovician tholeiitic, metavolcanic complex (Nacka Nacka Metabasic Igneous Complex), thought to represent a portion of the Molong arc, and dominantly pelitic, with less common psammitic, metasediments (Wagga-Omeo metamorphic belt). These rocks have been intruded by a Late Ordovician-Early Silurian S-type granodiorite, and a Middle to Late Silurian I-type granodiorite complex with a spatially related tholeiitic basic complex.

In terms of area, the Wagga-Omeo terrane is the largest of those described by Scheibner [1985] in the Tasmanides of southeastern Australia. The metasedimentary rocks of the terrane are, lithologically, more diverse than the often cited description of "quartz-rich flysch" [Gilligan and Scheibner, 1978; Basden, 1982; Scheibner, 1985; Basden et al., 1985] would suggest. Sedimentary structures indicate a dominantly northward-directed palaeocurrent flow throughout the terrane [Fagan, 1979; Cas et al., 1980; Powell, 1983]. The quartz-rich psammites in the southernmost portion of the terrane, become progressively less psammitic northward to the point where, in the Tumut district, the sediments are dominantly pelitic. Metamorphosed andesitic and dacitic volcaniclastics, interbedded with pelitic metasediments, are common along the eastern margin of the terrane in the vicinity of the Nacka Nacka Metabasic Igneous Complex. Based on both morphotectonic [Cas et al., 1980; Cas, 1983; Powell, 1983] and immobile trace element geochemical criteria, the Nacka Nacka Metabasic Igneous Complex is interpreted as part of an oceanic (Marianas-type) arc.

The Jindalee Terrane. The Jindalee terrane is composite and consists of a basement of ?Cambro-Ordovician schistose serpentinites and amphibolite schist, which may represent dismembered ophiolite, cherts and quartzites, collectively referred to as the Jindalee Group. Overlying Early to Late Silurian rocks form a 2000m thick sequence comprising andesitic volcaniclastics overlain by dacitic volcaniclastics. They are locally underlain by metabasalts of oceanic island or Marianas-type arc affinity. Early Devonian I-type plutons intrude the bounding faults, and co-magmatic volcanics with minor sediments unconformably overlie the trough sequence.

The Tumut Terrane. The Tumut terrane is also composite, but differs from the Jindalee terrane in having a basement which, although not exposed in the Tumut district, probably comprises Middle to Late Ordovician basic-intermediate volcanic rocks [Suppel et al., 1986]. This conclusion is based on the regional aeromagnetic pattern, and the presence of numerous gold deposits. Silurian A-type rhyolitic volcanics with associated terrestrial/ shoreline deposits, occur along the eastern edge of the terrane. Further west and dominating the terrane is a sequence of Silurian andesitic to dacitic volcaniclastics up to 8000m thick. In places, tholeiitic island arc basalt, chert, quartz-rich flysch and more dacitic volcaniclastics underlie this sequence. Disrupted ophiolite lithologies occur in fault-bound blocks within the terrane. As in the Jindalee terrane, Early Devonian I-type volcanics and minor sediments unconformably overlie the volcaniclastic sequence, and co-magmatic plutons intrude along bounding faults and shear zones.

The Mooney Mooney Terrane. The Mooney Mooney terrane comprises the Silurian Coolac ophiolite suite, a partially dismembered yet relatively complete ophiolite. It contains variously serpentinised and tectonised harzburgite, chromitite pods, layered dunite, wehrlite and clinopyroxenite cumulates, gabbros, diorites and plagiogranites, metabasalts of sea-floor origin, extremely rare boninites, and minor cherts, jaspers and pyritic sulphide-rich rocks (Fig. 3) [Franklin & Ashley, 1980; Ashley et al., 1979]. The various units in the Coolac ophiolite suite form well-defined bodies, often of considerable areal extent which display both tectonic and locally intrusive contacts.

The presence of numerous small bodies of intrusive and possibly extrusive plagiogranite with weak calc-alkaline tendencies [Ashley et.al., 1983], and of small-pillowed basalts, suggest the ophiolite comprises arc substrate rather than newly formed oceanic crust or scraped-off seamount. The Mooney Mooney terrane is entirely bounded by elements of the Mooney Mooney fault system (Fig. 2).

The Young Terrane. The Young terrane consists of a Middle Silurian tholeiitic basic complex, plus Middle Silurian S-type dacitic volcanics, supposedly co-magmatic with a Late Silurian S-type granodiorite. The terrane is distinguished from the Molong-Monaro terrane to the east by the absence of Ordovician volcanic rocks [E. Scheibner pers. comm.]. As with the Tumut and Jindalee terranes, Early Devonian I-type granites are present.

Accretionary History

The distribution throughout the region of Early Devonian I-type granites and volcanics is consistent with their respectively having stitching and overlapping relationships, and this suggests that all five terranes had accreted (or at least amalgamated) by the end of the Silurian. Subsequent movement on any of the bounding faults (e.g. the Cootamundra fault) probably only involved minor tectonism partitioning into strain-softened [White et al., 1980] fault and shear zones.

Basden [1982] and Basden et al. [1985] showed parts of the Tumut Trough sequence common to what are here recognised as three separate terranes (Tumut, Jindalee, Mooney Mooney) (Fig. 2). This suggests that the three were elements of a composite terrane, with the Tumut Trough sequence constituting an overlapping assemblage, before subsequent minor(?) dispersion and amalgamation.

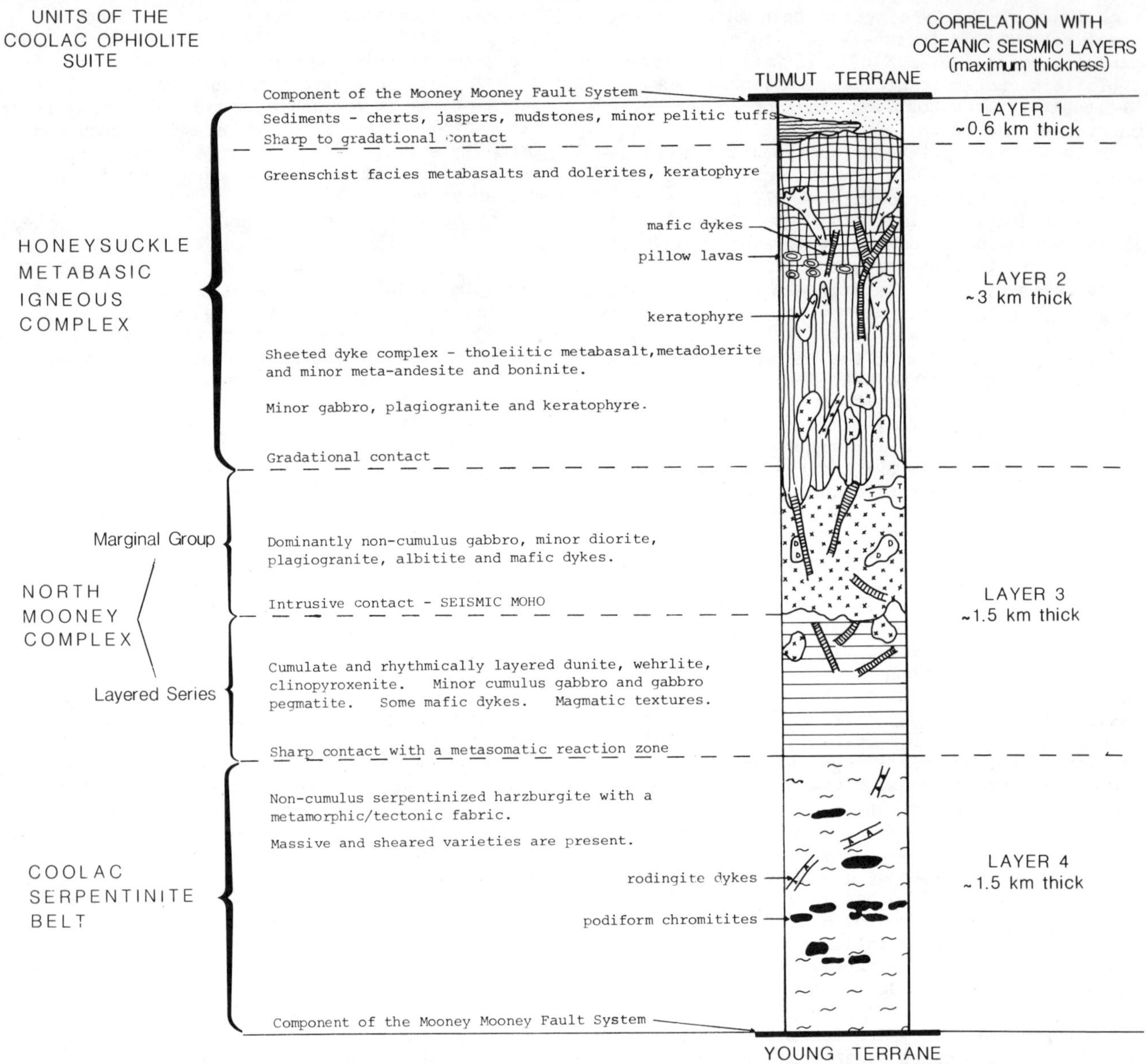

Fig. 3. Stratigraphy of the Coolac ophiolite suite, Mooney Mooney Terrane.

Based mainly on lithostratigraphic, geochemical, and very limited fossil and radiometric age dating evidence (Table 1), Basden has suggested that a case can be made for the terranes (other than the Mooney Mooney terrane) undergoing little change in relative position. The main pieces of supporting evidence invoked are:

(a) The possible provenance of the quartz-rich flysch of the Silurian Tumut Trough sequence being the Late Ordovician metasediments of the Wagga-Omeo terrane.

(b) The dacitic volcaniclastics of the Tumut and Jindalee terranes could be the deeper water equivalents of the dacitic volcanics of the Young terrane.

(c) The Middle Silurian tholeiitic basic complexes of the Wagga-Omeo and Young terranes could be related to the rifting which formed the Tumut Trough.

Unfortunately, none of the pieces of evidence is sufficiently conclusive to preclude other interpretations. Further, the "relative position" notion essentially relates to mapped (horizontal) distribution; and although this is generally incompatible with allochthonous terranes being amalgamated and accreted by dominantly strike-slip tectonics, it can accommodate thrust-tectonic amalgamation of allochthons, as well as parautochthonous interpretations.

Structure and Metamorphism Within the Terranes

Several of the recognised terranes contain rocks of substantially different ages. These different ages are expressed as contrasting intensities of deformation and metamorphism. Thus, the ?Cambro-Ordovician basement rocks of the Jindalee terrane have been multiply deformed, the Tumut Trough sequence has a relatively simple structural style and the Early Devonian volcanics are essentially undeformed. Structure and metamorphism within the terranes will therefore be examined in the context of the principal structurometamorphic units, some of which are not necessarily confined to a single terrane.

Based on mesoscale observations, six distinct structurometamorphic units can be delimited (Fig. 1). They are: the Wagga Anticlinorial Zone [Scheibner, 1972; Gilligan and Scheibner, 1978]; the Jindalee Group, Coolac ophiolite suite, and Bumbolee Creek Formation [Basden et al., 1985], all within the Tumut Synclinorial Zone; the remainder of the Tumut Synclinorial Zone; and the Young Anticlinorial Zone [op. cit.].

The Wagga Anticlinorial Zone comprises the Middle Ordovician Nacka Nacka Metabasic Igneous Complex, the Late Ordovician Wagga Metamorphic Belt metasediments, and younger (probable Silurian) granodiorites and basic intrusives. The high-temperature low-pressure metamorphism locally develops amphibolite facies assemblages and migmatites, but upper greenschist facies is more typical. The multiply deformed metasediments commonly exhibit two or three generations of mesoscale structures.

The Jindalee Group has two and usually three periods of penetrative deformation; the first event produced a transposition schistosity and the second a differentiated foliation. Metamorphic assemblages range from lower to upper greenschist facies.

The Coolac ophiolite suite comprises tectonised harzburgite (Coolac Serpentinite) overlain by a "knocker-terrane" mélange of constant polarity oceanic crust in a serpentinite matrix. In the superficially massive harzburgite, complex overprinting relations exist between layering and various types of foliation (including serpentine-vein foliations) and lineations. The somewhat massive oceanic crust, also displays layering and complex age relationships between variably spaced and oriented foliations defined by veining and ductile shear zones. Where the foliation is sufficiently intense the rocks have "retrograded" to the greenschist facies.

Quartz-rich flysch of the Tumut Trough sequence, the Bumbolee Creek Formation, although apparently conformable on a regional scale differs from the remainder of the Tumut Trough sequence within the Tumut Synclinorial Zone by locally showing complex mesoscale interference patterns. These relate to conjugate flexures superposed on the main cleavage-producing event. Metamorphism is of lowermost greenschist facies.

In the remainder of the Tumut Synclinorial Zone, pelitic rocks exhibit lowermost greenschist facies' assemblages and have a penetrative, north-northwest trending, subvertical cleavage. Minor kink folds locally overprint the cleavage and become far more intense in the vicinity of fault and shear zones.

The Young Anticlinorial Zone in the Tumut region consists of the Late Silurian Young Granodiorite, its marginal phases, and the Middle Silurian Goobarragandra Volcanics and Micalong Swamp Basic Igneous Complex. Away from ductile shear zones the rocks are essentially unfoliated.

With the possible exception of the Bumbolee Creek Formation, the degree and complexity of deformation, and intensity of metamorphism displayed by the structurometamorphic units, are consistent with age-relations within the recognised terranes.

Terrane Boundaries

The principal faults delimiting the terranes are the Gilmore fault zone, the Cootamundra-Gundagai-Killimicat fault system, and the Mooney Mooney fault system (Fig. 2).

<u>The Gilmore Fault Zone</u>. The Gilmore fault zone [Crook and Powell, 1976], referred to as the Gilmore suture by Scheibner [1985], forms the boundary between the Wagga-Omeo and Tumut terranes. On LANDSAT imagery, the fault zone corresponds with a broad, extremely prominent lineament, extending for several hundred kilometres from southeastern Victoria to central-western New South Wales. The fault zone crops out poorly in the Tumut region although our recent, detailed mapping has established a fault zone at least three kilometres wide. Within this narrow zone, several major fault systems delimit a number of moderately to intensely deformed and displaced blocks of Ordovician Wagga-Omeo terrane metasediments, intermediate volcanics and Middle to Late Silurian granodiorite. Individual faults and mylonite zones within the Gilmore fault zone form a braided pattern and are either slightly oblique or parallel to the penetrative, north-northwest trending, steeply east-dipping schistosity which characterises the metasediments of the Wagga-Omeo terrane in the Tumut district. The surface expression of the Gilmore fault zone resembles that of a weakly emergent or an eroded, tip-stick thrust front, as described by Morley [1986].

Based on a rift model and the thickness of the Tumut Trough sequence, Basden suggests a relative displacement during the Silurian of several thousand metres west-side up. Basden also suggests that the fault underwent right-lateral movement in the Late Silurian during the closure of the rift basin. The suggestions are, however, only as strong as the parautochthonous rift-basin model.

The Cootamundra-Gundagai-Killimicat Fault System. The Cootamundra-Gundagai-Killimicat fault system refers to the set of structures which collectively separate the Tumut and Jindalee terranes. It includes the Cootamundra fault [Basden, 1974], the Gundagai shear zone [Basden, 1982], and the Killimicat fault [Basden, 1982].

The Cootamundra fault is expressed as a zone of intense foliation, up to one kilometre wide, which narrows southward, and possibly becomes en echelon with the Gundagai shear zone and Killimicat fault southeast of Gundagai (Fig. 2). Basden et al. [1978] suggest a very steep easterly dip based on the attitude of the foliation. Although the fault is a boundary to terranes which amalgamated by the Early Devonian, a post Late Devonian movement of at least 3500m (west side down) can be substantiated to the north of the Tumut district [Basden, et al., 1978].

A subvertical zone of intense foliation and intermittent slivers of serpentinite characterise the Gundagai shear zone. The serpentinite is possibly of Cambro-Ordovician age but this is not unequivocal evidence. Although the northward extension of the shear zone has been mapped along the east side of the Jindalee Group in this region, it is likely that branches of the shear zone pass along both sides of the Jindalee Group.

The Killimicat fault is a faulted shear zone marked by a 200m-wide zone of intense foliation, kinking, and vein quartz. The foliation, and by inference the fault, dips steeply westward. Kink folds suggest a right-lateral component of movement which, because they fold the shear zone foliation which is form surface to the kinks, must have been late in the fault's history. Based on the rift-basin model and contrasts in thickness of the Tumut Trough sequence on either side of the fault, Basden has proposed a Silurian movement of up to 6000m west side down.

The Mooney Mooney Fault System. The Mooney Mooney fault system is a complex faulted shear zone (up to several kilometres wide) comprising two main faults which bound the Mooney Mooney terrane, and between which are numerous braided components. Immediately east of the main fault, the Young Granodiorite has been mylonitised over a zone ranging from zero to several hundred metres wide [Ashley and Chenhall, 1976]. Dips of 65^{0}E to 90^{0} on the eastern bounding fault are usual [Ashley et al., 1972; Basden et al., 1978], but the fault has been observed to be subhorizontal in some places. The S and C planes [Lister and Snoke, 1984] are subvertical and strike around 335^{0} and 350^{0} respectively. The elongation lineation (transport direction) plunges 50^{0}-70^{0} to the northwest, and kinematic indicators require a right-lateral, east side up motion. Basden also suggests a right-lateral movement, based on stratigraphic evidence. This portion of the Mooney Mooney fault system is therefore an east-dipping high-angle reverse fault with a right-lateral component. It is consistent with west-south-west thrusting of the Young terrane over the Mooney Mooney terrane.

We emphasise that parts of the boundary between the Mooney Mooney and Young terranes do not apparently conform with a simple high angle reverse fault [e.g. Basden et al., 1978]. We similarly emphasise that, although it is tempting to interpret the other terrane boundaries of the Tumut Trough as high angle reverse faults, more evidence is required before their nature can be stated with certainty.

Geophysics of the Terranes

A lack of detailed geophysical data, particularly palaeomagnetic determinations and seismic profiles, is one of the main factors hampering the recognition and definition of terranes in all Australian Palaeozoic mobile zones.

Available geophysical data in the Tumut district are largely restricted to regional surveys by the Bureau of Mineral Resources, Geology and Geophysics [Wyatt et al., 1980]. These data indicate the existence of two distinct zones separated by the Gilmore fault zone.

Most magnetic anomalies coincide with exposures of both mafic and felsic igneous units. The anomalies tend to be linear with trends of between 325^{0} and 025^{0} [Wyatt et al., 1980], reflecting the general trend of lithologic units in the Tumut region. Anomaly amplitudes reflect the bulk magnetic susceptibilities of rocks comprising each unit. The Gilmore fault zone corresponds with a marked change in anomaly amplitude. The Wagga-Omeo terrane is characterised by relatively low intensity northwest to north-northwest trending anomalies, with both the Ordovician intermediate volcanics and Silurian granodiorites of the terrane having, in general, low magnetic susceptibilities. The Tumut, Jindalee, Mooney Mooney and Young terranes, in contrast, are characterised by high amplitude, prominent, elliptical, north-northwest to north-northeast trending anomalies.

The Gilmore fault zone is marked by several coincident gravity and magnetic highs, thought to correspond with the Silurian Avenall Basic Intrusive Complex, a series of tholeiitic gabbro, norite and diorite stocks within the Wagga-Omeo terrane. The Coolac ophiolite suite and Honeysuckle Metabasic Igneous Complex of the Mooney Mooney terrane and the Jindalee Group, although associated with prominent magnetic anomalies, do not have significant gravity signatures. This suggests that these units have relatively little depth persistence, an interpretation that is consistent with a recent gravity survey over the Coolac ophiolite suite [P. Wellman, pers. comm.].

A detailed gravity survey [Watts, 1971] over the Mooney Mooney fault system near Cootamundra has been subject to differing interpretations [e.g. Watts, 1971; Basden et al., 1978]. All agree that the fault is east dipping; differences relate to whether it is a thrust or a high-angle reverse fault to a depth of about five kilometres. Since data obtained by surface mapping support a steep eastward dip, the high-angle reverse fault interpretation is preferred, although the available evidence is less than conclusive.

Detailed gravity data are not available for the Gilmore fault zone, major faults bounding the Tumut terrane, and the southern Mooney Mooney fault system. However, the small density contrasts between units on opposite sides of these fault systems could well preclude adequate resolution of fault orientation.

The nature of possible thrust systems into which the high angle reverse faults pass will only be resolved by detailed seismic reflection profiles over significant sections of the Tumut region. Nevertheless, we anticipate that at least the Tumut, Jindalee, Mooney Mooney and Young terranes are imbricate slices above a major plane of delamination.

Discussion and Conclusions

In view of the mediocre exposure and the paucity of palaeontologic, palaeomagnetic, and detailed structural data available from the region, lithostratigraphic correlation supplemented by geochemistry and radiometric age determinations have been the principal basis for tectonic interpretations. However, lithostratigraphy must be treated cautiously and, particularly where many age data are "soft", do not constitute a sound basis for plate tectonic reconstructions, such as the rift-basin model for the Tumut Trough and adjacent terranes. By recognizing terranes as suspect, lithostratigraphic imperfections are not overcome, but are at least placed in the context of the need for good, hard additional data.

In the current paper, we have summarised the basis for recognizing at least five terranes in the Tumut region and contend that they are suspect. Without palaeomagnetic and palaeobiogeographic control, we cannot discriminate between a parautochthonous rift-basin model for the Tumut Trough, an allochthonous thrust tectonic model, and an allochthonous strike-slip tectonic model. And this problem, some would contend [e.g. Scheibner, 1986], is common to much of the Lachlan Fold Belt. Nevertheless, on the strength of the present data, we believe that the Tumut Trough need never have a been rift-bound trough, and that the terrane distribution could well fit a thrust-tectonic, allochthonous model of amalgamation and presumably accretion.

Acknowledgements. Access to unpublished data of P. M. Ashley, G. M. Bradley, S. Dobos, P. Dupen, P. Harris, H. K. Jefferson, J. M. Kennard, J. D. Lightner, A. Lohan, S. Macklin, M. Murphy and P. Wellman assisted greatly in the preparation of this paper.

H. Basden publishes with the permission of the Secretary, New South Wales Department of Mineral Resources.

References

Ashley, P.M., Brown, P.F., Franklin, B.J., Ray, A.S. and Schneiber, E., Field and geochemical characteristics of the Coolac ophiolite suite and its possible origin in a marginal sea, J. Geol. Soc. Aust., 26, 45-60, 1979.

Ashley, P.M. and Chenall, B.E., Mylonitic rocks in the Young Granodiorite, southern New South Wales, J. Proc. Roy. Soc. N.S.W., 109, 25-33, 1976.

Ashley, P.M., Chenall, B.E., Cremer, B.E., Cremer, P.L., and Irving, A.J., The geology of the Coolac Serpentinite and adjacent rocks east of Tumut, New South Wales, J. Proc. Roy. Soc. N.S.W., 104(1-2), 11-29, 1972.

Barkas, J.P., Early Devonian igneous activity and some stratigraphic correlations in the Tumut region, New South Wales, Proc. Linn. Soc., N.S.W., 111(1), 13-26, 1976.

Basden, H., Preliminary report on the geology of the Cootamundra 1:100,000 sheet, Q. Notes Geol. Surv. N.S.W., 15, 7-18, 1974.

Basden, H., Preliminary report on the geology of the Tumut 1:100,000 sheet area, southern New South Wales, Q. Notes Geol. Surv., N.S.W., 46, 1-18, 1982.

Basden, H., Adrian, J., Clift, D.S.L., and Winchester, R.E., Geology of the Cootamundra 1:100 000 Sheet 8528, New South Wales Geological Survey, Sydney, 276 pp. 1978.

Basden, H., Franklin, B.J., Marshall, B. and Waltho, A.E., Tectonostratigraphy of the Tumut Trough and adjacent terranes. 3rd Circum-Pacific Terrane Conference, Sydney, Geol. Soc. Aust. Abstracts 14, 16-21, 1985.

Cas, R.A.F., Palaeogeographic and tectonic development of the Lachlan Fold Belt, southeastern Australia, Spec. Pub. 10, Geol. Soc. Aust., 104pp, 1983.

Cas, R.A.F., Powell, C.McA. and Crook, K.A.W., Ordovician palaeogeography of the Lachlan Fold Belt : a modern analogue and tectonic constraints, J. Geol. Soc. Aust., 27, 19-31, 1980.

Coney, P.J., Jones, D.L. and Monger, J.W.H., Cordilleran suspect terranes, Nature, 288, 329-333, 1980.

Crook, K.A.W., Forearc evolution in the Tasman Geosyncline : the origin of the southeastern Australian continental crust, J. Geol. Soc. Aust., 27, 215-32, 1980.

Crook, K.A.W. and Powell, C.McA., The evolution of the southeastern part of the Tasman Geosyncline, 25th Int. Geol. Congr., Field Guide, Excursion 17A, 122pp, 1976.

Degeling, P.R., Gilligan, L.B., Schneiber, E., and Suppel, D.W., Metallogeny and tectonic development of the Tasman Fold Belt system in New South Wales, Australia, in Schneiber, E. (ed.), Metallogeny and tectonic development of eastern Australia, Ore Geology Reviews, (in press), 1986.

Evernden, J.F. and Richards, J.R., Potassium-argon ages in eastern Australia, J. Geol. Soc. Aust., 9, 1-49, 1962.

Fagan, R., S-type granite genesis and emplacement in N.E. Victoria and its implications, Abstracts, Crust and Upper Mantle of Southeast Australia, Aust. Bur. Miner. Resour., Geol. Geophys., Rec., 1979/2, 29-30, 1979.

Franklin, B.J. and Ashley, P.M., Genesis of chromitite and sulphide-bearing deposits in ophiolitic rocks, New South Wales, Australia, Proc. Intl. UNESCO Symposium on Metallogeney of Mafic and Ultramafic Complexes, Athens, Greece, Oct. 1980, 37-50, 1980.

Gilligan, L.B. and Schneiber, E., The Lachlan Fold Belt in New South Wales, Tectonophysics, 48, 217-265, 1978.

Leitch, E.C. and Schneiber, E., A Tasman Fold Belt perspective on terrane analysis, 3rd Circum-Pacific Terrane Conference, Sydney. Geol. Soc. Aust., Abstracts 14, 136-139, 1985.

Lister, G.S. and Snoke, A.W., S-C mylonites, J. Struct. Geol., 6, 617-638, 1984.

Morley, C.K., A classification of thrust fronts, Bull. AAPG, 70, 12-25, 1986.

Owen, M. and Wyborn, D., Geology and geochemistry of the Tantangara and Brindabella 1:100,000 sheet areas, New South Wales and Australian Capital Territory, Aust. Bur. Miner. Resour. Bull. 204, 52pp, 1979.

Powell, C.McA., Tectonic relationship between the Late Ordovician and Late Silurian palaeogeographies of southeastern Australia, J. Geol. Soc. Aust., 30, 353-373, 1983.

Richards, J.R., Barkas, J.P. and Vallance, T.G., A lower Devonian point in the geological time scale, Geochem. J., 11, 147-153, 1977.

Scheibner, E., Tectonic concepts and tectonic mapping, N.S.W. Geol. Surv. Rec., 14, 37-83, 1972.

Scheibner, E., A plate tectonic model of the Palaeozoic tectonic history of New South Wales, J. Geol. Soc. Aust., 20, 405-426, 1973.

Scheibner, E., An outline of the tectonic development of New South Wales with special reference to mineralisation, in Markham, N.L. and Basden, H., (eds.), The Mineral Deposits of New South Wales, N.S.W. Geol. Surv., 1-39, 1975.

Scheibner, E., Suspect terranes in the Tasman Fold Belt System, eastern Australia, in Howell, D.G., (ed.), Tectonostratigraphic Terranes of the Circum-Pacific Region, Circum-Pacific Council for Energy and Mineral Resources, Earth Science Series, No. 1, 493-514, 1985.

Suppel, D.W., Warren, A.Y.E., Watkins, J.J., Chapman, J., Tenison Woods, K. and Barron, L., A reconnaissance study of the geology and gold deposits of the West Wyalong-Temora-Adelong District, Q. Notes Geol. Surv. N.S.W., 64, 1-30, 1986.

Watts, M.D., Geological interpretation of the Cootamundra gravity high, New South Wales, J. Geol. Soc. Aus., 18, 220-234, 1971.

White, S.H., Burrows, S.E., Carreras, J., Shaw, N.D. and Humphyeys, F.J., On mylonites in ductile shear zones, J. Struct. Geol., 2, 175-187, 1980.

Wyatt, B.W., Yeates, A.M. & Tucker, D.H., A regional review of the geological sources of magnetic and gravity anomaly fields in the Lachlan Fold Belt of New South Wales, B.M.R. Journal of Aust. Geol. Geophys. 5, 289-300, 1980.

THE EASTERN LACHLAN FOLD BELT OF SOUTHEAST AUSTRALIA: A POSSIBLE LATE ORDOVICIAN TO EARLY DEVONIAN SINISTRAL STRIKE SLIP REGIME

Gordon H. Packham

Ocean Sciences Institute and Department of Geology and Geophysics
The University of Sydney, Australia 2006

Abstract. The eastern part of the Lachlan Fold Belt has been interpreted as an Ordovician volcanic arc with a back-arc filled with quartz-rich turbidites. Using the present location of the arc as a datum two major dislocations have been identified. The western one has a displacement of 500km and the eastern one brings the forearc zone up against the back arc basin and has a minimum displacement of 350 km but is likely to be much larger. Ten other structures have been identified with smaller displacements. The sense of the displacements suggests a sinistral shear regime. A complex sequence of events followed the cessation of subduction and the onset of a strike-slip regime. The major features of the latest Ordovician to Early Devonian evolution of the belt can be explained within the framework of a sinistral shear model.

Introduction

The Lachlan Fold Belt of eastern Australia (the Lachlan Geosyncline of Packham, 1960) records a history of tectonic evolution from Cambrian to Early Carboniferous. This paper is mainly concerned with the development of the eastern part of the fold belt during its most active period from the Late Ordovician to the Early Devonian and in passing its relationship to the adjacent western part of the New England Fold Belt (the New England Geosyncline of Voisey [1959]) to the east. Although the oldest rocks in the New England Fold Belt are Cambrian, tectonic activity ended there in the early Triassic and most of its rocks are Devonian and younger [Leitch,1974]. The two fold belts are separated by the Permo-Triassic Sydney-Gunnedah Basin [Bembrick et al, 1980] and the northern part of the Lachlan Fold Belt is covered by the Triassic to Cretaceous Coonamble Embayment of the Surat Basin [Bembrick, 1976].

The tectonic framework of the eastern Lachlan Fold Belt in the Ordovician has been interpreted as a back-arc basin and volcanic arc regime [eg Oversby,1971, Packham and Falvey, 1971, Packham,1973, Scheibner,1974]. The cessation of arc volcanism at towards the end of the Ordovician was followed by a complex sequence of Silurian to Early Carboniferous events including folding, metamorphism, granitic intrusion, acid to intermediate volcanism and deposition of shallow to deep water sediments. During this interval the region became progressively cratonised [Crook,1974,1980] but as observed by Packham [1960] there were no further intervals dominated by andesitic volcanism of regional extent. Andesitic volcanism after this time such as in the Tumut Trough and on the northern Molong High is of local extent and of unknown tectonic significance. The change of tectonic style at the end of the Ordovician has been interpreted as the result of the cessation of subduction and the coupling of plate motions to the region occupied by the fold belt [Packham, 1973, Packham and Leitch, 1974]. Given that there was significant post-Ordovician deformation, the present configuration of the arc provides a clue to the style and magnitude of the subsequent deformation.

West of the volcanic arc is an extensive Ordovician quartz-rich turbidite facies derived from a southwesterly continental mass [Packham, 1973, Webby,1976] filling what has been supposed to be a back-arc basin. In the area south and west of Bathurst the volcanic and quartzose turbidite facies are intermixed on the 1:250 000 scale of mapping [Packham, 1967] and intimately interbedded in a single stratigraphic section [Scheibner, 1973] and can hence be regarded as parts of the same terrane. In the interpretation presented here this back arc-basin occupies two major areas referred to by Webby [1976] as the Wagga and Monaro Troughs (Figure 1). Scheibner [1974] regarded the southeastern highlands and the coastal strip as a fore-arc and trench complex that he called the Monaro Slope (Webby's Monaro Trough) and Basin. The tectonic interpretations of the eastern Lachlan Fold Belt made to date apart from that in a very recent paper by Ferguson et al [1986] have assumed only extension and compression with no significant changes in later-

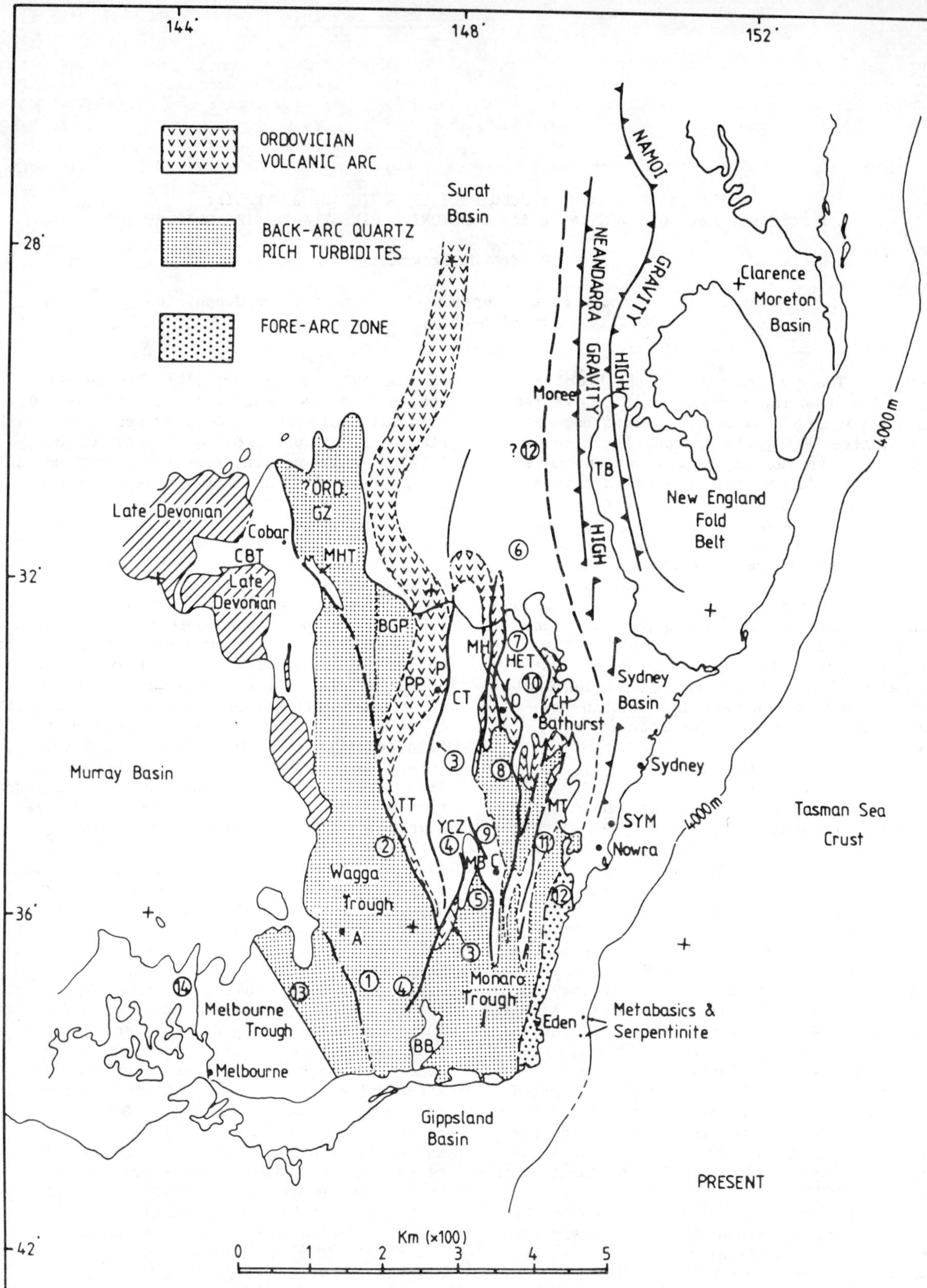

ORDOVICIAN VOLCANIC ARC
BACK-ARC QUARTZ RICH TURBIDITES
FORE-ARC ZONE
Surat Basin
NAMOI
NEANDARRA GRAVITY HIGH
GRAVITY HIGH
Clarence Moreton Basin
Moree
TB
New England Fold Belt
Late Devonian
Cobar
CBT
?ORD. GZ
MHT
BGP
PP
CT
MH
HET
CH
Bathurst
Sydney Basin
Sydney
Murray Basin
TT
YCZ
MT
SYM
Nowra
Tasman Sea Crust
Wagga Trough
A
Monaro Trough
Melbourne Trough
Melbourne
BB
Eden
Metabasics & Serpentinite
Gippsland Basin
PRESENT
4000m
Km (×100)

al geographic relationship of major tectonic elements [eg Packham,1973, Scheibner,1974, Cas et al,. 1980, Powell,1983a].

There has been a conflict of opinion about the nature of the unexposed basement under the Lachlan Fold Belt. The two alternative views put forward are that, as implied here, the crust was oceanic or that it was Precambrian continental crust. Unfortunately neither the deeper parts of the sediment sequence nor the basement are exposed. The arguments for a Precambrian basement are largely based on the geochemistry of the granites generated at depth while those for an oceanic regime are based on the outcropping rocks. The main points with appropriate references have been summarised by Powell [1984]. Apart from the circumstantial evidence of the geochemistry the only piece of evidence that may point to the presence of an older basement under the arc in at least the Molong High between the Cowra Trough and the Hill End Trough (Figure 1) is the occurrence of detrital grains in the lower part of the Gisbornian-Eastonian (Caradoc) Cliefden Caves Limestone on the arc that may have come from a metamorphic source [Webby and Packham, 1982].

The supply of quartz-rich sediment being fed into the region was not confined to the Ordovician. In the eastern part of the Lachlan Fold Belt there are intervals in which quartz-rich sediments are widespread for instance in the Early Silurian [Crook et al., 1973] and the Late Silurian as indicated in the outline of the tectonic history at the end of this paper. Whether this sediment came from outside the region or was recycled within the eastern part of the fold belt is not easy to assess but it is clear that the supply continued in the western part of the region, for instance in the Silurian to Early Devonian in the Melbourne Trough [VandenBerg and Garratt, 1976, Garratt, 1983] and in the Early Devonian in the Grampian Mountains of western Victoria [Spencer-Jones,1976] and the Cobar area of New South Wales [Glen,1982]. The eastern Lachlan Fold Belt was a sediment source in the Middle Devonian but received large volumes of quartz-rich sediment from the west in the Late Devonian [Powell,1984]. This establishes a prima faciae case for the fold belt having been adjacent to the Gondwanaland continental margin during its development but it does not preclude the possibility of terrane shuffling by movements essentially along strike as suggested here and in Fergusson et al. [1986].

Fergusson et al. [1986] pointed out that the region under discussion in this paper (their Benambra Terrane which "incorporates much of the Lachlan Fold Belt in New South Wales and eastern Victoria") has a contrasting geological history from the Melbourne Terrane to the west. They propose that the Wagga Trough (the western part of their Benambra Terrane) and the Melbourne Trough have been brought together some time in the Silurian to Early Devonian by an unknown amount of strike slip movement. They consider it to be dextral. They place the boundary between the two terranes at the eastern edge of the Mount Wellington "axis" of Thomas [1939] where an abrupt change of lithology occurs in the Late Ordovician strata from predominantly black shale on the west to thick quartzose turbidites on the east.

In this paper the argument will be developed that the Ordovician volcanic arc was initially a simple meridional belt and that the present distribution of the volcanics is the result of a number of sinistral strike-slip dislocations that took place between the Late Ordovician and the Early Devonian. The timing and sequence of these postulated dislocations is discussed and their development related to the complex tectonic evolution of the eastern part of the Lachlan Fold Belt.

Reconstruction of the Ordovician Arc

The Ordovician arc rocks that outcrop in the eastern Lachlan Fold Belt can be traced under the Mesozoic cover on regional gravity and aeromagnatic maps (Bureau of Mineral Resources) as gravity and magnetic highs (Figure 1). The western trend stands out clearly as a strong line of magnetic anomalies and a gravity high. On the basis of the character of the mangetic anomalies the northern end of the Molong High has been interpreted as swinging to the west and ending. An alternative interpretation is that it continues to the northnorthwest and meets the western volcanic trend between 30 and 31 degrees south following a gravity ridge. This second interpretation makes no material difference to the arguments presented here. The incorporation of the regional geophysics and field mapping by Suppel et al [1986] are the main reasons why Figure 1

Fig. 1. Present day tectonic elements. Major faults and sutures: (1) Keiwa Fault; (2) Gilmore Fault; (3) Kiandra-Narromine Structure; (4) Long Plain-Indi Fault; (5) Murrumbidgee Fault; (6) Columbine Thrust; (7) "Bell River" Fault; (8) Copperhannia-Lake George Fault; (9) Deakin Fault; (10) Wiagdon Thrust; (11) Lake Bathurst-Tarago Structure; (12) Eden-Nowra-?Moree Structure; (13) Mt. Wellington Fault; (14) Mt. Ida-McIvor Fault. Smaller subdivisions of the Lachlan Fold Belt: CBT, Cobar Trough; GZ, Girilambone Zone; MHT, Mineral Hill Trough; BGP, Bogan Gate Platform; PP, Parkes Platform; TT, Tumut Trough; CT, Cowra Trough; YCZ, Yass-Canberra Zone; MB, Murrumbidgee Basin; BB, Buchan Basin; MH, Molong High; HET, Hill End Trough; MT, Murruin Trough. Names: A, Albury; B, Bathurst; C, Canberra; Co, Cobar; O, Orange; P, Parkes; SYM, stockyard Mountain 1.

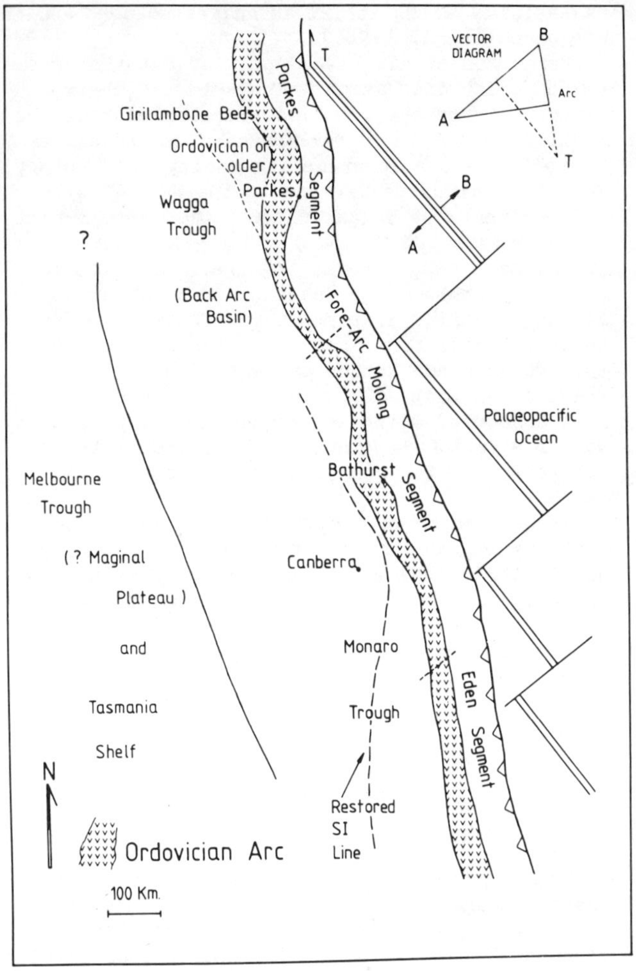

Fig. 2. Reconstruction of the eastern Lachlan Fold Belt at the end of the Ordovician.

differs significantly from the map of Cas et al [1980]. The pattern delineated is quite complex and is clearly not the outcrop of a simple arc. The assumption that an initially simple linear belt has been dislocated by subsequent deformations is the starting point of the present interpretation.

Thus, using the arc as a datum, two major dislocations can be postulated (Figure 1). The western one, the Kiandra - Narromine structure is a sinistral dislocation. The eastern one the Eden - Nowra - ?Moree structure is also possibly sinistral. Both are discussed below. When the displacement on the Kiandra - Narromine structure is restored the Monaro Trough is seen to be the southern extension of the Wagga Trough (Figure 2). Within the area east of the Melbourne Trough eleven other possible sinistral dislocations (Figure 1) have been identified. They appear to displace the arc on a smaller scale or juxtapose unlike stratigraphic successions. Sinistral shear also appears to have controlled the location and orientation of some of the major plutonic bodies that were emplaced in the belt in the Silurian and Early Devonian and which occupy about thirty five percent of its present erosion surface. The interpretation of the development of the region under the influence of sinistral shear is contrary to the views expressed by Cas et al. [1980], Powell [1983a] and Fergusson [1985] who argue that epi-Ordovician east-west folds that have been identified in a number of locations have been formed by dextral shear. These folds are overprinted at some of these locations by north-south folds that are related to the regional metamorphism of the Wagga Trough. Powell [1983a] goes on to develop the dextral shear argument to explain the formation of extensional Siluro-Devonian troughs within the region (e.g. the Hill End, Cowra and Cobar Troughs, Figure 1). Although in my opinion the observed features can be explained by sinistral shear the possibility of an interval or intervals of dextral motion is not precluded. The timing of such events is explored below.

Within the framework of my sinistral hypothesis, there are two ways in which the observed fold overprinting could occur: a) the folds were formed in a north-south orientation and rotated anticlockwise into an east-west direction by the shear before the imposition of a second set of folds or, b) the sinistral deformation that was responsible for the overprinting was preceded by a dextral phase.

The argument for the early east-west folds indicating dextral shear depends on the orientation of the folds at the time of their formation being the same as at present. However, all that can be inferred is the angular relationship at the time of overprinting. The scale of possible rotation could range from the mesoscopic to that of regional blocks. If as suggested above the early folds were rotated by the sinistral shear before a second phase of compression then the observed relationship would result. Furthermore if this were the correct explanation then neither the older nor the younger fold sets need to be the same age throughout the belt. The only requirement would be that at any locality there were two phases of folding.

In a region undergoing deformation by shear stress the sense of rotation of blocks will be controlled by the sense of the shear but the rotated blocks will be bounded by conjugate shears. This has been described recently by Nicholson at al. [1986] in the San Andreas Fault System. In the sinistral shear model proposed here for the Lachlan Fold Belt the direction of rotation of blocks should be anticlockwise as a response to motion on the major sinistral shear and conjugate strike-slip faults should be dextral. Dextral motion on shears in the region are suggested by Basden et al. [this volume]. Anticlockwise rotations may also affect the interpretation of paleoslope determinations from sedimentary structures. The compilation of determinations by Powell [1983a] indicates a reasonably consistent northerly sediment movement. If rotation had occurred at any of the measurement localities the movement indicated would be more easterly which would be more compatible with continent derived sediment filling the back-arc basin.

As an alternative to the single simple strike-slip model it may be supposed that that there was a period of dextral shear. It appears from the following discussion that it may a) have occurred before the displacement on the western dislocation or b) be associated with the formation of the eastern dislocation. These possibilities will be examined in the following discussion.

The Kiandra - Narromine Structure

This postulated structure separates the western outcrops of the Ordovician arc extending north and south through Parkes, from the eastern ones on the Molong High between the Cowra and Hill End Troughs and dislocated fragments further east. It is envisaged that this transform structure parallels the arc north of Parkes. In the south it cuts obliquely across the arc, passing into the back-arc basin where it loses its identity. At the southern limit of the arc in the vicinity of Kiandra part of the arc has been displaced to the northeast by the later Long Plain Fault (4 on Figure 1). From the mapping of Owen and Wyborn [1979] in this vicinity the Kiandra-Narromine structure is obscured by the overlapping Early Silurian Tantangara Formation except for a distance of 10 km where the volcanics of the Kiandra Group are separated from the quartz-rich turbidites of the Nungar Beds by an adamellite intrusion. The equivalence of age of the two sedimentary units has not been established. The youngest fossils in the Kiandra Group are Gisbornian (early Late Ordovician) and the only fossils identified from the Nungar Beds are Eastonian (middle Late Ordovician).The postulated displacement on the Kiandra - Narromine structure is about 500km. Along the Kiandra - Narromine structure are a number of ultramafic bodies , disrupted ophiolites, chert formations, basic volcanics and quartz-rich turbidites. These rocks it is suggested are a melange of elements from the fore-arc, arc and back-arc, together with minor new crust generated either during the main interval of transform motion or later along the pre-existing zone of weakness. The Tumut Trough is of profound significance in the understanding of the tectonic evolution of the fold belt but lack of detailed published stratigraphic and structural information make it impossible to arrive at an interpretation of its history.

The geological time scale of Harland et al. [1982] (Table 1) is used throughout this discussion, however a more recent scale for the Ordovician, Silurian and Devonian devised by McKerrow et al. [1985] is also quoted in the table. For the purposes of the present discussion the two scales are satisfactory. The McKerrow et al. scale has shorter Ashgill and Pridoli intervals.

Returning to the dating of a possible pre-sinistral dextral phase, the youngest fossils in the quartz-rich turbidites of the Wagga and Monaro Troughs are early Bolindian (early Ashgill) [Webby, 1981] dated at about 445 Ma on the scale of Harland et al. [1982] (Table 1).In the Kiandra area east of the Long Plain Fault the Ordovician arc volcanics are separated from the quartz-rich turbidites by part of the ?Late Silurian Kosciusko Batholith but both volcanics and turbidites are unconformably overlain by the Tantangara Formation which is in turn overlain by the Peppercorn Beds from which a late Llandovery (Telchyan) conodont fauna has been obtained [Owen and Wyborn,1979]. Since the unconformity beneath the Peppercorn Beds is a strong one the Tantangara Beds are considered to be early Llandovery. Thus there is possibly an interval of about 10 Ma (445 to 435 Ma) between the end of Ordovician sedimentation and the deposition of the Tantangara Beds. The sinistral displacement of the arc on the Kiandra - Narromine structure would have occurred in this interval but it is unlikely to be the same movement that was responsible for the folding at the time of the Wagga Trough metamorphism. Plate motion was probably taken up by strike slip first and by more areally extensive deformation when fault movement ceased. If this is correct then also relevant are the dates for the plutons that are associated with the metamorphism but post-date its maximum and cut across some of the metamorphic zones but are not themselves deformed. They put a minimum age on the deformation. Brooks and Leggo [1972] date them at 421+/-8 Ma. Richards and Singleton [1981] present a number of dates for minerals in plutonic rocks intruding the metamorphics including four results about 430 Ma but many are younger. They attribute this to "long duration heating or subsequent disturbance". On the radiometric evidence the aureole granites are unlikely to be younger than early Wenlock (base 428 Ma) and hence the longitudinal folds associated with the metamorphism are best dated as Early Silurian. The suggested transform displacement of 500km took place before the deposition of the Tantangara Beds. It is therefore likely that the major deformation of the Wagga Trough commenced after the cessation of strike-slip at about 435 Ma during the deposition of the Tantangara Beds and ended after their deformation; possibly by the start of the Wenlock. If there was dextral motion it most likely occurred prior to the sinistral strike-slip; that is in the late Ordovician.

The Eden - Nowra - ?Moree Structure

A second major dislocation is postulated onshore near the New South Wales south coast. The rocks of the Wagonga Beds outcropping along the coast have been interpreted as part of a subduction complex [Packham,1973, Scheibner, 1972,

TABLE 1. Time Scales in Ma

		a.	b.
Early Carboniferous		320-360	-354
Late Devonian	Famennian	360-374	354-365
	Frasnian	367-374	365-374
Middle Devonian	Givetian	374-380	374-383
	Eifelian	380-387	383-391
Early Devonian	Emsian	387-394	391-401
	Siegenian	394-401	401-406
	Gedinnian	401-408	406-412
Silurian	Pridoli	408-414	412-414
	Ludlow	414-421	414-420
Middle Silurian	Wenlock	421-428	420-425
Early Silurian	Llandovery	428-438	425-436
Late Ordovician	Ashgill	438-448	436-442
	Caradoc	448-458	442-454
Middle Ordovician	Llandeilo	458-468	454-461
	Llanvirn	468-478	461-470
Early Ordovician	Arenig	478-488	470-492
	Tremadoc	488-505	492-513

(a) Harland et al.[1982], and (b) McKerrow et al. [1985].

Jenkins et al, 1982]. Detailed studies of the coastal outcrops by C. Jenkins (pers comm.) and Powell [1983b] have provided evidence that the deformed sediments have the characteristics of an accretionary wedge. The rocks are cherts, basic volcanics, quartzo-felspathic greywackes and melange. In addition serpentinites, basic metavolcanics and deformed cherty sediments have been dredged by members of the Ocean Sciences Institute on the HMAS Cook from the continental slope at several localities up to 60 km east of the coast. These are probably an eastward extension of the coastal fore-arc subduction complex.

The sediments in this coastal strip about 20km wide have what Powell [1983b] calls stripy cleavage. Further west is the narrow Comerong Syncline of Late Devonian bimodal volcanics and overlying shallow water quartzose sediments [McIlveen, 1974]. West of the syncline are Ordovician sedimentary rocks with slaty cleavage, they are part of the Darriwillian (Middle Ordovician) [Jenkins, 1982] to early Bolindian [Webby, 1981] quartz-rich turbidite sequence of the Monaro Trough in which northerly-directed palaeocurrent measurements have been made at several localities [Powell, 1983a]. The volcanic arc and fore-arc basin are absent between the Monaro Trough which is interpreted as the back-arc basin and the accretionary wedge. It is suggested that the elongated Comerong Syncline has developed along the line of a transform structure that has brought together the rocks of these two tectonic elements.

Accepting the existence of this structure the next problem is to determine what has happened to the arc from which the accretionary wedge has been detached (the Eden segment). Only broad time constraints can be put on the time of movement but it is noteworthy that in the accounts of the Silurian stratigraphy of the area west of the Comerong Syncline [Felton and Huleatt, 1975] where the oldest beds are Ludlovian there is no mention of the presence of basic-intermediate volcanic detritus. If it is assumed that a phase of dextral movement dislocated the arc the fore-arc zone could have been carried south from the Parkes or Molong Segments displacing the arc that lay east of the Monaro Trough southwards. It could have only come from the Parkes Segment if this dextral dislocation preceded displacement on the Kiandra-Narromine structure. The minimum dextral displacement would have been 350km which would have had to occur before the development of the structurally complex pattern of deformation of the Molong segment of the arc under sinistral shear from the Middle Silurian to the Early Devonian. The timing of such dextral motion presents difficulties. Within the framework of the present discussion there are two possiblities, right at the end of the Ordovician or in the Middle Silurian. If it was at the end of the Ordovician it would correspond to the time that the early east-west folds were formed in the Wagga Trough but this interval was probably not longer than 5 Ma. The second possibility is that it was in the Middle Silurian after the sinistral movement of the Tumut - Narromine structure and the Wagga Trough deformation and prior to the sinistral deformation of the Molong segment of the arc and the formation of the Hill End Trough in the Late Silurian.

The other alternative is that the arc was

affected only by sinistral dislocation. This would mean that the Eden segment of the arc moved northward either doubling the arc, or if the displacement was greater, removing it from the region altogether.This movement could have taken place at any time from the early Wenlock to the Early Devonian as a single movement or as a succession of movements. There are two appropriate candidates for the lost arc. The first is the old volcanic arc of New England. The New England arc was east facing [Leitch,1975]. The outcropping lower Paleozoic fore-arc accretionary wedge (the Woolomin Formation) comprises cherts, meta-basalt, argillite, volcanogenic sediment and melange, an assemblage similar in a number of respects to the Wagonga Beds of the New South Wales south coast region. The cessation of subduction beneath the New England arc in the Silurian may be indicated by the hiatus recorded by Cawood [1976] between Middle Ordovician and Early Devonian strata of the arc west of the accretionary wedge. However a number of significant differences suggest that the New England arc, if part of the same Ordovician arc, was remote from the Lachlan Fold Belt from the Early Devonian to mid-Carboniferous. Thus a) quartzo-feldspathic greywackes abundant in the Wagonga Beds have not been described from the Woolomin Beds b) subduction recommenced in the New England arc in the Early Devonian, c) the widespread mid-Devonian Tabberabberan Orogeny of the Lachlan Fold Belt is absent from New England, d) the abundant quartz-rich Late Devonian sediment of the Lachlan Fold Belt is absent from the Tamworth Belt and e) the Early Carboniferous Kanimblan Orogeny of the Lachlan Fold Belt has not been identified in the Tamworth Belt where deposition is interrupted by minor discontinuities in the late Devonian and early Carboniferous [Korsch and Harrington, 1981].

An alternative location for the arc is represented on the 1:5 000 000 Gravity Map of Australia compiled by the Bureau of Mineral Resources in 1976 by a gravity high parallel to but the west of the outcrop of the New England fore-arc basin (the Tamworth Belt of Korsch and Harrington [1981]) and associated Namoi Gravity Ridge. The western gravity high (the Meandarra Gravity Ridge) parallels the Namoi Gravity Ridge for 500km to 27 degrees south (Figure 1). In the north the Meandarra High dies out while the Namoi High bends towards the east and joins a large north-south gravity high over the southern Queensland mid to late Palaeozoic volcanic arc. At the southern end the Namoi High trends eastwards towards the coast north of the Sydney basin paralleling the trend of the Tamworth Belt. The Meandarra High lies beneath the Surat and Gunnedah Basins for most of its length. At its southern end it appears to be broken by a 30km sinistral displacement at the northern edge of the Sydney Basin joining up with a gravity high under the basin parallel to the coast [Qureshi, 1984]. Gravity modelling by Qureshi suggests that the high under the Sydney basin correspondes to a thickness of 9-18km of basic material of density 2.95-2.85 t per cubic m. Drilling to the east of the gravity high at Stockyard Mountain 1 [Australian Oil and Gas et al,1965] in the southern Sydney Basin intersected Devonian quartz-rich sediments below the Permian thus locking the source of the Meandarra Gravity High into the Lachlan Fold Belt by at least the Late Devonian. If the missing arc is represented by the source of the gravity high, the exposed New England arc may have been at the north end of the Meandarra Gravity High and moved south to its present location during the Carboniferous. This could explain the similarities and differences between the New England and Lachlan Fold Belts. If the New England arc is the missing arc then the displacement on the Eden - Nowra -?Moree structure would have been at least 600km and the Meandarra High to the west of the Tamworth Belt would be part of the New England arc. If the New England arc had been attached to the northern end of the Meandarra High the sinistral strike-slip would have been at least 1200km. Qureshi [1984] in his discussion of the possible source of the anomaly points out that volcanics thought to be Early Permian or Late Carboniferous have been intersected in drill holes in the Gunnedah and Sydney Basins underlying the Permo-Triassic basin fills. They include basic and acid types. Qureshi points out that the width of the volcanics is greater than that of the anomaly and that the anomaly extends south of the limit of the volcanics. He considers that from the compositions of known Late Carboniferous volcanics they are unlikely to be the sole source of the anomaly. Although he favours igneous activity associated with extension preceding the formation of the Sydney Basin as an eplanation of the gravity high, Qureshi suggests that it might be caused by older volcanics or intrusives.

Strike-slip Rates

Taking the simplest model in which there is only sinistral shear, the total displacement on the 12 major dislocations is possibly 1600km plus perhaps 150km taken up by strain within the belt making a total of 1750km; or for the more complex New England model 2350km. If this took place between the latest Ordovician (445 Ma) and mid Early Devonian (395 Ma) the average rates would be 35 or 47 mm per year. The rate of movement on the Kiandra - Narromine structure prior to the Tantangara Beds may have been higher, perhaps 50 mm per year, with the rate being 31 or 46 mm per year for the remainder. None of these is excessive. If there was an early phase of dextral shear occupying say half the time between the end of deposition of the Ordovician quartz-rich turbidites and the Tantangara Bed deposition the rate of displacement on the Kiandra - Narromine structure would have had to have doubled to 100mm per year. While there is uncertainty about the exact date of commencement of deposition of the Tantangara Beds this rate is high, towards the top of plate velocities, and since the change from dextral to sinistral motion would presumably have resulted from a change in plate motions it would be expected that the change-over period would have taken some millions of years and the rates either side of such a change would be low. I therefore do not favour this scheme.

If the displacement on the Eden - Nowra - ?Moree structure was dextral moving the fore-arc zone of the Molong segment (and the Parkes segment prior to formation of the Kiandra-Narromine structure) southwards the displacement would have been at least 350km and additional strain would have been taken up in the deformation of the Wagga Trough. Assuming that the time available (445-435 Ma) was divided equally between sinistral and dextral motion this would again require large annual rates (>70 and 100mm per year respectively). These rates are again too high. A more complex solution that gives better rates is that the dextral movement on the eastern structure was a second dextral phase postdating the sinistral motion on the western structure. I find it difficult to fit this dextral interval into the geological history unless as discussed above it was in the Middle Silurian (428-421 Ma) after the deformation of the Wagga Trough but before the sinistral deformation of the Molong segment of the arc and the formation of the Hill End Trough. The dextral rate would be 50mm per year but the subsequent sinistral rate would have been around 15mm per year. The strongest argument against a dextral model for the displacement on the Eden - Nowra - ?Moree structure is that the Meandarra Gravity High appears to be part of the Lachlan Fold Belt and is in the right location for a sinistrally displaced arc segment.

The reconstructions presented here have two interesting paleogeographic implications: a) the Wagga Trough and the Monaro Troughs are segments of a single linear back-arc basin, and b) the former fore-arc zone occupies the topographically low region under the Coonamble embayment and the New South Wales south coast continental slope.

The plate tectonic model proposed is one in which subduction along an east facing trench interacted late in the Ordovician with an oblique spreading ridge trending south-south-west. The cessation of subduction resulted in transform motion. Only a small change in relative plate motions are required to result in sinistral or dextral motion (Figure 2). The change from one to the other can be accommplished with ease only if the rates of lateral slip are low. If there is a change in slip it should be preceded and followed by intervals of negligible motion. This does not seem to be the case in the models of alternating sinistral and dextral movements described above. The sinistral shear model is assumed in the summary of the tectonic evolution given at the end of this paper.

The Benambran and Bowning Orogenies

Orogenic activity in the Lachlan Fold Belt does not have a simple pattern in time and place. The view expressed by W.R.Browne in David [1950] was that there were a series of orogenic episodes that affected all or part of the fold belt. Movements were followed by peneplanation and marine transgression. This view was modified by Packham [1960] to define a more complex pattern of areas affected. Packham [1969] defined a number of orogenic phases within the orogenies of David [1950] to point out that the orogenies were not single dramatic episodes but that the record of orogenic activity was spread over much longer intervals. The epi-Ordovician Benambran Orogeny was extended from the end of the Ordovician to the middle of the Middle Silurian and the epi-Silurian Bowning Orogeny from the early Ludlow to the mid-Emsian. This paper has as its underlying argument that orogeny was continuous being, driven by sinistral shear and that therefore the Benambran-Bowning orogenic activity should be thought of as continuous, although locally the rates and styles of deformation varied considerably. The records of deformation indicated by unconformities are to be regarded in the first instance as of local significance and their interpretation has to be undertaken with much care.

After the Benambran-Bowning orogenic interval the region was affected by two later orogenic movements, the Tabberabberan (mid Devonian) and the Kanimblan (Early Carboniferous). Although it is likely that both of these movements were of considerable duration they differ in character from each other and the previous orogeny. The Tabberabberan Orogeny marked the end of deep water sedimentation in the fold belt and was followed by widespread non-marine and in the east some marginal marine deposits. The Kanimblan movement was the final stage of cratonisation of the fold belt. More details of these orogenies will be given in the outline of the tectonic history.

Although orogenic activity in the Benambran-Bowning interval was very intense in places it did not result in mountain building and as suggested above orogenesis was patchy. For example sedimentation was continuous from the Silurian into the Early Devonian in the Melbourne Trough [VandenBerg and Garratt,1976], the Hill End Trough [Packham, 1968] and the Cowra Trough [Savage, 1969] while the Tumut Trough was deformed [Crook and Powell, 1976] and the Late Silurian Kosciusko Batholith was unroofed and covered unconformably by the Snowy River Volcanics [Talent, 1965]. This patchiness could be attributed to two alternative hypotheses, namely that the fold belt is a collage of terranes with different geological histories that were brought together by the mid-Devonian or that the reaction of a particular region to the transpressional regime was variable depending on its crustal structure, composition and geological history.

There is no reason why both may not in part be the case.

In the scheme put forward here, where the region was under shear stress as a result of relative plate motions, the Ordovician back-arc - arc - trench system with an attendant high heat flow enhanced by the recent subduction of young crust and a spreading ridge, responded in a variety of ways. As well as the regional strain, mantle upwellings in extensional zones provided a driving mechanism for the complex tectonic, volcanic and plutonic episodes. Four contrasted responses are envisaged.

1. Folding and crustal thickening took place in regions of thicker sedimentary piles and relatively young oceanic crust and lower strength lithosphere (eg Monaro Trough).
2. Thick volcanic arc accumulations were more resistant to the effects of high heat flow and stress and in general were not so strongly deformed as regions of thick sediment accumulation (eg Molong High and Parkes Platform).
3. Upwelling of hot asthenospheric material occurred in regions in the arc or back-arc undergoing extensional strain; and where the crust was susceptible, melting took place that may have involved the basement and the sediment (for instance in thick piles of continent-derived sediments). Acid volcanics with subvolcanic granites were generated in regions of thinned crust or where the crust was less hydrous (eg Cowra and Tumut Troughs). Vertical and horizontal heat flow from extensional zones resulted in melting at depth of the deformed and thickened part of the complex (eg Wyangala Batholith environs). The rise of magma and the lowered strength of the crust and upper mantle allowed strong deformation to take place under regional stress resulting from plate motions. The elongation of a number of the granitic bodies that were emplaced is oblique to the structural trend and consistent with sinistral shear (eg Wyangala Batholith, most granite bodies in the Wagga Metamorphic Belt). In thick water-rich sediment piles hydrous low temperature melts that were generated could not reach the surface because of the rapid increase of melting point with fall in pressure (eg Wagga Metamorphic Belt). These remained asociated with their regional metamorphic aureoles.
4. Compression and deformation occurred where blocks of crust thickened by earlier deformations moved past each other; and where the relative strike-slip was sufficient, blocks with contrasting structural and stratigraphic histories were brought into juxtaposition (Long Plain Fault area).

Brief Outline of Tectonic Evolution

Based on the propositions enunciated above the following outline of the main features of the geological history of the eastern Lachlan Fold Belt is offered.

Early to Late Ordovician. A trench-arc system faced east with a narrow marginal sea comprising the Monaro and Wagga Troughs behind. The I-S line of Chappell and White [1974] that marks the eastern limit of granites that they consider to be formed from the partial melting of sediments (S-type granites) lies behind and sub-parallel to the arc when the arc is reconstructed (Figure 2). Granites formed from the partial melting of igneous rocks (I-type granites) occur on both sides of the line but principally to the east of it. The line may mark the western limit of the influence of arc crust.

Benambran Orogeny

1. Late Ordovician (early Bolindian). The change from a subduction regime to one of strike-slip can be explained by the subduction of a spreading ridge trending southeast. If the ridge was intersected by transform faults this could have been achieved in a few million years. The triple point would move south along the trench with sinistral transform motion north of it and subduction of young crust south (Figure 2). First evidence of tectonism occurred in the back-arc, eg gravity sliding east of Canberra [Crook et al, 1973] and possibly some of the early, now latitudinal folds, tabulated by Cas et al. [1980].

2 Late Ordovician to Early Silurian (mid-Bolindian to early Llandovery). The arc-marginal sea complex became a zone of strike-slip tectonics and rupturing of the arc occurred along the Kiandra - Narromine structure. This is the most likely time for the injection of mantle melts along this structure resulting in basic volcanism and the emplacement of ultramafic masses [Ashley et al, 1979] although Basden [1982] quoted potassium-argon dates on hornblendes from a gabbro from the complex as 425 and 426 +/-6 Ma. East of the Tumut Trough in the area now occupied by the Young Batholith the Monaro Trough rocks were brought up against the transform.

3. Early Silurian (Early to Late Llandovery, Figures 3,4). The uplift and deformation of the Wagga belt commenced. High temperature/low pressure metamorphism was accompanied by the generation of granites and their emplacement in an en echelon pattern in the Wagga Metamorphics under the influence of the sinistral shear. The granite melts had high water content and formed at low temperature. They solidified before reaching the surface and hence the lack of volcanics. Cooling ages of granites are as young as Ludlow. Less intense deformation took place along the southern extension of the belt (present southern highlands) where sediment thicknesses may have not been so great and therefore crustal temperatures were lower. Quartz-rich sediments were deposited in the Parkes area, and overlie volcanogenic sediment and limestone near Orange with little or no discordance on the underlying arc volcanics. By contrast quartz-rich proximal turbidites, presumably derived from the underlying Ordovician

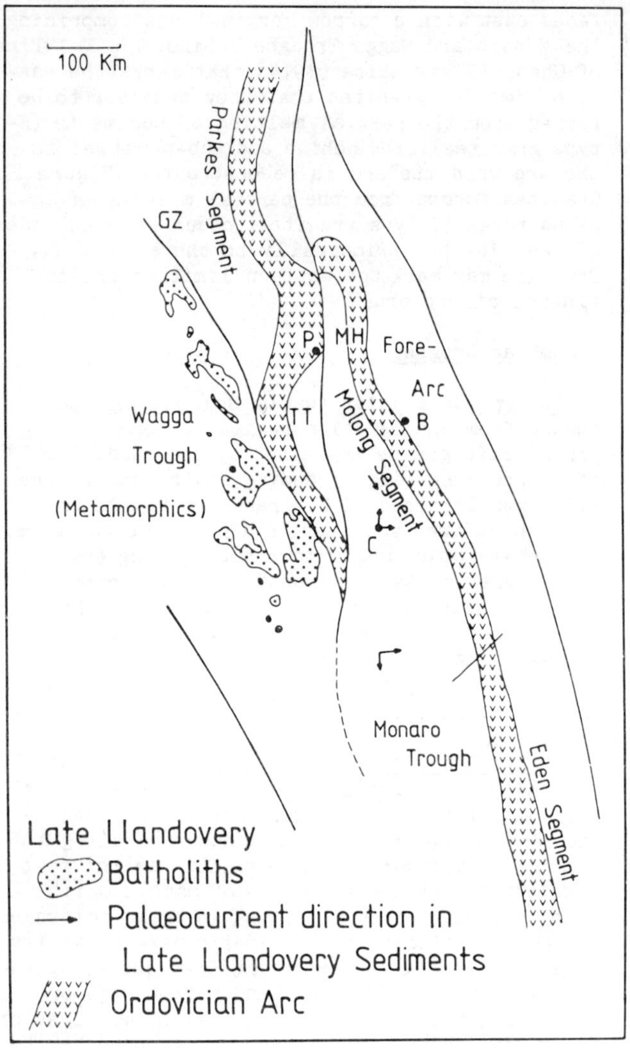

Fig. 3. Reconstruction showing the disposition of the major Ordovician tectonic elements of the eastern Lachlan Fold Belt in the late Llandovery.

turbidites, were deposited unconformably on the Ordovician sequence at a number of locations in the southern highlands [Crook et al., 1973, Pickett,1982]. The different relationship of Ordovician and Silurian rocks on the arc from that in the back-arc region turbidites reflects the different responses of these elements to the shear stress. The degree of deformation of the early Silurian sediments prior to the deposition of later Sulurian sediments was also variable. In the case of the Tantangara Beds there is a strong unconformity, but in other localities there is almost conformity.

4. Early Silurian (latest Llandovery). There was a widespread hiatus in deposition late in the Llandovery. Deposition commenced again in the Middle Silurian in most places but in others it did not start until the Late Silurian. Deformation continued as the metamorphic belt stiffened. Possibly movement occurred along the Gilmore Fault (Figure 1). Subsidence of the Cowra Trough commenced. Granite generation started in the Southern Highlands (Berridale, Murrumbidgee and Kosciusko Batholiths).

5. Middle Silurian to Late Silurian (Wenlock-Ludlow, Figures 5,6). The main subsidence phase of the Cowra Trough started. Granite emplacement commenced in the Southern Highlands (Berridale, Murrumbidgee and Kosciusko Batholiths - radiometric dates for the first two of these bodies are given in Roddick and Compston [1976] and Williams et al. [1975]). An outburst of acid/intermediate volcanism that was strongly concentrated in the Cowra Trough and its extension through Yass and Cooma mainly along the eastern side of the Kiandra-Narromine structure followed transform movement. Upwelling of hot mantle and crustal stretching along the transform may have triggered the volcanic episode by partial melting of the sediment pile or basement rocks. The genetic relationship between the granites and the volcanics has been established by Owen and Wyborn [1979] and Wyborn et al. [1981] although there may be a significant age difference between them. For instance the Goobragangra Volcanics east of the Tumut Trough are intruded by the Micalong Swamp Basic Igneous Complex which is in turn intruded by the granitic rocks chemically related to the volcanics [Owen and Wyborn, 1979]. Deposition of the dacitic volcanics and volcaniclastic sediment was in shallow water and thicknesses are some kilometres. The steady subsidence could have resulted from magma generated below the volcanic belt being transferred to the surface.

Two other zones of strong volcanic activity developed, east and southeast of Canberra in the Captains Flat-Woodlawn and Braidwood areas where

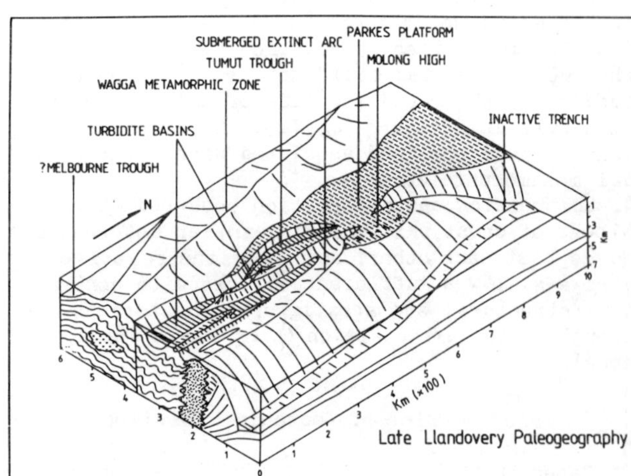

Fig. 4. Scaled block diagram of late Llandovery paleogeography.

deposition did not commence until Late Silurian [Felton and Huleatt,1975, Richardson,1979]. These later volcanics are I-types whereas the western ones are S-types [Wyborn et al., 1981]. Volcanics are not universal. Of particular interest is their absence from the Coolamon Plains sequence immediately east of the Goobragandra Volcanics, and from which it is separated by the Long Plain Fault. Volcanics are uncommon in the Yarrangobilly Caves area south of the Tumut Trough. They are also absent from the Mid to Late Silurian sections such as at Quidong [Pickett, 1982] that occur at scattered locations to the south, east of the Early Devonian Buchan Basin (Figure 1) but they do occur to the west of it in the Mitta Mitta River [VandenBerg,1976].

After the commencement of volcanic activity in the western Cowra-Yass volcanic belt, transform

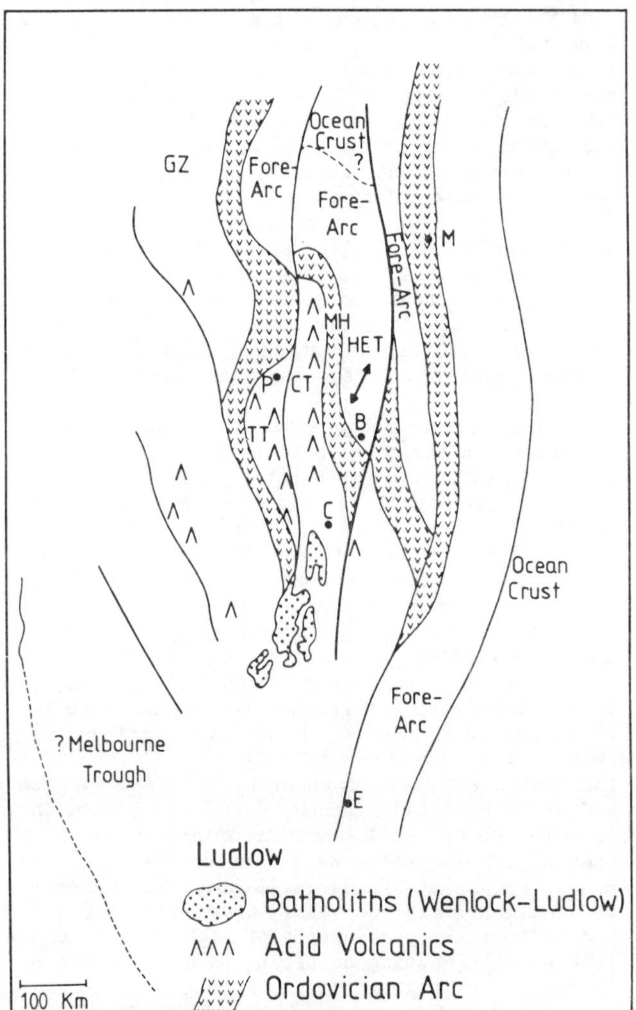

Fig. 5. Reconstruction showing the disposition of the major Ordovician tectonic elements of the eastern Lachlan Fold Belt in the Ludlow.

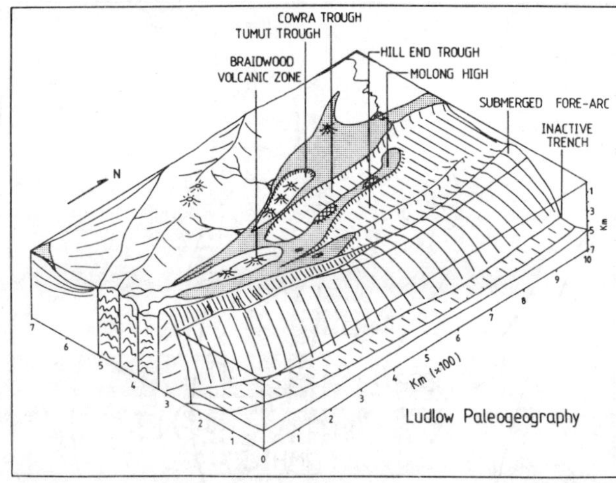

Fig. 6. Scaled block diagram of Ludlow paleogeography.

motion probably transferred to the east to produce the second major disruption to the arc (the Eden - Nowra - ?Moree structure) along the western edge of the Eden arc segment. A suggested second order dislocation responsible for the formation of the Hill End Trough [Packham,1968] may have formed at this time. It lay just west of the Wiagdon Thrust, the eastern structural boundary of the trough, and possibly along the line of the highly elongated Wologorong Granite [Shaw and Flood,1982] south of Bathurst. A significant amount of crustal extension must have occurred beneath the Hill End Trough so that the seven kilometre thick Siluro-Devonian turbidite sequence could accumulate.

Non-volcanic quartz-rich sedimentation commenced in the early Late Silurian eg the Bogan Gate Platform, the Hill End Trough, the Cowra Trough, the Taralga-Goulburn area and Yass. In the Hill End and Cowra Troughs it was interrupted by renewed acid volcanism.

6. Late Silurian (Pridoli). Quartz-rich sediments mainly of turbidite facies, were deposited at Taralga [Scheibner,1973] ,Woodlawn, Braidwood [Felton and Huleatt, 1975] and in the Cowra Trough [Packham, 1969]. Slope reversal in the eastern Hill End Trough [Packham,1968] and the deposition of shallow water carbonates and shales at Bungonia [Jones et al., 1984] may have accompanied movement on the suggested eastern transform. The Capertee High on the east of the Hill End Trough may have come into being by strike-slip motion along a structure concealed beneath the syncline of Late Devonian sediments some 10 to 15 km east of the Wiagdon Thrust. The Currawang Basalt in the Woodlawn area east of Canberra overlies Ludlow dacitic volcanics and is overlain by Pridoli turbidites. The extrusion of the basalt and the sudden increase in water depth are also possibly related to this movement. The relatively quiescent interval in the Pridoli in the

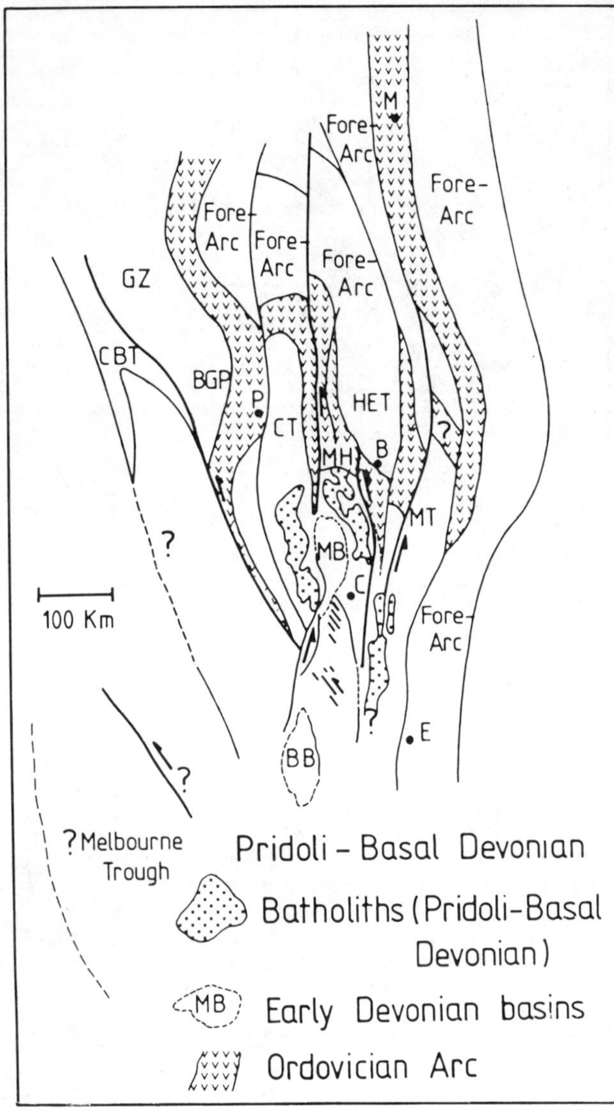

Fig. 7. Reconstruction showing the disposition of the major Ordovician tectonic elements in the Pridoli to basal Devonian.

east probably marks a switch in movement back to the west and major structures were reactivated. The Cheesmans Creek olistotrome on the Molong High comprises large blocks of Ordovician formations enclosed in Pridoli sediments [Sherwin, 1971]. The Ordovician blocks are up to 450m long and appear in reverse order to the stratigraphic sequence. They may have been derived from a local fault scarp, the precursor of the Columbine (Figure 1).

Bowning Orogeny

1. Latest Silurian to basal Devonian (Pridoli-Gedinnian) (Figures 7,8). Following the stiffening of the crust in the southern highlands and the Wagga Metamorphic Belt by metamorphism and intrusion it is suggested that the following events occurred close to the Silurian-Devonian boundary as a response to the more brittle nature of the crust.

(a) Movement along the Gilmore Fault resulted in the commencement of formation of the Cobar Trough on the northwest margin of the Wagga Metamorphic Belt.

(b) Sinistral displacement on the Long Plain Fault accompanied by the emplacement of the Young Granite brought the Coolamon Plains non-volcanic sequence against the Goobragandra Volcanics; the Tumut Trough [Crook and Powell,1976] was deformed and uplift and erosion unroofed the Kosciusko Batholith [VandenBerg,1976] .

(c) Extension took place in the Taemas area forming the Early Devonian Murrumbidgee Basin and northwards there was further extension in the Cowra Trough which deepened [Savage,1969].

(d) Northward movement of the Molong High resulting in slip on the Copperhannia-Lake George fault and displacement on the Columbine Thrust moved the Cheesemans Creek olistostrome north and emplaced the Wyangala Batholith in the zone of stretching south of the Molong High resulting from these movements. Rapid sedimentation occurred in the Cowra Trough.

(e) Slip occurred on the Copperhannia-Lake George Fault and the Lake Bathurst-Taralga structure allowing the Boro Granite to be emplaced. The volcanics that accompanied these movements are mostly unconformable on the underlying Ordovician or Silurian formations [Jones et al,1984]. The Braidwood and Wologorong Granites are also of this age.

(f) Northwest trending sinistral wrench faults developed in the more cratonised region of the southern highlands intersecting the Berridale and Murrumbidgee Batholiths and the Jerangle Granite Complex.

(g) The Bogong Granite and other small masses in the Tumut Trough were probably emplaced late in this series of events.

2. Early Devonian (Gedinnian-Siegenian). Short lived deep water quartz rich sediments in the axial region of the Cobar Trough [Glen, 1982] pass outwards into shallow water deposits to the west, and east onto the Bogan Gate Platform, [Webby,1972]. Dacitic volcanics were extruded in the southern Cobar Trough at Mt Hope and Nymagee and at Mineral Hill probably on fault lines. They are also found at Albury (?Silurian) and Jemba in the Wagga Metamorphic Belt [VandenBerg,1976] and on the Bogan Gate Platform [Webby,1972]. These volcanics are all presumably crustally derived and reflect the persistence of high mantle temperatures with melting occurring on a local scale in regions of higher strain. A last burst of strong volcanic activity in the Cowra Trough, and limited activity on the Molong High, is recorded by deposits conformable on Silurian beds [Packham,1968]. Siegenian carbonates follow on the Molong High. Volcanics and minor quartz-rich

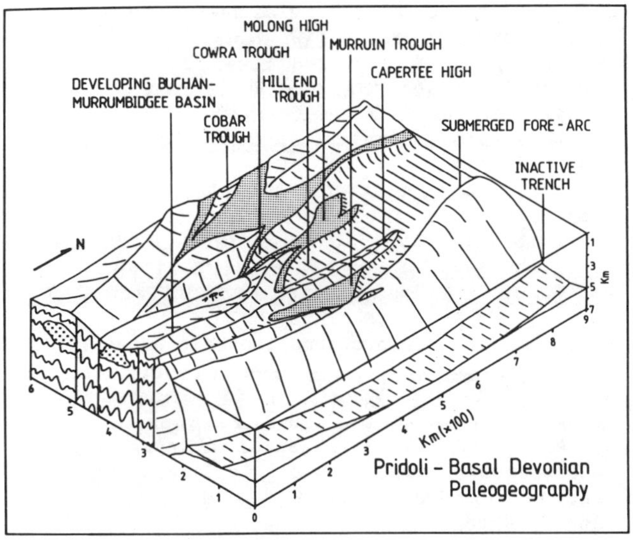

Fig. 8. Scaled block diagram of Pridoli-basal Devonian paleogeography.

sediments were deposited unconformably on Silurian rocks in the Tumut Trough [Richards et al., 1977, Moye et al.,1969]. Volcanics were also extruded unconformably on the Silurian in the Murrumbidgee Basin; and in the Coolamon area to the south they also overlie the Ordovician.(These volcanics and the later formations in the area are cut by the Long Plain Fault indicating its reactivation in Tabberabberan movements.) Acid volcanics were extruded unconformably on granite in eastern Victoria and on Silurian strata in southern NSW [VandenBerg,1976]. The most intense explosive activity is found in the northeast part of the Fold Belt in the Hill End Trough in the Merrions Tuff and Crudine Group [Packham,1968] and the Rockley-Burraga area [Packham,1969] and the Wollondilly Valley in the Bindook Volcanics etc, [Cas et al.,1981]. Although the strike-slip faulting was followed by widespread acid volcanism, granitic bodies of this age are not abundant. This might reflect the level of erosion. The Yeoval and Eugowra Granites in the northern part of the Cowra Trough and the Marulan Batholith in the northeast in the Goulburn Synclinal Zone of the Murruin Trough [Scheibner 1976] belong to this interval. The Bega Batholith inland from the New South Wales south coast was probably generated at this time since it has a cooling age after adjusting for new decay constants of 380 to 390 Ma [Chappell and White, 1976].

First Interlude

Early Devonian (Siegenian-Emsian). Sedimentation was predominantly non-volcanic with carbonates and shales accumulating at Buchan [Talent, 1956] and Taemas [Browne, 1959] but quartz-feldspatholithic turbidites derived from an eastern inactive volcanic source being deposited in the Hill End Trough [Packham,1968] and carbonates and muds were deposited on its western margin [Packham, 1969]. The flood of clastic debris into the Hill End Trough contrasts with the small clastic supply and hence subdued topopgraphy that must have existed in the vicinity of Taemas, Buchan and the western margin of the Hill End Trough. This indicates the limited affect of the Bowning Orogeny as a mountain building episode in the southern highlands. Carbonates muds and sands were deposited on the Parkes Platform on earlier Devonian volcanics. Clastic sediments were deposited in the Cobar and Melbourne Troughs and in the Tabberabbera area west of the Kiewa Fault where a very narrow north-south syncline contains a marine quartzose clastic sequence unconformable on the Ordovician [Talent,1963]. The Winburn Tuff [Packham, 1968] on the Capertee High may be an exception to the volcanic quiet episode.

It is possible that strike-slip along the Mount Wellington axis on the east of the Melbourne Trough took place during this interval of relative inactivity in the east.

Tabberabberan Orogeny

Middle Devonian. By the late early Devonian the Cobar Trough, the Melbourne Trough, the Tabberabbera sequence, the Cowra Trough, the Buchan and Taemas Basins and the Hill End Trough had not yet been through a major deformational cycle. They were deformed in the Tabberabberan episode [Glen et al,1986, Talent, 1965, VandenBerg and Garratt,1976, Packham,1968, 1969, 1978]. No upper limit on the deformation of the Taemas Basin can be established because of the absence of Late Devonian sediments. Powell et al [1976] favoured an Early Carboniferous (Kanimblan) age for the folding of the Hill End Trough. The Fold Belt became for a short time a sediment source area. Folds and faults in the Hill End Trough are suggestive of orthogonal compression. Further movement occurred on old major faults in the southern part of the belt during the folding of the Murrumbidgee and Buchan Basins which rest on a deformed and intruded basement.

The cause of the Tabberabberan orogenic event is not clear. It is uncertain whether the strike - slip regime was still operative at this time. Although the New England arc may have been somewhat remote at the time the cessation of transform motion may have been accompanied by the recommencement of subduction in the New England arc in the Early Devonian indicating that a new plate tectonic regime had evolved in the region. If it evolved from strike-slip to oblique subduction, sinistral shear may have continued taking up part of the inter-plate motion in the manner illustrated by Sarewitz and Karig [1986].

Second Interlude

Late Devonian. At this time volcanic activity and granite emplacement occurred in the former

Melbourne Trough [VandenBerg and Garratt,1976], and there was bimodal volcanism in the NSW south coastal belt [McIlveen,1974]. Regions uplifted in the Tabberabberan orogeny were eroded and subsided due to thermal contraction, becoming areas of sedimentation for essentially fluvial deposits most of which are quartz rich. The base of this sequence is diachronous [Webby,1972].

Kanimblan Orogeny

Early Carboniferous. The Kanimblan Orogeny was the terminal tectonic event in the Fold Belt. Its cause is unclear. Open folds with steep limbs near faults in general reflect the strength of much of the underlying basement resulting from previous deformations. Movement took place on some old sutures and deformation was of greater intensity in the northeast of the belt where a number of crosscutting granite bo dies were emplaced into the Hill End Trough, the Capertee High and the Currarong Syncline. Movements affected a region that extended from Broken Hill in the west to the western margin of the Sydney Basin but there is no evidence of a significant deformation in the sequence in the New England arc nor as mentioned previously is there any sign of the flood of quartz-rich detritus characteristic of the Late Devonian of the Lachlan Fold Belt. The New England arc was perhaps translated to its present position in the Carboniferous.

Acknowledgments. I have benefited from discussions with Evan Leitch, Chris Jenkins Kingsley Mills and Barry Webby in the preparation of this work. I would particularly like to thank Keith Crook, Evan Leitch and Erwin Scheibner for their careful reviews of the manuscript. Peter Schneider drafted the diagrams.

References

Ashley, P. M., P. L. Brown, B. J. Franklin, A. S. Ray, and E. Scheibner, Field and geochemical characteristics of the Coolac Ophiolite suite and its possible origin in a marginal sea, J. Geol. Soc. Aust., 26, 45-60, 1979.

Australian Oil and Gas Corp. Ltd., Farmout Drillers No Liability and Exoil (N.S.W.) Pty. Ltd., Drilling operations in the Sydney Basin, 1958-1962, Aust. Bur. Miner. Res. Geol. Geophys., Petroleum Search Subsidy Acts Publ., 12, 1965.

Basden, H., Preliminary report on the geology of the Tumut 1:100 000 Sheet area, southern New South Wales, Q. Notes Geol. Surv. N.S.W., 46, 1-18, 1982.

Basden, H., B. J. Franklin, B. Marshall, and A. E. Waltho, Terranes of the Tumut district, southeastern New South Wales, This volume.

Bembrick, C. S., Coonamble Embayment, Monogr. Australas. Inst. Min. Metall., 7(3), 302-306, 1976.

Bembrick, C. S., C. Herbert, E. Scheibner, and J. Stuntz, Structural subdivisions of the Sydney Basin, Bull. Geol. Surv., N.S.W., 26, 3-9.

Brooks, B., and M. D. Leggo, Local chronology and regional implications of a Rb-Sr investigation of granitic rocks from the Corryong district, Southeastern Australia, J. Geol. Soc. Aust., 19, 1-19, 1972.

Browne, I. A., Stratigraphy and structure of the Devonian rocks of the Taemas and Cavan areas, Murrumbidgee River south of Yass, N.S.W., J. Proc. R. Soc. N.S.W., 92, 115-128, 1959.

Cas, R., C. McA. Powell, and K. A. W. Crook, Ordovician palaeogeography of the Lachlan Fold Belt: modern analogue and tectonic constraints, J. Geol. Soc. Aust., 27, 19-31, 1980.

Cas, R. C., C. McA. Powell, C. L. Fergusson, J. G. Jones, W. D. Roots, and J. Fergusson, The Lower Kowmung Volcaniclastics: a deep-water succession of mass-flow origin, northeastern Lachlan Fold Belt, N.S.W., J. Geol. Soc. Aust., 28, 271-288, 1981.

Cawood, P. A., Cambro-Ordovician strata in northern New South Wales, Search, 7, 317-318, 1976.

Chappell, B. W., and A. J. White, Two contrasting granite types, Pac. Geol., 8, 173-174, 1974.

Chappell, B. W., and A. J. White, Plutonic rocks of the Lachlan Mobile Zone, Excur. Guide 25th Int. Geol. Congr., 13C, 40 pp., 1976.

Crook, K. A. W., Kratonization of West Pacific - type geosynclines, J. Geol., 82, 24-36, 1974.

Crook, K. A. W., Fore-arc evolution in the Tasman Geosyncline, J. Geol. Soc. Aust., 27, 215-232, 1980.

Crook, K. A. W., J. Bein, R. J. Hughes, and P. A. Scott, Ordovician and Silurian history of the southern part of the Lachlan Geosyncline, J. Geol. Soc. Aust., 20, 113-138, 1973.

Crook, K. A. W., and C. McA. Powell, The evolution of the southeastern part of the Tasman Geosyncline, Excur. Guide 25th Int. Geol. Congr.,17A, 122 pp., 1976.

David, T. W. E., The geology of the Commonwealth of Australia., 1, 747 pp., Arnold, London.

Felton, E. A. and M. B. Huleatt, M. B., Braidwood, New South Wales, scale 1:100 000 Geological Sheet Series, 8827, Geol Surv. N.S.W., Sydney, 1975.

Fergusson, C. L., Tectonic history of the Tabberabbera Belt, eastern Victoria, and implications for Ordovician reconstructions of southeastern Australia, Abstracts Geol. Soc. Aust., 14, 63-65, 1985.

Fergusson, C. L., D. R. Gray, and R. A. F. Cas, Overthrust terranes in the Lachlan Fold Belt, Geology, 14, 519-522, 1986.

Garratt, M.J., Silurian to Early Devonian facies and biofacies for the Melbourne Trough, central Victoria, Geol. Soc. Aust., J., 30, 121-148, 1983.

Glen, R. A., The Amphitheatre Group, Cobar, New South Wales: preliminary results of new mapping and implications for ore search, Q. Notes Geol. Surv. N.S.W., 49, 1-14, 1982.

Glen, R. A., R.D. Dallmeyer, and L. P. Black, Preliminary note on new isotopic evidence for, and implications of, an Early Devonian deformation at Cobar, Q. Notes Geol. Surv. N.S.W., 64, 26-29, 1986.

Harland, W. B., A. V. Cox, P, G. Llewellyn, C. A. G., Pickton, A. G., Smith, and R. A. Walters, A geological time scale, 131 pp., Cambridge University Press, Cambridge, 1982.

Jenkins, C. J., Darriwillian (Middle Ordovician) graptolites from the Monaro Trough sequence, east of Braidwood, Proc. N.S.W. Linn. Soc. N.S.W., 106, 173-179, 1982.

Jenkins, C. J., R.R. Kidd, and K. J. Mills, Upper Ordovician graptolites from the Wagonga Beds near Batemans Bay, New South Wales, J. Geol. Soc. Aust., 29, 367-373, 1982.

Jones, B.G., P. F. Carr, and C. G. Hall, The Early Devonian Tangerang Formation - Windellama region, N.S.W.- definition and paleoenvironmental significance, Aust. J. Earth Sci., 31, 75-90, 1984.

Korsch , R. J. and H. J. Harrington, H. J., Stratigraphic and structural synthesis of the New England Orogen, J. Geol. Soc. Aust., 28, 205-226, 1981.

Leitch, E. C., The geological development of the southern part of the New England Fold Belt, J. Geol. Soc. Aust., 21, 133-156, 1974.

Leitch, E. C., Plate tectonic interpretation of the Paleozoic history of the New England Fold Belt, Bull. Geol. Soc. Amer., 86, 141-144, 1975.

McIlveen, G. R., The Eden-Comerong-Yalwal rift zone and contained gold mineralization, Rec. Geol. Surv. N.S.W., 16, 247-277, 1974.

McKerrow, W. S., R. St. J. Lambert and L. R. M. Cocks, The Ordovician, Silurian and Devonian Periods, in The Chronology of the Geological Record, edited by N. J. Snelling, pp. 73-80, Mem. Geol. Soc., 10, 1985.

Moye, D. G., K. R. Sharp, and D. H. Stapleton, D.H., Lower and Middle Devonian Series - Snowy Mountains area, in Geology of New South Wales, edited by G. H. Packham , pp. 143146, J. Geol. Soc. Aust., 16, 1969.

Nicholson C., L. Seeber, and L. R. Sykes, Seismic evidence for conjugate slip and block rotation within the San Andreas Fault System, Southern California, Tectonics, 5, 629-648, 1986.

Oversby, B., Paleozoic plate tectonics in the southern Tasman Geosyncline, Nature Phys. Sci., 234, 45-48, 1971.

Owen, M. and D. Wyborn, D. Geology and geochemistry of the Tantangara and Brindabella 1:100 000 Sheet Areas, New South Wales and Australian Capital Territory, Bull. Aust. Bur. Miner. Resour. Geol. Geophys., 204, 1979.

Packham, G.H.(1960). Sedimentary history of the Tasman Geosyncline in southeastern Australia, 21st Int. Geol. Congr., 12, 74-83, 1960.

Packham, G. H., Bathurst, New South Wales, scale 1:250 000, Geological Series Sheet S1 55-8, Geol. Surv. N.S.W., 1967.

Packham, G H., The Lower and Middle Palaeozoic stratigraphy and sedimentary tectonics of the Sofala-Hill End-Euchareena region N.S.W., Proc. Linn. Soc. N.S.W., 93, 111-163, 1968.

Packham, G. H., The geology of New South Wales, J. Geol. Soc. Aust., 16(1), 654 pp., 1969.

Packham, G. H., A speculative Phanerozoic history of the south-west Pacific, in The Western Pacific: Island Arcs, Marginal Seas, Geochemistry, edited by P. J. Coleman, pp.369-388, Univ of W. A. Press, Perth, 1973.

Packham, G. H., Discussion: Timing of the deformation of the Hill End Trough, J. Geol. Soc. Aust., 25, 361-362, 1978.

Packham, G. H. and Falvey, D.A., An hypothesis for the formation of marginal seas in the Western Pacific, Tectonophysics, 11, 79-109, 1971.

Packham, G. H., and Leitch, E.C., The role of plate tectonics in the interpretation of the Tasman Orogenic Zone, in The Tasman Geosyncline, edited by A. K. Denmead, G. W. Tweedale, and A. F. Wilson, pp. 129-154.Geol. Soc. Aust., Qld. Div., Brisbane, 1974.

Pickett, J. W., The Silurian System in New South Wales, Bull. Geol. Surv. N.S.W., 29, 264 pp. 1972.

Powell, C. McA., Tectonic relationship between the Late Ordovician and Late Silurian palaeogeographies of southeastern Australia, J. Geol. Soc. Aust., 30, 353-373, 1983a.

Powell, C. McA., Geology of the New South Wales south coast and adjacent Victoria with emphasis on the pre-Permian structural history, Field Guide, Specialist Group Tectonics and Structural Geology, Geol. Soc. Aust., 1, 118 pp., 1983b.

Powell, C. McA., The Uluru Regime, in Phanerozoic earth history of Australia, edited by J. J. Veevers, pp. 290-340, Clarendon Press, Oxford, 1984.

Powell, C. McA., D. R. Edgecombe, N. M. Henry, and J. G. Jones, J. G., Timing of regional deformation of the Hill End Trough - a reassessment, J. Geol. Soc. Aust., 23, 407-422, 1976.

Qureshi, I.R., Tectonic significance of the Wollondilly-Blue Mountains gravity gradient zone, Aust. J. Earth Sci., 31, 293-302, 1984.

Richards, J. R., Barkas, J. P. and Vallance, T G. Lower Devonian point on the geological timescale, J. Geochem., 11, 147-153, 1977.

Richards, J. R. and O. P. Singleton, Palaeozoic Victoria, Australia: igneous rocks, ages and interpretation, J. Geol. Soc. Aust., 28, 395-421, 1981.

Richardson, S.J., Geology of the Michelago 1:100 000 sheet 8726, 253 pp., Geol. Surv. N.S.W., Sydney, 1979.

Roddick, J. C. and W. Compston, W., Radiometric evidence for the age of emplacement and cooling of the Murrumbidgee Batholith, J. Geol. Soc. Aust., 23, 223-233, 1976.

Sarewitz, D.R. and D. E. Karig, D.E., Processes

of allochthonous terrane evolution, Mindoro Island, Philippines, Tectonics, 5, 525-552, 1986.

Savage, N. M., Geology of the Manildra district, J. Proc. Roy. Soc. N.S.W., 101,159-178, 1969.

Scheibner, E. Actualistic models in tectonic mapping, Rpt. 24th Int. Geol. Cong., 3, 405-422, 1972.

Scheibner, E., Geology of the Taralga 1:100 000 Sheet, 8829, 79 pp., Geol. Surv. N.S.W., Sydney, 1972.

Scheibner, E., A plate tectonic model of the Palaeozioc tectonic history of New South Wales, J. Geol. Soc. Aust., 20, 405-426, 1974.

Scheibner, E., Explanatory notes on the Tectonic Map of New South Wales, 238 pp., Geol. Surv. N.S.W., Sydney, 1976.

Shaw, S.E., R. H. Flood, and G. H. Riley, The Wologorong Batholith, New South Wales, and the extension of the I-S line of the Siluro-Devonian granitoids, J. Geol. Soc. Aust., 29, 41-48, 1982.

Sherwin, L., Stratigraphy of the Cheeseman's Creek district, New South Wales, Rec. Geol. Surv. N.S.W., 13, 199-234, 1971.

Spencer-Jones, D., The Grampians Group of western Victoria, Geology of Victoria, edited by J. G. Douglas, and J. A. Ferguson, pp. 71-76, Geol. Soc. Aust. Spec. Publ., 5, 1976.

Suppel, D. W., A. Y. E. Warren, J. J. Watkins, J. Chapman, K. Tenison Woods, and L. Barron, A reconnaissance study of the geology and gold deposits of the West Wyalong - Temora - Adelong district, Q. Notes Geol. Surv. N.S.W., 64, 1-23, 1986.

Talent, J. A., Devonian brachiopods and pelecypods of the Buchan Caves Limestone,Proc. R. Soc. Vict., 68, 1-56, 1956.

Talent, J. A., The Devonian of the Mitchell and Wentworth Rivers, Mem. Geol. Surv. Vict., 24.

Talent, J. A. The stratigraphic and diastrophic evolution of central and eastern Victoria in middle Palaeozoic times, Proc. R. Soc. Vict., 78, 179-195, 1965.

Thomas, D. E., The structure of Victoria with respect to lower Palaeozoic rocks. Min. Geol. J. Vict., 1(4), 59-64, 1939.

VandenBerg, A. H. M., Eastern Victoria, Silurian - Middle Devonian, in Geology of Victoria edited by J. G. Douglas, and J. A. Ferguson, pp. 62-70,Geol. Soc. Aust. Spec. Publ., 5, 1976.

VandenBerg, A. H. M., and M. J. Garratt, Melbourne Trough, Silurian - Middle Devonian, in Geology of Victoria, edited by J. G. Douglas, and J. A. Ferguson, J. A., pp. 45-62, Geol. Soc. Aust. Spec. Publ., 5, 1976.

Voisey, A. H., Australian Geosynclines. Aust. J. Sci., 22, 188-198, 1959.

Webby, B. D., Devonian geological history of the Lachlan Geosyncline., J. Geol. Soc. Aust., 19, 99-123, 1972.

Webby, B. D., The Ordovician System in southeastern Australia, in The Ordovician System, edited by M. G. Bassett, pp. 417-446, Univ. Wales Press, Cardiff,1976.

Webby, B.D.(Ed.) The Ordovician System in Australia. Correlation chart and explanatory notes. Publ. Int. Union Geol. Sci., 6, 64 pp., 1981.

Webby, B.D., and G. H. Packham, Stratigraphy and regional setting of the Cliefden Caves Limestone Group (Late Ordovician), central-western New South Wales, . Geol. Soc. Aust., 29, 297-317, 1982.

Williams, I. S., W. Compston, B. W. Chappell, and T. Shirahase, Rubidium-strontium age determinations on micas from a geologically controlled composite batholith, J. Geol. Soc .Aust., 22, 497-505, 1975.

Wyborn, D., B. W. Chappell, and R. M. Johnston, Three S-type volcanic suites from the Lachlan Fold Belt, southeast Australia, J. Geophys. Res., 86, 10335-10348, 1981.

CONTRASTING STRUCTURAL STYLES IN THE PALEOZOIC SUBDUCTION COMPLEX OF THE SOUTHERN NEW ENGLAND OROGEN, EASTERN AUSTRALIA

Kenneth C. Cross

Western Mining Corporation Limited, Olympic Dam Project, P.O. Box 345, Unley, S.A. 5061, Australia

Christopher L. Fergusson

Department of Geology, University of Wollongong, P.O. Box 1144, Wollongong, N.S.W. 2500, Australia

Peter G. Flood

Department of Geology and Geophysics, University of New England, Armidale, N.S.W. 2351, Australia

Abstract. The New England Orogen of eastern Australia preserves a Paleozoic forearc with a volcanic chain in the west, a forearc basin in the centre and a subduction complex in the east. The western boundary of the subduction complex is the Peel Fault which in part is associated with ophiolitic serpentinite-matrix melange. The subduction complex has formed mainly by subduction accretion with the development of eight main tectonostratigraphic units. Most of these units have local younging directed towards the magmatic chain but the units themselves young in the opposite sense. One unit has formed by the accretion of an exotic island-arc fragment. The older portion of the subduction complex (?Silurian-Early Carboniferous) is dominated by tectonic melanges consisting of fault bound slivers of greywacke, mudstone, chert and greenstone arranged in an imbricate fault pattern. By contrast the younger portion has abundant relatively coherent units of monotonous folded and faulted turbidite successions. This marked change in structural style coincides with changing sedimentation patterns in the forearc basin and reflects an increased sediment supply to the trench accompanied by a relatively minor change in plate motion.

Introduction

The principal factors affecting the structural styles of subduction complexes are the supply of sediment to the trench and the rate of convergence [Karig and Sharman, 1975; von Huene, 1984]. Tectonic melanges, characterised by structures formed in a regime of high strain rates, dominate subduction complexes where convergence rates are fast and sediment supply is low. Folded and faulted units of deep-marine strata develop at subduction zones with high rates of sediment supply and low rates of convergence [e.g. Lesser Antilles, Westbrook et al., 1982]. The aim of this paper is to describe the contrasting structural styles in the Paleozoic subduction complex of the southern New England Orogen in eastern Australia (Figure 1), and to show that the structural variation reflects changes mainly in the sediment supply during the period of subduction.

Regional Setting

The main tectonic elements of the southern New England Orogen (Figure 1) in the Middle and Late Paleozoic were: a continental volcanic chain in the west, a forearc basin in the centre, and a subduction complex in the east [Leitch, 1975; Cawood, 1982a; Fergusson, 1984a]. Remnants of the volcanic chain are largely covered by the Permian-Triassic Gunnedah and Sydney Basins (Figure 2). The forearc basin is largely preserved in the Tamworth, Hastings and Emu Creek Blocks, and the subduction complex is exposed in a number of blocks east of the Peel Fault [Cawood and Leitch, 1985].

The Tamworth Block contains marine and terrestrial strata at least 9 km in thickness and is subdivided into three major successions, viz:

(1) Cambrian-Ordovician(?) strata consisting of mudstone, greywacke, conglomerate, and keratophyre that might be the exotic remnant of a palaeo-Pacific island arc (Cross, unpublished manuscript).

(2) Devonian strata consisting of mudstone, greywacke, conglomerate, limestone, dolerite,

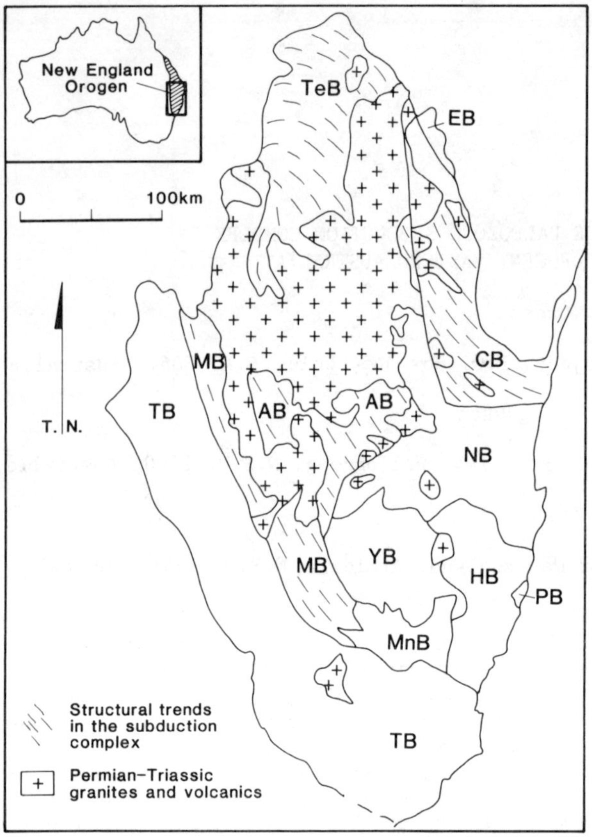

Fig. 1. Map of the southern New England Orogen showing the main blocks referred to in the text. The location of the New England Orogen is shown in the inset. Key to blocks: TB - Tamworth Block, MnB - Manning Block, MB - Macdonald Block, YB - Yarrowitch Block, HB - Hastings Block, PB - Port Macquarie Block, AB - Armidale Block, NB - Nambucca Block, CB - Coffs Harbour Block, TeB - Texas Block, EB - Emu Creek Block.

and spilite. Clastic rocks were derived from an intermediate to mafic volcanic source [Cawood, 1983]. Environments of deposition range from shallow marine reefs to apparently deeper marine settings with turbidite deposition [Leitch, 1974].

(3) Late Devonian-Early Carboniferous marine and terrestrial strata including mudstone, greywacke, conglomerate and minor oolitic limestone. The succession was derived from an intermediate volcanic source region [Leitch, 1974] and depositional environments change from deep marine turbidite fans in the east to shelf and terrestrial rocks in the west and the south [Leitch, 1974; Roberts and Engel, 1980].

(4) A Late Carboniferous-Early Permian predominantly conglomeratic succession with interbedded sandstone, mudstone, regionally extensive ignimbrite and basalt [Leitch, 1974; McPhie, 1983]. These rocks were derived from an intermediate to silicic volcanic source to the west, with a minor contribution from the Lachlan Orogen [McPhie, 1983], and were deposited in fluvatile and minor lacustrine environments [Leitch, 1974].

The Hastings Block consists of a Devonian to Early Permian succession similar to that in the Tamworth Block and is regarded by Cawood and Leitch [1985] as a sinistrally offset extension of the latter block. The Emu Creek Block contains Carboniferous and Permian rocks and is a dextrally offset portion of the northern continuation of the forearc basin preserved in the Tamworth Block [Murray et al., 1986].

The subduction complex constitutes much of the geology east of the Peel Fault (Figures 1 and 2). Other rocks in this region include highly deformed Early Permian sequences, the Early Permian Bundarra and Hillgrove Plutonic Suites, and Late Permian-Early Triassic granitoids and associated volcanics. In the north the subduction complex forms a z-shaped megafold (Figure 1) which formed in the Early Permian [Flood and Fergusson, 1982]. Farther south the subduction complex is truncated by the Hastings Block (Figure 2). The region between the z-shaped megafold and the Hastings Block is the Nambucca Block [Leitch, 1974]. It consists of multiply deformed Early Permian slate and phyllite which presumably overlie parts of the subduction complex.

Tectono-stratigraphy of the Subduction Complex

The subduction complex is subdivided into a number of tectono-stratigraphic units, which include both melanges and relatively coherent units. Cawood and Leitch [1985] recognize similar structural units and refer to these as terranes.

The ages of the tectono-stratigraphic units are not well determined as fossils are scarce. One potential constraint is lithological comparison of greywackes with those of the Tamworth Block. Systematic differences in the proportions of quartz, silicic and intermediate-mafic volcanic rock fragments and the presence or absence of ooliths have been used in correlations between these regions [Cawood, 1983; Fergusson, 1984b; Korsch, 1984]. These proposed correlations between the Tamworth Block forearc sequence and the associated New England subduction complex, and the lack of clearly defined truly exotic terranes, pose some problems for any attempts to apply terrane terminology in New England (see Discussion).

Within the subduction complex itself, three principal types of melanges are recognized [Cross and Fergusson, 1983]: (1) serpentinite-matrix melanges, which are associated with major faults, (2) melanges derived largely from the tectonic disruption of previously coherent marine strata and associated igneous rocks, and (3) tectonically disrupted olistostromes.

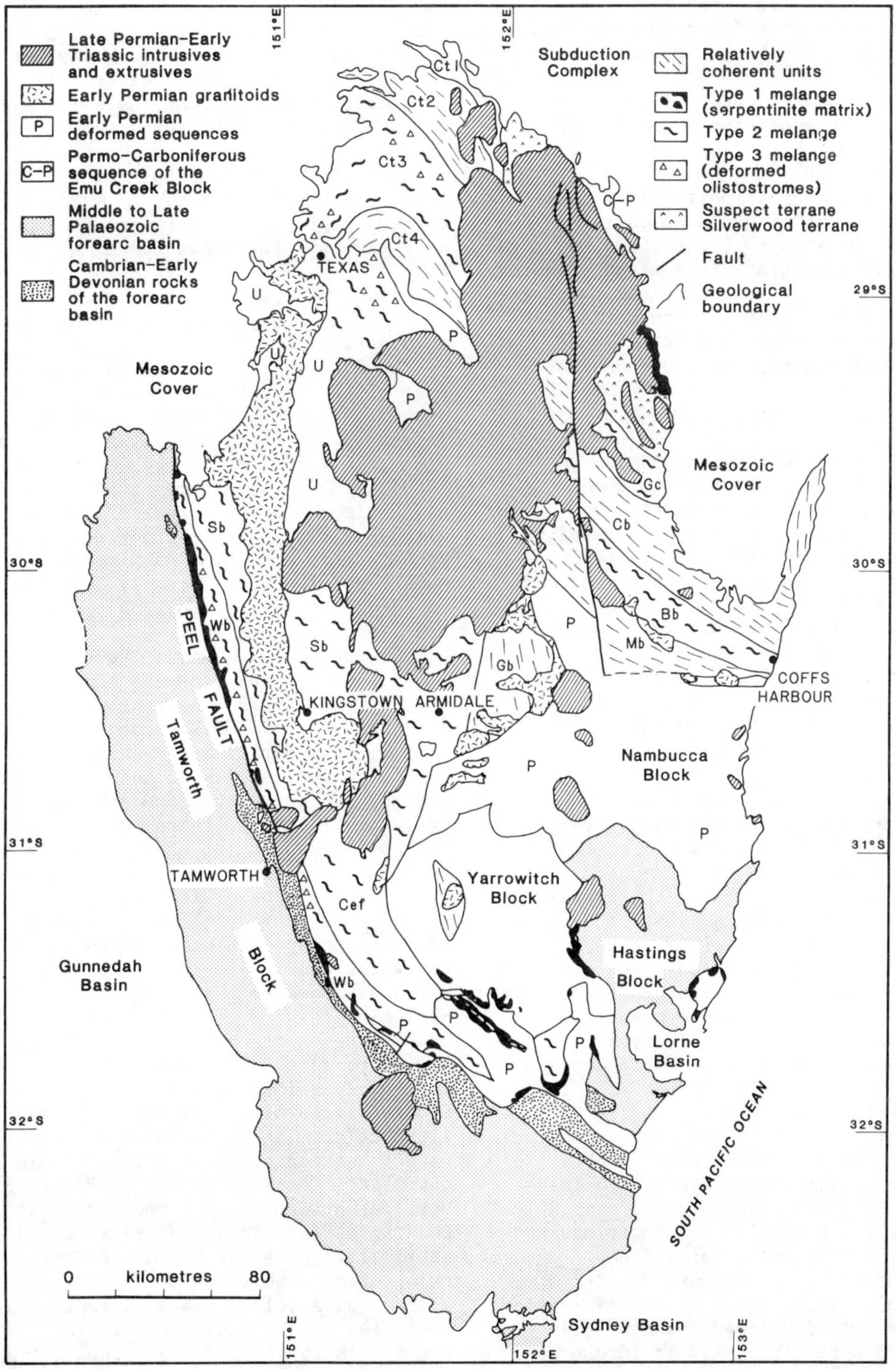

Fig. 2. Map of the southern New England Orogen showing the distribution of the three types of melange and relatively coherent units in the subduction complex. Units referred to in the text are: Wb - Woolomin beds, Cef - Cockburn Formation Sb - Sandon beds, Gb - Girrakool beds, Gc - Gundahl Complex, Cb Coramba beds, Bb - Brooklana beds, Mb - Moombil beds, Ct1, Ct2, Ct3, Ct4 - subunits of the Texas beds, U - undifferentiated rocks of the subduction complex.

Reconnaissance and detailed investigations indicate that such structurally incoherent rocks are more widespread throughout the subduction complex than previously recognized. This is especially true of type (1) and (2) melanges, whereas type (3) are restricted to localised zones in the vicinity of the Peel Fault [see Leitch and Cawood, 1980].

The tectono-stratigraphic units of the New England subduction complex are exposed in a number of blocks bordered by faults, and by intrusive contacts with granitic plutons (Figure 1). One of these, the Yarrowitch Block has been modified by significant Permian deformation and metamorphism [Leitch, 1974], and is excluded from the subsequent descriptions.

Macdonald, Manning and Armidale Blocks

The western boundary of the Macdonald and Manning Blocks is the Peel Fault. This structure is associated with discontinuous masses of type (1) serpentinite-matrix melange. Type (1) melange also occurs along the fault boundaries of the Manning and Yarrowitch Blocks and in the Port Macquarie Block. It has principally formed by the partial to intense tectonic disruption of serpentinized harzburgite, with lesser serpentinized dunite and wehrlite. (We refer to rocks such as these which have been tectonically mixed more or less in situ, as "autoclastic melange" [after Greenly, 1919].)

The proportion of sheared serpentinite matrix changes considerably from outcrop to outcrop, and within individual outcrops, and blocks of little-deformed serpentinite are commonly preserved. The shear foliation in the matrix is defined by mm-scale anastomosing shear fractures (the scaly foliation of others [e.g. Lash, 1985]). The foliation dips steeply and, for the most part, strikes sub-parallel to the general trend of most serpentinite masses.

Along the Peel Fault (Figure 2), type (1) melange predominantly consists of serpentinized harzburgite tectonite [cf. Nicolas et al., 1980] with minor serpentinized dunite. Locally this includes or is closely associated with combinations of highly serpentinized xenoblastic and cumulate peridotites, cumulate plagioclase-bearing peridotites, and assemblages of variably altered, typically subalkaline and sometimes low-Ti, olivine-bearing and olivine-free gabbroic and basaltic rocks (Figure 3a, Cross, unpublished data). Where these ultramafic-mafic associations are well developed the type (1) melange is termed ophiolitic melange [Cross and Fergusson, 1983]. Also locally abundant are exotic blocks including scattered high-pressure metamorphics [Barron et al., 1976; Offler, 1982a], and blocks of adjacent lithological units.

Type (1) melange at approximately 151°40'E 31°42'S (Figure 2) contains a lensoidal block (40 m in length) of fossiliferous Early Permian sedimentary serpentinite. This sandy to conglomeratic epiclastic rock was derived almost entirely from the rather distinctive host ultramafic unit and contains a fauna indicative of a relatively high-energy sub-littoral environment of deposition (Cross, unpublished manuscript). Although this indicates that at least this small part of the subduction complex was emergent, the occurrence of comparatively extensive olistostromes in the Wisemans Arm Formation (see below) suggests that the Early Devonian was a time of far more significant uplift and accompanying accretion-related deformation.

The Wisemans Arm Formation occurs immediately east of the Peel Fault (Figure 2). It is a westward younging coherent succession of interbedded turbiditic greywacke and mudstone with some type (3) melange [Leitch and Cawood, 1980]. The olistostromes consist of chert olistoliths up to hundreds of metres in length with smaller clasts of greenstone and limestone contained in a cleaved matrix of greywacke and mudstone. The unit is Early Devonian in age based on detailed petrographic correlations with the Tamworth Block and fossil content [Leitch and Cawood, 1980; Cawood, 1982b].

Adjacent to the Wisemans Arm Formation are the pre-Early Devonian Woolomin beds [Leitch, 1974] which extend for 300 km along strike and are up to 5 km in width. These consist of chert, impure chert, jasper, siliceous mudstone, mudstone, greenstone, green tuff and subordinate greywacke. Cawood [1982a] maintains that the Woolomin beds consist of a series of repeated fault slices each containing a sequence up to 1000 m in thickness with basalt at the base, overlain by chert with an increasing proportion of mudstone up-sequence. A few slices contain greywacke near their tops. Cross (unpublished mapping), however, has shown that large areas of Woolomin beds are type (2) tectonic melange consisting of steeply dipping mesoscopic to kilometre-sized slabs and blocks arranged in a more-or-less imbricate fashion in a matrix of moderately to highly disrupted bedded lithologies. In these rocks a complete spectrum exists between coherent strata within slabs, and autoclastic melange in the matrix (Figure 3). During the formation of the autoclastic melange, bedding surfaces were progressively destroyed until all boundaries became tectonic and many thin, often highly lecticular fragments are aligned parallel to the resultant shear foliation in the mudstone (Figure 3). Even in the more strongly deformed zones it is possible to confuse this lenticularity with bedding, especially in the extensive regions of limited and scattered exposure. Disrupted folds are locally present, especially in blocks of thin-bedded highly siliceous sediments (Figure 3d).

Overall, the Woolomin beds formed by the

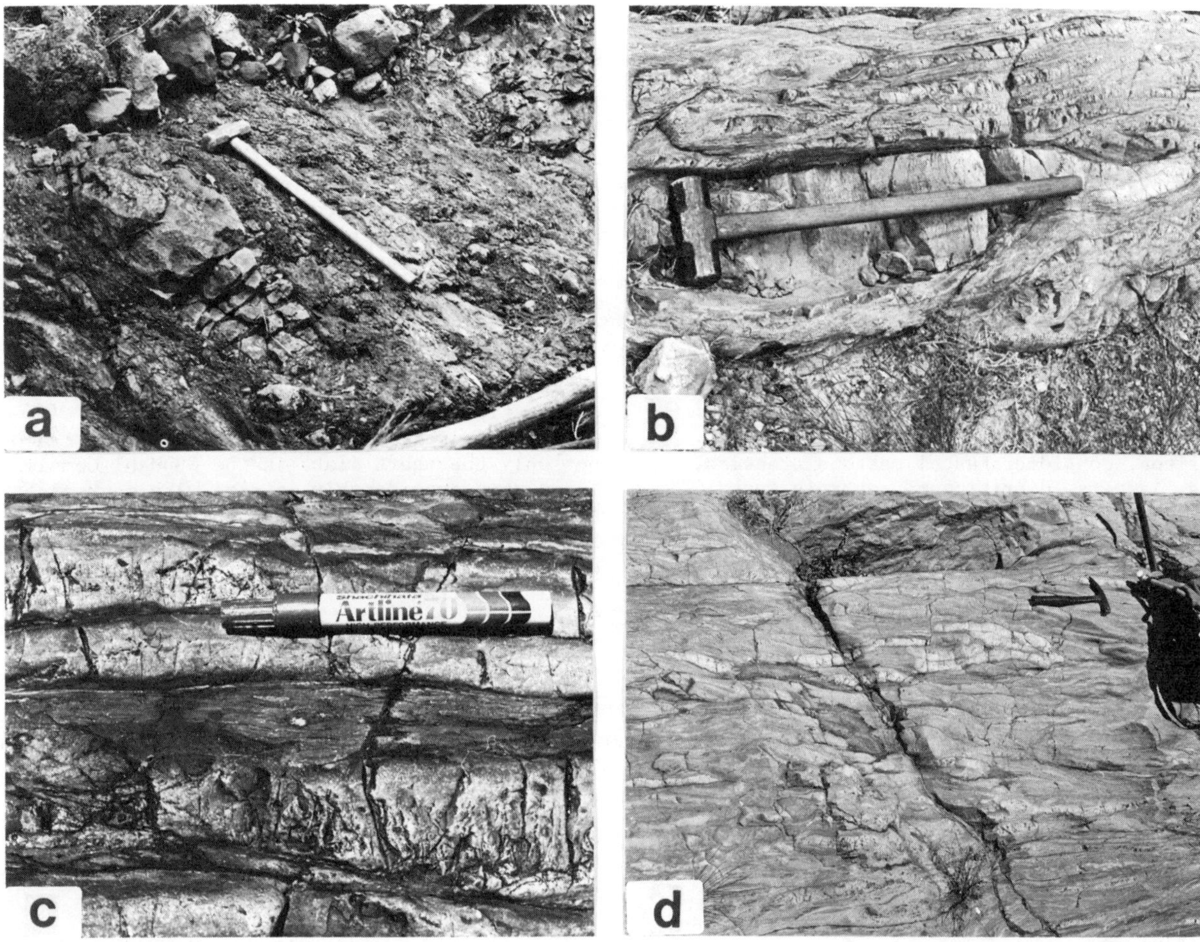

Fig. 3.(a) An olivine mela-norite inclusion in type 1 ophiolitic melange from the Peel Fault 100 km southeast of Tamworth. The matrix is highly sheared serpentinite with the hammer handle lying parallel to the shear foliation. (b)-(d) Examples of tectonic melange, with disrupted thin-bedded siliceous argillite beds contained in sheared argillite, from the Woolomin beds 100 km southeast of Tamworth. Note the disrupted fold hinge in (d) and the "miniraft" of siliceous argillite in (b) aligned parallel to the shear foliation.

imbrication of ocean-floor material which has been substantially disrupted by shear subparallel to the general trend of the larger tectonic slabs. Chert and greenstone are commonly spatially related and this reflects their pre-melange association. Throughout most of the Woolomin beds contacts between differing lithologies are faulted. The map interpretation of Cawood [1982a] is atypical of the Woolomin beds as a whole, especially as he did not describe the widespread structural dislocations in the "apparently coherent" chert, siliceous mudstone and mudstone lithologies.

East of the Woolomin beds is an extensive region of type (2) melange dominated by greywacke and mudstone with less abundant chert, jasper and greenstone. This tectonostratigraphic unit is called the Cockburn Formation in the Macdonald Block [Cawood, 1982a] and the Sandon beds in the Armidale Block. The Sandon beds and Cockburn Formation are probably Devonian in age, with some Early Carboniferous rocks in their eastern portion [based on petrographic comparison of greywackes with those of the Tamworth Block, see Cawood, 1982b; Korsch, 1984; P.G. Flood, unpublished data].

In the eastern part of the Armidale Block there is a multiply-deformed sequence of relatively coherent greywacke and mudstone [the Girrakool beds of Korsch, 1981a]. Much of the multiple deformation affecting these rocks is probably Early Permian in age [Korsch and Harrington, 1981]. The abundance of silicic volcanoclastic detritus in the greywackes of this unit suggests a correlation with the Late Carboniferous section of the Tamworth Block [Korsch, 1984].

Texas and Coffs Harbour Blocks

These two blocks each contain a major hinge of the z-shaped megafold delineated by the structural trends of the subduction complex (Figures 1 and 2). They also share a common tectono-stratigraphy which for the Carboniferous is matched with units further south.

The oldest unit is the Devonian Silverwood Group [Korsch, 1977]. It consists of a monotonous assemblage of fine-grained tuffaceous rocks and altered mafic to intermediate volcanics. Minor lithologies are greywackes, argillites, conglomerates, limestone breccias, cherts, jaspers, dioritic and gabbroic rocks. Intrusive relationships occur between the igneous and clastic rocks in the Coffs Harbour Block [Fergusson, 1984b]. Because of the abundance of massive lithologies and the lack of younging indicators the structure of the Silverwood Group is poorly understood. Unlike other units in the subduction complex, no consistent structural grain is evident. Although most of the unit appears to be coherent, autoclastic melange is locally developed. Limestone clasts in the Silverwood Group contain Early-Middle Devonian faunas [Day et al., 1978]. The Silverwood Group was probably deposited in an island arc-related setting, and may well be an exotic terrane [Flood and Fergusson, 1982].

To the southwest of the Silverwood Group is a type (2) tectonic melange up to 15 km in width. This is known as the Gundahl Complex in the Coffs Harbour Block [Fergusson, 1984a,b] and as subunit Ct1 of the Texas beds in the Texas Block [Fergusson and Flood, 1984]. The melanges consist of greywacke, mudstone, green tuff, bedded chert and greenstone. Greywackes are lithic-rich with abundant fragments of silicic and intermediate volcanics, and less common feldspar (mainly plagioclase) and quartz grains. Many of the greywackes contain ooliths and this, together with their overall petrographic character, is consistent with correlation with the Early Carboniferous of the Tamworth Block [e.g. Day et al., 1978]. On this basis they also correlate with the eastern part of the Sandon beds in the Armidale Block to the south.

The Gundahl Complex is the type example of type (2) melange. It consists of slabs and blocks contained in a deformed matrix (Figure 4). The matrix comprises mudstone with many striated and polished shear fractures that divide the rock into many small chips (scaly fabric, Figure 4a) and it is often poorly exposed. The slabs and blocks display varying intensities of internal deformation. Massive greywackes and greenstones are characterised by intricate networks of randomly oriented cataclastic veins, and the blocks themselves pinch and swell. Slabs containing well-bedded strata are disrupted by abundant cryptic bedding-parallel faults and contain numerous folds. In some areas strata were deformed when relatively unlithified [cf. Lash, 1985]. For example, pinch-and-swell structures with rounded greywacke fragments occur in incipiently to strongly deformed greywacke-argillite sequences. However, at any one locality most greywacke fragments are bounded by low-to-high angle shear fractures, indicating predominantly brittle behaviour (Figure 4b). Siliceous mudstones and cherts have abundant meoscopic brittle fractures indicating they were well-lithified prior to deformation (Figure 4).

On a map-scale it is often possible to trace out only the chert slabs in the Gundahl Complex. These vary considerably in size with many slabs tens of metres in length and fewer up to several kilometres in length. They overlap in an en echelon pattern, similar to the distribution of lecticular fragments in many outcrops of autoclastic melange. Detailed mapping shows that the Gundahl Complex has an imbricate structure with a repetition of a disrupted pre-melange trench-floor sequence [Fergusson, 1984a,b, 1985]. This sequence is similar to that proposed by Cawood [1982a] for the Woolomin beds, except that tectonic disruption has destroyed most of the original stratigraphic relations in the pre-melange sequence.

To the southwest of both the Gundahl Complex and subunit Ct1 of the Texas beds there are large regions dominated by mudstone and greywacke with an abundance of silicic volcaniclastic detritus [Fergusson, 1984b; Korsch, 1984; Fergusson and Flood, 1984]. These rocks may be correlated with the Late Carboniferous succession of the Tamworth Block and the Girrakool beds of the Armidale Block. Three main units are recognized : (1) subunit Ct2 of the Texas beds in the Texas Block and the Coramba beds of the Coffs Harbour Block, (2) subunit Ct3 of the Texas beds and the Brooklana beds, and (3) subunit Ct4 of the Texas beds and the Moombil beds (Figure 2).

The Coramba beds and subunit Ct2 consist of monotonous successions of greywacke and mudstone characterised by alternating intervals of "proximal" turbidites and "distal" turbidites in an association resembling mid- and outer-submarine fan desposits [Fergusson, 1985]. The northwest portions of these units are dominated by massive and/or laminated mudstone with minor interbedded greywacke and were probably deposited in a starved basin-plain environment. These rocks are deformed into tight to isoclinal macroscopic folds, with vergence to the southwest and the predominant sense of younging to the northeast [e.g. Fergusson, 1982]. Axial surfaces are either steeply dipping to the southwest or vertical, and are typically faulted. In addition to

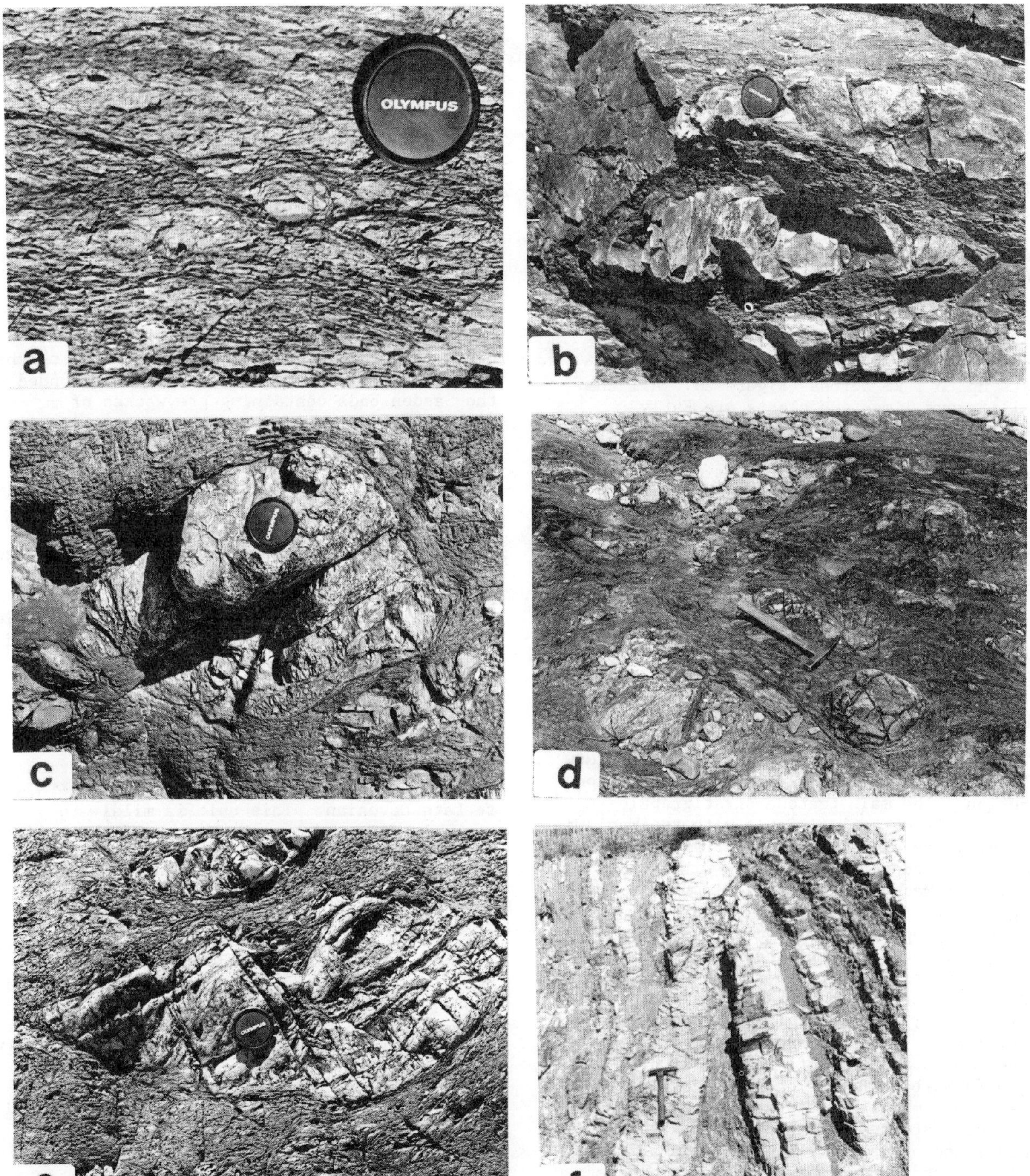

Fig. 4.(a)-(e) Examples of tectonic melange from the Gundahl Complex. Note the anastomosing shear foliation (scaly mudstone) in (a). Scaly mudstone wraps around a greywacke lenticular fragment in (b). In (c) an equant chert fragment is enclosed by scaly mudstone, whereas greenstone and chert fragments are both present in (d). The greenstone fragments in (e) may have originated from the tops of pillows as indicated by their lower concave upward boundaries. Lens cap is 5.5 cm across, and hammer is 33cm in length. (f) Example of bed repetition due to a bedding-parallel fault from subunit Ct4 of the Texas beds.

thickening related to macroscopic faulted folds, these beds have been tectonically thickened by repetition along cryptic bedding-parallel faults [Figure 4f; Korsch, 1981b]. Throughout the central Coffs Harbour Block cleavage is clockwise divergent from fold axial traces and macroscopic and mesoscopic folds are transected [Fergusson, 1982].

The Brooklana beds and subunit Ct3 contain less greywacke than the Coramba beds and, unlike the latter, they contain chert, jasper and greenstone. Reconnaissance mapping suggests that this unit consists of relatively coherent strata with an undetermined proportion of type (2) melange. Subunit Ct3 of the Texas beds contains a number of scattered occurrences of megabreccias (olistostromes), with limestone clasts up to 1 km across (Figure 2). All the fossiliferous limestone clasts are Early Carboniferous in age [Fergusson and Flood, 1984]. There are three feasible sources for these limestones: the forearc basin to the west, shallow marine portions of the trench-slope break and rafted-in caps to oceanic guyots on the oceanic plate.

The Moombil beds and subunit Ct4 are dominated by mudstone and are structurally massive.

Discussion

Subduction Accretion

The principal mechanism of growth in the New England subduction complex has been subduction accretion. The main tectono-stratigraphic units young away from the Peel Fault with the Woolomin beds being the oldest and the Girrakool beds and equivalents being the youngest. Within each unit in the Macdonald and Armidale blocks local younging directions are generally directed towards the magmatic chain to the west [Cawood, 1982a; R.G. Cuddy, unpublished mapping]. In the Texas and Coffs Harbour Blocks most local younging directions are towards the northeast [Korsch, 1981; Fergusson, 1984b; Fergusson and Flood, 1984]. The pattern differs from that farther south and reflects the location of Texas and Coffs Harbour Blocks on the central limb of the z-shaped megafold (Figures 1 and 2).

The consistent younging directions result from the addition of upward younging fault slices at the toe of the subduction complex, and these gradually rotate to steep orientations with outward growth [cf. Karig and Sharman, 1975]. In the type (2) melanges bedding-parallel extension has destroyed many original depositional contacts but, where preserved within slabs of coherent strata, younging directions still reflect the early accretion event. Accretion of the Coramba beds was accompanied by large scale folding with wavelengths up to 10 km.

Subduction Kneading

In uplifted subduction complexes contrasts in structural styles, lithofacies content and less commonly palaeobathymetric criteria have been used to delimit the extent of kneaded trench-slope basins [e.g. Moore et al.; 1980]. Fergusson [1984a] applied the first two of these criteria to identify several possible ancient trench-slope basin in the Gundahl Complex. The Wisemans Arm Formation may have formed in a trench-slope basin [Leitch and Cawood, 1980]. Elsewhere in the New England subduction complex trench-slope basins may be identified by out-of-sequence greywacke successions. For example, in the Kingstown area of the Armidale Block (Figure 2), greywackes containing ooliths (probably Early Carboniferous) are surrounded by the Sandon beds containing greywackes of probable Devonian age. Another example is in the Silverwood Group equivalent of the Coffs Harbour Block which contains greywackes consisting wholly of mafic volcanic detritus and less abundant out-of-sequence greywacke successions dominated by silicic volcanic detritus and probably Late Carboniferous in age [Fergusson, 1984b]. In general, kneaded trench-slope basins appear to constitute only a minor part of the New England subduction complex compared to the abundance of inferred accreted material.

Terrane Accretion

The Silverwood Group represents a fragment of relativly exotic island arc-related material accreted to the subduction complex in the Middle to Late Devonian. This unit is mildly to strongly deformed but shows none of the evidence of accretionary growth present in the units of the subduction complex. The Silverwood Group may have been originally separated from the Devonian convergent margin by a marginal sea [Day et al., 1978] and, along with the Woolomin beds, might well be considered an exotic terrane. Cawood and Leitch [1985] and Scheibner [1985] have liberally applied terrane terminology to tectono-stratigraphic units of the subduction complex. However, in view of the inferred "local" provenance for many of these rocks, we would consider it more prudent to restrict the term "terrane" to the more exotic elements of the complex, and to relate the characteristics of the other tectono-stratigraphic units to the nature of their various accretionary histories.

Contrasting Structural Styles: A Concluding Discussion

One of the most prominent features of the New England subduction complex is a change in structural style from the pre-Early Devonian-Early Carboniferous type (2) melanges to the Late Carboniferous relatively coherent units.

This change results from the deposition of large thicknesses of Late Carboniferous sediments in the trench and coincides with a major regression in the forearc basin [Roberts and Engel, 1980]. Increased sedimentation in the trench and forearc basin was caused by renewed volcanism and uplift along the volcanic chain [Leitch, 1974; McPhie, 1983].

In addition to the increased sediment supply there is also some evidence for a minor change in plate interaction. Clockwise cleavage divergences from axial traces in the Coffs Harbour Block are analogous to those in the Southern Uplands of Scotland which Stringer and Treagus [1981] interpreted to form from oblique subduction at the trench. Thus a change in relative plate motion may have occurred from normal subduction in the Early Carboniferous to oblique subduction in the Late Carboniferous.

The variation in structural style is also accompanied by a change in the nature of material accreted. The Late Carboniferous portion of the subduction complex has substantially less mudstone, tuff, chert and greenstone than the pre-Early Devonian to Early Carboniferous type (2) melanges. The absence of these lithofacies from the Late Carboniferous presumably reflects their dilution by voluminous relatively more proximal volcaniclastics and, to some extent, subduction and possible subcretion at depth beneath the subduction complex.

The overall gross characterics of the New England subduction complex appear to reflect the gradual transformation of the convergent margin from a Mariana-type to a Chilean-type; perhaps due to the subduction of progressively younger and therefore more buoyant oceanic crust [cf. Uyeda, 1982]. In particular, the presence of low-Ti ophiolitic rocks in type (1) melange, in the Woolomin beds, and near the base of the forearc succession [Offler, 1982b; Cross, unpublished data] is consistent with an early Mariana- type history [Crawford et al., 1981]. With the exception of scattered intraplate volcanic piles, representing seamounts and/or volcanic islands [Cawood, 1984; Offler, 1982; Cross, unpublished data] and the Silverwood Group, the oceanic crust subducted during the early history of the New England subduction complex was surprising lacking in transported terranes. Presumeably an open ocean somewhat akin to the central and eastern neo-Pacific is indicated.

Acknowledgements. The authors are grateful to Evan Leitch and Richard Norris for critical reviews of this paper. Therese Carmondy typed the mauscript onto the word processor. Draga Gelt of Monash University drafted the diagrams.

References

Barron, B. J., E. Scheibner, and E. Slansky, A dismembered ophiolite suite at Port Macquarie, New South Wales, Rec. Geol. Surv. N.S.W., 18, 69-102, 1976.

Cawood, P. A., Structural relations in the subduction complex of the Paleozoic New England Fold Belt, eastern Australia, J. Geol., 90, 381-392, 1982a.

Cawood, P. A., Correlation of stratigraphic units across the Peel Fault, in New England Geology, edited by P. G. Flood, and B. Runnegar, pp. 53-61, Department of Geology, University of New England and AHV Club, Armidale, 1982b.

Cawood, P. A., Modal composition and detrital clinopyroxene geochemistry of lithic sandstones from the New England Fold Belt (east Australia): a Paleozoic fore-arc terrain, Bull. Geol. Soc. Am., 94, 1199-1214, 1983.

Cawood, P. A., A geochemical study of metabasalts from a subduction complex in eastern Australia, Chem. Geol., 43, 29-47, 1984.

Cawood, P. A., and E. C. Leitch, Accretion and dispersal tectonics of the southern New England Fold Belt, eastern Australia, in Tectonostratigraphic Terranes of the Circum-Pacific Region, edited by D. G. Howell, pp. 481-492 Circum-Pacific Council for Energy and mineral Resources, Earth Sciences No.1, 1985.

Crawford, A. J., A. L. Beccaluva, and G. Serri, Tectono-magmatic evolution of the West Philippine-Mariana region and the origin of boninites, Earth Planet. Sci. Lett., 54, 346-356, 1981.

Cross, K. C., and C. L. Fergusson, Contrasting styles of melange development in the New England Orogen, Geol. Soc. Aust. Abs., 9, 35-36, 1983.

Day, R. W., C. G. Murray, and W. G. Whitaker, The eastern part of the Tasman Orogenic Zone, Tectonophysics, 48, 327-364, 1978.

Fergusson, C. L., Structure of the Late Paleozoic Coffs Harbour Beds, northeastern New South Wales, J. Geol. Soc. Aust., 29, 25-40, 1982.

Fergusson, C. L., The Gundahl Complex of the New England Fold Belt, eastern Australia: a tectonic melange formed in the Paleozoic subduction complex, J. Struct. Geol., 6, 257-271, 1984a.

Fergusson, C. L., Tectono-stratigraphy of a Paleozoic subduction complex in the Central Coffs Harbour Block of northeastern New South Wales, Aust. J. Earth Sci., 31, 217-236, 1984b.

Fergusson, C. L., Trench floor sedimentary sequences in a Paleozoic subduction complex, eastern Australia, Sediment. Geol., 42, 181-200, 1985.

Fergusson, C. L., and P. G. Flood, A Late Paleozoic subduction complex in the Border Rivers area of southeast Queensland, Proc. R. Soc. Queensl., 95, 47-55, 1984.

Flood, P. G., and C. L. Fergusson, Tectonostratigraphic units and structure of the Texas Coffs Harbour region, in New England Geology, edited by P. G. Flood, and B. Runnegar, pp. 71-78, Department of Geology, University of New England and AHV Club, Armidale, 1982.

Greenly, E., The Geology of Anglesey, Mem. Geol. Surv. U.K., 21, 1919.

Karig, D. E., and G. F. Sharman, Subduction and accretion in trenches, Bull. Geol. Soc. Am., 86, 377-389, 1975.

Korsch, R. J., A framework for the Paleozoic geology of the southern part of the New England Geosyncline, J. Geol. Soc. Aust., 23, 339-355, 1977.

Korsch, R. J., Structural geology of the Rockvale Block, northern New South Wales, J. Geol. Soc. Aust., 28, 51-70, 1981a.

Korsch, R. J., Deformational history of the Coffs Harbour Block, J. Proc. R. Soc. N.S.W., 114, 17-22, 1981b.

Korsch, R. J., Sandstone compositions from the New England Orogen, eastern Australia: implications for tectonic setting, J. Sediment. Petrol., 54, 192-211, 1984.

Korsch, R. J., and H. J. Harrington, Stratigraphic and structural synthesis of the New England Orogen, J. Geol. Soc. Aust., 28, 205-226, 1981.

Lash, G. G., Accretion-related deformation of an ancient (Early Paleozoic) trench-fill deposit, central Appalachian Orogen, Bull. Geol. Soc. Am., 96, 1167-1178, 1985.

Leitch, E. C., The geological development of the southern part of the New England Fold Belt, J. Geol. Soc. Aust., 21, 133-156, 1974.

Leitch, E. C., Plate tectonic interpretation of the Paleozoic history of the New England Fold Belt, Bull. Geol. Soc. Am., 85, 141-144, 1975.

Leitch, E. C., Crustal development in New England, in New England Geology, edited by P. G. Flood, and B. Runnegar, pp. 9-16, Department of Geology, University of New England and AHV Club, 1982.

Leitch, E. C., and P. A. Cawood, Olistoliths and debris flow deposits at ancient consuming plate margins: an eastern Australian example, Sediment. Geol., 25, 5-22, 1980.

McPhie, J., Outflow ignimbrite sheets from Late Carboniferous calderas, Currabubula Formation, New South Wales, Australia, Geol. Mag., 120, 487-503, 1983.

Moore, G. F., H. G. Billman, P. E. Hehanussa, and D. E. Karig, Sedimentology and palaeobathymetry of Neogene trench-slope deposits, Nias Island, Indonesia, J. Geol., 88, 161-180, 1980.

Murray, C. G., C. L. Fergusson, P. G. Flood, W. G. Whitaker, and R. J. Korsch, Plate tectonic model for the Carboniferous evolution of the New England Orogen, Aust. J. Earth Sci., in press, 1986.

Nicolas, A., F. Boudier, and J-L. Bouchez, Interpretation of peridotite structures from ophiolitic and oceanic environments, Am. J. Sci., 280A, 192-210, 1980.

Offler, R., The origin of exotic blocks in serpentinites, Peel Fault System, Glenrock Station, N.S.W., in New England Geology, edited by P. G. Flood, and B. Runnegar, pp.43-51, Department of Geology, University of New England and AHV Club, Armidale, 1982a.

Offler, R., Geochemistry and tectonic setting of igneous rocks in the Glenrock Station area, N.S.W., J. Geol. Soc. Aust., 29, 443-445, 1982b.

Roberts, J., and B. A. Engel, Carboniferous palaeogeography of the Yarrol and New England Orogens, eastern Australia, J. Geol. Soc. Aust., 27, 167-186, 1980.

Scheibner, E. C., Suspect terranes in the Tasman Fold Belt System, eastern Australia, in Tectonostratigraphic Terranes of the Circum-Pacific Region, edited by D. G. Howell, pp.493-514, Circum-Pacific Council for Energy and Minerals Resources, Earth Sciences Series, No. 1, 1985.

Stringer, P., and J. E. Treagus, Asymmetrical folding in the Hawick Rocks of the Galloway area, Southern Uplands, Scott. J. Geol., 17, 129-148, 1981.

von Huene, R., Tectonic processes along the front of modern convergent margins - research of the past decade, Annual Rev. Earth Planet. Sci., 12, 359-381, 1984.

Uyeda, S., Subduction zones: an introduction to comparative subductology, Tectonophysics, 81, 133-159, 1982.

Westbrook, G. K., M. J. Smith, J. H. Peacock, and M. J. Poulter, Extensive underthrusting of undeformed sediment beneath the accretionary complex of the Lesser Antilles subduction zone, Nature, 300, 625-628, 1982.

CARBONIFEROUS FAUNAS: THEIR ROLE IN THE RECOGNITION OF TECTONOSTRATIGRAPHIC TERRANES IN THE TASMAN BELT, EASTERN AUSTRALIA

John Roberts

Department of Applied Geology, University of New South Wales, Kensington, N.S.W. 2033, Australia

Abstract. Two major marine invertebrate assemblages are present in Carboniferous shelfal successions of Australia: a high diversity, warm water, Early Carboniferous Cosmopolitan Assemblage; and a low diversity, cold water, Middle-Late Carboniferous Gondwanan Assemblage. A comparison between Cosmopolitan Assemblages of the Tasman Belt and intracratonic basins of Western Australia indicates that there are a large number of genera but few species common to both areas: this is interpreted as indicating geographic separation of the two regions during the Early Carboniferous. The first evidence of close proximity of the two regions is the occurrence of late Visean Western Australian brachiopod species in the *Marginirugus barringtonensis* Zone in Queensland, in the latter region a zone of a transition between the Cosmopolitan and Gondwanan Assemblages. The likelihood of the Yarrol-New England portion of the Tasman Belt constituting a separate terrane located north of the Australian continent during the Early Carboniferous is indicated by palaeomagnetic data of equatorial aspect from New England, but a more southerly aspect from the cratonic region of northeastern Australia. During latest Visean-early Namurian time the Yarrol-New England belt moved rapidly southwards and docked with the Australian craton, and for the first time cratonic material was deposited in shelf regions of the belt. A deterioration of climate caused by the southerly movement, as well as the Hercynian reorganisation of continents and oceanic circulation, led to the replacement of the Cosmopolitan Assemblage by the Gondwanan Assemblage.

The eastern Tasman Belt initially consisted of a volcanic arc, a shelf and a subduction complex. Structural dislocation, particularly dextral translation of the northern part of the belt in the latest Carboniferous, divided the belt into the Yarrol and New England Orogens. Deep water sediments of the subduction complex contain fossil debris, ooliths and intermediate to silicic volcanic detritus that indicate provenance linkage of the arc, shelf and subduction complex during the Early Carboniferous.

The only Carboniferous rocks allochthonous to the Tasman Belt are found in a small sliver on the eastern margin of the Yarraman Block near Murgon, and in the northern part of the D'Aguilar Block. These may belong to the Gympie Terrane.

Introduction

Carboniferous marine invertebrate faunas of the Tasman Belt in eastern Australia are found almost exclusively in shallow marine shelf sediments, and the succession of faunas has been subdivided into a number of biostratigraphic zones (Fig. 1). Some fossil debris, as well as ooliths, are also present in deep water sediments of an adjacent subduction complex, providing evidence of provenance linkage.

This paper examines the Carboniferous faunal record of the Tasman Belt and intracratonic basins elsewhere in Australia to determine the extent to which marine invertebrates, particularly brachiopods, can be used to demonstrate that segments of the present Australian continent were once disjunct tectonostratigraphic terranes. Supporting stratigraphic, sedimentological, biogeographic, palaeomagnetic and radiometric data are used to enhance the palaeontological evidence.

Carboniferous Faunas

Two distinct marine invertebrate assemblages are present within the Tasman Belt of eastern Australia, an older (Early Carboniferous) Cosmopolitan Assemblage and a younger (Middle-Late Carboniferous) Gondwanan Assemblage (Fig. 1). The Cosmopolitan Assemblage contains representatives of virtually all preservable groups and is diverse in species and genera. Groups with high levels of diversity include brachiopods [Maxwell, 1954; Campbell and McKellar, 1969; Roberts, 1975], bryozoans [Engel, 1975, 1978] trilobites, [Engel and Morris 1975, 1980, 1983, 1984, 1985] and rugose and tabulate corals [Pickett, 1969; Jull 1969, 1974a, b]; those with a medial rate of diversity include the bivalved molluscs, gastro-

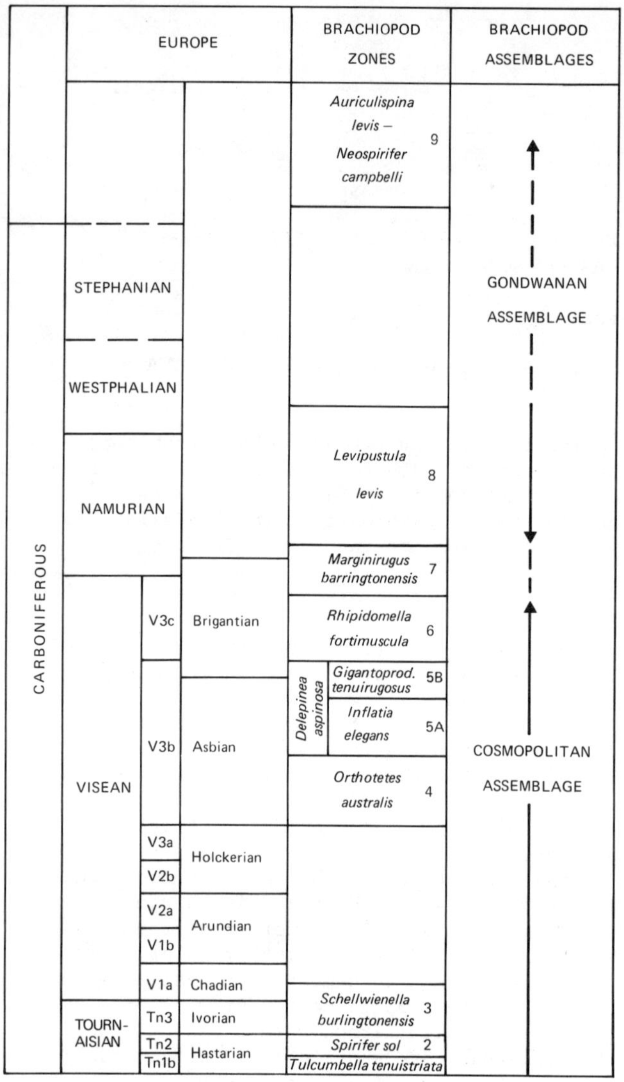

Fig. 1. Carboniferous brachiopod assemblages of eastern Australia.

pods [Campbell and Engel, 1963], echinoderms, particularly crinoids [Campbell and Bein, 1971; Lindley, 1979], and conodonts [Jenkins, 1974; Mory and Crane, 1982]; and those with low rates of diversity include the ammonoids [Campbell et al., 1983], algae [Mamet and Roux, 1983] and foraminifera. The assemblage has few endemic genera [Roberts, 1981].

The Gondwanan Assemblage, on the other hand, has an extremely low diversity [Campbell and McKellar, 1969; Roberts, 1981], and fewer of the major groups of organisms are represented. Brachiopods, bryozoans, bivalves and gastropods are the most abundant of the marine organisms [Campbell, 1961, 1962; Maxwell, 1964; McKellar, 1965; Fleming 1969, 1972; Roberts et al., 1976], but their level of diversity is far lower than that of the previous assemblage [Roberts, 1981]. Conulariids [Campbell, 1961], trilobites [Amos et al., 1960] and conodonts are poorly represented, and the rugose and tabulate corals and ammonoids are absent.

Affinities of the Cosmopolitan Assemblages

Australia. Early Carboniferous marine invertebrate faunas in the Bonaparte, Canning and Carnarvon Basins in the western part of the Australian continent are also cosmopolitan in nature, but differ from those of the Tasman Belt at either the generic or specific level. Using the brachiopods, probably the most abundant and best known group, Roberts [1971] maintained that this difference was caused by isolation rather than by climatic contrast; Runnegar and Campbell [1976] suggested that environmental factors, particularly the type of sediment, could also have been responsible for faunal differences, successions in the west being dominated by limestones and quartz sandstones whereas those in the Tasman Belt were largely volcanogenic. The virtual absence of foraminifera from eastern Australia, ascribed to the abundance of volcanogenic detritus (B. L. Mamet, University of Montreal, personal communication, 1977), may support their view, but recent palaeomagnetic work by Klootwijk [1985] suggests that the two regions could have been well separated during the Early Carboniferous.

A comparison between the Early Carboniferous brachiopod faunas of the Bonaparte Basin and the Yarrol-New England Orogen indicates a reasonable degree of similarity at the generic level, but little in common at the specific level (Table 1). Early Carboniferous brachiopod faunas of the Bonaparte Basin [Roberts, 1971; Thomas, 1971] are present in two successions separated by a postulated erosional break and a transgressive marine black shale which generally lacks brachiopod faunas [Roberts, 1985]. Tournaisian faunas contain 44 species belonging to 33 genera, 5 of which are endemic to Western Australia; 20 of these genera also occur in the Yarrol-New England Orogen. Only three species belonging to *Schizophoria*, *Rhipidomella* and *Leptagonia* are common to both areas, and these have a world-wide distribution. Visean faunas (all late Visean) contain 40 species belonging to 35 genera, one of which is endemic to Western Australia; 23 of these genera are also present in the Yarrol-New England Orogen. The same three species mentioned above are present in eastern Australia together with *Echinoconchus gradatus*, *Dorsoscyphus spinulosus*, *Hustedia paula* and *Punctospirifer pauciplicatus*, the last three species occurring in a younger zone (*Marginirugus barringtonensis*) than in the Bonaparte Gulf Basin. Overall, the entire Early Carboniferous brachiopod fauna from the Bonaparte Basin contains 78 species belonging to 55 genera, 6 of which are endemic to Western

TABLE 1. Summary of Numbers of Taxa in Australian Carboniferous Brachiopod Assemblages

Age	Western Australia	Common	Eastern Australia
Post-Visean	0	0	19(5)** / 26
Late Visean	35(1) / 40	23 / 7	64(3) / 94
Tournaisian	33(5) / 44	20 / 2	38(2)* / 52

genera (endemics) / species

* Contains early Visean assemblages.

** Figures refer only to the *L. levis* Zone. The late Visean figures for eastern Australia do not include the *Marginirugus barringtonensis* Zone.

Australia and 32 of which are also present in faunas of the same age in the Yarrol-New England Orogen.

Early Carboniferous brachiopod faunas from the Yarrol-New England Orogen contain 134 species belonging to 77 genera, five of which are endemic. The fauna can be subdivided into assemblages, separated by a middle Visean hiatus [Roberts 1981, 1985], similar to those in the Bonaparte Basin but containing a far more extensive Visean fauna. The Tournaisian-early Visean assemblage contains 52 species belonging to 38 genera, and the Late Visean assemblage 94 species belonging to 64 genera.

Apart from the cosmopolitan species of *Schizophoria*, *Rhipidomella* and *Leptagonia* and *Echinoconchus gradatus*, the only area in which eastern and western faunas occur together is in the Yarrol Orogen, where 3 species from the *Punctospirifer pauciplicatus* Zone [Roberts, 1971] are found with those of the *Marginirugus barringtonensis* Zone in the Killala Creek Limestone [Nazer, 1977]. Although differing sedimentary facies may have been a contributing factor, I believe that the virtual absence of species common to both areas throughout the Early Carboniferous, a period of strong cosmopolitanism, points to geographic isolation of the Yarrol-New England Orogen from the Western Australian basins. The significance of the first appearance of Western Australian species in eastern Australia is dealt with below.

Overseas. A strong degree of cosmopolitanism existed in Early Carboniferous faunas [Boucot and Gray, 1979; Ross and Ross, 1985], probably as a result of a low level of climatic differentiation which began in the Late Devonian and continued into the Carboniferous. At a generic level Australian Early Carboniferous shallow marine invertebrates have affinities with faunas in western Europe, central and western North America, and Asia, particularly Japan [Campbell and McKellar, 1969; Jully, 1969; Roberts, 1975; Campbell et al., 1983; see Roberts et al., 1985 for information on brachiopods, trilobites, bryozoans, conodonts,ammonoids, spores and pollen and plant fossils]. Whilst these faunas cannot be used to indicate relative positions in a palaeogeographic sense, they nevertheless contrast markedly with faunas of the Gondwanan Assemblage which developed over a short span of time between the latest Visean and early Namurian.

Affinities of the Gondwanan Assemblage

Australia. Gondwanan Assemblages are not recognised in Western Australia because most basins underwent uplift and regression late in the Visean. Marine sediments of Middle to Late Carboniferous age are present in the subsurface of the offshore Bonaparte Basin and possibly the Carnarvon Basin, but macrofaunas have not been collected.

Overseas. The Gondwanan Assemblage is represented by the *Levipustula levis* Zone in Argentina [Amos et al., 1973], indicating close faunal links and the existence of similar climatic conditions in South America and eastern Australia. At least 8 species comparable with those in the *L. levis* assemblage, including a species of *Levipustula*, are present in a fauna in the Lake Baikal region of USSR [Kotlyar and Popeko, 1967; Roberts et al., 1976], but are accompanied by other forms not represented in either Australia or South America. One or two elements of the assemblage reported from Westphalian faunas in northwestern Europe [Boger and Fiebig, 1963] and Spain [Winkler Prins, 1968] are now considered distinct from the Gondwanan Assemblage (C.F. Winkler Prins, Rijksmuseum van Geologie en Mineralogie, Leiden, personal communication,1986). There are no other close faunal ties with elsewhere in the world.

The Change from Cosmopolitan to Gondwanan Assemblages

The change between the two assemblages was relatively abrupt and took place during the latest Visean to earliest Namurian within the *Marginirugus barringtonensis* Zone. Campbell and McKellar [1969] and Roberts [1981] believed the faunal change was caused by a deterioration in the climate due to rapid southward movement of the Australian continent, but new palaeomagnetic information provided by Klootwijk [1985] now suggests that much of that southerly movement may have taken place within the Yarrol-New England Orogen prior to docking with the Australian craton. Klootwijk [1985] indicated that volcanics in the Early Carboniferous Isismurra Formation at Rouchel in the New England Orogen were extruded in an equatorial latitude, whereas contemporaneous volcanics from the Newcastle Range Volcanics near Georgetown in North Queensland (on the Precambrian craton of Australia) were extruded at a latitude 30°S of the equator. If both magnetisations are Early Carboniferous in age, it is possible that the entire Yarrol-New England Orogen was a separate terrane during the earlier part of its geological history. Attachment to the Australian craton took place in the latest Visean-earliest Namurian, coinciding with an upsurge of volcanism, uplift and generation of a widespread disconformity, and the imposition of a new depositional pattern throughout the orogen [Roberts and Engel, in press]. Although there is little evidence of structural dislocation within the Yarrol-New England Orogen at this time, docking with the Australian craton appears to coincide with widespread deformation in the Lachlan Orogen and central Australia [Powell, 1984]. Clasts from a cratonic region, which had been absent from Early Carboniferous and older successions, were shed into the Yarrol-New England Orogen from the west throughout the remainder of the Carboniferous. The change coincides with the commencement of glaciation and postulated generation of a widespread ice cap [Powell and Veevers, 1984]. The deterioration of climate in Australia may to some extent have been triggered by southward movement of the continent as suggested by Irving [1966], but would also have been influenced by a worldwide Hercynian reorganisation of continents and oceanic circulation patterns [Ross and Ross, 1985].

The abrupt nature of the faunal change in eastern Australia was highlighted by analysis of the brachiopod faunas [Roberts, 1981] which demonstrated that 94% of all species became extinct at the end of the *Rhipidomella fortimuscula* Zone. The change from the Cosmopolitan to Gondwanan Assemblage took place abruptly in the southern part of the Tasman Belt, but in the north some of the cosmopolitan genera lingered on into the *Marginirugus barringtonensis* Zone [Nazer, 1977]; these 'stragglers' had all disappeared by the *Levipustula levis* Chron. The *M. barringtonensis* fauna in Queensland is particularly important because it contains the only Western Australian species, apart from the entirely cosmopolitan species of *Schizophoria*, *Rhipidomella* and *Leptagonia*, to penetrate into the Yarrol-New England Orogen. Their presence may indicate that the two regions were, for the first time, in reasonably close proximity. The same explanation may also account for the first appearance of *Echinoconchus gradatus* in the latest Visean or Namurian of the Bonaparte Basin [Roberts, 1971], well after the species first appeared in eastern faunas [Roberts, 1975].

Palaeogeographic Setting of the Faunas

Shelly faunas were virtually confined to a shelf area which became divided tectonically into the Yarrol and Tamworth Shelves of the Yarrol and New England Orogens (Fig. 2). A volcanic arc flanked the shelf to the west, and in the east the shelf edge appears to have been defined by the Yarrol-Peel Fault System. Sediments deposited on the shelf consisted of volcanogenic lithic sandstone, siltstone and mudstone; oolitic limestone was more widespread in the Yarrol Shelf than in the Tamworth Shelf.

East of the Yarrol-Peel Fault System deep water sediments (pelagic sediments and turbidites) as well as melanges [Fergusson, 1984] accumulated as an accretionary prism or subduction complex. Earlier workers, for example Day et al. [1978] and Roberts and Engel [1980], termed these areas slope and basin provinces.

Dextral movement of the Yarrol Orogen southwards along a transcurrent fault in the latest Carboniferous was responsible for the generation of the Texas-Coffs Harbour Megafold, and the present configuration of Carboniferous outcrops (Fig. 2) [Murray et al., in press].

Fossil Debris and Ooliths of the Subduction Complex

In the Yarrol Orogen deep water sediments are present in a number of fault-bounded blocks east of the Yarrol Fault including the Coastal, Yarraman, D'Aguilar and Beenleigh Blocks (Fig. 2). Sediments of the Coastal Block in the northern Yarrol Orogen have been subdivided into the Doonside, Wandilla and Shoalwater Formations and are referred to the Curtis Island Group [Kirkegaard et al., 1970]. Although the latter authors considered the Shoalwater Formation as the oldest unit within the group, Murray [1974] identified oceanic deposits of the Doonside Formation as the basal unit; this formation is possibly Devonian or older in age. The overlying Wandilla Formation consists of rhythmically interbedded mudstone and turbiditic greywacke, and minor chert. The greywackes contain mainly volcanic rock fragments but in places have a significant quantity of allochthonous ooliths and occasional elliptical crinoid stems resembling *Platyplateium*. Fleming et al. [1975] suggested that the main source of sediment for the Wandilla Formation was the adjacent Yarrol Shelf which contained carbonate-rich volcanogenic sediments. Since ooliths were abundant on the shelf throughout the Early Carboniferous but were absent from the end of the Visean, an Early Carboniferous age was assigned to the Wandilla Formation. The overlying Shoalwater Formation, a thick turbiditic sequence of interbedded ortho- and protoquartzite, lacks fossil and volcanogenic debris and is considered to be Late Carboniferous in age. Although the post-Visean sandstones of the Yarrol Shelf are more quartzose than those of Early Carboniferous age, they contain far less quartz than the turbiditic sands in the Shoalwater Formation. Fleming et al. [1975] attempted to explain this anomaly by suggesting

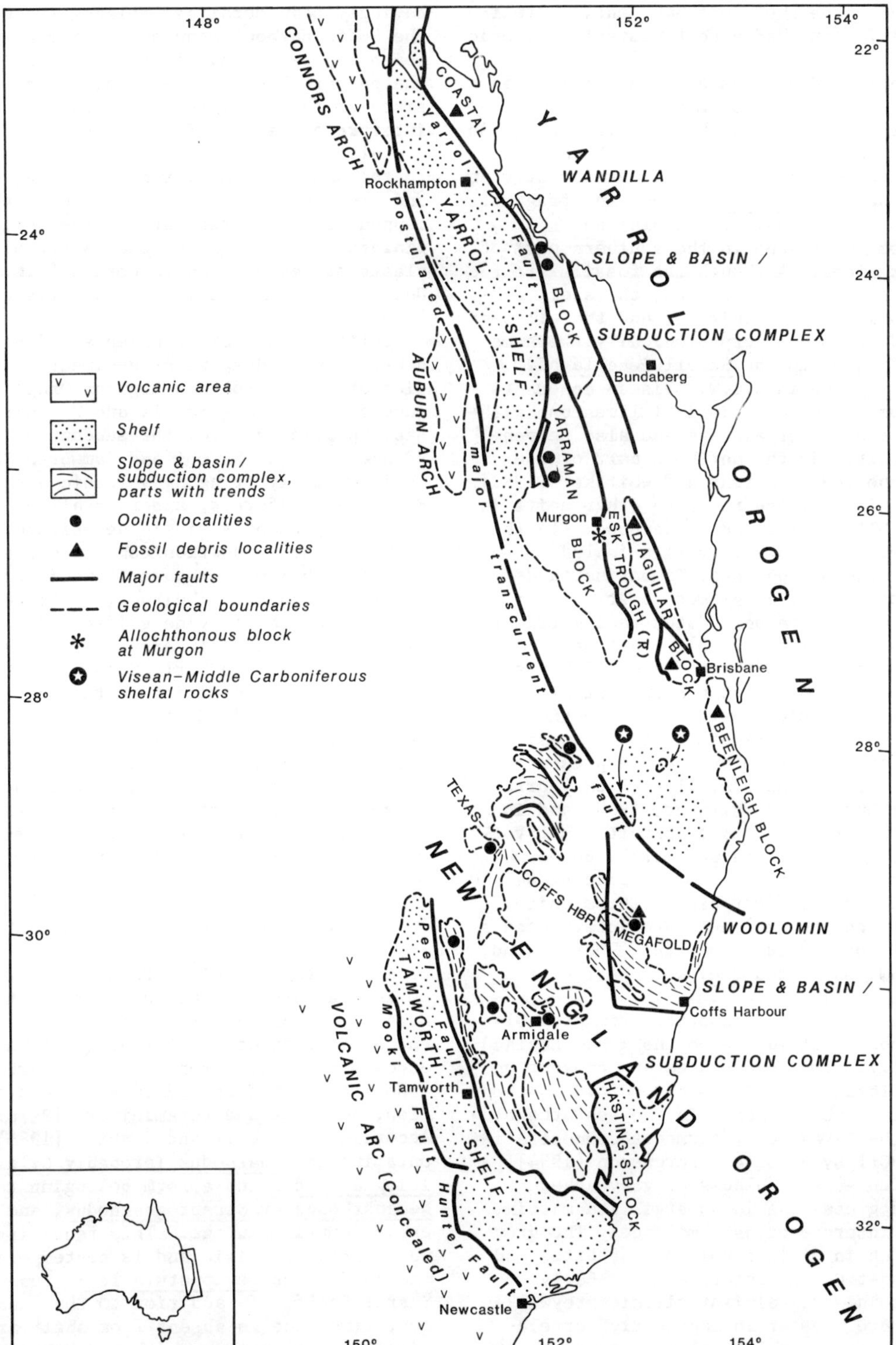

Fig. 2. Carboniferous rocks of the Yarrol and New England Orogens within the Tasman Belt of eastern Australia showing the occurrence of ooliths and fossil debris within deep water sediments, and the position of a possibly allochthonous block at Murgon, Queensland.

deposition while the shelf was emergent, whereas Harrington [1974] invoked a northwesterly cratonic source.

Fossil debris and ooliths are also present in turbiditic greywackes in the medial part of the Neranleigh-Fernvale beds in the D'Aguilar Block in the southern Yarrol Orogen (Fig. 3). Fleming et al. [1974] reported crinoid stems, particularly an ovoid form resembling Platyplateium, and fragments of punctate spiriferoids, productoids and bryozoans from Mount Nebo in the southern part of the D'Aguilar Block. Although the fossils are too fragmentary to give a precise age, the shell debris has Carboniferous affinities, and the elliptical crinoid stems are morphologically the same as stems restricted to the Schellwienella cf. burlingtonensis Zone in N.S.W. Visean conodonts extracted from blocks of crinoidal limestone within diamictites of uncertain age and also serpentinite at Kilkivan in the northern part of the block were reported by Murray and Whitaker [1982]. The relationship of these rocks to other units in the northern D'Aguilar Block is unknown. In the Beenleigh Block fragmentary fossils including Paraconularia are present near Tamborine, and granitic clasts from a conglomerate at Tallebudgera Creek have been dated radiometrically as 344 and 340 Ma [Green, 1973]. Part of Neranleigh-Fernvale beds therefore could be as old as Early Carboniferous, though Murray et al. [1979] suggest a possible extension into the Devonian. Other stratigraphic units of possibly Late Carboniferous to Permian age are present in the northern part of the D'Aguilar Block [Murray et al., 1979; Murray and Whitaker, 1982] and include the Rocksberg Greenstone (basaltic metavolcanics with minor chert and phyllite), Booloumba beds (slate and phyllite with minor greenstone and chert), and unnamed metamorphics. These units lack fossils, and the sediments are interpreted as having been deposited in deep water. Murray and Whitaker [1982] suggested they were part of the allochthonous Gympie terrane.

Ooliths have also been reported from probably correlative units within the northern New England Orogen (Fig. 2). Two occurrences are present on limbs of the Texas-Coffs Harbour megafold in the northern part of the orogen. In the west the oldest part of the Texas beds, termed tectonostratigraphic unit Ct1 by Flood and Fergusson [1984], consists of lens-shaped blocks of greywacke, argillite and greenstone in an argillite matrix, and has been interpreted as a melange. The greywackes are rich in silicic volcanic detritus and ooliths and unit Ct1 is interpreted as Early Carboniferous in age. Similar silicic greywackes with ooliths are present in apparently correlative rocks in the Gundahl Complex [Fergusson, 1984] on the eastern limb of the megafold. Higher parts of the Texas beds (units Ct2-4) are believed to be Late Carboniferous in age [Fergusson and Flood, 1984], though other workers [Lucas, 1960; Korsch and Harrington, 1981] suggest they are much older. One unit (Ct3) contains allochthonous limestones having Visean coral assemblages [Strusz, 1974]. The Coffs Harbour sequence of Fergusson [1984], which apparently overlies the Gundahl Complex on the eastern limb of the fold, lacks ooliths, consists of felsitic greywacke and argillite, and is interpreted as Late Carboniferous in age. Ooliths are also present in greywackes generally referred to the Sandon beds in two localities near Armidale, and are also present in greywackes previously mapped as undifferentiated Palaeozoic rocks near Bingara, though some of the latter also contain clasts of Early Devonian corals [P.G. Flood, University of New England, unpublished data, 1985].

Oolith and fossil fragments and volcanogenic debris within deep water sediments on the eastern side of the Yarrol-New England Orogen therefore suggest that the Wandilla and Woolomin Slope and Basin provinces (i.e. the subduction complex) were adjacent to the Yarrol and Tamworth Shelves with their westerly fringing volcanic arcs during the Early Carboniferous, and that all of these provinces were part of a single depositional system. Although the subduction complex has been divided into a number of terranes [Cawood and Leitch, 1985; Leitch and Scheibner, 1985] the fossil debris and ooliths provide evidence for provenance linkage between shelf areas and the subduction complex during the Early Carboniferous. Dextral faulting and deformation in the latest Carboniferous and additional episodes of faulting in the Permian and Triassic may well have influenced the relative positions of these provinces because of lateral movement along the Peel and Yarrol Fault systems. The relationship between Late Carboniferous shelf areas and subduction complexes cannot be ascertained from present evidence.

Possible Allochthonous Fossils at Murgon

A sliver of Middle Carboniferous rocks located on the eastern edge of the Yarraman Block at Murgon, Queensland (Fig. 2), appears to be different from other Carboniferous shelf and deep water successions. Limestone lenses interbedded with andesitic volcanics and mudstone contain conodonts, fish, annelids and foraminifera [Palmieri, 1969]. According to Nicoll and Jenkins [1985] the conodonts include Gnathodus (probably G. glaber), Idiognathoides and a form belonging to either Neognathodus or Streptognathodus, and are probably early Namurian in age. This fauna is regarded as warm water in origin and is contemporaneous with the cold water Levipustula levis Zone fauna on the Yarrol Shelf. In addition to the faunal differences, limestone is absent from shelf areas at this time, and volcanism, which had been essentially dacitic during the Early Carboniferous, had ceased in the Yarrol Orogen at the end of the Visean. The rocks from Murgon therefore appear to be allochthonous. Adjacent rocks of the possibly Carboniferous to Permian Maronghi Creek Beds in the Yarraman Block [Murphy et al., 1976] consist of chert, mudstone, greywacke and minor basalt. Poor exposure prevents an understanding of their relationship with the Middle Carboniferous sediments.

If the northern D'Aguilar Block, which also contains allochthonous Carboniferous limestones, can be proved distinct from the remainder of the Carboniferous subduction complex, it is possible that it and the allochthonous block at Murgon constitute the western part of the Gympie Terrane [Murray and Whitaker, 1982; Harrington and Korsch, 1985].

Hastings Block

In a tectonic sense, the Hastings Block in the eastern part of the New England Orogen presently appears to be out of place. Geologically, the block has features similar to parts of the Tamworth Shelf [Leitch, 1974] but is located east of subduction complex sediments (Fig. 2). The Carboniferous sedimentary succession commences with turbiditic sediments containing fragmentary Early Carboniferous shelly fossils [Voisey, 1934], followed by shallow water sandstone containing the *L. levis* fauna, and ends with sandstone, conglomerate and multi-coloured mudstone also with *L. levis* faunas. These faunas belong to the Cosmopolitan and Gondwanan Assemblages respectively. Structurally, the Hastings Block is gently folded, moderately faulted and has undergone fewer and less intense deformations than the adjoining region. Its present position on the eastern side of the subduction complex can be explained by sinistral movement during the Late Permian in a manner similar to that proposed by Cawood [1982], and it is interpreted here as a fragment of the Tamworth Shelf moved northward from a position on strike with sediments of similar age in the southeastern part of the shelf.

Conclusions

1. During the Early Carboniferous there were virtually no species of marine invertebrates common to the Yarrol-New England Orogen and the intracratonic basins of Western Australia, despite this period being one of cosmopolitanism. Although the two regions have contrasting shelfal successions (volcanogenic clastics and oolitic limestone versus calcarenite and quartzose clastics), faunal differences at the species level point towards geographic isolation of the two regions.

2. The first appearance of Western Australian Visean faunas in the *Marginirugus barringtonensis* Zone of the Yarrol Orogen coincides with the change from the Cosmopolitan to the Gondwanan Assemblages. In eastern Australia this change was almost certainly induced by a deterioration of climate probably caused by rapid movement of the Yarrol-New England Orogen from equatorial to more southern latitudes [Klootwijk, 1985]. The occurrence of Western Australian species in the Yarrol Orogen, and one eastern Australian species in the Bonaparte Basin, W.A., may indicate that the regions were in close proximity for the first time at the end of the Visean. It is significant that shortly afterwards (during the Namurian) the first influx of cratonic clasts appeared in shelf sediments of the New England Orogen.

3. Fossil debris, ooliths, and intermediate to silicic volcanic detritus in sediments of the subduction complex indicate that the volcanic arc, shelf and subduction complex provinces of the Yarrol-New England Orogen were part of a single depositional system during the Early Carboniferous.

4. Carboniferous rocks allochthonous to the Yarrol-New England Orogen have been identified on the basis of faunal or lithological evidence on the eastern edge of the Yarraman Block, and in the northern D'Aguilar Block.

5. Although tectonically 'out of place', the Hastings Block is part of the Tamworth Shelf.

Acknowledgments. Dr P.G. Flood, Department of Geology, University of New England, generously provided unpublished maps and written material on the occurrence of fragmentary fossils and ooliths in the New England region. The manuscript was critically reviewed by Dr J.W. Pickett, Geological Survey of New South Wales.

References

Amos, A.J., B. Antelo, C.R. Gonzalez, M.P. Marinelarena, and N. Sabattini, Sintesis sobre el conocimiento bioestratigratico del Carbonico y Permico de Argentina, *Actas del Quinto Congresso Geologico Argentino*, *3*, 3-20, 1973.

Amos, A.J., K.S.W. Campbell, and R. Goldring, *Australosutura* gen. nov. (Trilobita) from the Carboniferous of Australia and Argentina, *Palaeontology*, *3*, 227-236, 1960.

Boger, H. and H. Fiebig, Die Fauna des Westdeutchen Oberkarbons II. Die articulaten brachiopoden des Westdeutchen Oberkarbons, *Palaeontographica, A 122*, 111-165, 1963.

Boucot, A.J., and J. Gray, Epilogue: A Paleozoic Pangaea?, in *Historical biogeography, plate tectonics, and the changing environment*, edited by A.J. Boucot and J. Gray, 465-482, Oregon State University Press. 1979.

Campbell, K.S.W., Carboniferous fossils from the Kuttung rocks of New South Wales, *Palaeontology*, *4*, 428-474, 1961.

Campbell, K.S.W., Marine fossils from Carboniferous glacial rocks of New South Wales, *J. Paleont.*, *36*, 38-52, 1962.

Campbell, K.S.W. and J. Bein, Some Lower Carboniferous crinoids from New South Wales, *J. Paleont.*, *45*, 419-436, 1971.

Campbell, K.S.W., D.A. Brown and A.R. Coleman, Ammonoids and the correlation of the Lower Carboniferous rocks of eastern Australia. *Alcheringa*, *7*, 75-123, 1983.

Campbell, K.S.W., and B.A. Engel, The faunas of the Tournaisian Tulcumba Sandstone and its members in the Werrie and Belvue Synclines,

New South Wales. J. Geol. Soc. Aust., 10, 55-122, 1963.

Campbell, K.S.W., and R.G. McKellar, Carboniferous invertebrates: sequence and affinities, in Stratigraphy and Palaeontology, Essays in Honour of Dorothy Hill, edited by K.S.W. Campbell, 79-119, Australian National University Press, Canberra, 1969.

Cawood, P.A., Tectonic reconstruction of the New England Fold Belt in the Early Permian: an example of development at an oblique slip margin, in New England Geology, edited by P.G. Flood and B. Runnegar, pp. 25-34, Department of Geology, University of New England and AHV Club, Armidale, 1982.

Cawood, P.A. and E.C. Leitch, Accretion and dispersal tectonics of the southern New England Fold Belt, eastern Australia, in Tectonostratigraphic terranes of the Circum-Pacific region, edited by D.G. Howell, pp. 481-492, Circum-Pacific Council for Energy & Mineral Resources, Earth Science Series No. 1, 1985.

Day, R.W., C.G. Murray, and W.G. Whitaker, The eastern part of the Tasman Orogen Zone, Tectonophysics, 48, 327-364, 1978.

Engel, B.A., Fenestrate bryozoans with large apertural form in the Carboniferous of eastern Australia, Proc. Linn. Soc. N.S.W., 103, 135-170, 1978.

Engel, B.A., A new ?bryozoan from the Carboniferous of eastern Australia, Palaeontology, 18, 571-605, 1975.

Engel, B.A. and N. Morris, Linguaphillipsia (Trilobita) in the Carboniferous of eastern Australia, Senck. leth., 56, 147-189, 1975.

Engel, B.A. and N. Morris, New Cyrtosymbolinae (Trilobita) from the Lower Carboniferous of eastern Australia, Senck. leth., 60, 265-289, 1980.

Engel, B.A. and N. Morris, Phillipsia-Weberiphillipsia in the Early Carboniferous of eastern New South Wales, Alcheringa, 7, 223-251, 1983.

Engel, B.A. and N. Morris, Conophillipsia (Trilobita) in the Early Carboniferous of eastern Australia, Alcheringa, 8, 23-64, 1984.

Engel, B.A. and N. Morris, The biostratigraphy of Carboniferous trilobites in eastern Australia, 10e Congr. Int. Stratigr. Geol. Carbonifere, Compte Rendu, Madrid 1983, 2, 491-499, 1985.

Fergusson, C.L., Tectono-stratigraphy of a Palaeozoic subduction complex in the central Coffs Harbour Block of north-eastern New South Wales, Aust. J.Earth Sci., 31, 217-236, 1984.

Fleming, P.J.G., Fossils from the Neerkol Formation of central Queensland, in Stratigraphy and palaeontology, essays in honour of Dorothy Hill, edited by K.S.W. Campbell, pp. 264-275, Australian National University Press, Canberra, 1969.

Fleming, P.J.G., Redescription of fenestellid species from the Upper Carboniferous of eastern Australia, Geol. Surv. Qd, Publ., 354, Palaeont. Pap., 29, 1-8, 1972.

Fleming, P.J.G., R.W. Day, C.G. Murray, and W.G. Whitaker, Late Palaeozoic invertebrate macrofossils in the Neranleigh-Fernvale Beds, Qd Govt Min. J., 75, 105-107, 1974.

Fleming, P.J.G., C.G. Murray and W.G. Whitaker, Late Palaeozoic invertebrate fossils in the Wandilla Formation, and the deposition of the Curtis Island Group, Qd Govt Min. J., 76, 416-422, 1975.

Flood, P.G. and C.L. Fergusson, The geological development of the northern New England province of the New England Fold Belt, 1984 Field Conference. Volcanics, granites and mineralisation of the Stanthorpe-Emmaville-Drake region, edited by H.K. Herbert and J.M.W. Rynn, pp. 1-19, Geological Society of Australia, Queensland Division, Brisbane, 1984.

Green, D.C., Radiometric evidence of the Kanimblan Orogeny in south eastern Queensland and the age of the Neranleigh-Fernvale Group, J. Geol. Soc. Aust., 20, 153-160, 1973.

Harrington, H.J., The Tasman Geosyncline in Australia, in The Tasman Geosyncline - a symposium, edited by A.K. Denmead, G.W. Tweedale and A.F. Wilson, pp. 383-407, Geological Society of Australia, Queensland Division, Brisbane, 1974.

Harrington, H.J., and R.J. Korsch, Deformation associated with the accretion of the Gympie Terrane in eastern Australia, in Third Circum-Pacific Terrane Conference, Extended Abstracts. Geol. Soc. Aust., Abs., 14, 104-108,

Irving, E., Paleomagnetism of some Carboniferous rocks from New South Wales and its relation to geological events, J. Geophys. Res., 71, 6025-6051, 1966.

Jenkins, T.B.H., Lower Carboniferous conodont biostratigraphy of New South Wales, Palaeontology, 17, 909-924, 1974.

Jull, R.K., The Lower Carboniferous corals of eastern Australia: a review, in Stratigraphy and Palaeontology, Essays in Honour of Dorothy Hill, edited by K.S.W. Campbell, pp. 120-139, Australian National University Press, 1969.

Jull, R.K., Aphrophyllum and allied genera of rugose corals from Lower Carboniferous (Visean) beds in Queensland. Proc. R. Soc. Qd, 85, 1-26, 1974a.

Jull, R.K., The rugose corals Lithostrotion and Orionastraea from Lower Carboniferous (Visean) beds in Queensland. Proc. R. Soc. Qd, 85, 57-76, 1974b.

Kirkegaard, A.G., R.D. Shaw and C.G. Murray, Geology of the Rockhampton and Port Clinton 1:250,000 sheet areas. Geol. Surv. Qd, Rep., 38, 55 pp. 1970.

Klootwijk, C., Palaeomagnetism of the Tasman Fold Belt: indication for mid-Carboniferous large-scale southward displacement of the New

England region, in Third Circum-Pacific Terrane Conference, Extended Abstracts, Geol. Soc. Aust., Abs, 14, 124-127, 1985.

Korsch, R.J., and H.J. Harrington, Stratigraphic and structural synthesis of the New England Orogen, J. Geol. Soc. Aust., 28, 205-266, 1981.

Kotlyar, G.V. and L.I. Popeko, Biostratifrafia, Mshanki i Brakhiopodi verchnogo paleozoya Zabaikalya. (Biostratigraphy, bryozoa and brachiopods of the Upper Palaeozoic of the Zabaikal region, Zap. Zabaikal. Fil., Geogr. O-va, S.S.S.R., 28, 323 pp., 1967.

Leitch, E.C., The geological development of the southern part of the New England Fold Belt, J. Geol. Soc. Aust., 21, 133-156, 1974.

Leitch, E.C. and E. Scheibner, E. A Tasman Fold Belt perspective on terrane analysis, in Third Circum-Pacific Terrane Conference, Extended Abstracts. Geol. Soc. Aust., Abs, 14, 136-141, 1985.

Lindley, I.D., An occurrence of a camerate crinoid genus Eumorphocrinus in the Early Carboniferous of New South Wales, J. Proc. R. Soc. N.S.W., 112, 121-124.

Lucas, K.G., The Texas area, J. Geol. Soc. Aust., 7, 229-341,

McKellar, R.G., An Upper Carboniferous brachiopod fauna from the Monto district, Queensland, Geol. Surv. Qd, Publ., 328, 15 pp., 1965.

Mamet, B. and A. Roux, Algues Devono-Carboniferes de l'Australia, Revue de Micropaleontologie, 26, 63-131, 1983.

Maxwell, W.G.H., Upper Palaeozoic formations in the Mt. Morgan district - faunas, Pap. Dep. Geol. Univ. Qd, 4(5), 1-69, 1954.

Maxwell, W.G.H., The geology of the Yarrol region. Part 1. Biostratigraphy, Pap. Dep. Geol. Univ. Qd, 5 (9), 1-79, 1964.

Mory, A.J. and D.T. Crane, Early Carboniferous Siphonodella (Conodonta) faunas from eastern Australia, Alcheringa, 6, 275-303, 1982.

Murphy, P.R., H. Schwarzbock, L.C. Cranfield, I.W. Withnall, and C.G. Murray, Geology of the Gympie 1:250,000 sheet area, Geol. Surv. Qd, Rep., 96, 1-157, 1976.

Murray, C.G., Alpine-type ultramafics in the northern part of the Tasman Geosyncline - possible remnants of Palaeozoic ocean floor, in The Tasman Geosyncline - a symposium, edited by A.K. Denmead, and G.W. Tweedale and A.F. Wilson, pp. 161-181, Geological Society of Australia, Queensland Division, Brisbane, 1974.

Murray, C.G., C.L. Fergusson, P.G. Flood, W.G. Whitaker and R.J. Korsch, Plate tectonic model for the Carboniferous evolution of the New England Fold Belt, Aust. J. Earth Sci., in press, 1986.

Murray, C.G., G.R. McClung and W.G. Whitaker, Early Permian fossils from the Amamoor Beds and the age of the Booloumba Beds - a new stratigraphic unit of the North D'Aguilar Block. Qd Govt Min. J., 80, 71-78, 1979.

Murray, C.G. and W.G. Whitaker, A review of the stratigraphy, structure and regional tectonic setting of the Brisbane Metamorphics, in New England Geology, edited by P.G. Flood and B. Runnegar pp. 79-94, Dept. Geol. Univ. New England and AHV Club, Armidale, N.S.W., 1982.

Nazer, R., Late Visean brachiopods with Western Australian affinities from the Yarrol Shelf, Qd Govt Min. J., 78, 126-131, 1977.

Nicoll, R.S. and T.B.H. Jenkins, Conodont biostratigraphy, in The Carboniferous of the World, Vol. 2, edited by C.M. Diaz, pp. 114-117, IUGS Publication 20, Instituto Geologico y Minero de Espana and Empressa Nacional Adaro de Investigaciones Mineras S.A., Spain, 1985.

Palmieri, V., Upper Carboniferous conodonts from limestones near Murgon, south-east Queensland, Geol. Surv. Qd, Publ., 341, Palaeont. Pap., 17, 1-13, 1969.

Pickett, J.W., Lower Carboniferous coral faunas from the New England district of New South Wales, Palaeont. Mem. Geol. Surv. N.S.W., 15, 1-38, 1966.

Powell, C. McA. Terminal fold-belt deformation: relationship of mid-Carboniferous megakinks in the Tasman Fold Belt to coeval thrusts in cratonic Australia, Geology, 12, 546-549, 1984.

Powell, C. McA. and J.J. Veevers, Termination of the Uluru Regime: the mid-Carboniferous lacuna, Phanerozoic Earth History of Australia, edited by J.J. Veevers, pp. 348-350, Oxford Geological Sciences Series, 2. Clarendon Press, Oxford, 1984.

Roberts, J., Devonian and Carboniferous brachiopods from the Bonaparte Gulf Basin, northwestern Australia, Bull. Bur. Miner. Resour., Geol. Geophys. Aust., 122, 319 pp., 1971.

Roberts, J., Early Carboniferous brachiopod zones in eastern Australia, J. Geol. Soc. Aust., 22, 1-32, 1975.

Roberts, J., Control mechanisms of Carboniferous brachiopod zones in eastern Australia, Lethia, 14, 123-134, 1981.

Roberts, J., Carboniferous sea level changes derived from depositional patterns in Australia, 10e Congr. Int. Stratigr. Geol. Carbonifere, Compte Rendu, Madrid 1983, 4, 43-64, 1985.

Roberts, J. and B.A. Engel, Carboniferous palaeogeography of the Yarrol and New England Orogens, eastern Australia, J. Geol. Soc. Aust., 27, 167-186, 1980.

Roberts, J., and B.A. Engel, Depositional and tectonic history of the southern New England Orogen, Aust. J. Earth Sci., in press, 1986.

Roberts, J., J.W. Hunt and D.M. Thompson, Late Carboniferous marine invertebrate zones of eastern Australia, Alcheringa, 1, 197-225,1976.

Roberts, J., et al., Australia, in The Carboniferous of the World, Vol. 2, edited by

C.M. Diaz, pp. 7-145, IUGS Publication 20, Instuto Geologico y Minero de Espana and Empressa Nacional Adaro de Investigaciones Mineras S.A., Spain, 1985.

Ross, C.A. and J.R.P. Ross, Carboniferous and Early Permian biogeography, Geology, 13, 27-30, 1985.

Runnegar, B. and K.S.W. Campbell, Late Palaeozoic faunas of Australia, Earth Sci. Rev., 12, 235-257, 1976.

Strusz, D.L., Report on collections of Carboniferous corals from the Warwick, Goondiwindi, and Inverell sheet areas, Queensland, in Palaeozoic geology of the Warwick and Goondiwindi 1:250 000 sheet areas, Queensland and New South Wales, by F. Olgers, P.G. Flood and A.D. Robertson, Rep.Bur. Miner. Resour., Geol. Geophys. Aust., 164, 101-104, 1974.

Thomas, G.A., Carboniferous and Early Permian brachiopods from Western and northern Australia, Bull. Bur. Miner. Resour., Geol. Geophys. Aust., 56, pp. 276, 1971.

Voisey, A.H., A preliminary account of the geology of the middle north coast district of New South Wales, Proc. Linn. Soc. N.S.W., 59, 333-347, 1934.

Winkler Prins, C.F., Carboniferous Productidina and Chonetidina of the Cantabrian Mountains (NW Spain): systematics, stratigraphy and palaeoecology, Leid. geol. Meded., 43, 41-126, 1968.

BIOGEOGRAPHIC SIGNIFICANCE OF SOME ORDOVICIAN FAUNAS IN RELATION TO EAST AUSTRALIAN TASMANIDE SUSPECT TERRANES

Barry D. Webby

Department of Geology & Geophysics, University of Sydney, N.S.W., 2006, Australia

Abstract. The continental platform-shelf remnants of the Australian sector of Gondwana, for example, the Georgina Basin and Tasmanian Shelf, have varied Ordovician marine invertebrate faunas, especially of shallow-water benthonic organisms. Biogeographically they are low-latitude faunas. The suspect terranes of the adjacent, major Tasmanide fold belt of eastern Australia also contain some biogeographically useful Ordovician faunas, in particular the trilobites, nautiloids, brachiopods, stromatoporoids and corals of localized Late Ordovician shallow-water carbonate deposits. These are also low-latitude faunas, and are well represented in the remnants of the island arc (Parkes Platform and Molong High) of central New South Wales, in the Tamworth terrane of NE New South Wales and the Broken River Embayment of northern Queensland. They provide low-latitudinal constraints for these structural blocks. The corals of the most easterly-placed Tamworth terrane seem to have the closest affinities to North America. Major strike-slip movements are not detectable on faunal evidence, though one earliest Ordovician trilobite fauna is anomalous in having an influx of 'higher latitude' Argentinian immigrants. This isolated occurrence is an offshore 'high' at Waratah Bay, central Victoria, and may be explained by major strike-slip displacement or a changing pattern of ocean currents. The Late Ordovician, low-latitude island-arc faunas of central New South Wales are regarded as biogeographically distinct from equivalents in the continental shelf remnant of Tasmania. This seems consistent with a separation of the two regions by a marginal sea (Wagga Trough) like the present-day Tasman Sea.

Introduction

The Ordovician faunas of the Tasmanide belt of Eastern Australia include a wide variety of benthonic and pelagic (planktonic and nektonic) organisms. The benthonic forms are the most useful for biogeographic analysis because such organisms must adapt to local temperature and substrate conditions as well as rely on larval stages for their dispersal. However these forms have the most limited geographic distribution, being confined mainly to the continental margin of Gondwana and to the fringes of islands within the former Tasmanide ocean. In contrast the pelagic forms have a widespread distribution within the various terranes and other crustal blocks (Fig. 1) but are much less sensitive indicators of provincialism, usually only registering broad-scale temperature differences.

The planktonic Ordovician graptolite assemblages of the Tasmanides have been interpreted by Skevington [1974] and Jaanusson [1979] as belonging to the Pacific province. Skevington has shown the province as straddling the equator. However, Fortey [1984] has proposed a different subdivision based on biofacies. The graptolite assemblages of the Tasmanides are part of a widely distributed, deep-water 'oceanic' biofacies which apparently does not have marked latitudinal control [Fortey and Cocks, 1986]. The conodonts seem to comprise two depth-related associations. One is characterized by elements of the North American Midcontinent province [Sweet and Bergstrom, 1974], apparently adapted to shallow equatorial waters of the remnants of the continental shelf in Tasmania and the offshore arc-type islands of central New South Wales. The second is typified by representatives of the North Atlantic province, which occupied deeper, cooler waters of the outer shelf or slope [Burrett et al. 1984].

Other faunas discussed in the foregoing account include the nautiloids of both nektonic and nekto-benthonic kinds, the pelagic and benthonic trilobites, and the sessile benthonic groups such as the brachiopods, corals, stomatoporoids, sphinctozoan sponges and edrioblastoid echinoderms. The benthonic forms prove to be the most useful for biogeographic analysis.

In discussing the biogeographic relationships of these faunas it has proved most convenient to use the Australian tripartite subdivision of the Ordovician time scale (Fig. 2) - Lower (or Early) Ordovician for the Lancefieldian-Yapeenian stages (equivalent to the British Tremadoc-Arenig excluding the uppermost Arenig, or American Canadian), Middle Ordovician for the Darriwilian (i.e., British uppermost Arenig-Llanvirn, or American Whiterockian-lowest Chazyan), and Upper

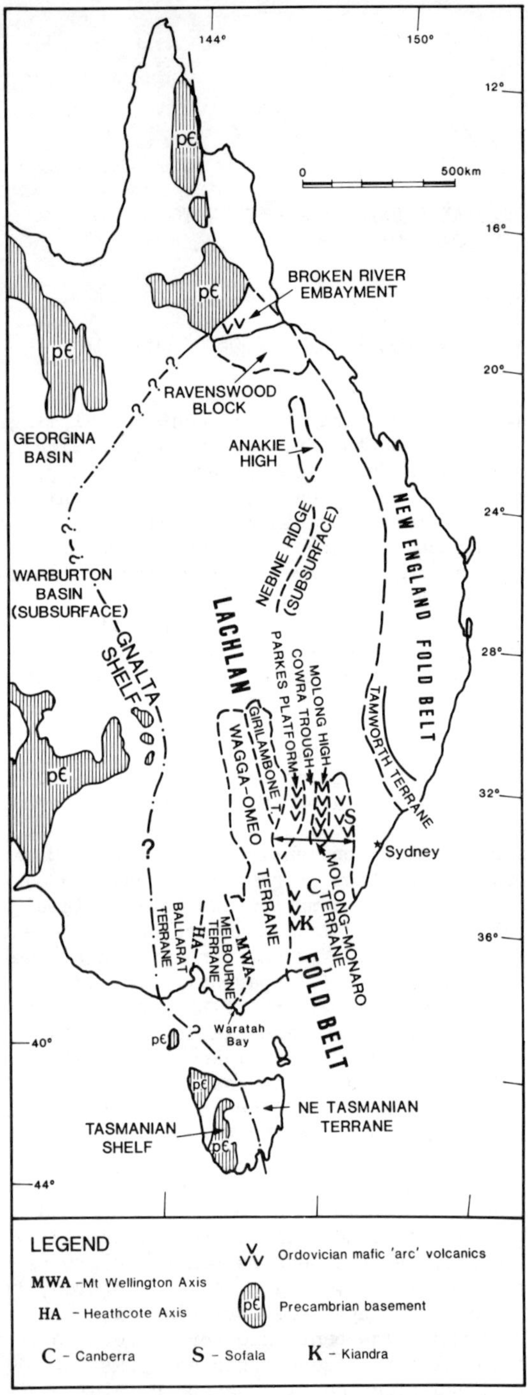

Fig. 1. Map of eastern Australia showing the location of Ordovician platform sequences of the Gondwana continental margin (Georgina Basin, Warburton Basin, Gnalta Shelf and Tasmanian Shelf), and the various suspect terranes and other crustal blocks with Ordovician rocks of the Tasmanides (Lachlan Fold Belt and New England Fold Belt).

(or Late) Ordovician for the Gisbornian-Bolindian stages (i.e., British Llandeilo-Ashgill, or American Chazyan-Richmondian).

Embleton [1973,in Veevers, 1984] has used palaeomagnetic data to establish that the Australian continent, as a part of east Gondwana, occupied an equatorial-low latitude position throughout the 75 million-year period of the Ordovician. The presence of Early Ordovician evaporites in the Daly River Basin of NW Australia and the Middle Ordovician evaporites of the Amadeus Basin in Central Australia, along with the widespread development of warm-water platform carbonates, for instance in the Early Ordovician of the Georgina Basin, and in the Late Ordovician of the Tasmanian Shelf with rich coral and stromatoporoid faunas, testifies to the general climatic picture of the Australian craton occupying low latitudes, and parts of it being invaded periodically by warm, shallow seas [Webby, 1978].

This paper focuses initially on the nature of biogeographic relationships in the Australian (= Gondwanan) continental margin with a view to providing an essential background for the wider discussion and analysis of the biogeography of the discrete terranes or crustal blocks of the Tasmanides (Fig. 1).

Faunal Patterns of the Continental Margin

Georgina Basin

The Ordovician continental margin of Australia is somewhat difficult to define (Fig. 1) because of the very limited areas of Ordovician outcrop and known subsurface expression. The Georgina Basin has one of the more complete shallow, continental platform sequences with fossiliferous carbonates and siliciclastics of Early-Middle Ordovician age [Shergold and Druce, 1980]. The basin has been represented, along with the adjoining Amadeus Basin of Central Australia, as a part of a large, shallow, gently shelving embayment of the continental platform, opening to the south and east towards the deeper, oceanic settings of the Tasmanides, and lying in the warm-water, equatorial zone [Webby, 1978].

The interbedded carbonate and siliciclastic successions of the Georgina Basin contain rich faunas, like the trilobites documented by Fortey and Shergold [1984] from the mid-late Arenig equivalent Nora Formation. Half the eleven genera of this fauna are regarded as endemic, though some of these generic components have links with other 'northern' basinal sequences, most notably with the Canning Basin of Western Australia. The other half includes a few genera which are also found in other Gondwanan regions like SW China, the Himalayas and Argentina, and a few others (at least one of which has a pelagic habit) have North American links, presumably by way of the circum-equatorial belt. The abundance of endemics (especially peculiar asaphid trilobites with inflated, tuberculate frontal glabellar lobes) has suggested to Fortey and Shergold [1984] that some degree of isolation existed between this region and other cratonic areas.

Another group which is well represented in the Georgina Basin is the nautiloids. Those of the Coolibah and Nora Formations have been claimed to exhibit conspicuous endemism [Wade 1977a, 1977b], most notably the actinoceratoid family, the Georginidae. However Burrett and Stait [1985] have identified the two genera (Georgina and Mesaktoceras) comprising the family as occurring also in Thailand and Tibet, and these are just two of the Georgina Basin nautiloids to show close links with Thailand and Malaysia (see discussion of Tasmanian Shelf, below).

Warburton Basin

Ordovician deposits have been recorded from numerous wells in the north-eastern corner of South Australia (Fig. 1). Poorly preserved Early and Middle (or early Late) Ordovician graptolites have been reported from two widely separated wells and, along with the occurrences of black shales [Gatehouse, 1986], suggest the local development of a deeper-water basinal facies, possibly a tongue-like extension from the Tasmanide ocean. The Early Ordovician development of deeper water deposition seems to have been linked in a much broader transgression of the sea (Larapintine Sea) across the Australian craton to the Amadeus and Canning Basins [Webby, 1978]. Possibly for a period of time during the mid-late Arenig this seaway acted as a deep-water barrier to migration between the shallow on-craton areas of the Georgina Basin, and the other shelf areas of the continental margins like the Gnalta Shelf and Tasmanian Shelf to the south (Fig. 1).

Gnalta Shelf

Ordovician sequences and faunas of the Gnalta Shelf in western New South Wales (Fig. 1), especially those at Mt Arrowsmith, have most frequently been compared with their counterparts in the Amadeus Basin. The conodont assemblages of the Tabita Formation at Mt Arrowsmith are similar to those of the Horn Valley Siltstone of the Amadeus Basin [Warris, unpublished data, 1967; Kennedy, 1975; Cooper, 1981] and like them also appear to have mixed North American Midcontinental and North Atlantic affinities. The nautiloids have some resemblances, with Anthoceras from contemporaneous horizons at Mt Arrowsmith and from the Amadeus and Canning Basins being very similar [Crick and Teichert, 1983]. Now, however, following Stait and Laurie's [1985] revision, the Mt Arrowsmith material is grouped as a different, albeit related, species.

Whittington and Hughes [1974] have noted that in earliest Ordovician (Tremadoc) times the Australian margin of Gondwana lay in presumed warm tropical waters represented by the Tsinaniid trilobite faunal province. However, one recent discovery, a species of Hysterolenus from a possible earliest Ordovician siltstone succession near Koonenberry Mountain (Wang Qizheng, pers. comm., Oct., 1986) may be interpreted as suggesting the presence of the cooler Olenid-Ceratopygid faunal province or, more likely, a cooler, deeper water, environment on the outer edge of the Gnalta Shelf. Elsewhere Hysterolenus has only been recorded from Sweden, Wales, Asiatic parts of the Soviet Union and China, as well as apparently from New Zealand (= Ruapyge Wright [1979]).

Tasmanian Shelf

The only other major area of Ordovician outcrop with close links to a known Precambrian basement and consequently either a remnant of the Gondwanan continental margin or a microcontinent is the Tasmanian Shelf (Fig. 1). A number of divergent viewpoints have been expressed concerning whether the present position of Tasmania is close to its original Ordovician position or that it formerly lay as much as 1000 km farther to the west; see summary in Webby [1976, 1978]. The shelf region has a more complete record of Ordovician deposits, including Late Ordovician strata, than other continental margin areas to the north but, like the Gnalta Shelf, the lower part of the Early Ordovician succession has major breaks and thick accumulations of non-marine molasse-type conglomerates.

Banks and Burrett [1980] have noted that the conodont assemblages of the Tasmanian carbonate platform are mainly of the warm-water North American Midcontinental type. There is also at least one multielement conodont genus (Tasmanognathus) from Blackriveran (Early Caradoc) equivalents formerly thought to be endemic [Burrett, 1979] but now found in N China [Lin et al. 1984]. In the coeval, deeper, thin-bedded carbonate 'slope' environments of southern Tasmania a different assemblage is represented, with typical elements of the cooler North Atlantic province [Burrett et al. 1984]. It seems likely therefore that the so-called provincial differences here reflect a depth (and temperature) stratification, thus limiting the usefulness of conodonts in establishing relationships in space.

Graptolites have been recorded from a number of horizons throughout the Ordovician successions of the Tasmanian Shelf but not from the typical massive Gordon Subgroup carbonates [Burrett, in Webby et al. 1981]. The Early Ordovician Florentine Valley Formation and Caroline Creek Sandstone, and the Middle-Late Ordovician Gordon Subgroup carbonates have reasonably varied benthonic faunas, some with striking examples of provincialism.

The nautiloid faunas from the lower part of the

Gordon Subgroup (late Early-Middle Ordovician) include a few endemics like Allocotoceras but also a number of forms which have a much wider distribution [Teichert and Glenister, 1953; Stait, 1984a; Burrett and Stait, 1985] . Wutinoceras, Manchuroceras and Armenoceras are also known from the Georgina Basin, Thailand, N China and North America, with the latter two additionally recorded from the Siberian Platform, and Manchuroceras from S China. Chaohuceras is another genus with links to N China. On the other hand the nautiloid associations of the upper part of the Gordon Subgroup (Benjamin Limestone and equivalents) have a more restricted distribution with a higher proportion of endemic genera and component of nektobethonic forms, like the members of the Oncocerida and Discosorida [Stait, 1982, 1984b].

Other molluscs from the Early Ordovician of Tasmania show interesting patterns of distribution. The macluritid gastropod Teichiispira is also represented in the Georgina and Canning Basins and in Thailand and Malaysia [Gilbert-Tomlinson, 1973; Burrett and Stait, 1985], and the bellerophontid gastropod Peelerophon is known from the Canning Basin, Thailand, Argentina and southern Europe [Jell et al. 1984].

Early-Middle Ordovician trilobite faunas include comparatively few endemic forms. The Florentine Valley Formation (mainly siltstones) of approximately late Tremadoc-early Arenig age has a fauna which, at the generic level, seems to have generally closer affinities with Eastern Asia than to other regions [Jell and Stait, 1985a], and even includes the species Dikelokephalina asiatica Kobayashi and Hystricurus penchiensis Lu in common with occurrences in S Korea and NE China, respectively. Dikelokephalina also occurs in the Canning Basin [McTavish and Legg, 1976]. On the other hand, the Caroline Creek Sandstone of probable early-middle Arenig age includes an assemblage with at least two genera exhibiting a restricted distribution [Jell and Stait, 1985b]. Etheridgaspis appears to be endemic, though Burrett [1973] has noted its occurrence also in Central Asia. Tasmanocephalus has additionally been reported by Shergold [1971] from the Mt Arrowsmith area of the Gnalta Shelf. Others in the Caroline Creek Sandstone like the pelagic Carolinites have more cosmopolitan relationships.

The Late Ordovician successions of Tasmania exhibit two distinctly different trilobite assemblages [Banks and Burrett 1980; Burrett et al. 1983, 1984]. In the peritidal to subtidal carbonate platform environments of the Gordon Subgroup, the assemblages are characterized by occurrences of benthonic forms such as Pliomerina, Amphilichas, Remopleurides, Eokosovopeltis and other genera. Closest biogeographic relationships of this fauna are with the central New South Wales 'island arc' (see below), SE Asia, Korea and Kazakhstan, within the Pliomerina province of the much wider 'band-like' circum-equatorial Remopleurided realm [Webby, 1971a, 1974]. In contrast the deeper water calcareous shale, micrite and carbonate turbidite succession of the slope environments in southern Tasmania exhibit a benthonic component with blind forms (a trinucleid resembling Eirelithus and the raphiophorids Nanshanaspis and Bulbaspis) and sighted forms (Nileus and Pseudobasilieus); also a pelagic component with a blind (Shumardia) and a sighted (Telephina) form. This assemblage has mixed cosmopolitan and Chinese affinities.

Further discussions of the Late Ordovician faunal relationships on the Tasmanian Shelf, especially of the brachiopod, coral and stromatoporoid faunas, are given in the discussion of the contemporaneous 'island-arc' faunas of central New South Wales (see below).

Summary of Continental Margin Biogeography

A number of basic problems emerge from this brief review of Ordovician biogeographic relationships in the continental margin. Firstly in defining the margin because of the very limited areas of Ordovician outcrop and known surface expression (Fig. 1), and secondly in achieving meaningful comparisons from region to region because the particular assemblages are seldom of precisely the same age or represent exactly the same sedimentary environment. Thirdly, apart from Tasmania, there is no record of Late Ordovician cratonic deposits in Australia, so region-by-region comparisons of the continental margin are restricted to the Early-Middle Ordovician interval. Only the Georgina Basin, Gnalta Shelf and Tasmanian Shelf exhibit an adequately varied Early-Middle Ordovician fossil record, but in the Georgina Basin the faunas are preserved mainly in carbonates, in the Gnalta Shelf chiefly in siliciclastics, and in Tasmania Early Ordovician siliciclastics, passing up into Middle Ordovician carbonates. Many of the differences between the various continental margin faunas already outlined seem to reflect age and/or facies control rather than to represent significant biogeographic distinctions.

Furthermore it is possible that the continental margin may have been disrupted during the late Early Ordovician with the rifting associated with the opening of the east-west Larapintine Sea across Australia [Webby, 1978, fig.8], resulting in the separation of the 'Carpentaria' block (including the Georgina Basin) from the rest of the Gondwanan continental mass. There is also the issue of whether Tasmania was displaced in post-Ordovician times, as mentioned previously. Baillie [1984] has suggested that it may have been much farther to the west, in a straight line between western Victoria and the north-eastern part of Victoria Land in Antarctica. Perhaps an

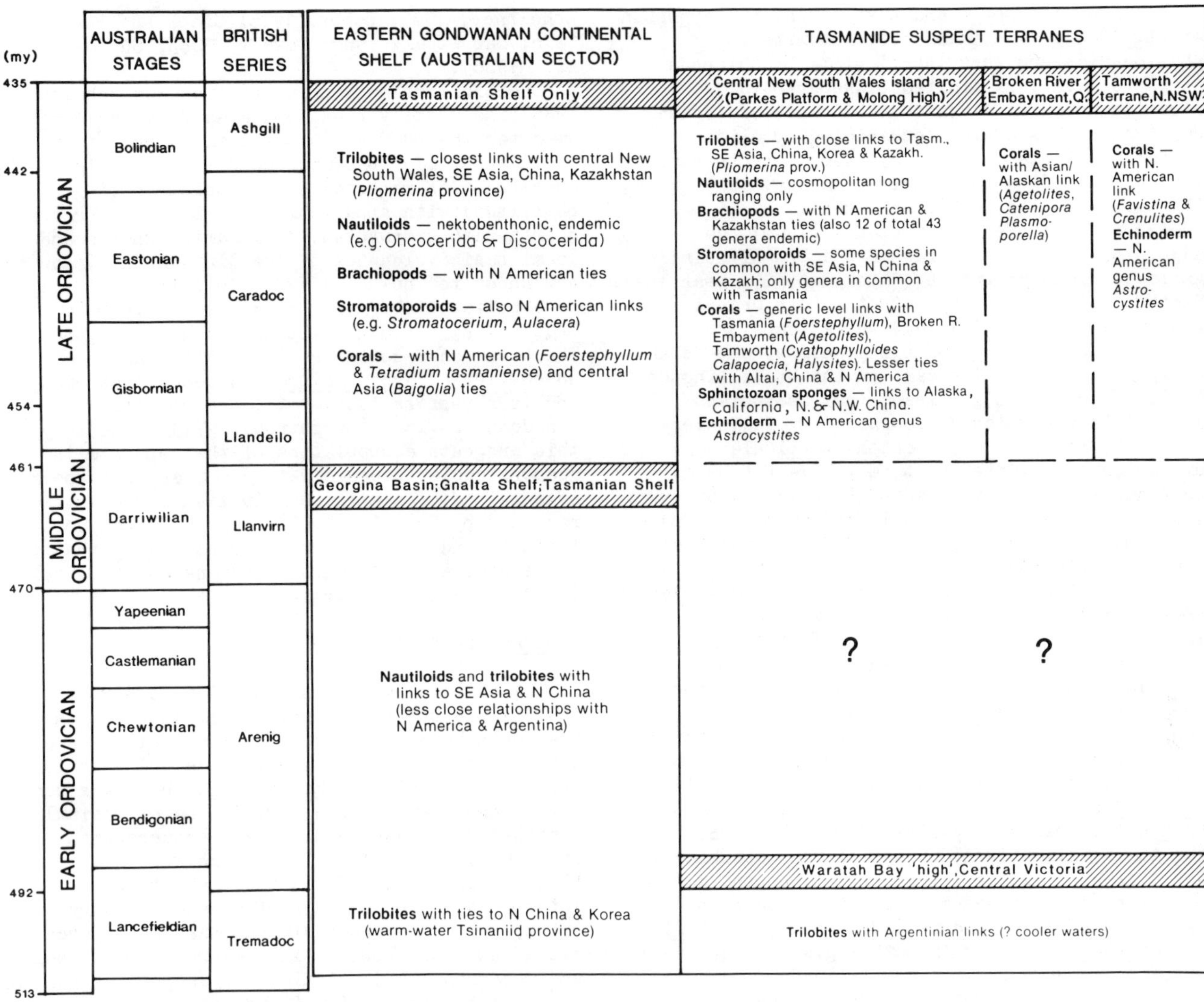

Fig. 2. Diagram to show the more significant biogeographic patterns of the Ordovician sequences in eastern Australia, both in the Gondwana continental margin and in the Tasmanide suspect terranes.

active programme of palaeomagnetic work may help to resolve some of these problems.

Taking the generalized biogeographic data from the different regions of the Australian continental margin it may be concluded that in the Tremadoc the closest links were with Eastern Asia (N China and Korea), within the equatorial Tsinaniid trilobite province of Whittington and Hughes [1974], and during the Arenig, with SE Asia and N China, and to a less extent, Argentina and North America (Fig. 2). What can be said about the Late Ordovician is based on Tasmania, and it confirms earlier patterns of close biogeographic relationships with SE Asia, and also important links with Kazakhstan and North America probably along the circum-equatorial belt.

Faunal Patterns in the Suspect Terranes of the Tasmanides

Ballarat, Melbourne and NE Tasmanian Terranes

The Ballarat terrane (Fig. 1) contains a thick Early-Middle Ordovician (Lancefieldian-Darriwilian) graptolite succession of turbidites and black shales [VandenBerg, in Cooper and Grindley, 1982]. The strata are thought to have been deposited in a 'continental margin sediment prism' from the south and west [Cas, in Fergusson et al. 1986]. The fauna of graptolites, phyllocarids and rare inarticulate brachiopods seems to have very limited biogeographic significance, but given the abundance of typical representatives of the 'oceanic' graptolite biofacies [Fortey and Cocks, 1986], the terrane may be identified as a part of an oceanic basin at the margin of the former continent.

According to Fergusson et al. [1986] the adjoining Melbourne terrane has a broadly similar geologic history, but the thick deposits near the Heathcote Axis range up into the Late Ordovician (Gisbornian and Bolindian), while there are condensed successions of black shales, cherts and phosphatic rock to the east on the Mt Wellington Axis. It may be significant that Eastonian graptolite assemblages recognized by VandenBerg [in Webby et al. 1981] - or the particular biofacies represented by them - have not been recorded from near the Heathcote Axis but are very well developed to the east in the Mt Easton Shale.

At Waratah Bay (fig. 1), on the extension of the Mt Wellington Axis to the south, there is a localized offshore 'high' of earliest Ordovician (Early Tremadoc) age, with shallow water limestones and mudstones (Digger Island Formation). Jell [1985] identified the trilobites of this fauna as comprising eighteen taxa (four possibly endemic genera) with a significant component of Argentinian species (three identical) and representatives of three other genera being very similar. The presence of only one species (*Pilekia* sp.) as resembling an occurrence in adjacent Tasmania is explained by Jell [1985] as due to the slightly older age of the assemblage. There is also little in common between the Waratah Bay fauna and the assemblages of the Georgina Basin. Only two non-agnostid genera (*Leiostegium* and *Onychopyge*) occur in both regions. The latter genus is known also from New Zealand.

Little of biogeographic significance can be said about the Early Ordovician 'basinal' graptolitic slates (Mathinna Beds of Banks and Smith [1986]) of the NE Tasmanian terrane (Fig. 1).

Wagga-Omeo and Girilambone Terranes

The Wagga-Omeo terrane [Leitch and Scheibner, 1985] extends through a large part of east-central Victoria and central New South Wales (Fig. 1). It includes a thick complex of quartz-rich greywackes and mudstones with rare graptolitic horizons, ranging from Early Ordovician (Bendigonian) to Late Ordovician (Bolindian) in age [Webby, 1976; Kilpatrick and Fleming, 1980; Sherwin, 1983]. Extensive parts of the complex have been regionally metamorphosed [Vallance, in Packham 1969; Crook and Powell, 1976; VandenBerg, 1978]. The sequences have been interpreted as those of a marginal sea (Wagga Trough) [Packham, 1973; Scheibner, 1973; Webby, 1976; Crook and Powell, 1976; Cas et al. 1980; Powell, 1983; Suppel et al. 1986].

The Girilambone terrane [Leitch and Scheibner, 1985] has usually been separated from the Wagga-Omeo terrane because it has been thought to include an older basement complex, but this succession of metasediments (schists and phyllites) with apparently associated lower grade quartz-rich greywackes, radiolarian cherts and local basic volcanics of the Girilambone Group is now known to include diagnostic late Darriwilian to earliest Gisbornian (Middle to early Late Ordovician) conodonts [Stewart and Glen, 1986]. The occurrence comes from cherts of the succession. *Pygodus serra* has been identified, and is a representative of the North Atlantic conodont province [Bergstrom, 1971]. Possibly this suggests accumulation of the deposits in deeper, cooler oceanic waters. Graptolite-bearing successions of similar Darriwilian age have been recorded by Sherwin [1983] from the Wagga-Omeo terrane, and this raises the possibility that the Wagga-Omeo and Girilambone terranes were deposited in the same sedimentary basin.

Molong-Monaro Terrane

The eastern half of the Lachlan Fold Belt from the Gilmore suture [Suppel et al. 1986] to the New South Wales south coast region comprises a complex terrane with remnants of a volcanic island arc (Parkes Platform, Molong High and areas near Sofala and Kiandra; see Fig. 1) and of basinal quartz-rich greywacke and shale successions previously referred to the Monaro Trough [Scheibner, 1972]. The latter have been interpreted as components of an accretionary prism, part fore-arc basin and part outer arc slope deposits [Powell, in Veevers, 1984; Powell and Conaghan 1984]. Again, the deep-water basinal successions provide few faunas of use in analysing biogeographic relationships. There are some distinctive elements of the Late Ordovician graptolite faunas in areas of the 'volcanic' Parkes Platform and Molong High, such as *Orthograptus apiculatus* and *Amplexograptus inuiti* [Sherwin, 1971, 1973] which do not occur in coeval Victorian successions and may suggest the existence of separate biofacies representing circulation of localized shallower or laterally equivalent water masses [Finney, 1984]. In the Canberra area, Middle Ordovician conodonts (*Pygodus serra* and *Periodon aculeatus*) occur in the deeper-water graptolite shales of the Pittman Formation [Nicoll, 1980]. Like the occurrence in the cherts of the Girilambone Group, the conodonts are typical representatives of the cooler North Atlantic province.

Shallow-water benthonic assemblages are only well represented in the volcanic associations of this terrane - biotas that lived on the fringes of offshore volcanic islands or upon their submerged remnants [Webby, 1985]. The Parkes Platform and Molong High are the main areas of outcrop of the volcanic-limestone-shale facies, and of diversified associated shelly and coralline faunas (see discussion below). Smaller areas of volcanics near Sofala and Kiandra (Fig. 1) have virtually no record of associated shallow-water deposits or accompanying benthonic faunas. Only known exceptions are a species of coral Plasmoporella in limestone boulders of an agglomerate in the Sofala Volcanics at Sofala [Pickett, 1978] and a solitary streptelasmatid rugose coral in a tuff of the Nine Mile Volcanics near Kiandra [Legg, in Owen and Wyborn, 1979].

Given the similarities of the Ordovician geologic histories of the Parkes Platform and Molong High with the coincidence of Early and Late Ordovician phases of volcanicity [Webby, 1976], it seems likely that the two regions originally abutted each other. The volcanics may have erupted from a single major centre in the Early Ordovician, then again in the Late Ordovician prior to rifting of the Cowra Trough [Scheibner, 1973]. Alternatively the Molong High may have originated in a single linear feature, in continuity with and on the prolongation of the Parkes Platform to the south. Only later was it displaced to its present position along major transform faults [Packham, 1985]. Or, as a third alternative, the volcanics originated in separate centres of the Parkes Platform and Molong High, with the Cowra Trough already existing as an interarc basin.

Island-arc Faunas

Relationships between the Parkes Platform and the Molong High. Although the contemporaneous Ordovician successions of the Parkes Platform and Molong High have similar facies they exhibit certain faunal differences which suggest the two areas may not have been in the closest possible juxtaposition. Some genera of trilobites have markedly different species like, for instance, the occurrence of Amphilichas nasutus from the Billabong Creek Limestone of the Parkes Platform and A. encyrtos from the correlative shallow 'shelf' limestones of the Quondong Formation to the west side of the Molong High [Webby, 1974]. Similarly Parkesolithus gradyi is represented in graptolitic siltstones of the basal Goonumbla Volcanics of the Parkes Platform in a closely equivalent facies and stratigraphic level to the occurrence of P. dictyotos of the graptolitic basal Malongulli Formation on the eastern 'slope' of the Molong High [Webby, 1974; Webby et al. 1981]. There are similarities between contemporaneous brachiopod faunas especially the articulates of the 'island shelf' carbonate environments of the Parkes Platform and Molong High, with at least five out of thirteen species in common. The inarticulates of equivalent, deeper 'slope' settings are also similar, but the articulates are different, for example, the species of Sericoidea and Durranella [Percival, 1979a].

The coral Foerstephyllum and the stromatoporoid Stromatocerium have been recorded by Pickett [1985] from the Billabong Creek Limestone of the Parkes Platform, but possibly from a lower horizon than the horizons containing species of Tetradium, Hillophyllum and Ecclimadictyon which have common occurrences with equivalent limestones on the Molong High [Webby and Semeniuk, 1971; Webby, 1971b; Webby and Morris, 1976]. Comparing the assemblages in the limestones of the lower part of the Goonumbla Volcanics of the Parkes Platform with counterparts on the Molong High there are again some differences, for instance, between species of Palaeophyllum [McLean and Webby, 1977], and in lacking a similar record of early appearances of favositids and halysitids as seen on the Molong High [Webby et al. 1981]. These faunal differences suggest some sort of separation or barrier, but this may not be more significant than representing 'eastern' and 'western' sides of a single linear volcanic arc either prior to rifting [Scheibner, 1973] or before major transform motion [Packham, 1985], especially given the more marked faunal contrasts in the 'slope' environs.

General island-arc biogeography. The Late Ordovician faunas of the island-arc setting (Parkes Platform and Molong High) show significant differences in biogeographic relationships when compared with other regions like the Tasmanian Shelf (Fig. 2). At the more generalized level of relationships there are the nautiloid assemblages, mainly members of the Michelinoceratida and Tarphycerida, which invaded the shallow seas around the offshore islands and over their submerged counterparts during a relatively short period of predominant carbonate depositional conditions. They mainly include representatives of long ranging, near cosmopolitan, entirely nektonic stocks [Stait et al. 1985], in striking contrast to the patterns seen in Tasmania where the coeval faunas comprised a significant endemic component and representation of nektobenthonic forms such as the Oncocerida and Discosorida [Stait, 1982, 1984b]. The absence of nektobenthonic forms from the offshore volcanic 'highs' suggests some sort of deep-water barrier preventing their dispersal from the Tasmanian (=Gondwanan) continental margin. Hewitt and Stait [1985] have suggested that at least one nektonic member of the New South Wales island-arc fauna may have been capable of swimming to depths of at least 800 m. This survivor of the predominantly Early Ordovician group of ellesmeroceratids is the genus Bactroceras which occurs only in the deeper 'slope' environs of the Late Ordovician Malongulli Formation of the Molong High [Stait et al. 1985].

It has a strengthened skeleton of connecting rings and septa which has led Hewitt and Stait [1985] to suggest by analogy with implosion calculations of Recent Nautilus that it probably lived as a 'pelagic predator'.

Late Ordovician trilobite assemblages in the shallow carbonate island-shelf facies of the Parkes Platform and Molong High are characterized by occurrences of Pliomerina and Amphilichas [Webby, 1971a, 1974]. A possible new harpid genus is the only likely endemic form in this association which includes a total of some eleven genera. The fauna has strongest biogeographic affinities with Tasmania, SE Asia and Kazakhstan, belonging to the Pliomerina province of the much larger, band-like equatorial Remopleuridid realm. The carbonate platform deposits of Tasmania have a similar biofacies with common occurrences of Pliomerina, Amphilichas, Remopleurides and Eokosovopeltis but with few species in common [Banks and Burrett, 1980; Burrett et al. 1983, 1984]. Many of the members of the island shelf biofacies are derived from stable, conservative, widely dispersed stocks with relatively low rates of evolutionary activity - with few genera diversifying to produce more than one temporally distinct species [Webby, 1985].

A second community type is represented in the island slope by occurrences of Parkesolithus and Malongullia. The assemblage is associated with graptolites and includes seven genera. Some of the trilobites exhibit well developed eyes (Encrinuraspis and Remopleurides) and some have poorly formed eye tubercles (Parkesolithus and Malongullia); they probably lived in depths no greater than 200 m. Parkesolithus and Malongullia appear to be short-lived endemics. The trinucleid Parkesolithus has uncertain, possibly hangchungolithine affinities [Hughes et al. 1975]. It differs from Tasmanian and New Zealand occurrences - the Tasmanian forms are related to the South American genus Guandacolithus and to Eirelithus [Burrett, el al. 1984] and the South American genus Incaia occurs in New Zealand [Hughes and Wright, 1970]. Malongullia has links with other sighted raphiophorids of the Subfamily Ampyxinellinae with a restricted distribution in the Asian region, specifically Kazakhstan and China [Chugaeva, 1958; Koroleva, 1959; Chang, 1980]. The slope facies in the southern part of Tasmania has produced a distinctly different assemblage of blind raphiophorids (Bulbaspis and Nanshanaspis), trinucleids, and pelagic Telephina of mixed cosmopolitan and Chinese affinities [Burrett et al. 1984]. This leads to the suggestion that there are more marked biogeographic contrasts in the slope than the shelf environments of the two regions. In the respective slope areas, the differences are mainly at the generic level, whereas in the respective shelves the differences are chiefly at the species level.

A third type of assemblage, of Triarthrus, Geragnostus? and Shumardia, from the basinal graptolite facies adjoining the Molong High, is mainly composed of pelagic, cosmopolitan and long-ranging holdovers from the Cambrian.

In studies of the Late Ordovician brachiopod faunas of the Parkes Platform and Molong High, Percival [1978, 1979a, 1979b; unpublished data 1980] has shown that between one quarter and one-third of the total forty-three genera are endemic and that nearly all the forty-nine species have a restricted distribution. Percival has also recognized the fauna as broadly belonging to the equatorial American realm [Williams, 1973] with closest provincial affinities to the Scoto-Appalachian province in the late Gisbornian-middle Eastonian (early-middle Caradoc) to a closer faunal connection with Kazakhstan in the late Eastonian (late Caradoc). Typical North American genera include Hesperorthis, Doleroides, Rhynchotrema, Sowerbyella, Protozyga, Christiania, Trigrammaria, Sowerbyites, Ptychopleurella, Eridorthis, Dinorthis, Plectorthis and Skenidioides. Characteristic Kazakh genera include Anoptambonites, Kassinella, Durranella, and Dulankarella. It should be noted that these latter genera, which occur in lithic sandstones of the Goonumbla Volcanics on the Parkes Platform, have a lower diversity and more restricted distribution than the North American elements which characterize the shallow carbonates of the island shelf, both of the Parkes Platform and Molong High. The shallow waters of the offshore island arc not only provided staging posts for the dispersal of some fifteen North American genera and eight Kazakh genera but also acted as an important centre for the origin of new endemic genera such as Anomaloglossa, Infurca and Gunningblandella [Percival, 1978, 1979a, 1979b; Webby, 1985], and the formation of distinctive shell banks composed of the large inarticulate Eodinobolus. These are the earliest known aggregations of large shells in the geological record [Webby and Percival, 1985]. In contrast the deeper island slope environments include relatively fewer endemics (one out of seven genera) and a much higher proportion of inarticulates (four out of seven genera).

Relationships with contemporaneous Tasmanian brachiopod faunas have not yet been fully clarified. Laurie, in Webby et al. [1981] and Burrett et al. [1981, 1984] have listed broadly equivalent early-middle Caradoc elements as including Strophomena, Rhynchotrema, Hesperorthis, Murinella, Skenidioides, Clitambonites, Sowerbyella and Sowerbyites with several new genera, and a middle-late Caradoc fauna with the addition of Dinorthis, Holtedahlina, Protozyga, Hallina and Ptychopleurella. Taken together these occurrences similarly suggest a strong affinity to the American realm of Williams [1973]. Indeed about 50% or more of the generic components of this fauna are known to have common occurrences with the New South Wales island shelf, but it is less markedly endemic, it exhibits a less strong connection with Kazakhstan and there are no Eodinobolus shell banks. Only one species,

Sowerbyites vesciseptus is known to have a common occurrence [Burrett et al. 1981]. The Tasmanian platform assemblage may therefore be regarded as provincially distinct from the New South Wales island shelf.

Significant provincial differences also exist between respective Late Ordovician stromatoporoid faunas of the New South Wales island shelf and Tasmania. Cystistroma and Cliefdenella, this latter now regarded as a sphinctozoan sponge [Webby, 1985], are typical representatives of the N.S.W. fauna but have not been recorded from the Tasmanian Shelf. In contrast Pachystylostroma and Aulacera are found in Tasmania but not known from coeval successions of the N.S.W. island shelf [Webby, 1980]. Clathrodictyon and Ecclimadictyon colonized the Tasmanian Shelf after earlier appearances (possibly origins) in the N.S.W. island arc but none of the five Tasmanian species [Webby and Banks, 1976] is the same as the seven N.S.W. species [Webby 1969; Webby and Morris, 1976]. Closer relationships exist between the stromatoporoids of the N.S.W. island arc and the SE Asia-China-Kazakhstan region. Rosenella woyuensis, Labechiella regularis and Ecclimadictyon amzassensis occur also in Kazakhstan and, additionally, the first species is also recorded from Malaysia and N China, with the second also known from N China. Labechia variabilis is another found in Malaysia, N China and N Korea [Webby, 1975, 1980; Webby et al. 1985].

Again there are few close similarities between the coral assemblages of the New South Wales island arc and the Tasmanian Shelf. The fauna I assemblage of late Gisbornian-early Eastonian (early Caradoc) age [Webby et al. 1981] is dominated by species of Tetradium and Bajgolia [Webby and Semeniuk, 1971; Webby, 1977], with occurrences of bank-forming T. cribriforme being especially distinctive [Webby and Packham, 1982] and by the appearance of the apparently endemic rugosan Hillophyllum [Webby, 1971b]. Of the eight species of Tetradium only one is probably conspecific with a Tasmanian species, while there is no evidence of the distinctive ramose T. tasmaniense (with its affinities to North American Paleoalveolites) in the N.S.W. fauna. Pickett [1985] has established a link by finding the North American-Tasmanian tabulate genus Foerstephyllum in the Billabong Creek Limestone of the Parkes Platform, but corresponding New South Wales faunal elements like the rugose genus Hillophyllum and bank-forming T.cribriforme have not apparently been confirmed from Tasmania. The family Bajgolidae which diversified first in the N.S.W. island arc has a distribution which suggests it next spread to Tasmania and the Altai fold belt and then to Scotland [Yang, 1984].

There are similarities between the fauna II and III assemblages of middle Eastonian-early Bolindian (middle Caradoc-early Ashgill) age of the N.S.W. island arc and the Tasmanian Shelf except that the N.S.W. heliolitine corals of fauna II are much more diverse [Webby and Kruse, 1984] and the genus Halysites of fauna III - the earliest halysitine with well developed coenenchymal tubuli [Webby and Semeniuk, 1969] - has no exact counterpart in Tasmania. Moreover at the species level (see, for example, the species of rugosan Favistina [Webby, 1987]) there is little in common between the N.S.W. and Tasmanian assemblages.

Ordovician sphinctozoan sponges like Cliefdenella have been recorded from the New South Wales island arc [Webby and Rigby, 1985] and from other fold belts in the Palaeo-Pacific rim, localities in the Klamath Mountains of California and Alaska [Rigby and Potter, 1980, 1986]. Cliefdenella has also recently been found in N China and in the Altai fold belt of NW China (Lin Baoyu, pers. comm., Nov., 1986).

The echinoderm genus Astrocystites in the Late Ordovician fringing carbonate succession of the Molong High [Webby 1968] is the sole representative of the class Edrioblastoidea. It has a fairly restricted distribution being most widely reported from Caradoc-Ashgill equivalents in North America [Mintz, 1970; Sprinkle, 1973, 1982] but it is also known from an isolated locality in the Late Ordovician Trelawney Beds of north-eastern New South Wales [Webby, 1968] and apparently from the Ashgill of Sweden [Paul, 1976].

Tamworth Terrane

Cawood and Leitch [1985] have referred to the Tamworth terrane of the New England Fold Belt in northeastern New South Wales (Fig.1) as including rocks of Ordovician age. Conodonts of late Darriwilian-Gisbornian (late Llanvirn-early Caradoc) age are recorded by Cawood [1976] from limestone clasts of the debris-flow brecias and limestone succession and, like the other Middle Ordovician conodont occurrences from the Molong-Monaro and Girilambone terranes of the Lachlan Fold Belt [Nicoll 1980; Stewart and Glen, 1986], proved to be a cooler North Atlantic-type association.

Other fault blocks within the Tamworth terrane have produced diverse Late Ordovician faunas, especially corals [Hall, 1975]. In one of these the fauna appears to occur in allochthonous blocks (the Trelawney Beds of Philip [1966]) within the Early Devonian succession. The assemblage is dominated by species of Palaeophyllum, Favistina, Cyathophylloides, Crenulites, Palaeofavosites, Plasmopora and Plasmoporella [Hall, 1975; Webby 1987]. In another fault block near the major Peel Fault System, a second coral assemblage is represented in apparently allochthonous shallow water limestones of a deeper bedded chert and mudstone sequence. It includes a similar coral content but with the addition of the halysitines (Quepora, Catenipora and Halysites) and the genus Calapoecia. The presence of these halysitines and Calapoecia might suggest links with the Molong High, but closer inspection of the faunas shows no common occurrences at the species level [Webby, 1986]. In fact, there are a number of typical North American 'Red River'-type coral components, especially forms like *Favistina*, *Palaeophyllum*, *Cyathophylloides* and *Crenulites* [Flower, 1961], at least one identical species of *Favistina* - *F. neminghensis* [Etheridge, 1918] of the Tamworth terrane - probably being identical with the eastern Canadian *F. honoratensis* Bolton 1979 - an occurrence of the mainly North American edrioblastoid genus *Astrocystites* [Webby, 1968] and conodonts of North American Midcontinental aspect [Philip, 1966]. These suggest the limestone remnants near Tamworth may be more closely linked to North American sequences than to counterparts in the Lachlan Fold Belt (Fig. 2). The coral biogeographic relationships do not entirely support the suggestion that these Ordovician remnants of the New England Fold Belt originated from a position to the south of the Parkes Platform and Molong High and in the same linear volcanic arc [Packham, 1985].

Anakie High

Another coral fauna of similar age is reported by Anderson and Palmieri [1977] from an isolated occurrence on the eastern flank of the Anakie High in Central Queensland (Fig. 1). Unfortunately however the corals have not been studied to establish their relationships. They occur in shallow-water limestones of a predominantly flysch-type siltstone, sandstone and shale succession (Fork Lagoons Beds). Richly diverse conodont assemblages of North American Midcontinental and endemic aspects have been described by Palmieri [1978]. C.R. Barnes (pers. comm., 1983) has noted that the endemic component of this fauna is also seen in the Late Ordovician conodont assemblages of the Molong High.

Ravenswood Block

A very different type of association has been reported by Henderson [1983] from the Early-Middle Ordovician volcanic-sedimentary terrane of the Ravenswood Block in northern Queensland (Fig. 1). The varied assemblages of graptolites in the succession are of slender, many stiped dichograptids of the Subfamily Sigmagraptinae of the oceanic biofacies of Fortey and Cocks [1986]. The associated trilobites comprise the pelagic genera *Carolinites* and *Opipeuter*, and representatives of the deep water, basinal 'olenid' community type - the genera *Acanthopleurella* and *Hypermecaspis*. These faunas however, while of considerable biostratigraphic and palaeoenvironmental use do not provide a great deal of biogeographic interest, and do not help in establishing the relative position of the block in the Ordovician.

Broken River Embayment

In addition to the basinal quartz-rich 'flysch' successions (Wairuna Formation, Judea Formation and equivalents) of possible Late Ordovician age, with their associated spilites, jaspers and radiolarian cherts [Arnold and Henderson 1976; Arnold and Fawckner, 1980], there is an andesitic volcanic-limestone arc-type association (Everetts Creek Volcanics and Carriers Well Limestone). The limestone has been assigned a Late Ordovician age based on the conodonts (Telford, unpublished data, 1972), but also includes occurrences of favositid, halysitine, heliolitine and solitary rugose corals [Hill, in White, 1965]. Of special interest are the records of *Agetolites*, *Catenipora* and *Plasmoporella* [Hill et al. 1969]. The Asian-Alaskan tabulate genus *Agetolites* has not been reported previously from other East Australian Late Ordovician successions, but it has recently been found in the limestone unit at the top of the Malachi's Hill Beds [Semeniuk, 1970] on the western side of the Molong High, central New South Wales, along with the endemic rugosans *Bowanophyllum* and *Rhabdelasma*, and the cerioid rugosan *Cyathophylloides* [McLean and Webby, 1977; Webby, 1987]. They belong to the fauna IV assemblage of middle Bolindian (middle Ashgill) age [Webby et al. 1981]. Jenkins [1978] has also recorded *Agetolites* from the earliest Silurian of the Panuara area, on the eastern side of the Molong High. More details of the Carriers Well faunas are needed to establish the biogeographic affinities, but for the present they seem clearly linked by the presence of *Agetolites*, the abundant heliolitine element and the associated volcanics to the central N.S.W. island arc (Fig. 2).

Conclusions

1. The varied, shallow-water Ordovician faunas of the Georgina Basin, Gnalta Shelf and Tasmanian Shelf are interpreted as representing assemblages which lived in low latitudes along parts of the former Gondwana continental margin. Similarly the widely dispersed, discrete blocks (suspect terranes) of the Tasmanides, more especially those with Late Ordovician shallow-water deposits - the 'island arc' (Parkes Platform and Molong High) of central New South Wales, the Tamworth terrane of northeastern N.S.W., the Anakie High and the Broken River Embayment of central and northern Queensland - all exhibit faunas of warm water aspect, and they too are characteristic of low latitudes. There is no clear-cut evidence of large-scale strike-slip movements like those involving displacements of low latitude faunas to positions of high latitude as discussed by Hallam [1986] for the Permian-Jurassic interval of the Pacific margin. The only sort of major displacements which remain undetected are those in directions more-or-less parallel to palaeo-

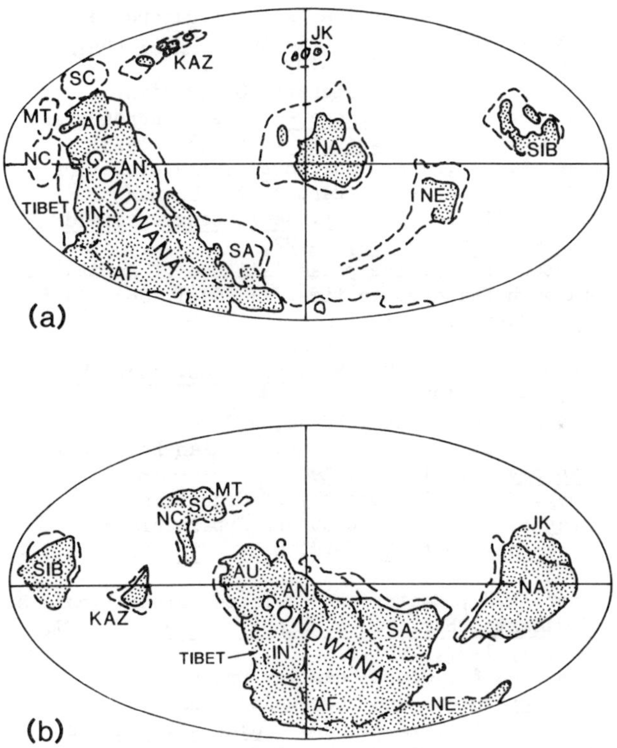

Fig. 3 Ordovician world palaeogeographic map reconstructions; (a) of Burrett and Stait [1985] and (b) of Scotese and others [1979]. Note that the eastern Gondwana margin has a strongly oblique across-latitude relationship in (a), and has a slightly oblique, almost subparallel, low-altitude orientation in (b). AU, Australia; AN, Antarctica; SA, South America; IN, India; AF, Africa; NA, North America, NE, Northern Europe; NC, North China; SC, South China; MT, Malaysia-Thailand, KAZ, Kazakhstan; SIB, Siberian Platform; JK, Jano-Kolymian Block.

latitude. The Ordovician palaeogeographic map reconstruction of Burrett and Stait [1985] shows the eastern margin of Gondwana aligned essentially across the latitudes in a NNW-SSE direction (Fig. 3a). In such an orientation it is doubtful whether large-scale strike-slip movements parallel to continental margin would go undetected. In contrast where the original eastern margin of the Gondwana continent was more oblique or aligned east-west (Fig. 3b), as Ziegler et al. [1979] and Scotese et al.[1979] have shown, then such movements would be faunally difficult to detect [Fortey and Cocks, 1986].

2. There is only one known faunal anomaly in the Tasmanide belt. The general pattern of Early Ordovician biogeographic relationships of the Australian continental margin faunas is with North China and Korea (Tsinaniid trilobite province of Whittington and Hughes [1974]), whereas, the offshore 'high' at Waratah Bay, central Victoria, has closer faunal connections with Argentina (Fig. 2). Jell [1985] has noted that the Waratah Bay fauna has little in common with the trilobite assemblages of adjacent Tasmania, adding that this may be explained by the fauna being slightly older. Such an interpretation implies a marked change in general circulation patterns with an earlier influx of the ?cooler-water Argentinian immigrants, then replacement by warm-water forms with Tasmanian-E Asian links. Alternatively, there was no significant change in circulation, but a major strike-slip displacement of the Waratah Bay 'high' from a higher-latitude position off South America.

3. The Late Ordovician coral assemblages of the Tamworth terrane of northeastern N.S.W. have closer links with North America than the other Tasmanide coral-bearing associations. The terrane is now in the most easterly part of the Tasmanides (Fig. 1), and may have originally occupied a position in the equatorial belt relatively nearer to North America. Possibly the shallow-water assemblages come from remnants of oceanic islands of the Palaeo-Pacific Ocean, between the continental blocks of North America and the Australian sector of Gondwana.

4. The Late Ordovician island-arc faunas of central New South Wales are provincially distinct from equivalent faunas of the Tasmanian Shelf. While at the generic level they exhibit similar relationships, especially among the trilobites, brachiopods and corals, there are very few species in common. From the nautiloid distribution patterns it seems that the island arc occupied an offshore position with a deep barrier separating it from the Gondwana continental margin of Tasmania [Stait et al. 1985; Webby, 1985]. It appears unlikely that the two regions occupied markedly different latitudinal positions but they were probably separated by a major marginal sea (the Wagga Trough), as suggested by Packham [1973], Scheibner [1973], Webby [1976] and others.

Acknowledgements This study has been supported by funds from the Australian Research Grants Scheme (grants nos. E73/15102, E79/15763 and E82/15297).

References

Anderson, J.C., and V. Palmieri, The Fork Lagoons Beds, an Ordovician unit of the Anakie Inlier, central Queensland. Qd Govt. Min. J., 78, 260-263, 1977.

Arnold, G.O., and R.A. Henderson, Lower Palaeozoic history of the southwestern Broken River Province, North Queensland. J.Geol.Soc. Aust., 23, 73-93, 1976.

Arnold, G.O., and J.F. Fawckner, The Broken River and Hodgkinson Provinces. In R.A. Henderson and P.J. Stephenson, (eds). The Geology and

Geophysics of Northeastern Australia. Geol. Soc. Aust., Qd Divn, 175-189, 1980.

Baillie, P., A Palaeozoic suspect terrane in Southeastern Australia and north Victoria Land, Antarctica. Abstr.Geol. Soc. Aust., 12, 43, 1984.

Banks, M.R., and C.F. Burrett, A preliminary Ordovician biostratigraphy of Tasmania. J. Geol. Soc. Aust., 26, 363-376, 1980.

Banks, M.R., and E.A. Smith, A graptolite from the Mathinna Beds, north-eastern Tasmania. Aust. J. Sci., 31, 118-119, 1968.

Bergström S.M., Conodont biostratigraphy of Middle and Upper Ordovician of Europe and Eastern North America. Mem. Geol. Soc. Am. 127, 83-162, 1971.

Burrett C.F., Ordovician biogeography and continental drift. Palaeogeogr.,Palaeoclimatol., Palaeoecol., 13, 161-201, 1973.

Burrett, C.F., Tasmanognathus: a new Ordovician conodontophorid genus from Tasmania. Geol. et Palaeont., 13, 31-38, 1979.

Burrett C.F, and B. Stait, South East Asia as a part of an Ordovician Gondwanaland - a palaeobiogeographic test of a tectonic hypothesis. Earth Planet. Sci. Letters, 75, 184-190, 1985.

Burrett, C.F., Stait, B., and J. Laurie, Gordon Subgroup (Ordovician) carbonates at Precipitous Bluff and Point Cecil, southern Tasmania, Australia. Pap. Proc. R. Soc. Tasm., 115, 93-99, 1981.

Burrett, C.F., Stait, B., and J. Laurie, Trilobites and microfossils from the Middle Ordovician of Surprise Bay, southern Tasmania, Australia. Mem. Ass. Australas. Palaeontols,1, 177-193, 1983.

Burrett, C.F., Stait, B., Sharples, C., and J. Laurie Middle-Upper Ordovician shallow platform to deep basin transect, southern Tasmania, Australia. Palaeont. Contrib. Univ. Oslo, 295, 149-157, 1984.

Cas, R., Powell, C.McA., and K.A.W. Crook, Ordovician palaeogeography of the Lachlan Fold Belt: modern analogue and tectonic constraints. J. Geol. Soc. Aust., 27, 19-31, 1980.

Cawood, P.A., Cambro-Ordovician strata, northern New South Wales. Search, 7, 317-318, 1976.

Cawood P.A., and E.C. Leitch, Accretion and dispersal tectonics of the southern New England Fold Belt, eastern Australia. In Howell, D.G. (ed.), Tectonostratigraphic Terranes of the Circum-Pacific, Circum-Pacific Council Earth Sciences Series, No. 1, 481-492, 1985.

Chang, W.T., On the Miomera and Polymera (Trilobita). Scientia Sinica, 23, 223-234, 1980.

Chugaeva M.N., Trilobites of the Ordovician of the Chu-Ili Mountains. Trudy geol. Inst. Akad. Nauk. SSSR, 9, 5-138, 1958.

Cooper, B.J., Early Ordovician conodonts from the Horn Valley Siltstone, Central Australia. Palaeontology, 24, 147-183, 1981.

Cooper R.A., and G.W. Grindley, (eds.), Late Proterozoic to Devonian sequences of southeastern Australia, Antarctica and New Zealand and their correlation. Spec. Publ. Geol. Soc. Aust. 9, 1-103, 1982.

Crick, R.E., and C. Teichert, Ordovician endocerid genus Anthoceras: its occurrence and morphology. Alcheringa 7, 155-162, 1983.

Crook, K.A.W., and C.McA.Powell, The evolution of the southeastern part of the Tasman Geosyncline. 25th Int. Geol. Congr., Field Guide Excursion 17A, 1-122, 1976.

Embleton, B.J.J, The palaeolatitude of Australia through Phanerozoic time. J. Geol. Soc. Aust. 19, 475-482, 1973.

Etheridge, R. Jr., Two remarkable corals from the Devonian of New South Wales. Rec. Aust. Mus. 12, 49-51, 1918.

Fergusson, C.L., Gray, D.R., and Cas, R.A.F., Overthrust terranes in the Lachlan Fold Belt. Geology 14, 519-522, 1986.

Finney, S.C., Biogeography of Ordovician graptolites in the southern Appalachians. Palaeont. Contrib. Univ. Oslo, 295, 167-176 1984.

Flower, R.H., Montoya and related colonial corals. New Mexico Bur. Mines, Miner. Resour. Mem. 7, part 1, 1-97, 1961.

Fortey, R.A., Global earlier Ordovician transgressions and regressions and their biological implications. Palaeont. Contrib. Univ. Oslo, 295, 37-50, 1984.

Fortey, R.A., and L.R.M. Cocks, Marginal faunal belts and their structural implications, with examples from the Lower Palaeozoic. J. Geol. Soc. Lond., 143, 151-160, 1986.

Fortey, R.A., and J.H. Shergold, Early Ordovician trilobites Nora Formation, central Australia. Palaeontology 27, 315-366, 1984.

Gatehouse, C.G., The geology of the Warburton Basin in South Australia. Aust. J. Earth Sci. 33, 161-180, 1986.

Gilbert-Tomlinson, J., The Lower Ordovician gastropod Teiichispira in northern Australia. Bull. Bur. Min. Resour. Aust. 126, 65-88.

Hall, R.L., Upper Ordovician coral faunas from north-eastern New South Wales. J. Proc. R. Soc. N.S.W., 108, 75-93, 1975.

Hallam, A., Evidence of displaced terranes from Permian to Jurassic faunas around the Pacific margins. J. Geol. Soc. Lond. 143, 209-216, 1986.

Henderson, R.A., Early Ordovician faunas from the Mount Windsor Subprovince, northeastern Queensland. Mem. Ass. Australas. Palaeontols, 1, 145-175, 1983.

Hewitt, R.A., and B. Stait, Phosphatic connecting rings and ecology of an Ordovician ellesmerocerid nautiloid. Alcheringa 9, 229-243, 1985.

Hill, D., Playford, G., and J.T. Woods, Ordovician and Silurian Fossils of Queensland. Queensland Palaeontogr. Soc., Brisbane, 1-18.

Hughes, C.P., and A.J. Wright, The trilobites Incaia Whittard 1955 and Anebolithus gen.nov. Palaeontology, 13, 677-690, 1970.

Hughes, C.P., Ingham, J.K., and R. Addison, The morphology, classification and evolution of the Trinucleidae (Trilobita). Phil. Trans. R. Soc. Lond., B,272, 537-607, 1975.

Jaanusson, V., Ordovician. In R.A. Robison and C. Teichert, (eds). Treatise on Invertebrate Paleontology, Part A. Geol. Soc. Am. and Univ. Kansas Press, A136-166. 1979.

Jell P.A., Tremadoc trilobites of the Digger Island Formation, Waratah Bay, Victoria. Mem. Mus. Vict. 46, 53-88, 1985.

Jell, P.A., and B. Stait, Tremadoc trilobites from the Florentine Valley Formation, Tim Shea area, Tasmania. Mem. Mus. Vict., 46, 1-34, 1985a.

Jell, P.A., and B. Stait, Revision of an early Arenig trilobite faunule from the Caroline Creek Sandstone, near Latrobe, Tasmania. Mem. Mus. Vict., 46, 35-51, 1985b.

Jell, P., Burrett, C.F., Yochelson, E., and B. Stait, The Early Ordovician bellerophontid Peelerophon oehlerti (Bergeron 1889) from Argentina, Australia and Thailand. Alcheringa 8, 169-176, 1984.

Jenkins, C.J., Llandovery and Wenlock stratigraphy of the Panuara area, central New South Wales. Proc. Linn. Soc. N.S.W., 102, 109-130, 1978.

Kennedy, D.J., Conodonts from a Lower-Middle Ordovician formation, Mt. Arrowsmith, northwest New South Wales, Australia. Geol. Soc. Am., Abstr. with Programs 7, 796, 1975.

Kilpatrick, D.J., and D.P. Fleming, Lower Ordovician sediments in the Wagga Trough : discovery of early Bendigonian graptolites near Eskdale, northeast Victoria. J. Geol. Soc. Aust., 27, 69-73, 1980.

Koroleva M.N., New genera of trilobites from the Middle and Upper Ordovician of northern Kazakhstan. Dokl. Akad. Nauk. SSSR, 124, 1313-1316, 1959.

Leitch, E.C., and E. Scheibner, A Tasman Fold Belt perspective on terrane analysis. Abstr.Geol. Soc. Aust. 14, 136-141, 1985.

Lin, Baoyu, Qiu, Hongrong and Changcheng Xu, New Observations of Ordovician strata in Shetai district of Urad Front Bannor, Nei Mongol (Inner Mongolia). Geol. Review. 30, 95-105, 1984.

McLean, R.A. and B.D. Webby, Upper Ordovician rugose corals of central New South Wales. Proc. Linn. Soc. N.S.W. 100, 231-244, 1976.

McTavish, R.A., and D.P. Legg, The Ordovician of the Canning Basin, Western Australia. In Bassett M.G. (ed.), The Ordovician System. Univ. Wales Press & Nat. Mus. Wales Cardiff, 447-478, 1976.

Mintz, L.W., The Edrioblastoidea: re-evaluation based on a new specimen of Astrocystites from the Middle Ordovician of Ontario. J. Paleont. 44, 872-880, 1970.

Nicoll, R.S., Middle Ordovician conodonts from the Pittman Formation, Canberra, ACT. BMR Journ. 5, 150-153, 1980.

Owen, M. and D. Wyborn, Geology and geochemistry of the Tantangara and Brindabella area. Bull. Bur. Min. Resour. Aust., 204, 1-52, 1979.

Packham G.H., The Geology of New South Wales. J. Geol. Soc. Aust. 16, 1-654, 1969.

Packham, G.H., A speculative Phanerozoic history of the south-west Pacific. In Coleman P.J. (ed.), The Western Pacific: Island Arcs, Marginal Seas, Geochemistry. Univ. W. Aust. Press, Perth, 369-388, 1973.

Packham, G.H., Transforming the eastern Lachlan Fold Belt. Abstr. Geol. Soc. Aust., 14, 174-177, 1985.

Palmieri V. Late Ordovician conodonts from the Fork Lagoons Beds, Emerald area, Central Queensland. Publ. Geol. Surv. Qd, 369, 1-31, 1978.

Paul, C.R.C., Palaeogeography of primitive echinoderms in the Ordovician . In M.G. Bassett, (ed.), The Ordovician System, Univ. Wales Press & Nat. Mus. Wales, Cardiff, 553-574, 1976.

Percival, I.G., Inarticulate brachiopods from the Late Ordovician of New South Wales, and their palaeoecological significance. Alcheringa 2, 114-141, 1978.

Percival, I.G., Ordovician plectambonitacean brachiopods from New South Wales. Alcheringa 3, 91-116, 1979a.

Percival, I.G., Late Ordovician articulate brachiopods from Gunningbland, central western New South Wales. Proc. Linn. Soc. N.S.W., 103, 175-187, 1979b.

Philip, G.M., The occurrence and palaeogeographic significance of Ordovician strata in northern New South Wales, Aust. J. Sci., 29, 112-113, 1966.

Pickett, J., Further evidence for the age of the Sofala Volcanics. Q. Notes Geol. Surv. N.S.W., 31, 1-14, 1978.

Pickett J., Correlation of a mid to late Ordovician section near Parkes, New South Wales. Rec. Geol. Surv. N.Z., 9, 77, 1985.

Powell, C.McA., Geology of the N.S.W. South Coast and adjacent Victoria with emphasis on the Pre-Permian structural history. Geol. Soc. Aust., Spec. Group Tectonics & Struct. Geol., Field Guide 1, 1-118, 1983.

Powell, C.McA., and P.J. Conaghan, Assessment of the accretionary prism model for the eastern Lachlan Fold Belt in the late Ordovician. Abstr. Geol. Soc. Aust., 11, 59-61, 1984.

Rigby, J.K., and A.W. Potter, Ordovician sphinctozoan sponges from the Klamath Mountains, California, the first known early Paleozoic occurrence. Geol. Soc. Am. Abstr. with Programs, 12, 509, 1980.

Rigby, J.K., and A.W. Potter, Ordovician sphinctozoan sponges from the eastern Klamath Mountains, northern California. Paleont. Soc. Mem. 20, 1-45, 1986.

Scheibner E., Actualistic models in tectonic mapping. 24th Int. Geol. Congr. 3, 405-422, 1972.

Scheibner, E., A plate tectonic model of the Palaeozoic tectonic history of New South Wales. J. Geol. Soc. Aust., 20, 405-426, 1973.

Scotese, C.R., Bambach, R.K., Barton, C., Van der Voo, R., and A.M. Ziegler, Paleozoic base maps. J. Geol. 87, 217-277, 1979.

Semeniuk, V., The Lower-Middle Palaeozoic stratigraphy of the Bowan Park area, central-western New South Wales. J. Proc. R. Soc. N.S.W., 103, 15-30, 1970.

Shergold J.H. Resume of data on the base of the Ordovician in northern and central Australia. Mém. Bur. Rech. Géol. Minièr. 73, 391-402, 1971.

Shergold, J.H. and E.C. Druce, Upper Proterozoic and Lower Palaeozoic rocks of the Georgina Basin. In R.A. Henderson and P.J. Stephenson, (eds), The Geology and Geophysics of Northeastern Australia, Geol. Soc. Aust., Qd. Divn, 149-174, 1980.

Sherwin, L., Stratigraphy of the Cheesemans Creek district, New South Wales. Rec. Geol. Surv. N.S.W., 13, 199-237.

Sherwin, L., Stratigraphy of the Forbes-Bogan Gate district. Rec. Geol. Surv. N.S.W. 15, 47-101,1973.

Sherwin, L., New occurrences of Ordovician graptolites from central New South Wales. Q. Notes Geol. Surv. N.S.W., 53, 1-4, 1983.

Skevington, D., Controls influencing the composition and distribution of Ordovician graptolite faunal provinces. Spec. Pap. Palaeont. 13, 59-73, 1974.

Sprinkle, J., Morphology and evolution of Blastozoan echinoderms. Mus. Comp. Zool. Harv. Univ. Spec. Publ., 1-284, 1973.

Sprinkle, J., (ed.), Echinoderm faunas from the Bromide Formation (Middle Ordovician) of Oklahoma. Univ. Kansas, Paleont. Contrib. Monogr. 1, 1-369, 1982.

Stait, B., Ordovician Onoceratida (Nautiloidea) from Tasmania, Australia. N. Jb. Geol. Palaont. Mh. 1982, 607-618, 1982.

Stait, B., Re-examination and redescription of the Tasmanian species of Wutinoceras and Adamsoceras (Nautiloidea, Ordovician). Geol. et Palaeont. 18, 53-57, 1984a.

Stait, B., Ordovician nautiloides of Tasmania - Gouldoceratidae Fam. Nov. (Discosorida). Proc. R. Soc. Vict. 96, 187-207, 1984b.

Stait, B., and J. Laurie, Ordovician nautiloids of central Australia, with a revision of Madiganella Teichert and Glenister. BMR Journ. 9, 261-266, 1985.

Stait, B., Webby, B.D., and I.G. Percival, Late Ordovician nautiloids from central New South Wales, Australia. Alcheringa 9, 143-157, 1985.

Stewart, I.R., and R.A. Glen, An Ordovician age for part of the Girilambone Group at Yanda Creek, east of Cobar. Q. Notes Geol. Surv. N.S.W., 64, 23-25, 1986.

Suppel, D.W., Warren, A.Y.E., Watkins, J.J., Chapman, J., Tenison Woods, K., and L. Barron, A reconnaissance study of the geology and gold deposits of the West Wyalong-Temora-Adelong district. Q. Notes, Geol. Surv. N.S.W., 64, 1-23, 1986.

Sweet, W.C. and S.M. Bergstrom, Provincialism exhibited by Ordovician conodont faunas. Spec. Publ. Soc. Econ. Paleont. Min., 21,189-202, 1974.

Teichert, C., and B.F. Glenister, Ordovician and Silurian cephalopods from Tasmania, Australia. Bull. Am. Paleont., 34, 187-248, 1953.

VandenBerg, A.H.M., The Tasman Fold Belt system in Victoria. Tectonophysics, 48, 267-297, 1978.

Veevers, J.J., (ed.), Phanerozoic earth history of Australia. Clarendon Press, Oxford, i-xv, 1-418, 1984.

Wade, M., Georginidae, new family of actinoceratoid cephalopods, Middle Ordovician, Australia. Mem. Qd. Mus. 18, 1-15, 1977a.

Wade, M., The siphuncle in Georginidae and other Ordovician actinoceroid cephalopods. Lethaia, 10, 303-315, 1977b.

Webby, B.D. Astrocystites distans, sp. nov., an edrioblastoid from the Ordovician of eastern Australia. Palaeontology 11, 513-525, 1968.

Webby, B.D., Ordovician stromatoporoids from New South Wales, Palaeontology, 12, 637-662, 1969.

Webby, B.D., The trilobite Pliomerina Chugaeva from the Ordovician of New South Wales. Palaeontology, 14, 612-622, 1971a.

Webby, B.D., The new Ordovician genus Hillophyllum and the early history of rugose corals with acanthine septa. Lethaia, 4, 158-168, 1971b.

Webby, B.D., Upper Ordovician trilobites from central New South Wales. Palaeontology 17, 203-252, 1974.

Webby, B.D., Succession of Ordovician coral and stromatoporoid faunas from central-western New South Wales, Australia. In Sokolov, B.S. (ed.), Drevniye Cnidaria, Vol. II. Nauka, Novosibirsk, 57-68, 1975.

Webby, B.D., The Ordovician System in south-eastern Australia. In Bassett, M.G. (ed.) The Ordovician System. Univ. Wales Press & Nat. Mus. Wales, Cardiff, 417-466, 1976.

Webby, B.D., Upper Ordovician tabulate corals from central-western New South Wales. Proc. Linn. Soc. N.S.W. 101, 167-183, 1977.

Webby, B.D., History of the Ordovician continental platform and shelf margin of Australia. J. Geol. Soc. Aust. 25, 41-63, 1978.

Webby, B.D., Biogeography of Ordovician stromatoporoids. Palaeogeogr. Palaeoclimatol. Palaeoecol. 32, 1-19, 1980.

Webby, B.D., Influence of a Tasmanide island-arc on the evolutionary development of Ordovician faunas. Rec. Geol. Surv. N.Z., 9, 99-101, 1985.

Webby, B.D., The Ordovician genus Favistina Flower and a related colonial coral from New South Wales, Australia. New Mexico Bur. Mines, Miner. Resour., Mem. in press, 1987.

Webby, B.D., and M.R. Banks, Clathrodictyon and Ecclimadictyon (Stromatoporoidea) from the Ordovician of Tasmania. Pap. Proc. R. Soc. Tasm., 110, 129-137, 1976.

Webby, B.D., and P.D. Kruse, The earliest heliolitines: a diverse fauna from the Ordovician of New South Wales. Palaeontogr. Americana, 54, 164-168, 1984.

Webby, B.D., and D.G. Morris, New Ordovician stromatoporoids from New South Wales. J. Proc. R. Soc. N.S.W. 109, 125-135, 1976.

Webby, B.D. and G.H. Packham, Stratigraphy and regional setting of the Cliefden Caves Limestone Group (Late Ordovician), central-western New South Wales. J. Geol. Soc. Aust., 29 297-317,1982.

Webby, B.D., and J.K. Rigby, Ordovician sphinctozoan sponges from central New South Wales. Alcheringa 9, 209-220, 1985.

Webby, B.D., and V. Semeniuk, Ordovician halysitid corals from New South Wales. Lethaia 2, 345-360, 1969.

Webby, B.D., and V. Semeniuk, The Ordovician coral genus Tetradium Dana from New South Wales. Proc. Linn. Soc. N.S.W., 95, 246-259, 1971.

Webby, B.D., Wyatt, D., and C.F. Burrett, Ordovician stromatoporoids from the Langkawi Islands, Malaysia. Alcheringa 9, 159-166, 1985.

Webby, B.D., VandenBerg, A.H.M., Cooper, R.A., Banks, M.R., Burrett, C.F., Henderson, R.A., Clarkson, P.D., Hughes, C.P., Laurie, J., Stait, B., Thomson, M.R.A., and G.F. Webers, The Ordovician System in Australia, New Zealand and Antarctica : correlation chart and explanatory notes. Int. Union Geol. Sci., Publ. 6, 1-64, 1981.

White, D.A., The geology of the Georgetown/Clarke River area, Queensland. Bull. Bur. Min. Resour. Aust., 71, 1-165, 1965.

Whittington, H.B., and C.P. Hughes, Geography and faunal provinces in the Tremadoc Epoch. Spec. Publ. Econ. Paleont. Min. 21, 203-218, 1974.

Williams, A., Distribution of brachiopod assemblages in relation to Ordovician palaeogeography. Spec. Pap. Palaeontology 12, 241-269, 1973.

Wright, A.J., Evaluation of a New Zealand Tremadocian trilobite. Geol. Mag., 116, 353-364, 1979.

Yang, Shengwu, Ordovician tabulate coral assemblages of Britain and their zoogeographical relationships. Palaeontogr. Americana 54, 169-173, 1984.

Ziegler, A.M., Scotese, C. R., McKerrow, W.S., Johnson, M.E., and R.K. Bambach, Paleozoic paleogeography. Ann. Rev. Earth Planet. Sci. 7, 473-502, 1979.

SOUTH WESTERN PACIFIC PERMIAN AND TRIASSIC MARINE FAUNAS: THEIR DISTRIBUTION AND IMPLICATIONS FOR TERRANE IDENTIFICATION

N. W. Archbold

Department of Geology, University of Melbourne, Parkville, Victoria, 3052, Australia

Abstract. Permian and Triassic marine faunas of the Southwest Pacific are assessed with a view to identifying allochthonous terranes. The highly distinctive and provincial nature of Permian faunas, particularly the Brachiopoda, provides strong support for the Gondwanan affinities of many regions now in Asia. These have been interpreted as having been rifted off the Indo-Australian sector of Gondwanaland during the Late Palaeozoic and Early Mesozoic. Possible allochthonous terranes in New Zealand are indicated by the occurrence of Permian tropical fusulinid limestones and distinctive Tethyan Triassic faunas in related rock associations, although alternative explanations of these faunas have been proposed. The identification of an allochthonous terrane in the Gympie area, Queensland, of the eastern Australian continental margin is not confirmed by faunal data. Nevertheless, the identification of allochthonous terranes that have their origin and termination within a single faunal province cannot be clearly established by palaeontological assessments.

Introduction: Fossil Faunas and Allochthonous Terranes

The Southwest Pacific, incorporating for this analysis the regions of southern Asia, Southeast Asia, New Guinea, Australia, New Zealand and New Caledonia, displays a variety of Permian and Triassic marine faunas. Some exhibit endemism whilst others are transitional with faunas of adjacent regions. Distinct faunal provinces can be recognised for the Permian and are shown in Figure 1. Analysis of the faunas provides critical data for the determination of palaeogeographical relationships and palaeoclimates during the Permo-Triassic. Such data has previously enabled the recognition of allochthonous terranes from the occurrence of adjacent faunal and deduced palaeoclimatological anomalies. Proposals for a plexus of microplates and terranes now constituting the western portion of the North American continent draw heavily on palaeontological as well as palaeomagnetic and structural data [Chamberlain and Lambert, 1985; Irving, 1985]. These data from the western Cordillera of North America suggest that the constituent microplates travelled considerable latitudinal and longitudinal distances in order to juxtapose contrasting tropical and temperate marine faunas [Ross and Ross, 1985]. Nevertheless such data must be used with care as has been demonstrated by a recent analysis of the Permo-Triassic of Japan [Choi, 1985], which has indicated that microplate tectonics such as applied to Japan [Nur and Ben-Avraham, 1982; Stevens, 1983] may not be the only way to interpret the available data.

This paper analyses Permo-Triassic faunas of the Southwest Pacific in the light of known faunal provincialism in an attempt to indicate allochthonous terranes within that region. Movements of disjunct terranes within a single faunal province cannot be demonstrated from palaeontological data.

Permian Faunas and Climates

Permian marine faunas of the Southwest Pacific fall within three broad faunal provinces [Archbold, 1983a], specifically the Cimmerian, Westralian and Austrazean Provinces (Figure 1). Brachiopoda appear to be a particularly useful group for elucidating provinces, apparently because of their sensitivity to water temperatures. Brachiopoda are given primacy in this study although they are by no means the only group to be utilised. Additional brachiopod studies indicate that Cimmerian faunas were relatively uniform during the earliest Permian but that subsequent assemblages were considerably differentiated leading to the subdivision of the Cimmerian Province during the later Permian [Zhan and Lee, 1977; Nakamura et al., 1985], into the Iranosin and Himalayan Subprovinces [Archbold and Thomas, in press].

Permian palaeoclimates of the regions covered by the three faunal provinces have been progressively outlined by Dickins [1977, 1978, 1984, 1985a] from which the following general

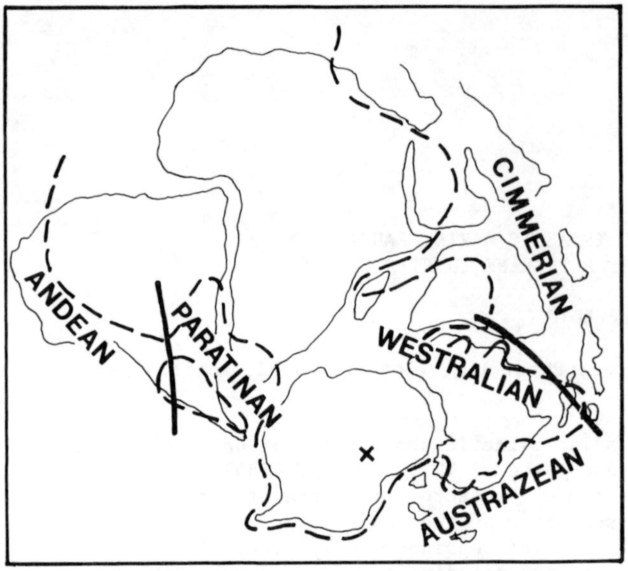

Fig. 1. Marine Permian provinces of Gondwana [from Archbold, 1983a]. Star indicates generalised position of Permian south pole.

trends are summarised. Cimmerian regions from Arabia in the west to Irian Jaya in the east were characterised by cold climates early in the Permian (Asselian to Early Sakmarian) which were replaced by warm to tropical climates later in the Early Permian. Temperate faunas, established in the interval between the cold and tropical faunas, are better developed in the eastern portion of the Cimmerian Province. The Westralian Province demonstrates a short lived cold stage (Asselian to Early Sakmarian) followed by extensive development of temperate faunas with tropical elements becoming apparent only in the Late Permian. The Austrazean Province was dominated by cold and cool temperate conditions throughout the Permian with possible warming in the New Zealand sector in the Late Permian [Dickins, 1985a]. Against this generalised backdrop of climatic interpretations, the faunas of the three provinces and their tectonic implications are discussed below.

Cimmerian Faunas

The large Cimmerian Province stretches from Arabia in the west to Irian Jaya in the east. Many components of the province have been interpreted as having been rifted off from the northern margin of Gondwana during the Late Palaeozoic and Mesozoic [e.g. Sengor, 1979; Sengor et al., 1980; Nur and Ben-Avraham, 1982]. Evidence of Early Permian glaciation is well documented in regions of the Province such as the Arabian Peninsula [Braakman et al., 1982; Dickins, 1985a] and north western Tibet in the North Tibetan continent [Liang et al., 1983; Sun, 1984]. Other regions of the Province lack irrefutable evidence of glaciation, but have pebbly mudstones, pebble slate horizons or boulder beds occupying comparable stratigraphical positions, as in Thailand [Waterhouse, 1982a] and the Himalayan regions [Acharyya, 1979]. Although Timor and Irian Jaya have yet to yield any direct lithological evidence of Early Permian glaciation at comparable stratigraphical horizons, marine faunas from these regions independently indicate Gondwanan affinities.

The cold water bivalve Eurydesma in Asselian to Early Sakmarian faunas of the North Tibetan continent [Lui and Cui, 1983] and Afghanistan [Termier et al., 1974] provides good evidence for a Gondwanan origin of these regions. However, the widespread Sakmarian faunas with the brachiopods Punctocyrtella and/or Subansiria which occur stratigraphically above glacial horizons and in many other regions are equally indicative of a Gondwanan affinity. Such faunas are known from the Arabian Peninsula [Hudson and Sudbury, 1959], Iran [Waterhouse, 1976], the Pamirs [Grunt and Dmitriev, 1973], Afghanistan [Plodowski, 1968; Termier et al., 1974], the North Tibetan continent [Hu, 1983; Sun, 1984], Kashmir [Bion, 1928], Aranachal Pradesh [Singh, 1978] and Irian Jaya [Archbold, 1983b]. In addition, a newly discovered Sakmarian brachiopod fauna from Timor, collected by Mr S.T. Barkham, Royal Holloway and Bedford New College, University of London, also includes Punctocyrtella and other forms shared with Westralian Sterlitamakian (Sakmarian) faunas. Faunas from Thailand that are apparently correlative with the Eurydesma faunas [Waterhouse, 1982a] are less clear in their affinities because they contain fewer genera in common with Gondwanan faunas. Other Sakmarian faunas from Cimmerian regions also include genera such as Globiella, Taeniothaerus and Brachythyrinella which are useful Gondwana indicators [Waterhouse and Gupta, 1977; Archbold, 1982, 1984a; Zhan and Lee, 1984].

During Artinskian and younger Permian times climatic warming eliminated the strong Gondwanan element and increased the Tethyan aspect of Cimmerian faunas. Stronger links with Gondwanan (i.e. Westralian) faunas remained in eastern parts of the province such as in Irian Jaya where Late Artinskian/Early Kungurian faunas demonstrate significant links with faunas of Thailand, probably Timor and the Westralian Province [Archbold, 1981; Archbold et al., 1982]. At this time these links extended into the Himalayan region, where faunas of this age are poorly known, as indicated by the occurrence of Retimarginifera [Archbold, 1984b]. Other noteworthy eastern Cimmerian elements include the distinctive ammonoid Pseudoschistoceras as yet only known from Timor, Irian Jaya and Western Australia [Glenister et al., 1984].

By later Permian times full differentiation of the Cimmerian Province into the western Iranosin Subprovince (characterised by such brachiopods as Tyloplecta, Ogbinia, Araxathyris and Edriosteges) and the eastern Himalayan Subprovince, including Timor (characterised and dominated by Leptodus, Costiferina, Derbyia, large Neospirifer and Spiriferella) had taken place [Nakamura et al., 1985, p. 195]. This differentiation may have reflected a tropical climate in the west and a subtropical climate in the east. Also supporting the Gondwanan affinity of various Cimmerian continental elements, are the occurrences of mixed Glossopteris - Cathaysian floras in such areas as Kashmir, the Gangdise (=Lhasa) block of southern Tibet, Thailand and Irian Jaya [e.g. Archbold et al., 1982; Li et al., 1985; Dickins, 1985a].

All of the regions constituting the Cimmerian Province were therefore peripheral to Gondwana during the Permian. It is likely that the short duration larval stages of marine, shallow-water, benthonic organisms (especially Brachiopoda) were able to migrate throughout the Cimmerian Province (full migration being restricted later in the Permian by climatic differentiation) and those with a somewhat greater temperature range tolerance were able to penetrate the Westralian Province. Mixed floras in continental sequences of the Cimmerian Province also argue for a peripheral Gondwanan position, the floras occupying an intermediate geographical position between the cold areas of Gondwana and the tropical Cathaysian continent of south China. Such conclusions have been increasingly used [e.g. Crawford, 1979] to demonstrate that during the Permian, the Tethys cannot have been a wide oceanic gap, as shown on classical Permian reconstructions [e.g. Smith et al., 1973; Bambach et al., 1980]. Earlier palaeontological and lithological investigations [e.g. Ridd, 1971; Audley-Charles et al., 1972] appeared to be at variance with limited palaeomagnetic data [McElhinny et al., 1974] which placed continental elements, now located in the Cimmerian Province, considerably north of the equator during the Permian.

However, data obtained during the last decade from Asia have reversed the picture. Palaeontological and stratigraphical information [e.g. Ahmad, 1978, 1982; Archbold, 1981; Waterhouse, 1983; Wang, 1984; Dickins, 1985a; Rowley et al., 1985], supported by palaeomagnetic [e.g. Zhu and Teng, 1984; Jin et al., 1985; Chamalaun, 1977; Soffel and Forster, 1980; Wensink, 1982] and structural data have been used in syntheses to indicate the complex break- up of the northern margin of Gondwana, and the progressive rift-drift northward of numerous microcontinents during the Late Palaeozoic and Mesozoic [e.g. Ridd, 1980; Boulin, 1980; Mitchell, 1981; Stauffer, 1983; Audley-Charles, 1983, 1984; Huang et al., 1984; Wang and Sun, 1985; Burrett and Stait, 1985]. Whilst there are differences in details between the syntheses, the overall picture of progressive rifting of the northern margin of Gondwana during the Late Palaeozoic and the Early Mesozoic is becoming accepted. The rifting of Peninsula India from Gondwana took place subsequently during the Late Mesozoic and, having drifted north, was re-united with earlier fragments of Gondwana [e.g. Zhu and Teng, 1984]. Australia (including New Guinea) has converged northwards on Asia since the Neogene [Audley-Charles, 1986], so that adjacent blocks during the Permian that were subsequently split apart are again potentially going to become part of a single continental unit.

Westralian Faunas

Western Australia, is regarded as being a typical rift-drift continental margin [e.g. Hamilton, 1979; Veevers, 1984] although differences of opinion exist as to the continental region adjacent to the western margin. Generally either Peninsular India, Greater India (i.e. Tibet) or elements of southeast Asia have been placed adjacent to Western Australia (see references quoted above for typical reconstructions).

Marine faunas of the western Australian basins have, since early this century, been known for their Tethyan aspect, and their close faunal links with the classic Salt Range, Himalayan and Timor faunas indicate that western Australia was at the southeastern extremity of the warm Tethyan seas characteristic of the Cimmerian Province. The faunas began with characteristic cold water Gondwanan assemblages, generally of Early Sakmarian age [Archbold, 1982] containing the bivalve Eurydesma and brachiopods such as Brachythyrinella, Punctocyrtella and the endemic Lyonia. Late Sakmarian, Artinskian and Kungurian faunas include the distinctive Cimmerian genera Demonedys, Comuquia, Dyschrestia and Costiferina as minor elements. However, the faunas are marked by the endemic development of many species of Neochonetes (Sommeriella), Streptorhynchus, Permorthotetes, Taeniothaerus, small Tomiopsis and the large number of Neospirifer species (related to species of Neospirifer in the eastern part of the Cimmerian Province - see Archbold and Thomas, 1986) and large spiriferid genera such as Latispirifer and Crassispirifer which exhibit some links with Late Permian Himalayan faunas. Cool to warm temperate conditions are indicated for these faunas [Dickins, 1978].

The distinctive Late Permian (Chhidruan) faunas of the Canning and Bonaparte Gulf Basins have been compared with Himalayan and Salt Range faunas [Thomas, 1969] because of the occurrence of Derbyia, Megasteges, Waagenoconcha, Costiferina, Leptodus, large Neospirifer and small Tomiopsis. A faunal diversity slightly

TABLE 1. Permian Brachiopoda From the Westralian and Austrazean Provinces of the Gondwanan Realm

Westralian Province	Austrazean Province
Lingula	*Lingula*
Streptorhynchus	*Orbiculoidea*
Arctitreta	*Rhipidomella*
Kiangsiella	*Schuchertella*
Derbyia	*Streptorhynchus*
Permorthotetes	*Notostrophia*
Neochonetes	*Neochonetes*
Svalbardia	*Tivertonia*
Quinquenella	?*Svalbardia*
?*Chonetinella*	*Capillonia*
Tornquistia	*Licharewiella*
Demonedys	*Strophalosia*
Mingenewia	*Echinalosia*
Strophalosia	*Wyndhamia*
Heteralosia	*Pseudostrophalosia*
Echinalosia	*Taeniothaerus*
Wyndhamia	*Aulosteges*
Lialosia	*Megasteges*
strophalosiid genus A	*Lethamia*
strophalosiid genus B	*Waagenoconcha*
Taeniothaerus	*Azygidium*
Aulosteges	*Filiconcha*
Megasteges	*Anidanthus*
Leptodus	*Megousia*
Waagenoconcha	*Terrakea*
Stictozoster	*Auriculispina*
Comuquia	*Costatumulus*
?*Lethamia*	*Cancrinella*
Dyschrestia	*Linoproductus*
Retimarginifera	*Pseudosyrinx*
Callytharrella	*Subansiria*
Costiferina	*Permasyrinx*
Cancrinella	*Sulcicosta*
Lyonia	*Punctocyrtella*
Globiella	*Trigonotreta*
Punctocyrtella	*Fusispirifer*
?*Pseudosyrinx*	*Aperispirifer*
Myodelthyrium	*Neospirifer*
licharewiid genus A	*Brachythyrinella*
licharewiid genus B	*Cancellospirifer*
Brachythyrinella	*Spiriferella*
Neospirifer	*Sulciplica*
Fusispirifer	*Pustuloplica*
Latispirifer	*Licharewia*
Cratispirifer	*Attenuatella*
Costatispirifer	*Biconvexiella*
Crassispirifer	*Martinia*
neospiriferid genus A	*Spinomarginia*
Trigonotreta	*Ingelarella*
Spiriferella	*Glendonia*
Elivina	*Notospirifer*
Spirelytha	ingelarelliid genus A
Phricodothyris	*Lamnaespina*
Tomiopsis	*Punctospirifer*
?*Fredericksia*	*Pustulospiriferina*
Gjelispinifera	*Cleiothyridina*
Lamnaespina	*Spirigerella*
Callispirina	*Pugnoides*
Cleiothyridina	*Wairakiella*
Stenoscisma	*Stenoscisma*
Camerisma	*Plekonella*
Rhynchopora	*Psilocamara*
Hustedia	*Fletcherithyris*
Fletcherithyris	*Gilledia*
Hoskingia	*Maorielasma*
Yochelsonia	*Marinurnula*
Gilledia	*Amygdalocosta*
?*Hemiptychina*	*Glossothyropsis*
	?*Pseudodielasma*

lower than Himalayan faunas suggests that the faunas inhabited subtropical to tropical waters [Dickins, 1978]. By placing India off Western Australia, the Himalayan Front is located at approximately the same latitude as, and relatively close to, either the Canning Basin (Figure 1) or the Bonaparte Gulf Basin [Sastri et al., 1981]. The major Late Permian transgression in the Himalayan region characterised by the Late Perian *Lamnimargus himalayensis* fauna is of the same age as the Chhidruan transgression in the Canning and Bonaparte Basins. The similarity of these faunas argues for a closer relative position, during the Permian, of the Himalayan Front and northwestern Australia than permitted by the Greater India model of Gondwana [c.f. Veevers, 1984].

Austrazean Faunas

Striking differences between the Permian brachiopod faunas of the Westralian and Austrazean Provinces (Table 1) appear to reflect the absence of direct marine communication between the two regions except around the top of the continent. Diversity data and generic distributions for eastern Australian Permian brachiopod faunas indicate cold temperatures for Tasmanian faunas from Asselian to Kazanian times with more temperate conditions progressively farther north from the Sydney basin to the Bowen Basin and Gympie area. Faunal links are strongest during the Early Permian (Asselian to Tastubian) with both *Punctocyrtella* and *Brachythyrinella* being present in the eastern faunas [Archbold, 1982], presumably reflecting pan Gondwanan cold marine waters and hence the ability of such forms to migrate around the north of Australia. Migration of forms across Australia through the central intracratonic basins at the time of maximum marine transgression (Tastubian) appears unlikely in view of postglacial palaeogradient directions which suggest a ridge separating the eastern and western basins [Wopfner, 1981]. The shallow seas of the ancestral southern Australian-Antarctica depression [Veevers, 1984] transgressed into the Arckaringa Basin and Victorian region at this time but links with the Perth Basin are unlikely as indicated by non-marine sedimentation in the southern Perth and Collie Basins at this time.

From Late Sakmarian times onward, Brachiopoda are particularly distinctive between the Westralian and eastern Australian faunas. For the entire Permian, there are 69 genera in the Westralian faunas and 55 genera in eastern Australian faunas. The 16 shared genera invariably demonstrate a considerable difference in species diversity between the two regions and only two or three species may be shared. Post Tastubian links, although minor, are strongest between Westralian and Queensland faunas. Other groups such as Bivalvia, Gastropoda and Bryozoa show stronger links between the two regions at the generic level but few species are shared. Approximately half of the known Foraminifera species are shared between the Westralian Province and eastern Australian faunas.

New Zealand faunas demonstrate a strong endemic eastern Australian aspect and hence both areas are placed in the Austrazean Province [Archbold, 1983a; Grant-Mackie, 1985]. These faunal links are further enhanced by the recent discovery of the bivalve *Eurydesma* and the chonetid brachiopod *Tivertonia* in New Zealand faunas [Begg and Ballard, 1985]. About 50 brachiopod genera are present in New Zealand faunas [Waterhouse, 1982b] of which 25 are shared with eastern Australian faunas. Eighteen are shared with Westralian faunas and only four of these are absent from eastern Australian assemblages. Ten genera have not yet been identified in Australia [Archbold and Thomas, 1985] but it is noteworthy that eastern Australian brachiopods are poorly known and require modern description. The New Zealand faunas are closest to those of Queensland (Bowen Basin and Gympie area). Meagre faunas from New Caledonia are Austrazean in character. Palaeogeographical positions adjacent to north east and east of Queensland have been suggested for the New Caledonia faunas [Waterhouse, 1969]. No reasons exists to exclude New Caledonia faunas from the Austrazean Province [Grant-Mackie, 1985].

The structural complexity of New Zealand has provided ample scope for the application of the terrane concept of tectonics [e.g. Howell, 1980; Bishop et al., 1983]. Anomalous New Zealand Permian faunas consisting of fusulinids and corals in limestone are interpreted as tropical in aspect and have created interest for some time [Hornibrook, 1951; Leed, 1951]. Recently, these faunas and rocks have been regarded as being indicative of allochthonous terranes [Sporli and Gregory, 1981; Ozawa and Kanmera, 1983], Only one brachiopod species, *Martiniopsis woodi*, has been recorded from these limestone faunal assemblages of Northland, North Island [Waterhouse, 1968]. Unfortunately, the critical specimens have never been figured or fully described. *Martiniopsis woodi* is a characteristic species of a diverse brachiopod assemblage from terrigenous clastic sequences of the south island of New Zealand. A detailed stratigraphical and geochemical analysis of lavas associated with the Northland fusulinid limestones has indicated that the successions may not be allochthonous to New Zealand [Ramsay and Moore, 1985]. Subtropical offshore currents and a position near Queensland for much of New Zealand during the later Permian may account for the occurrences. Alternatively, local climatic warming has been proposed for New Zealand [Waterhouse, 1982b; Dickins, 1985a] during the times that the limestone was being deposited. Nevertheless, palaeontological data from the Triassic of the same terrane adds further data in support of the allochthonous model [Grant-Mackie, 1985].

The Gympie Terrane of Queensland may also have been tectonically transported to its present position [Harrington, 1983]. Stratigraphical and structural data of the Gympie Terrane are comparable with those of the Brook Street and Maitai Terranes of New Zealand, and geochemical data from the Gympie volcanics are also partly comparable [Sivell and Waterhouse, 1986]. However, Permian faunas from the Gympie Terrane [Runnegar and Ferguson, 1969] appear to be closer to those of the Bowen Basin, Queensland, than to those of New Zealand. Nevertheless, distinctions are minor and palaeontological evidence alone does not disprove relative displacement of the Gympie Terrane within the northern portion of the Austrazean Province to its present position.

Triassic Faunas

Marine Triassic faunas of the Southwest Pacific fall into two distinctive provinces - specifically faunas of Tethyan (and wider) aspect, and those of the Maorian Province, which may represent a cool water remnant of the extensive Gondwanan Realm of the Permian [Fleming, 1979]. The distinctive Maorian faunas occur in New Zealand and New Caledonia and investigations continue to demonstrate the distinctiveness of the faunas and the similarities of the two regions [Campbell, 1984; Grant-Mackie, 1985]. Benthonic Brachiopoda and Bivalvia are generally endemic at the generic level in the province. Ammonites, as free-swimming organisms and pterioid bivalves with longer duration pelagic larvae, migrated freely into the Maorian Province, as did Foraminifera and conodonts, and are of Tethyan and cosmopolitan affinities. The bulk of the Maorian Triassic faunas are of Middle and Late Triassic age but several Early Triassic ammonites are known. Not included within the Maorian Province are the faunas of the Torlesse Zone (Canterbury Suite), which include significant Tethyan elements and only a small proportion of the Maorian genera [Grant-Mackie, 1985]. The Torlesse Zone also includes the Permian fusulinid limestones discussed above and hence the combined evidence of Permian and Triassic faunas provides strong evidence for the allochthonous nature of the zone [Grant-Mackie, 1985].

Early Triassic faunas are known from two onshore areas of the Australian mainland (the Perth Basin, Western Australia and the Gympie area, Queensland) and a Middle Triassic fauna is known from the north-western Austalian Sahul Shoals No. 1 Well on the continental shelf [Skwarko and Kummel, 1974]. Ammonites and benthonic elements of these faunas are essentially Tethyan in character, although the Queensland faunas chiefly consist of endemic species of Tethyan and cosmopolitan genera [Runnegar, 1969]. No late Triassic faunas are known from Australia.

Middle and Late Triassic ammonite and benthonic faunas of Papua New Guinea are Tethyan and cosmopolitan in character, although, as with the Queensland Early Triassic, benthonic elements are generally distinct at the species level. The distinctive brachiopod Clavigera in the Kuta Formation of Papua New Guinea may indicate a minor link with the Late Triassic faunas of the Maorian Province [Skwarko et al., 1976].

Conclusions

Triassic world climates were essentially hot [Dickins, 1985b] and the marine faunas essentially cosmopolitan (except for the Maorian Province). As a result it is the provincial nature of the Permian faunas of the south west Pacific that exhibit the most potential for indicating rifted elements of Gondwana now in Asia and allochthonous terranes within the Austrazean Province. High provinciality caused by major latitudinal climatic zones reduces the size of provinces and increases their number. Under such conditions, allochthonous terranes become more apparent from palaeontological data.

Permian faunas of the southwest Pacific, particularly from areas such as northwest Australia, Timor, Irian Jaya and Thailand, demonstrate close links with faunas from the Middle East and Southern Asia. Regions such as Iran, the Pamirs, Afghanistan and Tibet represent a mosaic of continental blocks that were part of the northern margin of Gondwana during the Permian and now represent allochthonous terranes welded onto the southern margin of central Asia. Both Permian and Triassic palaeontological evidence points to an allochthonous origin for the Torlesse Zone of New Zealand and, as a result, the Permian fusulinid and coral limestone faunas are excluded from the Permian Austrazean Province.

Acknowledgements. The author has appreciated discussions with Dr G.A. Thomas, University of Melbourne and Dr J.M. Dickins, Bureau of Mineral Resources, Geology and Geophysics on issues discussed above, but is alone responsible for the opinions expressed. L. Archbold and I. Munro typed the manuscript.

References

Acharyya, S.K., India and southeast Asia in Gondwanaland fit, Tectonophysics, 56, 261-276, 1979.

Ahmad, F., "Gondwanaland", the concept that failed. Seventh Birbal Sahni Memorial Lecture, 27 pp., Birbal Sahni Institute of Palaeobotany, Lucknow, India, 1978.

Ahmad, F., The myth of oceanic Tethys, Boll. Soc.Paleont.Italiana, 21, 153-168, 1982.

Archbold, N.W., Permian brachiopods from Western Irian Jaya, Indonesia, Geol. Res. Dev. Centre, Paleont. Ser., 2, 1-25, 1981.

Archbold, N.W., Correlation of the Early Permian faunas of Gondwana: Implications for the Gondwanan Carboniferous-Permian boundary, J. Geol. Soc. Aust., 29, 267-276, 1982.

Archbold, N.W., Permian marine invertebrate provinces of the Gondwanan Realm, Alcheringa, 7, 59-73, 1983a.

Archbold, N.W., The Permian brachiopod faunas of Irian Jaya, Indonesia: Gondwanan or South East Asian?, Abstracts Geol. Soc. Aust., 9, 226-227, 1983b.

Archbold, N.W., Early Permian marine faunas from Australia, India and Tibet: An update on the Gondwanan Carboniferous-Permian boundary, Bull. Ind. Geol. Assoc., 17, 133-138, 1984.

Archbold, N.W., Western Australian occurrences of the Permian brachiopod genus Retimarginifera, Alcheringa, 8, 113-122, 1984.

Archbold, N.W., C.J. Pigram, N. Ratman and S. Hakim, Indonesian Permian brachiopod fauna and Gondwana-South-East Asia relationships, Nature, 296, 556-558, 1982.

Archbold, N.W., and G.A. Thomas, Western Australian Permian Brachiopoda: Their biogeography and biostratigraphy, Record N.Z. Geol. Surv., 9, 8-10, 1985.

Archbold, N.W., and G.A. Thomas, Neospirifer and Trigonotreta (Spiriferida, Brachiopoda) from the Permian of Western Australia, Alcheringa, 10, 125-161, 1986.

Archbold, N.W., and G.A. Thomas, Permian Brachiopoda from Western Australia: A review in time and space, in Proceedings of 1st International Congress on Brachiopods, Brest, France, September 9-13, in press, 1985.

Audley-Charles, M.G., Reconstruction of eastern Gondwanaland, Nature, 306, 48-50, 1983.

Audley-Charles, M.G., Cold Gondwana, warm Tethys

and the Tibetan Lhasa block, Nature, 310, 165, 1984.
Audley-Charles, M.G., Rates of Neogene and Quaternary movements in the Southern Banda Arc based on micropalaeontology, J. Geol. Soc. London, 143, 161-175, 1986.
Audley-Charles, M.G., D.J. Carter, and J.S. Milson, Tectonic development of Eastern Indonesia in relation to Gondwanaland dispersal, Nature Physical Science, 239, 35-39, 1972.
Bambach, R.K., C.R. Scotese, and A.M. Ziegler, Before Pangea: the geographies of the Paleozoic world, American Scientist, 68, 26-38, 1980.
Begg, J.G., and H.R. Ballard, The first Eurydesma fauna from New Zealand, its correlation and depositional environment, Record N.Z. Geol. Surv., 9, 14-15, 1985.
Bion, H.S., The fauna of the Agglomeratic Slate Series of Kashmir, Mem. Geol. Surv. India, Palaeont. Indica, New Series, 12, 1-42, 1928.
Bishop, D.G., J.D. Bradshaw, and C.A. Landis, Provisional Terrane map of South Island, New Zealand, Stanford University Publications, Geological Sciences, 18, 24-31, 1983.
Boulin, J., Afghanistan structure, Greater India concept and eastern Tethys evolution, Tectonophysics, 72, 261-287, 1981.
Braakman, J.H., B.K. Levell, J.H. Martin, T.L. Potter, and A. van Vliet, Late Palaeozoic Gondwana glaciation in Oman, Nature, 299, 48-50, 1982.
Burrett, C., and B. Stait, South East Asia as a part of an Ordovician Gondwanaland - a palaeogeographic test of a tectonic hypothesis, Earth Planet. Sci. Letters, 75, 184-190, 1985.
Campbell, H.J., New records and taxa of Permian and Triassic fossils from New Caledonia and New Zealand, Alcheringa, 8, 151-167, 1984.
Chamalaun, F.H., Palaeomagnetic evidence for the relative positions of Timor and Australia in the Permian, Earth Planet. Sci. Letters, 34, 107-112, 1977.
Chamberlain, V.E., and R. St.J. Lambert, Cordillera, a newly defined Canadian microcontinent, Nature, 314, 707-713, 1985.
Choi, D.R., Late Permian - Early Triassic paleogeography of northern Japan: Did Pacific microplates accrete to Japan?, Geology, 12, 728-731, 1984.
Crawford, A.R., The myth of a vast oceanic Tethys, the India-Asia problem and earth expansion, J. Petrol. Geol., 2, 3-9, 1979.
Dickins, J.M., Permian Gondwanan climate, Chayanica Geologica, 3, 11-22, 1977.
Dickins, J.M., Climate of the Permian in Australia: the invertebrate faunas, Palaeogeog. Palaeoclimat. Palaeoecol., 23, 33-46, 1978.
Dickins, J.M., Evolution and climate in the Upper Palaeozoic, in Fossils and Climate, edited by P. Brenchley, pp. 317-327, John Wiley & Sons, Chichester, United Kingdom, 1984.
Dickins, J.M., Palaeobiofacies and palaeobiography of Gondwanaland from Permian to Triassic, in The Tethys, edited by K. Nakazawa and J.M. Dickins, pp. 83-92, Tokai University Press, Tokyo, Japan, 1985a.
Dickins, J.M., Climate of the Triassic, Record N.Z. Geol. Surv., 9, 34-36, 1985b.
Fleming, C.A., The Geological History of New Zealand and its life, 141 pp, Auckland University Press, Auckland, New Zealand, 1979.
Glenister, B.F., L.M. Glenister, and S.K. Skwarko, Lower Permian Cephalopods from Western Irian Jaya, Indonesia, Geol. Res. Devel. Centre, Pal. Ser., 4, 74-85, 1983.
Grant-Mackie, J.A., New Zealand - New Caledonian Permian - Jurassic faunas, biogeography and terranes, Record N.S. Geol. Surv., 9, 50-52, 1985.
Grunt, T.A., and V.Yu. Dmitriev, Permskie brakhiopody Pamira, Akad. Nauk SSSR, Paleont. Inst., Trudy, 136, 1-212, 1973.
Hamilton, W., Tectonics of the Indonesian Region, United States Geol. Surv. Prof. Pap., 1078, 1-346, 1979.
Harrington, H.J., Correlation of the Permian and Triassic Gympie Terrane of Queensland with the Brook Street and Maitai Terranes of New Zealand, in Permian Geology of Queensland, pp. 431-436, Geological Society of Australia, Queensland Division, Brisbane, Australia, 1983.
Hornibrook, N.deB., Permian fusulinid foraminifera from the North Auckland Peninsula, New Zealand, Trans. Roy. Soc. N.Z., 79, 319-321, 1951.
Howell, D.G., Mesozoic accretion of exotic terranes along the New Zealand segment of Gondwanaland, Geology, 8, 487-491, 1980.
Hu, C.-M., New genera and species of spiriferid brachiopods in the Upper Carboniferous and Lower Permian of Duoma, Rutog, Xizang, China, Earth Science, J. Wuhan College Geology, 1983, No. 1, 105-117, 1983.
Huang, J.-Q., G.-M. Chen, and B.-W., Chen, Preliminary analysis of the Tethys - Himalayan Tectonic Domain, Acta Geol. Sinica, 58, 1-17, 1984.
Hudson, R.G.S., and M. Sudbury, Permian Brachiopoda from South-East Arabia, Notes et Mem. Moyen-Orient, 7, 19-55, 1959.
Irving, E., Whence British Columbia, Nature, 314, 673-674, 1985.
Leed, H., Permian reef-building corals from the North Auckland Peninsula, New Zealand, N.Z. J. Sci. Tech., B33, 126-128, 1951.
Li, X.-X., Y.-M. Wu, and Z.-B. Fu, Preliminary study on a mixed Permian flora from Xiagangjiang of Gerze District, Xizang and its palaeobiogeographic significance, Acta Palaeont. Sinica, 24, 150-170, 1985.
Liang, D.-Y., Z.-T. Nie, T.-Y. Guo, B.-W. Xu, Y.-Z. Zhang, and W.-P. Wang. Permo-

Carboniferous Gondwana-Tethys facies in southern Karakorum Ali, Xizang (Tibet), Earth Science, J. Wuhan College Geology, 1983, No. 1, 9-27, 1983.
Lin, J.-L., M. Fuller, and W.-Y. Zhang, Preliminary Phanerozoic polar wander parths for the North and South China blocks, Nature, 313, 444-449, 1985.
Liu, B.-P., and X.-S. Cui, Discovery of Eurydesma - fauna from Rutog, Northwest Xizang (Tibet), and its biogeographic significance, Earth Science, J. Wuhan College Geology, 1983, No. 1, 79-92, 1983.
McElhinny, M.W., N.S. Haile, and A.R. Crawford, Palaeomagnetic evidence shows Malay Peninsula was not a part of Gondwanaland, Nature, 252, 641-645, 1974.
Mitchell, A.H.G., Phanerozoic plate boundaries in mainland S.E. Asia, the Himalayas and Tibet, J. Geol. Soc. London, 138, 109-122, 1981.
Nakamura, K., D. Shimizu, and Z.-T. Liao, Permian palaeobiogeography of brachiopods based on the faunal provinces, in The Tethys edited by K. Nakazawa and J.M. Dickins, pp. 185-198, Tokai University Press, Tokyo, Japan, 1985.
Nur, A., and Z. Ben-Avraham, Oceanic plateaus, the fragmentation of continents, and mountain building, J. Geophys. Res., 87, 3644-3661, 1982.
Ozawa, T., and K. Kanmera, Tectonic terranes of Late Palaeozoic rocks and their accretionary history in the Circum-Pacific region viewed from fusulinacean paleobiogeography, Stanford University Publications, Geological Sciences, 18, 158-160, 1983.
Plodowski, G., Neue spiriferen aus Afghanistan, Senckenb. Lethaea, 49, 251-259, 1968.
Ramsay, W.R.H., and P.R. Moore, Mineralogy and chemistry of a pillow lava, Northland, New Zealand, and its tectonic significance, N.Z. J. Geol. Geophys., 28, 471-485, 1985.
Ridd, M.F., South-east Asia as a part of Gondwanaland, Nature, 234, 531-533, 1971.
Ridd, M.F., Possible Palaeozoic drift of S.E. Asia and Triassic collision with China, J. Geol. Soc. London, 137, 635-640, 1980.
Ross, C.A., and J.R.P. Ross, Carboniferous and Early Permian biogeography, Geology, 13, 27-30, 1985.
Rowley, D.B., A. Raymond, J.T. Parrish, A.L. Lottes, C.R. Scotese, and A.M. Ziegler, Carboniferous paleogeographic, phytogeographic and paleoclimatic reconstructions, Int. J. Coal Geology, 5, 7-42, 1985.
Runnegar, B.N., A Lower Triassic ammonoid fauna from southeast Queensland, J. Paleont., 43, 818-828, 1969.
Runnegar, B.N., and J.A. Ferguson, Stratigraphy of the Permian and Lower Triassic marine sediments of the Gympie District, Queensland, Pap. Dept. Geol. Univ. Qd, 6, 247-281, 1969.
Sastri, V.V., B.S. Venkatachala, V. Narayanan, The evolution of the east coast of India, Palaeogeog., Palaeoclimat., Palaeoecol., 36, 23-54, 1981.
Sengor, A.M.C., Mid-Mesozoic closure of Permo-Triassic Tethys and its implications, Nature, 279, 590-593, 1979.
Sengor, A.M.C., Y. Yilmaz, and I. Ketin, Remnants of a pre-Late Jurassic ocean in northern Turkey: Fragments of Permian-Triassic Paleo-Tethys?, Bull. Geol. Soc. Amer., 91, 599-609, 1980.
Singh, T., Brachiopods from the Permian Formation of Siang District, Arunachal Pradesh, Contrib. Himal. Geol., 1, 171-188, 1978.
Sivell, W., and J.B. Waterhouse, Petrogenesis of the Gympie Group volcanics and their relation to Early Permian volcanism in eastern Australia and New Zealand, Abstract Geol. Soc. Aust., 15, 179, 1986.
Skwarko, S.K., and B. Kummel, Marine Triassic molluscs of Australia and Papua New Guinea, Bull. Bur. Miner. Resour. Geol. Geophys. Australia, 150, 111-139, 1974.
Skwarko, S.K., R.S. Nicoll, and K.S.W. Campbell, The Late Triassic molluscs, conodonts and brachiopods of the Kuta Formation, Papua New Guinea, B.M.R. J. Aust. Geol. Geophys., 1, 219-230, 1976.
Smith, A.G., J.C. Briden, and G.E. Drewry, Phanerozoic world maps, Palaeontol. Assoc. London, Spec. Paps Palaeontol., 12, 1-42, 1973.
Soffel, H.C., and H.G. Forster, Apparent Polar Wander Path of Central Iran and its geotectonic interpretation, J. Geomag. geoelectr., 32, suppl. 3, 117-135, 1980.
Sporli, K.B., and M.R. Gregory, Significance of Tethyan fusulinid limestones of New Zealand, in Gondwana Five, edited by M.M. Cresswell and P. Vella, pp. 223-229, A.A. Balkema, Rotterdam, The Netherlands, 1981.
Stauffer, P.H., Unravelling the mosaic of Paleozoic crustal blocks in Southeast Asia, Geol. Rund., 72, 1061-1080, 1983.
Stevens, C.H., Corals from a dismembered Late Paleozoic paleo-Pacific plateau, Geology, 11, 603-606, 1983.
Sun, T., Late Carboniferous - Early Permian strata and fauna in North Ali District, Xizang (Tibet), China, Exploration of Nature, 3, 71-83, 1984.
Termier, G., H. Termier, A.F. de Lapparent, and P. Marin, Monographie du Permo-Carbonifere de Wardak (Afghanistan Central), Doc. Lab. Geol. Fac. Sci. Lyon, Hors Serie, 2, 1-167, 1974.
Thomas, G.A., The Permian brachiopod faunas of Western Australia, in Gondwana Stratigraphy, pp. 217-234, UNESCO, Paris, France, 1969.
Veevers, J.J. (editor), Phanerozoic Earth History of Australia, 418 pp, Clarendon Press, Oxford, United Kingdom, 1984.
Wang, N.-W., Le paleocontinent qingzangindien et sa saturation avec le paleocontinent

cathaysian, in Mission Franco-Chinoise au Tibet 1980, edited by J.L. Mercier and Li Guangcen, pp. 33-54, Editions du Centre National de la Recherche Scientifique, Paris, France, 1984.

Wang, Y.-G., and D.-L. Sun, The Triassic and Jurassic paleogeography and evolution of the Qinghai-Xizang (Tibet) Plateau, Can. J. Earth Sci., 22, 195-204, 1985.

Waterhouse, J.B., The classification and description of Permian Spiriferida (Brachiopoda) from New Zealand, Palaeontographica Abt., A, 129, 1-94, 1968.

Waterhouse, J.B., A new Permian fauna from New Caledonia, and its relationships to Gondwana and the Tethys, in Gondwana Stratigraphy, pp. 249-272, UNESCO, Paris, France, 1969.

Waterhouse, J.B., World correlations for Permian marine faunas, Pap. Dept. Geol. Univ. Qd, 7, 2, 1-232, 1976.

Waterhouse, J.B., An early Permian cool-water fauna from pebbly mudstones in south Thailand, Geol. Mag., 119, 337-354, 1982a.

Waterhouse, J.B., New Zealand Permian brachiopod systematics, zonation and paleoecology, Paleont. Bull. N.Z. Geol. Surv., 48, 1-158, 1982b.

Waterhouse, J.B., How wide was the Tethyan "Ocean", Contrib. Himal. Geol., 2, 246-249, 1983.

Waterhouse, J.B., and V.J. Gupta, Permian zones and correlations of the Himalayas, Bull. Ind. Geol. Assoc., 10, 2, 1-19, 1977.

Wensink, H., Tectonic inferences of paleomagnetic data from some Mesozoic formations in central Iran, J. Geophys., 51, 12-23, 1982.

Wopfner, H., Development of Permian intracratonic basins in Australia, in Gondwana Five, edited by M.M. Cresswell and P. Vella, pp. 185-190, A.A. Balkema, Rotterdam, The Netherlands, 1981.

Zhan, L.-P., and L. Lee, The distribution of Permian brachiopods in China, 15 pp, Science Press, Peking, People's Republic of China, 1977.

Zhan, L.-P., and L. Lee, Some new understanding concerning the study of the Permian system in China, Bull. Chinese Acad. Geol. Sci., 1984, 9, 169-183, 1984.

Zhan, Z.-W., and J.-W. Teng, Preuves paleomagnetiques de la derive vers le Nord de fragments de la plaque indienne apres la separation du continent de Gondwana et de la collision avec la plaque eurasiatique, in Mission Franco-Chinoise au Tibet 1980, edited by J.L. Mercier and Li Guangcen, pp. 15-20, Editions du Centre National de la Recherche Scientifique, Paris, France, 1984.

OROCLINAL BENDING, FRAGMENTATION AND DEFORMATION OF TERRANES IN THE NEW ENGLAND OROGEN, EASTERN AUSTRALIA

R.J. Korsch[1] and H.J. Harrington[2]

Division of Continental Geology[1] and Special Projects & Geoscience Services Branch[2]
Bureau of Mineral Resources, GPO Box 378, Canberra, ACT 2601, Australia

Abstract. The New England Orogen in eastern Australia consists essentially of two pre-Permian terranes which form a subduction-related forearc basin - accretionary wedge couple. The orogen has been disrupted by major oroclinal bending; this is a mechanism of terrane dispersion in that the Coffs Harbour Block (orocline) was transported from the north by bending, and docked with the southern part of the orogen. One effect was to increase the overall width of the orogen. A second effect of the oroclinal bending was to cause metamorphism, intense mesoscopic crenulation, and structural shredding of rocks in the Nambucca Block. A third effect was the fragmentation of rocks on the concave side of the Manning orocline to produce several crustal blocks with complicated geological histories.

Introduction

The recognition that tectonostratigraphic terranes in western North America can be exotic, allochthonous, displaced, accreted, or suspect [e.g. Jones et al., 1977; Coney et al., 1980; Alvarez et al., 1980] and the development of the concept of terrane analysis [Jones et al., 1983] have led to the suggestion that the Paleozoic Australian Tasman Orogen [Harrington, 1974] consists of several terranes [e.g. Howell, et al., 1983; Scheibner, 1984; Schermer et al., 1984]. Workers have begun to apply terrane analysis concepts to the New England Orogen section (Fig. 1) of the Tasman Orogen [e.g. Harrington, 1983; Cawood and Leitch, 1985; Scheibner, 1984, 1985; and several papers in this volume].

The New England Orogen is an ideal region in which to examine the relationship between tectonostratigraphic terrane development and the deformational features in the rocks. The latter range from intense penetrative deformation to macroscopic folding over very large regions, and can be related to events such as terrane docking or oroclinal bending. Thus, this paper discusses the recognition of terranes in the New England Orogen (Fig. 1), the development of major oroclines, and some of the effects that resulted from the formation of the oroclines. Only events in the Permian are considered here. This paper relies heavily on published literature, including reviews of the geology of New England by Leitch [1974], Korsch [1977] and Korsch and Harrington [1981].

Pre-Permian Paleogeography of the New England Orogen

During the Devonian and Carboniferous the New England Orogen consisted essentially of three distinct parallel belts, which were, from west to east, (1) a volcanic arc, most of which is now not exposed, (2) the Tamworth Belt - a thick sequence of terrestrial to marine volcaniclastic sediments derived from the arc, and (3) the Tablelands Complex - predominantly deepwater marine turbidites, also derived from the volcanic arc [e.g. Leitch, 1974; Day et al., 1978]. In plate tectonic terms these belts have been interpreted as volcanic arc, forearc basin, and accretionary wedge (trench and ocean basin deposits), respectively, [e.g. Solomon and Griffiths, 1972; Scheibner, 1973; Leitch, 1975a; Crook, 1980; Cawood, 1982; Fergusson, 1984a; R.J. Korsch, University of New England, unpublished thesis, 1975].

The accretionary wedge was formed by the progressive addition of at least three tectonostratigraphic units (Figs 1 and 2). As defined by Korsch [1977], these are the Woolomin association (? Silurian to Early Devonian), the Sandon association (Early Devonian to Early Carboniferous) and the Coffs Harbour association (? Early to Late Carboniferous).

The Terrane Concept in the New England Orogen

The New England Orogen has been described by numerous workers in terms of structural blocks, most of which are fault bounded [e.g. Voisey, 1959; Pogson and Hitchins, 1973; Scheibner, 1973; Rod, 1974; Leitch, 1974; Korsch, 1977; Crook,

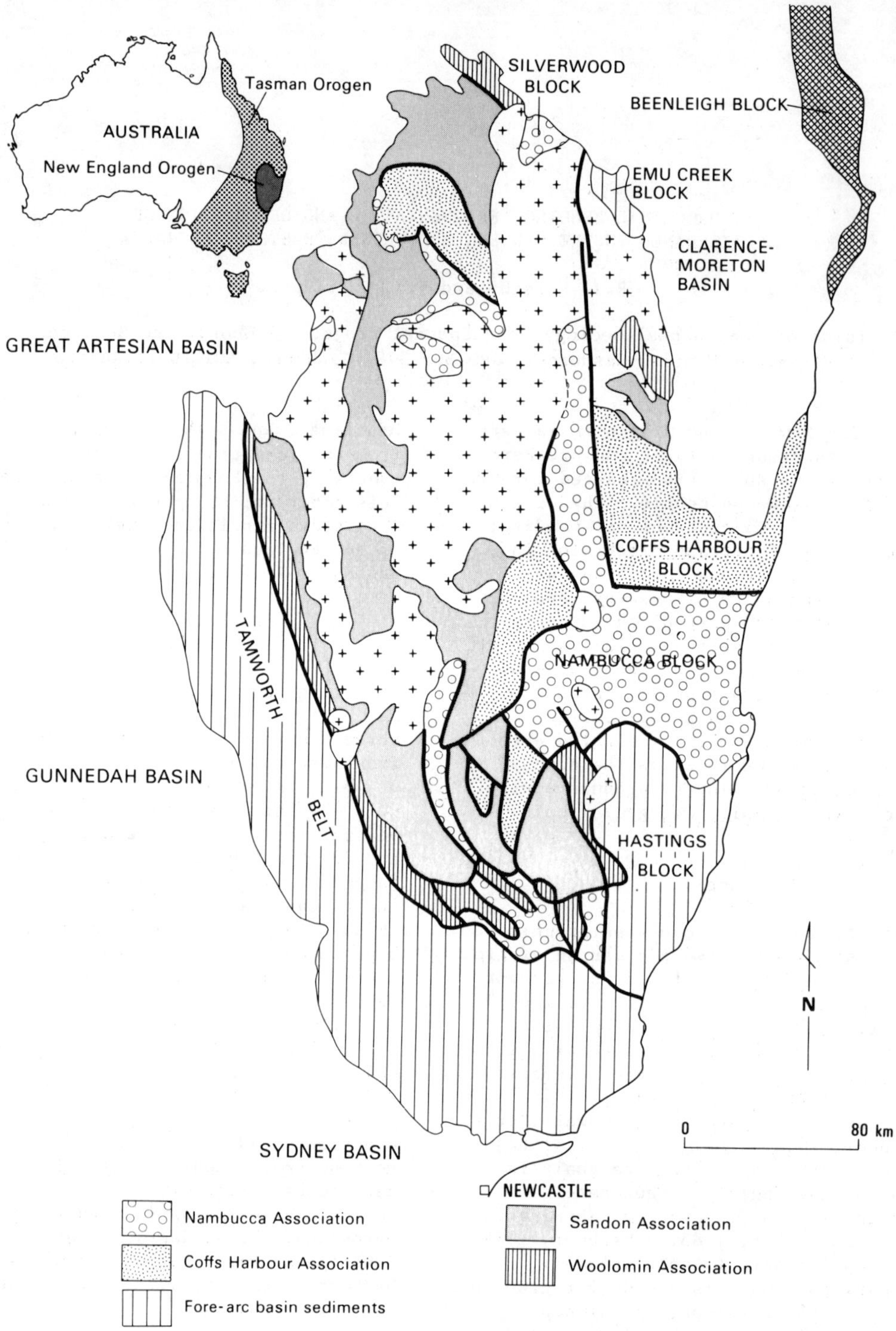

Fig. 1. Simplified geological map showing the distribution of lithological associations in the New England Orogen [after Korsch, 1977]. Note that in the northern central part of the Orogen we correlate the informal units of Flood and Fergusson [1982] as follows : Ct_1 = Woolomin association; Ct_2 and Ct_3 = Sandon association; and Ct_4 = Coffs Harbour association. Late Paleozoic and Mesozoic granitoids and felsic volcanic rocks have been combined (crosses). Tertiary and Quaternary rocks have been omitted for clarity.

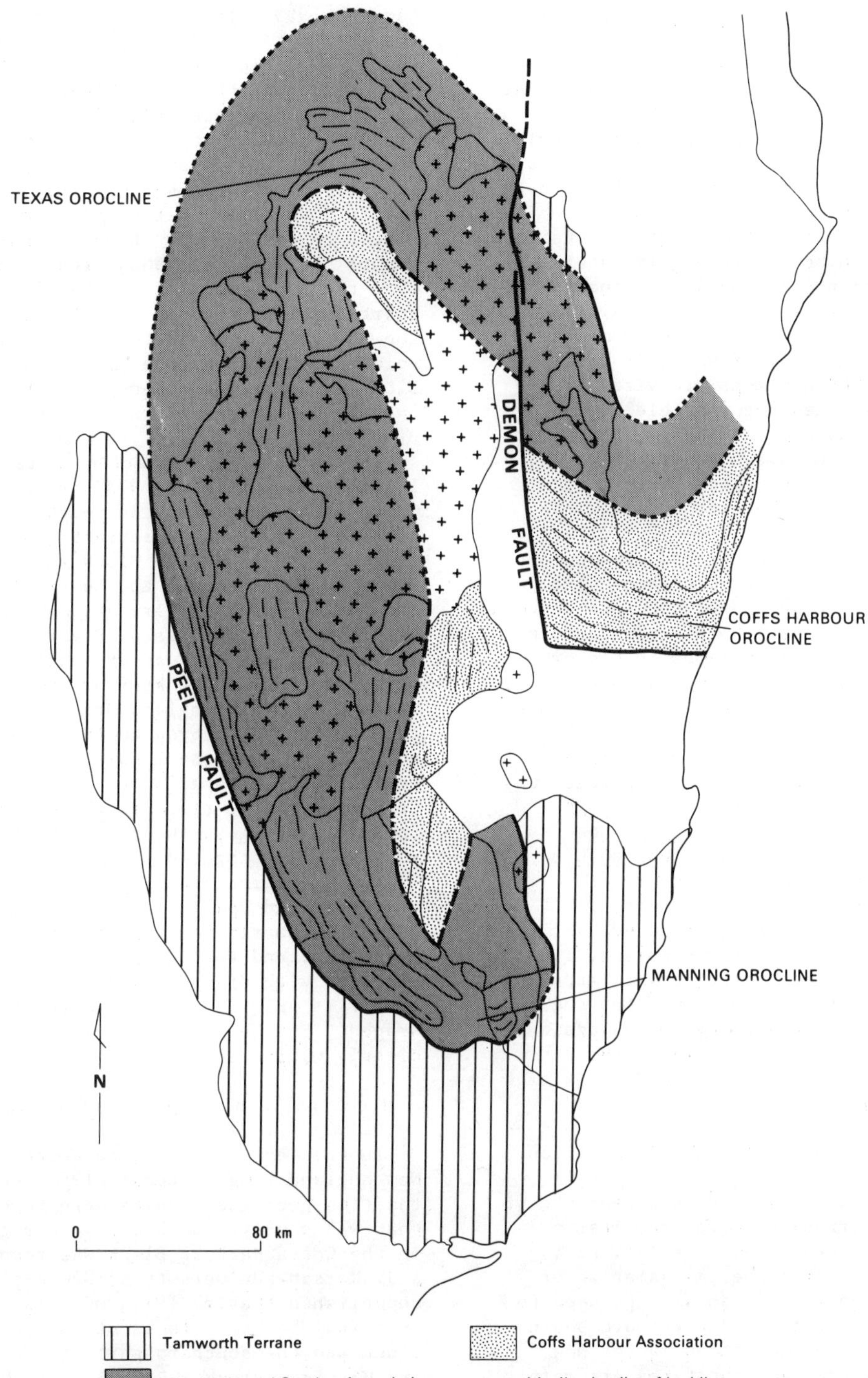

Fig. 2. Present day distribution of the Texas, Coffs Harbour and Manning oroclines in the New England Orogen, including interpretation beneath younger cover. Heavy solid line separates the Tamworth terrane from the accretionary wedge sequences, and heavy dashed line separates the Sandon association from the Coffs Harbour Association. In places, particularly in the south, the stipple pattern for the Woolomin and Sandon associations also covers areas of presently exposed Permian sequences.

1980; Cawood, 1984]. There is, however, only limited agreement as to the definition and boundaries of the various blocks. If the concepts of terrane analysis are applied on the basis of the rigorous procedures defined by Jones et al. [1983], most of the blocks should be accorded "terrane" status, and several would consist of two or more terranes. In fact, over 25 terranes could be postulated. This is because much of the Tablelands Complex consists of turbidites which became intensely deformed during incorporation into the accretionary wedge and during post-accretion tectonic processes, producing a mudrock-matrix melange in places [eg. Fergusson, 1984b]. As well, stratigraphic marker horizons are virtually absent, and it is often not possible to correlate between adjacent blocks. The blocks thus appear to have had different histories, even though the rocks were deposited initially in the same general environment.

Cawood [1984], Scheibner [1984], Cawood and Leitch [1985] and Leitch and Scheibner [1985] have divided the New England Orogen into between three and seven terranes. Rather than attempt to describe all possible terranes in the orogen, we divide it into two units: the Tamworth terrane, which is a stratigraphic terrane in the terminology of Jones et al. [1983], and the pre-Permian accretionary wedge, which is a disrupted composite terrane. This approach was also adopted by Fergusson et al. [1985] who referred to the accretionary wedge as the Gwydir terrane, a name we prefer not to use because we used the term Gwydir to refer to an early Paleozoic to Carboniferous tectonic period (TP1) in the orogen [Korsch and Harrington, 1981; Harrington and Korsch, 1985].

In the southern part of the coeval Yarrol Orogen (located to the north of the New England Orogen and separated from it by the Mesozoic Clarence-Moreton Basin) we recognise three accreted terranes (the Curtis Island, Beenleigh and Gympie terranes).

Provenance Linkage and Terrane Recognition

Provenance linkage, the sudden appearance of detritus in one terrane derived from its neighbour, was defined by Jones et al. [1983] as one method for dating the amalgamation or accretion of terranes. It can also be used to document a common source to the forearc basin and accretionary wedge sequences in the New England Orogen, where the sandstones are mainly quartz-poor lithic types. Using petrography of sandstones, Korsch [1984] showed that the Tamworth Belt had a distinctive petrographic evolution, and also that most of the associations in the Tablelands Complex correlate with those in the Tamworth Belt [Korsch, 1984, figs 7, 12]. This provenance linkage was also argued by Cawood [1983] for a small area straddling the Peel Fault to the south of Tamworth.

One exception to the quartz-poor lithic composition of the New England sandstones is the Beenleigh Block (Fig. 3), where the sandstones are distinctly more quartzose compared to other associations of the accretionary wedge [Korsch, 1977, 1984] and have a much higher K-feldspar content. E.M. Lohe [University of Queensland, unpublished thesis, 1980] clearly demonstrated that the sandstones of the Beenleigh Block are quartzo-feldspathic in composition and he also showed that the oolitic-bearing felsic volcanolithic sandstones of the South D'Aguilar Block (and the Sandon Association of the New England Orogen) were absent from the Beenleigh Block.

The Shoalwater Formation [Kirkegaard et al., 1970; Fleming et al., 1975] of the Curtis Island Group about 500 km to the north (Fig. 3) contains quartzose sandstones distinctly different from those of the Beenleigh Block and the New England Orogen. In the past, the Curtis Island Group and Beenleigh Block were grouped under the term Wandilla Slope and Basin [Fleming et al., 1975]. The slope and basin was extended by Day et al. [1978] to include rocks from the Coffs Harbour Block of the New England Orogen. Because of their different petrographic characteristics, we conclude that the Beenleigh Block and at least the Shoalwater part of the Curtis Island Group are exotic terranes which formed elsewhere (possibly along strike to the north) and were subsequently accreted to the New England Orogen.

The Beenleigh Block and Curtis Island Group are separated spatially by the Gympie terrane and it is considered that the two units are unrelated, and that they developed as separate terranes with discrete and separate geological histories prior to their accretion to the orogen. Further work on these terranes is clearly warranted.

Oroclinal Bending in the New England Orogen

In the Texas area in the northern part of the New England Orogen, Lucas [1960] recognised that the geological units were folded in an arc that was concave southwards. A regional syncline in the Coffs Harbour Block was recognised by R.J. Korsch [University of New England, unpublished thesis, 1975] who also postulated oroclinal bends in rocks of the Coffs Harbour Block and the southern part of the Tamworth Belt and Hastings Block. Rotation of the Hastings Block was also proposed in the model of Scheibner [1976]. This led to the Texas arc and Coffs Harbour regional syncline being considered parts of the same large z-shaped megafold by Flood and Fergusson [1982] and Fergusson and Flood [1984] which they referred to as the Texas - Coffs Harbour mega - fold

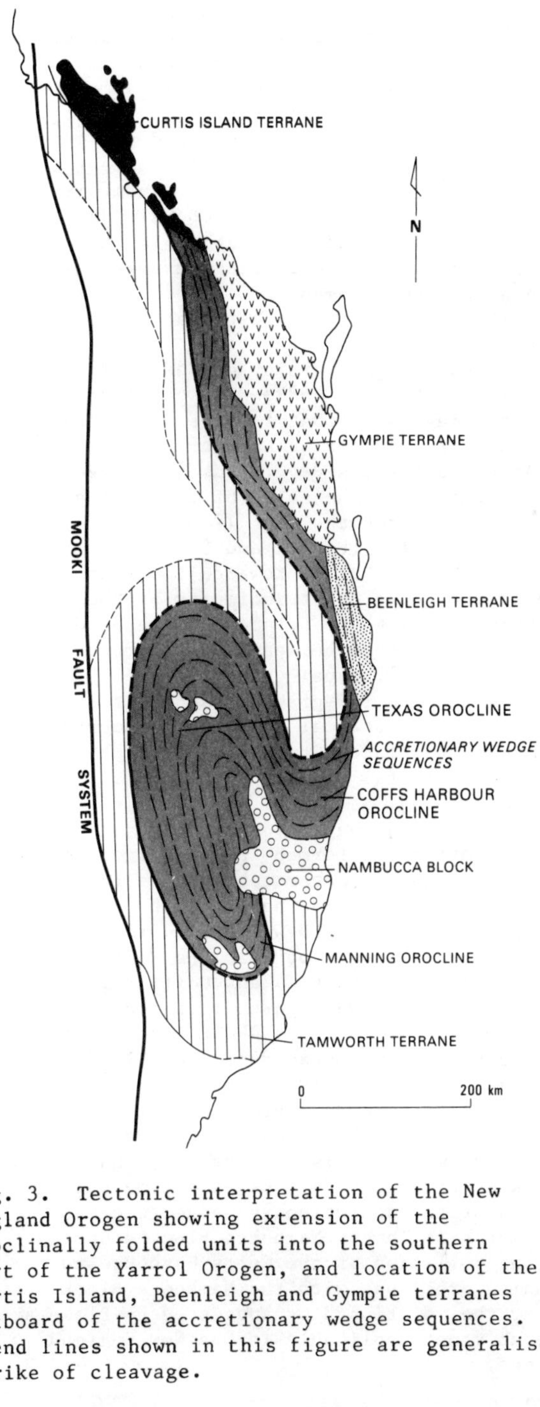

Fig. 3. Tectonic interpretation of the New England Orogen showing extension of the oroclinally folded units into the southern part of the Yarrol Orogen, and location of the Curtis Island, Beenleigh and Gympie terranes outboard of the accretionary wedge sequences. Trend lines shown in this figure are generalised strike of cleavage.

[see also Murray et al., 1986]. As almost all the tectonostratigraphic units in the New England Orogen are involved in the megafold, it can be regarded as a double orocline [Carey, 1955], and we extend the concept here to include oroclinal bending of the Manning area and the southern part of the Tamworth Belt, referring to the bends as the Texas, Coffs Harbour and Manning oroclines (Fig. 2). An interesting point is that the Manning orocline has the opposite sense of vergence to the Texas and Coffs Harbour oroclines. Although the mechanism for this is unknown, similar opposed senses of vergence have been described for deformed ribbons in the Alpine-Mediterranean orogen by Van der Linden [1985].

Structural trends of bedding and cleavage follow the oroclinal bends. The shapes of the oroclines have been modified by subsequent tectonism, and the width of the Woolomin and Sandon associations on the limb between the Manning and Texas oroclines has been increased by extension associated with emplacement of plutons of the New England Batholith. However, the total width of the orogen has been increased by the oroclinal bending.

The recognition of oroclines in the New England Orogen has illuminated some geological problems in the eastern part of the orogen, such as the significance of the Baryulgil Serpentinite, the Emu Creek Block, and Silurian fossils described by Fletcher [in Whiting, 1954]. The Baryulgil Serpentinite has been transported southsoutheastwards during the oroclinal bending and can now be regarded as a former northern equivalent of the serpentinites along the Peel Fault [see distribution in fig. 1 of Korsch, 1977]. This was also proposed by Runnegar [1974], although he inferred that the serpentinites represented the sole of a single shallow spoon-shaped thrust sheet under almost the entire New England Orogen.

We consider that the Emu Creek Block is equivalent to the Tamworth Belt and that it was once a part of the continuous quasilinear forearc basin that included the Tamworth Belt and the Yarrol Belt of the Yarrol Orogen. A similar correlation was proposed by Day et al. [1978] and Murray and Whitaker [1982], although they invoked strike-slip faulting rather than oroclinal bending to transport the Emu Creek Block (part of their Demon Block) south.

The timing of the oroclinal bending in the orogen is constrained by several geological events. Bedding and D1 cleavage in the Coffs Harbour Block are deformed by the bending [Korsch, 1981]. Metamorphism associated with this cleavage development has a Rb-Sr whole rock isochron of 318 ± 8 Ma [Graham and Korsch, 1985] and hence the bending is late Carboniferous or younger. Murray et al. [1986] consider that the oroclinal bending occurred at about 300 Ma (late in the late Carboniferous). However, bedding and faulted margins of many of the Early Permian sedimentary slivers in the Texas-Inverell region described by Olgers et al. [1974] have been deformed parallel to the curve of the orocline (e.g. Glenlyon, Alum Rock, Terrica sequences). Complex deformation occurred in the core of the orocline at Silver Spur [see Olgers et al., 1974]. We infer that this deformation occurred at the same time as the formation of the oroclines. It is even feasible that the formation of these small basins, their filling and deformation all occurred during development of the oroclines.

The oroclines formed prior to intrusion of the Late Permian plutons of the New England Supersuite, because these plutons truncate the oroclinal structures. The supersuite has isotopic ages ranging from 265±13 Ma [Hensel et al., 1985] to 242 Ma [Evernden and Richards, 1962; Cooper et al., 1963]. Hence, we consider that the oroclines developed at about 280-265 Ma in the late Early Permian [time scale of Harland et al., 1982].

The normal concept of terrane accretion [e.g. Jones et al., 1983] revolves around the transportation of terranes via sea-floor spreading or strike-slip faulting prior to the actual accretion event at a continental margin. We present here a third method of terrane transportation, namely oroclinal bending of two originally quasilinear terranes. This led to the arrival of the Coffs Harbour block (orocline) at its present site from one located somewhere to the north.

We infer that the Beenleigh terrane arrived at its present site either during or following development of the oroclines although arrival prior to oroclinal bending cannot be conclusively ruled out. It is possible that the Beenleigh terrane excised part of the orocline in the northeastern part of the orogen (as inferred in Fig. 3). Recent deep crustal seismic reflection profiling by the Bureau of Mineral Resources across the Beenleigh terrane has shown it to be thin-skinned, less than 5 km thick [Korsch et al., 1986]. The minimum time of arrival of the terrane is constrained by deposition of the Ipswich Basin, which is mostly covered by the Clarence-Moreton Basin (Fig. 1). The Ipswich Basin is an overlap assemblage, depositionally overlying and completely covering the suture of the juxtaposed Beenleigh terrane and oroclines of New England. The oldest formation in the Ipswich Basin is Ladinian-Carnian (about 235 Ma, using the time scale of Harland et al. [1982]).

Deformation in the Nambucca Block Resulting From Oroclinal Bending

The Nambucca Block (Fig. 1) consists of an Early Permian succession [Runnegar, 1970] of mainly fine-grained clastic sediments that have been intensely crenulated and tectonically shredded [Leitch, 1978]. Accompanying metamorphism reached pumpellyite-actinolite and greenschist facies belonging to an intermediate pressure facies series [Leitch, 1975b]. K-Ar ages of 268-249 Ma for slates from the block were obtained by Leitch and McDougall [1979], who considered that the ages dated the metamorphism which accompanied deformation of the sedimentary sequence. Following recent interpretations of K-Ar ages in metamorphic belts [e.g. Adams et al., 1985], we interpret the K-Ar ages as cooling ages, and the maximum age [268 Ma] thus provides a minimum age for the regional metamorphism which therefore occurred between 286 Ma and 268 Ma. Hence, the metamorphism and deformation in the Nambucca Block were occurring at the same time as the development of the oroclines (280-265 Ma), and we suggest that the two events are intimately related.
C.L. Fergusson [University of New England, unpublished thesis, 1982] also concluded that the megafold (that is, the oroclines) must have formed during the Early Permian.

As the oroclines developed, relative movement towards the north of the Hastings Block which forms the east limb of the Manning orocline (Figs 1 and 2), and of the Coffs Harbour orocline towards the south, caused the young, soft sediments of the Nambucca Block to be trapped by the bending and to be squeezed between the two more rigid blocks (Fig. 3), causing the intense deformation and low-grade metamorphism. This led to the symmetrical pattern of isograds and deformational intensity, both of which increase towards a central high in the Nambucca Block [Leitch, 1978].

Blake [1985] suggested that high P/T metamorphic belts formed in subduction zones, but that high T/P metamorphic belts formed during crustal thickening caused by accretion of terranes. In the New England Orogen, Korsch [1978] suggested that the Nambucca Block (intermediate P/T) and Coffs Harbour Block (high T/P) were parts of paired metamorphic belts, but a Rb-Sr isochron for the Coffs Harbour metamorphism of 318±8 Ma [Graham and Korsch, 1985] indicates that this metamorphism occurred before deposition of the sediments of the Nambucca Block. We suggest here that metamorphism in the Coffs Harbour Block is related to processes acting during growth of the accretionary wedge, whereas metamorphism in the Nambucca Block occurred during collision of the Coffs Harbour and Manning oroclines with the Nambucca Block. Thus, the relationships suggested by Blake are probably not applicable in the New England Orogen.

Fragmentation of the Southern Part of the New England Orogen

One problem which hindered recognition of the oroclines is the fragmented character of the southern part of the orogen (Fig. 4). Nevertheless, recognition of the orocline (Fig. 2) might account for the fragmentation of the southern part into several fault-

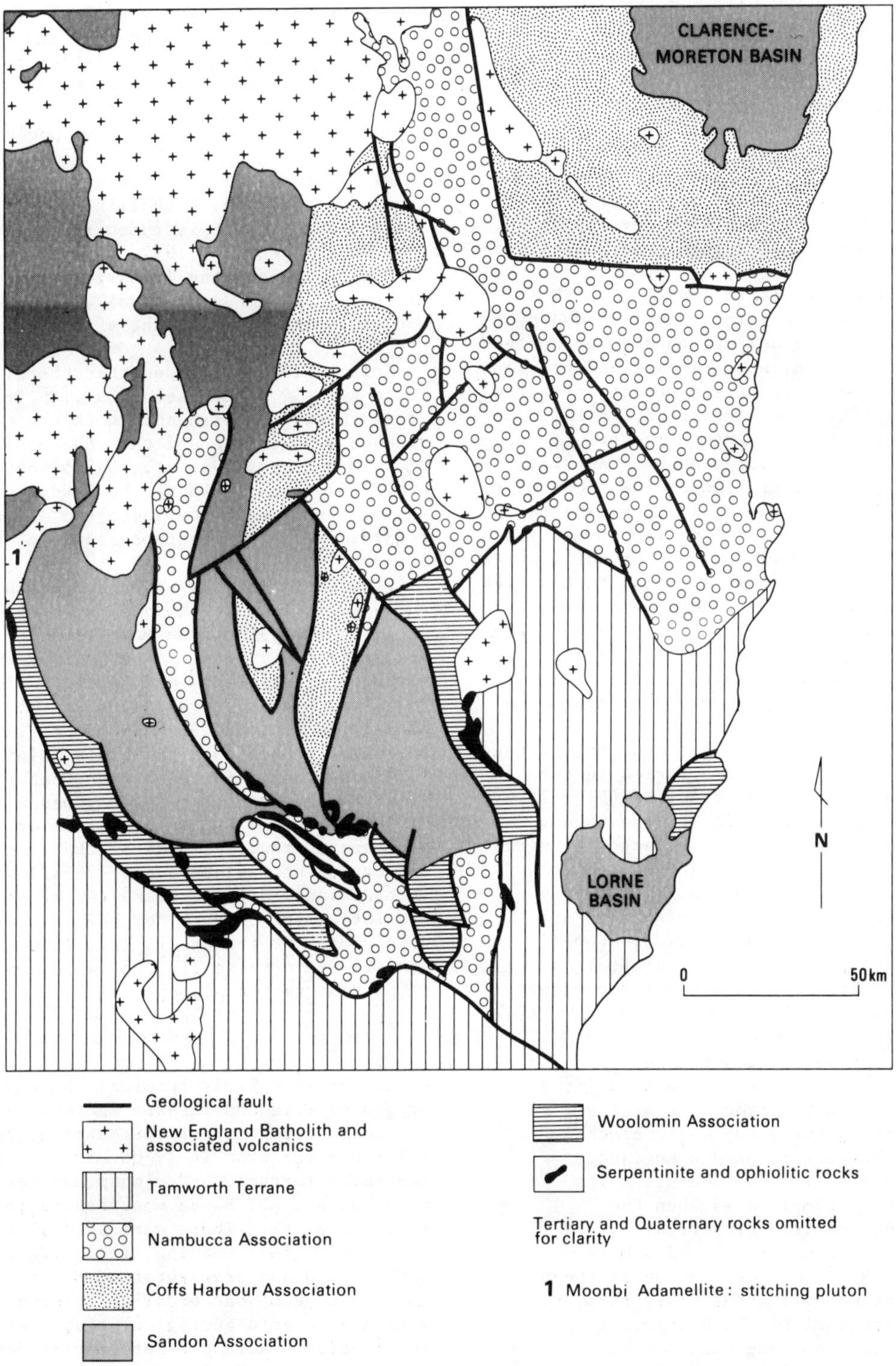

Fig. 4. Geological map of the southern part of the New England Orogen showing many crustal blocks produced by fragmentation during oroclinal bending. Map after Pogson and Hitchins [1973] and R.J. Korsch [University of New England, unpublished thesis, 1975].

bounded blocks or schollen [Dewey and Sengor, 1979]. Some of the rocks forming these schollen are parts of the accretionary wedge and some are parts of the Early Permian Nambucca association [Korsch, 1977]. Thus the southern part can be regarded as a terrane that has been fragmented, but only partially dispersed, and this has had the effect of complicating the structural pattern.

During the Late Carboniferous to Early Permian the tectonic regime changed, with development of the Mooki transform fault system [Harrington and Korsch, 1985], the accompanying formation of the Sydney-Bowen transform basin system, and the cessation of subduction. This was followed by the formation of the oroclines in the New England Orogen. The change in tectonism was accompanied by a change in the style of sedimentation in the Tablelands Complex. The widespread, thick, turbidite sequences that were characteristic of the Carboniferous were succeeded by scattered early Permian fault-controlled basins [Scheibner, 1976; Korsch, 1982] in which great thicknesses (up to several kilometres) of marine diamictites were deposited. The basins vary from pull-aparts associated with major strike-slip faults to rifts and small half-grabens associated with normal faults and strike-slip faults [see Korsch, 1982]. The subduction-related volcanic arc was replaced by rift-related bimodal basalt-rhyolite suites [Harrington, 1982; Korsch, 1982; Harrington and Korsch, 1985].

It is interesting to note that the most intense fracturing and fragmentation occurred in the concave side of the Manning orocline (Figs 2, 4), which is the tightest orocline of the three. After fragmentation, although the schollen were to some extent confined, they interacted in a complex fashion, with extensional, compressional, and strike-slip processes all operating [see Korsch, 1982]. Jones et al. [1982], Crowell [1985] and others have shown how crustal schollen can develop and rotate between major strike-slip faults. Rotation of a scholle has implications for the geological processes operating between adjacent schollen. Rapidly changing processes can be associated with the same boundary segment or with a scholle rotating between two major blocks [R.J. Korsch, University of New England, unpublished thesis, 1975]. The situation becomes extremely complicated when the fragmented zone consists of several schollen such as those in the model of Jones et al. [1982] or those in the southern New England Orogen during the Early Permian. To the north and west of the Hastings Block (Fig. 1) the schollen have complicated histories. Some schollen consist of deformed Nambucca association (Fig. 4) suggesting that rifting was followed by convergence [see Harrington and Korsch, 1985, fig. 5]. Nevertheless, it is likely that fragmentation into several blocks and oroclinal bending were both occurring at the same time and the spatial and temporal relationships cannot be regarded as fortuitous.

Stitching Plutons in the New England Orogen

At least six distinctive suites form the New England (granitoid) Batholith [Hensel et al., 1985] and they were intruded in four distinct time slices [Korsch, 1977] between the Late Carboniferous and the Triassic. In a few cases pluton stiching [Jones et al., 1983] can be used to help define timing of movements in the orogen.

Korsch [1977] recorded that the Hillgrove Suite (312±10 Ma; Hensel et al., [1985]) had a close spatial relationship with the Coffs Harbour association and that the Bundarra Suite (287±10 Ma; Hensel et al., [1985]) occurs entirely within rocks of the Sandon association, and hence these two suites can be regarded as being "terrane specific" [Irwin, 1985].

Plutons of the New England Supersuite (in which we include the Nundle Suite of Hensel et al., [1985]) intrude most lithological associations, including the Early Permian Nambucca association, and have a maximum age of 265±13 Ma [Hensel et al., 1985]. The Moonbi Adamellite cuts the Peel Fault and is a stitching pluton that provides a minimum age for cessation of fault movement between the Tamworth Belt and accretionary wedge sequences. Elsewhere in New England, plutons of the supersuite truncate rocks of the oroclines and hence the maximum age of 265±13 Ma (Artinskian, Early Permian) can be used to provide a minimum date for the cessation of development of the oroclines. Thus in a general sense these plutons can be regarded as post-accretion plutons.

Conclusions

In the late Early Permian, the New England Orogen experienced intense tectonic activity, including formation of the Manning, Texas and Coffs Harbour oroclines, docking of the Beenleigh terrane, and widespread mesoscopic deformation, which was most severe in the Nambucca Block. There were several effects of the oroclinal bending, including intense deformation and fragmentation of rocks in the concave side of one orocline to produce schollen with complicated geological histories. The penetrative mesoscopic deformation was very different from the Triassic macroscopic deformation that accompanied the arrival of the Gympie terrane. Although plutons of the New England Batholith occupy much of the New England

Orogen, only a few were stitching plutons that can be used to date the minimum age of accretionary events.

Analysis of terranes in the New England Orogen has suggested a further mechanism of terrane dispersion, namely by oroclinal bending. During bending, the Coffs Harbour block (orocline) was transported from the north, docking with the remainder of the orogen, and roughly doubling its overall width.

Acknowledgments. We wish to thank P.J. Cook, I.M. Hodgson, C. Klootwijk and J.F. Lindsay for their careful reviews of the manuscript, Vera Ashby for drafting the figures, and Rosemary Nicholls for typing the camera-ready copy. Published with permission of the Director, Bureau of Mineral Resources.

References

Adams, C.J., Bishop, D.G. and Gabites, J.E., Potassium-argon age studies of a low-grade, progressively metamorphosed greywacke sequence, Dansey Pass, South Island, New Zealand, J. Geol. Soc. London, 142, 339-349, 1985.

Alvarez, W., Kent, D.V., Premoli-Silva, I., Schweickert, R.A., and Larson, A., Franciscan Complex limestone deposited at 17° south paleolatitude, Geol. Soc. Am. Bull., 91 (Pt.1), 476-484, 1980.

Blake, M.C., Jr., Significance of metamorphic tectonites in terrane accretion, Geol. Soc. Australia Abstracts, 14, 23-26, 1985.

Carey, S.W., The orocline concept in geotectonics, Pap. Proc. Royal Soc. Tasmania, 89, 255-287, 1955.

Cawood, P.A., Structural relations in the subduction complex of the Paleozoic New England Fold Belt, Eastern Australia, J. Geol., 90, 381-392, 1982.

Cawood, P.A., Modal composition and detrital clinopyroxene geochemistry of lithic sandstones from the New England Fold Belt (east Australia): a Paleozoic forearc terrane, Geol. Soc. Am. Bull., 94, 1199-1214, 1983.

Cawood, P.A., Accretionary tectonics and terrane dispersal within the New England Fold Belt, Eastern Australia, Stanford Univ. Pub. Geol. Sci., 18, 50-52, 1984.

Cawood, P.A. and Leitch, E.C., Accretion and dispersal tectonics on the southern New England Fold Belt, eastern Australia, in D.G. Howell, ed., Tectonostratigraphic terranes of the circum-Pacific region, Circum-Pacific Council Ener. Min. Res. Earth Sci. Series, 1, 481-492, 1985.

Coney, P.J., Jones, D.L. and Monger, J.W.H., Cordilleran suspect terranes, Nature, 288, 329-333, 1980.

Cooper, J.A., Richards, J.R. and Webb, A.W., Some potassium-argon ages in New England, New South Wales, J. Geol. Soc. Australia, 10, 313-316, 1963.

Crook, K.A.W., Fore-arc evolution in the Tasman Geosyncline: the origin of the southeast Australian continental crust, J. Geol. Soc. Australia, 27, 215-232, 1980.

Crowell, J.C., The recognition of transform terrane dispersal within mobile belts, in D.G. Howell, ed., Tectonostratigraphic terranes of the circum-Pacific region, Circum-Pacific Council Ener. Min. Res. Earth Sci. Series, 1, 51-62, 1985.

Day, R.W., Murray, C.G. and Whitaker, W.G., The eastern part of the Tasman orogenic zone, Tectonophysics, 48, 327-364, 1978.

Dewey, J.F. and Sengor, A.M.C., Aegean and surrounding regions: complex multiplate and continuum tectonics in a convergent zone, Geol. Soc. Am. Bull., 90 Part I, 84-92, 1979.

Evernden, J.F. and Richards, J.R., Potassium-argon ages in eastern Australia, J. Geol. Soc. Australia, 9, 1-49, 1962.

Fergusson, C.L., Tectono-stratigraphy of a Paleozoic subduction complex in the central Coffs Harbour Block of north-eastern New South Wales, Australian J. Earth Sci., 31, 217-236, 1984a.

Fergusson, C.L., The Gundahl Complex of the New England Fold Belt, eastern Australia: a tectonic melange formed in a Paleozoic subduction complex, J. Struct. Geol., 6, 257-271, 1984b.

Fergusson, C.L. and Flood, P.G., A Late Paleozoic subduction complex in the Border Rivers area of southeast Queensland, Proc. Royal Soc. Queensland, 95, 47-55, 1984.

Fergusson, C.L., Flood, P.G. and Cross, K.C., The Gwydir terrane; a Paleozoic subduction complex in the southern New England Fold Belt of eastern Australia. Geol. Soc. Australia Abstracts, 14, 66-70, 1985.

Fleming, P.J.G., Murray, C.G. and Whitaker, W.G., Late Paleozoic invertebrate fossils in the Wandilla Formation and the deposition of the Curtis Island Group. Queensland Govt. Mining J., 76, 416-422, 1975.

Flood, P.G. and Fergusson, C.L., Tectono - stratigraphic units and structure of the Texas - Coffs Harbour region, in P.G. Flood and B.N. Runnegar, eds., New England Geology, Univ. New England, Armidale, 71-78, 1982.

Graham, I.J. and Korsch, R.J., Rb-Sr geochronology of coarse-grained greywackes and argillites from the Coffs Harbour Block, eastern Australia, Isotope Geosci., 4, 45-54, 1985.

Hall, R.L., Late Ordovician coral faunas from northeastern New South Wales, J. Proc. Royal Soc. New South Wales, 108, 75-93, 1975.

Harland, W.B., Cox, A.V., Llewellyn, P.G., Pickton, C.A.G., Smith, A.G., and Walters, R.,

A geologic time scale, Cambridge Univ. Press, Cambridge, 131p, 1982.

Harrington, H.J., The Tasman Geosyncline in Australia, in Denmead, A.K., Tweedale, G.W. and Wilson, A.F., eds., The Tasman Geosyncline - a Symposium, Geol. Soc. Australia Queensland Div, Brisbane, 383-407, 1974.

Harrington, H.J., Tectonics and the Sydney Basin, Univ. Newcastle Adv. Study Sydney Basin, 16th Symp., 15-19, 1982.

Harrington, H.J., Correlation of the Permian and Triassic Gympie Terrane of Queensland with the Brook Street and Maitai terranes of New Zealand, in Permian Geology of Queensland, Geol. Soc. Australia Queensland Div., Brisbane, 431-436, 1983.

Harrington, H.J. and Korsch, R.J., Tectonic model for the Devonian to middle Permian of the New England Orogen, Australian J. Earth Sci., 32, 163-179, 1985.

Hensel, H.D., McMulloch, M.T. and Chappell, B.W., The New England Batholith: constraints on its derivation from Nd and Sr isotopic studies of granitoids and country rocks, Geochim, Cosmochim. Acta, 49, 369-384, 1985.

Howell, D.G., Jones, D.L. and Schermer, E.R., Tectonostratigraphic terranes of the frontier Circum-Pacific region, Am. Assoc. Petrol. Geol. Bull., 67, 485-486, 1983.

Irwin, W.P., Age and tectonics of plutonic belts in accreted terranes of the Klamath Mountains, California and Oregon, in D.G. Howell, ed., Tectonostratigraphic terranes of the circum-Pacific region, Circum-Pacific Council Ener. Min. Res. Earth Sci. Series, 1, 187-200, 1985.

Jones, D.L., Silberling, N.J. and Hillhouse, J.W., Wrangellia - a displaced terrane in north-western North America, Canadian Jour. Earth Sci., 14, 2565-2577, 1977.

Jones, D.L., Cox, A., Coney, P. and Beck, M., The Growth of North America, Sci. Am., 247, 70-84, 1982.

Jones, D.L., Howell, D.G., Coney, P.J. and Monger, H.W.H., Recognition, character and analysis of tectonostratigraphic terranes in western North America, J. Geol. Ed., 31, 295-303, 1983.

Kirkegaard, A.G., Shaw, R.D. and Murray, C.G., Geology of the Rockhampton and Port Clinton 1:250 000 sheet areas, Geol. Surv. Queensland Rep., 38, 155p, 1970.

Korsch, R.J., A framework for the Paleozoic geology of the southern part of the New England geosyncline, J. Geol. Soc. Australia, 23, 339-355, 1977.

Korsch, R.J., Regional-scale thermal metamorphism overprinting low-grade regional metamorphism, Coffs Harbour Block, northern New South Wales, J. Proc. Royal Soc. New South Wales, 111, 89-96, 1978.

Korsch, R.J., Deformational history of the Coffs Harbour Block, J. Proc. Royal Soc. New South Wales, 114, 17-22, 1981.

Korsch, R.J., Early Permian tectonic events in the New England Orogen, in P.G. Flood and B.N. Runnegar, eds., New England Geology, Univ. New England, Armidale, 35-42, 1982.

Korsch, R.J., Sandstone compositions from the New England orogen, eastern Australia: Implications for tectonic setting, J. Sedim. Petrol., 54, 192-211, 1984.

Korsch, R.J. and Harrington, H.J., Stratigraphic and structural synthesis of the New England Orogen, J. Geol. Soc. Australia, 28, 205-226, 1981.

Korsch, R.J., Lindsay, J.F., O'Brien, P.E., Sexton, M.J. and Wake-Dyster, K.D., Deep crustal seismic reflection profiling, New England Orogen, eastern Australia: Telescoping of the crust and a hidden, deep, layered sedimentary sequence, Geology, in press, 1986.

Leitch, E.C., The geological development of the southern part of the New England Fold Belt, J. Geol. Soc. Australia, 21, 133-156, 1974.

Leitch, E.C., Plate tectonic interpretation of the Paleozoic history of the New England Fold Belt, Geol. Soc. Am. Bull., 86, 141-144, 1975a.

Leitch, E.C., Zonation of low grade regional metamorphic rocks, Nambucca Slate belt, northeastern New South Wales, J. Geol. Soc. Australia, 22, 413-422, 1975b.

Leitch, E.C., Structural succession in a late Paleozoic slate belt and its tectonic significance, Tectonophysics, 47, 311-323, 1978.

Leitch, E.C. and McDougall, I., The age of orogenesis in the Nambucca Slate Belt: A K-Ar study of low-grade regional metamorphic rocks, J. Geol. Soc. Australia, 26, 111-119, 1979.

Leitch, E.C. and Scheibner, E., A Tasman Fold Belt perspective on terrane analysis, Geol. Soc. Australia Abstracts, 14, 136-141, 1985.

Lucas, K.G., The Texas Area. J. Geol. Soc. Australia, 7, 229-235, 1960.

Murray, C.G., Fergusson, C.L., Flood, P.G., Whitaker, W.G. and Korsch, R.J., Plate tectonic model for the Carboniferous evolution of the New England Fold Belt, Australian Jour. Earth Sci., in press, 1986.

Murray, C.G., and Whitaker, W.G., A review of the stratigraphy, structure and regional tectonic setting of the Brisbane Metamorphics, in P.G. Flood and B.N. Runnegar, eds., New England Geology, Univ. New England, Armidale, p. 79-94, 1982.

Olgers, F., Flood, P.G. and Robertson, A.D., Palaeozoic geology of the Warwick and Goondiwindi 1:250 000 sheet areas, Queensland and New South Wales, Bureau Mineral Res. Geol. Geophys. Report, 164, 109p., 1974.

Pogson, D.J. and Hitchins, B.L., New England 1:500 000 geological sheet, Geol. Surv. New South Wales, Sydney, 1973.

Rod, E., Structural interpretation of the New

England region, J. Proc. Royal Soc. New South Wales, 107, 90-99, 1974.
Runnegar, B.N., The Permian faunas of northern New South Wales and the connection between the Sydney and Bowen basins, J. Geol. Soc. Australia, 16, 697-710, 1970.
Runnegar, B.N., The geological framework of New England, Geol. Soc. Australia Queensland Div. Field Conf. New England Area, 9-19, 1974.
Scheibner, E., A plate tectonic model of the Paleozoic tectonic history of New South Wales, J. Geol. Soc. Australia, 20, 405-426, 1973.
Scheibner, E., Explanatory notes on the tectonic map of New South Wales, Geol. Surv. New South Wales, Sydney, 283p., 1976.
Scheibner, E., Suspect terranes in the Tasman Fold Belt System (eastern Australia), Stanford Univ. Pubs. Geol. Sci., 18, 170-174, 1984.
Scheibner, E., Suspect terranes in the Tasman Fold Belt system, eastern Australia, in D.G. Howell, ed., Tectonostratigraphic terranes of the circum-Pacific region, Circum-Pacific Council Ener. min. Res. Earth Sci. Series, 1, 493-514, 1985.
Schermer, E.R., Howell, D.G., and Jones, D.L., The origin of allochthonous terranes: Perspectives on the growth and shaping of continents, Ann. Rev. Earth Planet. Sci., 12, 107-131, 1984.
Solomon, M. and Griffiths, J.R., Tectonic evolution of the Tasman Orogenic Zone, eastern Australia, Nature, 237, 3-6, 1972.
Van der Linden, W.J.M., Looping the loop; the geotectonics of the Alpine-Mediterranean region, Geol. Mijnbouw, 64, 281-295, 1985.
Voisey, A.H., Tectonic evolution of northeastern New South Wales, Australia, J. Proc. Royal Soc. New South Wales, 92, 191-203, 1959.
Whiting, J.W., Limestone deposits, Parish Braylesford, County Gresham, New South Wales Mines Dept. Annual Rep. 1950, p.87, 1954.

EVIDENCE FOR SINISTRAL MOVEMENT ON THE PEEL FAULT SYSTEM IN SERPENTINITES, GLENROCK STATION, N.S.W.

Robin Offler

Department of Geology, University of Newcastle, New South Wales, 2308, Australia

Adrian Williams

Bond Corporation Holdings Ltd., International House, 26 St. Georges Terrace, Perth, 6000, Australia

Abstract. The serpentinites in the Glenrock Station area contain mesoscopic and microscopic structures resembling those observed in mylonites. These structures have been developed during movement on the Peel Fault System, a major suture extending for over 270 km in the Southern New England Fold Belt. The earliest structure to develop is a fine, penetrative schistosity (S_1); textural evidence suggests that S_1 is a plane of flattening and is parallel to the XY plane of the finite strain ellipsoid. Overprinting it at an angle of 45° or less, are a set of microshears (S_2) which show a sinistral sense of movement according to vergence relationships and kinematic indicators. Less commonly developed at this time was another set of microshears with a dextral displacement of the first foliation. Superposed on all early foliations is a second set of microshears (S_3) which folds and displaces in a left lateral sense, earlier foliations. A chrysotile fibre lineation was produced during both shearing events.

Microscopic features indicate that the serpentine minerals behaved in a plastic fashion during the formation of S_1 and in a brittle-plastic fashion while the microshears were being formed, suggesting that shearing took place at shallower depths in the crust.

The geometry of the foliations and the chrysotile lineations reveals that the movement on the Peel Fault System is dominantly strike-slip. Movement on this zone appears to have been short lived occurring between 262 and 270 Ma. This fact and the observation that shear strain has not been high, suggests that the strike-slip displacement was not substantial. Thus the serpentinites in the Glenrock Station area record evidence for a sinistral, dominantly strike-slip movement on the Peel Fault System.

Introduction

In the last decade, major advances have been made in the unravelling of the tectonic history of the continents bordering the Circum-Pacific region, particularly North America. Studies have shown that the present shape of these continents is the result of both accretion and dispersion of terranes of various sizes and age. These terranes are fault bounded and characterised by a geological history different from their neighbours [Howell et al, 1985]. Displacement along the faults may vary from a few kilometres to many thousands of kilometers.

In Eastern Australia, several terranes have been recognised in the New England Fold Belt (see inset, Figure 1), a north trending, Palaeozoic structure extending approximately 1600 km in N.S.W. and Queensland [Cawood and Leitch, 1985]. The southern portion of this belt is divided into the Tablelands Complex and the Tamworth Belt [Runnegar, 1974], and Zone A and B by Leitch [1974]. Distinctive rock sequences, structural styles and types of metamorphism occur in these zones. Scheibner [1985] recognises five suspect terranes in the Tablelands Complex; Cawood and Leitch [1985], however, recognise only four.

Dividing the Tablelands Complex and the Tamworth Belt is the Peel-Manning Fault System, a major suture extending for approximately 270 km (Figure 1). It is delineated by lenticular masses of serpentinite of varying size and shape, containing exotic blocks which have been carried from the deeper levels of the crust and mantle. Geophysical studies, relationships between the serpentinites and the associated sediments, and cleavage orientation within the serpentinites suggest that the Peel Fault System dips vertically or steeply to the east [Crook, 1963; Ramsay and Stanley, 1976; Leitch, 1980; Glen and Butt, 1981; Offler and Williams, 1985]. Little is known of the attitude of the Manning River Fault System.

This suture occurs close to a long-lived, Palaeozoic convergent plate margin and separates forearc basin from accretionary-subduction complex rocks. At the beginning of the Permian, the plate vector orientation changed resulting in the development of a strike-slip margin [Roberts and Engel, 1986]. Strike-slip movement taken up by the Peel Fault System at this time was later terminated

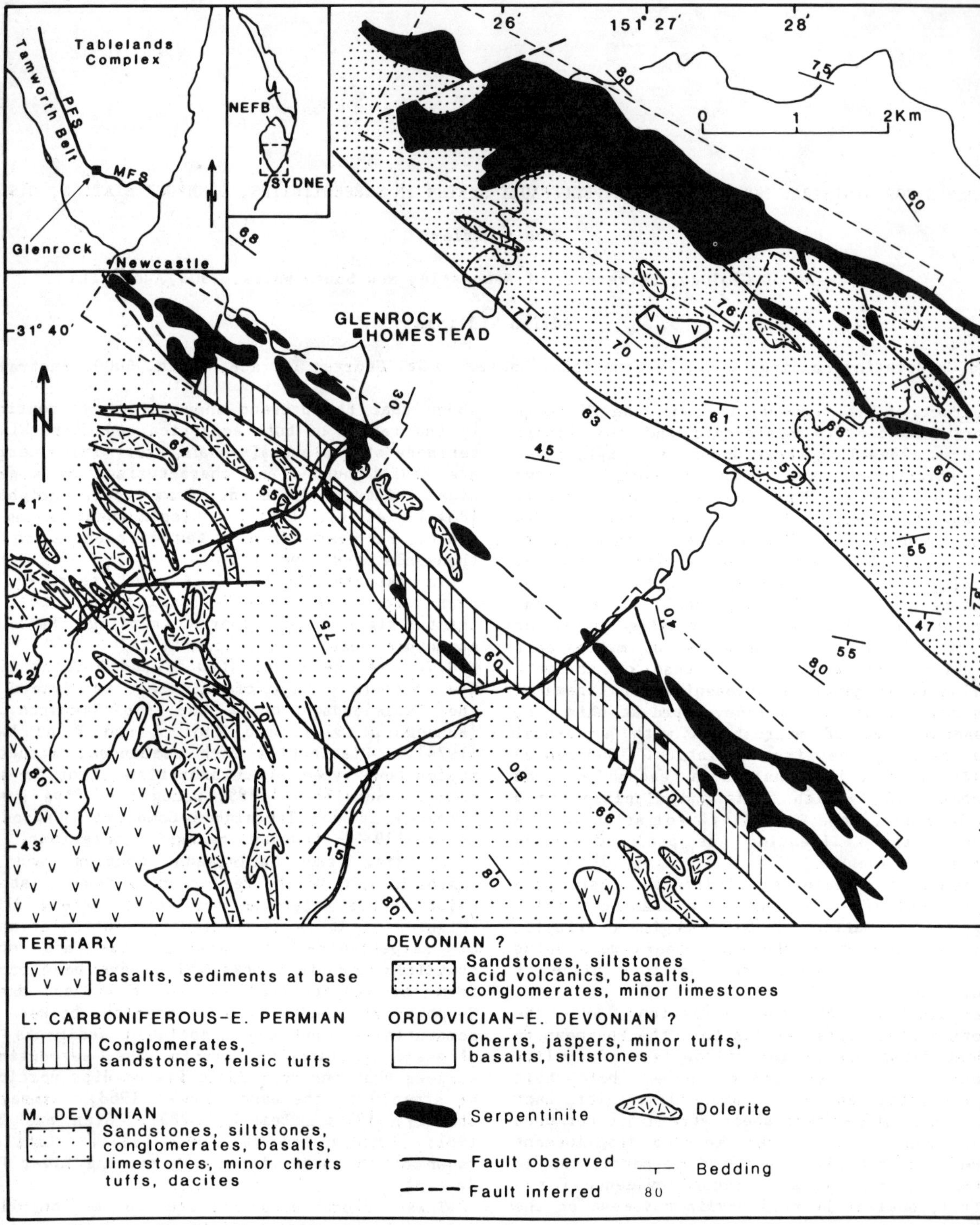

Fig. 1 Simplified geological map of the Glenrock Station area; serpentinites enclosed by dashed lines are those shown in Fig. 4. Large inset shows the position of the Peel-Manning Fault System within the Southern New England Fold Belt, smaller inset the position of the New England Fold Belt (NEFB). PFS - Peel Fault System, MFS - Manning Fault System.

TABLE 1. Nature of Movement--Peel Fault System

Study	Comment
	Dextral
Proud and Osborne (1952)	
Crook (1963)	also had an upthrust component
Scheibner and Glen (1972)	up to 150 km transcurrent movement; earlier an obduction zone
Korsch (1982)	
Shaw and Flood (1982)	
	Sinistral
Rod (1974)	
Corbett (1976)	earlier overthrusting
Cawood (1982)	
Offler and Williams (1985)	
	Thrust
Benson (1913)	
Runnegar (1974)	moved 150 km to W
Scheibner (1976)	120 km of thrust movement
Evans and Roberts (1979)	
Shaw and Flood (1981)	W thrusting
Harrington and Korsch (1985)	E dipping underthrusting

either before or with the intrusion of the Moonbi Adamellite and Inlet Monzonite in the Late Permian [250 Ma; Cooper et al, 1963; see later discussion].

There is general agreement that thrusting occurred on the Peel Fault System at some stage in its history, however, opinions vary as to the amount of thrusting that took place and whether the strike-slip movement was dextral or sinistral (Table 1). A better understanding of this area evolved as a result of the extensive field studies of Leitch, Korsch and others [Cawood and Leitch, 1985 and references therein]. Further interpretations were proposed but in none of them was there any attempt to account for the presence of exotic blocks in the serpentinites such as blueschists [Leitch, 1980; Offler, 1982] and eclogites [Shaw and Flood, 1974] which clearly indicate that thrust movements were substantial, nor to determine the movement patterns from the mesoscopic structures in the serpentinites. Several sets of cleavages were recognised by the investigators [Crook, 1963; Scheibner and Glen, 1972] and some were aware that these were a record of deformation on this major structure [Scheibner, 1976; Williams, 1979; Glen and Butt, 1981], but none realised their kinematic significance. This important aspect became apparent to the authors during a re-examination of specimens obtained from the Glenrock Station area, in an earlier study [Williams, 1979]. It showed that the textures and the cleavage relationships in the serpentinites resemble those in S-C mylonites [Berthé et al, 1979; Simpson and Schmid, 1983].

These features are described in this paper as well as the geometry and kinematic significance of the planar structures in the serpentinites. It will be shown that a sinistral strike-slip movement is recorded by the serpentinites in this area. These results provide the incentive for others to look more carefully at serpentinites, particularly those that border possible tectonostratigraphic terranes, since the information gained will help to elucidate the movements responsible for bringing the terranes in juxtaposition.

Geologic Setting

The serpentinites occur in the Glenrock Station area 240 km north of Sydney (Figure 1). They occur as lenticular cleaved and massive bodies along fault zones which form part of the Peel Fault System (Williams, 1979; Figure 1). Within them are exotic blocks of amphibolite, greenschist, quartzite, blueschist, lherzolite and a variety of ultramafic to mafic cumulate rocks [Offler, 1982]. Between the fault zones are folded and faulted rocks of varying age. Most belong to the Tamworth Group, a forearc basin sequence of Devonian age or to the Woolomin Formation, a subduction complex sequence of Early Palaeozoic age. A fault silver containing Late Carboniferous to Early Permian age rocks occurs in the central part of the area and to the north-east, a forearc basin sequence of unkown age. Assemblages in these rocks indicate prehnite-pumpellyite to prehnite-actinolite facies metamorphic conditions.

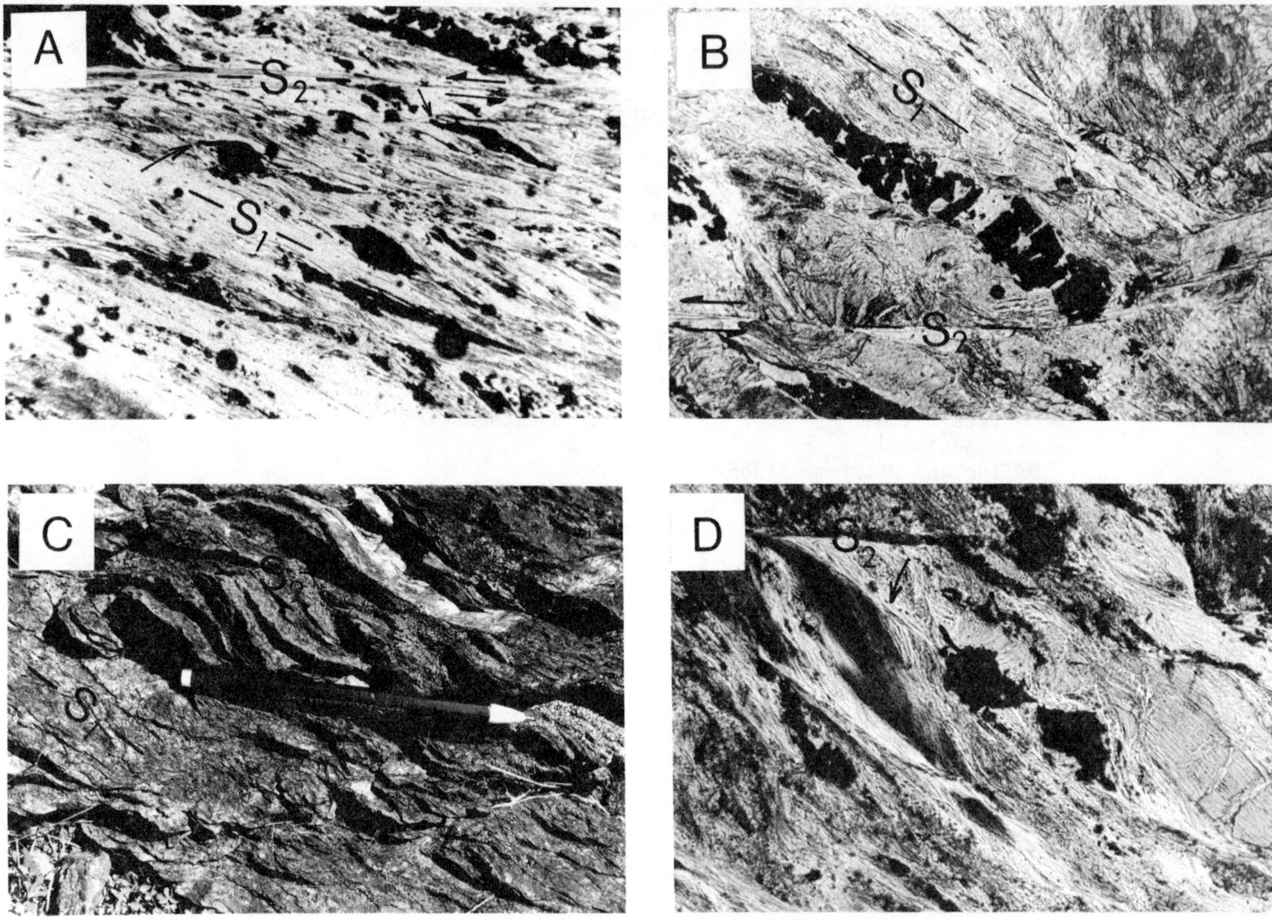

Fig. 2(A) Augen and elongate aggregates of magnetite parallel to S_1. Note drag of some aggregates into S_2 (arrows). AW 207 Photo length 0.45 mm. (B) Extended magnetite aggregates parallel to S_1 truncated by S_2 defined by chrysotile. RO 86.45 Photo length 0.9 mm. (C) Outcrop showing S_1 - S_2 relationships in serpentinite. S_2 is parallel to the pen. Note dragging of S_1 into S_2 (arrow). (D) Serpentine "fish" localised between S_2 planes. AW 14B Photo length 0.9 mm.

Structure

The serpentinite lenses contain two and sometimes three cross-cutting foliations and minor folds. These foliations are well developed in the smaller, lenticular bodies and at the margins of the larger bodies. The foliations also occur in schistose bands within the massive serpentinites. The earliest foliation to develop is a closely spaced penetrative, fine schistosity (S_1) which is more easily recognised in areas where later phases of deformations are less intense. Microscopically, it appears as a wispy, parallel alignment of lizardite and orthochrysotile which wrap around inclusions of massive serpentine and relict spinel aggregates; magnetite aggregates strung out parallel to S_1, or defining augen,often occur (Figure 2A). Many of the inclusions are bastites which consist of lizardite or antigorite surrounding trails of minute platy spinels. These spinels often define folds which have an axial plane orientation parallel to S_1 [Williams, 1979]. This feature and the observation that other inclusions have symmetrically oriented pressure fringes and are commonly extended parallel to S_1 (Figure 2B), suggests that S_1 is a plane of flattening and corresponds to the XY plane of the finite strain ellipsoid.

In addition to the lizardite bearing bastites, there are those consisting entirely of matted, decussate aggregates of antigorite plates which are less deformed and exhibit pressure fringes of lizardite and chrysotile. The lower strain shown by the bastites formed of antigorite indicates that this mineral has a mechanical behaviour different to that of lizardite, a characteristic noted previously by Williams [1979] and more recently by Wicks [1984] in serpentinites from

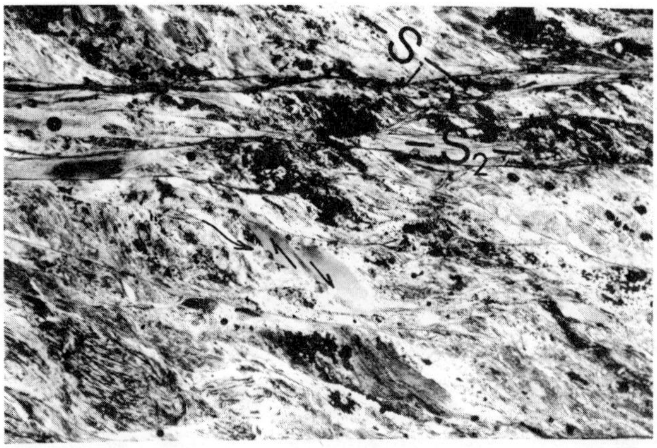

Fig. 3 Anthithetic microfaulting in serpentinised inclusion (arrow) oriented parallel to S_1. AW 14B Photo length 3.5 mm.

other regions. The textural features of the serpentine minerals suggest that the chrysotile and lizardite behaved in an incompetent manner and antigorite in a competent fashion. The experimental work on a lizardite 1T with chrysotile $-2M_{c1}$ + minor brucite mixture by Raleigh and Paterson [1965], suggests that S_1 formed when temperatures were less than 300 C and confining pressures over 3 Kb. However, these experiments were carried out on specimens from which the pore water had been driven off. Since pore water lowers the compressive strength of lizardite-chrysotile bearing serpentinites, the confining pressures proposed above must be considered maximum.

In both serpentinite belts, S_1 is overprinted by a set of micro shears (S_2 and S_2') which intersect the earlier schistosity at an angle up to 45°. S_2 is the dominant shear plane and shows a sinistral sense of movement according to S_1 - S_2 vergence relationships and drag of S_1 into S_2 whilst S_2' exhibits a dextral sense of movement. The age relationship between the two is not known. In some outcrops, S_1 is crenulated between S_2 or S_2' planes, but in most, it is dragged into the later foliation (Figure 2C). Both S_2 and S_2' are more widely spaced than S_1 (< 0.5 - 30 cm) and are axial plane to open mesoscopic folds.

Evidence for shearing is readily apparent in thin section; S_1 is clearly dragged into S_2 (Figure 2A), bastites are bent and fractured (see Figure 9, Williams, 1979) and serpentine "fish" resembling the mica fish seen in mylonites (Figure 2D; Lister and Snoke, 1984) are apparent. Displacement of bastites, serpentinised peridotites and magnetite aggregates is also common (Figure 2B); antithetic faults in serpentinised inclusions rarely occur (Figure 3). In some serpentinised pyroxenes containing cleavage favourably oriented to allow slip to take place, strain is taken up by the formation of kinks.

A second set of microshears (S_3) which also has a sinistral sense of movement, displaces and refolds the earlier foliations in the northern serpentinite belt but is less well developed in the southern belt (Figure 4). Movement on the S_3 planes resulted in the formation of abundant, open, mesoscopic folds defined by S_1 and S_2. They very in wavelength from 4cm to 30cm and have their axial planes oriented at 45° or less to S_3. Mesoscopically and microscopically, these shear sets show features similar to S_2 but are more widely spaced (1cm to 50cm). In some parts of the northern belt, a lineation defined by chrysotile fibres can be observed on S_2 and S_3.

In the previous study, Williams [1979] concluded that "some type of brittle-ductile transitional behaviour" should be invoked to explain the textures produced during events responsible for the shear sets. The present study substantiates this conclusion. However, it would be more appropriate to describe the behaviour of the serpentine minerals during the later events as brittle-plastic [Rutter, 1985] and plastic when S_1 formed. Such a change in behaviour suggests that the foliations were formed at different crustal levels. This observation is significant because it implies that there was at least some dip-slip component on the fault zones during the deformation of the serpentinites.

Geometry

The orientation of the main structural elements is illustrated in Figures 4, 5 and 6. S_1, the earlier foliation recognised in the serpentinites, strikes meridionally and dips steeply east or west. However, there is some variation in the geometry of S_1, particularly in the southern belt. Much of this can be attributed to variation in shear strain that appears to have existed during the formation of S_2 because the angle between this plane and S_1 varies throughout the serpentinite belts. It is also in part due to folding associated with later shearing events and to local strain heterogeneities built up in proximity to the mechanically strong phacoids of massive serpentinites and exotic blocks that are common in the serpentinites. Shear strain would be higher adjacent to the blocks resulting in the formation of S_1 and S_2 in a slightly different orientation to the normal trend [cf Sibson, 1980]. Such a situation has been noted next to massive serpentinite phacoids on both a micro- and mesoscopic scale where S_1 and S_2 show a marked deflection away from the normal trend.

S_2 trends north-west and dips steeply north-east or south-west substantiating the near vertical attitude of the Peel Fault System note elsewhere by other authors. However, in the northern belt, many of the S_2 planes have a north-north-westerly strike. A similar picture is shown by the axial planes of the F_2^1 folds which have a relatively constant orientation in the southern belt but are highly variable in attitude in the northern belt, even on a small scale (Figures 7A,

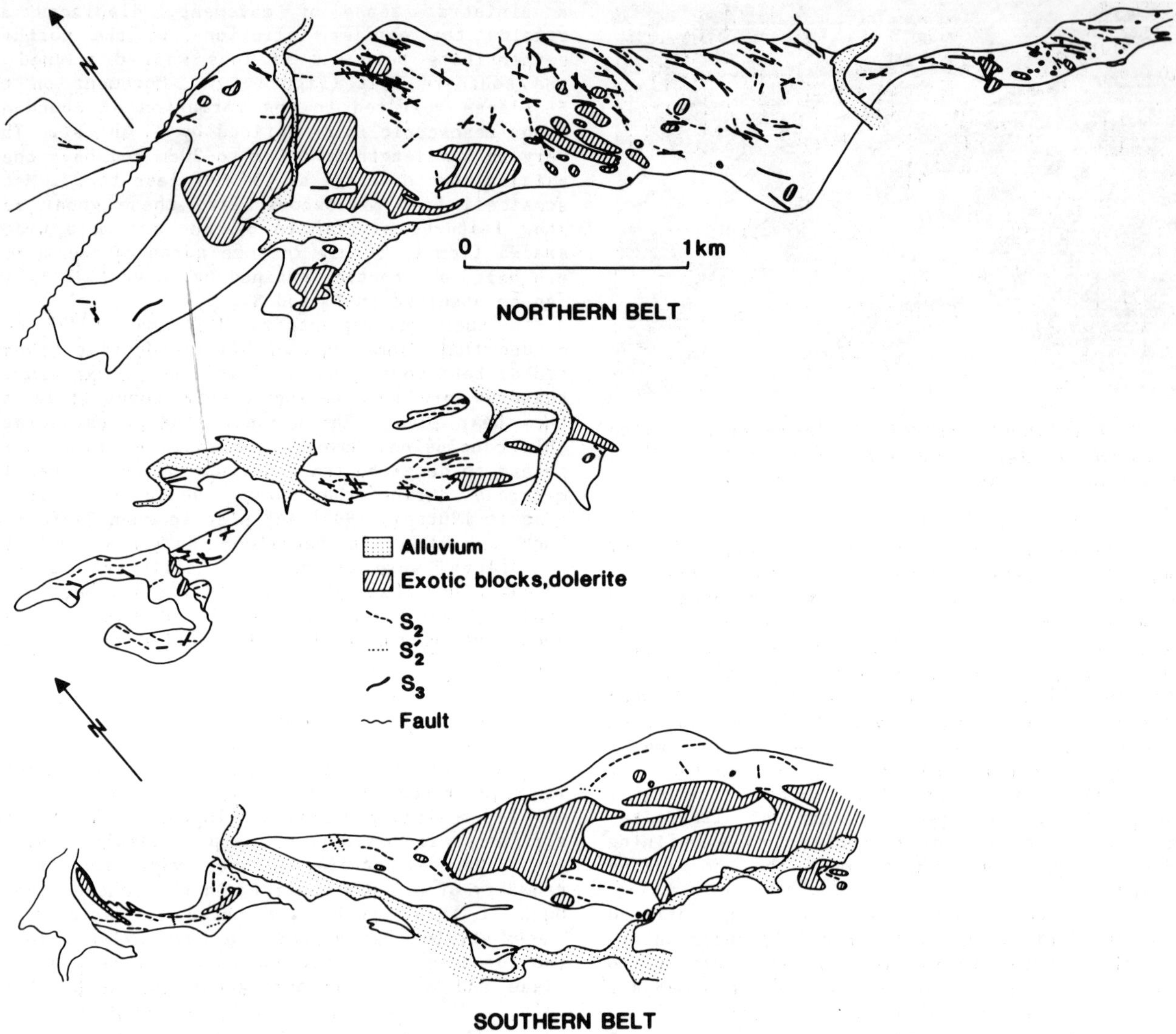

Fig. 4 Trends of S_1, S_2, S_2' and S_3 in serpentinites defining the two main belts in the area.

B,E). This is a result of the more intense folding associated with the formation of S_3 that has occurred in this belt. S_1 - S_2 intersections (L_2^1; Figures 5, 6) and plunges of F_2^1 folds (Figures 7A, B) in general, are near vertical, although there is some spread of the poles along a great circle corresponding to mean orientation of S_2.

S_3 is clearly discordant to S_2 in the northern belt, trending north-westerly and dipping steeply to the south-west. This discordance is even more pronounced in the north-western part of the belt where relationships between all foliations are well defined (Figure 8). However, in the southern belt this relationship is only apparent in plan (Figure 4) and not in the contoured stereographic projection of poles to S_3 (Figure 6). This seemingly anomalous situation has been brought about by the open folding that has taken place subsequent to the formation of S_3.

S_2 - S_3 (L_3^2) intersections and F_3^2 have steep plunges similar to those defined by S_1 - S_2, however, they show less variation in the magnitude of the plunge (Figures 5, 6, 7B). In contrast to the lineations defined by intersecting shear surfaces, those defined by chrysotile fibres (L_2^2, L_3^3) plunge at moderate to shallow angles in several directions (Figure 7D).

The chrysotile fibres that occur in S_2 cannot be used for slip analysis because they have been recorded in areas where this foliation is folded. Those in S_3 indicate a predominantly strike-slip component on the serpentinite belt (Figure 7D). Further evidence in support of this proposed movement comes from the cleavage intersections which

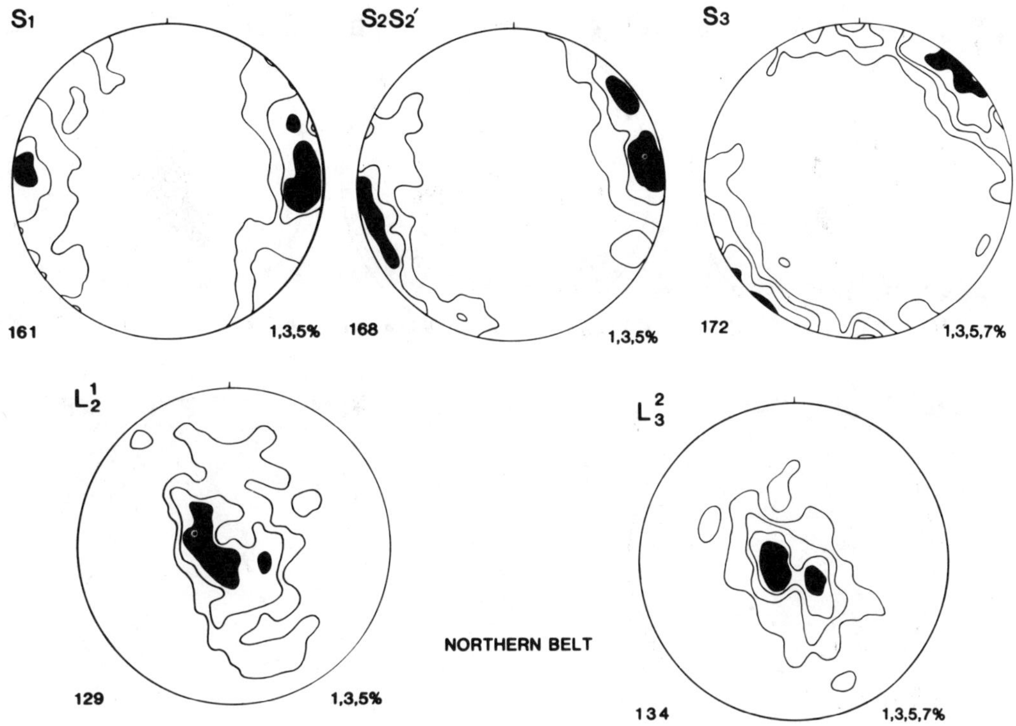

Fig. 5 Synoptic diagrams of planar and linear elements in the northern serpentinite belt. Terminology after Bell and Duncan, 1978.

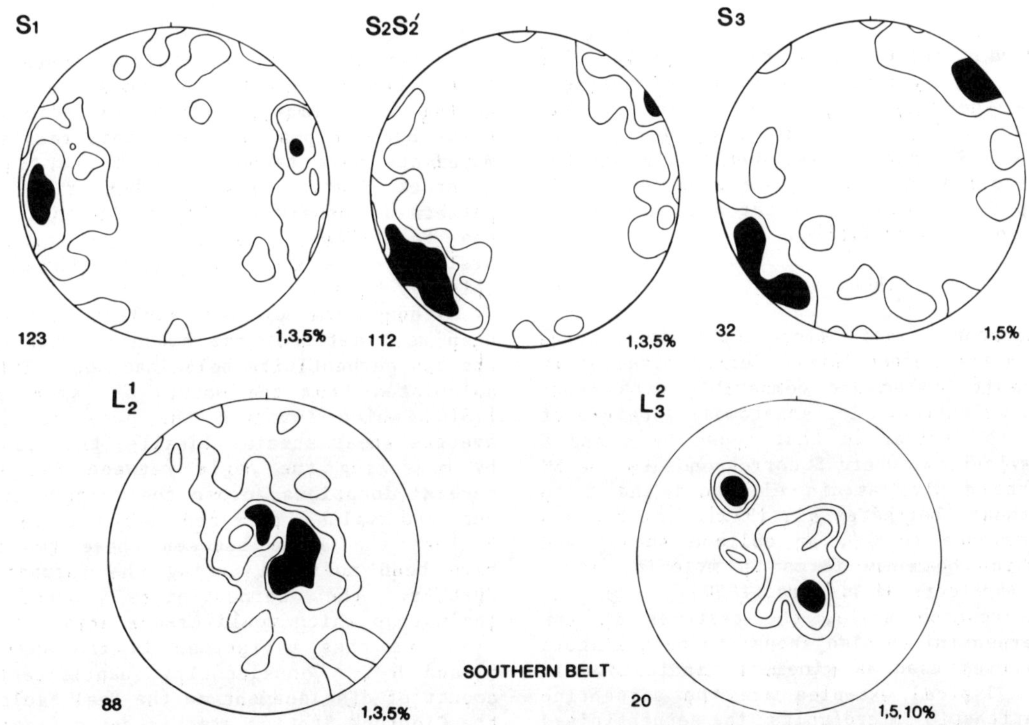

Fig. 6 Synoptic diagrams of planar and linear elements in the southern serpentinite belt.

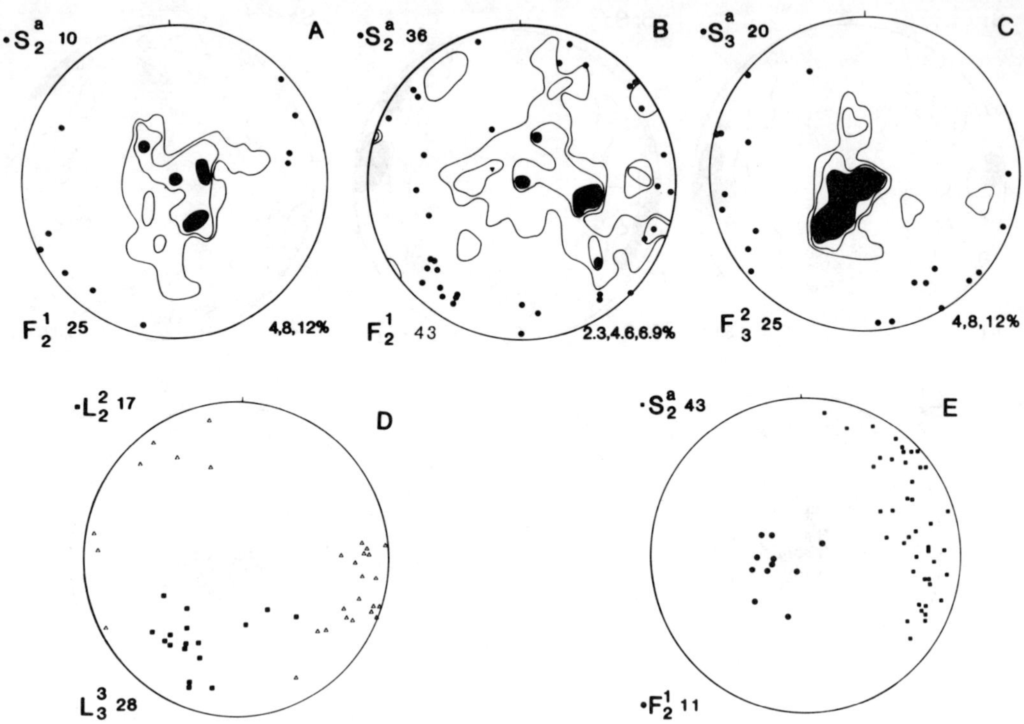

Fig. 7 Synoptic diagrams of (A) fold axes F_2^1 and axial plane orientations S_2^a in the southern belt (B) F_2^1, S_2^a (C) F_3^2 and S_3^a in the northern belt (D) chrysotile lineations L_2^2 and L_3^3 in S_2 and S_3 planes in the northern belt (E) F_2^1 and S_2^a in one location of the northern belt.

plunge near vertically (Figure 9). Since the slip direction is 90° away from the cleavage intersections along the shear plane this indicates that the movement on the serpentinites during the production of S_2 and S_3 was dominantly strike-slip. This is indicated in Figure 9 where it will be noted that the majority of the slip directions plot on or near the primitive.

Discussion

The mesoscopic and microscopic structures developed in the serpentinites during movement on the Peel Fault System are comparable with those observed in mylonites. The anastomosing nature of S_1 and S_2 is similar to that shown by S and C planes in mylonites, where S corresponds to the XY plane of the finite strain ellipsoid and C to planes of shear [Berthé et al, 1979]. S_3 bears a close resemblance to C′, an oblique shear band cleavage which commonly forms in mylonite zones [White, 1979; Platt and Vissers, 1980].

On a microscope scale, the textures in the deformed serpentinites also appear to be analagous to the textures used as kinematic indicators in mylonites. Typical examples are the serpentine "fish", antithetic microfaults in serpentinised inclusions and the deflection of S_1 into S_2 [cf Simpson and Schmid, 1983; Lister and Snoke, 1984]. These textures and the different types of deformational behaviour shown by the serpentine minerals during the formation of the foliations, indicate that both dip-slip and sinistral strike-slip movements have taken place on the faults in the Glenrock Station area. Whether this movement pattern is consistent along the entire length of the Peel Fault System is not yet known but preliminary studies further north suggest that this may be so.

Although the movement pattern in this area has been well defined, the amount of displacement on the two serpentinite belts has not. This could be calculated from the method of Ramsay and Graham [1970] which involves the determination of the average shear strain. Ideally, this could be done by measuring the angle between S_1 and S_2 at several locations across the serpentinite bodies, but the values obtained may not be meaningful because the angle between these two planes may have been modified during the formation of S_3. Further, the serpentinites contain numerous inclusions which would create strain heterogeneities and thus variations in the angles between S_1 and S_2. Consequently, quantification of the amount of displacement on the Peel Fault System in the Glenrock Station area is not a feasible proposition. It is likely that displacement has not been great because in outcrop and for the area as

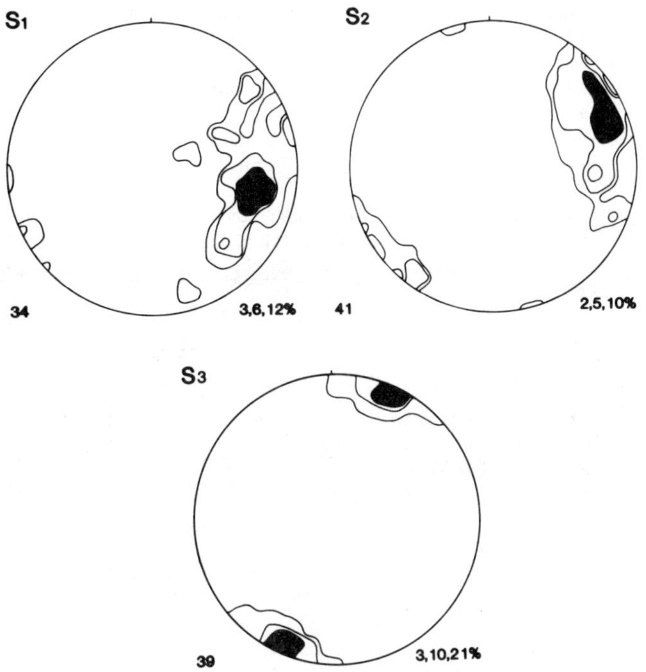

Fig. 8 Synoptic diagrams of planar elements in small area in the north-west section of northern serpentinite belt.

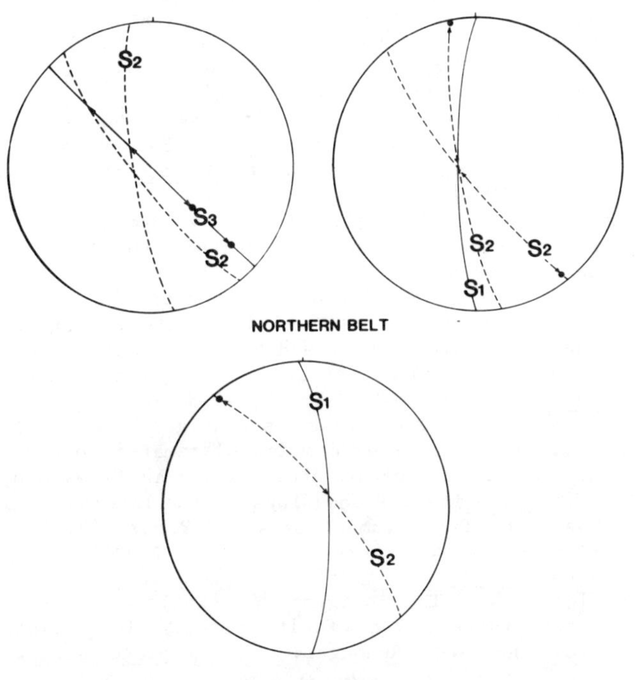

Fig. 9 Great circles represent mean S_1, S_2 and S_3 for the northern and southern belts and large dot, slip direction. Mean directions taken from maxima in Figs. 5 and 6.

a whole, the angle between S_1 and S_2 is relatively high, indicating that shear strains have not been great.

The timing of the movements that produced the structures in the serpentinites that delineate the Peel Fault System is difficult to determine. Current opinion has it that subduction ceased at the beginning of the Permian (~ 285 Ma; Roberts and Engel, [1986]). This implies that transcurrent movement which replaced the earlier convergent motion, commenced at about this time. Evidence in support of this comes from the dating of nephrites in the serpentinites (273 ± 5.8 Ma, 280 ± 5.6 Ma, Lanphere and Hockley, [1976]) and from the occurrence of fauna of Artinskian-Sakmarian age (~ 270 Ma) in sediments cut by the Peel Fault System [Price, 1973]. However, the timing can be further constrained if folding of the forearc basin sequence and transcurrent movement on the zone were contemporaneous, a situation that could arise if plate convergence was oblique [Ave Lallemant et al, 1985]. If this is correct, then movement must have been terminated before the late Early Permian, because the first post-tectonic granitoid in the New England Fold Belt, the Barrington Tops Granodiorite (262 Ma, Hensel et al, [1985]; 265 ± 2 Ma, Roberts et al, [1986]) intruded folded and faulted Devonian to Early Permian forearc basin sediments. Thus transcurrent movement on the Peel Fault System must only have occurred for a few million years implying that displacement along the margin was not excessive. What movements took place prior to the emplacement of the serpentinites along the fault zones and their subsequent deformation, is not known. Only further investigations will provide the answers.

Conclusions

The serpentinites in the Glenrock Station area contain several generations of foliations, lineations and mesoscopic folds which have been produced during movements on the Peel Fault System. The earliest foliation to form, S_1, is a plane of flattening and corresponds to the XY plane of the finite strain ellipsoid. This is overprinted by a set of microshears S_2 and S_2' which intersect this plane at 45 ° or less. Kinematic indicators resembling those in S-C mylonites and vergence relationships, reveal that S_2, the dominant shear plane, has a sinistral sense of movement whilst S_2' has a dextral sense of movement. A second set of microshears S_3 with a sinistral sense of movement, displaces and refolds the earlier foliations.

Textural features indicate that during the formation of S_1, the serpentine minerals behaved in a plastic fashion. Their behaviour changed to brittle-plastic when the microshears were produced suggesting that the shearing occurred at shallower crustal levels.

The orientation of chrysotile lineations in S_3 and geometrical relationships between the foliations, show that movement on the Peel Fault System in this area, is dominantly strike-slip. This movement appears to have been relatively short-lived occurring after deposition of sediments of Artinskian-Sakmarian age (~ 270 Ma), and before the emplacement of the first post tectonic granitoid, the Barrington Tops Granodiorite, in the New England Fold Belt at 262-265 Ma. This fact and the observation that shear strain has not been high suggests that the strike-slip displacement was not substantial.

This study has shown that serpentinites within fault zones develop mesoscopic and microscopic structures resembling those observed in mylonites. Like those in mylonites they indicate both direction and sense of fault movement.

Since these rocks border tectonostratigraphic terranes, the elucidation of their structural history would lead to a better understanding of the dispersal or accretion patterns responsible for bringing the terranes to their present position. If the structures in them are ignored, as has happened so often in the past, then less soundly based tectonic syntheses will result.

Acknowledgements. We acknowledge the constructive comments of Paul Lennox, Peter Flood and Evan Leitch who reviewed the manuscript and also thank Wal Crebert for drafting the figures, Barbara Sakac for the typing and Jan Walker for the plates. This study was supported by Tertiary Education Commission Funds.

References

Avé Lallemant, W.J. Schmidt and J.L. Kraft, Major Late-Triassic strike-slip displacement in the Seven Devils terrane, Oregon and Idaho: a result of left-oblique plate convergence. Tectonophysics, 119, 299-328, 1985.

Bell, T.H. and A.C. Duncan, A rationalized and unified shorthand terminology for lineations and fold axes in tectonites. Tectonophysics, 47, T1-T5, 1978.

Benson, W.N., The geology and petrology of the Great Serpentine Belt of N.S.W., Part 1: Introduction. Proc. Linn. Soc. N.S.W., 38, 490-517, 1913.

Berthé, D., P. Choukroune, and P. Jegouzo, Orthogneiss, mylonite and noncoaxial deformation of granites: the example of the South Amorican Shear Zone. J. Struct. Geol., 1, 31-42, 1979.

Cawood, P.A., Tectonic reconstruction of the New England Fold Belt in the Early Permian: and example of development at an oblique-slip margin. In: Flood, P.G. and B. Runnegar. New England Geology. Proceedings of a symposium on the Geology of the New England region. University of New England, Armidale, 25-34, 1982.

Cawood, P.A. and E.C. Leitch, Accretion and dispersal tectonics of the Southern New England Fold Belt, Eastern Australia. In: Howell, D.G. Tectonostratigraphic terranes of the Circum-Pacific Region. Circum-Pacific Council for Energy and Mineral Resources Earth Science Series, No 1, 481-492, 1985.

Corbett, G.J., A new fold structure in the Woolomin Beds suggesting a sinistral movement on the Peel Fault. J. Geol. Soc. Aust., 23, 401-406, 1976.

Crook, K.A.W., Structural geology of part of the Tamworth Trough. Proc. Linn. Soc. N.S.W., 87, 397-409, 1963.

Evans, P.R. and J. Roberts, Evolution of central eastern Australia during the late Palaeozoic and early Mesozoic. J. Geol. Soc. Aust., 26, 325-340, 1979.

Glen, R.A. and B.C. Butt, Chrysotile Asbestos at Woodsreef, New South Wales. Econ. Geol., 76, 1153-1169, 1981.

Harrington, H.J. and R.J. Korsch, Tectonic model for the Devonian to middle Permian of the New England Orogen. Aust. J. Earth Sci., 32, 163-179, 1985.

Hensel, H.D., M.T. McCulloch, and B.W. Chappell, The New England Batholith: constraints on its derivation from Nd and Sr isotopic studies of granitoids and country rocks. Geochimica et Cosmochimica Acta, 49, 369-384, 1985.

Howell, D.G., D.L. Jones, and E.R. Schermer, Tectonostratigraphic terranes of the Circum-Pacific region. In: Howell, D.G., Tectonostratigraphic terranes of the Circum-Pacific Region. Circum-Pacific Council for Energy and Mineral Resources Earth Science Series, No 1, 3-30, 1985.

Korsch, R.J., Early Permian tectonic events in the New England Orogen. New England Geology. Proceedings of the symposium on the Geology of the New England region. University of New England, Armidale, 35-42, 1982.

Lanphere, M.A. and J.J. Hockley, The age of nephrite occurrences in the Great Serpentine Belt of New South Wales. J. Geol. Soc. Aust., 23, 15-17, 1976.

Leitch, E.C., The geological development of the southern part of the New England Fold Belt. J. Geol. Soc. Aust., 21, 133-156, 1974.

Leitch, E.C., Structural succession in a Late Palaeozoic slate belt and its tectonic significance, in Burns, K.L. and R.W.R. Rutland., eds., Structural characteristics of tectonic zones. Tectonophysics, 47, 311-323, 1978.

Leitch, E.C., The Great Serpentine Belt of New South Wales: diverse mafic-ultramafic complexes set in a Palaeozoic arc. In: A. Panayiotou, ed., Ophiolites. Proceedings of the International Ophiolite Symposium, Cyprus, 1979,637-648, 1980.

Lister, G.S. and A.W. Snoke, S-C Mylonites, J. Struct. Geol., 6, 617-638, 1984.

Offler, R., The origin of exotic blocks in serpentinites, Peel Fault System. In: Flood, P.C. and B.N. Runnegar. New England Geology. Proceedings of a symposium on the Geology of the New England region. University of New England, Armidale, 43-51, 1982.

Offler, R. and A.J. Williams, Evidence for sinistral movement on the Peel Fault System in serpentinites, Glenrock Station, N.S.W. 3rd

Circum-Pacific Terrane Conference, Sydney. 171-173, 1985.

Platt, J.P. and R.L.M. Vissers, Extensional structures in anisotropic rocks. J. Struct. Geol., 2, 397-410, 1980.

Price, I., A new Permian and Upper Carboniferous(?) succession near Woodsreef, N.S.W., and its bearing on the Palaeogeography of western New England. Linn. Soc. N.S.W. Proc., 97, 202-210, 1973.

Proud, J.S. and G.D. Osborne, Stress-environment in the genesis of chrysotile, with special reference to the occurrence of Woodsreef, near Barraba, New South Wales. Econ. Geol., 47, 23-37, 1952.

Raleigh, C.B. and M.S. Paterson, Experimental deformation of serpentinite and its tectonic implications. J. Geophys. Res., 70, 3965-3985.

Ramsay, J.G. and R.H. Graham, Strain variation in shear belts. Can. J. Earth Sci., 7, 786-813, 1970.

Ramsay, W.R.H. and J.M Stanley, Magnetic anomalies over the western margin of the New England Fold Belt, northeast New South Wales. Geol. Soc. Am. Bull., 87, 1421-1428, 1976.

Roberts, J. and B.A. Engel, Depositional and tectonic history of the South New England Orogen. Aust. J. Earth Sci., in press, 1986.

Roberts, J., B.A. Engel and N. Morris, Geology of the Hunter-Myall region, New South Wales. Camberwell 9133, Dungog 9233 and Bulahdelah 9333 1 : 100000 sheets. Geol. Surv. N.S.W., Explan. Notes, in press, 1986.

Rod, E., Structural interpretation of New England Region. J. Proc. Roy. Soc. N.S.W., 107, 90-99, 1974.

Runnegar, B.N., The geological framework of New England. Geol. Soc. Aust., Qd. Div. Field Conf. New England area, 9-19, 1974.

Rutter, E.H., On the nomenclature of mode of failure transitions in rocks. Tectonophysics, 122, 381-387, 1985.

Scheibner, E. and R.A. Glen, The Peel Thrust and its tectonic history (New South Wales). Q. Notes Geol. Surv. N.S.W., 8, 2-14, 1972.

Scheibner, E., Explanatory notes on the tectonic map of New South Wales, scale 1:1 000000. Mines Dept., Geol. Survey New South Wales, 283p, 1976.

Scheibner, E., Suspect terranes in the Tasman Fold Belt System, Eastern Australia. In: Howell, D.G., Tectonostratigraphic terranes of the Circum-Pacific Region. Circum-Pacific Council for Energy and Mineral Resources. Earth Science Series, No. 1, 493-514, 1985.

Shaw, S.E. and R.H. Flood, Eclogite from Serpentinite near Attunga, New South Wales. J. Geol. Soc. Aust., 21, 377-385, 1974.

Shaw, S.E. and R.H. Flood, The Bundarra Plutonic Suite: A summary of new data. In: Flood, P.C. and B.N. Runnegar. New England Geology. Proceedings of a symposium on the Geology of the New England region. University of New England, Armidale, 183-192, 1982.

Shaw, S.E. and R.H. Flood, The New England Batholith, eastern Australia: geochemical variations in time and space. J. Geophys. Res., 86, 10530-10544, 1981.

Sibson, R.H., Transient discontinuities in ductile shear zones. J. Struct. Geol., 2, 165-171, 1980.

Simpson, C. and S.M. Schmid, Microstructural indicators of sense of shear in shear zones. Geol. Soc. Am. Bull., 94, 1281-1283, 1983.

Wicks, F.J., Deformation histories as recorded by serpentinites. II. Deformation during and after serpentinization. Can. Mineral., 22, 197-203, 1984.

Williams, A.J., Foliation development in serpentinites, Glenrock, New South Wales. Tectonophysics, 58, 81-95, 1979.

LATE PALAEOZOIC AND EARLY MESOZOIC EVOLUTION OF THE NEW ENGLAND OROGEN IN SOUTHEASTERN QUEENSLAND - A PHOTOGEOLOGICAL STUDY

C.R. Nash

Hunting Geology and Geophysics (Australia) Pty. Ltd., PO Box 365, Fyshwick ACT 2609, Australia

Abstract. Possible constraints upon late Palaeozoic and early Mesozoic evolution of the New England Orogen in southeastern Queensland are suggested by systematically oriented megascopic structures interpreted from satellite imagery and small-scale aerial photography. The region investigated includes portions of the Yarrol-Silverwood-Tamworth, Wandilla and Gympie terranes and reflects the imprint of late Palaeozoic crustal shortening followed by extension in Mesozoic and Cainozoic times. Interpreted stress trajectories associated with these events are oriented approximately normal to the axis of the orogen.

Comparisons with transcurrent fault regimes of the North American Cordillera do not support the concept of early Mesozoic oblique compression of the New England Orogen. Instead, it is proposed that NNW trending grabens and parallel belts of epizonal granites which formed at this time were the result of early Mesozoic taphrogenesis which preceded the ultimate late Mesozoic and Cainozoic continental splitting along the eastern margin of Australia.

Introduction

Satellite imagery and small-scale aerial photographs provide an optimum base for the interpretation of megascopic structure (10-1000 km). As pointed out by Gold [1980], the role of remotely-sensed data in structural geology increases proportionately with the scale of structure under consideration. Interpretation of large structures depends upon the recognition of regional lithologic inhomogeneities, expressed as differential landforms, rather than the direct measurement of discontinuities in the field which forms the basis of mesoscopic structural analysis.

The application of these methods is constrained by the two-dimensional nature of most photogeological data. Although the attitudes of faults and dipslopes are often visible in stereomodels, in most cases the traces of penetrative planar structures and lineaments are defined by the single line of their intersection with the Earth's surface. It is consequently desirable to combine the results of photogeological interpretation with all available geological and geophysical data in structural studies [Nash et al., 1980].

Little consensus exists with regard to the origin of the pervasive fracture-controlled lineament patterns which characterise continental regions, and which invariably result from photogeological studies. It has been recognised for over a century that tensile strain within the brittle carapace of the crust is an important cause of pervasive fracturing [Hodgson, 1976], and that fractures of this type may be generated by modest tectonic uplifts of 0.5-1 km, extending over regions of 10-1000 km [Nur, 1978]. This mechanism is also invoked by Nur to explain the asymmetric frequency distributions of observed lineament sets, which are often a function of lithospheric strength and pre-existing planes of weakness.

Not all fractures however are generated in this manner, particularly in orogenic belts. Recent investigations of the Appalachian foreland region of the northeastern United States have shown that compression may produce systems of fracturing normal to crustal shortening fabrics, thus providing indicators of palaeostress trajectories [Engelder and Geiser, 1980]. In the Cordillera of western Canada, continental-scale strike-slip faults are related to transpressive strain imposed by the northward motion of the Pacific Plate and its antecedents in late Mesozoic and Cainozoic times [Gabrielse, 1985]. Systems of smaller faults detected by field mapping [Seraphim and Hollister, 1976] and thrust, fold and lineament trends interpreted from Landsat imagery [Nash, 1985] all suggest that strike-slip faulting of the northern Cordillera was accompanied by possible conjugate fracture sets and by crustal shortening fabrics oriented normal to palaeostress directions induced by plate convergence. In many instances these subsidiary structural fabrics provide better evidence of regional stress directions than do the large strike-slip dislocations themselves, which are frequently the sites of subsequent tensional faulting.

Sets of fractures formed by compression or

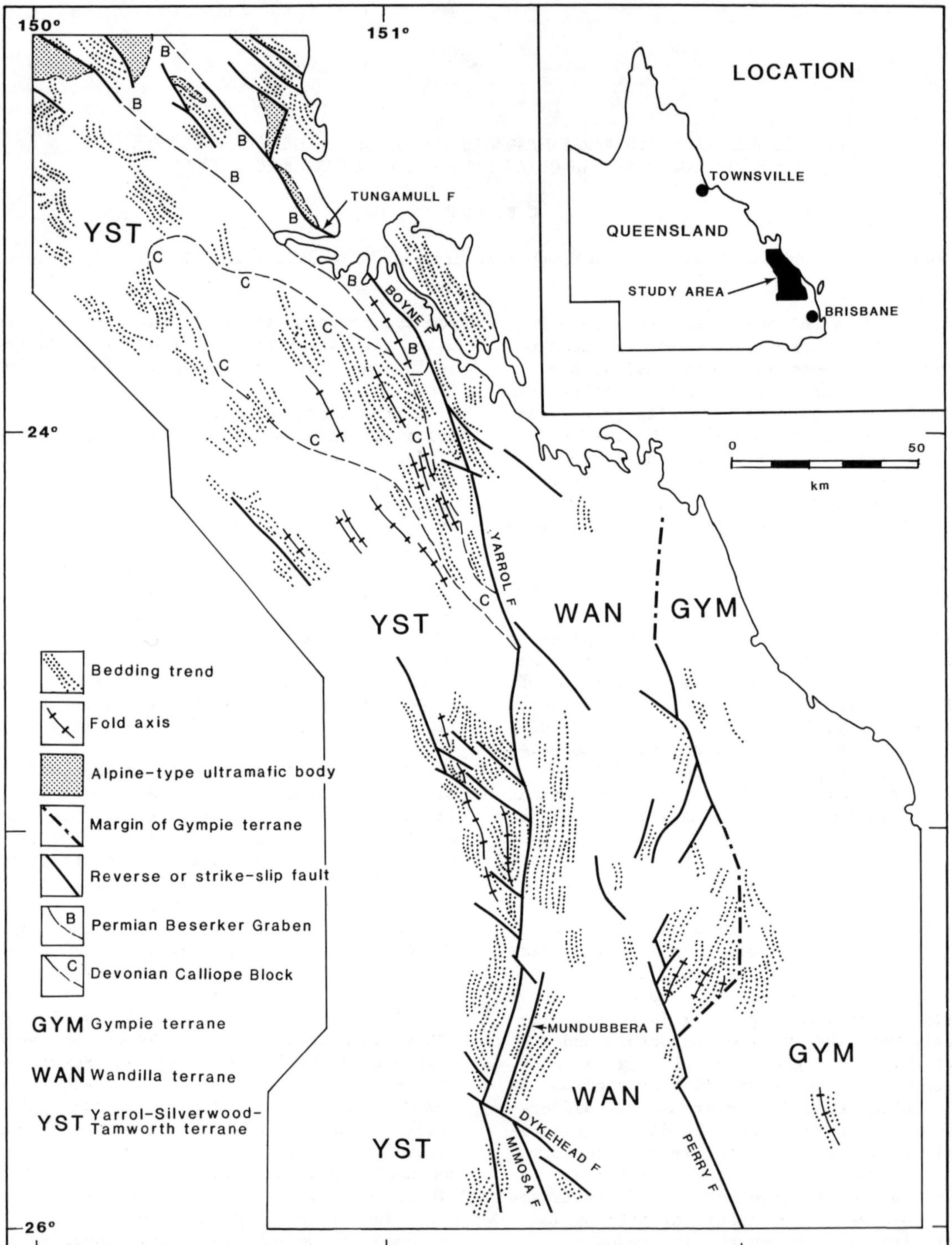

Fig. 1. Megascopic planar structural elements of the Rockhampton-Maryborough area (location shown by inset diagram).

transpression will tend to remain recognisable even if later tensile fracturing does occur, since preferential reactivation of the earlier structures may be anticipated. The utility of these structures as palaeostress indicators is however largely dependent upon the recognition of temporal relations between fracturing and unequivocal structural fabrics related to synchronous crustal shortening or extension.

These concepts have been applied to the structural investigation of a 41,000 km^2 region of the New England Orogen, located between Rockhampton and Maryborough in southeastern Queensland (Figure 1). The region includes portions of the Yarrol-Silverwood-Tamworth (YST), Wandilla (WAN) and Gympie (GYM) terranes [Howell et al., 1985; Scheibner, 1985]. The stratigraphy and inferred tectonic evolution of the region, based upon mapping by the Geological Survey of Queensland, are described by Day et al., [1974; 1978; 1983] and by Cranfield [1986]. Subsequent photogeological studies [Nash, 1984; 1985; 1986] have revealed distinctive regional patterns of megascopic structures which are apparently related to crustal shortening followed by post-orogenic tensile fracturing.

Megascopic Structure of the New England Orogen in Southeastern Queensland

The study area falls mainly within the eastern highlands physiographic province of Australia and is consequently characterised by deep fluvial dissection and minimal development of Cainozoic superficial cover which hinders photogeological interpretation over much of the continent. Analysis of small-scale aerial photography and satellite imagery over this area [Nash, 1986] suggests that several sets of megascopic structures are present, respectively generated during the Permo-Triassic transition of the New England Orogen from an active to a passive margin and during the late Mesozoic-early Cainozoic opening of the Tasman Sea.

Late Palaeozoic Structures

The traces of planar structural elements seen on remotely-sensed data mainly correspond to bedding traces within Palaeozoic volcaniclastic sequences of the constituent terranes of the New England Orogen (Figure 1). The primary bedding structures are deformed into meridional to NW trending open folds in the structural domain to the west of the Yarrol Fault and associated high-angle reverse faults. The western structural domain (YST terrane) corresponds to the Yarrol Shelf and Calliope Block of Day et al., [1978]. The domain to the east of the Yarrol Fault (WAN terrane) is largely composed of Devonian-Carboniferous flysch intruded by Permo-Triassic batholiths and plutons, and corresponds to the Coastal Block of Day et al., [1978]. Field observation reveals a ubiquitous mesoscopic slaty cleavage in the central domain, which is generally vertical and related to tight folding. The cleavage is locally superimposed upon an earlier bedding foliation [Nash, unpub. data]. The GYM terrane or Gympie Block [Day et al., 1978] is composed of deformed Carboniferous marine clastic rocks, undeformed basaltic volcanics and sediments of Permian age, and Early Triassic sediments [Cranfield, 1986]. The Gympie terrane is largely covered by soils of the Queensland coastal plain and consequently provides a poor subject for photogeological analysis.

The distribution of megascopic planar structural elements in the YST terrane reflects a change from meridional trends in the southern part of the study area to NW trends in the north. These variations are broadly mirrored in the shape of the adjacent Bowen foreland basin to the west (Figure 2), in which 5 km of paralic-paludal sediments accumulated in the 260-240 Ma interval [Veevers, 1984]. The trends of structures in the YST terrane and of the Bowen Basin are generally compatible with westerly-directed compression, possibly associated with a convergent margin near the present position of the Queensland coast during Late Permian times, as envisaged by Harrington and Korsch [1985a]. The variation from meridional to NW trends may be a reflection of compression of the YST terrane against an earlier continental margin. The inferred suture is manifested in published gravity data [BMR, 1976] as a prominent paired Bouguer anomaly.

Planar structural elements in the western part of the WAN terrane are parallel to trends in the YST terrane. NNE trends in the eastern portion of the WAN terrane may represent earlier structures than the regional N-S trend.

The late Palaeozoic compression of the Rockhampton area coincided temporally with the emplacement of a belt of batholiths and plutons with isotopic ages of 260-230 Ma [Day et al., 1983]. Plutonism occurred in both the YST and WAN terranes. The plutonic belt displays a broad sigmoidal form with trends similar to those displayed by planar structures [Nash, 1986].

Early to Middle Triassic Structures

Day et al., [1974] drew attention to a major change in tectonic style which took place in southeastern Queensland around the Permo-Triassic interval and which is prominently reflected in the photogeological data. The formation of the linear NNW trending Esk Trough (Figure 2) is indicated by isotopic ages of 236-242 Ma from andesitic volcanics within the structure [Webb, 1981]. Directions of rift propagation transect earlier Palaeozoic compressional structures in the southern part of the study area [Nash, 1986]. In the northern part of the study area, rifting appears to have followed earlier structural trends which display local northwesterly orientations. After formation of the Esk Trough, a linear NNW trending belt of epizonal plutons was emplaced along a major linea-

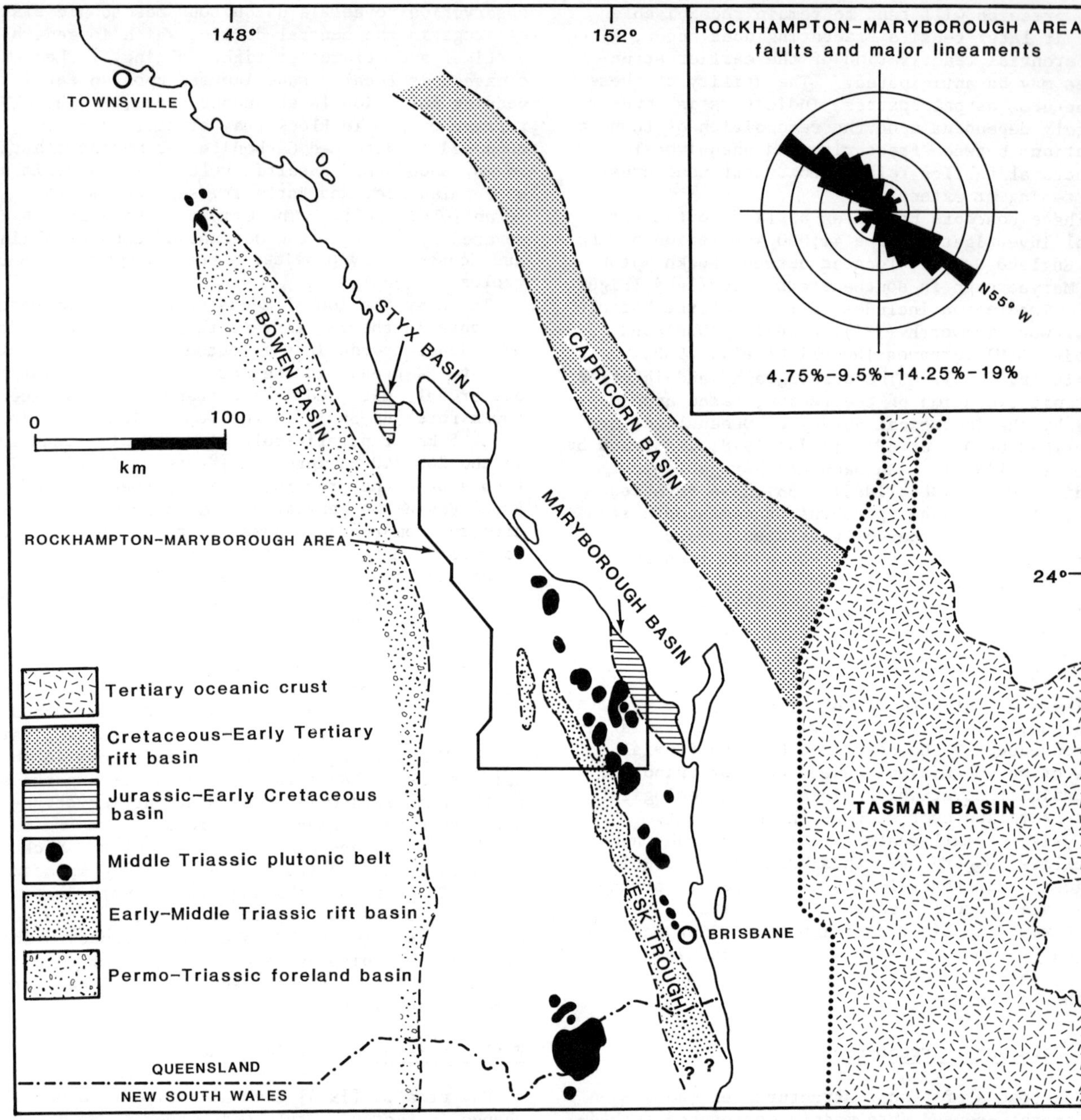

Fig. 2. Eastward migration of late Palaeozoic to Early Tertiary basins in southeastern Queensland and location of NNW trending belt of Triassic epizonal plutons. Inset diagram shows strong preferred orientations of lineaments and faults of inferred early Mesozoic and late Mesozoic-early Cainozoic origin.

ment zone (the Mount Perry lineament) which represents the northern extension of the Esk Trough [Nash, 1985; 1986]. These plutons are predominantly granophyric and form easily recognisable resistant massifs on aerial photographs. Available isotopic data [Day et al., 1983] would suggest Middle-Late Triassic age for the plutons.

The regional fault and fracture patterns of the Rockhampton-Maryborough area (Figure 2-inset) are represented by NNW to WNW trending photolineaments and Landsat lineaments [Nash, 1984] which transect late Palaeozoic plutonic rocks and planar structures of the YST, WAN and GYM terranes. These structures and NW oriented silicic dykes which are spatially associated with the mid-Triassic plutons [Nash, 1986] suggest Early to Middle

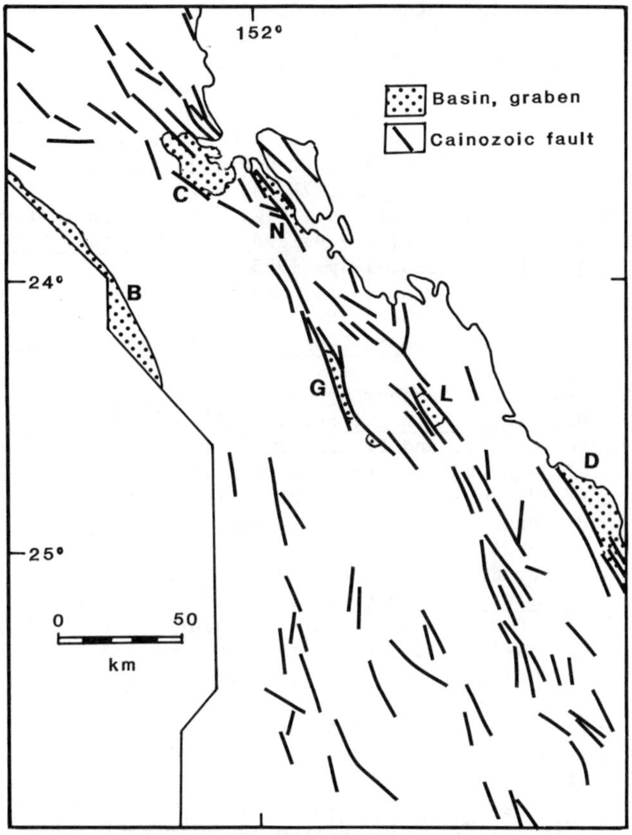

Fig. 3. Cainozoic faults and Early Tertiary basins and grabens of the Rockhampton-Maryborough area. (B-Biloela Basin; C-Casaurina Basin; N-Narrows Graben; L-Lowmead Graben; D-Bundaberg Trough).

Triassic extension directions between N20°E and N65°E. Subsequent crustal shortening is revealed by folding and thrusting of the contents of the Esk Trough [Murphy et al., 1976] and by NW oriented stylolite fabrics and minor reverse faults in dykes of inferred Triassic age [Nash, unpub. data]. Triassic plutonism and deformation largely preceded the terminal eruption of flat-lying Late Triassic ignimbrites in the central part of the area, although Stephens [1986] suggests a comagmatic relationship between volcanics and plutonic rocks in the south.

Late Cretaceous to Early Tertiary Structures

Southeastern Queensland was subjected to tensile stresses during Late Cretaceous to Early Tertiary opening of the Tasman Sea between 76 and 55 Ma [Shaw, 1978]. Deposition in the NNW trending Capricorn rift basin (Figure 2) began with a Late Cretaceous rift sequence, followed by Oligocene and Miocene marine deposits [Ericson, 1976]. Isostatically induced epeirogenic movements in southeastern Queensland at this time produced a series of NNW oriented normal faults and grabens [McConnochie and Henstridge, 1985], shown in Figure 3. Many of the photogeologically mapped Cainozoic structures are reactivated faults of earlier generation. Eocene sedimentation filled the grabens with fluviatile and lacustrine sediments; subsequent movements along the bounding faults of the grabens are revealed by different levels of sediment within the structures relative to their shoulders. Repeated reactivation of Tertiary block faults caused the formation of prominent NNW trending fault-line scarps observed at the present time.

Discussion

Photogeologically interpreted megascopic structural data suggest that late Palaeozoic deformation within the New England Orogen of southeastern Queensland resulted from westerly-directed tectonic transport. Large paired Bouguer anomalies along the western margin of the YST terrane [BMR, 1976] probably reflect accretion of the terrane to the Australian continent prior to the Early Permian volcanism which marked the early development of the Bowen Basin along this suture, and which has been variously attributed to back-arc spreading [Henderson and Johnson, 1985] and to a calc-alkaline volcanic arc [Murray, 1985]. The prominent N-S structural trends interpreted from remotely-sensed data are regarded as evidence of compression during late Palaeozoic times, coincident with the evolution of the Bowen foreland basin between 260 and 240 Ma. The broadly concordant plutonic belt emplaced during this period may represent a region of reduced stress within the orogen. The western portion of the WAN terrane appears to bear the imprint of the same deformation which characterises the YST terrane, possibly superimposed upon an earlier set of structures. Accretion of the WAN terrane presumably predated the formation of the Early Permian Beserker Graben (Figure 1), inferred to be of similar age to the basal volcanic sequence of the Bowen Basin [Day et al., 1983]. The Beserker Graben occupies a position between the WAN and YST terranes and is deformed along similar trends to the terranes on either side.

No evidence is forthcoming from the present study to support the concept of oblique-slip movements in Early and Middle Triassic times suggested by Evans and Roberts [1980], nor is there photogeological evidence to support the concept of mid-Triassic accretion of the GYM terrane proposed by Harrington and Korsch [1985b]. Stress transmission across convergent margins [Uyeda, 1982] might be expected to generate transpressive shortening structures in orientations near to those predicted by Wilcox et al [1973] and which are observed at various locations in the Cordillera of North America [Harding, 1976; Nash, 1985], notwithstanding the variable tangential components implicated in stress transmission across obliquely convergent plate boundaries [Henry, 1986]. In place of these structures, photogeological data suggest that extension and associated rift volcanism prevailed during early Mesozoic times in the study area. Short-lived episodes of Late Triassic compression may however have induced local strike-slip movements along large tensional faults of earlier generation [Stephens, 1986; Flood and Garces, 1986].

It is argued instead that the New England Orogen has behaved essentially as a divergent margin since early Mesozoic times. It has been suggested that much of Pangaea was subjected to anomalous tensile stresses at this time [Bott and Mithen, 1983]; similar conclusions were reached by Kent [1976; 1977] and Dingle [1982] from studies of the margins of Africa, where rifts formed as early as Permo-Triassic times, long before the final opening of the Atlantic and Indian Oceans. Similar examples are found in the Triassic-Liassic rift basins of the Appalachian region and their analogues in Morocco, formed on either side of the central zone of fracturing along which Jurassic opening of the North Atlantic took place [van Houten, 1977].

It is suggested that these processes were active in the accreted terranes forming the New England Orogen of southeastern Queensland from early Mesozoic times onward. Following Middle-Late Triassic plutonism along the initial NNW axes of rifting, the latter became effectively cratonised except for relatively minor Late Cretaceous-Early Tertiary tectonism in the Maryborough Basin to the east [Ellis, 1976] and epeirogenesis in the study area [McConnochie and Henstridge, 19ε5]. The locations of the Jurassic-Early Cretaceous Maryborough Basin and the Late Cretaceous-Tertiary Capricorn Basin (Figure 2) indicate progressive migration of basin axes to the east of the initial Triassic zone of rifting, and may reflect rift jumping as described by Wood [1983].

Acknowledgements. The research described in this paper forms part of a PhD project undertaken by the writer at Macquarie University, Sydney. The writer wishes to express his gratitude to David Durney for his discussion of structural and tectonic concepts, and to Hunting Geology and Geophysics (Australia) Pty. Ltd., who kindly provided facilities for research.

References

BMR, Gravity map of Australia, Bur. Miner. Resour. Geol. Geophys. Aust., 1976.

Bott, M.H.P., and D.P. Mithen, Mechanism of graben formation - the wedge subsidence hypothesis, Tectonophysics, 94, 11-22, 1983.

Cranfield, L.C., The geology of the South Burnett district, in Willmott, W.F., editor, 1986 Field Conference Guidebook, Brisbane, Geol. Soc. Aust. Qld. Div., 1-12, 1986.

Day, R.W., L.C. Cranfield, and H. Schwarzböck, Stratigraphy and structural setting of Mesozoic basins in southeastern Queensland and northeastern New South Wales, in Denmead, A.K., G.W. Tweedale and A.F. Wilson, editors, The Tasman Geosyncline - A Symposium, Brisbane, Geol. Soc. Aust. Qld. Div., 319-362, 1974.

Day, R.W., C.G. Murray, and W.G. Whitaker, The eastern part of the Tasman Orogenic Zone, Tectonophysics, 48, 327-364, 1978.

Day, R.W., W.G. Whitaker, C.G. Murray, I.H. Wilson, and K.G. Grimes, Queensland Geology, Geol. Surv. Qld. Publ., 383, 194p, 1983.

Dingle, R.V., Continental margin subsidence: a comparison between the east and west coasts of Africa, Am. Geophys. Union, Geodynam. Ser., 6, 59-71, 1982.

Ellis, P.L., Maryborough Basin, Aust. Inst. Min. Metallurgy, Monogr. Ser., 7, 447-450, 1976.

Engelder, T., and P. Geiser, On the use of regional joint sets as trajectories of palaeostress fields during development of the Appalachian Plateau, New York, Jl. Geophys. Res., 85, 6319-6341, 1980.

Ericson, E.K., Capricorn Basin, Aust. Inst. Min. Metallurgy, Monogr. Ser., 7, 464-473, 1976.

Evans, P.R., and J. Roberts, Evolution of central Australia during the late Palaeozoic and early Mesozoic, Jl. Geol. Soc. Aust., 26, 325-340, 1980.

Flood, P.G., and B. Jun Garces, The Tarong Basin, Queensland, in Willmott, W.F., editor, 1986 Field Conference Guidebook, Brisbane, Geol. Soc. Aust. Qld. Div., 77-81, 1986.

Gabrielse, H., Major dextral transcurrent displacements along the northern Rocky Mountain Trench and related lineaments in north-central British Columbia, Bull. Geol. Soc. Amer., 96, 1-14, 1985.

Gold, D.P., Structural geology, in Siegal, B.S., and A.R. Gillespie, editors, Remote Sensing in Geology, New York, John Wiley, 419-483, 1980.

Harding, T.P., Tectonic significance and hydrocarbon trapping significance of sequential folding synchronous with San Andreas faulting, San Joaquin Valley, California, Am. Assoc. Petrol. Geol. Bull., 60, 356-378, 1976.

Harrington, H.J., and R.J. Korsch, Late Permian to Cainozoic tectonics of the New England Orogen, Aust. Jl. Earth Sci., 32, 181-203, 1985a.

Harrington, H.J., and R.J. Korsch, Deformation associated with the accretion of the Gympie terrane in eastern Australia, Geol. Soc. Aust. Abs. Ser., 14, 104-108, 1985b

Henderson, R.A., and D.P. Johnson, Summary of the Permian-Quaternary development of the Mackay-Ayr region, coastal North Queensland, in Johnson, D.P., and A.W. Stephens, editors, 1985 Field Conference Guidebook, Brisbane, Geol. Soc. Aust. Qld. Div., 1-8, 1985.

Henry, C.D., East-northeast-trending structures in western Mexico: evidence for oblique convergence in the Mesozoic, Geology, 14, 314-317, 1986.

Hodgson, R.A., Review of significant early studies in lineament tectonics, Utah Geol. Assoc. Publ., 5, 1-10, 1976.

Howell, D.G., D.L. Jones, and E.R. Schermer, Tectonostratigraphic terranes of the circum-Pacific region, in D.G. Howell, editor, Tectonostratigraphic Terranes of the Circum-Pacific Region, Houston, Circum-Pacific Council for Energy and Mineral Resources, 3-30, 1985.

Kent, P.E., Major synchronous events in continental shelves, Tectonophysics, 36, 87-91, 1976.

Kent, P.E., The Mesozoic development of aseismic continental margins, Jl. Geol. Soc. Lond., 134, 1-18, 1977.

McConnochie, M.J., and D.A. Henstridge, The Lowmead Graben - geology, Tertiary oil shale genesis and regional tectonic implications, Aust. Jl. Earth Sci., 32, 205-218, 1985.

Murphy, P.R., H. Schwarzböck, L.C. Cranfield, I.W. Withnall, and C.G. Murray, Geology of the Gympie 1:250,000 Sheet area, Geol. Surv. Qld. Rep., 96, 1976.

Murray, C.G., Tectonic setting of the Bowen Basin, Geol. Soc. Aust. Abs. Ser., 17, 5-16, 1985.

Nash, C.R., Tectonic interpretation of the northern Tasman Orogenic Zone from Landsat structural data, Proc. 3rd Australas. Remote Sensing Conf., Gold Coast, Qld., 433-441, 1984.

Nash, C.R., Comparative evolution of fault systems in the Yarrol Fold Belt and the Canadian Cordillera - a Landsat study, Geol. Soc. Aust. Abs. Ser., 14, 164-167, 1985.

Nash, C.R., Permo-Triassic evolution and metallogenesis of the Rockhampton-Maryborough area, Queensland - a photogeological investigation, Ore Geol Rev., 1, (in press), 1986.

Nash, C.R., P.R. Boshier, M.M. Coupard, A.C. Theron, and J.G. Wilson, Photogeology and satellite image interpretation in mineral exploration, Minerals Sci. Engng., 12, 216-244, 1980.

Nur, A., The tensile origin of fracture-controlled lineaments, Proc. 3rd International Conf. on Basement Tectonics, Denver, Basement Tectonics Committee, 155-167, 1981.

Scheibner, E., Suspect terranes in the Tasman Fold Belt System, eastern Australia, in D.G. Howell, editor, Tectonostratigraphic Terranes of the Circum-Pacific Region, Houston, Circum-Pacific Council for Energy and Mineral Resources, 493-514, 1985.

Seraphim, R.H., and V.F. Hollister, Structural settings, Can. Inst. Min. Metallurgy Spec. Vol., 15, 30-43, 1976.

Shaw, R.D., Sea floor spreading in the Tasman Sea: a Lord Howe Rise - eastern Australia reconstruction, Bull. Aust. Soc. Explor. Geophys., 9, 75-81, 1978.

Stephens, C., Late Triassic silicic volcanism near Gayndah, in Willmott, W.F., editor, 1986 Field Conference Guidebook, Brisbane, Geol. Soc. Aust. Qld. Div., 32-38, 1986.

Uyeda, S., Subduction zones: an introduction to comparative subductology, Tectonophysics, 81, 133-159, 1982.

Van Houten, F.B., Triassic-Liassic deposits of Morocco and eastern North America: comparison, Am. Assoc. Petrol. Geol. Bull., 61, 79-99, 1977.

Veevers, J.J., editor, Phanerozoic Earth History of Australia, Oxford Press, 418p, 1984.

Webb, J.A., A radiometric time scale of the Triassic, Jl. Geol. Soc. Aust., 28, 107-121, 1981.

Wilcox, R.E., T.P. Harding, and D.R. Seely, Basic wrench tectonics, Am. Assoc. Petrol. Geol. Bull., 57, 74-96, 1973.

Wood, C.A., Continental rift jumps, Tectonophysics, 94, 529-540, 1983.

PROVISIONAL SUBDIVISION OF NORTHEASTERN NORTH ISLAND, NEW ZEALAND, INTO TECTONOSTRATIGRAPHIC TERRANES

J. Duane Gibson[1]

Geology Department, University of Auckland, Auckland, New Zealand

Abstact. Three tectonostratigraphic terranes can be provisionally defined for the northeastern portion of North Island, New Zealand. The Koranga and Oponae-Waitahaia terranes are interpreted as subduction-accretion complexes that were emplaced during the latest Jurassic-Early Cretaceous and Albian-Cenomanian, respectively. The basal portion of the Matakaoa terrane is either a seamount or an island arc sequence with discrete intervals of Cretaceous and Lower Tertiary magmatic activity. The accretionary complexes of the Koranga and Oponae-Waitahaia terranes differ in the following aspects: 1) melange emplacement during the Albian of Early Jurassic and Albian strata is unique to the Oponae-Waitahaia terrane, 2) a regional Albian-Cenomanian (Ngaterian-Arowhanan) angular unconformity that is present within the Oponae-Waitahaia terrane is absent within the Koranga terrane, and 3) the oldest known units of the Koranga terrane consist of Lower and Upper Jurassic flysch sequences while Lower Jurassic chert and associated spilite, representing the uppermost portions of oceanic crust, are the only definite pre-Aptian lithologies of the Oponae-Waitahaia terrane. Both terranes have similar Cenomanian to Santonian strata, a widespread middle Late Cretaceous unconformity, and correlative Maastrichtian and younger strata. This overlap sequence linked the Koranga and Oponae-Waitahaia terranes by the late Albian-Cenomanian.

The Matakaoa terrane accreted to the Koranga-Oponae-Waitahaia composite terrane during the middle Tertiary. Detritus thought to be derived from the Matakaoa Volcanic Group forms a distinct igneous conglomerate lithofacies of the late Early Miocene (Otaian-Altonian) Ihungia Formation. At its type locality, this formation is separated from older units by a low-angle decollement. To the south, similar units are autochthonous.

The middle Cretaceous to Tertiary sequences of the Oponae-Waitahaia and Matakaoa terranes, as well as an unknown portion of the Koranga accretionary complex, have been involved in a middle to late Tertiary fold-and-thrust system. This system is related both to the emplacement of the Matakaoa terrane and to continued Neogene subduction off the eastern margin of North Island. All three of these terranes are currently separating from the rest of North Island by combined rigid-body rotation that results from extension across the Taupo Volcanic Zone and dextral strike-slip(?) displacement along regional faults in the western Raukumara Peninsula.

Introduction

During the last 20 years, major advances have been made in establishing stratigraphic and structural relationships in the Gisborne-East Cape region of northeastern North Island [Black, 1980; Crook and Feary, 1982; Feary, 1979; Feary and Hill, 1978; Feary and Pessagno, 1980; Isaac, 1977; Kenny, 1984; Moore, 1978; Moore et al., 1982; Phillips, 1985; Pirajno, 1979, 1980; Speden, 1972, 1973, 1975, 1976a, 1976b, 1977; Stoneley, 1968; Strong, 1976a, 1980]. Three tectonostratigraphic terranes are suggested from these relationships: the Koranga, Oponae-Waitahaia, and Matakaoa terranes (Figs. 1 and 2). The Koranga terrane lies in the southwest of the Raukumara Peninsula and continues southward for an unknown distance. Near East Cape, the Matakaoa terrane forms massifs that contain a variety of mafic and associated sedimentary units. This terrane is well exposed in the shore platform between Cape Runaway and Matakaoa Point and underlies the broad continental shelf to the north [Gillies, 1983]. The Oponae-Waitahaia terrane lies between these two terranes, extending northward from the central Raukumara Peninsula to the eastern Bay of Plenty coast and northeastward to at least the Waitahaia River catchment (Fig. 1). This study summarizes the internal stratigraphy, amalgamation, and subsequent disruption of the three proposed tectonostratigraphic terranes.

[1]Now at Tech. Reps., Inc., Albuquerque, New Mexico 87110 U.S.A.

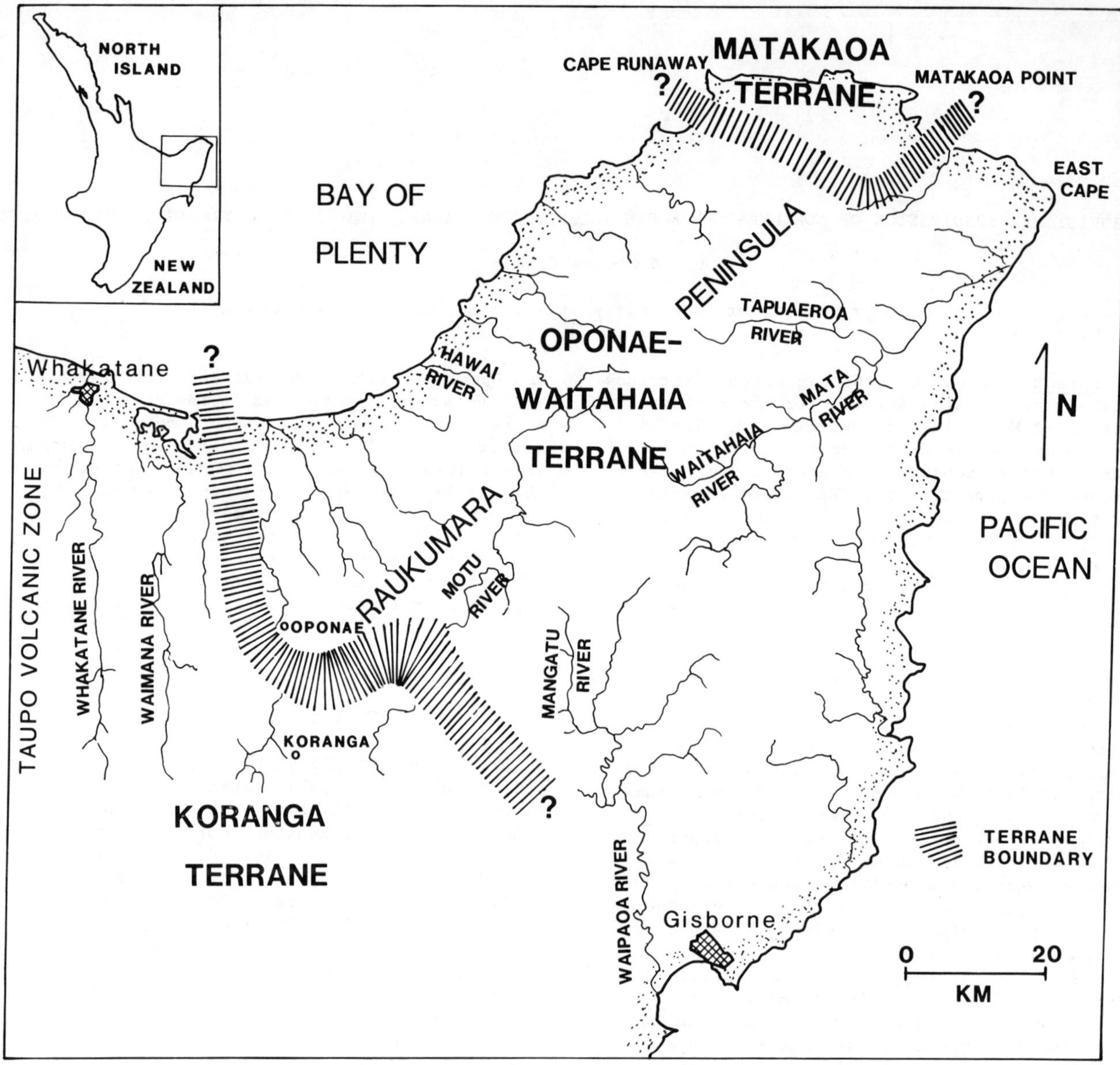

Fig. 1. Provisional tectonostratigraphic terranes of the Gisborne - East Cape (Raukumara Peninsula) region, northeastern North Island, New Zealand. Hachured lines show approximate terrane boundaries. Southeast continuation of the terrane boundary between the Koranga and Oponae-Waitahaia terranes is covered by Upper Tertiary strata.

Terranes

Koranga Terrane

The Koranga terrane is named for the Koranga Valley where tectonostratigraphic relationships within the terrane are best documented [Moore, 1978]. It contains a Jurassic accretionary complex unconformably overlain by a mildly deformed, discontinuous, fining-upwards, Aptian to Tertiary passive margin sequence (Fig. 2). The portion of the terrane below the unconformity consists of Jurassic flysch mapped as "Urewera Greywacke" [Grindley, 1960; Kingma, 1965; Feary, 1979], "Urewera Group" [Moore, 1978], "Torlesse Supergroup" [Speden, 1972; Feary and Pessagno, 1980], and "Torlesse terrane" [Sporli, 1978]. This unit contains dominantly volcaniclastic massive sandstone, alternating sandstone/ argillite, and massive argillite [Speden, 1972;

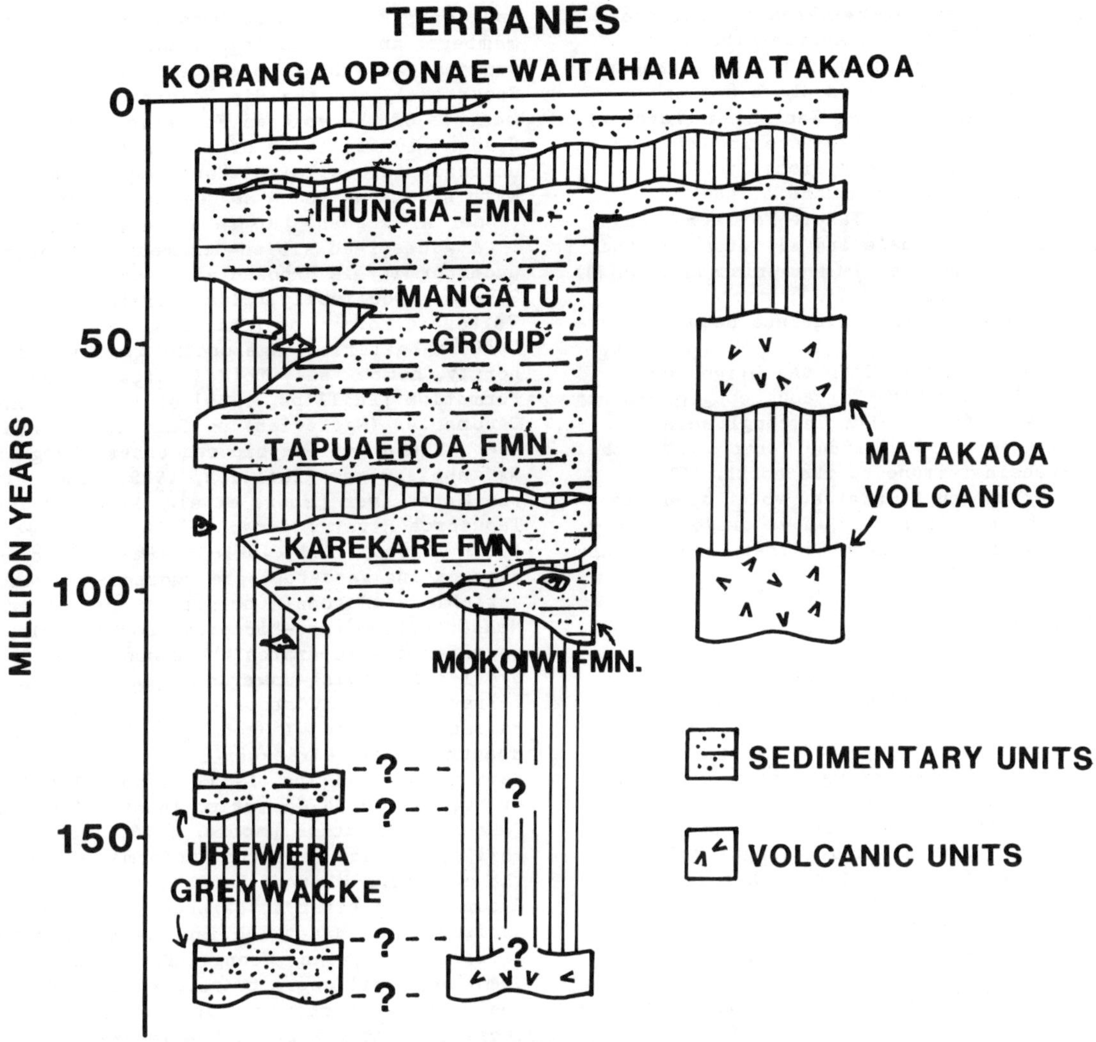

Fig. 2. Correlation of units of the Koranga, Oponae-Waitahaia, and Matakaoa terranes. Koranga terrane, after Isaac [1977], Moore et al. [1982], Speden [1972, 1973], Stoneley [1968]; Oponae-Waitahaia terrane, after Black [1980], Feary [1979], Feary and Pessagno [1980], Kenny [1984], Moore [1957], Moore et al. [1982], Phillips [1985], Speden [1976b], Stoneley [1968]; Matakaoa terrane, after Brothers and Delaloye [1982], Pirajno [1980], Strong [1976a, 1976b, 1980]. ("Fmn." = Formation).

Feary, 1979] with subordinate conglomerate that indicates a complex continental source. Speden [1976a] recognized two ages within the sparsely fossiliferous Urewera Greywacke, an Early Jurassic (Ururoan) sequence and a Late Jurassic (Heterian, Ohauan, and Puaroan) sequence. Neocomian strata (118-137 Ma) are apparently completely missing from this terrane as well as the rest of the New Zealand landmass [Sporli, 1978].

The unconformably overlying Aptian (Korangan) Koranga and Aptian-Albian (Urutawan) Te Wera Formations [Isaac, 1977; Moore, 1978; Speden, 1975] are fossiliferous lithic sandstone and conglomerate with rapid, lateral transitions in lithofacies. Locally, the conglomerate contains clasts derived from the immediately underlying Urewera Greywacke. Elsewhere, silicic plutonic and volcanic lithologies are abundant. A nearshore marine depositional environment is suggested for these formations. The Koranga Formation is locally absent and the Te Wera Formation lies directly on the Urewera Greywacke.

The Te Wera Formation is areally restricted and, where it is present, grades conformably into the widespread lower Albian (Urutawan) to locally Coniacian-Santonian (Teratan) Karekare Formation [Isaac, 1977; Moore, 1978; Speden, 1975]. The base of the Karekare Formation is time-transgressive. The age of strata unconformably overlying the Urewera Greywacke and other pre-Karekare units varies from Aptian-Albian (Urutawan) to Turonian-Coniacian (Mangaotanean).

All of the units mentioned above and the areally restricted Coniacian-Santonian (Teratan) Moanui Formation [Moore, 1978] are unconformably overlain by the thick-bedded, locally glauconitic, quartzose sandstone of the Maastrichtian (Haumurian) Tahore Formation and siliceous, micaceous shale and siltstone of the Maastrichtian-Paleocene(?) (Haumurian-Teurian(?)) Rakauroa Formation [Isaac, 1977; Moore, 1978].

Paleocene, Eocene, and Oligocene sequences are laterally restricted [Isaac, 1977; Moore, 1978; Stoneley, 1968]. Included in the Oligocene sequences are distinctive "igneous conglomerates" that are equated to the Ihungia conglomerate lithofacies [Kenny, 1980, 1984] found within the Oponae-Waitahaia terrane to the north. The matrix-supported conglomerates, which occur at various stratigraphic levels, have predominantly mafic igneous clasts with subordinate sedimentary and metamorphic lithologies.

Oponae-Waitahaia Terrane

This terrane extends from the Oponae melange [Feary, 1979; Feary and Pessagno, 1980] of the central Raukumara Peninsula to the Waitahaia River catchment [Phillips, 1985; Kenny, 1984] of the northeastern Raukumara Peninsula. "Exotic" Early Jurassic cherts [Feary, 1979; Feary and Pessagno, 1980] and "non-exotic" Albian (Motuan) strata are present within the melange, which was emplaced during the Albian (Motuan). The Oponae melange is overlain unconformably by an Albian (Motuan) marine sequence.

Descriptions by Hill [1974], Hoolihan [1977] and Moore [1957] suggest that units and relationships along the eastern Bay of Plenty are similar to those at Oponae. The Aptian-Albian (Korangan) fossils at Hawai River are the oldest obtained from the eastern Bay of Plenty [Speden, 1973]. These fossils are derived from concretions within the disrupted and tightly to isoclinally folded Tokaroa Zone of Hoolihan [1977]. That author believed that this tectonic zone is unconformably overlain by Albian (undifferentiated Clarence) age strata, although he did not observe the contact in his field area. Wellman [1959] considered the Aptian units at Hawai River to be stratigraphically continuous with sequences further inland and mapped much of the eastern Bay of Plenty as Taitai Series (late Early Cretaceous) strata. Speden [1973, 1975] interpreted the same region as a mosaic of Jurassic strata with outliers of middle Cretaceous and younger rocks. It is possible that much of the sequence in the eastern Bay of Plenty is similar to the Jurassic flysch units of the Koranga terrane.

The sequences in the eastern Bay of Plenty are similar to the Mokoiwi Formation [Speden, 1976b; Kenny, 1984; Gibson, 1986] of the northeastern Raukumara Peninsula that contains three distinct members: an alternating sandstone/mudstone sequence, a sandstone sequence (the Taitai Sandstone), and the Rip Volcanics. All three members of the Mokoiwi Formation are believed by Speden [1976b] to be of Albian age, although the unfossiliferous Taitai Sandstone is associated with strata as young as Cenomanian-Turonian (upper Arowhanan) [Moore et al., 1982].

A widespread Albian-Cenomanian (Ngaterian) unconformity is exposed along the Waitahaia River. Isoclinally folded strata of the Waitahaia Formation [Phillips, 1985] of Albian (Motuan-Ngaterian) age unconformably underlie gently, homoaxially folded units of the Karekare Formation [Phillips, 1985] of late Albian-Cenomanian (Ngaterian) to Santonian (Teratan) age. The stratigraphic sequences along the Waitahaia River [Phillips, 1985; Kenny, 1984], the Mata River [Moore et al., 1982], and the Tapuaeroa River [Speden, 1976b; Moore et al., 1982; Gibson, 1986] include similar carbonaceous alternating sandstone and mudstone sequences with variable sandstone/mudstone ratios. A pre-Maastrichtian (pre-Piripauan to Haumurian) unconformity separates these sequences from overlying, fining-upwards sequences of sandstone, siltstone, and mudstone that are locally carbonaceous and glauconitic. The Campanian-Maastrichtian (Piripauan-lower Haumurian) Owhena Formation [Phillips, 1985] is correlated to parts of the Tapuaeroa Formation [Kenny, 1984] and the Rakauroa Formation [Moore, 1978; Isaac, 1977]. Phillips [1985] maps the conformable contact with the overlying Whangai Formation of the Mangatu Group by the first appearance of a concretionary horizon. The Maastrichtian (Haumurian) to Paleocene (Teurian) Whangai Formation is dominantly siliceous to calcareous mudstone, but the unit has highly variable lithofacies. Locally, it may represent a lateral equivalent of the upper portion of the Tapuaeroa Formation. The remainder of the Maastrichtian (Haumurian) to Oligocene (Landon) Mangatu Group has a poorly defined internal stratigraphy of flysch, calcareous mudstone, limestone, greensand, and swelling mudstone lithofacies [Phillips, 1985; Kenny, 1984]. These units are in fault contact with the sandy mudstones and subordinate igneous conglomerate units of the lower Miocene (Otaian-Altonian) Ihungia Formation [Kenny, 1980, 1984; Phillips, 1985]. By analogy with studies by Stoneley [1968] to the S, the Ihungia Formation is interpreted as having been originally in depositional contact with the underlying Mangatu units.

Matakaoa Terrane

Exposed near East Cape and extending to the N beneath the adjacent continental shelf is the Matakaoa Volcanic Group [Pirajno, 1980; Rutherford, 1980; Gifford, 1970; Moore, 1957; Gillies, 1983]. This group contains basalt, dolerite, gabbro, and other mafic plutonic lithologies with minor associated sedimentary units. The Matakaoa Volcanic Group is dominantly tholeiitic, but a secondary alkali-enriched tholeiitic or calc-alkaline trend may be present [Pirajno, 1980]. Discrimination diagrams of immobile trace elements suggest two contrasting magmatic settings, one with intra-plate oceanic affinity and the other with island arc affinity [Pirajno, 1980].

Microfossils from the intercalated sedimentary units suggest two ages of magmatic activity [Katz, 1976; Strong, 1976a, 1976b, 1980]: a late Early Cretaceous (Barremian-Cenomanian, probably Albian) age and an early to middle Eocene (Dannevirke to earliest Arnold) age. Three K-Ar age determinations from hornblende separates obtained from the Matakaoa Volcanic Group range from 27 to 43 Ma [Brothers and Delaloye, 1982]. A third early Miocene age, perhaps reflecting the latest deformational overprinting, is suggested from whole-rock K-Ar analysis [Strong, 1980].

The Matakaoa Volcanic Group is overlain unconformably by the lower Miocene (Waitakian-Otaian) Whakai Formation [Chapman-Smith and Grant-Mackie, 1971]. This unit consists of mudstone with sandstone, grit, and conglomerate that is locally rich in mafic igneous clasts. It has been correlated with the co-eval Ihungia Formation [Chapman-Smith and Grant-Mackie, 1971]. Locally, various lithofacies of the Mangatu Group are faulted against the Matakaoa Volcanic Group [Chapman-Smith and Grant-Mackie, 1971]. This contact is interpreted in this study as a terrane boundary.

Terrane Amalgamation and Accretion

Koranga+Oponae-Waitahaia Amalgamation

Amalgamation of these two terranes occurred in the late Albian to Cenomanian, following a widespread angular unconformity within the Oponae-Waitahaia terrane (Fig. 2). The overlying Cenomanian to Santonian strata of the Oponae-Waitahaia terrane are locally coarser than coeval strata of the Koranga terrane. The sequences are tentatively interpreted as an overlap sequence with amalgamation of the Koranga and Oponae-Waitahaia terranes occurring within the late Albian to Cenomanian (Fig. 2). No unique fault or suture zone has been mapped between the Koranga and Oponae-Waitahaia terranes. The existence and location of this terrane boundary is based on regional stratigraphic and structural relationships.

Additional evidence is found within the Upper Cretaceous to support pre-Upper Cretaceous amalgamation of the two terranes. Both terranes contain a terrane-wide, pre-Maastrichtian (pre-Mata) angular unconformity. Similar Upper Cretaceous to Lower Tertiary units, including the Whangai Formation, overlie this unconformity. Depositional contacts have been shown between all known members of the Upper Cretaceous to Lower Tertiary overlap sequences.

Matakaoa+Koranga-Oponae-Waitahaia Amalgamation

A middle Tertiary accretion of the Matakaoa terrane to the Koranga-Oponae-Waitahaia composite terrane is suggested by two features: 1) a similar age of deformation within both terranes, and 2) provenance linking of material derived from the Matakaoa terrane that was deposited on the Koranga-Oponae-Waitahaia composite terrane.

Early Miocene whole rock K-Ar ages [Strong, 1980] probably represent the age of prevasive deformation and argon loss within the Matakaoa terrane, rather than the age of magmatic emplacement. This age is similar to the latest Oligocene-Early Miocene age that is favored for emplacement of thrust nappes by Kenny [1984], Mazengarb [1982], Black [1980], and Stoneley [1968] within the composite Koranga-Oponae-Waitahaia terrane.

The Early Miocene (Upper Otaian) to Middle Miocene (lower Lillburnian) Ihungia Formation consists dominantly of massive sandy mudstone and flysch sequences. The formation includes a distinctive igneous conglomerate with clasts of various mafic and associated chert lithologies [Kenny, 1984]. These clasts are believed to be derived from the eroding Matakaoa terrane that was adjacent to the Koranga-Oponae-Waitahaia composite terrane at the time of deposition of this unit.

Terrane Disruption

Tertiary Fold-and-thrust Belt

Middle Tertiary deformation. The Cretaceous and Tertiary overlap sequence of the Koranga-Oponae-Waitahaia composite terrane has been partially disrupted by a fold-and-thrust belt (Fig. 3). The basal decollement or frontal thrust [Morley, 1986] has been mapped by Stoneley [1968] and Mazengarb [1982] in the upper Mangatu and Waipaoa catchments (Fig. 1). Stoneley [1968] supported an overall southwest-directed transport for this fold-and-thrust system. A somewhat analogous contact between a Tertiary "autochthon" and "allochthon" has been mapped by Kenny [1984] and Phillips [1985] in the Waitahaia area. Elsewhere, the basal decollement may be buried, and its location and geometry remains debated. The possibility that the Tertiary fold-and-thrust belt extends into the Urewera Greywacke within the central Raukumara Peninsula as well as Jurassic(?)-Cretaceous flysch units of the eastern Bay of Plenty cannot be ruled out. Widespread Tertiary nappe transport would be difficult to recognize in these areas [Hoolihan, 1977; Hill, 1974].

Late Tertiary deformation. The early Miocene Ihungia Formation is interpreted in its type locality as a thrust sheet that was probably

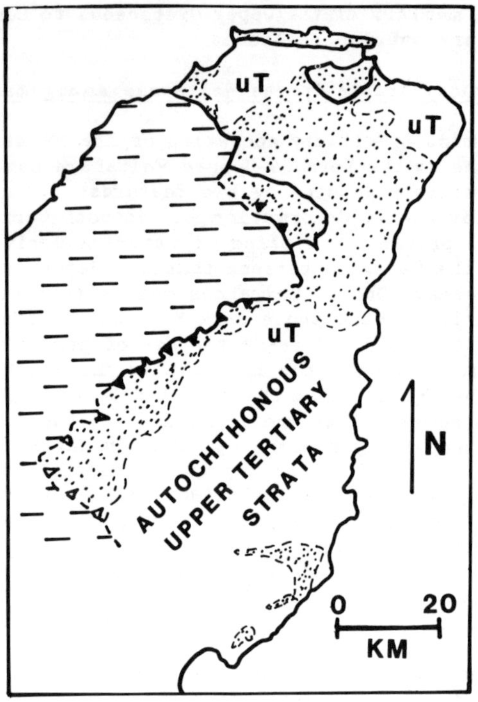

Fig. 3. Areal extent (stippled) of Tertiary fold-and-thrust belt of northeastern North Island [after Moore, 1985]. Hachured area to west and southwest was apparently not affected by this fold-and-thrust system. ("uT" - autochthonous Upper Tertiary strata).

emplaced from the north in post-middle Miocene (post-Lillburnian) time [Kenny, 1984; Phillips, 1985]. The relative extent of nappe tectonics within the latest Oligocene-early Miocene and following the middle Miocene is difficult to determine. Much of the deformation and tectonic transport of Jurassic through Lower Tertiary strata may be young, even less than 10 Ma, resulting from continued subduction along the Hikurangi Trough.

Quaternary Deformation

Ongoing deformation of the Koranga-Oponae-Waitahaia-Matakaoa composite terrane results from movement along regional strike-slip(?) faults along the western margin of the composite terrane, such as the Whakatane and Waimana faults [Speden, 1975], and from regional uplift along the eastern margin of the Raukumara Peninsula [Berryman and Hull, 1984; Yoshikawa et al., 1980]. In addition, ongoing extension along the Taupo Volcanic Zone is rotating this composite terrane away from the remainder of North Island [Walcott and Mumme, 1982].

Comparison With the Rest of New Zealand

The tectonostratigraphic sequences found in the northeastern North Island are similar to those of the Wairarapa/southern Hawkes Bay [Moore and Speden, 1979], northeastern South Island [Bishop et al., 1985], and Northland [Ballance and Sporli, 1979]. Bishop et al. [1985] include Late Jurassic to Early Cretaceous strata within the Pahau terrane, a younger division of the Torlesse terrane. These authors recognize a localized(?) intra-Albian disturbance within the Pahau terrane that is apparently also present in the Wairarapa area [Moore and Speden, 1979]. Early Jurassic units, such as the flysch of the Koranga terrane and the chert of the Oponae-Waitahaia terrane, are apparently not present within the Pahau terrane. Dating of chert within the Pahau terrane of South Island is limited, however.

Similarities and differences between tectonostratigraphic sequences in Northland and the East Coast have been summarized by Ballance and Sporli [1979]. Both Northland and the Raukumara Peninsula contain ophiolitic seamount bodies [Brothers and Delaloye, 1982; Brothers, 1983], the Tangihua Volcanics in Northland and the Matakaoa Volcanics of the Raukumara Peninsula. Allochthonous Cretaceous-Tertiary sedimentary sequences are comparable, although minor variations in lithofacies and stratigraphic sequences are found. The major difference between Northland and the East Coast is within the autochthonous units overlying Jurassic greywacke. Autochthonous middle to upper Cretaceous units are not present in Northland. The oldest autochthonous units overlying greywacke basement are thin Lower Tertiary sequences that resemble coeval sequences on the Raukumara Peninsula. Autochthonous, pre-Eocene (Arnold) strata were either never deposited on the Jurassic greywacke units or, more probably, were removed prior to the Upper Eocene.

Summary

Three tectonostratigraphic terranes are defined within northeastern North Island, two of which are accretionary complexes while the third is an accreted seamount complex. The older of the accretionary complexes, the Koranga terrane, amalgamated to the younger accretionary complex, the Oponae-Waitahaia terrane, during the middle Cretaceous. The oldest units of the Koranga terrane are Early and Late Jurassic flysch sequences that are overlain unconformably by a late Early Cretaceous (Aptian) and younger marine sequence. This terrane was apparently not deformed while the Oponae-Waitahaia terrane was emplaced during the Albian-Cenomanian. The oldest units of the Oponae-Waitahaia terrane are Early Jurassic chert and associated spilite, presumably the uppermost portions of an Early Jurassic oceanic crust, and middle Cretaceous flysch-like units. At Oponae, these units were incorporated into a melange that was emplaced during the middle Albian (intra-Motuan). This melange is unconformably overlain by an Albian and younger overlap sequence that stitched this terrane to the Koranga terrane within the Albian. Albian flysch within the Oponae-Waitahaia terrane are folded into upright, isoclinal folds along the Waitahaia and Mata Rivers. These folds are unconformably overlain by openly folded Cenomanian and younger strata. Both the isoclinal folding of Albian strata and emplacement of melange during the Albian-Cenomanian are thought to have occurred within a subduction-accretion complex.

Magmatic activity that formed the Matakaoa Volcanics occurred in two discrete episodes in the middle Cretaceous and the Lower Tertiary. This seamount terrane accreted to the composite Koranga-Oponae-Waitahaia terrane during the middle Tertiary. Detritus from the Matakaoa seamount terrane is included within the Ihungia Formation and correlative units of Early Miocene (Altonian) age found within all three terranes.

All of these terranes have undergone post-accretion deformation within an oblique convergent margin. The fold-and-thrust system that began with accretion of the Matakaoa terrane continued into the late Cenozoic. Regional, N-S trending, dextral strike-slip(?) faults have offset the western portion of these terranes while the eastern portion has been uplifted.

During the last few million years, the Koranga-Oponae-Waitahaia-Matakaoa composite terrane has rotated away from the rest of North Island owing to spreading across the Taupo Volcanic Zone.

Acknowledgments. The author is grateful to Peter Ballance and Bernhard Sporli for discussion during this study. Comments by Evan Leitch and an anonymous reviewer were appreciated.

References

Ballance, P.F., and K.B. Sporli, Northland Allochthon, J. R. Soc. N.Z., 9, 259-275, 1979.

Berryman, K., and A. Hull, Guidebook to the North Island scientific excursions, International symposium on recent crustal movements of the Pacific region, R. Soc. Misc. Ser., R. Soc. N.Z., 8, Wellington, New Zealand, 1984.

Bishop, D.G., J.D. Bradshaw, and C.A. Landis, Provisional Terrane Map of South Island, New Zealand, in Tectonostratigraphic Terranes of the Circum-Pacific Region, edited by D.G. Howell, pp. 515-522, Circum-Pacific Council for Energy and Mineral Resources, 1985.

Black, R.D., Upper Cretaceous and Tertiary geology of the Mangatu State Forest, Raukumara Peninsula, New Zealand, N.Z. J. Geol. Geophys., 23, 293-312, 1980.

Brothers, R.N., Tertiary accretion of ophiolit seamounts, North Island, New Zealand, in Accretion Tectonics in the Circum-Pacific Regions, edited by M. Hashimoto and S. Uyeda, pp. 307-318, 1983.

Brothers, R.N., and M. Delaloye, Obducted ophiolites of North Island, New Zealand: origin, age, emplacement, and tectonic implications for Tertiary and Quaternary volcanicity, N.Z. J. Geol. Geophys., 25, 257-274, 1982.

Chapman-Smith, M., and J.A. Grant-Mackie, Geology of the Whangaparaoa area, Eastern Bay of Plenty, N.Z. J. Geol. Geophys., 14, 3-38, 1971.

Crook, K.A.W., and D.A. Feary, Development of New Zealand according to the fore-arc model of crustal evolution, Tectonophysics, 87, 65-107, 1982.

Feary, D.A., Geology of the Urewera Greywacke in the Waioeka Gorge, Raukumara Peninsula, New Zealand, N.Z. J. Geol. Geophys., 22, 693-708, 1979.

Feary, D.A., and P.H. Hill, Mesozoic Radiolaria from cherts in the Raukumara Peninsula, New Zealand, N.Z. J. Geol. Geophys., 21, 363-373, 1978.

Feary, D.A., and E.A. Pessagno, Jr., An Early Jurassic age from chert within the Early Cretaceous Oponae Melange (Torlesse Supergroup), Raukumara Peninsula, New Zealand, N.Z. J. Geol. Geophys., 23, 623-628, 1980.

Gibson, J.D., Tectonic history of the Cretaceous Mokoiwi Formation and adjacent units, northeastern North Island, New Zealand, Ph.D. thesis, Univ. Auckland, Auckland, N.Z., 165 pp., 1986.

Gifford, W.G.R., The igneous geology of the Mangaroa Range, Hicks Bay area, M.Sc. thesis, Univ. Auckland, Auckland, N.Z., 107 pp., 1970.

Gillies, P.H., A marine geophysical study of the junction of the Kermadec and Hikurangi subduction systems, Ph.D. thesis, Univ. Auckland, Auckland, N.Z., 1983.

Grindley, G.W., Sheet 8 Taupo (1st ed.), Geological Map of New Zealand 1:250000, N.Z. Dept. Scientific and Industrial Research, Wellington, 1960.

Hill, P.H., Taitai Series rocks at Te Kaha, M.Sc. thesis, Univ. Auckland, Auckland, N.Z., 1974.

Hoolihan, K., Basement geology of the Motu River area, Raukumara Peninsula, M.Sc. thesis, Univ. Auckland, Auckland, N.Z., 1977.

Isaac, M.J., Mesozoic geology of the Matawai District, Raukumara Peninsula, Ph.D. thesis, Univ. Auckland, Auckland, N.Z., 1977.

Katz, H.R., Cretaceous foraminifera from the Matakaoa Volcanic Group - letters to the editor, N.Z. J. Geol. Geophys., 19, 943-947, 1976.

Kenny, J.A., Geology of the Ihungia catchment, Raukumara Peninsula, M.Sc. thesis, Univ. Auckland, Auckland, N.Z., 1980.

Kenny, J.A., Stratigraphy, sedimentology and structure of the Ihungia decollement, Raukumara Peninsula, North Island, New Zealand, N.Z. J. Geol. Geophys., 27, 1-19, 1984.

Kingma, J.T., Sheet 6 East Cape, Geological Map of New Zealand 1:250000, N.Z. Dept. Scientific and Industrial Research, Wellington, 1965.

Mazengarb, C., Structure and stratigraphy at the front of the Maungahaumi nappes, Mangatawa, Raukumara Peninsula, M.Sc. thesis, Univ. Auckland, Auckland, N.Z., 1982.

Moore, P.R., Geology of the western Koranga Valley, Raukumara Peninsula, New Zealand, N.Z. J. Geol. Geophys., 21, 1-20, 1978.

Moore, P.R., Distribution, age, and relationships of volcanic rocks, ("East Coast volcanics") in the Gisborne - East Cape region, N.Z. Geol. Surv. rec. no. 8, 16-23, 1985.

Moore, P.R., and I.G. Speden, Stratigraphy, structure, and inferred environments of deposition of the Early Cretaceous sequence, eastern Wairarapa, New Zealand, N.Z. J. Geol. Geophys., 22, 417-433, 1979.

Moore, P.R., I.G. Speden, M.J. Isaac, and C.P. Strong, Notes on Cretaceous and Early Tertiary stratigraphic sections in the Gisborne Region, East Coast workshop, Informal publication of the N.Z. Geol. Surv., 44 pp., 1982.

Moore, W.R., Geology of the Raukokore Area, Raukumara Peninsula, North Island, M.Sc. thesis, Victoria University of Wellington, Wellington, N.Z., 1957.

Morley, C.K., A Classification of Thrust Fronts, Am. Assoc. Pet. Geol. Bull., 70, 12-25, 1986.

Phillips, C.J., Upper Cretaceous and Tertiary geology of the upper Waitahaia River, Raukumara Peninsula, New Zealand, N.Z. J. Geol. Geophys., 28, 595-607, 1985.

Pirajno, F., Geology, geochemistry, and mineralisation of a spilite-keratophyre association in Cretaceous flysch, East Cape area, New Zealand, N.Z. J. Geol. Geophys., 22, 307-328, 1979.

Pirajno, F., Subseafloor mineralization in rocks of the Matakaoa Volcanics around Lottin Point, East Cape, New Zealand, N.Z. J. Geol. Geophys., 23, 313-334, 1980.

Rutherford, P.G., Geology of the Matakaoa Volcanic Group, Cape Runaway area, M.Sc. thesis, Univ. Auckland, Auckland, N.Z., 1980.

Speden, I.G., New fossil localities in the Torlesse Supergroup, western Raukumara Peninsula, New Zealand, N.Z. J. Geol. Geophys., 15, 433-445, 1972.

Speden, I.G., Distribution, stratigraphy and stratigraphic relations of Cretaceous sediments, western Raukumara Peninsula, New Zealand, N.Z. J. Geol. Geophys., 16, 243-268, 1973.

Speden, I.G., Cretaceous stratigraphy of the Raukumara Peninsula, N.Z. Geol. Soc. Bull., 91, 69 pp., 1975.

Speden, I.G., Fossil localities in Torlesse rocks of the North Island, New Zealand, J. R. Soc. N.Z., 6, 73-91, 1976a.

Speden, I.G., Geology of Mt. Taitai, Tapuaeroa Valley, Raukumara Peninsula, N.Z. J. Geol. Geophys., 19, 71-119, 1976b.

Speden, I.G., Taitai Series (Early Cretaceous) and the elimination of the Mokoiwian Stage, N.Z. J. Geol. Geophys., 20, 537-562, 1977.

Sporli, K.B., Mesozoic tectonics, North Island, New Zealand, Geol. Soc. Am. Bull., 89, 415-425, 1978.

Stoneley, R., A Lower Tertiary decollement on the East Coast, North Island, New Zealand, N.Z. J. Geol. Geophys., 11, 128-156, 1968.

Strong, C.P., Cretaceous foraminifera from the Matakaoa Volcanic Group, N.Z. J. Geol. Geophys., 19, 140-143, 1976a.

Strong, C.P., Cretaceous foraminifera from the Matakaoa Volcanic Group-Reply, N.Z. J. Geol. Geophys., 19, 945-947, 1976b.

Strong, C.P., Early Paleogene foraminifera from Matakaoa Volcanic Group (Note), N.Z. J. Geol. Geophys., 23, 267-272, 1980.

Walcott, R.I., and T.C. Mumme, Paleomagnetic study of the Tertiary sedimentary rocks from the East Coast of the North Island, New Zealand, Geophys. Div. Rep. no. 189, N.Z. Dept. Scientific and Industrial Research, Lower Hutt, New Zealand, 44 pp., 1982.

Wellman, H.W., Divisions of the New Zealand Cretaceous, Trans. R. Soc. N.Z., 87, 99-163, 1959.

Yoshikawa, T., Y. Ota, M. Yonekura, A. Okada, and N. Iso, Marine terraces and their tectonic deformation on the northeast coast of the North Island, New Zealand, Geogr. Rev. Jap., 53, 235-262 (In Japanese with English abstract), 1980.

ASPIRING TERRANE: AN OCEANIC ASSEMBLAGE FROM NEW ZEALAND AND ITS IMPLICATIONS FOR TERRANE ACCRETION IN THE SOUTHWEST PACIFIC

R. J. Norris and D. Craw

Department of Geology, University of Otago, Dunedin, New Zealand

Abstract. Recent interpretations of the Mesozoic Rangitata orogen of New Zealand recognise a series of accreted terranes. The terrane most generally accepted as allochthonous is the quartzo-feldspathic Torlesse terrane. Its suture with the neighbouring volcanoclastic Caples terrane is enigmatic, occurring somewhere in an extensive zone of highly deformed schists. Recent work has recognised a lithological association of oceanic character, the Aspiring lithological association, within the schists between the two terranes. The margins are complex but presently dominated by large, nappe-like folds and shear zones belonging to D2/D3 phase of deformation. The presence of an oceanic suite between the two clastic terranes adds further support to the allochthonous view of the Torlesse. The Aspiring association is here considered as a terrane itself, although it probably represents part of an accretionary complex originally developed between the Torlesse and Caples terranes. Disrupted terranes in general are likely to contain segments of original suture zones and may have greater displacements within them than along their margins.

Introduction

The Rangitata Orogen of New Zealand has an orogenic history extending back to at least early Permian time with final consolidation and uplift occurring in mid-Cretaceous [Coombs et al., 1976]. Early attempts at interpretation [e.g. Fleming, 1970] recognised a volcanic arc (Brook Street terrane in Figure 1) bordering the western foreland (Tuhua terrane) and bounded to the east by a miogeosynclinal (Murihiku terrane) and a eugeosynclinal suite (Caples and Torlesse terranes). The latter was envisaged to have accumulated in an elongate geosyncline and been deformed and metamorphosed during Cretaceous time by westward subduction of Pacific seafloor. Problems with such a model include, firstly, the mainly early Permian age of the Brook Street terrane compared with the Triassic and Jurassic age of the volcanogenic sediments and tuffs of the Brook Street and Murihiku terranes [Waterhouse, 1964; Coombs, 1954; Campbell and Coombs, 1966; Houghton, 1981], and secondly, the quartzo-feldspathic continent-derived detritus which characterises the Torlesse terrane lying oceanward of the contemporaneous volcanogenic belt [Landis and Bishop, 1972; Force, 1972]. Landis and Bishop [1972] addressed the second problem by invoking "leap-frogging" of quartzo-feldspathic sediment across the volcanic belt, while Spörli [1978] suggested a lateral supply into a trench. Bradshaw and Andrews [1973] argued for an easterly source for the Torlesse, an argument accepted by Coombs et al. [1976] in their major synthesis of the Rangitata orogen in South Island.

Further evidence for the separate nature of the Hokonui and Torlesse terranes came from the recognition of the Caples terrane as a suite of redeposited volcanogenic sediments, originally thought of as part of the Torlesse terrane [Campbell and Campbell, 1970] but totally different in provenance [Kawachi, 1974; Turnbull, 1979, 1980; Carter et al., 1978; McKinnon, 1983]. While the Caples terrane is at least partly contemporaneous with the Torlesse terrane and has a similarly complex structural history, the contrasted provenance, much more akin to that of the Brook Street and Murihiku terranes, provides a strong argument in favour of an allochthonous Torlesse.

The Dun Mountain ophiolite belt was recognised by Coombs et al. [1976] as a major suture separating Brook Street/Murihiku and Caples terranes, the Murihiku being interpreted as a frontal-arc basin and the Caples as a fore-arc/trench accretionary complex. The Torlesse was envisaged as an allochthonous sedimentary prism rafted in and accreted sometime between Triassic and Cretaceous [Blake, Jones and Landis, 1974]. The source of the Torlesse was not identified. Norris et al. [1977] first proposed an origin of the Torlesse as accumulating at an active continental margin continuous with the ensimatic Hokonui/Caples arc, and being emplaced by oblique-slip displacement. This model was refined by Bradshaw et al. [1981] and further developed

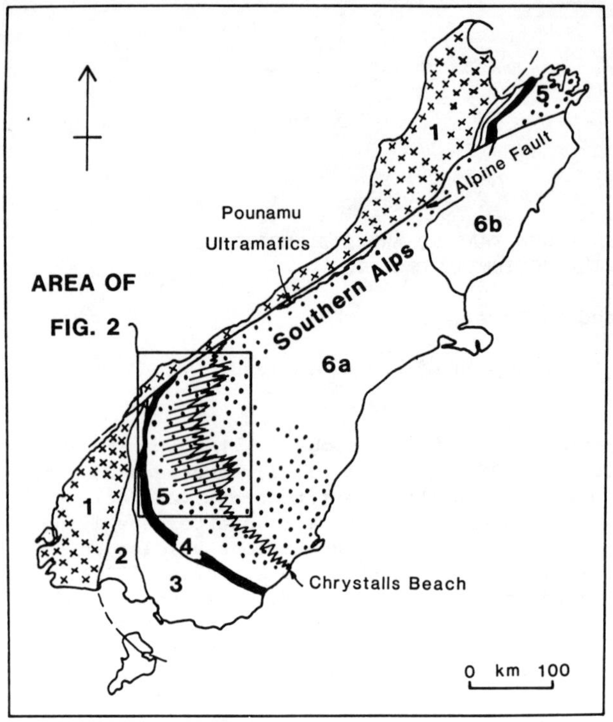

Fig. 1. Tectonic map of South Island, largely based on Bishop et al. [1983], and Coombs [1985]. 1 - Tuhua terrane [undifferentiated]; 2 - Brook Street terrane; 3 - Murihiku terrane; 4 - Dun Mountain terrane [ophiolite belt]; 5 - Caples terrane; 6 - Torlesse terrane, [6a: older Torlesse, 6b: younger Torlesse]. Haast Schist zone is shown stippled and area underlain by Aspiring association is cross-hatched.

by McKinnon [1983], although Kamp [1984] preferred to invoke a mythical Pacific continent as a source.

Howell [1980] argued that the absence of a Triassic/Jurassic volcanic arc as a source of the Murihiku terrane requires major post-Jurassic displacements between the latter and the Tuhua terrane, an argument accepted by Bishop et al. [1983] and supported by recent work [Coombs, 1985; Mattinson et al., 1986]. Howell also argued that the western part of the Caples terrane represents a disrupted fore-arc basin sequence, displaced from its parent arc and accreted against the Dun Mountain ophiolite, while the eastern part of the Caples is an accretionary complex formed beneath a trench slope.

The development of ideas on the evolution of the Rangitata orogen range from the early fixist geosynclinal models through more mobilist views of accretionary margins [Landis and Bishop, 1972; Coombs et al., 1976] to those embodying a complex of accreted terranes [Howell, 1980; Bishop et al., 1983]. Most of the later authors accept a major break between the volcanogenic Caples terrane and the voluminous quartzo-feldspathic sediments of the Torlesse terrane. Between the two lies the Haast Schist, a zone of complexly deformed, mainly greenschist facies, schists which has gradational contacts with the Caples and Torlesse rocks on either flank. The Caples-Torlesse boundary is simply said by most authors to lie "somewhere in the Haast Schist". The lack of an obvious major fault or suture has led to criticisms of the treatment of Caples and Torlesse as separate terranes, and to some workers [e.g. Korsch and Wellman, 1986] returning to viewing the Caples-Torlesse as a single accretionary wedge.

This paper reviews recent mapping in northwest Otago that lends further weight to the terrane model and, in addition, examines the definition of terranes and the nature of terrane boundaries.

The Aspiring Association

The Haast Schist is the name given to all rocks possessing a cleavage or foliation lying between the Caples and Torlesse metasediments. In their lower textural and metamorphic grades, the schists are only weakly cleaved sediments with sedimentary features still preserved. At high textural grades, bedding has largely been destroyed by intense transposition, multiple refolding and extensive metamorphic segregation. Nevertheless, even within medium and high grade schist, broad lithologic variations reflect changes in parent lithology and allow recognition of different lithologic associations.

Torlesse rocks are dominated by thick sequences of quartzo-feldspathic flysch varying between sand-dominated and mud-dominated [Carter et al. 1978]. These give rise to distinctive quartzo-feldspathic schists with well developed veining and quartz-albite segregation laminae. Metavolcanic horizons tend to be rare, thin, isolated features [e.g. Bishop 1974; Spörli et al., 1974; Ward and Spörli, 1979]. Caples-derived schists on the other hand tend to be less quartzose and less veined and segregated. Their distinction in outcrop rather than regionally, however, is far from easy.

In the Mt. Aspiring region of northwest Otago (Figure 2), the schists are dominated by pelitic lithologies, basic metavolcanic schists and associated cherts [Hutton, 1940; Henley and Norris, 1974; Craw, 1984]. These lithologies are very minor constituents of the flanking Caples and Torlesse terranes. This association (termed the Aspiring lithologic association by Craw [1984]) is most readily interpreted as an oceanic suite of lithologies [Henley and Norris, 1974; Cooper, 1976; Craw, 1984] situated between the dominantly coarse clastic flysch sequences of the two neighbouring terranes.

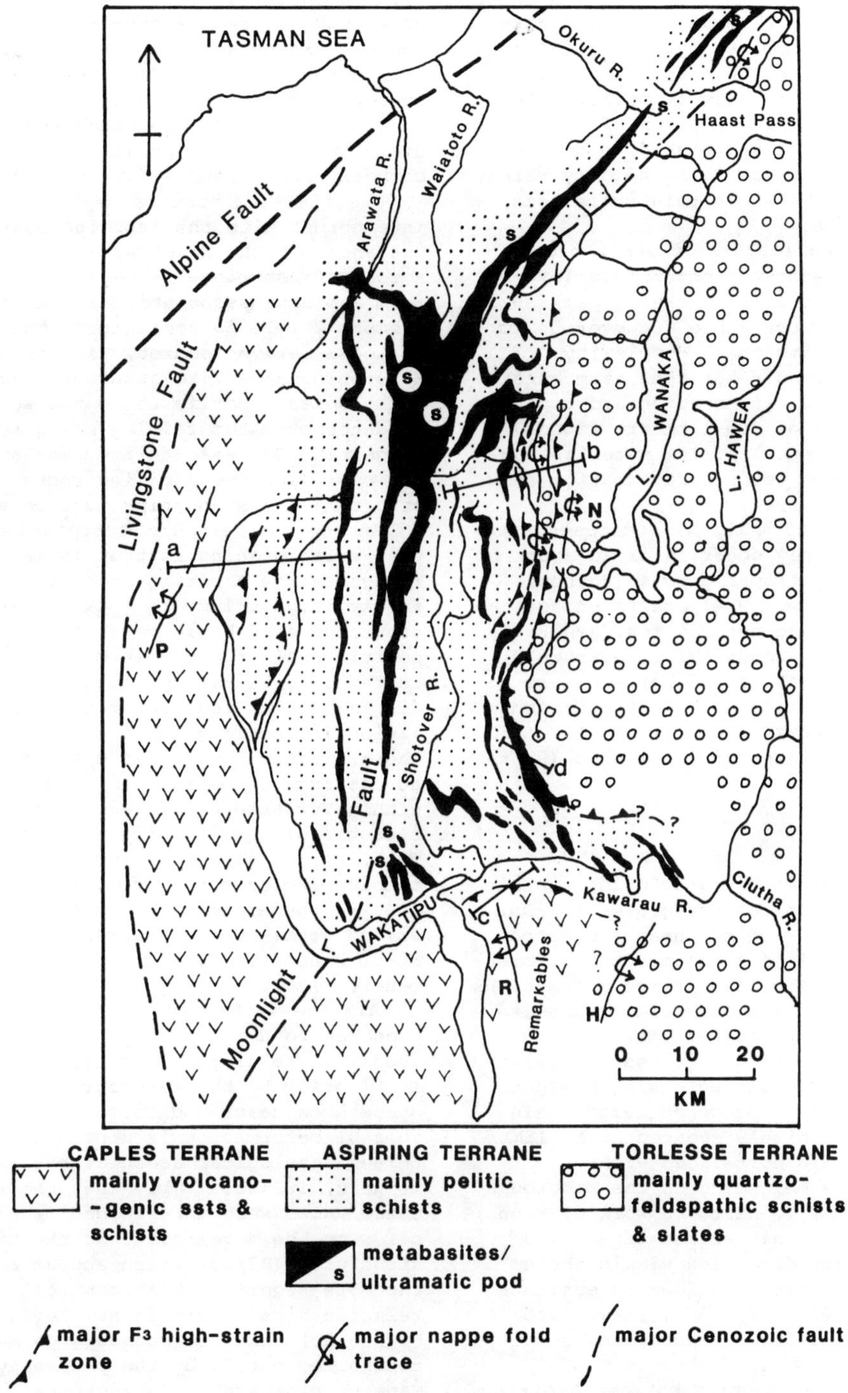

Fig. 2. Simplified geological map of northwest Otago showing probable structural relationships and distribution of Caples, Torlesse and Aspiring terranes. Locations of cross-sections in Figure 3 (a, b, c, d) are indicated. Major nappe folds are: P - Poseidon anticline [Bishop et al. 1976]; N - Niger nappe [Craw, 1985]; R - Remarkables nappe; H - Horn Range antiform [Turnbull, 1981].

Description

Detailed descriptions of Aspiring association rock-types have been presented by Craw [1984] and are only summarised here. The association is dominated (c. 70%) by a variety of pelitic rock-types, including thick (>1 km) piles of moderately coarse-grained, non-segregated, micaceous schist. The latter lithology is found only rarely elsewhere in Otago and then in relatively thin, discontinuous horizons.

Substantial volumes (c.20%) of basic meta-volcanic rocks (greenschists) occur interlayered with pelitic schist and are remarkably persistent, some zones being traceable for over 50 km along strike despite structural complexity. Thicknesses of metavolcanic horizons vary considerably due to structural attenuation and repetition, but thicknesses of 0.5-1 km are found. Pillow lavas and fragmental pillow material can be recognised in places despite deformation and recrystallisation.

Metacherts (up to 100 m thick) and associated Fe-Mn-rich rocks commonly occur in or adjacent to meta-volcanic units. Ultramafic pods [Hutton, 1940; Cooper, 1976; Craw, 1984] are found scattered through the Aspiring association, generally within the metavolcanic zones. Minor quartz-feldspar-rich (psammitic) schist horizons are interlayered with pelitic schist, but constitute only a very small proportion (<10%) of the Aspiring association. Rare pods of marble (up to 50 m thick) are also found.

Structural Relations

The rocks of the Aspiring association exhibit at least five generations of structures (D1-D5; Craw, [1985]). The last two of these are associated with relatively late fault movements, probably of Cenozoic age. The first three generations produce penetrative fabrics and large-scale isoclinal folds (see Figure 3).There is a strong possibility that D2 and D3 structures are related and formed consecutively during a single major deformational event which produced large-scale nappe-like folds with highly sheared lower limbs [Craw, 1985]; these are correlated with structures defined as belonging to the Manorburn generation [Norris, 1977, based on work by Means, 1966]. Rotation of fold hinges and lineations towards the stretching direction within these high-strain zones (Figure 4) allows an estimate to be made of the shear direction [Craw, 1985; S. Cox, University of Otago unpublished report, 1985].

In many places, the junction between Aspiring association schist and schists more typical of Torlesse derivation appears to be a high-strain zone, or series of such zones, associated with the lower limb of a Manorburn generation nappe, the Torlesse-derived schist being thrust westward over the Aspiring association (Figure 3), [Craw, 1985; Norris and Craw, unpub. observations]. The relationship with the Caples terrane schists, at least locally, also seems to be one of a major shear zone beneath a nappe-like structure of Caples terrane schists, in this case thrust eastward [Craw, 1985; S. Cox, University of Otago unpublished report,1985]. The field relationships in general are more complicated than this however, and it is not possible to map a single boundary shear zone (cf. Figure 3).

The flanking schists may be considerably interthrust with the Aspiring rocks - for instance, in the Haast River area [Cooper, 1974], a zone of Aspiring-like schists lies along strike from Aspiring association in the type area (Figure 2). It is overlain to the east by Torlesse-derived schist, also folded into a large Manorburn nappe-like structure. However, similar lithologies underlie it to the west (without any evidence of later folding being responsible), suggesting Torlesse-derived schist both underlies and over-lies the Aspiring rocks. Such complex structural relationships must be expected.

Another problem in interpreting the structural relationships is that it is difficult to recognise the Aspiring association unless a fairly large volume of rock is represented. Small occurrences of basic meta-volcanics, cherts and pelitic schists occur in both Torlesse and Caples terranes [e.g. Bishop, 1974; Turnbull, 1979]; the distinguishing feature of the Aspiring association is the large area of schist dominated by these lithologies. It is not possible at present to distinguish metavolcanic or chert horizons within Torlesse or Caples terrane schists from tectonic slices of Aspiring association rocks. In fact, narrow zones of highly sheared mélange containing inclusions of oceanic material similar to that in the Aspiring association have been described within Caples terrane [Kawachi, 1974; K. S. Pound, University of Otago unpublished report, 1982].

All the rocks have suffered extensive penetrative deformation including large scale isoclinal folding and imbrication [Bishop et al., 1976] prior to the Manorburn generation structures described above [Craw, 1985]; whether this D1 deformation is continuous between the three lithological associations is not clear. If it were, it would imply that the rocks of the three units were interfolded on a regional scale prior to the development of the Manorburn structures (D2/D3) which appear at present to be the more significant in controlling the contact relationships. There is some evidence (e.g. Figure 3b) that the boundary geometry is not controlled purely by the presently mapped Manorburn generation structures. Either the Manorburn structures are more complex than mapped or they are imposed on a pre-existing (?D1) contact. Folding and shear zones together with a regionally penetrative foliation of Manorburn generation completely dominate the margins of the Aspiring association, making any pre-existing relationships very difficult to ascertain.

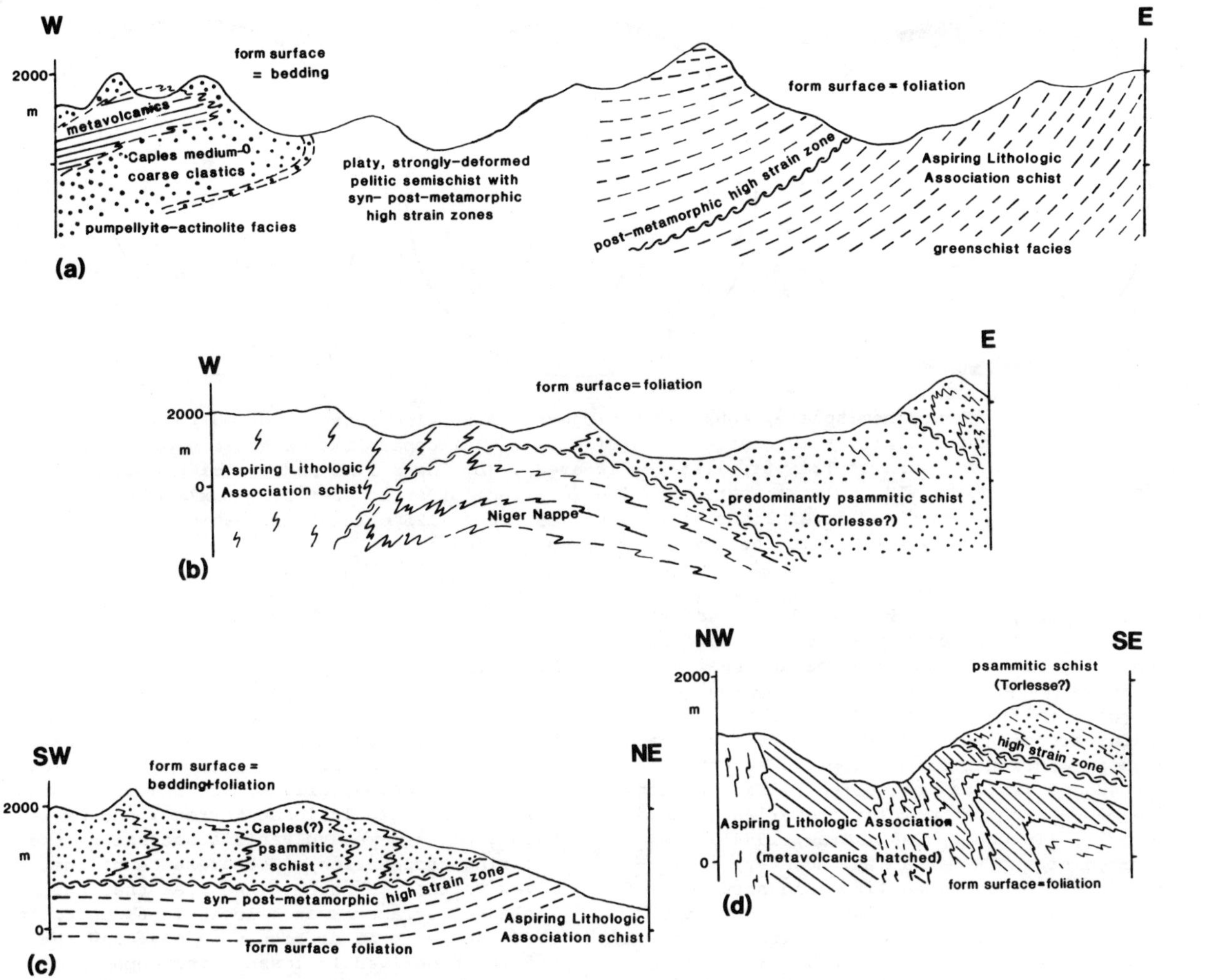

Fig. 3. Cross-sections (partially schematic) on lines indicated in Figure 2. (N.B. The scale is not the same in each section; however, in each section, the horizontal and vertical scales are equal).

Metamorphism

Most of the Aspiring association rocks so far mapped exhibit greenschist facies metamorphism, including chlorite and biotite zones [Hutton, 1940; Craw, 1981]. Estimated P/T conditions for the chlorite zone rocks are around 4.5 kb and 400°C [Yardley, 1982; Kawachi et al.,1983]. The general metamorphic facies sequence within the Haast Schists can be considered as Barrovian in type [Cooper, 1972; Turner, 1981, p.428-438], although Brown [1975, 1977] presents evidence for relatively high P/T conditions. In addition, remnant cores of blue amphibole are occasionally found within the more usual green actinolitic amphibole of the Aspiring association schists or as inclusions in albite porphyroblasts [Hutton, 1940] and have been used to argue for an early high P/T phase of metamorphism [Yardley, 1982]. Similar rare blue amphibole is found in the Caples terrane which also locally exhibits lawsonite-albite-chlorite facies [Kawachi, 1975].It is not clear whether this higher P/T event affected all the Haast Schists or was confined to the Caples and Aspiring units. In any case, it appears to have pre-dated the Manorburn generation structures which currently appear to control the margins of the Aspiring association.

Regional Distribution

The main area of outcrop of the Aspiring association is in the type area of north-west Otago. To the north, similar lithologies continue as far as the Haast River area [Craw, 1984; cf. Cooper, 1974] and may relate to the ophiolitic fragments and associated rocks of the Pounamu Ultramafic Belt in the central Southern Alps

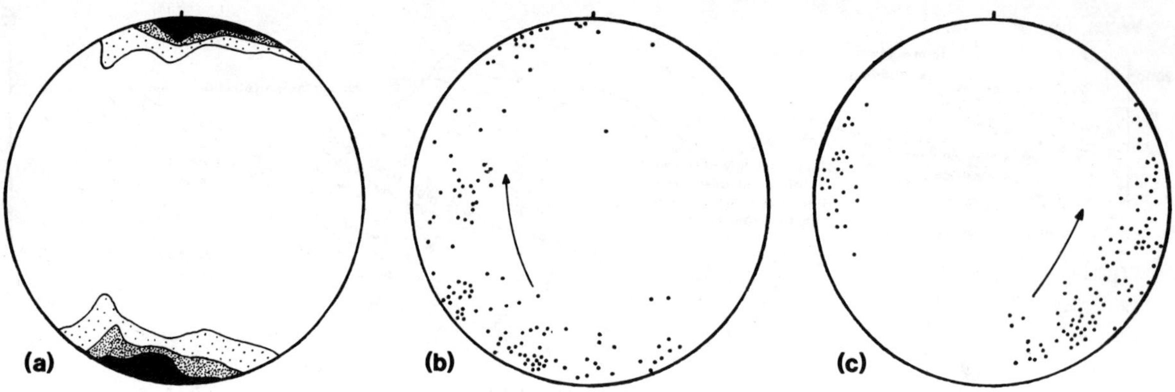

Fig. 4. Lower hemisphere, equal area projections showing rotation of D3 fold hinges and lineations into stretching direction. (a) upper limb of Niger nappe (245 points, contours at 12%, 6%, 1.2% per 1% area); (b) lower limb and associated high strain zones (145 points); note rotation (arrowed) of structures from maximum in (a); (c) upper and lower limbs of Remarkables nappe (125 points); lineations are progressively rotated in the lower limb in direction of the arrow.

[Cooper and Reay, 1983]. Southward, the Aspiring association lithologies appear to terminate in the vicinity of the Kawarau Gorge and the Remarkables Mountains. The rocks cropping out in the gorge itself and on the lower slopes of the mountains are of Aspiring type, and are separated by high strain zones from psammitic schists above, suggesting that the gorge forms a tectonic window below a thrust complex of Torlesse schist (Figure 2), [Turnbull, 1981; Craw, 1984]. The main ridge of the Remarkables, however, is located within a large nappe-like fold structure of probable Caples-derived schist verging to the east and also overlying Aspiring lithologies along a high strain zone [S. Cox, University of Otago unpublished report, 1985]. The relationship between this east-verging structure of Caples derivation and the west-verging structures of Torlesse derivation such as the Horn Range antiform immediately to the east [Turnbull, 1981] is not known at present. The Aspiring association, however, appears to be entirely covered by these overthrusting units to the south.

On the coast south of Dunedin (Figure 1), a 20 km long section through low-grade metagreywackes contains large thicknesses of melange and broken formation containing abundant radiolarian chert in a deformed argillitic matrix [Campbell and Campbell, 1970; Nelson, 1982]. Occasional meta-volcanic horizons also occur. Rocks of the Caples terrane are structurally involved with the melanges and also outcrop extensively to the south and west. It is possible that this more oceanic suite of lithologies on the northeastern margin of the Caples terrane represents the Aspiring association reappearing from beneath its cover. Apart from the coastal strip, however, exposure is relatively poor in east Otago and little is known about the detailed structure between the Remarkables and the sea.

Discussion

Regional Relationships

In northwest Otago, we can recognise three broad lithological associations within the Haast Schists: two clastic associations, one volcanogenic (the Caples terrane) and the other quartzofeldspathic (the Torlesse terrane) and between them a pelagic-metavolcanic association (the Aspiring association). The presence of the latter gives further support to the view that the Torlesse terrane is an allochthonous unit, in that, while it is difficult to envisage the continent-derived Torlesse being deposited alongside the Caples rocks in a trench outboard of an ensimatic volcanic arc, it is even more difficult to arrange for the deposition of a pelagic suite of oceanic sediments between them.

A major uncertainty in any argument, however, is the age relationships between the three units. The age of the Torlesse is reasonably well established as Permian to Jurassic, although large gaps in the range of biostratigraphic dates occur [Campbell and Warren, 1965; McKinnon, 1983]. The age of the Caples terrane is much less certain, but definitely includes Permian and probably Triassic strata [Campbell and Campbell, 1970; Turnbull, 1980], so that it is at least in part coeval with the Torlesse. The age of the Aspiring association is totally unknown, so it is possible that it is entirely older than either of the other two and even formed a basement on which they were deposited.

The original relationships of the Aspiring association with the two flanking terranes are enigmatic. The Caples rocks become more and more pelitic eastwards [Bishop et al., 1976] and there is some suggestion that they may grade into, or interdigitate with, rocks of the Aspiring association [Craw, 1984], (e.g. Figure 3), although as described earlier, any such relationship has now been greatly modified by later thrusting. Cooper and Reay [1983] consider that basal Torlesse-derived schists may stratigraphically overlie the disrupted ophiolite bodies of the Pounamu Ultramafic Belt; while the stratigraphic relationship is far from proven and there is no clear correlation between the Pounamu rocks and the Aspiring association, the possibility of a similar relationship remains. In Otago, the Aspiring-Torlesse boundary is dominated by Manorburn generation structures to the extent that any original relationship is undecipherable at present.

Is the Aspiring Lithological Association a Terrane?

The question arises as to whether the Aspiring lithological association should be redefined as the "Aspiring terrane". Terranes have been defined as "fault-bounded geologic entities of regional extent, each characterized by a geological history that is different from the histories of contiguous terranes" [Jones et al., 1982]. The present boundaries of the Aspiring association are largely tectonic, although they are certainly more complex than single faults. While the age relationships are uncertain, the lithologies indicate an oceanic suite as distinct from the adjacent clastic suites. The present spatial relations are not directly compatible with inferred geologic histories. As we accept the terrane concept as essentially being a nongenetic way of subdividing an orogen into tectono-stratigraphic units [e.g. Coney et al., 1980], we therefore propose that the Aspiring association be considered as a suspect terrane.

Problems may arise, however, if we start pushing terranes around the globe as fault-bounded, discrete entities. The Aspiring terrane is most reasonably interpreted as an accretionary complex of oceanic material and therefore only became a discrete entity at its site of accumulation, which may well have been its present site. At its south-western margin, it may have graded into or underlain the Caples terrane. Similarly, at its other margin it may have graded into or underlain the Torlesse terrane. It represents the closure of a piece of ocean, maybe originally separating the Caples and Torlesse terranes. Taking another view, the Aspiring terrane could itself be considered as a complex suture zone between the Caples and Torlesse terranes; it may be allochthonous with respect neither to the Caples nor the Torlesse, yet these two could still be allochthonous with respect to each other.

In terms of the classification presented by Jones et al. [1982], the Aspiring association is a metamorphosed disrupted terrane. Until such time as its original relationships with Caples and Torlesse can be elucidated, it is best viewed as a terrane in its own right. It should be borne in mind, however, that the displacements within such a terrane may be far greater than the displacements along its margins, as it probably contains segments of an imbricate suture zone originally developed by the approach of the two adjacent terranes.

Acknowledgements. Field work was supported by grants from the N. Z. University Grants Committee and the Otago University Research Committee. We thank D. G. Bishop, D. S. Coombs, A. F. Cooper, C. A. Landis, and I. M. Turnbull for valuable discussion and comment, and Simon Cox for information on the Remarkables.

References

Bishop, D. G., Stratigraphic, structural and metamorphic relationships in the Dansey Pass area, Otago, New Zealand. N. Z. J. Geol. Geophys., 17, 301-335, 1974.

Bishop, D. G., Bradshaw, J. D., Landis, C. A., and Turnbull, I. M., Lithostratigraphy and structure of the Caples terrane of the Humboldt Mountains, New Zealand, N. Z. J. Geol. Geophys., 19, 827-848, 1976.

Bishop, D. G., Bradshaw, J. D., and Landis, C.A., A provisional terrane map of the South Island, New Zealand, in Proceedings of the Circum-Pacific Terrane Conference, edited by D. G. Howell, D. L. Jones, A Cox , and A. Nur, pp. 24-31, Stanford University Publications, Geological Sciences, XVIII, Stanford , California, 1983.

Bradshaw, J. D., and Andrews, P. B., Geotectonics and the New Zealand Geosyncline, Nature, 241, 14-16, 1973.

Bradshaw, J. D., Adams, C. J., and Andrews, P. B., Carboniferous to Cretaceous on the margin of Gondwana, the Rangitata phase of New Zealand, in Proceedings of the Fifth International Gondwana Symposium, Wellington, New Zealand, edited by M. M. Cresswell, and P. Vella, pp. 217-221, A.A. Balkema, Rotterdam, The Netherlands, 1980.

Brown, E. H., Comparison of the mineralogy and phase relations of blueschists from the north Cascades, Washington, and greenschists from Otago, New Zealand, Geol. Soc. Am. Bull., 85, 333-344, 1975.

Brown, E. H., The crossite content of Ca-amphibole as a guide to pressure of metamorphism, J. Petrol., 18, 53-72, 1977.

Campbell, J. D., and Warren, G., Fossil localities of the Torlesse Group in the South Island, Trans. R. Soc. N. Z., Geol., 3, 99-137, 1965.

Campbell, J. D. and Coombs, D. S., Murihiku Supergroup (Triassic-Jurassic) of Southland and South Otago, N. Z. J. Geol. Geophys., 9, 393-398, 1966.

Campbell, J. K., and Campbell, J. D., Triassic

tube fossils from Tuapeka rocks, Akatore, South Otago, N. Z. J. Geol. Geophys., 13, 392-399, 1970.

Carter, R. M., Hicks, M. D., Norris, R. J., and Turnbull, I. M., Sedimentation patterns in an ancient arc-trench-ocean basin complex: Carboniferous to Jurassic Rangitata Orogen, New Zealand, in Sedimentation in submarine canyons, fans and trenches, edited by D. J. Stanley and G. Kelling, 340-361, Dowden, Hutchinson and Ross, Stroudsberg, Virginia, 1978.

Coney, P. J., Jones, D. L., and Monger, J. W. H., Cordilleran Suspect Terranes, Nature, 288, 329-333, 1980.

Coombs, D. S., The nature and alteration of some Triassic sediments from Southland, New Zealand, Trans. R. Soc. N. Z., 82, 65-109, 1954.

Coombs, D. S., New Zealand Terranes (abstract), in Third Circum-Pacific Terrane Conference Extended Abstracts, edited by E. C. Leitch, Geological Society of Australia Abstracts, 14, 45-48, 1985.

Coombs, D. S., Landis, C. A., Norris, R. J., Sinton, J. M., Borns, D. J., Craw, D., The Dun Mountain Ophiolite Belt, New Zealand, its tectonic setting, constitution, and origin, with special reference to the southern portion, Am. J. Sci., 276, 561-603, 1976.

Cooper, A. F., Progressive metamorphism of meta-basic rocks from the Haast Schist Group of southern New Zealand, J. Petrol., 13, 347-492, 1972.

Cooper, A. F., Multiphase deformation and its relationship to metamorphic crystallisation at Haast River, south Westland, New Zealand, N. Z. J. Geol. Geophys., 17, 855-880, 1974.

Cooper, A. F., Concentrically zoned ultramafic pods from the Haast Schist zone, South Island, New Zealand, N. Z. J. Geol. Geophys., 19, 603-623, 1976.

Cooper, A. F., and Reay, A., Lithology, field relationships and structure of the Pounamu Ultramafics from the Whitcombe and Hokitika Rivers, Westland, New Zealand, N. Z. J. Geol. Geophys., 26, 359-379, 1983.

Craw, D., Oxidation and microprobe-induced potassium mobility in iron-bearing phyllo-silicates from the Otago schists, New Zealand, Lithos, 14, 49-57, 1981.

Craw, D., Lithologic variations in Otago Schist, Mount Aspiring area, northwest Otago, New Zealand. N. Z. J. Geol. Geophys., 27, 151-166, 1984,

Craw, D., Structure of schist in the Mt. Aspiring region, northwestern Otago, New Zealand, N. Z. J. Geol. Geophys., 28, 55-75, 1985.

Fleming, C. A., The Mesozoic of New Zealand, chapters in the history of the Circum-Pacific mobile belt, Q. J. Geol. Soc. London, 125, 125-170, 1970.

Henley, R. W., Metamorphism of the Moke Creek lode, Otago, New Zealand, N. Z. J. Geol. Geophys., 18, 229-237, 1975.

Henley, R. W., and Norris, R. J., The Central Otago metavolcanics - remnants of a spreading axis? (abstract), Geol.Soc. N.Z. Annual Conference, Wellington, Abstr., 1974.

Houghton, B. F., Lithostratigraphy of the Takitimu Group, central Takitimu Mountains, western Southland, New Zealand, N. Z. J. Geol. Geophys., 24, 3389-348, 1981.

Howell, D. G., Mesozoic accretion of exotic terranes along the New Zealand segment of Gondwanaland, Geology, 8, 487-491, 1980.

Hutton, C. O., Metamorphism in the Lake Wakatipu region, western Otago, New Zealand, N. Z. Dept. Sci. Indust. Res. Mem., 5, 90 pp., 1940.

Jones, D. L., Howell, D. G., Coney, P. J., and Monger, J. W. H., Recognition, character and analysis of tectonostratigraphic terranes in western North America, in Accretion tectonics in the circum-Pacific regions, edited by M. Hashimoto and S. Uyeda, 21-36, D. Riedel Publishing Co., Boston, 1982.

Kamp, P. J. J., Pacifica and New Zealand, proposed eastern elements in Gondwanaland's history, Nature, 288, 659-664, 1980.

Kawachi, Y., Geology and petrochemistry of weakly metamorphosed rocks in the upper Wakatipu district, southern New Zealand, N. Z. J. Geol. Geophys, 17, 169-208, 1974.

Kawachi, Y., Pumpellyite-actinolite and contig-uous facies metamorphism in part of upper Wakatipu district, South Island, New Zealand, N. Z. J. Geol. Geophys, 18, 401-441, 1975.

Kawachi, Y., Grapes, R. H., Coombs, D. S., and Dowse, M., Mineralogy and petrology of a piemontite-bearing schist, western Otago, New Zealand, J. Metamorph. Geol, 1, 353-372, 1983.

Korsch, R. J., and Wellman, H. W., The geological evolution of New Zealand and the New Zealand region, in The Ocean Basins and their Margins, Vol. 7B, The Pacific, edited by A. E. M. Nairn, F. G. Stehli, and S. Uyeda, Plenum Press, New York, in press, 1986.

Landis, C. A., and Bishop, D. G., Plate tectonics and regional stratigraphic-metamorphic relations in the southern part of the New Zealand Geosyncline, Geol. Soc. Am. Bull., 83, 2267-2284, 1972.

MacKinnon, T. C., Origin of the Torlesse terrane and coeval rocks, South Island, New Zealand, Geol. Soc. Am. Bull., 94, 967-985, 1983.

Mattinson, J. M., Kimbrough, D. L., and Bradshaw, J. Y., Western Fiordland orthogneiss, early Cretaceous arc magmatism and granulite facies metamorphism, New Zealand, Contrib. Mineral. Petrol., 92, 383-392, 1986.

Means, W. D., A macroscopic recumbent fold in schist near Alexandra, Central Otago, N. Z. J. Geol. Geophys, 9, 169-208, 1966.

Nelson, K. D., A suggestion for the origin of mesoscopic fabric in accretionary mélange, based on features observed in the Chrystalls Beach Complex, South Island, New Zealand, Geol. Soc. Am. Bull., 93, 625-634, 1982.

Norris, R. J., Fold generations in the Otago schist - their definition and correlation (abstract), Geol. Soc. N. Z. Annual Conference, Queenstown, Abstr., 1977.

Norris, R. J., Landis, C. A., and Ward, C. M., The relationship of the Rangitata orogen to possible plate configurations in the south Pacific, (abstract), Geol. Soc. N. Z. Annual Conference, Queenstown, Abstr., 1977.

Spörli, K. B., Mesozoic tectonics, North Island, New Zealand, Geol. Soc. Am. Bull, 89, 415-425, 1978.

Spörli, K. B., Stanaway, K. J., and Ramsay, W. R. H., Geology of the Torlesse Supergroup in the southern Liebig and Burnett ranges, Canterbury, New Zealand, J. R. Soc. N. Z., 4, 177-192, 1974.

Turnbull, I. M., Petrography of the Caples terrane in the Thomson Mountains, northern Southland, New Zealand, N. Z. J. Geol. Geophys., 22, 709-727, 1979.

Turnbull, I.M., Structure and interpretation of the Caples terrane of the Thomson Mountains, northern Southland, New Zealand, N. Z. J. Geol. Geophys., 23, 43-62, 1980.

Turnbull, I. M., Contortions in the schists of the Cromwell district, Central Otago, New Zealand, N. Z. J. Geol. Geophys., 24, 65-86, 1981

Turner, F. J., Metamorphic Petrology (2nd edition), McGraw-Hill, New York, 524 pp., 1981.

Ward, C. M., and Spörli, K. B., Exceptionally large steeply plunging folds in the Torlesse terrane, New Zealand, J. Geol., 87, 187-193,1978.

Waterhouse, J. B., Permian stratigraphy and faunas of New Zealand, Bull. N. Z. Geol. Surv., n.s.72, 101 pp., 1964.

Yardley, B. W. D., The early metamorphic history of the Haast Schists and related rocks of New Zealand, Contrib. Mineral. Petrol., 81, 317-327, 1982.

TECTONOSTRATIGRAPHIC TERRANES OF THE CROISSILLES HARBOUR REGION, SOUTH ISLAND, NEW ZEALAND

C. A. Landis

University of Otago, Dunedin, New Zealand

M. C. Blake, Jr.

U.S. Geological Survey, Menlo Park, California

Abstract. The boundary between Hokonui and Te Anau assemblages is flanked by a broad (10-20 km) zone of imbricated slabs of late Paleozoic and Mesozoic lithostratigraphic terranes. Five terranes are mapped, three of predominantly sedimentary character (Dun Mountain-Maitai, Rai, Pelorus) are separated by two consisting of ophiolitic melange (Patuki, Croisilles). A regional stratigraphy is mapped within the Late Permian Maitai Group, but formational subdivision of Rai and Pelorus rocks is not attempted because of degree of deformation and the more monotonous aspect of these graywacke-dominated suites. Sandstone petrography and petrochemistry indicate provenance of sands mainly from dissected and undissected arc sources. Each of the three sedimentary belts possesses its own lithic character and we can see no evidence for original stratigraphic interrelations. The two ophiolitic melanges are similar, but not identical. They probably formed through tectonic disruption of the Dun Mountain ophiolite. Correlation seems more likely for these but must await further evidence. Each terrane possesses distinctive structural style, metamorphic grade, and probably metamorphic timing. The Croisilles area represents a complex zone of rocks formed under both convergent and divergent conditions; their evolution began early in the Permian and continued until achieving the present relationship, probably by middle Mesozoic time. Original relations between the five terranes remain equivocal and they should be regarded as suspect terranes until further constraints are available.

Introduction

The area surrounding Croisilles Harbour in the Nelson-Marlborough Sounds, northern South Island, New Zealand (Fig. 1), contains rocks spanning the boundary between the Hokonui and Te Anau assemblages [Carter et al., 1974; Blake et al., 1974; Turnbull, 1977] of the Rangitata Orogen [Waterhouse, 1975]. These two lithotectonic assemblages, each comprising two or more terranes characterized by voluminous late Paleozoic-Mesozoic graywackes, are of contrasting lithological, structural, and metamorphic character, and the boundary separating them is a tectonic contact of first-order importance [Landis and Bishop, 1972]. An irregular coast line, 500 m relief, and wave-washed coastal exposures permit detailed mapping of these rocks.

In summary, the Hokonui assemblage is characterized by very thick, orderly sequences of sedimentary rock underlain to the west by thick arc volcanic and volcaniclastic rocks, and to the east by thinner ophiolitic basement. Hokonui rocks are preserved in gently plunging synclinal folds with intervening anticlines either poorly developed or lacking. Sediment is volcanogenic, containing abundant vitric tuff, and is commonly fossiliferous. Metamorphism is of burial type (zeolite, prehnite-pumpellyite, lawsonite-albite-chlorite) with local development of a weak metamorphic fabric.

The Te Anau assemblage, in contrast, is characterized by thick but complexly deformed sedimentary rocks. Again, the rocks are predominantly volcanogenic but they are generally coarser grained and lacking in vitric tuffs while containing cherts. Fossils are extremely rare. Metamorphism is of low grade regional character with penetrative deformation features widespread. These rocks pass gradually into a regional schist zone, the Haast Schist. The base, at its only exposure, is a thick pile of basaltic volcanics. Correlation between Hokonui and Te Anau assemblages is not recognized.

Previous work includes regional mapping of the Nelson-Marlborough Sounds at 1:250,000 scale by the New Zealand Geological Survey [Beck, 1964], showing gross structure and distribution of rock types. Stratigraphy of Permian rocks is described by Waterhouse [1964, 1967], while lith-

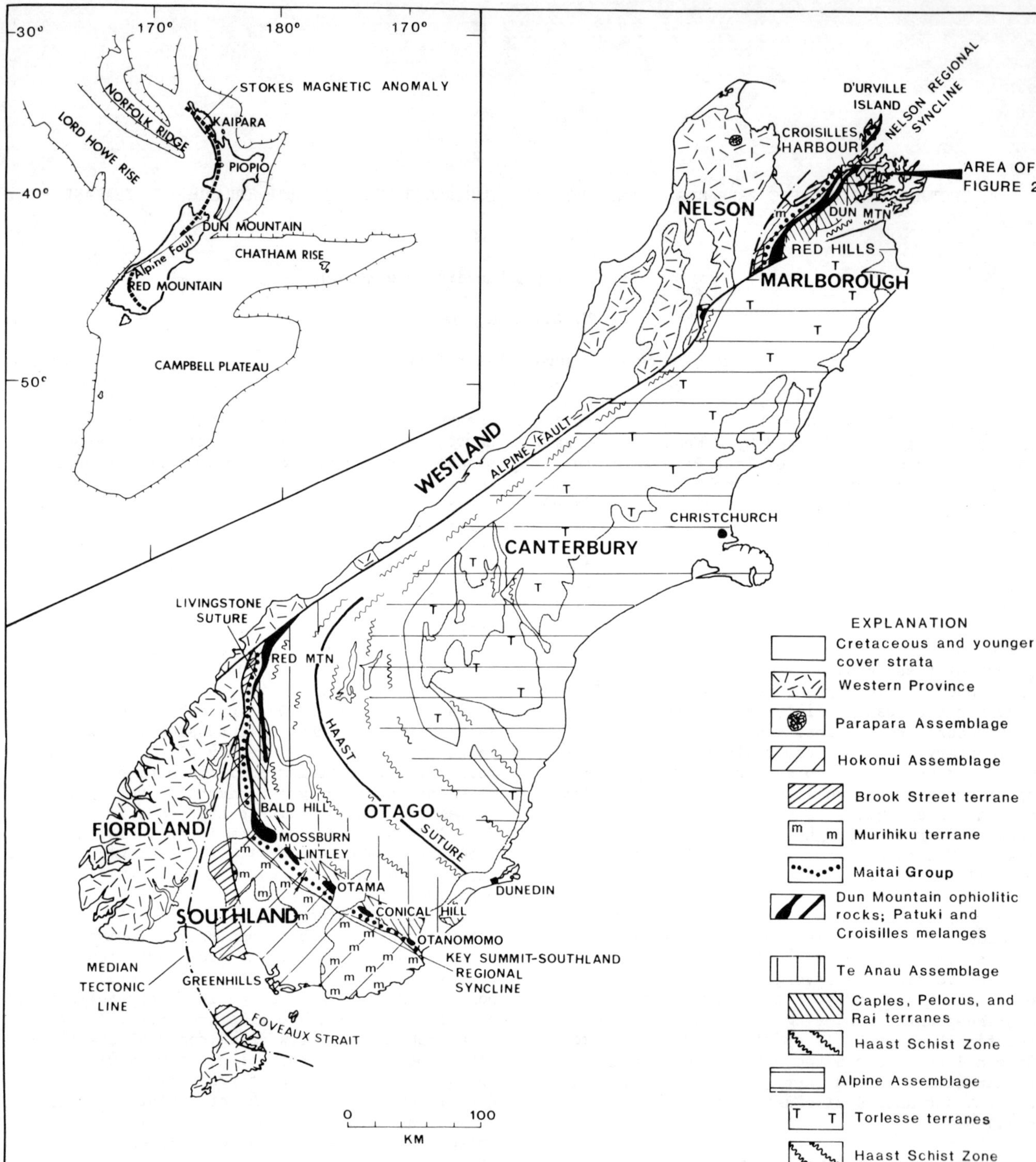

Fig. 1. Geological sketch map of South Island, New Zealand, showing location of Croisilles Harbour area; modified after Coombs and others (1976). Inset shows location with respect to Southwest Pacific bathymetric features (2000 m isobath) and structural trends as defined by Alpine fault and Stokes magnetic anomaly. The highly indented coastline in the area east of Croisilles Harbour is referred to as the Marlborough Sounds.

ology and structure of unfossiliferous graywackes and semi-schists underlying the eastern part of the area are described by Vitaliano [1968]. Ophiolitic rocks cross the area in two well defined belts originally defined as lithostratigraphic units, e.g., the Dun Mountain Formation of Beck [1964], similarly the Dun Mountain Ultramafics of Waterhouse [1964] in the northwest, and the Croisilles Volcanics [Beck, 1964; Waterhouse, 1964] in the southeast. Our previous work [Blake and Landis, 1973] showed that the rocks of the Dun Mountain belt comprise a dismembered ophiolite suite [see also Hunt, 1974; Johnston, 1977, and Davis et al., 1980]. Other aspects of these rocks in the Croisilles Harbour area have been described by Coleman [1966], Coombs et al. [1976], and Dickins et al. [1987]. The regional geology, petrology, and tectonic setting of the Dun Mountain ophiolite belt is described by Coombs et al. [1976].

The present study was undertaken to investigate the nature of the boundary between the contrasting Hokonui and Te Anau assemblages, the significance of ultramafic rocks occurring at this boundary, and, in particular, to determine the nature and origin of the Croisilles Volcanics and their relations to adjoining sedimentary rocks and rocks of the Dun Mountain ophiolite belt. The Croisilles Volcanics have previously been regarded as a regionally extensive volcanic formation interstratified within a thick geosynclinal sedimentary sequence [Beck, 1964; Waterhouse, 1964, 1967].

Our work and that of Dickins et al. [1987], reveals that the Croisilles Volcanics comprises an ophiolitic melange, rather like the melanges at Dun Mountain [Blake and Landis, 1973; Johnston, 1981], and is not to be regarded as a stratigraphic unit. The Croisilles melange is tectonically bounded on both sides by thick arc-derived volcanogenic sandstones and siltstones. The western of these, the Rai Sandstone, separates the Croisilles from the penecontemporaneous Patuki ophiolitic melange [Johnston, 1981], which is sometimes regarded as a portion of the Dun Mountain ophiolite belt. The eastern sequence, generally known as the Pelorus Group, passes gradationally eastward into a regional schist terrane, part of the Haast Schist.

Sample and outcrop locations are indicated by six-digit grid reference (G.R.) numbers. These relate to the 1949 New Zealand grid. As applied to Figure 2, the first two easterly and northerly numbers are given on the margins; the third is interpolated.

Tectonic and Stratigraphic Units

Five tectonostratigraphic terranes [Jones et al., 1983] of Permian and probable Mesozoic age are mapped (Fig. 2). From northwest to southeast the terranes are Dun Mountain-Maitai terrane, Patuki melange, Rai terrane, Croisilles melange, and Pelorus terrane. Although an internal stratigraphy is recognized in some terranes, stratigraphic linkages between terranes are unknown. A summary of terrane characteristics is given in Table 1.

Previous terrane subdivision [Blake et al., 1974; Coombs et al., 1976; Howell, 1980; Bishop et al., 1985] has involved only 2 or 3 units in this area, reflecting assumed lithostratigraphic interrelations and plausible linkages in conceptual models. Although recognizing the risks of terrane proliferation, we prefer a five-fold subdivision, because each of the belts is separated by a regional fault across which original structural and stratigraphic relations have not been demonstrated.

Dun Mountain-Maitai Terrane

Within the map area (Fig. 2), the Dun Mountain-Maitai terrane comprises the Maitai and Lee River Groups [Waterhouse, 1964], dominantly volcanogenic rocks deposited in a Late Permian arc-flank environment. In areas to the south, Maitai strata rest with conformity or near conformity on the underlying Dun Mountain ophiolite [Coombs et al., 1976]. Elsewhere an unconformable relationship is proposed [Davis et al., 1980; Johnston, 1981]. Four constituent formations are mapped in the Croisilles Harbour region.

Maitai Group. Wooded Peak Limestone. Well-bedded, rather pure, light- to medium-gray, fine-grained limestones, correlated with the Wooded Peak Limestone [Waterhouse, 1964] are well exposed at two cliffed coastal localities within the mapped area (Fig. 2)--Onetea and Taipare bays (Pahakorea), and similar rocks also crop out at the base of the Maitai Group on the ridge northwest of the lagoon between Otarawao and Pakiaka points. In addition to the predominant light- to medium-gray limestone, other Wooded Peak rock types include dark gray, foetid limestone, recrystallized (phyllitic) greenish-gray calcareous volcanogenic sandstone and siltstone, and occasional pink and green marbles which occur near contacts with ultramafites. Thickness of the unit is approximately 200 m at Pahakorea [Coleman, 1966]. At Onetea Bay, the sequence, somewhat deformed, is of comparable thickness.

Petrographically, representative limestone samples consist almost entirely of silt to sand size prisms of comminuted shell derived from atomodesmatinid bivalves [cf. Wood, 1956]. Terrigenous debris, generally comprising less than 10% of these rocks, includes plagioclase (albite), quartz, clinopyroxene, and volcanic rock fragments. Authigenic epidote, pyrite, sericite, and chlorite are widespread, while stilpnomelane and lawsonite occur in a few samples. Thin, planar bedding predominates. Argillaceous beds tend to be rhythmically interlayered with limestone, and some limestones are graded. Bedding features and overall lithology suggest a turbidite origin for the Wooded Peak Limestone [see also Coleman, 1966;

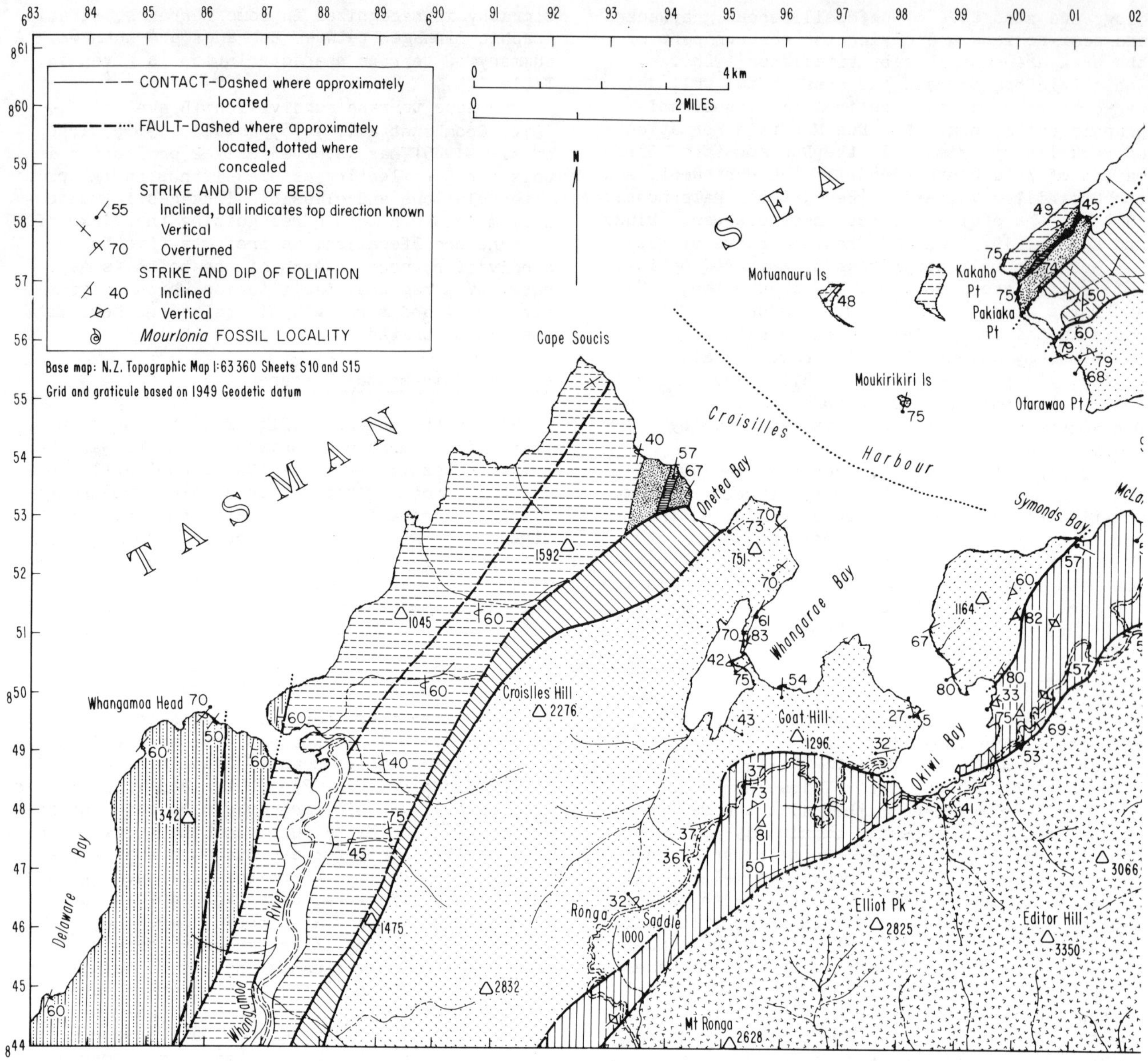

Fig. 2. Geologic map of Croisilles Harbour

Figures 28, 29, Fronticepiece; Landis, 1974a, p. 232, 236, 1980].

Maitai Group. "Tramway-Little Ben" Formation. Rocks of the Tramway and Little Ben Formations [Waterhouse, 1959] overlie the Wooded Peak Limestone and underlie the Greville Formation in the Dun Mountain area. Sedimentary facies comparable to each formation in its type section are also recognized here--the Tramway facies is typified by thin-bedded, very fine sandstones and siltstones which contain atomodesmatinid fragments; the Little Ben facies is characterized by thicker bedded, coarser grained, green, characteristically unfossiliferous volcanogenic sandstone. However, consistent stratigraphic relations are not recognized in the study area and clear division into two formations was not achieved. For the present study we prefer to combine rocks of both facies into one "compound formation."

Petrographically, rocks of the Tramway facies are characterized by detrital albite, quartz (10-15%), volcanic rock fragments and atomodesmatinid prisms. In contrast, the Little Ben facies is characteristically a quartz-poor to quartz-free mixture of volcanic rock fragments and mineral grains. Pyroxene and hornblende are abundant and several samples from Onetea and Otarawao contain

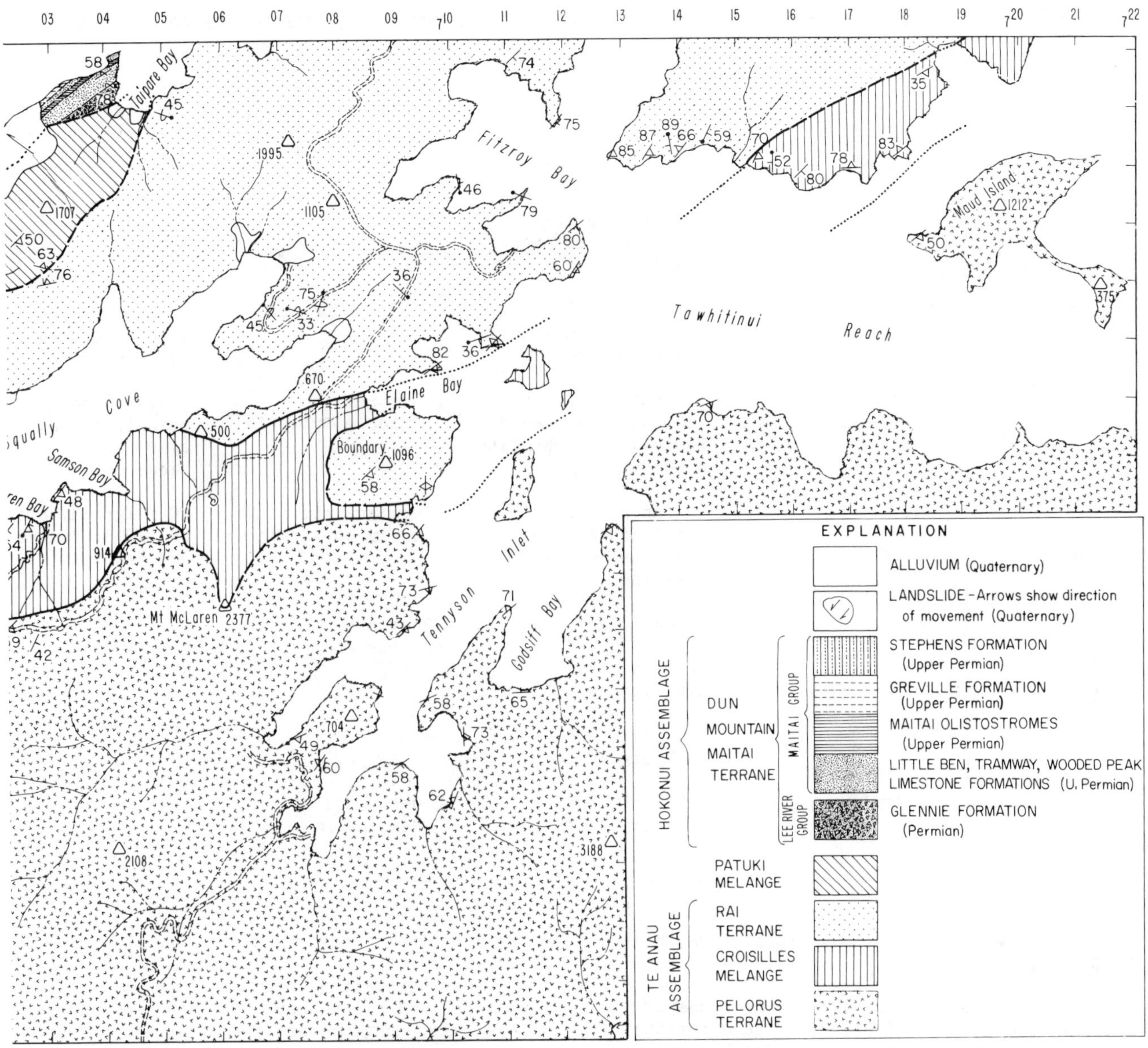

Nelson-Marlborough Sounds. Elevations in feet.

ultramafic debris (see discussion of Sandstone Petrography). Authigenic minerals, especially chlorite and pumpellyite, are widespread in the Little Ben facies, and to a lesser extent in rocks of the Tramway facies.

Maitai Group. Greville Formation. The Greville Formation [Wellman in Rigg et al., 1957; Waterhouse, 1964], characterized by gray, rhythmically interlaminated and frequently graded, fine-grained sandstones and siltstones, comprises the most widespread formation mapped. Penetrative axial plane cleavage is widely developed. These rocks are well exposed in coastal outcrops west of Onetea Bay, at Kakaho Point and on Moukirikiri and Motuanauru islands. In addition to the distinctive gray laminated rocks, Greville strata also include thicker bedded, greenish-gray volcanic sandstones, nodular limestone bodies and rare lensoidal masses of carbonaceous material. Fossils are rare in Greville strata, but recent discoveries, particularly of ammonoids, have given a measure of control on age [Waterhouse, 1973; Furnish et al., 1976; Johnston and Stevens, 1978]. These ammonoids (collected outside the present area), are regarded as late Middle Permian by Waterhouse [1973], whereas Furnish et al. [1976] regard the ammonoids as Early Triassic. On the basis of regional structure and

TABLE 1. Summary of Terrane Characteristics, Croisilles Harbour Region

Amalgamated Terrane Assemblage	Tectonostratigraphic Terrane	Characteristic Lithology	Age of Igneous and Sedimentary Rocks	Structural and Stratigraphic Characteristics
Hokonui	Brook Street	pyroclastics, lavas, dikes; deep-sea; tholeiitic to calc-alkaline	Lower Permian	thick (15 km) homoclinal sequence
	Murihiku	volcaniclastic sands and muds; deep-sea; calc-alkaline	Triassic-Jurassic	thick (10 km) open synclinal sequence
	Dun Mountain-Maitai	volcaniclastic sands and muds; deep-sea; tholeiitic; ophiolite basement	Lower Permian to Middle Triassic	thick (6 km) near-isoclinal synclinal sequence
Te Anau	Rai	volcaniclastic sands and muds; deep sea; calc-alkaline	?Permian-Triassic (poorly constrained)	stratigraphy not resolved; folded and faulted
	Pelorus	volcaniclastic sands and muds; deep-sea; calc-alkaline	?Permian-Triassic (poorly constrained)	stratigraphy not resolved; folded and faulted
Suture Zone Terranes	Patuki	ophiolite and dark argillite	Permian	Melange
	Croisilles	ophiolite and dark argillite	Lower-Middle Permian	Melange

stratigraphy, we provisionally support Waterhouse's age designation.

Sedimentary features suggest that Greville sediments are largely distal [in the sense of Bouma and Hollister, 1973] turbidite deposits: predominance of laminated and graded fine sandstones and siltstones, absence of shallow-water sedimentary structures, and of fossils. A similar conclusion was reached by Waterhouse [1964]. Rapid transitions to interlayered, coarse-grained, massive redeposited sandstones and along-strike persistence of facies and thickness, suggest a possible origin of the formation as elongate coalescing submarine fans.

Petrographically, the gray laminated rocks consist predominantly of angular grains of quartz and albite and highly altered volcanic rock fragments floating in a dense chlorite-sericite matrix. Tiny tablets of lawsonite (0.1 mm) are characteristic of Greville rocks. The lawsonite, along with chlorite and sericite, is frequently oriented parallel to slaty cleavage.

Maitai Group. Stephens Formation. Stephens Formation, the uppermost unit of the Maitai Group [Wellman in Rigg et al., 1957; Waterhouse, 1964] comprises redeposited green and gray volcanogenic sandstones and siltstones with lesser amounts of vitric tuff, conglomerate and limestone. These rocks are well exposed along the rocky coast in the southwestern part of the area. Although not studied in detail, they do bear a general resemblance to well documented Stephens strata from D'Urville Island and Stephens Island [Waterhouse, 1964; Campbell et al., 1984].

Maitai Group. Maitai olistostromes. Lenticular bodies of fine-grained, characteristically calcareous, serpentinite containing angular to well-rounded blocks of spilite, dolerite, gabbro, peridotite and serpentinite, and argillite occur within the Wooded Peak Formation at Onetea and Pahakorea, and within Tramway-Little Ben rocks at Otarawao. Much of this rock can be described as ophicalcite, resembling similar rocks described from the Pennine nappes [Bernoulli and Weissert, 1985]. These lenses range from conformable to slightly disconformable with respect to enclosing

TABLE 2. Selected Electron Microprobe Analyses of Detrital Grains From Little Ben-Tramway Sandstone (O.U. 23629)

	Amphibole			Clinopyroxene			Chromite
SiO_2	47.5	42.2	42.0	51.8	52.1	50.6	0.1
TiO_2	1.6	2.0	2.8	0.6	0.5	0.8	0.1
Al_2O_3	8.0	14.3	13.2	2.5	2.4	5.1	16.8
FeO*	12.5	9.6	11.1	10.2	9.8	6.1	20.9
MgO	14.8	15.5	14.4	16.1	15.9	15.0	10.3
CaO	10.6	11.7	11.1	18.4	18.1	21.9	0.0
Na_2O	1.5	2.5	2.4	0.3	0.3	0.3	0.0
K_2O	0.3	0.6	0.5	0.0	0.0	0.0	0.0
MnO	0.5	0.1	0.2	0.3	0.3	0.1	0.3
Cr_2O_3	0.0	0.1	0.0	0.0	0.1	0.0	51.6
Total	97.3	98.6	97.7	100.2	99.5	99.9	100.1

Maitai strata. Thickness at Pahakorea is 90-150 m while at Onetea the body is 240-300 m thick. Similar bodies may also occur in lower Maitai strata at Graham Creek and at Ohaua, D'Urville Island, 10 km to the south and north, respectively, of the mapped area [Coleman, 1966].

The mechanism by which these masses were emplaced remains unclear: they closely resemble portions of the underlying Dun Mountain ophiolite belt, and indeed they may be diapiric intrusions of this unit. However, their overall conformable nature, coupled with evidence presented below, strongly suggests olistostromal (e.g., submarine mass flow) emplacement, perhaps from an already deformed Dun Mountain ophiolite. At Pakiaka Point, where the termination of one body is well exposed, Maitai strata spanning the interval directly along strike from the pinch-out, show no evidence of significant shearing or tectonism. Furthermore, immediately overlying sandstones contain ophiolite-derived clastic detritus (Table 2), which is not found in most Maitai rock. Low temperature metasomatic rocks (rodingites) are extensively developed at the contacts between the olistostromes and enclosing rock [Coleman, 1966].

A modern, and probably similar, occurrence of submarine debris flow of serpentinite emanating from a diapiric seamount in the forearc region of the Mariana island arc is documented by Fryer et al. [1985].

Lee River Group. The Lee River Group was proposed by Waterhouse [1964, p. 23] to embrace a diverse suite of rocks distinguished by an "abundance of volcanic and gabbroic and ultramafic igneous rocks" and lying above the Pelorus Group and below the Maitai Group. Four subdivisions were recognized (youngest to oldest): Dun Mountain Formation or Ultramafics, Patuki Volcanics, Rai Sandstone, Croisilles Volcanics. Basalts, dolerites, and volcanic breccias west and east (above and below) Dun Mountain rocks were all included in the Patuki Volcanics, but stratigraphic continuity was not established. Subsequent mapping in the Red Hills region [Walcott, 1969] distinguished a new volcanic unit, the Glennie Formation, lying above the main ultramafic mass and below the Maitai sediments. The Glennie has been subsequently mapped at Dun Mountain [Johnston, 1977, 1981; Davis et al., 1980] and we recognize it in the Croisilles Harbour area as well. We map Glennie rocks as underlain by sheared serpentinite marking the boundary of the Patuki melange. All other contacts within the Lee River Group as originally proposed are tectonic, and original stratigraphic relations are unknown. Furthermore, a significant proportion of the original Lee River Group is characterized by melange structure. Thus, we propose to restrict the Lee River Group to those dominantly mafic plutonic, volcanic, and interbedded sedimentary rocks lying in sequence with and stratigraphically beneath Maitai Group sedimentary rocks, and also including similar and originally continuous strata now displaced by faulting. This usage is compatible with that of Johnston [1981]. Constituent units comprise the Glennie Formation and any underlying mafic or ultramafic rock which can be confidently regarded as originating in stratigraphic sequence with Glennie or Maitai strata [e.g., Tinline Formation of Davis et al., 1980]. Rai, Croisilles, and Patuki rocks are to be excluded from the Group until demonstrable stratigraphic relations are established.

Lee River Group. Glennie Formation. A suite of stratified volcanogenic rocks--pillow lavas, pillow-bearing hyaloclastite, massive basaltic flows, dolerite sills, and volcanic breccia--which underlies the Wooded Peak Limestone at Taipare Bay is correlated with the Glennie Formation [Walcott, 1969; Johnston, 1977]. These rocks comprise a well exposed 200 m thick west-facing sequence. They are bounded on the east by sheared serpentinite of the Patuki melange. The western contact, with the Maitai Group is not exposed. Relations are broadly conformable; however, a gully at the contact suggests minor faulting. Elsewhere (e.g., Dun Mountain and Red Hills), unfaulted sequences from Glennie into Maitai rocks are recognized [Walcott, 1969; Johnston, 1977, 1981; Davis et al., 1980]. Volcanogenic breccias intervening between Maitai limestone and underlying basalts have been included within Maitai Group strata at Dun Mountain [Lauder, 1965] and in correlative strata (Upukerora Formation) in Southland [Landis, 1974a, 1980].

Pillow basalts are well exposed in oldest Glennie strata at Taipare Bay. They contain conspicuous epidote-filled vesicles and are highly altered, being appropriately regarded as spilites. Minor copper mineralization is locally evident (e.g. at Taipare Bay). Overlying Glennie breccias and hyaloclastites are petrographically similar to the pillow basalts; interstitial material includes hematitic volcanic detritus and dense jasperoid. Still younger rocks comprise sheets of basalt and dolerite, and these, in turn, are conformably overlain by poorly stratified volcanic breccias. The youngest Glennie rocks consist of angular clasts (2.5-30 cm) of underlying rock types set in a matrix of hematitic debris and fine-grained limestone. Although included within the Glennie Formation at the present map scale, these breccias are provisionally regarded as correlatives of the Upukerora Formation (basal Maitai Group) as described by Landis [1980].

Similar breccias occur widely at the Maitai-Lee River contact elsewhere in Nelson [Lauder, 1965; Walcott, 1969; Carman, 1968] and in Otago and Southland [Grindley, 1958; Landis, 1974a, 1980; Coombs et al., 1976]. They characteristically rest upon basalts and dolerites of the Dun Mountain ophiolite belt and are regarded as redeposited submarine talus deposits.

Patuki Melange

A belt of melange comprising disrupted ophiolite plus variably rounded blocks and tectonic slices of sedimentary rock extends more or less continuously 120 km from the Alpine Fault near Red Hills north to D'Urville Island (Fig. 1). In the Croisilles Harbour area, the belt is characteristically less than 1 km in width, but elsewhere it may be 3 km or wider. Distribution corresponds essentially to the area mapped as Dun Mountain Formation by Beck [1964]. Previous work and summary of melange features are presented by Blake and Landis [1973] and Coombs et al. [1976]. The name Patuki melange was proposed to describe the ophiolitic melange lying east of Lee River plutonic rocks in the Dun Mountain area [Davis et al., 1980, Johnston, 1981].

In the area mapped, melange matrix is dominantly sheared serpentinite; however, sheared spilitic basalt and argillite are also recognized. Tectonic inclusions can be divided into two main types--an igneous group of ophiolitic parentage and a sedimentary suite of varied parentage. Basalts and dolerites similar to rocks described by Waterhouse [1964] and Sivell and Waterhouse [1984a] as Patuki volcanics and Patuki intrusive suite, and serpentinized peridotite, gabbro, and rodingite similar to those in the Dun Mountain region comprise the main igneous inclusions. Chromitite occurs at Onetea Bay.

The most characteristic sedimentary inclusions in the melange are argillites of uncertain provenance. These are characteristically indurated, dark grey mudstones, many of which have undergone metasomatic alteration [Reed, 1959; Coleman, 1966] and which appear to comprise a distinctive inclusion type in melanges from Dun Mountain to D'Urville Island. These tough rocks, frequently referred to (incorrectly) as baked argillite, were used extensively for tools and weapons by the pre-European Maori [Duff, 1946; Reed, 1959].

Many argillites contain thin graded silt laminae [e.g., Coleman, 1966, Fig. 19]; others appear to be massive. Although some blocks may have been derived from Rai and Maitai strata, the characteristic fine grain size and dark color, the great predominance of argillite over sandstone, and the widespread absence of medium to coarse grained volcanic sandstone, argue against an origin by tectonic erosion from adjacent rocks of either Rai or Maitai terranes as presently exposed. "Phantom stratigraphy" reconstructed from inferred original relations between tectonic blocks suggests that the argillite inclusions were derived from distal turbidite-hemipelagite deposits originally capping ophiolitic sea floor. In addition, blocks of distinctive siliceous vitric tuff which occur between Pakiaka Point and Taipare Bay suggest proximity of the ophiolitic basement to an island arc.

It is tempting to regard the Patuki melange as a disrupted remnant of the relatively intact ophiolite sequence which occurs at Red Hills [Walcott, 1969] and Red Mountain [Coombs et al., 1976; Sinton, 1977, 1980] and in fact our earlier studies do just that [Blake and Landis, 1973; Coombs et al., 1976]. However, the general rarity and widespread absence in melange of ophiolitic breccia blocks (compare Glennie and Upukerora formations), and conversely the absence of thick dark argillite and siliceous tuffs (compare melange blocks) from the top of the intact ophiolite and overlying basal Maitai Group strata suggest that such specific correlations are tenuous. Thus, the Patuki melange of the area described is best regarded as a tectonic unit containing blocks probably derived from the Lee River Group but also elements of unknown stratigraphic affinity.

Rai Terrane

The Rai Sandstone was introduced by Waterhouse [1964] as a formation overlying the Croisilles

Volcanics and comprising "bands of massive green and blue sandstone, some with channelled bases and some with black argillite chips. Bands of black argillite are up to 20 ft thick, and bands of alternating sandstone and mudstone are also present in beds 1/4 to 1 in. thick, rarely with calcareous concretions." The Rai Formation was proposed and mapped regionally by Beck [1964] who described it as "mainly a fine-grained dark argillite." Additional rock types including green tuff and serpentinite were mentioned. Both Waterhouse and Beck record the presence of the bivalve Atomodesma; however, no specific localities are cited. Davis et al. [1980] and Johnston [1981] include the Rai within the Pelorus Group (see below).

Stratigraphy of Rai rocks remains poorly known. Neither the base nor the top of the unit is recognized, the mapped boundaries being tectonic zones. Indeed, the real unity of this diverse suite of rocks lies not in stratigraphy but in gross lithologic character and tectonic position between the Patuki and Croisilles melanges. Although a lithostratigraphically based Rai unit may be eventually warranted, we prefer to consider it as a tectonostratigraphic terrane in the meantime.

The rocks we map as Rai terrane are generally similar to the Rai Sandstone described by Waterhouse [1964]. Athough sandstone is the most conspicuous rock type, mudstones are prevalent in many areas. Basaltic rocks, serpentinites, and exogenous conglomerates were not recognized. The most conspicuous Rai rocks mapped are gray, greenish gray or grayish red (hematitic), massive, volcanogenic graywackes. In hand specimen, fine-grained fuzzy-bordered lithic grains, detrital augite, and hematitic lithic fragments are conspicuous. Some graywackes show graded bedding, others appear quite massive. In some rocks outsized angular rip-off clasts of red, green, and gray mudstone are widely dispersed. Rhythmically layered, very thin bedded to laminated gray and grayish red volcanogenic sandstones and mudstones are also widespread. Bedding in these rocks is parallel-sided, sharp at bases, frequently graded and locally internally cross-laminated. Deposition was from turbidity currents and other mass flow types, probably in a submarine fan environment. Other rock types mapped within the Rai include fine-grained felsic vitric-crystal tuff and red chert. These rocks were found only in the area immediately north of Elaine Bay.

Rai rocks are structurally complex [Fig. 2; see also Beck, 1964]. Several coastal outcrops reveal mesoscopic folding within finer-grained sequences and larger scale structures are inferred (Fig. 2). Folds range from broad open structures along the west side of Okiwi Bay to tight nearly isoclinal features at Whangarae Bay.

Age and correlation of Rai rocks remain major problems. Apart from a few poorly preserved radiolaria and some carbonaceous material, we found no fossils in the Rai. Beck [1964] and Waterhouse [1964, 1967, 1982] regarded Rai as early Permian in age, on the basis of assumed stratigraphic position and presence of the atomodesmatinid bivalves; however, no locality data were cited. We failed to find macrofossils in any of the rocks we mapped as Rai and in view of the absence of stratigraphic continuity with known fossiliferous rocks, we are reluctant to accept a precise age designation. A fault-bounded sliver of sandstone and limestone correlated with Rai and containing Atomodesma sp? aff. trechmanni (Marwick) of probable Later Permian age is described [Johnston, 1977] from 2 km southeast of Dun Mountain. This identification and age is challenged by Waterhouse [1982] who argues that the fossil is allied to Aphanais otamaensis and is of Early Permian age. Johnston describes the fossiliferous sliver as more than 200 m thick and becoming progressively more sheared on both margins before passing into melanges. These rocks, included by Beck [1964] with the Croisilles, are of a facies quite unlike any Rai rocks mapped by us, or those defined as Rai by Waterhouse [1964], and we are reluctant to accept their stratigraphic correlation with Rai rocks of the Croisilles Harbour area.

On balance, we favor a Permian age for at least some Rai strata, but are hesitant to propose detailed correlations. Furthermore, a Mesozoic age should not be ruled out (see discussion of chemical data later in this paper). Rocks of comparable lithology and tectonic position [Blake et al., 1974] in North Island [Kear, 1971; Schofield, 1974; Sporli, 1978] are of known Jurassic age, while in southeastern Otago strata in a similar tectonic position are of probable Late Triassic age [Campbell and Campbell, 1970].

Chemical analyses of seven Rai volcanic sandstones (Table 3) reveal composition comparable to andesite and basaltic andesite. Certain anomalies, e.g., high $FeO:Fe_2O_3$, low $CaO:SiO_2$ ratios probably reflect diagenetic and perhaps predepositional processes, while metasomatic processes have probably affected the alkali ratio in specimen no. 8, a large block interpreted in the field to be Rai sandstone lying within the Croisilles melange. Petrochemical affinities are discussed in a later section of this paper.

Petrographically, Rai sandstones are weakly metamorphosed volcanic sandstones, feldspathic volcanic sandstones, and volcanic-sedimentary lithic sandstones. Volcanic rock fragments are predominantly andesitic although palagonite and fine-grained felsic and basic volcanic detritus occur in moderate amounts. Sedimentary and metasedimentary rock fragments, mainly argillite and slate, are also numerous. Granitic and ultramafic sand-grains are widespread but uncommon. Mineral grains such as feldspar (albitized plagioclase and very rare potassium feldspar), quartz, pyroxene, hornblende, mica and epidote occur in varying amounts. Except for feldspar, mineral

TABLE 3. Chemical Analyses, Pelorus and Rai Metasedimentary Rocks, Croisilles Harbor Area

	Pelorus					Rai						
	1	2	3	4	5	6	7	8	9	10	11	12
Oxides, wt %												
SiO_2	69.0	54.4	66.8	60.9	59.0	58.1	56.9	64.7	52.4	56.9	53.7	53.7
Al_2O_3	15.4	18.2	16.1	16.6	18.0	15.9	15.9	17.2	16.3	17.6	16.1	17.2
Fe_2O_3	1.1	1.4	1.2	3.7	1.7	4.0	1.9	0.9	1.0	1.7	1.3	1.7
FeO	3.2	7.2	2.4	2.0	4.8	3.8	5.0	6.1	5.0	7.7	6.4	7.2
MgO	1.4	3.9	1.4	2.0	3.0	4.3	5.1	4.9	2.0	4.7	5.0	4.8
CaO	0.9	5.4	1.5	3.2	3.9	4.0	5.2	2.3	1.1	6.3	4.8	5.7
Na_2O	3.5	2.7	5.0	3.8	4.5	3.0	3.2	6.0	4.2	2.7	2.7	4.9
K_2O	1.8	1.4	2.2	2.2	2.1	2.2	1.5	0.3	1.4	1.2	1.8	0.3
H_2O	2.4	4.0	1.8	2.9	2.6	3.0	3.3	3.0	2.8	4.4	3.4	3.6
TiO_2	0.8	1.1	0.7	0.9	1.0	1.0	1.0	1.0	0.9	1.2	1.1	1.3
P_2O_5	0.2	0.2	0.1	0.2	0.2	0.2	0.1	0.2	0.1	0.2	0.2	0.2
MnO	0.1	0.1	0.1	0.1	0.1	0.1	0.1	0.1	0.1	0.2	0.1	0.1
CO_2	0.0	0.1	0.0	0.0	0.0	0.0	0.0	0.0	0.0	0.1	0.0	0.1
Total	99.8	100.1	99.3	98.5	100.9	99.6	99.2	100.1	99.6	100.4	99.8	100.8
Trace elements, ppm												
B	50	10	30	10	30	15	20	0	20	15	20	10
Ba	300	200	500	300	700	150	200	70	200	150	200	70
Be	1	0	1.5	1.5	0	0	0	0	0	0	0	0
Co	7	20	10	10	15	20	20	20	10	30	20	20
Cr	50	150	20	50	30	200	200	150	70	100	200	70
Cu	30	50	5	30	30	70	30	30	30	50	20	30
Ga	10	15	15	15	20	15	15	15	10	15	15	10
Nb	0	7	0	0	0	0	0	0	0	0	0	0
Ni	15	30	10	15	15	50	50	50	50	20	50	20
Pb	7	0	20	20	0	7	0	10	0	0	0	0
Sc	15	30	10	15	20	30	20	20	20	30	20	30
Sr	150	700	200	300	700	500	300	150	150	500	200	300
V	70	150	50	100	100	200	150	150	100	200	100	150
Y	20	30	15	30	20	30	20	30	20	20	15	30
Yb	2	3	2	3	2	3	2	3	3	3	2	3
Zr	200	100	150	150	150	100	100	150	150	70	70	100

Chemical analyses by Hezekiah Smith and Lowell Smith, U.S.G.S. Method used was a single solution procedure (Shapiro and Brannock, 1962).

Semiquantitative spectrographic analysis by Chris Heropoulous, U.S.G.S.

Description of Samples

1. Sheared gray sandstone, Pelorus terrane coastal exposure southern side of Croisilles melange, Okiwi Bay. G.R. 989484. 2. Weakly sheared gray sandstone, Pelorus terrane, southeast side of Croisilles melange; road cutting 100 m southwest Stratford Farm turnoff. G.R. 019509. 3. Strongly sheared gray sandstone showing transposed bedding, Pelorus terrane; coastal outcrop west end of Maud Island. G.R. 180577. 4. Weakly sheared massive gray sandstone, Pelorus terrane, coastal headland east of Harveys Bay. G.R. 089487. 5. Massive gray sandstone, Pelorus terrane, coastal outcrop, point at mouth of Tawa Bay. G.R. 127528. 6. Massive hematitic sandstone, Rai terrane, cuttings in French Pass road. G.R. 943473. 7. Massive sandstone, Rai terrane, coastal outcrop, headland north of boulder beach east side Okiwi Bay. G.R. 986509. 8. Greenish gray sandstone containing conspicuous grains of detrital augite, Rai terrane, hillslope outcrop southwest of Elaine Bay. G.R. 080534. 9. Poorly sorted, gray sandstone, Rai terrane, coastal promontory, Whangarae Bay. G.R. 967507. 10. Massive greenish gray volcanic graywacke, Rai terrane, blocks on side of road to Otarawao, near Matarau. G.R. 029555. 11. Massive weakly sheared sandstone, Rai terrane, coastal outcrop east of Sheep Point, Tawhitinui Reach. G.R. 133590. 12. Massive greenish gray sandstone; Bush covered bluffs along ridge crest, northeast of Otarawao. G.R. 029569.

grains seldom exceed a few percent in abundance. Matrix is abundant but is partially authigenic [Cummins, 1962]. Low-grade metamorphic effects are widespread; feldspars are extensively sericitized and saussuritized and volcanic glass is devitrified. Authigenic prehnite, pumpellyite, epidote and chlorite are widespread. Many lithic fragments are replaced (to varying degrees) by hematite, and textural relations (the presence of side-by-side texturally identical grains of hematitized and nonhematitized lithic clasts) suggest that most of the hematite formed prior to deposition.

Croisilles Melange

The Croisilles Volcanics [Waterhouse, 1964, p. 23] was defined as a formation 130-1,000 m thick consisting of a "complex of spilitic pillow lava, red and green tuff, serpentinite, and black argillite." Persistent bands of conglomerate [Bell et al., 1911] were also noted. The regional extent of the unit was demonstrated by Beck [1964]. Beck's description of the rocks is similar to that of Waterhouse (see above) except that distinctive conglomerate bands were included by Beck within the adjoining Pelorus Group.

Croisilles rocks are well exposed at several localities within the area mapped. Intense tectonic mixing of widely different rock types is obvious at these localities, and we found no evidence to suggest that the unit is stratified either internally or with respect to the adjoining units. In addition, aerial-photo studies reveal a classic hummocky texture characteristic of tectonic blocks in sheared matrix ("knocker topography") within the less well-studied bush-covered inland areas underlain by Croisilles rock. Consequently, like Coombs et al. [1976], Johnston [1981] and Dickins et al. [1987], we have mapped these rocks as a tectonic unit, the Croisilles melange.

A wide variety of rock types occur as tectonic blocks (fist-size up to 2 km long slabs) within sheared serpentinite and sheared mudstone matrix: greenstone (abundant), argillite (abundant), volcanic sandstone, siltstone, chert, peridotite, pyroxenite, blocky serpentinite, steatite, amphibolite, gabbro, dolerite, volcanic breccia and agglomerate, roundstone conglomerate, and albite granite. Shearing and granulation occasionally penetrate tectonic blocks [compare Coombs et al., 1976, plate 1b]; however, most blocks are conspicuously tough, massive and not pervasively tectonized.

Mineralogically the rocks are variably reconstituted. Most Ca-plagioclase is replaced by albite and pumpellyite (and other phases), while some relict pyroxene and hornblende is altered to chloritic phases and pumpellyite. Argillite and siltstone tend to be strongly indurated, and to contain metamorphic actinolite, pumpellyite and stilpnomelane. Conglomeratic rocks are conspicuous in many Croisilles outcrops. Volcanic detritus comprises the bulk of pebbles, some of which show clear evidence of pre-depositional epidote- and pumpellyite-forming metamorphism. Some second-cycle conglomerate clasts are also present.

In general, the Croisilles melange bears a close resemblance to the Patuki melange. However, blocks of chert, amphibolite, fossil-bearing mudstone and sandstone, and conglomerate are more abundant in the Croisilles.

Several good exposures of the Croisilles melange are reasonably accessible. Roadcuts in the saddle south of Goat Hill (G.R. 962489) reveal a representative sequence:

red chert (southeast)	
shear zone	
greenstone	15-20 m
shear zone	
amphibolite and serpentinite	5-6 m
shear zone	6-7 m
peridotite (variably crushed) (northwest)	

A roadcut near the head of the track to Whangarei, 1 km southwest of the Goat Hill saddle, reveals easterly-dipping sheared mudstone containing lenticular bodies of greenstone. Other good Croisilles exposures occur along the east side of Okiwi Bay (G.R. 991487), in natural and roadside exposures in the vicinity of Trig 1524 (G.R. 009499), at MacLaren Bay (G.R. 031532) and along the coast around Picnic Bay (G.R. 152591).

Some melange blocks contain fossils. Prismatic holes, probably molds of atomodesmatinid shell fragments, occur in siltstone immediately adjacent to a zone of sheared talc-serpentinite near the head of the Ronga Valley (G.R. 933446). Volcanic sandstone containing atomodesmatinid fragments and the Permian gastropod *Mourlonia* were collected and described from a roadside exposure between Okiwi Bay and MacLaren Bay [Waterhouse, 1966]. This locality (G.S. 9477) was relocated. We note that the grid reference given by Waterhouse (S15-042326) is incorrect; the correct reference is S15-057534). The fossiliferous rock is a slumped melange block; a likely source area has been located by M. R. Johnston near the ridge crest 500 m SSE of the road [Dickins et al., 1987].

Although previously regarded as being of Early Permian age [Waterhouse, 1964], re-examination of the Croisilles *Mourlonia* by Dr. J. M. Dickins indicates a Late Permian age (Queensland Fauna 4). Thus the Croisilles fossiliferous sandstone is comparable in age to the Maitai group [Dickins, et al., 1987].

Pelorus Terrane

Graywacke and semischist southeast of the Croisilles melange are mapped as Pelorus terrane. These rocks, referred to by Beck [1964], Waterhouse [1961], Davis et al. [1980], and Johnston [1981] as the Pelorus Group, are re-

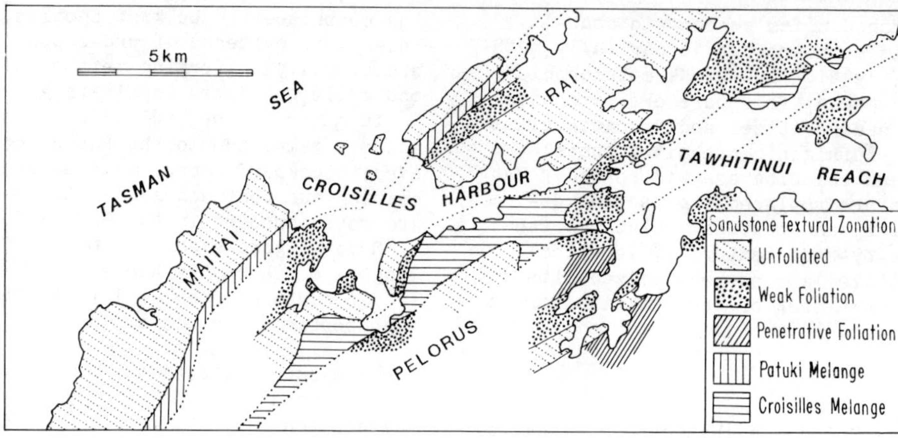

Fig. 3. Metamorphic textural zones within Croisilles Harbour-Tawhitinui Reach area. Weak and penetrative foliation correspond respectively to textural zones 2A and 2B of Bishop (1972).

garded here as a tectonic unit. No stratigraphic type section has been defined, the rocks being bounded on the northwest by the Croisilles melange, while to the southeast they grade through metamorphic transitions into the Haast Schist. Furthermore, internal stratigraphic continuity remains to be demonstrated.

Pelorus rocks within the map area are predominantly lithic volcanogenic sandstone, and at many localities they resemble rocks of the Rai terrane. In general, however, Pelorus sandstones are slightly more quartzose and tend to show a greater degree of reconstitution. Near the northwestern contact, medium to coarse sandstones are widely exposed, usually occurring in beds 30-100 cm thick. These are best seen in the vicinity of Okiwi Bay, where they show well-developed graded bedding and sole marks. Farther southeast, slates and fine-grained sheared sandstones comprise the dominant rocks. Rare horizons of tuffaceous siltstone and pale green siliceous shale (? chert, tuff) are also present. A large area of Pelorus rock lies southeast of the area described here; lithology and structure of these rocks are described by Vitaliano [1968].

Much of the rock we map as Pelorus is quite distinctly metamorphosed with pervasive cleavage and local development of axial plane quartz foliation. The distribution of schists and non-schistose rock is quite irregular (Fig. 3), and the transition to Haast Schist rocks is very complex, suggesting post-metamorphic deformation.

Five new chemical analyses of largely volcanogenic Pelorus graywackes are presented in Table 3. These along with five analyses of Pelorus rocks [Vitaliano, 1964] collected 1-5 km southeast of the area shown in Figure 2 show compositions resembling andesite, although decidedly more silicic than volcanic graywackes of the Rai.

Structure

Throughout most of the 150 km outcrop distance (Red Hills to D'Urville Island) in Nelson and Marlborough, the rocks of the Dun Mountain-Maitai terrane, Patuki melange, Rai terrane, Croisilles melange, and Pelorus terrane form a series of northeast-trending subparallel belts (Fig. 1). Within the Croisilles Harbour area (Fig. 2) our mapped contacts locally diverge from the less complicated pattern seen on the small scale map, but the overall arrangement remains. As shown by both outcrop pattern and structural measurements on bedding surfaces, the sedimentary and volcanic rocks of the Dun Mountain-Maitai terrane are folded into a tight, axially faulted syncline [Fig. 2; Beck, 1964]. The southeastern contact of this terrane is marked by sheared and serpentinized ultramafic rock, the margin of the Patuki melange. In general, basal Maitai strata, and underlying ophiolitic rocks are vertical to overturned, dipping steeply southeast parallel to the serpentinite marked fault zone contact.

To the east of the Patuki melange the westernmost sedimentary rocks of the Rai terrane also generally dip steeply southeast and are frequently overturned. Within the Rai beds, however, the trend of bedding attitudes is quite variable with a significant number of attitudes trending northwest, markedly askance to the regional northeast-southwest outcrop pattern. In addition, observations of facing directions and mesoscopic folds indicate a much more complex style of deformation in the Rai than in Maitai Group rocks, a situation recognized by Beck [1964].

The southeastern margin of the Rai terrane with the Croisilles melange structurally resembles the southeastern margin of the Dun Mountain-

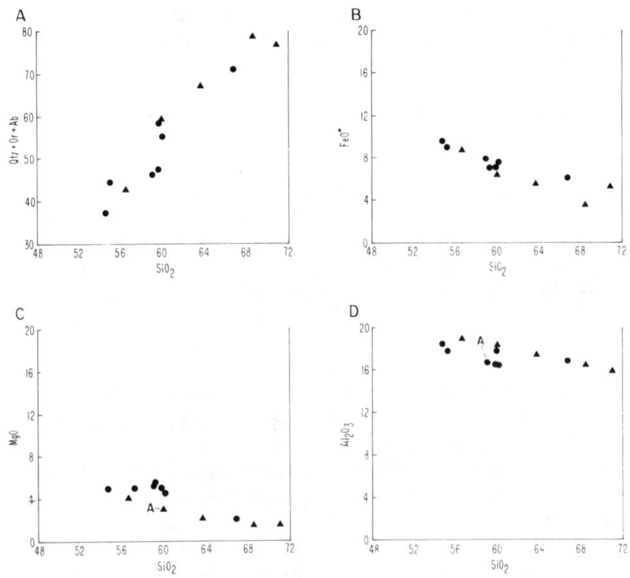

Fig. 4. Chemical diagrams for Rai (circles) and Pelorus (triangles) sandstones from study area. Analyses plotted on anhydrous basis. A = average andesite of Chayes (1965).

Maitai terrane (described above). The contact is marked by the appearance of sheared argillite and serpentinite followed by melange containing abundant blocks of altered volcanic rock, gabbro, and ultramafics. In the vicinity of Okiwi Bay shearing in the Rai sandstone and siltstone increases toward the contact with the Croisilles melange. Within the latter unit there is no recognizable sequence of rock types, however, it is notable that the Croisilles melange includes all the rock types of classical ophiolite sequences. It is our impression that shearing and mixing of rock types is more pervasive within the Croisilles melange than within the Patuki.

Southeast of Elaine Bay contacts of the Croisilles melange with adjacent Rai and Pelorus terranes diverge strongly from the more nearly linear trends found elsewhere. Thus a pronounced "hook" to the south is mapped near Mt. McLaren, while an apparent bifurcation to the east encloses, at Boundary triangular station, a large inlier of relatively coherent Rai sandstone (Fig. 2). Although structural data are inconclusive, it is probable that these diversions are the result of rather complex folding.

Southeast of the Croisilles melange lie the metasedimentary rocks of the Pelorus terrane which show extensive development of cleavage and of bedding transposed into the plane of cleavage (s_1). In inferred fold hinge regions, sandstone beds may be reduced to phacoids floating in siltstone. Development of schistosity varies widely, with distinctly schistose rocks belonging to textural zones IIA and IIB of Bishop [1972] being particularly widespread (Fig. 3). Some schistose rocks, particularly near the Croisilles contacts, contain incipient quartz segregation folia (textural zone IIIA of Bishop, 1972), as well as local development of 5-25 cm thick quartz and quartz-pumpellyite veins. Rather abrupt changes from schist to graywacke are mapped. Whether these are related to folding is unknown, but it is clear that the schist-graywacke transition is a complex one.

In conclusion, regional structure can be described as a sequence of five imbricate slabs, each independently deformed internally. In general, slab boundaries dip steeply southeasterly, so that the Pelorus terrane, the most highly metamorphosed unit, can also be described as the structurally highest.

Sandstone Petrochemistry

Chemical analyses of 12 Rai and Pelorus sandstone samples (Table 3) are consistent with petrographic observations that the rocks consist mainly of first cycle andesitic detritus. Thus, despite their sedimentary origin and diagenetic and low grade metamorphic alterations, we have analyzed chemical data using procedures commonly applied in igneous petrology. Several variation diagrams (Fig. 4) show clear-cut igneous trends comparable to those characteristic of Cenozoic arc volcanic suites. Comparison with chemographic plots of Miyashiro [1974, 1975] indicates overlap between calc-alkalic and tholeiitic series. Available data, including trace element contents (Ba, Ni, Cr, V) are more consistent with island arc tholeiites rather than abyssal tholeiites.

Other variation diagrams (Fig. 5) show greater scatter than would be expected from a single volcanic suite, and these are taken to indicate that the alkali elements behave rather erratically during diagenesis and low grade metamorphism. Thus, although a trend is evident in the K_2O-SiO_2 plot, two samples are anomalously low. The Na_2O-SiO_2 plot reveals much greater scatter,

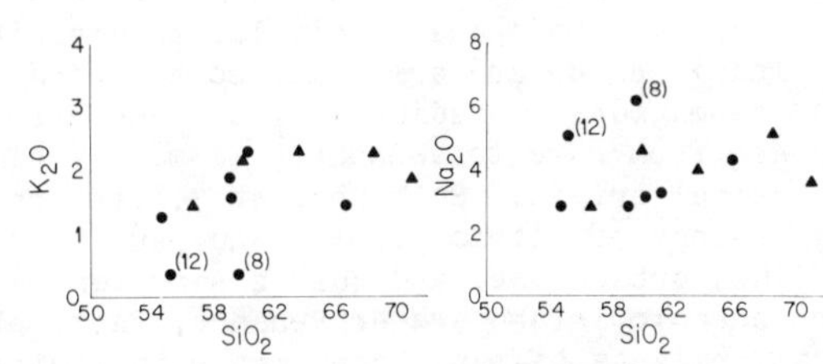

Fig. 5. Alkali-silica variation diagrams for Rai (circles) and Pelorus (triangles) sandstones. Analyses plotted on anhydrous basis. Samples 8 and 12 show evidence of post-depositional alkali re-distribution.

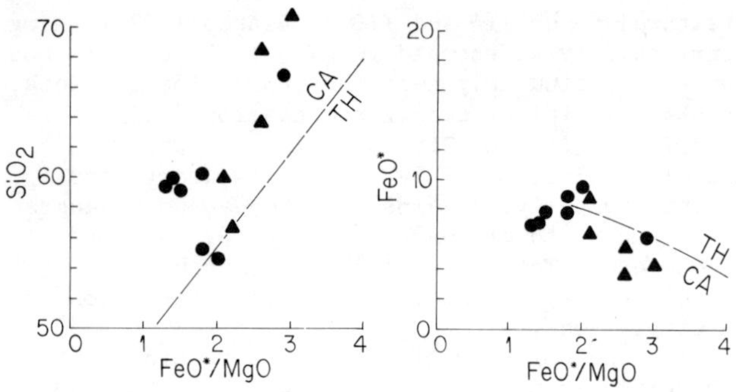

Fig. 6. Chemical variation diagram for Rai (circles) and Pelorus (triangles) sandstone. CA = calc-alkaline, TH = tholeiitic fields after Miyashiro (1974).

an effect which we relate to metasomatic processes associated with albitization. In contrast, a CaO-SiO_2 plot (not included in paper) shows a well defined trend, but of a one which lies on the low calcium side of the trend obtained by plotting average compositions from several island arc and continental margin calc-alkalic and tholeiitic series [Miyashiro, 1974]. This may suggest a uniform loss of CaO during alteration of arc-derived detritus.

Chemical data also provide evidence that the superficially similar first cycle arc-volcanic sandstone suites flanking the Croisilles melange (i.e., Rai and Pelorus) comprise genetically and compositionally different units. Thus, Figure 6 shows Rai and Pelorus rocks occupying distinct fields of SiO_2:FeO*/MgO and FeO*: FeO*/MgO plots. Furthermore, several trace element contents (e.g., B, Ba, Be, Cr, Ni, Co, Pb) confirm this contrast. Although some overlap is evident, Rai sandstones are decidely less evolved in an igneous sense than comparable Pelorus sandstones.

Sandstone Provenance

Petrographic study reveals a dominantly volcanic source for Maitai, Rai, and Pelorus sandstones. Small amounts of plutonic-hypabyssal material are mineralogically similar to associated volcanic sands and are regarded as being broadly co-magmatic. Sedimentary-derived detritus ranges from rare to abundant, is mainly of intra-basinal origin. Bulk chemistry, texture and mineralogy of lithic clasts, and paucity of microcline, orthoclase, and quartz sand generally suggest oceanic island arc provenance. Attempts at modal analyses of sandstones met with limited success. Degree of low grade metamorphic reconstitution, including albitization, generally obscures original grain boundaries and matrix-cement relations; fine grained authigenic prehnite, pumpellyite, chlorite, sericite and other minerals are abundant. Thus, only two Rai samples were point-counted with confidence, these yielded Q:F:R ratios of 3:5:72 and 0:10:90. They resemble Te Anau sandstones described by Mackinnon [1983] from Caples terrane, but are slightly richer in lithics.

In addition to the predominantly arc-volcanic detritus, one example, Tramway sample (S15-181 = G.R. 007580) is distinctly richer in detrital quartz and poorer in volcanic lithic debris than most sandstones studied. A similar relationship is noted in Tramway samples from elsewhere in the Nelson area [Watters in Waterhouse, 1964] and in correlative strata southeast of the Alpine fault [Landis, 1974b]. This sand is of "dissected arc" [Dickinson et al., 1983] provenance. Ophiolitic clasts have been recorded from several rocks of Little Ben lithology. One sample contains detrital grains of serpentinite, pyroxene, hornblende and chromite in addition to basic volcanic debris. Chemical analyses (Table 2) reveal that clinopyroxenes and hornblendes resemble analyzed materials from gabbros and mafic dikes at Red Mountain [Sinton, 1980] while chromites resemble those from the Red Mountain ultramafites [Sinton, 1977]. The iron-rich nature of pyroxene and its characteristic association with brown hornblende, together with associated mafic volcanic and ultramafic debris, suggest erosion of an arc-root ophiolite. The close stratigraphic association of these sandstones indicates penecontemporaneous provenance of undissected arc, quartzo-feldspathic crystalline, and ophiolitic sediment.

The location of source terranes for these sedimentary rocks remains enigmatic. Accepting a Permian age for Maitai and possibly for Rai and Pelorus strata, it appears that more than one Permian volcanic arc is required. The Brook Street terrane [Blake et al., 1974; Coombs et al., 1976], a largely middle Permian volcanic arc, lies to the west of the Maitai rocks (Fig. 1) and comprises one possible source. It is characterized by basic pyroclastic and epiclastic volcanic sediments with interlayered dacites, andesites and basaltic lavas of tholeiitic to calc-alkaline nature [Waterhouse, 1964; Williams, 1978; Houghton, 1981, 1985; Sivell and Rankin, 1983]. In addition, it contains local suites of mafic-ultramafic plutonic, hypabyssal, and volcanic rocks some of possible ophiolitic affinities [Mössman, 1973; Nauman, 1973; Sivell and Rankin, 1983], and is invaded by granitic-dioritic plutons [Williams and Harper, 1978; Price and Sinton, 1978], mainly of Mesozoic age. In view of their more mafic nature, it is unlikely that Brook Street volcanic rocks comprise a substantial source for Maitai sedimentary rocks (Fig. 7).

The source of the clastic detritus of the Rai and Pelorus terranes is even more problematical. They have been widely regarded as deeper water distal correlatives of Maitai and Murihiku strata [Blake et al., 1974; Coombs et al., 1976; Carter et al., 1978; Davis et al., 1981; Mackinnon, 1983], and also of Brook Street

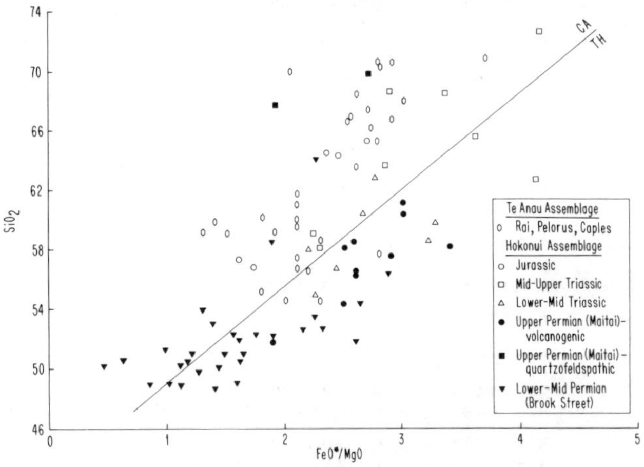

Fig. 7. Chemographic relations between Te Anau and Hokonui assemblage rocks. All data refer to sandstones and siltstones except for Brook Street which also includes lavas, pyroclastics and dike rocks. Samples showing evidence of extensive metasomatism are omitted. CA = calc-alkaline and TH = tholeiitic fields after Miyashiro (1974). Note tendency for rock suites to occupy distinctive fields. Data from Boles (1974), Coombs (1954), Kawachi (1974), Landis (1969, 1974b, and unpublished analyses), Reed (1957), Challis (1971), Houghton (1977) Williams (1978), Williams and Smith (1979), Sivell (1984), Sivell and Rankin (1983) and the present paper.

derivation. However, chemical data argue against this interpretation (Fig. 7). In particular, rocks of the Te Anau assemblage (Rai, Pelorus, Caples) are of calc-alkalic character and chemically distinct from characteristically tholeiitic Maitai and the less silicic, tholeiitic to calc-alkaline, Brook Street rocks. The analyzed Te Anau rocks do bear a distinct resemblance to Mesozoic volcaniclastic strata of the Hokonui assemblage. We conclude that Te Anau rocks were derived from an arc terrane now removed by tectonic processes, and furthermore, that they are not distal equivalents of the Maitai strata.

Metamorphism

All rocks studied have undergone some degree of low-grade metamorphism. Using field criteria [Bishop, 1972], we have mapped three textural zones in metasedimentary rocks (Fig. 3). The resulting map shows an irregular distribution.

A relationship appears to exist between degree of textural reconstitution and structural boundaries. For example, in the Rai terrane the more completely reconstituted rocks [compare textural Zone IIA of Bishop, 1972] are localized along the contact with Patuki melange, and a similar relation can be seen in Pelorus rocks along the Croisilles melange boundary. A conspicuous exception to this generalization is seen along the Pelorus-Croisilles contact near the road to McLaren Bay where massive unsheared graywacke crops out adjacent to the contact. These rocks resemble Rai sandstones, being anomalously basic (see Sandstone Petrochemistry) and rich in clastic augite compared to representative Pelorus samples.

Mineralogical indicators of metamorphism were studied in approximately 200 samples using routine petrographic and X-ray techniques. The distribution of several key metamorphic minerals is shown in Figure 8. Nearly all rocks contain metamorphic pumpellyite. Representative Maitai metasediments contain abundant lawsonite with lesser prehnite and epidote and no stilpnomelane or actinolite. In contrast, Maitai rocks in narrow metasomatic contact zones adjacent to serpentinite bodies at Pahakorea and Onetea contain prehnite, wollastonite, idocrase, hydrogrossular, actinolite and aragonite while lacking lawsonite [Coleman, 1966; writers' observations]. Rai metasediments contain moderately abundant prehnite and some epidote (particularly along the southeastern contact with the Croisilles melange), while lawsonite, stilpnomelane and actinolite are rare. Metasedimentary rocks of the Pelorus terrane contain actinolite and abundant stilpnomelane, with less common epidote, one sample with lawsonite and none with prehnite.

In terms of metamorphic facies, Maitai rocks belong to the lawsonite-albite-chlorite facies, Rai to prehnite-pumpellyite facies locally transitional to lawsonite-albite-chlorite facies, while Pelorus rocks are predominantly in the pumpellyite-actinolite facies, passing eastward into greenschist facies [Vitaliano, 1968]. Melange blocks, particularly in the more intensely studied Croisilles melange, contain assemblages assigned to a variety of facies--prehnite-pumpellyite, pumpellyite-actinolite and amphibolite. In addition, trace amounts of blue amphibole occur in several metavolcanic rocks and may indicate localized blueschist facies conditions. In a general way these relations indicate a NW-SE increase in temperature:pressure gradient prevailing in intact sedimentary terranes, while melange blocks include a wide range of metamorphites ranging from high T:P early formed ocean floor metamorphites and metasomatites [compare Coombs et al., 1976, p. 585; Sivell and Waterhouse, 1984b] to later low T:P burial and regional metamorphosed arc-volcanic sediments.

Geochronology

Radiometric dates and regional considerations indicate that deformation and metamorphism were polyphase and extended over a prolonged period of time. Detrital low-grade metamorphic rock fragments are widespread in sandstone; pebbles of metamorphosed and veined volcanic rock, sand-

TABLE 4. K-Ar Dates From Metamorphic Rocks, Croisilles Harbour Region

Number	Sample	K_2O(Wt%)	Ar^{40} cc/gm	% Rad.	Age (m.y.)
S15-1A	Pelorus slate (WR)	1.81	1.384×10^{-5}	87.3	187
S15-47E	Pelorus slate (WR)	0.92	7.922×10^{-6}	85.9	209
S15-136A	Rai slate (WR)	1.81	1.093×10^{-5}	89.6	149
S15-155A	Rai slate (WR)	2.32	1.446×10^{-5}	89.3	154
S15-2A	Croisilles amphibolite (WR)	0.49	5.438×10^{-6}	87.1	265
S15-2A	Croisilles amphibolite (H)	0.57	6.518×10^{-6}	88.1	272
22293	Maitai slate (WR)	1.66	7.87×10^{-6}	81.7	115

Decay Constants: $\lambda_\varepsilon+\lambda'_\varepsilon = 0.581\times10^{-10}yr^{-1}$
$\lambda_\beta = 4.962\times10^{-10}yr^{-1}$
$^{40}K/K_{total} = 1.167\times10^{-4}$mol/mol

S15-1A = G.R. 989484
S15-47E = G.R. 054527
S15-136A = G.R. 106558
S15-155A = G.R. 096534
S15-2A = G.R. 961489
2293 reported in Harper and Landis, 1967

WR = Whole Rock
H = Hornblende
Pelorus and Rai samples collected less than 200 m from Croisilles melange contact.
Analyst C. T. Harper.

stone, and conglomerate, occur in conglomerates (including blocks in melange). Amphibolite inclusions in Croisilles melange record an early Permian phase of sea floor metamorphism not recognized in metasedimentary rock.

K-Ar dates from metasedimentary and metaigneous rocks (Table 4) suggest recrystallization and cooling episodes ranging from Permian to Cretaceous. Thus, dates from an amphibolite block in the Croisilles melange reflect ophiolite recrystallization during Early Permian time. This is consistent with U-Pb dating of zircon from a plagiogranite block from Croisilles rocks at Elaine Bay which indicates 275-280 Ma age of magmatic crystallization [Dickins et al., 1987]. Two Pelorus slates yielded latest Triassic-Early Jurassic and Middle Jurassic whole rock ages while two Rai samples yielded Late Jurassic dates [Jurassic geochronology after Kent and Gradstein, 1985]. Four Maitai slates from the Nelson-Marlborough Sounds, including one from the present area, were dated previously at 110-130 Ma (Early Cretaceous) [Harper and Landis, 1967]. On the basis of available data, we suggest that each of the tectonic units experienced distinctive metamorphic and thermal histories.

Recent K-Ar dating of Torlesse (Alpine assemblage) slates and schists [Adams et al., 1985] from northern Otago (Fig. 1) reveals ages ranging between 154-199 Ma. Taken in conjunction with data presented here, a good case can be made for early to mid-Jurassic collisional suturing of Te Anau and Alpine assemblages [Bishop et al., 1985]. The actual boundary (the Haast Suture in Figure 1) is cryptic, being obscured by collision-related metamorphism.

Conclusions - Tectonic Evolution

Geology of the Croisilles Harbour region reflects a complex history of processes associated with active plate margins. Sedimentary and volcanic rocks show evidence for deep sea accumulation, the sandstones being redeposited by mass flow phenomena along the flanks of volcanic arcs, the massive dark mudstones accumulating as distal turbidites and hemipelagic deposits, while volcanic rocks of the ophiolitic melanges represent slices of crust generated in submarine environments. Structural and metamorphic features are consistent with processes at both divergent and convergent margins. Locations of implied spreading centers, subduction zones, and island arcs remain speculative, and it is probable that these have been largely removed during Mesozoic orogenesis. Key points which must be considered in any reconstruction are summarized below:

Arc-derived Upper Permian Maitai strata rest in depositional contact upon Lower Permian oceanic crust of the Lee River Group, in some areas with little or no angular unconformity. Basal Maitai rocks comprise sedimentary breccias of volcanic arc and ophiolitic parentage and overlying redeposited limestones composed of bivalve prisms. Intervening pelagic and chemical sediments are lacking. Overlying sediments are predominantly of andesitic to basaltic arc-volcanic provenance. However, volumetrically minor interbedded rocks include quartzofeldspathic sandstones, suggesting a secondary provenance of continental margin character and also olistostromes and sandstones of ophiolitic provenance.

Croisilles and Patuki rocks comprise ophio-

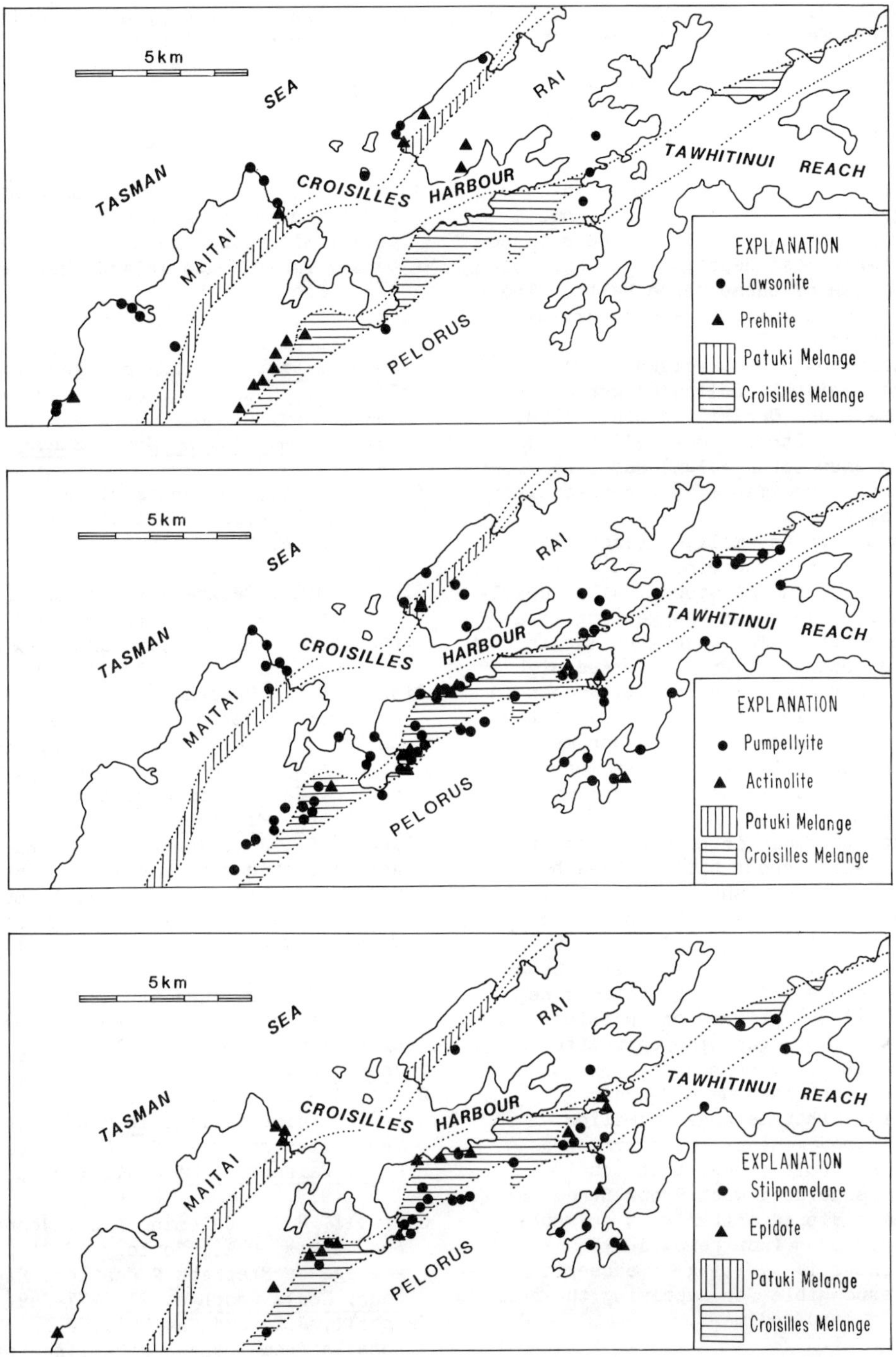

Fig. 8. Distribution of diagnostic lower-grade metamorphic minerals.

litic melanges which resemble each other in lithologic content, age, and style of deformation; however their original continuity cannot be demonstrated. Blocks of dark gray argillite comprise the predominant sedimentary rock type in both Croisilles and Patuki ophiolitic melanges. "Phantom stratigraphy" assembled from melange blocks suggests ophiolitic sequences overlain by hemipelagic gray muds with little or no intervening chemical or pelagic sediment. Comparable sequences of argillite are not known from intact Dun Mountain-Maitai, Rai, or Pelorus terranes.

Petrographic data indicate derivation of Te Anau assemblage (Rai and Pelorus) sandstones from a volcanic arc. Maitai sands are also of arc-volcanic provenance but differ in composition and are generally finer-grained and thinner-bedded than Te Anau rocks. Petrochemical and sedimentological data indicate that Maitai, Rai, and Pelorus rocks cannot be regarded as shallow-, intermediate- and deep-water portions of one contemporaneous sedimentary system. There is no evidence for interrelated deposition of the volcanic arc sandstones of these three belts. The only clear-cut Permian volcanic arc assemblage presently exposed in New Zealand is the Brook Street terrane. Brook Street volcanism was active during or immediately following generation of the Dun Mountain and Croisilles ophiolites. Geochemistry of Brook Street rocks eliminates them as a likely source for voluminous late Permian and Mesozoic sediments of the Hokonui or Te Anau assemblages.

Low T:P metamorphic assemblages (lawsonite-albite-chlorite facies) predominate in Maitai rocks and are widely, though sporadically, developed in Rai and Pelorus rocks, suggesting metamorphism of these strata in convergent margin environments. In contrast, earlier-formed high T:P assemblages characterize ophiolitic rocks in the Patuki and Croisilles melanges. The latter are regarded as forming during sea-floor metamorphism, while the lawsonite-forming metamorphism is associated with Mesozoic deformation and regional metamorphism.

Numerous plate tectonic interpretations of New Zealand's Permian-Cretaceous history have been proposed and many of these show broad agreement [e.g., Landis and Bishop, 1972; Coombs et al., 1976; Carter et al., 1978; Wood, 1978; Mackinnon, 1983]. In view of the points summarized above, we believe that future models need to be more complex; Howell [1980] offered one possible step in this direction. We regard the ophiolites as having been generated in, or near to, one or more island arcs (i.e., not at true mid-ocean spreading centers), that melange formation probably reflects movement in transcurrent as well as convergent fault zones, that the subduction zone associated with Brook Street arc volcanism may have dipped either east or west, and that the other arcs (now removed by subduction, strike-slip faulting or subaerial erosion) were responsible for supplying the bulk of Hokonui and Te Anau sediment.

Acknowledgments. Discussions with many colleagues, particularly D.S. Coombs, D.L. Kimbrough, and M.R. Johnston have proven valuable. Reviews by D.G. Howell, N.K. Huber, and E.C. Leitch and an anonymous reviewer substantially improved the manuscript. C.T. Harper is thanked for radiometric data. The final manuscript was completed while one of the authors (C.A.L.) was a visitor at the Department of Geology, University of Colorado, Denver. Financial support was provided by a University of Otago research grant (C.A.L.) and research grants from the Fullbright Commission and Auckland University (M.C.B.).

References

Adams, C.J., Bishop, D.G., and Gabites, J.E., Potassium-argon age studies of low-grade progressively metamorphosed graywacke sequence, Dansey Pass, South Island, New Zealand. Jour. GeoL. Soc. London, 142, 339-349, 1985.

Beck, A.C., Sheet 14 Marlborough Sounds Geological Map of New Zealand 1:250,000, Dept. Sci. Indust. Res., Wellington, New Zealand, 1964.

Bell, J.M., Clarke, E. de C., and Marshall, P., The geology of the Dun Mountain Subdivision, Nelson, New Zealand Geol. Survey Bull. 12, 71, 1911.

Bernoulli, D., and Weissert, H., Sedimentary fabrics in Alpine ophicalcites, South Pennine Arosa Zone, Switzerland, Geology, 13, 755-758, 1985

Bishop, D.G., Progressive metamorphism from prehnite-pumpellyite to greenschist facies in the Dansey Pass area, Otago, New Zealand, Geol. Soc. America BuLl., 83, 3177-3197, 1972.

Bishop, D.G., Bradshaw, J.D., and Landis, C.A., Provisional terrane map of South Island, New Zealand, in Howell, D.G., ed., 1985, Tectonostratigraphic terranes of the Circumpacific region, Earth Sci. Series, Circumpacific Council Energy Min. Res., 515-521, 1985.

Blake, M.C., Jr., Jones, D.L., and Landis, C.A., Active Continental Margins: contrasts between California and New Zealand, in Burk, C.A., and Drake, C.L., eds., The Geology of Continental Margins, New York, Springer-Verlag, 853-872, 1974.

Blake, M.C., Jr., and Landis, C.A., The Dun Mountain ultramafic belt--Permian oceanic crust and upper mantle in New Zealand, U.S. Geol. Survey Jour. Res., 1, 529-534, 1973.

Bouma, A.H., and Hollister, C.D., Deep ocean basin sedimentation, in Middleton, G.V., ed., Turbidites and Deep Sea Sedimentation, Soc. Econ. Paleontol. Mineral. (short course), 79-118, 1973.

Campbell, H.J., Coleman, A.C., Johnston, M.R., and Landis, C.A., Geology of Stephens Island and age of Stephens Formation, New Zealand Jour. Geol. Geophys. 27, 277-289, 1984.

Campbell, J.K, and Campbell, J.D., Triassic tube fossils from Tuapeka rocks, Akatore, South Otago, New Zealand Jour. Geol. Geophys., 13, 392-399,1970.

Carman, M.F., A comparison of some Permian rocks on opposite sides of the Alpine fault, South Island, New Zealand, Royal Soc. New Zealand Trans. Geol., 6, 91-130, 1968.

Carter, R.M., Landis, C.A., Norris, R.J., and Bishop, D.G., Suggestions towards a high-level nomenclature for New Zealand rocks, Royal Soc. New Zealand Jour. 4, 5-18, 1974.

Carter, R.M., Hicks, M.D., Norris, R.J., Turnbull, I.M., Sedimentation patterns in an ancient arch-trench-ocean basin complex: Carboniferous to Jurassic Rangitata Orogen, New Zealand, in Stanley, D.J., and Kelling, G., eds., Sedimentation in submarine canyons, fans, and trenches, Stroudsburg, Pennsylvania, Dowden, Hutchinson and Ross, 340-361, 1978.

Challis, G.A., Chemical analyses of New Zealand rocks and minerals with C.I.P.W. norms and petrographic descriptions, 1917-1957, Part 1: Igneous and pyroclastic rocks, New Zealand Geol. Survey Bull., 5, 94, 1971.

Chayes, F., The chemical composition of Cenozoic andesite: Oregon Dept. Geol. Mineral Ind. Bull. 65, 1-11, 1965.

Coleman, R.G., New Zealand serpentinites and associated metasomatic rocks, New Zealand Geol. Survey Bull. 76, 102, 1966.

Coney, P.J., Jones, D.L., and Monger, J.W.H., Cordilleran suspect terranes, Nature, 288, 329-333, 1980..

Coombs, D.S., Landis, C.A., Norris, R.J., Sinton, J.M., Borns, D.J., and Craw, D., The Dun Mountain ophiolite belt, New Zealand, its tectonic setting, constitution and origin, with special reference to the southern portion, Am. Jour. Sci., 276, 561-603, 1976.

Craw, D., Melanges and associated rocks, Livingstone Mountains, Southland, New Zealand, New Zealand Jour. Geol. Geophys., 22, 443-454, 1979.

Cummins, W.A., The graywacke problem, Liverpool Manchester Geol. Jour., 38, 51-72, 1972.

Davis, T.E., Johnston, M.R., Rankin, P.C., and Stull, R.J., The Dun Mountain ophiolite belt in east Nelson, New Zealand, Proc. Internat. Ophiolite Symposium, Cyprus, 480-496, 1980.

Dickins, J.M., Johnston, M.R., Kimbrough, D.L., and Landis, C.A., The stratigraphic and structural position and age of the Croisilles melange, East Nelson, in press, 1987.

Dickinson, W.R., Beard, L.S., Brakenridge, G.R., Erjavec, J.L., Ferguson R.C., Inman, K.F., Knapp, R.S., Lindberg, F.A., and Ryberg, T.P., Provenance of North American Phanerozoic sandstones in relation to tectonic setting, Geol. Soc. America Bull., 94, 222-235, 1983.

Dickinson, W.R., Interpreting provenance relations from detrital modes of sandstone, in Zuffa, G.G., ed., Provenance of arenites, D. Reidel Publ. Co., 333-361, 1985.

Duff, R., 1946, Native quarries of baked argillite, Records Canterbury (New Zealand) Museum, 5, 115-124, 1946.

Furnish, W.M., Glenister, B.F., Kummel, B., Spinosa, C., Sweet, W., and Teichert, C., Reinterpretation of ceratitic ammonoids from the Greville Formation, New Zealand, Geol. Mag., 113, 39-46, 1976.

Fryer, P., Ambos, E.L., Hussong, D.M., Origin and emplacement of Mariana forearch seamounts, Geology, 13, 774-777, 1985.

Harper, C.T., and Landis, C.A., K-Ar ages from regionally metamorphosed rocks, South Island, New Zealand, and some tectonic implications, Earth Planet. Sci. Letters, 2, 419-429, 1967.

Houghton, B.F., Geology of the Takitimu Group and associated intrusive rocks, central Takitimu Mountains, western Southland, New Zealand, Ph.D. thesis, Univ. Otago, Dunedin, New Zealand, 1977.

Houghton, B.F., Lithostratigraphy of the Takitimu Group, central Takitimu Mountains, western Southland, New Zealand, New Zealand Jour. Geol. Geophys., 24, 333-348, 1981.

Houghton, B.F., Petrology of the calcalkaline lavas of the Permian Takitimu Group, southern New Zealand, New Zealand Jour. Geol. Geophys., 28, 649-665, 1985.

Howell, D.G., Mesozoic accretion of exotic terranes along the New Zealand segment of Gondwanaland, Geology, 8, 487-491, 1980.

Hunt, T.M., Rock magnetism in the alpine-peridotite body at Dun Mountain, New Zealand, Royal Soc. New Zealand Jour., 4, 433-445, 1974.

Johnston, M.R., The Rai Sandstone in the Dun Mountain area, east Nelson, Royal Soc. New Zealand Jour., 7, 113-117, 1977.

Johnston, M.R., Sheet 027AC Dun Mountain. Geological map of New Zealand, 1:50,000, 1st ed., Wellington, New Zealand, Dept. Sci. Ind. Res. 1981.

Johnston, M.R., and Stevens, G.R., New fossil localities in the Maitai Group, Nelson, and comments on its age, New Zealand Jour. Geol. Geophys., 21, 113-115, 1978.

Jones, D.L., Howell, D.G., Coney, P.J., and Monger, J.W.H., Recognition, character and analysis of tectonostratigraphic terranes in western North America, in Hashimoto, M., and Uyeda, S., eds., Accretion Tectonics in the Circum-Pacific Region, Terra Scientific Publ. Co., Tokyo, p. 21-35, 1983.

Kawachi, Y., Geology and petrochemistry of weakly metamorphosed rocks in the upper Wakatipu District southern New Zealand, New Zealand Jour. Geol. Geophys., 17, 169-208, 1974.

Kear, D., Basement rock facies--northern North Island, New Zealand Jour. Geol. Geophys., 14, 275-283, 1971.

Kent, D.V., and Gradstein, F.M., A Cretaceous and Jurassic geochronology, Geol. Soc. America Bull., 96, 1419-1427, 1985.

Landis, C.A., Stratigraphy, lithology, structure, and metamorphism of Permian, Triassic, and Tertiary rocks between the Mararoa River and Mount Snowden, western Southland, New Zealand, Royal Soc. New Zealand Jour. 4, 229-251, 1974a.

Landis, C.A., Little Ben Sandstone, Maitai Group (Permian): nature and extent in the Hollyford-Eglinton region, South Island, New Zealand, New Zealand Jour. Geol. Geophys., 23, 551-567, 1980.

Landis, C.A., and Bishop, D.G., Plate tectonics and regional stratigraphic-metamorphic relationships in the southern part of the New Zealand Geosyncline, Geol. Soc. America Bull., 83, 2266-2284, 1972.

Mackinnon, T.C., Origin of the Torlesse terrane and coeval rocks, South Island, New Zealand, Geol. Soc. America Bull. 94, 967-985, 1983.

Miyashiro, A., 1974, Volcanic rock series in island arcs and active continental margins, Am. Jour. Sci., 274, 321-355, 1974.

Miyashiro, A., Volcanic rock series and tectonic setting, Ann. Rev. Earth Planet. Sci., 3, 251-269, 1975.

Mossman, D.J., Geology of Greenhils Ultramafic Complex, Bluff Peninsula, Southland, New Zealand, Geol. Soc. America Bull., 84, 39-62, 1973.

Nauman, C.R., Alabaster Group rocks (Lower Permian) in southern Skippers Range, northwest Otago, New Zealand, Royal Soc. New Zealand Jour. 3, 527-544, 1973.

Price, R.C., and Sinton, J.M., Geochemical variations in a suite of granitoids and gabbros from Southland, New Zealand, Contr. Mineral. Petrol., 67, 267-278, 1978.

Reed, J.J., Soda-metasomatized argillites associated with the Nelson ultramafic belt, New Zealand Jour. Geol. Geophys., 2, 906-915, 1959.

Rigg, T., Chittenden, E., and Hodgson, I., A survey of soils, vegetation and agriculture of the Eastern Hills, Waimea County, Nelson, Cawthorn Institute Nelson, New Zealand, 42,1957.

Schofield, J.C., Stratigraphy, facies, structure and setting of the Waiheke and Manaia Hill Groups, east Auckland, New Zealand Jour. Geol. Geophys., 17, 807-838, 1974.

Shapiro, L., and Brannock, W.W., Rapid analysis of silicate, carbonate, and phosphate rocks: U.S. Geol. Survey Bull. 1144-A, 56, 1962.

Sinton, J.M., Equilibration history of basal alpine-type peridotite, Red Mountain, New Zealand, Jour. Petrology, 18, 216-246, 1977.

Sinton, J.M., Petrology and evolution of the Red Mountain ophiolite complex, New Zealand, Am. Jour. Sci., 280, 296-328, 1980.

Sivell, W.J., Low-grade metamorphism of the Brook Street Volcanics, D'Urville Island, New Zealand, New Zealand Jour. Geol. Geophys., 27, 167-190, 1984.

Sivell, W., and Rankin, P., Arc-tholeiite and ultramafic cumulate, Brook Street Volcanics, west D'Urville Island, New Zealand, New Zealand Jour. Geol. Geophys., 26, 239-257, 1983.

Sivell, W.J., and Waterhouse, J.B., The Patuki intrusive suite: closed system fractionation beneath a slow-spreading ridge, Lithos, 17, 1-18, 1984a.

Sivell, W.J., and Waterhouse, J.B., Oceanic ridge metamorphism of the Patuki Volcanics, D'Urville Island, New Zealand, Lithos, 17, 19-36, 1984b.

Sporli, K.B., Mesozoic tectonics, North Island, New Zealand, Geol. Soc. America Bull., 89, 415-425, 1978.

Turnbull, I.M., Nomenclature in the Rangitata Geosyncline (note), New Zealand Jour. Geol. Geophys., 20, p. 803-809, 1977.

Vitaliano, C.J., Petrology and structure of the southeastern Marlborough Sounds, New Zealand, New Zealand Geol. Surv. Bull., 74, 40, 1968.

Walcott, R.I., Geology of the Red Hill Complex, Nelson, New Zealand, Royal Soc. New Zealand Trans. Earth Sci., 7, 57-88. 1969.

Waterhouse, J.B., Stratigraphy of the lower part of the Maitai Group, Nelson, New Zealand Jour. Geol. Geophys., 2, 944-953, 1959.

Waterhouse, J.B. Permian stratigraphy and faunas of New Zealand, New Zealand Jour. Geol. Geophys., 72, 101, 1984.

Waterhouse, J.B., The age of the Croisilles Volcanics, eastern Nelson, Royal Soc. New Zealand Trans. Geol., 3, 175-181, 1966.

Waterhouse, J.B., An ophicerated ammonoid from the New Zealand Permian and its implications for the Permian-Triassic boundary, Geol. Mag., 110, 305-329, 1973.

Waterhouse, J.B., The Rangitata Orogen, Pacific Geol. 9, 35-73, 1975.

Waterhouse, J.B., The Early Permian age of the Rai Sandstone, east Nelson, New Zealand, New Zealand Jour. Geol. Geophys., 25, 359-362, 1982.

Williams, J.G., Eglinton Volcanics--stratigraphy, petrography and metamorphism, New Zealand Jour. Geol. Geophys., 21, p. 713-732, 1978.

Williams, J.G., and Harper, C.T., Age and status of the Mackay Intrusives in the Eglinton-upper Hollyford area, New Zealand Jour. Geol. Geophys., 21, 733-742, 1978.

Williams, J.G., and Smith, I.E.M., Geochemical evidence for paired arcs in the Permian volcanics of southern New Zealand, Contr. Mineral. Petrol., 68, 285-291, 1979.

Wood, B.L., The Geology of the Gore Subdivision, New Zealand Geol. Survey Bull., 53, 128, 1956.

Wood, B.L., The Otago Schist megaculmination: its possible origins and tectonic significance in the Rangitata Orogen of New Zealand, Tectonophysics, 47, 339-368, 1978.

TERRANE BOUNDARIES AND TERRANE DISPLACEMENT IN NORTHERN VICTORIA LAND, ANTARCTICA: SOME PROBLEMS AND CONSTRAINTS

J.D. Bradshaw

University of Canterbury, Christchurch, New Zealand

Abstract. In recent years, northwest trending belts of contrasted geology in northern Victoria Land have been interpreted as terranes. Numerous tectonic hypotheses for their origin and relationship have been advanced, involving either strike slip or convergence with east and/or west dipping subduction zones. The eastern boundary of the central Bowers terrane is a zone of thrusting and tectonic mixing of probable early Ordovician age, and marked by the "Millen Schists" or Millen shear zone. The Leap Year Fault, previously regarded as the terrane boundary, is a younger high angle reverse dip slip or oblique slip fault which juxtaposes different tectonic levels. The tectonics of the western boundary are less clear. Significant thrusting has recently been demonstrated on this boundary too, but it is likely that this documents only the last major movement. Thrusting alone is inadequate to explain features including: the loss of the eastern part of the Wilson terrane, the probable loss of the western part of the Bowers terrane, the provenance of some Bowers terrane rocks, and patterns of granitoid geochemistry. A complex movement history extending over a period of 150 Ma is indicated.

Introduction

Geologically contrasted northwest-southeast trending belts of Early Paleozoic rocks have been mapped in northern Victoria Land for over twenty years and more recently have been shown to extend for 400km from the Ross Sea to the Southern Ocean [Stump et al.,1983]. The central Bowers terrane [Weaver et al., 1984, Bradshaw et al., 1985], which is crucial to the understanding of the relationship between these belts, is between 15 and 35 km wide and comprises at least 10km of Cambrian to Ordovician sediments and volcanics locally cut by Late Paleozoic granites and overlain by younger volcanics (Figure 1). The Early Paleozoic rocks are termed the Bowers Supergroup which is subdivided into the Sledgers Group (oldest), the Mariner Group, and the Leap Year Group (Figure 2). The Sledgers Group comprises at least 3.5km of basic to acid volcanics and interfingering sediments, and is concordantly overlain by the Mariner Group. The latter is at least 2.5km thick and consist of sandstone, fossiliferous calcareous mudstone and carbonates, which form a largely non-volcanogenic regressive sequence with marginal marine rocks at the top. The Leap Year Group rests on a marked erosion surface which truncates the Mariner Group towards the north and west, and comprises the local polymict Carryer Conglomerate overlain by at least 4km of fluviatile and marginal marine Camp Ridge Quartzite. Belts of more deformed and metamorphosed rocks lying to the east and west were initially interpreted as Precambrian which led to the use of the terms "Bowers Trough" and "Bowers Graben" [Gair et al., 1969; Laird et al., 1982]. Further research has shown that these notions are inappropriate because the rocks to the east, the Robertson Bay terrane, are much the same age [Wright and Findlay, 1984], and those to the west were also involved in Cambro-Ordovician metamorphism and magmatism.

A turning point in the investigations was the geochemical study of the Early to Middle Cambrian volcanics of the Bowers terrane [Weaver et al.,1984] which clearly indicated a primitive island arc setting for the early Bowers rocks, inconsistent with their current position between contemporaneous belts of continental character and continental provenance. The proposal [Weaver et al., 1984, Bradshaw et al., 1985] that the belts represented discrete terranes juxtaposed tectonically has been widely accepted, but the tectonic explanations have been criticised and obviously require modification in the light of new data. Alternative tectonic models have been proposed by Gibson and Wright [1985], Kleinschmidt and Tessensohn [in press,1986], and Gibson [this volume], involving convergent tectonics and east or west dipping subduction zones.

A map and a synoptic terrane diagram are presented in Figure 1. East of the Bowers terrane, a single Robertson Bay terrane is recognised; the narrow Millen terrane [see Bradshaw et al., 1985] has proved to be a zone of mechanical mixing of Bowers and Robertson Bay terrane rocks, and is unwarranted. West of the Bowers terrane the term Wilson terrane has been widely used although the

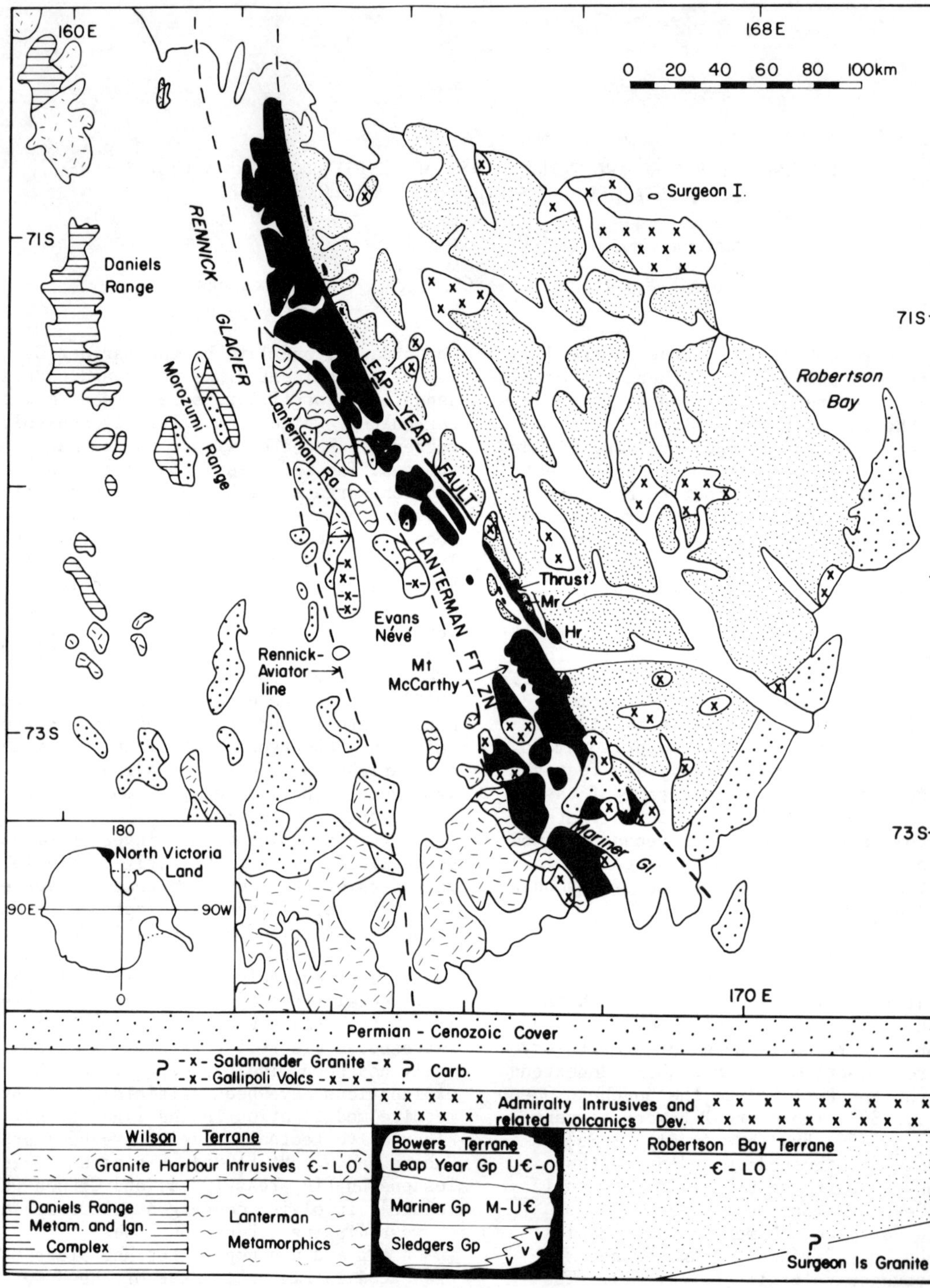

Fig 1. Terrane Map and diagram for north Victoria Land. Mr - Millen Range. Hr - Handler Ridge.

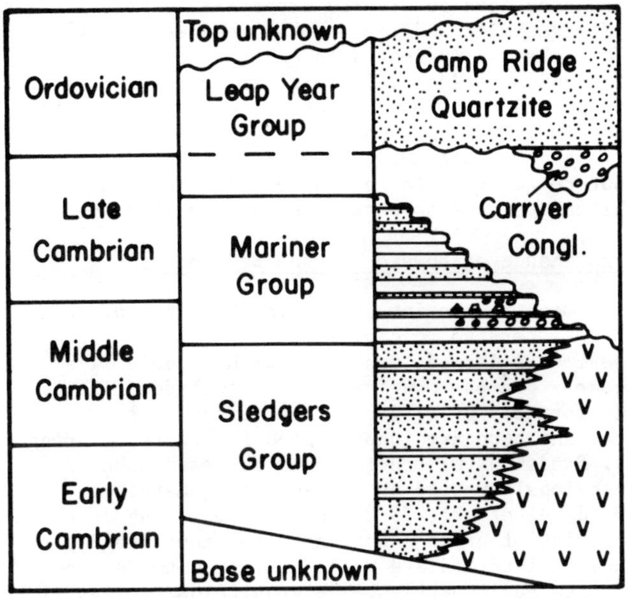

Fig 2. Subdivision of the Bowers Supergroup showing only names employed in text.

unity of this belt prior to the intrusion of the Granite Harbour Intrusives is not firmly established.

The purpose of this paper is to assess tectonic models in the light of constraints from new work. Let me emphasise that the Cambrian Period was as long as the Cenozoic Era during which there have been major changes in plate configuration in the southwest Pacific sector, and that the geological timespan under discussion, about 200 Ma, in equal in length to the history of the whole Pacific floor and longer than the history of other oceans.

Eastern Boundary of the Bowers Terrane

The original identification [Weaver et al., 1984] of the Leap Year Fault as the terrane boundary is incorrect and this fault largely juxtaposes different structural levels near the terrane boundary. In the Bowers terrane to the west, upright very gently plunging folds in the youngest unit of the Bowers Supergroup, the Camp Ridge Quartzite, are typical. To the east, recumbent and reclined folds closely related to major thrusts involve both Robertson Bay and Bowers terrane rocks in a complex tectonic contact zone which reaches a width of 25km in the Millen Range -Mt McCarthy area. Schistosity and mineral lamination are variably developed, sometimes overprinted by upright chevron folds and crenulation cleavage. East of the Millen Range, at Handler Ridge, a syncline in Robertson Bay rocks immediately east of the thrust zone contains, a bouldery olistostrome with large carbonate blocks from which latest Cambrian to earliest Ordovician macro and microfossils have been obtained [Wright et al., 1984, Burrett and Findlay, 1984]. The rocks present a contrast to the normally monotonous fine clastics of the Robertson Bay terrane, and suggest a perturbation of the typical Robertson Bay submarine fan environment [Field and Findlay, 1983] at the beginning of the Ordovician. K-Ar geochronology [Adams and Kreuzer, 1984] which suggests post metamorphic cooling also commenced in the early Ordovician, constrains the age of thrusting, metamorphism, and terrane amalgamation to a short interval, probably in Tremadoc or Arenig times.

The Leap Year Fault juxtaposes the simpler higher structural levels of the eastern Bowers terrane with the recumbent fold-thrust zone, leading to the omission of 3-5 km of strata. The remarkable parallelism of the Leap Year Fault and the Camp Ridge Syncline for over 100km suggests a similar underlying structural control. In the southern Evans Neve area both the syncline and the fold-thrust zone are offset dextrally (15km) by a north-south fault which may merge with the Leap Year Fault [Bradshaw, 1985]. A final point of interest is that paleocurrents in the youngest formation of the Bowers terrane, the 4+km Camp Ridge Quartzite, are parallel to the terrane boundary in the southern and central sectors. It is possible that the same basic tectonic control influenced patterns of subsidence, recumbent folding and thrusting, upright folding and high angle faulting. The widespread unconformity below the Leap Year Group and the striking change in lithology from the underlying Mariner Group suggest this influence started in the Late Cambrian and, as outlined above, amalgamation was completed in the Early Ordovician. The distribution of the Admiralty Intrusives [Weaver et al., 1984, Tessensohn, 1984, and Borg et al., in press, 1986] confirm that the Robertson Bay terrane and the Bowers terrane were a single entity in the Devonian. Geochemical gradients within the Admiralty Intrusives support a decreasing crustal influence on granitoid petrogenesis towards the south-southwest, consistent with a lack of thick continental crust below the Bowers terrane but inconsistent with the existence of the Wilson terrane in that direction. Contrary to previous interpretations, Borg et al., [1986] maintain that the Admiralty Intrusives are restricted to the Robertson Bay and Bowers terranes, and that the western boundary of the latter is a fundamental discontinuity in the crust.

Western Boundary of the Bowers Terrane

The western boundary of the Bowers terrane, the Lanterman Fault and an accompanying narrow strip of rocks showing high strain and greenschist metamorphism, has proved difficult to interpret. Recent comprehensive structural studies [Gibson, 1984, Gibson et al., 1984, and Gibson, this volume] clearly demonstrate the importance of thrusting and the emplacement of the high grade

Lanterman Metamorphics over the Bowers terrane. However, important questions remain over the timing and total tectonics of this boundary. Salient points for consideration are discussed below.

Igneous Rocks of the Wilson and Bowers Terranes

The Granite Harbour Intrusives (Figure 1) are confined to the Wilson Terrane and comprise predominantly "S" type granitoids with minor "I" types near the eastern margin. Intrusion ages (whole rock Rb/Sr) range from 515-480 Ma [Vetter et al., 1983], consistent with a protracted period of syn-and post-tectonic emplacement [Plummer et al., 1983, Roland et al., 1984]. Interpretations of isotope geochemistry, [Borg et al., in press, 1986] suggest that the age of magma generation was not greater than 550 Ma. Geochemical gradients suggest decreasing crustal influence towards the east. Trends are similar to those of continental margin plutonic provinces elsewhere, but are truncated, suggesting that only the western "inboard" portion is preserved. A similar conclusion is indicated by the recognition within the Wilson terrane of a broad western low pressure metamorphic belt and narrow and incomplete eastern medium pressure belt [Grew et al., 1984]. Metamorphism is thought to be a consequence of crustal thickening in excess of 60 km by continental collision [Grew and Sandiford,1985], and granite magma genesis is probably a related event. The development of this margin was contemporaneous with the erruption of the intra-oceanic Bowers arc. The polarity of subduction related to the arc is uncertain, but the presence of boninitic rocks in the west and lavas with backarc geochemistry at Mt McCarthy and Handler Ridge (S. D. Weaver, personal communication, 1985) favours the suggestion of eastward directed subduction [see also Wodzicki and Robert, 1984]. It is difficult to tie both sets of events to a single convergent margin regime.

Metamorphism in the Eastern Wilson Terrane

The Lanterman metamorphics form the eastern zone of the Wilson terrane. Studies by Grew and Sandiford [1984,1985] suggest a three stage development, (i) early high pressure metamorphism at 7-10 Kb and 700°C, followed by (ii) isothermal decompression to 5.5-6.4 Kb, equivalent to 8 km of uplift, and finally (iii) cooling to 300-370°C and decompression equal to a further 3-5 km of uplift. The early high pressure phase is based on the interpretation of a staurolite-talc assemblage in ultramafic pods as autochthonous relics within the dominant amphibolite grade metasediments of the second phase. Alternatively the ultamafic lenses may have been tectonically emplaced within the amphibolites, although in this circumstance the isothermal character of their emplacement is difficult to explain. The ultramafic rocks occur close to, and parallel with, the terrane boundary and have been considered important in that context [Kleinschmidt and Tessensohn, in press, 1986]. They clearly suggest deep crustal involvement in tectonic processes. It is clear however, that the pods relate to an event at least as early as the amphibolite metamorphism, and older than the greenschist metamorphism of the highly deformed suture zone. Again the eastern margin of the Wilson terrane appears to have a complex history.

Provenance of Bowers Terrane Sediments

Despite the intra-oceanic character of the Bowers terrane volcanics, the interbedded sediments include sandstone with granule size quartz grains. Conglomerates interbedded in the conformably overlying Mariner Group near the terrane boundary (unpublished fieldwork, J, Begg, J. Bradshaw, W. Buggisch and F. Tessensohn, previously mapped as Carryer Conglomerate by Bradshaw et al., 1982), are polymict and contain cobbles of green sandstone, limestone, feldsparphyric intermediate volcanics, granitoids and sparse basic pebbles. The granitoids are two mica granites of "S" type chemistry which plot in the continent collision fields of Pearce et al., [1984] diagram (S. D. Weaver, personal communication, 1986). Associated Mariner Group sediments include quartzose fine sandstone, mudstone, limestone, and olistostromes with carbonate mega-blocks, derived from a shallow platform probably across a contemporaneous fault line to the west [Bradshaw et al., 1982]. The Mariner Group becomes consistently finer towards the east. The Carryer conglomerate, a local valley fill or canyon fill conglomerate developed at the base of the Leap Year Group near the terrane boundary, is similarly polymict with pebbles of acid and basic volcanics, quartz sandstone, red mudstone, and red chert. The remainder of the Leap Year Group is overwhelmingly quartzose.

Bowers sediments clearly show that the arc had access to a continental source terrane and that from Late Middle Cambrian a carbonate platform and a mixed sedimentary, volcanic and plutonic continental terrane lay closeby to the west. Marked convergence between the arc and a continent occurred within Middle Cambrian times.

Terrane Boundary Zone Conglomerates

The interpretation of the two deformed conglomerates adjacent to the Lanterman Fault is difficult and involves two unresolved problems, the relative age of the conglomerates, and their relationship to other units of the Bowers terrane. The more easterly, predominantly volcanic derived Husky Conglomerate has always been included with the Bowers rocks, but the western polymict and typically more deformed Lanterman Conglomerate was initially included with the Wilson rocks [Bradshaw et al 1982]. Recently more detailed work [Gibson 1984] has suggested that both conglomerates are part of the Bowers terrane with increasing deformation and metamorphism towards the Lanterman Fault. With regard to relative age, Laird and Bradshaw [1983,p.123] cite a possible unconformity of Husky Conglomerate on Lanterman Conglomerate and Gibson [1984 fig 8 and p.139] reports blocks of possible Lanterman Conglomerate within Husky Conglomerate. Nonetheless Gibson [1985] now appears to regard the Husky Conglomerate as the older.

The affinities of the clasts in the Husky Conglomerate are clearly with the volcanics of the Sledgers Group, and as conglomerates with basaltic pebbles occur in the latter, correlation with the Sledgers Group has been suggested [Bradshaw and Laird 1983]. It is significant however that in detail the geochemistry of many clasts is distinctly more boninitic in character than the main Bowers volcanics [Weaver et al 1984] suggesting derivation from a more "outboard" sector, not currently recognised in north Victoria Land. Such a region is unlikely to have been exposed so close to the arc during the main arc building phase and therefore a post Sledgers Group age is indicated. The coarseness, angularity, and oligomict character of Husky Conglomerate contrast with the Lanterman and other Bowers terrane conglomerates, and suggest deposition close to source. The suggestion [Gibson and Wright, 1985] that the conglomerate is derived from obducted Cambrian Volcanics presents some difficulties. Either the Husky Conglomerate is younger than they propose and post-dates the Mariner and Leap Year Groups, or is allochthonous and derived from some more remote source, as it is clear that the area to the west was not buried by obducted volcanics during Mariner and Leap Year Group sedimentation.

The Lanterman Conglomerate and the comparable Black Spider Conglomerate [Gibson et al 1984] contain a range of lithologies comparable with but not identical to those of the Mariner Group or Carryer Conglomerate including sandstone, mudstone, limestone, volcanic, and rare granitoids. This suggests that they may be contemporaneous with late Middle and Upper Cambrian conglomerates of similar mixed continental provenance.

Provenance and Granitiod Suites

Provenance studies also have a bearing on the interpretation of granitoid geochemistry. Borg et al., [in press, 1986] propose that the Lanterman Fault coincides with a major discontinuity in the lower crust across which two granitoid suites of opposite polarity face each other. They further conclude that the chemical gradients within the younger Admiralty Intrusives are inconsistent with the presence of continental crust west of the amalgamated Bowers-Robertson Bay terrane before the Devonian. If this is the case then the continental source west of the Bowers terrane in the Cambrian must, at least temporarily, have been removed during the Late Ordovician and Silurian. This may have been achieved by the opening of a marginal sea or by strike slip tectonics which may have led to the removal of the origin source and the substitution of the present Wilson terrane in a Californian style continental margin. The absence of evidence of crustal thinning and the indications of truncation along the terrane boundary, favour the second alternative. In either case the separation of the north Victoria Land terranes from the Precambrian East Antarctic Shield is indicated.

Nature of Boundary Fault

Unlike the eastern boundary, the western boundary fault is steep and relatively straight and the Bowers terrane rocks retain upright structures, low strain, and low metamorphic rank to within a few hundred metres of the Lanterman Fault except in the Mariner Plateau area. The Bowers terrane shows no evidence of having been extensively overridden by Lanterman Metamorphics. Although the structural case for thrusting is convincing, comparison with the Alpine Fault of New Zealand, a modern active plate boundary, suggests that preserved fabrics reflect either the latest or most compressive structural regime. Fabric studies of Alpine Fault tectonites [Sibson et al., 1981] reflect the latest oblique dextral compressive regime with horizontal to vertical movement ratios in the range 0.8 to 2.7. The fabrics do not indicate the 480km of dextral offset indicated by the displacement of older terrane boundaries and analysis of southwest Pacific plate dynamics. Thus while thrusting on the Lanterman Fault has been strongly supported, it need not be expressing total movement or tectonic history of the terrane boundary.

Geochronology

The contribution of geochronology is equivocal. In a discussion of K-Ar data, Adams and Kreuzer [1984] note the following points, that rock and mineral ages in the Wilson terrane increase eastwards slightly to a maximum of 500 Ma, but that ages in the Bowers terrane show an enigmatic scatter with a similar 500Ma maximum near the eastern boundary but a wide range down to less than 300 Ma.

Adams and Kreuzer also note a relationship between age and sample site altitude in the Bowers terrane. It is also noticeable however, that whereas ages in or near the eastern terrane boundary are commonly high, those on or near the western terrane boundary are low, usually less than 400 Ma. There is also a distinct steepening of their cumulative frequency plot in the 385-420 Ma range. Thus geochronology does not rule out activity in the Late Silurian-Devonian about the time of generation of Admiralty intrusive magmas [Borg et al., in press, 1986].

Conclusions

1. The early preference for strike slip models for north Victoria Land proposed by Weaver et al., [1984] and Bradshaw et al., [1985] must be modified in the light of evidence of thrusting on both terrane boundaries.

2. Thrusting was probably the dominant process on the eastern boundary of the Bowers terrane and involved tectonic interlayering and greenschist metamorphism. Amalgamation was probably completed between 510-490 Ma, in the earliest Ordovician. There is no known source for the carbonate blocks which suggests some excision or telescoping. On the other hand, the distribution of detrital chromite [Wyborn, 1983] suggests an original proximity of the Bowers and Robertson Bay terranes.

3. The western boundary is much more complex and simple orthogonal convergence models do not satisfy the data. The salient points are:

a) Crustal thickening and magma generation in the Wilson terrane from about 550 Ma overlaps the generation of the Bowers terrane arc. The margin of the Wilson terrane continental crust is unusually abrupt but the scale of truncation is not easily explained by thrusting, and there is no evidence that the Bowers terrane was overridden by a very extensive slab of Wilson rocks.

b) There is apparently conflicting evidence from geochemistry and sedimentary geology about the timing of amalgamation between the Bowers terrane and continental crust. Geochemical trends within the Devonian "I" type Admiralty Intrusives cannot easily be reconciled with the presence of thick continental crust west of the Bowers terrane. Composition of sediments within the Bowers terrane to the west indicates access to a continental source from the Middle Cambrian. The abundance of carbonate and volcanic clasts and the paucity of metamorphics (despite the presence of granites) points to possible loss of intervening zones. Rates of uplift consequent to crustal thickening to more than 60 km would be expected to produce more immature sediments than the bulk of the Mariner and Leap Year Groups if deposition was so close to the Wilson terrane.

Thus while thrusting has been demonstrated on both boundaries of the Bowers terrane, there is no reason to relate them to a single period of compression. Timing of movement on the eastern boundary is well constrained, but evidence from igneous, sedimentary, and metamorphic geology points to a complex history of movement extending back at least to the Middle Cambrian on the western boundary. Evidence of early proximity to continental crust does not rule out subsequent movements which could include the opening of a marginal basin or terrane substitution by strike slip movement before final accretion in early Devonian. The strong compression and thrusting shown by the Lanterman Fault zone tectonites probably records only the last major phase of strong compression during which the Wilson terrane acted as a rigid block and retained its undisturbed isotopic age pattern. Similar features are seen along the Alpine Fault in New Zealand where disturbed isotopic age patterns and high strain tectonites characterize one side only.

Although the combination of sedimentological and geochemical data suggest a great deal more mobility than previously proposed, the rates of movement involved are consistent with modern ones.

Acknowledgements. I wish to thank the organisers of the 3rd Circum Pacific Terrane Conference for their help and encouragement, and particularly Dr Evan Leitch. Dr Stephen Weaver advised me on matters geochemical and made significant improvements to the manuscript. Dr George Gibson's comments were also most helpful.

References

Adams, C. J. and H. Kreuzer, Potassium-Argon age studies of slates and phyllites from the Bowers and Robertson Bay terranes, north Victoria Land, Antarctica. Geol. Jahrbuch B60, 265-288, 1984.

Borg, S. G., E. Stump, B. W. Chappell, M. T. McCulloch, D. Wyborn, R. L. Armstrong, and J. R. Holloway, Granitoids of northern Victoria Land, Antarctica, Implications of chemical and isotopic variations to regional crustal structure and tectonics, Am J. Sci., in press, 1986.

Bradshaw, J, D., Terrane boundaries and terrane displacement in northern Victoria Land, Antarctica, Geol. Soc. Australia, Abstracts 14, 30-33, 1985.

Bradshaw, J. D. and M. G. Laird, Pre-Beacon geology of northern Victoria Land: a review , (pp.98-101) in Oliver R. L., P. R. James and J. B. Jago, editors, Antarctic Earth Science. Aust. Acad. Sci. Canberra, 1983,

Bradshaw, J. D., M. G. Laird, and A. Wodzicki, Structural style and tectonic history in northern Victoria Land, (pp.809-816) in C. Craddock, editor, Antarctic Geoscience, Univ. Wisconsin Press, Madison, 1983.

Bradshaw, J. D., S. D. Weaver, and M. G. Laird, Suspect terranes and Cambrian tectonics in northern Victoria Land, (pp.467-479) in Howell, D., editor, Tectonostratigraphic terranes of the circum Pacific region, Circum Pacific Council for Energy and Mineral Resources, Earth Sci. Ser. 1. Houston, 1985.

Burrett, C. and R. H. Findlay, Cambrian and Ordovician conodonts from the Robertson Bay Group, Antarctica and their tectonic significance, Nature 307, 723-725,1984.

Field, D. B. and H. H. Findlay, Sedimentology of the Robertson Bay Group, northern Victoria Land, (pp.102-106) in Oliver, R. L., James, and J. B. Jago, editors, Antarctic Earth Science, Aust. Acad. Sci., Canberra, 1983.

Gair, H. S., A. Sturm, S. J. Caryer, and G. W. Grindley, The Geology of northern Victoria Land, Sheet 13, in Bushnell, V. C., and C. Craddock, editors, Geologic maps of Antarctica. Antarct. Map Folio Ser., Folio 12, Pl.1, 1969.

Gibson. G. M., Deformed conglomerates in the eastern Lanterman Range, north Victoria Land, antarctica, Geol. Jahrbuch B60, 117-143, 1984.

Gibson, G. M., Lanterman Fault: Boundary between allochthonous terranes, northern Victoria Land, Antarctica. Geol. Soc. Aust. Abstract 14, 79-82, 1985.

Gibson, G. M., F. Tessensohn, and A. J. Crawford, Bowers Supergroup rocks west of the Mariner Glacier and possible greenschist facies equivalents, Geol. Jahrbuch B60, 289-318, 1984.

Gibson, G. W. and T. O. Wright, The importance of thrust faulting in the tectonic development of northern Victoria Land, Antarctica, Nature 315, 480-483, 1985.

Grew E. S. and M. Sandiford, Staurolite-talc assemblage in tourmaline-phlogopite-chlorite schist from northern Victoria Land, Antarctica, and its petrogenetic significance. Contrib. Min. Petrol. 87, 337-350, 1984.

Grew, E. S. and M. Sandiford, Staurolite in garnet-hornblende-biotite schist from the Lanterman Range, northern Victoria Land, Antarctica, N. Jb. Miner. Mh. H.9, 396-410, 1985.

Grew, E. S., G. Kleinschmidt, and W. Schubert, Contrasted metamorphic belts in north Victoria Land, Geol. Jahrbuch B60, 253-263, 1984.

Kleinschmidt, G. and F. Tessensohn, Tectonic development of northern Victoria Land, Antarctica, 6th Gondwana Symposium, Columbus, Ohio, in press, 1986.

Laird, M. G. and J. D. Bradshaw, New data on the Lower Palaeozoic Bowers Supergroup, northern Victoria Land, (pp.123-126) in Oliver, R L, P. R. James, and J. B. Jago, Antarctic Earth Sciences, Aust. Acad. Sci. Canberra, 1983.

Laird, M. G., J. D. Bradshaw, and A. Wodzicki, Stratigraphy of the Upper Precambrian and Lower Paleozoic Bowers Supergroup, northern Victoria Land, Antarctica, (pp.535-542) in Craddock,C., editor, Antarctic Geoscience. Univ. Wisconsin Press, Madison, 1982.

Pearce, J. A., N. B. W. Harris, and A. G. Tindale, Trace element discrimination diagrams for the tectonic interpretation of granitic rocks, J. Petrol. 25, 956-983, 1984.

Plummer, C. C., R. S. Babcock, J. W. Sheraton, C. J. Adams, and R. L. Oliver, Geology of the Daniels Range, northern Victoria Land, Antarctic, (pp.113-117) in Oliver, R. L., P. R. James, and J. B. Jago, editors, Antarctic Earth Science, Aust. Acad. Sci., Canberra, 1983.

Sibson, R. H., S. H. White, and B. K. Atkinson, Structure and distribution of fault rocks in the Alpine Fault Zone, New Zealand, (pp. 197-210) in McClay, K. R. and N. J. Price, editors, Thrust and Nappe Tectonics, Geol. Soc. Lond. Sp. Pb. 9,1982.

Stump, E., M. G. Laird, J. D. Bradshaw, J. R. Holloway, S. G. Borg, and F. E. Lapham, Bowers graben and associated tectonic features cross northern Victoria Land, Antarctica, Nature 304, 334-6.

Tessensohn, F., Geological and tectonic history of the Bowers structural zone, northern Victoria Land, Antactica, Geol. Jarhbuch B60, 371-96. 1984.

Vetter, U., N. W. Roland, H. Kreuzer, A. Hohndorf, H. Lenz, and C. Besang, Geochemistry, petrogaphy, and geochronology of the Cambro-Ordovician and Devonian-Carboniferous granitoids of northern Victoria Land, Antarctica, (pp.140-143) in Oliver, R. L., P. R. James, and J. B. Jago, editors, Antarctic Earth Science, Aust. Acad. Sci., Canberra, 1983.

Weaver, S. D., J. D. Bradshaw, and M. G. Laird, Geochemistry of Cambrian volcanism in northern Victoria Land, Antarctica, Earth and Planet. Sci. Letters 68, 128-40 1984.

Wodzicki, A., and R. Robert, Geology and geological history of the Bowers Supergroup, northern Victoria Land, Antarctica Journal United States 19, (5), 29-30, 1984.

Wright, T. O. Sedimentology of the Robertson Bay Group, north Victoria Land, Antarctica, Geol. Jarhbuch B41, 127-38, 1984.

Wright, T. O., and R. H. Findlay, Relationships between Robertson Bay Group and Bowers Supergroup - new progress and complications from the Victory Mountains, north Victoria Land, Geol. Jahrbuch B60, 105-116, 1984.

Wright, T. O., R. J. Ross, and J. E. Repetski, Newly discovered youngest Cambrian or oldest Ordovician fossils from Robertson Bay terranes (formerly Precambrian) northern Victoria Land, Antarctica, Geology 12, 301-5, 1984 .

Wyborn, D., Chemical control on stratigraphic relations in northern Victoria Land and some possible relations with southeast Australia, (pp. 145) in Oliver, R. L., P. R. James, and J. B. Jago, editors Antarctic Earth Science, Aust. Acad. Sci., Canberra, 1983.

METAMORPHISM AND DEFORMATION IN THE BOWERS SUPERGROUP: IMPLICATIONS FOR TERRANE ACCRETION IN NORTHERN VICTORIA LAND, ANTARCTICA

George M. Gibson

School of Applied Science, Darling Downs Institute, Toowoomba, Queensland, Australia 4350

Abstract. L-S tectonites are widely developed in greenschist facies rocks along the western margin of the Bowers terrane. They are spatially and genetically related to deformation along the adjacent terrane boundary (Lanterman Fault) which is interpreted here as a major overthrust with crystalline basement rocks (Wilson terrane) emplaced over the Bowers terrane. Stretching fabrics within the tectonites plunge steeply down-dip within the regional foliation and indicate tectonic transport was primarily directed towards the northeast, at high angles to the dominant northwest trend of both the Lanterman Fault and the Bowers terrane. These relations are inconsistent with normal dip-slip or pure strike-slip displacement along the terrane boundary although reverse-slip combined with some component of right lateral transcurrent motion (i.e. transpression) could explain some of the structural data reported here. The Lanterman Fault possibly originated as a back-thrust following collision between a west-facing island arc and the Gondwana (Antarctic) margin.

Introduction

Whilst it is generally agreed that northern Victoria Land can be subdivided (Figure 1) into three major tectonostratigraphic units (Wilson, Bowers and Robertson Bay terranes), there is still considerable debate about the origin and sense of displacement along the boundary faults - the Lanterman and Leap Year Faults. In recent years these two structures have been variously interpreted as high angle normal faults [Stump et al., 1983], transcurrent faults [Weaver et al., 1984; Bradshaw et al., 1985], regional overthrusts [Gibson and Wright, 1985; but also see Bradshaw, 1985] or in terms of transpression [Gibson, 1985]. These very different opinions emphasise the uncertainties surrounding the main cause of terrane amalgamation and accretion in northern Victoria Land during the Lower Paleozoic Ross Orogeny. Resolution of the problem requires more detailed information on the minor structures associated with each boundary. This paper presents such information for the western margin of the Bowers terrane.

Lanterman Fault

The Lanterman Fault [Dow and Neall, 1974], or more strictly the Lanterman Fault Zone [Bradshaw et al., 1982], forms the greater part of the present boundary between the Wilson and Bowers terranes (Figure 1) and has a history of movement extending back over several hundred million years; its northern extension is the Bowers Fault [Dow and Neall, 1974]. The Lanterman Fault has been mapped from its type locality in the Lanterman Range southwards into the Mountaineer Range [Gibson et al., 1984] and is only one of several major subparallel faults that affect the western margin of the Bowers terrane. These faults generally dip steeply southwest and have so disrupted the geology of this important boundary that the western margin of the Bowers terrane has the characteristics of an imbricate zone in which rocks of varying degrees of metamorphism and structural development are juxtaposed. Indeed, at least one fault-bounded sliver of more highly metamorphosed Wilson terrane occurs within this zone and thus lies east of the Lanterman Fault [Gibson et al., 1984; Tessensohn, 1984].

Wilson Terrane

The Wilson terrane contains mainly multiply-deformed amphibolite facies gneisses which in the Lanterman Range have undergone at least two phases of medium to high grade metamorphism: an early medium-high pressure event in which kyanite was stable and a later, lower pressure event in which sillimanite and sillimanite-K feldspar assemblages were produced [Grew and Sandiford, 1984]. Metamorphism was accompanied by the emplacement of granitoids collectively known as the Granite Harbour Intrusives [Gunn and Warren, 1962; Sturm and Carryer, 1970]. The granitoids have predominantly S-type characteristics [Wyborn, 1983; Vetter et al.,

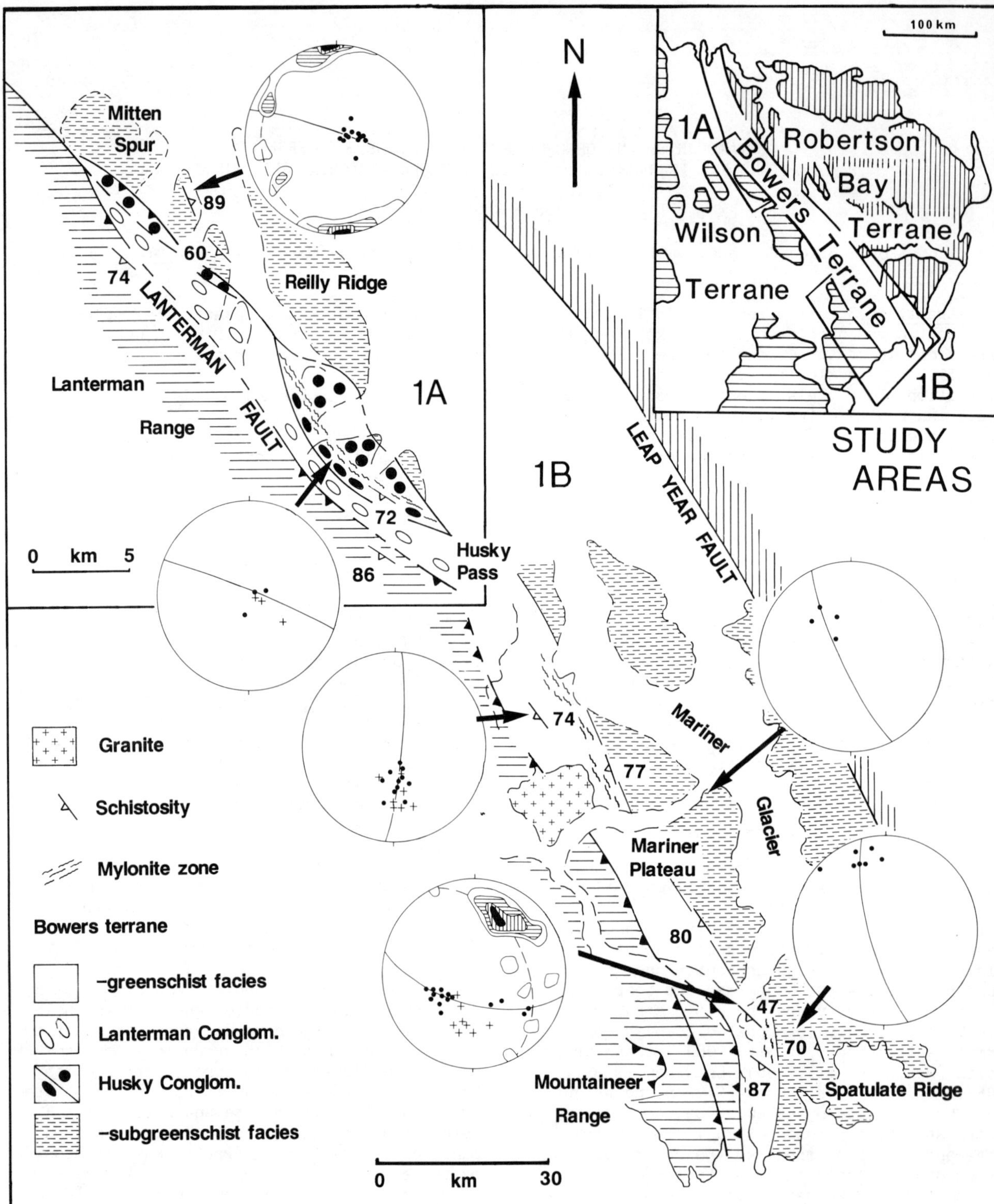

Mitten
Spur
89
60
74
LANTERMAN
Reilly Ridge
Lanterman
Range
FAULT
1A
72
86
Husky
Pass
0 km 5
N
1A
Robertson
Bowers
Bay
Wilson
Terrane
Terrane
Terrane
Terrane
1B
100 km
STUDY
AREAS
1B
LEAP YEAR FAULT
74
Mariner
77
Glacier
Mariner
Plateau
80
47
70
87
Mountaineer
Range
Spatulate Ridge
0 km 30
Granite
Schistosity
Mylonite zone
Bowers terrane
-greenschist facies
Lanterman Conglom.
Husky Conglom.
-subgreenschist facies

1983] and in the Lanterman Range were largely intruded penecontemporaneously with deformation [Roland et al., 1984; Sandiford, 1985]. A Cambro-Ordovician age has been determined for Granite Harbour Intrusives in the Daniels and Morozumi Ranges [Vetter et al., 1983] although elsewhere in the Wilson terrane even older granites have been reported [Kreuzer et al., 1985].

At least four phases of folding (F1-F4) are recognised [Gibson, 1984; Roland et al., 1984; Kleinschmidt et al., 1984] in the Wilson terrane, the first of which persists mainly as rare, rootless isoclines. Second generation structures include tight to isoclinal folds (F2) as well as a conspicuous stretching lineation which plunges steeply down-dip within the F2 axial-surface schistosity, S2. The schistosity generally strikes northwest parallel to both the Lanterman Fault and the regional trend.

The third and fourth phases of deformation were associated with retrogressive metamorphism to the greenschist facies and the widespread development of crenulation structures. The F3 deformation produced several macroscopic folds and was largely responsible for the reorientation of S2 into its present steep attitudes. F4 folds are largely confined to the eastern Lanterman Range and developed in response to late movements on the Lanterman Fault.

Bowers Terrane

Stratigraphy

The Bowers terrane is largely underlain by rocks of the 10km thick, Cambrian-early Ordovician Bowers Supergroup. At the base of this sequence is the Sledgers Group (Figure 2) which comprises two interdigitating units: the Molar Formation consisting mainly of marine volcanogenic sediments and the Glasgow Volcanics made up of pillow lavas, basaltic and andesitic breccias, and subordinate dacite-rhyolite [Laird and Bradshaw, 1983; Bradshaw et al., 1985]. Basic rocks in the Glasgow Volcanics have a chemistry consistent with an intra-oceanic origin [Weaver et al., 1984]. Overlying the Sledgers Group (Figure 2) are the non-volcanogenic, shallow to marginal marine Mariner Group and the fluviatile, dominantly conglomeratic Leap Year Group [Laird, 1981; Laird and Bradshaw, 1983]. A marked angular unconformity usually separates these two units although locally the Mariner Group is absent altogether, and the Leap Year Group rests directly on the underlying Sledgers Group [Laird and Bradshaw, 1983; Gibson et al., 1984].

Also included in the Bowers Supergroup are the Lanterman and Husky Conglomerates (Figure 2), both of which occur in the zone of interest along the western margin of the Bowers terrane and for this reason are given special consideration here. However, because both units are entirely fault-bounded (Figure 1), their stratigraphic affinities are still uncertain. Correlation of the Lanterman Conglomerate with the Leap Year Group is suggested here (Figure 2) because the two units appear little different in terms of either their compositions or depositional environments. Both have a significant quartz or quartzofeldspathic component and both contain cross-bedded quartz sandstones [cf. Laird, 1981 and Sandiford, 1985]. They both have characteristics indicative of a fluviatile origin.

The Husky Conglomerate may also in part be of shallow water origin. It too contains cross-bedded sandstones as well as channel conglomerates and coarsely graded pebble beds [Gibson, 1984] although Laird et al. [1982] also report thick mudstone units with scattered boulders in the Husky Conglomerate and these are less likely to represent shallow water conditions.

The Husky Conglomerate contains mainly mafic detritus consistent with derivation from the Glasgow Volcanics [Gibson, 1984]. On this basis it is either a lateral equivalent of volcanic breccias recently described from the Sledgers Group [Laird and Bradshaw, 1983] or else it is younger than the Glasgow Volcanics and may once have rested directly upon them [Gibson, 1984, 1985]. The younger age is preferred here and the Husky Conglomerate is shown on Figure 2 as a correlative of the Carryer Conglomerate, currently interpreted [Laird and Bradshaw, 1983] as a basal facies of the Leap Year Group. Alternatively, the Husky Conglomerate could be a lateral equivalent of the Mariner Group although this is considered less likely given the shallow water characteristics exhibited by some parts of the Husky Conglomerate.

In addition to problems about their stratigraphic affinities, there are also uncertainties about the mutual age relations of

Fig. 1. (opposite) Map of the western margin of the Bowers terrane showing the distribution of greenschist facies rocks and orientation of D1 structural elements. Fold axes (F1 in the subgreenschist facies areas and F2 in the higher grade rocks) are shown as dots whereas F2 stretching lineations are represented by crosses. Contour interval for poles to foliation (S0 or S1) is 3%, 6%, 9% and 12% per 1% area. Inset map of northern Victoria Land shows regional setting of the Bowers terrane.

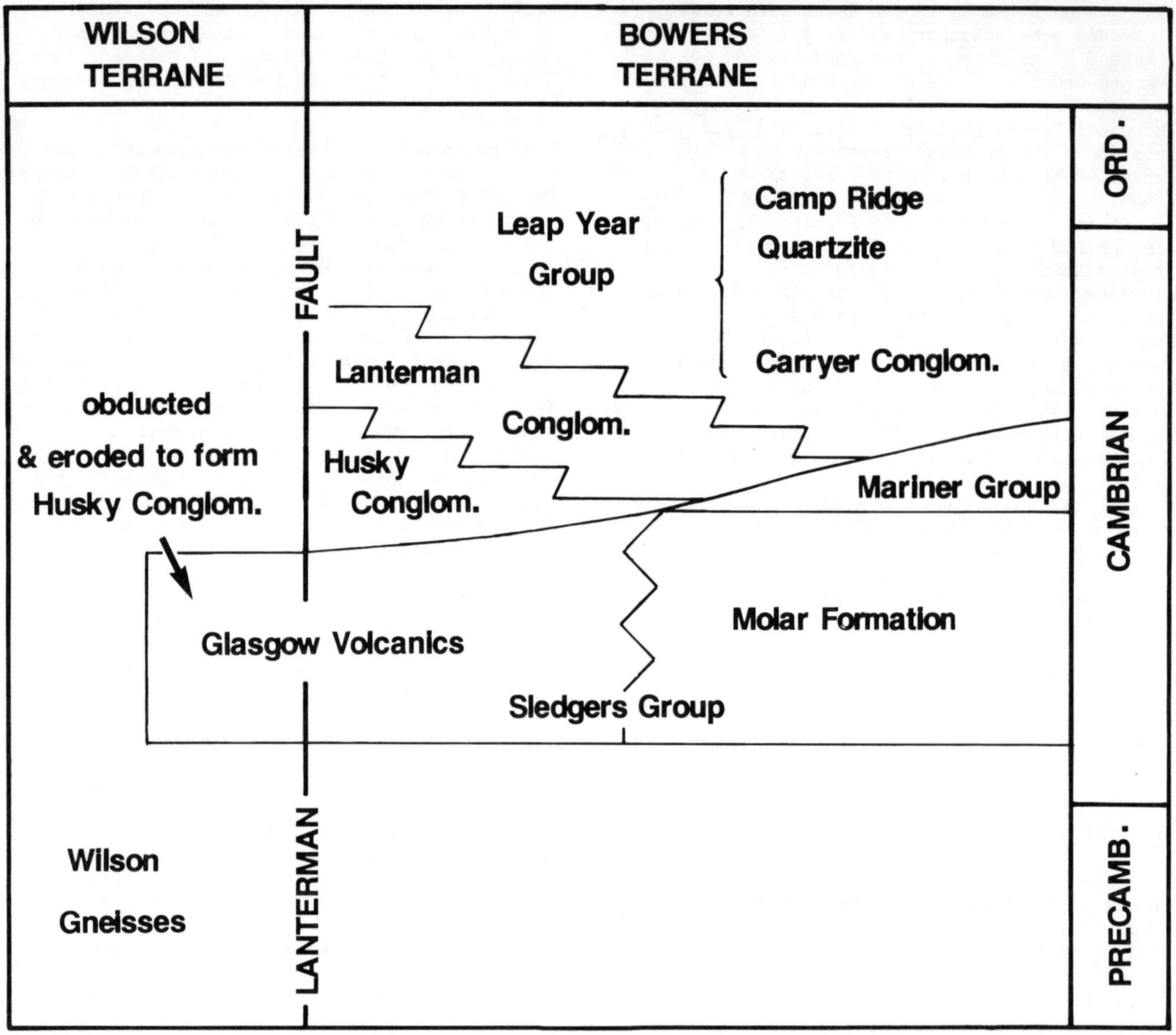

Fig. 2. Stratigraphy of the Bowers terrane [modified after Laird and Bradshaw, 1983].

the Husky and Lanterman Conglomerates. Rare boulders of inferred Lanterman Conglomerate have been reported in the Husky Conglomerate and this has led to the conclusion that the Husky Conglomerate unconformably overlies the former [Laird and Bradshaw, 1983; Gibson, 1984]. Thin-sections cut from these various lithologies have since permitted a more detailed comparison to be made between the boulders and Lanterman Conglomerate. Contrary to expectations the boulders (of pebbly psammite) have a slightly more mafic matrix than typical Lanterman Conglomerate and contain considerably less potash feldspar. Moreover, the pebbles within the psammite are compositionally similar to clasts in other parts of the Husky Conglomerate. It is concluded that the boulders represent blocks of reworked Husky Conglomerate rather than Lanterman Conglomerate. Earlier interpretations about the contact relations of the two conglomerates may have to be revised. Indeed, in this context it is worth noting that light green metaconglomerates having the characteristics of both the Lanterman and Husky Conglomerates occur on Indecision Spur in the eastern Lanterman Range [Gibson, 1984, p. 129]. A sedimentary transition from one conglomerate unit into the other now seems more likely. Because the Lanterman Conglomerate shows greater affinities with the Leap Year Group rather than the Sledgers Group, and as such is unlikely to have been derived from the latter,

TABLE 1. Summary of Lower Paleozoic Deformation in the Bowers Terrane

Deformational Event	D1		D2
Folding Phase	F1	F2	F3
Fold Style	Open to tight in Molar Formation; not recognised in H.C. or L.C.	Tight to isoclinal in H.C. and L.C. Kink folds in weakly metamorphosed Sledgers Group.	Open crenulations and kink folds; major open synform in eastern Lanterman Range.
Axial Surface Foliation	Slaty cleavage (S1) in Molar Formation; flattened pebble fabric in H.C. and L.C.	Penetrative schistosity (S2) or flattened pebble fabric in H.C. and L.C.; crenulation cleavage in slates and phyllites on Spatulate Ridge.	None or weakly developed crenulation cleavage.
Associated Lineations	S1/S0 intersection lineations.	S1/S2 intersection lineations; stretched pebbles and vesicles, elongated mineral aggregates.	-
Trend & Plunge of Axes	NW; gentle to steep.	Dominantly SSE or SW; Moderate to steep.	NW-SE; subhorizontal.
Attitude of Axial Surface	NW; Subvertical.	NW; dipping steeply SW.	WNW; subvertical.
Associated Metamorphism	Subgreenschist facies in Sledgers Group; lower greenschist facies in H.C. and L.C.	Greenschist facies in H.C. and L.C.; lower greenschist facies on Spatulate Ridge.	Retrograde; lower greenschist facies.
Associated Structures/ Developments	S1 in mylonite zones?	Main phase of mylonitisation; development of sheath folds; reorientation of F2 folds parallel to stretching lineations.	Shear zones; reorientation of S2 into steeper attitudes.

HC = Husky Conglomerate; LC = Lanterman Conglomerate.

it is inferred to be younger than the Husky Conglomerate.

Deformation and Metamorphism

The earliest phase of deformation (F1) in the Bowers Supergroup was generally associated with prehnite-pumpellyite or pumpellyite-actinolite facies metamorphism [Wodzicki et al., 1982] and produced mainly open, gently plunging, northwest-trending folds with a prominent subvertical axial-surface slaty cleavage [Bradshaw et al., 1982; Gibson et al., 1984]. Nearer the contact with the Wilson terrane, these folds become progressively tighter and more variably plunging, sometimes steepening quite markedly along strike, although their axial surfaces show no significant deviation from the regional northwest trend. Fold axes in the Molar Formation in the eastern Lanterman Range vary from subhorizontal to subvertical (Figure 1).

Such abrupt changes in plunge without any corresponding change in the orientation of the axial surfaces suggests non-cylindrical folding rather than the reorientation of the F1 structures through superposed folding. Non-cylindrical folds have also been observed at Mt Belolikov in the northern Bowers Mountains [Jordan et al., 1984] and demonstrate that this style of deformation is more widespread throughout the Bowers terrane than hitherto recognised.

With increasing proximity to the Wilson terrane metamorphism reaches the greenschist facies. Greenschist facies rocks occupy a 2-5km wide zone along the western margin of the Bowers terrane. Both the Husky and Lanterman Conglomerates lie within this zone as well as a substantial slice of Sledgers Group rocks further south in the Mountaineer Range (Figure 1). In all these rocks there is evidence for at least two, and locally up to three generations of structures. These structures are summarised in Table 1.

The earliest structures in the Lanterman and Husky Conglomerates are a flattened pebble fabric and layer-parallel schistosity, S1, neither of which appears to be directly related to folding. F1 folds are unknown in either formation and the earliest recognisable folds are second generation structures (=F2) developed in the earlier S1 fabrics. Second generation folds usually trend northwest-southeast (Figure 1) and occur at all scales from centimetre-size crenulations (Figure 3a) up to major structures with wavelengths of 100m. Fold styles in the Husky Conglomerate are typically open and upright but westwards the second generation folds become increasingly more tightly appressed. F2 folds in the Lanterman Conglomerate and more severely deformed parts of the Husky Conglomerate are tightly asymmetric or more rarely isoclinal, and possess a well-developed axial-surface foliation, S2. This foliation is commonly a crenulation cleavage (Figure 3a) but with increasing deformation westwards develops into a fully penetrative axial-surface schistosity. It is primarily a flattened pebble fabric in the more highly deformed parts of the Husky Conglomerate but also includes a significant component of orientated biotite, actinolite and epidote (Figure 3b).

In comparison, S1 contains more chlorite and biotite and lesser amounts of actinolite (Figure 3c). S2 evidently formed at higher temperatures than S1 although both probably evolved during the one episode of prograde greenschist facies metamorphism. A similar conclusion can be made for the Lanterman Conglomerate where the S1 and S2 surfaces are defined by essentially the same greenschist facies phases, principally biotite and muscovite (Figure 3d). It follows that the first and second phases of deformation are not entirely independent of each other. Rather they form a continuum and are themselves the result of a single progressive deformation (D1) which was responsible for both the initial development of the S1 fabrics and their subsequent folding about the F2 axes.

Clast shape varies considerably throughout the Lanterman and Husky Conglomerates and ranges from virtually undeformed pebbles, through triaxial ellipsoids into flattened, oblate spheroids, some with axial ratios in excess of 10:10:1 [Gibson, 1984; see also Sandiford, 1985]. Flattened pebbles are more typical of the Lanterman Conglomerate whereas pebbles in the Husky Conglomerate some distance from the Lanterman Fault are only weakly orientated within the regional foliation (?=S2) and at best are triaxial ellipsoids with their long axes (X dimension of the finite-strain ellipsoid) inclined steeply ESE. With increasing deformation the pebbles become progressively flattened and/or elongated within the regional foliation. In the more severely deformed parts of the Husky Conglomerate just north of Husky Pass (Figure 1) the long axes of pebbles are parallel to the axial direction of steeply plunging, isoclinal folds (Figure 1). Such folds are uncommon but have S2 as the axial-surface foliation (Figure 3c). Pebble shape at this locality is evidently the result of F2 strain. Pebble long axes plunge steeply ESE (Figure 4).

Second generation structures are also widely developed in Sledgers Group rocks metamorphosed to the greenschist facies in the Mountaineer Range (Figure 1). They include tight to isoclinal mesoscopic folds as well as one major structure mapped [Gibson et al., 1984] in deformed conglomerates at the northern extremity of Spatulate Ridge. Pebbles within these conglomerates are flattened parallel to the axial surface of this fold and have their long dimensions inclined at various angles to the

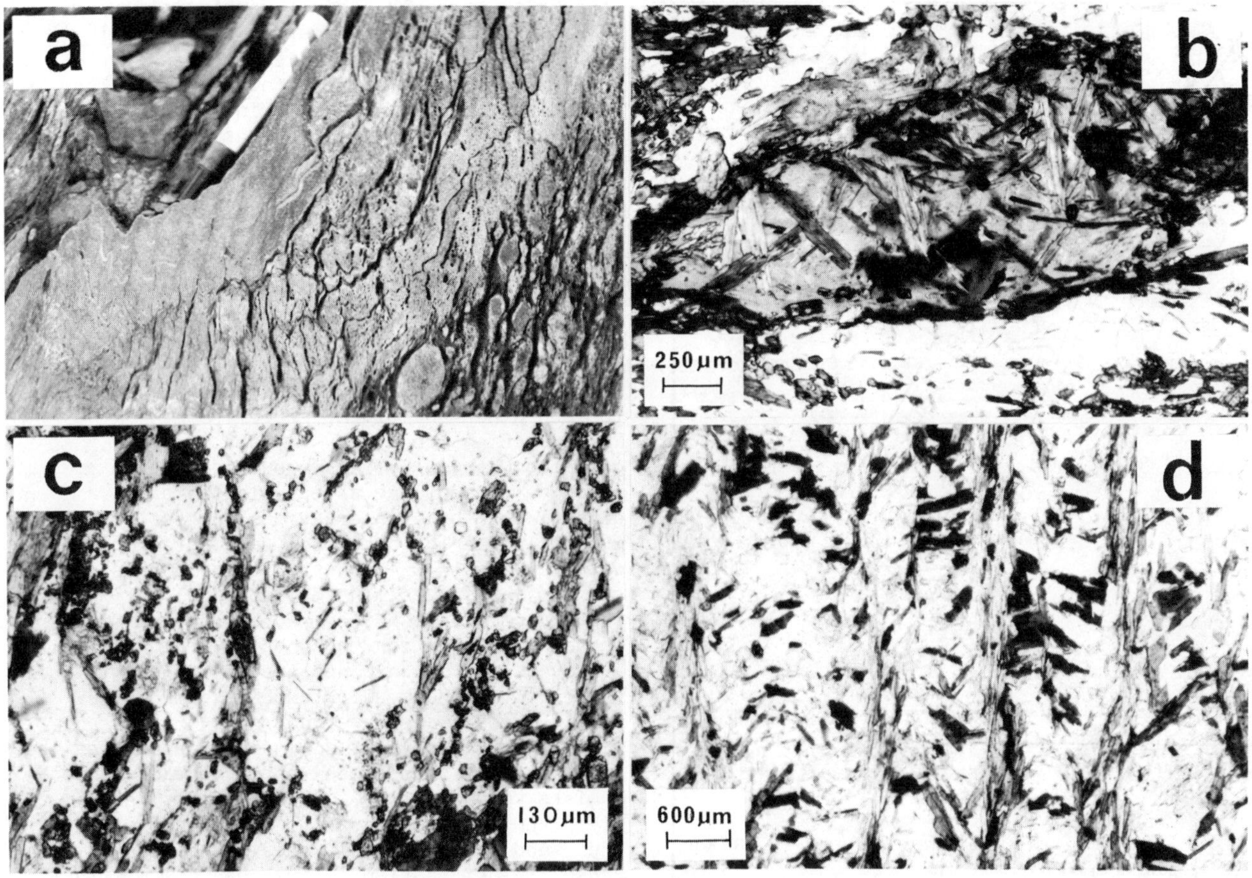

Fig. 3(a). F2 crenulations in metapsammite. Note quartz veins parallel to earlier S1 schistosity and the alignment of pebbles parallel to S2. (b). Ultramafic clast flattened parallel to S2 in Husky Conglomerate. The clast contains chlorite (matrix) and biotite (dark color) orientated at high angles to the external S2 schistosity which is defined here by the parallel alignment of actinolite, biotite and epidote. (c). S2 (vertical) defined by orientated actinolite, biotite and epidote (small, high relief prisms) has virtually obliterated the earlier S1 fabric (inclined from top right to bottom left) from a metapsammite in the Husky Conglomerate. S1 formed at slightly lower metamorphic grades than S2 and contains greater amounts of biotite and chlorite. (d). S2 crenulation cleavage in the Lanterman Conglomerate. Note the greater amounts of biotite orientated parallel to S1 in the microlithons.

fold axis (Figure 1). Elsewhere in the region these same conglomerates are characterised by stretched, rather than flattened, pebble fabrics. Stretched pebble fabrics are typically developed in areas where the F2 folds are both isoclinal and steeply plunging.

S2 is axial planar to the F2 folds and intensifies westwards, sometimes to the extent that the earlier slaty cleavage (=S1) is almost completely obliterated from metasediments close to the Lanterman Fault. In phyllites west of Spatulate Ridge the S2 cleavage domains are so closely spaced (<50 microns) that S1 is barely perceptible even with the aid of a microscope; it persists only as occasional relicts within the intervening microlithons. However, sufficient remains of the S1/S2 intersection relationships to demonstrate that this region is underlain by several extremely tight F2 folds even though individual closures were not identified. The intersection lineations plunge steeply to the SE.

Second generation fabrics in the metabasites (Glasgow Volcanics) on the Mariner Plateau include stretched vesicles and elongated pyroclastic fragments. They plunge steeply SSE, parallel or subparallel to the local F2 fold axes (Figure 1).

In some parts of the greenschist facies belt, a third generation of structures is

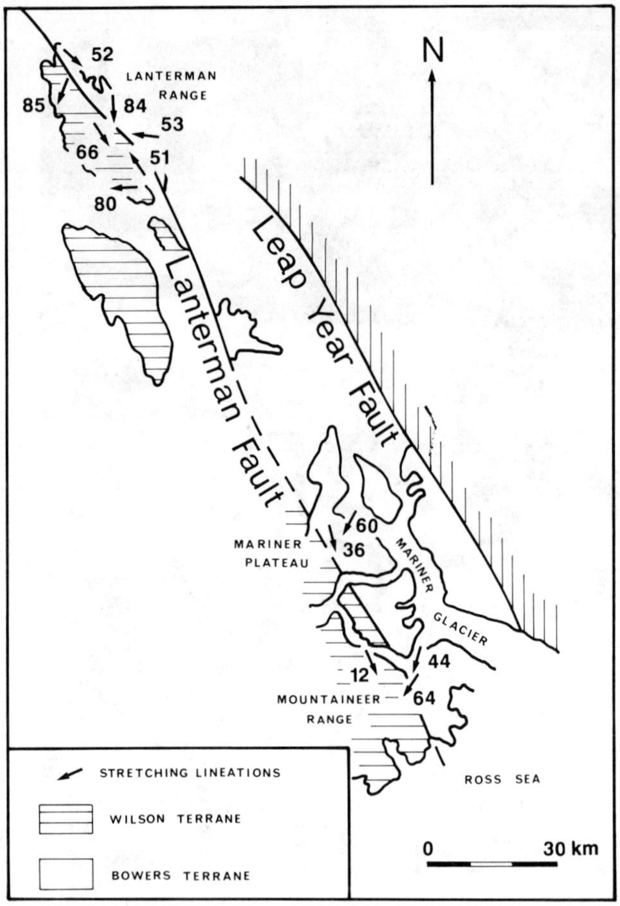

Fig. 4. Generalised distribution and orientation of measured stretching lineations in the Wilson and Bowers terranes.

developed. They originated during the second phase of deformation, D2 (Table 1) and include several shear zones as well as many small scale kink folds and crenulations [Gibson, 1984]. A minor amount of retrogressive metamorphism accompanied this deformation although its most obvious effects occur in the eastern Lanterman Range where S2 in the Husky and Lanterman Conglomerates is deformed into a major synform trending WNW. D2 fold axes and lineations are subhorizontal.

Mylonites

Mylonitic rocks occur at several localities along the western margin of the Bowers terrane. They are typically developed in zones of intense F2 strain and for the most part dip steeply southwest parallel to the regional foliation. One such zone is 200m wide and occurs in greenschist facies metabasites on Spatulate Ridge (Figure 1). It has undergone extensive recrystallisation and possesses a streaky to strongly laminated mylonitic fabric (=S2) which is axial planar to intrafolial folds developed in an earlier schistosity (=S1).

Tight to isoclinal intrafolial folds also occur in mylonites on the Mariner Plateau (Figure 1). They deform the earlier S1 (cataclastic) foliation and have themselves developed a mylonitic layering parallel to their axial surfaces. Chlorite and flattened quartz grains are contained within this surface and define S2. In addition S2 commonly contains a penetrative lineation defined by elongated mineral aggregates and stretched porphyroclasts. In one mylonite the stretched porphyroclasts, second generation fold axes and S1/S2 intersection lineations are all parallel to each other and plunge steeply SSE. F2 stretching lineations in less strongly deformed Glasgow Volcanics outside of the mylonite zone plunge steeply in the same direction. These observations suggest that the mylonitic fabrics developed penecontemporaneously with folding in the country rocks and support the earlier suggestion [Gibson et al., 1984] that mylonitisation occurred during the first phase of deformation, D1.

Several mylonite zones are recognised in the Lanterman and Husky Conglomerates. Pebbles within these zones have undergone extremes of D1 flattening, so much so in some areas that the conglomerates take on the appearance of banded schists [Gibson, 1984, Figure 3]. In fact given the degree of grain size diminution in these and other conglomerates in the Lanterman Range, mylonitisation clearly had an important influence on fabric development throughout this entire region. Mylonitic fabrics are much more extensively developed throughout the Lanterman Conglomerate than has hitherto been recognised.

Discussion

A preliminary model for the deformation and metamorphism along the western margin of the Bowers terrane has already been proposed [Gibson, 1985; Gibson and Wright, 1985] and this section is primarily concerned with a detailed presentation of the evidence upon which it is based.

In essence, deformation and metamorphism throughout this zone are believed to be the direct result of overthrusting along the western boundary of the Bowers terrane. Overthrusting occurred during, or immediately after, regional metamorphism of the Wilson gneisses so that these rocks were still at elevated temperatures when emplaced over the Bowers terrane. This initiated cooling and widespread retrogressive metamorphism in the Wilson gneisses whilst in the underlying Bowers terrane an inverted thermal gradient was established. Bowers rocks immediately beneath the thrust contact underwent greenschist facies metamorphism. The model not only explains why greenschist facies rocks occur along the western margin of the Bowers terrane but why the metamorphic grade decreases so rapidly with increased distance from the Lanterman Fault. In this context it is worth noting that some of the most intensely metamorphosed units in the Bowers Supergroup (e.g. Husky and Lanterman Conglomerates) may also be among its stratigraphically highest members (Figure 2). If so, then the greenschist facies rocks cannot have been dragged up from depth (lower stratigraphic levels) along the Lanterman Fault.

Higher temperatures and a greatly increased superincumbent load also facilitated the ductile deformation and flattening observed in greenschist facies rocks along the western margin of the Bowers terrane. Flattening produced both the mylonites and the penetrative S2 schistosity as well as the isoclinal, tightly appressed style of the F2 folds. Rocks with S2 as the dominant foliation also contain a variety of extensional lineations (stretched pebbles, etc). These rocks conform to the description of L-S tectonites and are usually developed in areas of intense D1 strain where the F2 folds are both strongly flattened and plunge steeply down-dip within the S2 surface. Folds and other features of the D1 deformation are best accounted for by a model of progressive simple shear [Escher and Watterson, 1974]. Thus linear fabrics are transverse extension (stretching) lineations analogous to those described elsewhere in the literature [e.g. Bryant and Reed, 1969; Williams, 1978; Lisle, 1984; Shackleton and Ries, 1984] whereas folds now plunging down-dip within the regional foliation (Figure 1) originally had horizontal or sub-horizontal fold axes and were subsequently rotated into the direction of maximum extension (X direction of finite-strain ellipsoid) during overthrusting. This would account for the non-cylindrical nature of many D1 folds in the Bowers terrane and in extreme cases could have led to the development of sheath folds [Gibson and Wright, 1985]. Sheath folds have not been directly observed along the western margin of the Bowers terrane although their presence in some of the mylonites and more intensely deformed rocks adjacent to the Lanterman Fault is suspected on the basis of parallelism between the stretching lineations and F2 fold axes. In areas dominated by sheath folds, fold axial directions would be expected to lie close to the principal axis of the finite strain ellipsoid, as defined by the stretching lineations [Cobbold and Quinquis, 1980].

Pebble shapes in the Lanterman Conglomerate are typically oblate spheroids with their short axes perpendicular to the S2 schistosity (XY plane of finite strain ellipsoid). Deformation in this unit was dominantly one of flattening with equal extension in both the X and Y directions. This fabric cannot be explained in terms of simple shear alone. More likely deformation resulted from a combination of simple shear and superimposed pure shear flattening [cf. Hossack, 1968; Chapman et al., 1979]. No quantitative data are yet available to support this contention although this is the type of deformation expected to occur beneath an advancing nappe or thrust sheet [Ramberg, 1975]. Moreover, not all flattening need have occurred during the course of overthrusting; some may have followed the final emplacement and settling of the Wilson terrane. It is perhaps significant that the most intensely flattened conglomerates occur immediately adjacent to the Lanterman Fault.

Tectonic Significance of Stretching Lineations

Linear fabrics along the western margin of the Bowers terrane are of D1 age and generally plunge steeply southwest or SSE (Figure 4). Some degree of reorientation may have occurred during the later D2 deformation although given the localised nature of this event, significant deviations of the lineations from their original trend are considered unlikely. The one exception is in the eastern Lanterman Range north of Husky Pass where the D1 fabrics have been rotated into steep attitudes about a major D2 synform. Lineations in this region originally had somewhat shallower attitudes.

Stretching lineations are also widely developed in amphibolite facies schists and gneisses along the eastern margin of the Wilson terrane (Figure 4). They originated during the second phase of deformation (F2) in these rocks and are equated here with the D1 fabrics in the Bowers terrane. They include stretched mineral aggregates as well as elongated pressure shadows around garnet porphyroblasts and for the most part are parallel to the local F2 fold axes [Roland et al., 1984; Kleinschmidt et al., 1984]. These lineations, together with those in the Bowers terrane, define the direction of regional extension (X direction of the finite-strain ellipsoid) and indicate that thrusting was largely directed towards the northeast, consistent with crustal shortening at high angles to the trend of both the Lanterman Fault and the Bowers terrane. These relations are inconsistent with either normal dip-slip or pure strike-slip displacement along the western boundary of the Bowers terrane. This boundary is better interpreted in terms of thrust faulting although reverse-slip combined with some component of right-lateral transcurrent motion (i.e. transpression) could explain the orientation of the stretching lineations described here.

The case for thrust faulting is further strengthened by recent observations on Wilson Group rocks in the Mountaineer Range. A stack of thrust sheets occurs in this region with the highest grade rock on top [Kleinschmidt et al.,

1984; Tessensohn, 1984]. Sillimanite gneisses (Murchison Formation) are thrust northeast over kyanite- and staurolite-bearing metasediments (Dessent Formation) which in turn are thrust over greenschist facies rocks (Bowers terrane). The thrusts dip moderately southwest, parallel or subparallel to the regional foliation and the inferred contact between the Wilson and Bowers terranes.

Whilst the structural data reported in this paper give strong support to the case for thrust faulting along the western margin of the Bowers terrane, they do not entirely preclude the possibility of earlier strike-slip displacement. Fabric development in any fault zone may simply reflect the later stages of deformation. Stretching lineations developed in mylonites along the Alpine Fault of New Zealand are a case in point. There, the lineations preserve only the record of late Cenozoic oblique compression and provide no information about the sense of displacement during the earlier stages of fault development [Sibson et al., 1981]. Yet for most of its history, displacement along the Alpine Fault was essentially pure strike-slip, a conclusion for which there is much independent support [e.g. Carter and Norris, 1976]. Independent evidence in support of strike-slip motion along the western margin of the Bowers terrane has proved more elusive. Furthermore, if strike-slip faulting has occurred on this boundary, then it must have happened very early in the tectonic evolution of northern Victoria Land, certainly before the development of S2 in the Bowers terrane and probably before deposition of the Lanterman and Husky Conglomerates.

Origin of the Lanterman Fault

Several plate tectonic models have now been proposed for northern Victoria Land [e.g. Weaver et al., 1984; Tessensohn, 1984; Gibson, 1985; Gibson and Wright, 1985; Kleinschmidt and Tessensohn, in press, 1986]. Almost all portray the Middle Cambrian Sledgers Group as an island arc although there is still no general agreement about the exact mechanism by which the Wilson and Bowers terranes were amalgamated. Weaver et al. [1984] favoured juxtaposition through transcurrent faulting and drew attention to the fact that Lower Ordovician S-type granitoids now occur within 6km of island arc-tholeiites (Glasgow Volcanics). They also noted that no fragment of a subduction complex had been reported along the contact zone between the Wilson and Bowers terranes although Kleinschmidt and Tessensohn [in press, 1986] have recently reinterpreted the Husky Conglomerate as part of one such fragment [cf. Gibson, this paper]. However, even if such fragments do exist, they still cannot alter the fact that there has been significant crustal loss along the contact zone between the two terranes. Virtually all tectonic models proposed for northern Victoria Land also require the excision of the early Gondwana continental margin and its associated miogeoclinal assemblage.

Whilst strike-slip faulting offers one possible solution to the problem of missing crust between the Wilson and Bowers terranes, evidence in support of such a model has proved hard to find. In fact much of the detailed structural data presented in this paper would seem to preclude significant strike-slip displacement altogether. Stretching fabrics developed along the western margin of the Bowers terrane are usually steeply plunging and in some cases orientated at high angles to the required direction (Figure 4). They indicate vertical, rather than lateral, displacement. Horizontal or subhorizontal lineations consistent with a predominantly strike-slip regime are unknown.

Gibson and Wright [1985] argued that amalgamation of the Bowers and Wilson terranes occurred through thrust faulting following an island arc-continent collision. In this model, the Lanterman Fault represents a former suture which subsequently evolved into a major regional overthrust (Figure 5a). Alternatively (Figure 5b) the Lanterman Fault may have originated as a post-collision back-thrust [Gibson, 1985]. Both models satisfy the structural data presented here although the second interpretation has several inherent advantages. First, obduction leads to the uplift and erosion of arc-related materials which were then available to produce the Husky Conglomerate and other sediments in the Bowers terrane. Concomitant with obduction the original Gondwana continental margin is tectonically buried beneath the overriding island arc assemblage. Second, unlike the first model, there is no need to explain the absence of a subduction complex along the Lanterman Fault [cf. Weaver et al., 1984]. Any such complex would have been removed by erosion along with the arc rocks. Third, with continued uplift and erosion of the Wilson terrane, increasingly greater amounts of continentally-derived detritus became available leading to the deposition of the Leap Year Group (including the Lanterman Conglomerate) and eventually to the complete burial of the arc. By this stage Lower Ordovician Granite Harbour Intrusives were unroofed and contributed granitic detritus to a rapidly filling sedimentary basin to the east.

Age of Thrusting

Thrust faulting in the Bowers terrane was essentially synchronous with the D1 deformation although only the broadest time limits can be placed on either event. K-Ar whole rock ages are available [Adams et al., 1982; Adams and

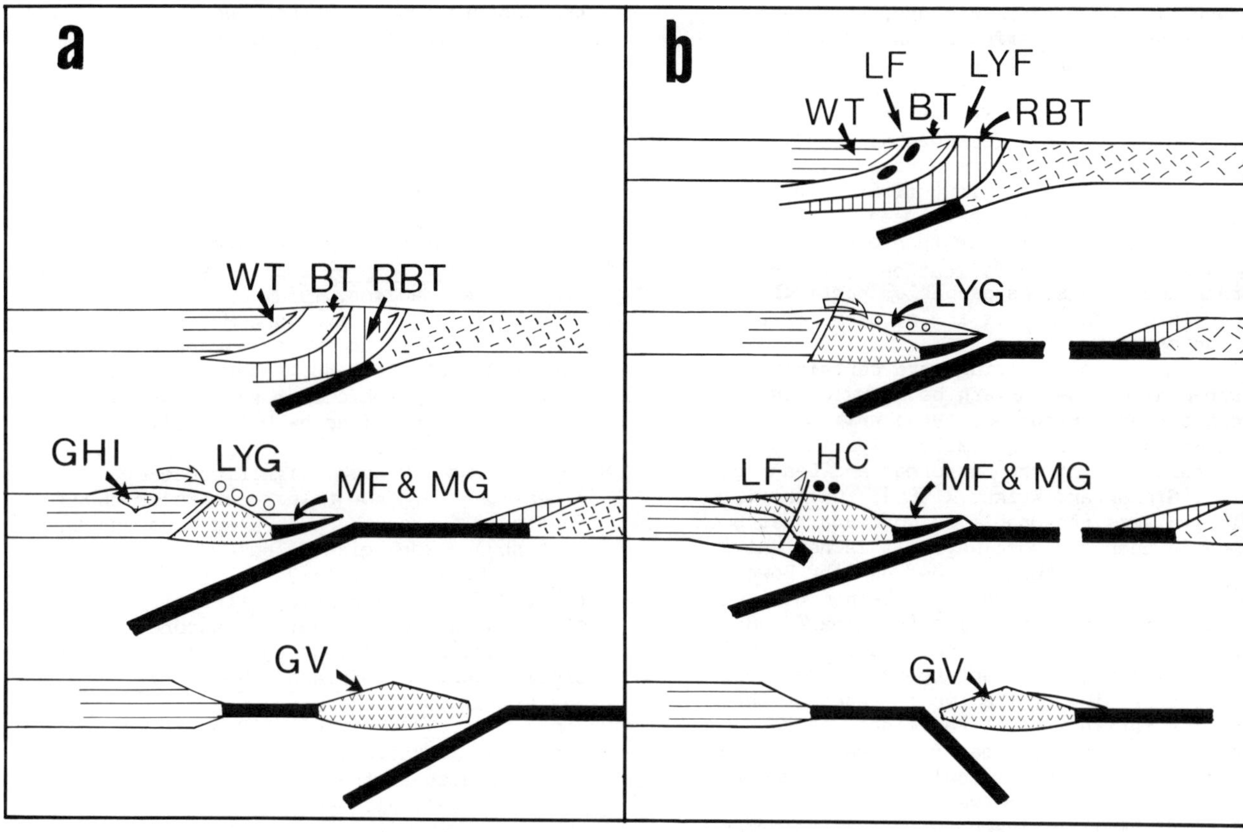

Fig. 5. Alternative tectonic models for the Lower Paleozoic Ross Orogeny in northern Victoria Land [after Gibson, 1985; Gibson & Wright, 1985]. Both depict an island arc - continent collision but offer a different interpretation for the Lanterman Fault. In Fig. 5a, the Lanterman Fault evolved from a former suture between the arc (east-facing) and Gondwana margin (Wilson terrane) whereas in Fig. 5b, it represents a post-collision back thrust (west-facing arc). In Fig. 5b, faulting occurred penecontemporaneously with erosion of the newly obducted and uplifed Glasgow Volcanics leading to the deposition of the Husky Conglomerate. Following uplift and erosion of the Wilson terrane (both models), the arc rocks were buried beneath the Leap year Group. Finally, the Leap Year Fault was formed following accretion of the Robertson Bay terrane from the east. Abbreviations: BT = Bowers terrane; GV = Glasgow Volcanics; GHI = Granite Harbour Intrusives; HC = Husky Conglomerate; LF = Lanterman Fault; LYF = Leap Year Fault; MF = Molar Formation; MG = Mariner Group; RBT = Robertson Bay terrane; WT = Wilson terrane.

Kreuzer, 1984] for low grade slates and phyllites in the Bowers terrane but the bulk of these fall in the range 510-275 m.y. Moreover, it is uncertain whether the maximum ages come from rocks in which S1 or S2 is the predominant schistosity. At best the radiometric data give a minimum age of about 500 m.y. for the D1 deformation and hence a minimum latest Cambrian age for the initiation of thrusting. If this interpretation is correct, then thrusting may have already been underway prior to deposition of the Leap Year Group. Indeed vertical displacement along the Lanterman Fault may have been a major factor in controlling sedimentation patterns throughout the Bowers terrane. In the tectonic model (Figure 5) presented here, uplift along the Lanterman Fault first exposed obducted island arc materials and then continental basement rocks. This interpretation certainly accords well with the localisation of the Husky and Lanterman Conglomerates along the Lanterman Fault (Figure 1). Deposition of the Leap Year Group may even have proceeded penecontemporaneously with deformation in the underlying Sledgers Group.

Reactivation of the Lanterman Fault

Late-stage shear zones and faults are widespread throughout the Bowers and Wilson terranes in the Lanterman Range. At one locality near Mt Lugering Lower Ordovician granitic rocks (Granite Harbour Intrusives) have been thrust northeast over Permian sediments and Jurassic Ferrar dolerite [Roland et al., 1984; Tessensohn, 1984]. Deformation in the Lanterman Range evidently extended into the Mesozoic [cf. Grindley and Oliver, 1983] and there have been two periods of thrust faulting, rather than one. Reactivation and/or modification of the original terrane boundary seems probable, particularly as the late-stage faults are subparallel to the Lanterman Fault.

Two periods of faulting have certainly occurred along the eastern boundary of the Bowers terrane [Bradshaw, 1985; Wright, 1985] and there are striking similarities in the field relations of the Lanterman and Leap Year Faults [Gibson and Wright, 1985]. On this basis it seems reasonable to conclude that both formed in much the same way. It is now known that the original boundary between the Bowers and Robertson Bay terranes has been modified by later high-angle faulting [Bradshaw, 1985; Wright, 1985]. The high-angle Leap Year Fault is effectively the present boundary for 200km but juxtaposition of the two terranes is the result of earlier thrust faulting. These observations indicate that reactivation and/or modification of the original terrane boundaries was not confined to the western side. The geological similarities between the Lanterman and Leap Year Faults are thus reaffirmed.

Acknowledgements. The author is indebted to Dr F Tessensohn and the Bundesanstalt fur Geowissenschaften und Rohstoffe for the invitation to join Ganovex III.
Drs H J Harrington and M G Laird are thanked for their constructive reviews of the manuscript whilst stimulating discussions on the structural problems of northern Victoria Land with Drs J D Bradshaw, W Laing and fellow members of the Ganovex expedition are fondly remembered. The support of the Australian Antarctic Division and Darling Downs Research Services is also gratefully acknowledged. Finally, I thank the organisers of the Third Circum-Pacific Terrane Conference, Drs E Leitch and E Scheibner, for the opportunity to visit Sydney.
Miss M Schulz typed the manuscript.

References

Adams, C J, J E Gabites, M G Laird, A Wodzicki and J D Bradshaw, Potassium-argon geochronology of the Pre-cambrian-Cambrian Wilson and Robertson Bay Groups and Bowers Supergroup, northern Victoria Land, Antarctica, in Antarctic Geoscience, edited by C Craddock, pp. 543-548, Madison, 1982.

Adams, C J and H Kreuzer, Potassium-argon studies of slates and phyllites in the Bowers and Robertson Bay terranes, northern Victoria Land, Antarctica, Geol. Jb., B60, 265-288, 1984.

Bradshaw, J D, Terrane boundaries and terrane displacement in northern Victoria Land, Antarctica, Geol. Soc. Aust. Abstracts (extended), Vol. 14, 30-33, 1985.

Bradshaw, J D, M G Laird and A Wodzicki, Structural style and tectonic history in northern Victoria Land, Antarctica, in Antarctic Geoscience, edited by C Craddock, pp 809-816, Madison, 1982.

Bradshaw, J D, S D Weaver and M G Laird, Suspect terranes and Cambrian tectonics in northern Victoria Land, Antarctica, in Tectonostratigraphic terranes of the Circum-Pacific region edited by D G Howell, pp 467-479, 1985.

Bryant, B and J C Reed, Significance of lineation and minor folds near major thrust faults in the southern Appalachians and the British and Norwegian Caledonides, Geol. Mag., 106, 412-429, 1969.

Carter, R M and R J Norris, Cainozoic history of southern New Zealand: an accord between geological observations and plate-tectonic predictions, Earth Planet. Sci. Lett., 31, 85-94, 1976.

Chapman, T J, N J Milton and G D Williams, Shape fabric variations in deformed conglomerates at the base of the Laksefjord Nappe, Norway, Jl. Geol. Soc., 136, 683-691, 1979.

Cobbold, P R and H Quinquis, Development of sheath folds in shear regimes, Jl. Struct. Geol., 2, 119-126, 1980.

Dow, J A S and V E Neall, Geology of the lower Rennick Glacier, northern Victoria Land, Antarctica, N.Z. Jl. Geol. Geophys., 17, 659-714, 1974.

Escher, A and J Watterson, Stretching fabrics, folds and crustal shortening, Tectonophys., 22, 223-231, 1974.

Gibson, G M, Deformed conglomerates in the eastern Lanterman Range, northern Victoria Land, Antarctica, Geol. Jb., B60, 117-141, 1984.

Gibson, G M, Lanterman Fault: boundary between allochthonons terranes in northern Victoria Land, Antarctica, Geol. Soc. Aust. Abstracts (extended) 14, 79-82, 1985.

Gibson, G M, F Tessensohn and A R Crawford, Bowers Supergroup south of the Mariner Glacier and possible equivalents of greater deformation and higher metamorphic degree, Geol. Jb., B60, 289-318, 1984.

Gibson, G M and T O Wright, Importance of thrust faulting in the tectonic development of northern Victoria Land, Antarctica, Nature, 315, 480-483, 1985.

Grew, E S and M Sandiford, A staurolite-talc assemblage in tourmaline-phlogopite-chlorite schist from northern Victoria Land, Antarctica, and its petrogenetic significance, Contrib. Mineral. Petrol., 87, 337-350, 1984.

Grindley, G W and P J Oliver, Post-Ross Orogeny cratonisation of northern Victoria Land, in Antarctic Earth Science edited by R L Oliver, P R James and J B Jago, pp 133-137, Canberra, 1983.

Gunn, B M and G Warren, Geology of Victoria Land between Mawson and Mulock Glaciers, Antarctica, N.Z. Geol. Surv. Bull., 71 pp, 1962.

Hossack, J R, Pebble deformation and thrusting in the Bygdin area, S Norway, Tectonophys., 5, 315-339, 1968.

Jordan, H, A R Crawford, R H Findlay, G Mortimer, P Muller and M Schmidt-Thome, Geology of the northern Bowers Mountains, north Victoria Land, Geol. Jb., B60, 57-81, 1984.

Kleinschmidt, G, N W Roland and W Schubert, The metamorphic basement complex in the Mountaineer Range, north Victoria Land, Antarctica, Geol. Jb., B60, 213-251, 1984.

Kleinschmidt, G and F Tessensohn, Tectonic development of northern Victoria Land, Antarctica in Proceedings of the Sixth Gondwana Symposium, Ohio State University, Columbus, in press, 1986.

Kreuzer, H, G Delisle, K From, A Horndorf, H Lenz, P Muller and U Vetter, Paleomagnetic and radiometric results on pre-Mesozoic rocks from northern Victoria Land, Abstracts Volume, Sixth Gondwana Symposium, Ohio State University Miscell. Publ. 231, 56, 1985.

Laird, M G, Lower Palaeozoic rocks of the Ross Sea area and their significance in the Gondwana context, Jl. Royal Soc. N.Z., 11, 425-438, 1981.

Laird, M G and J D Bradshaw, new data on the Lower Palaeozoic Bowers Supergroup, northern Victoria Land, in Antarctic Earth Science edited by R L Oliver, P R James and J B Jago, pp 123-126, Canberra, 1983.

Lisle, R J, Strain discontinuities within the Sere-Koli Nappe Complex, Scandinavian Caledonides, Jl. Struct. Geol., 6, 101-110, 1984.

Ramberg, H, Particle paths, displacement and progressive strain applicable to rocks, Tectonophys., 28, 1-37, 1975.

Roland, N W, G M Gibson, G Kleinschmidt and W Schubert, Metamorphism and structural relations of the Lanterman Metamorphics, Geol. Jb., B60, 319-361, 1984.

Sandiford, M, Structural evolution of the Lanterman Metamorphic Complex, northern Victoria Land, Antarctica, N.Z. Jl. Geol. Geophys., 28, 443-458, 1985.

Shackleton, R M and A C Ries, The relation between regionally consistent stretching lineations and plate motions, Jl. Struct. Geol., 6, 111-117, 1984.

Sibson, R H, S H White and B K Atkinson, Structure and distribution of fault rocks in the Alpine Fault Zone, New Zealand, in Thrust and Nappe Tectonics edited by K R McClay and N J Price, pp 197-210, London, 1981.

Stump, E, M G Laird, J D Bradshaw, J R Holloway, S G Borg and K E Lapham, Bowers graben and associated tectonic features cross northern Victoria Land, Antarctica, Nature, 304, 334-336, 1983.

Sturm, A and S J Carryer, Geology of the region between the Matusevich and Tucker Glaciers, northern Victoria Land, Antarctica, N.Z. Jl. Geol. Geophys., 13, 408-435, 1970.

Tessensohn, F, Geological and tectonic history of the Bowers Structural zone, north Victoria Land, Antarctica, Geol. Jb., B60, 371-396, 1984.

Vetter, U, N W Roland, H Kreuzer, A Hohndorf, H Lenz and C Bessang, Geochemistry, petrography and geochronology of the Cambro-Ordovician and Devonian-Carboniferous granitoids of northern Victoria Land, Antarctica, in Antarctic Earth Science edited by R L Oliver, P R James and J B Jago, pp 140-143, Canberra, 1983.

Weaver, S D, J D Bradshaw and M G Laird, Geochemistry of Cambrian volcanics of the Bowers Supergroup and implications for the early Palaeozoic tectonic evolution of northern Victoria Land, Antarctica, Earth Planet. Sci. Lett., 68, 128-140, 1984.

Williams, G D, Rotation of contemporary folds into the X direction during overthrust processes in Laksefjord, Finmark, Tectonophys., 48, 29-40, 1978.

Wodzicki, A, J D Bradshaw and M G Laird, Petrology of the Wilson and Robertson Bay Groups and Bowers Supergroup, northern Victoria Land, Antarctica, in Antarctic Geoscience edited by C Craddock, pp 549-554, Madison, 1982.

Wright, T O, Thrust tectonics in northern Victoria Land, Abstracts Volume, Sixth Gondwana Symposium, Ohio State University Miscell. Publ. 231, 107, 1985.

Wyborn, D, Chemistry of Palaeozoic granitoids of northern Victoria Land, in Antarctic Earth Science edited by R L Oliver, P R James and J B Jago, p 144, Canberra, 1983.

THE MESOZOIC TECTONOSTRATIGRAPHIC TERRANES AND ACCRETIONARY HERITAGE OF SOUTH-EASTERN MAINLAND ASIA

Michael P. Klimetz

2115 Shore Parkway, Brooklyn, New York 11214

Abstract. The extant continental configuration of south-eastern mainland Asia, a complex collage of Tethyan and Circum-Pacific suture-bounded tectonostratigraphic terranes, reveals a predominantly Mesozoic accretionary heritage. Permo-Triassic accretion of the Tarim and Qaidam-Alashan Blocks and the Inner Mongolia-Great Khingan Arc of northern China and Mongolia was followed by the Late Triassic to Early Jurassic accretion of terranes of the Songban Ganzi Accretionary Complex and South China Superterrane including the Indosinia and Shan-Thai-Malaya Blocks. The Medial Jurassic to Early Cretaceous accretion of the Sino-Korea/Songliao-Bureya Superterrane and Jiangfudong Terrane of eastern China, the Mongol-Okhotsk Terrane of eastern Mongolia and Transbaikalia and the West Burma Terrane were in turn followed by the Cretaceous accretion of the Sikhote-Alin Terrane of the southern Soviet Far East, the Ogcheon Terrane of southern Korea and the Fujian Terrane of south-eastern coastal China. The accretion of each terrane or groups of terranes was accompanied by intense silicic to intermediate igneous activity, regional metamorphism, widespread penetrative deformation, ophiolite emplacement, and regional unconformity. Incisement by Mesozoic-Cenozoic marginal-Pacific transcurrent faulting and lithospheric extension and Paleogene to Recent intracontinental deformation have profoundly altered original accretionary relationships and created seemingly contradictory architectonic geometries.

Introduction

Several extinct convergent plate margins recognized within the complex, anastomosing network of orogenic belts that transect eastern China, Southeast Asia, Mongolia and the southern Soviet Far East strikingly reveal the composite architecture of the south-east Asian continental firmament. Evaluations of the dispersal, extent, lithospheric character and accretionary framework of component blocks/terranes, style of convergence at their respective margins, chronology and sequence of accretion, syn- and subsequent post-accretionary deformation/reorganization have been presented by a variety of researchers for China and Korea [Bally et al., 1980; Li et al., 1980; Yin, 1980; Cui, 1982; Huan et al., 1982: Klimetz, 1983a, b; Lee et al., 1983: Li and Tang, 1983; Yang and Chen, 1983: Cui, 1984; Ji, 1984; Schermer, 1984; Tang and Feng, 1984; Z. M. Zhang et al., 1984; Ji and Coney, 1985], Southeast Asia [Ridd, 1980; Mitchell, 1981; Bunopas, 1982; Bunopas and Vella, 1983; Helmcke, 1985; Stauffer, 1985], Mongolia and the southern Soviet Far East [Parfenov, et al., 1983; Kovalenko et al., 1984; Mazarovich, 1985; Parfenov and Natal'in, 1985; Kimura and Tamaki, 1986; Klimetz, 1986] and the entire region [Burrett, 1974; Roy, 1978: Li et al., 1982; Klimetz, 1983c; Sengor, 1984; Terman, 1984; Howell et al., 1985]. Despite differing yet compelling interpretative schemes, all have demonstrated a clear conceptual understanding of the complexity and developmental longevity of a region characterized by essentially continuum Phanerozoic accretion and deformation. The absence of detailed information on many regions of critical importance, a prolonged history of complex accretion and remobilization, suspect isotopic age determinations, successive over-printing of deformation and metamorphism, concealment beneath vast expanses of Neogene fluvial clastics, plateau basalt flows and Quaternary loess, and obscurement of many regional pre-Cenozoic structural relationships through uplift, subsidence and incisement as a consequence of India's collision and continuing penetration into Tibet [Molnar and Tapponnier, 1975; Molnar and Chen, 1978] have precluded a simple, straight-forward tectonic interpretation.

An understanding of the great "Tertiary paroxysm" [Argand, 1924] and its implications for the evolution of orogenic belts, the rheological behavior of immense tracts of stressed continental lithosphere and the accretionary expansion of continents constitutes one of the most significant fundamental problems in contemporary tectonics. Despite the plethora of information which has recently emerged from numerous geologic and geophysical investigations as well as detailed satellite imagery interpretation [Tapponnier and Molnar, 1977, 1979; Molnar and Tapponnier, 1978], slip-line field analysis [Tapponnier and Molnar,

1976], imaginative rheological modeling [Tapponnier et al., 1982] and complex finite strain calculations [England and Houseman, 1986] of the development of the regional Cenozoic tectonic infrastructure, a satisfactory evaluation of the pre-Tertiary positions, shapes, orientations, and subsequent displacement histories of component blocks/terranes nonetheless remains elusive. Obviously, an appraisal of Mesozoic (and older) tectonic relationships and events is requisite to an understanding of the development of the post-Mesozoic continental edifice.

Accretionary Framework of South-Eastern Mainland Asia

The loci of Late Proterozoic and Paleozoic terrane amalgamation appear to be organized in a roughly concentric, progressively expansive arrangement about the southern and eastern periphery of the Archean-Proterozoic nucleus of the Siberian Craton. These in turn are confined by tracts of Mesozoic (and Cenozoic) accretion which comprise an immensely complex, broad and remarkably continuous network of orogenesis which persists virtually uninterrupted from central Europe through Tibet, Cathaysia and Indochina [Sengor, 1984] to its inevitable merger with Circum-Pacific orogenic belts. It is at this juncture, however, where Mesozoic (as well as Cenozoic) tectonic relationships are extremely deceptive and difficult to interpret.

The extant continental configuration of the region, a chaotic collage of terranes of Tethyan and Circum-Pacific paleogeographic heritage but uncertain parentage, reveals a predominantly Mesozoic rather than Paleozoic accretionary establishment [Klimetz, 1983a, b, c]. Despite recent discussion of the relative merits of a Mesozoic [Klimetz, 1985] as opposed to a Paleozoic [Z. M. Zhang et al., 1984, 1985] assembly scheme for eastern China and adjacent regions for example, this notion has attracted support from a breadth of tectonics and paleomagnetics investigators [see Haile, 1981; McElhinny et al., 1981; Achache, 1983; Chan et al., 1984; Kent et al., 1984; Schermer, 1984; Sengor, 1984; W. Zhang et al., 1984; Zhao and Coe, 1984; Achache and Courtillot, 1985; Bai et al, 1985; Ji and Coney, 1985; Lin, 1985; Lin et al, 1985; Opdyke et al., 1986] and offers vital tectonic and paleogeographic constraints for the evolution of both the Tethyan and Circum-Pacific realms.

Mesozoic Tectonostratigraphic Terranes of South-Eastern Mainland Asia

A simplified illustration of the Mesozoic tectonostratigraphic terranes and principal tectonic features of mainland south-eastern Asia, presented in Figure 1, is an interpretation compiled primarily from Bogolepov et al. [1973], Ermikov [1973], Nagibina [1973, 1981], Reedman and Um [1975], Nagibina et al. [1977], Zonenshain [1977], Huang [1978], Fontaine and Workman [1978], Yin [1980], Mitchell [1981], Klimetz [1983c], Sengor [1984], Howell et al. [1985], Ji and Coney [1985], Mazarovich [1985] and Parfenov and Natal'in [1985]. Component blocks/terranes have been subdivided in accordance with a variety of tectonic and stratigraphic criteria and whose boundaries are essentially coincident with satisfactorily established suture zones which comprise belts of penetratively deformed bathyal marine deposits, subduction zone mélange, paired metamorphism and intermediate to silicic magmatism in association with ultramafic rocks of presumed oceanic character. Contrary to published tectonostratigraphic terrane maps in which lithotectonic character is employed as the basis of subdivision and illustration, these terranes have been grouped into chronotectonic domains in which the timing of accretion of individual or groups of terranes or superterranes to Eurasia appeared to be approximately synchronous. This was accomplished primarily for the purpose of elucidating the compelling geometrical peculiarities and seemingly contradictory architectonic relationships extant in continental south-eastern Asia.

Permo-Triassic

The Tarim and Qaidam-Alashan Blocks and Inner Mongolian-Great Khingan Arc comprise the principal Permo-Triassic accreted terranes of the region. Despite the fact that terranes with an identical docking age are suspected in eastern China, northern Thailand and Vietnam, for simplicity and the express purpose of highlighting the expansive evolutionary development of the south-east Asian margin only Mongolian and adjacent northern Chinese terranes have been outlined.

The crystalline plinth of the Tarim Block, confined by the north-vergent Southern Tienshan (STS) suture to the north and the north-vergent Kunlunshan-Altyntagh (KLA) suture to the east and south, comprises some of the oldest rocks in China. The similarity of its Archean (?) to Proterozoic basement complex and Lower Paleozoic shallow water marine cover sequences to those of Sino-Korea is conspicuous. The Tarim Block appears to have behaved as a stable continental terrane throughout the Phanerozoic with deformation restricted to its extant margins. The most complete Paleozoic sections are exposed in the marginal Kelpintagh and Kuruktag uplifts and in the Kunlunshan, are ubiquitously deformed and metamorphosed and often intruded by silicic to intermediate plutons of Carboniferous to Early Triassic age. Subordinate exposures and decidedly less complete successions described mainly from drill core are known from the interior, are mildly metamorphosed and display a characteristic Late Ordovician through Devonian hiatus that is also observed in peripheral uplifts. Carboniferous platform carbonates and clastics are succeeded by Permian and Lower Triassic flysch which comprise thick, metamorphosed, imbricate thrust assemblages

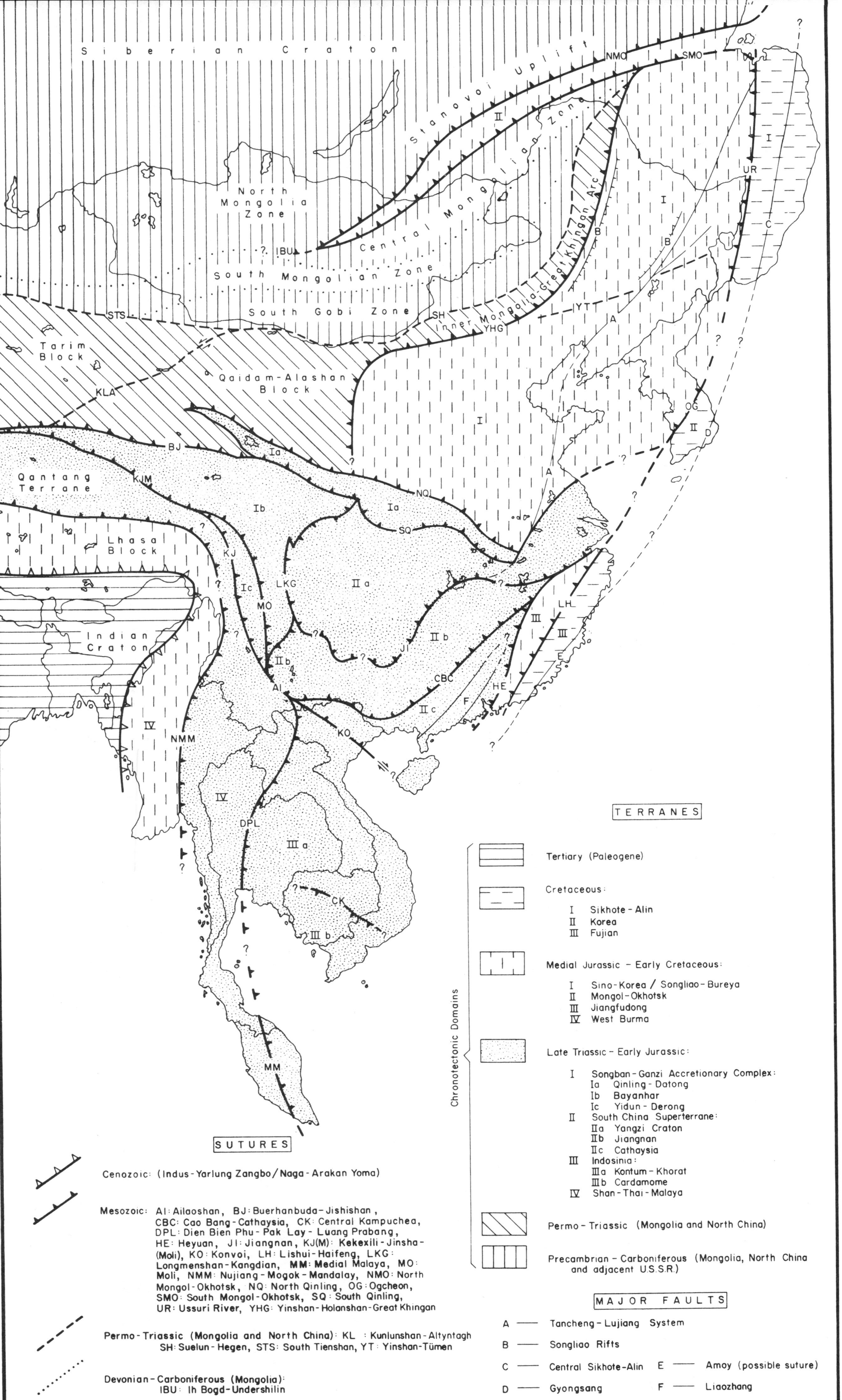

Fig. 1. Mesozoic tectonostratigraphic terranes of south-eastern mainland Asia.

of the Tienshan and Kunlunshan-Altyntagh orogenic belts. Upper Triassic through Pliocene terrestrial, molasse facies clastics are devoid of volcanics and contain minor intercalations of marine deposits, overlie Precambrian as well as Paleozoic strata with marked unconformity, and have accumulated in a series of marginal faulted and flexural basins directly attributable to Mesozoic to Recent thrust and transcurrent deformation. An approximately 1150-meter-thick sequence of Cenomanian through Early Senonian paralic and shallow marine deposits from the Kashgar Foredeep in the westernmost part of the Tarim is the only marine Cretaceous established in this part of China [Meyerhoff and Willums, 1976]. Recent paleomagnetic studies support a Permo-Triassic docking of the Tarim Block to Eurasia [Bai et al, 1985].

The Qaidam-Alashan Block, a superterrane comprised of three major Paleozoic accreted terranes, the Qaidam, Nanshan and Alashan respectively from southwest to northeast, which amalgamated during the Late Ordovician to Early Silurian, docked against Eurasia in the Permo-Triassic. Ji and Coney [1985] further recognize three terranes in the Nanshan. It is bracketed by the north-vergent Kunlunshan-Altyntagh (KLA)--eastern Suelun-Hegen (SH) sutures to the west and north, the north-vergent Buerhanbuda-Jishishan (BJ)-North Qinling (NQ) sutures to the south and the west-vergent Yinshan segment of the Yinshan-Holanshan-Great Khingan (YHG) suture to the east. The Paleozoic Qaidam and Alashan Terranes are essentially of continental character, comprised of a Precambrian crystalline core overlain by strongly deformed and metamorphosed Lower Paleozoic platform and rise sequences and Middle Paleozoic metacarbonates extensively intruded by Devonian through Permian granite. Thick Upper Paleozoic, Mesozoic and Cenozoic deposits are primarily terrestrial, molasse facies clastics, also devoid of volcanics which, in the Qaidam Terrane, have been deformed into a series of northwest-southeast trending anticlines. An imbricate, thrust-faulted assemblage of thick Triassic-Jurassic clastics [Liu and Kong, 1984] form the eastern margin of the Alashan Terrane at the Yinshan. The intervening Nanshan Terrane is entirely of oceanic character, containing thick sequences of Late Proterozoic through Ordovician deep water marine clastics, ophiolites, island-arc volcanics, subduction zone mélange and glaucophane schists intruded by Silurian-Early Carboniferous granite. Metamorphosed Siluro-Devonian flysch deposits are unconformably overlain by Permo-Carboniferous island-arc volcanics, shallow marine clastics and volcaniclastics which are in turn intruded by Late Permian-Early Triassic granite and granodiorite. Unconformably overlying Upper Permian-Lower Triassic terrestrial facies clastics and Triassic marine deposits are also deformed and metamorphosed and intruded by Late Triassic-Jurassic granite, especially near the boundary of the peculiarly shaped Qinling-Datong Terrane, the North Qinling (NQ) suture. The Nanshan Terrane is abruptly truncated at its northwesternmost termination by the Kunlunshan-Altyntagh (KLA) suture and at its southeasternmost termination by the Yinshan-Holanshan-Great Khingan (YHG) suture and are loci of Recent sinistral transcurrent and dip-slip faulting, respectively [Tapponnier and Molnar, 1977].

The Inner Mongolian-Great Khingan Arc, a northeastern prolongation and appendage of the Qaidam-Alashan Block comprises a narrow tract of Proterozoic (?) crystalline basement unconformably overlain by strongly deformed and metamorphosed Lower and Middle Cambrian marine platform and rise deposits intruded by Medial to Late Cambrian granite. An unconformably overlying mix of metamorphosed Carboniferous and Permian platform deposits and flysch are intruded by Carboniferous, Permian and Triassic granite and collectively comprise an imbricate, south-vergent thrust-faulted assemblage in association with ophiolite, glaucophane schists and subduction zone mélange [Feng and Tang, 1984; Q. Wang, 1986] along its northern periphery, the Suelun-Hegen (SH) suture. Weakly metamorphosed Jurassic shallow marine (?) to paralic facies clastics, associated with thick flows of andesite, are unconformably overlain by Cretaceous terrestrial and paralic facies clastics and volcaniclastics that are intercalated with coal seams and thick flows of andesite and rhyolite. Jurassic and Cretaceous granites intrude the entire sequence. Overlying Cenozoic sequences consist of terrestrial facies clastics and occasional basalt flows. The Inner Mongolian-Great Khingan Arc is confined to the south and east by the south- and east-vergent Holanshan-Great Khingan segment of the Yinshan-Holanshan-Great Khingan (YHG) suture of Late Jurassic-Early Cretaceous age.

Late Triassic-Early Jurassic

The Late Triassic-Early Jurassic tectonostratigraphic terranes of south-eastern mainland Asia occupy most of central and southern China and the entire of adjacent Southeast Asia from the Qinlingshan to the Malay Peninsula. They comprise a complex collage of Tethyan and Circum-Pacific terranes of both continental and oceanic aspect, woven by a remarkably sinuous yet continuous network of sutures and whose original accretionary framework appears to have been drastically different from their contemporary architectonic arrangement. A subdivision of terranes based on lithotectonic character, paleogeographic heritage and other tectonic criteria is presented in Figure 1 and establishes a basis of classification and further discussion of other Mesozoic terranes as well.

The Songban-Ganzi Accretionary Complex (I) consists of at least three distinct Tethyan terranes of mixed continental and oceanic/island-arc character, the Qinling-Datong (Ia), Bayanhar (Ib) and Yidun-Derong (Ic).

The Qinling-Datong Terrane (Ia) extends eastwards as a narrow embayment from the Datongshan through the Qinlingshan and Dabieshan to the Huaiyang massif, an uplift of Proterozoic and Lower Paleozoic rocks akin to the basement complex of the Yangzi Craton. The Qinlingshan and Dabieshan (central and easternmost) segments appear to be continental and island arc in character, and are comprised of a metamorphosed Proterozoic through Silurian basement complex overlain by strongly deformed and metamorphosed Devonian through Lower Permian platform and paralic deposits and Upper Permian-Lower Triassic flysch which have been intruded by island-arc type Triassic through Cretaceous granite and granodiorite [Xiang et al., 1984]. The Datongshan (westernmost) segment appears to be strictly oceanic in character, perhaps the remnants of a restricted marine embayment, and consists of Late Permian-Triassic flysch intruded by Triassic-Jurassic granite and granodiorite. The Qinling-Datong Terrane is confined by the south-vergent North Qinling (NQ), South Qinling (SQ) and Buerhanbuda-Jishishan (BJ) sutures. Ophiolites [Zhang and Zhu, 1984], subduction zone mélange [Feng and Zhu, 1980; Feng and Tang, 1984], a paired metamorphic belt [Wu, 1980; Z. Z. Zheng et al., 1981] and a profound Late Triassic-Early Jurassic unconformity [Jiang and Chu, 1964] have been observed along the North Qinling (NQ) suture. Slivers of dismembered ophiolite and sheared Permo-Triassic flysch and Mesozoic granodiorite have also been observed in the South Qinling (SQ) suture.

The Bayanhar Terrane (Ib) is roughly triangular in outline and is confined by the south-vergent Buerhanbuda-Jishishan (BJ) suture of the north, the south- and east-vergent Longmenshan-Kangdian (LKG) suture to the southeast and the northeast-vergent Kekexili-Jinsha-Moli (KJM) and Moli (MO) sutures to the southwest. The oldest rocks in the Bayanhar Terrane are exposed west of the Longmenshan and consist of a strongly deformed and metamorphosed Proterozoic through Ordovician basement complex unconformably overlain by Silurian marine clastics which are in turn unconformably overlain by Devonian through Permian platform carbonates, clastics and a few intercalations of paralic, coal-bearing shales. Thick sequences of strongly deformed and regionally metamorphosed Triassic carbonates and flysch [Q. Zhang, 1981] unconformably overlain by Jurassic terrestrial facies clastics characterize the interior of the terrane [Y. G. Wanq et al., 1983], and are intruded by Late Triassic-Jurassic granite and granodiorite and Early Cretaceous granite. The east-west trending Buerhanbuda-Jishishan (BJ) suture contains dismembered ophiolite [Yin, 1980], subduction zone mélange [Feng and Tang, 1984], Late Permian calc-alkaline volcanics, strongly deformed Triassic flysch and Late Triassic granodiorite and extends from the Kunlunshan to the western Qinlingshan. Similarly, the Kekexili-Jinsha-Moli (KJM) and Moli (MO) sutures also contain dismembered ophiolite [Yin, 1980], subduction zone mélange [Li and Xiao, 1984; Li and Zhang, 1985], Late Permian-Early Triassic calc-alkaline volcanics, strongly deformed Triassic flysch and Late Triassic-Jurassic granite and granodiorite and extends from the Kunlunshan through the Kekexilishan and Bayankelashan to the Ailaoshan of central Yunnan. The Longmenshan-Kangdian (LKG) suture contains sporadic exposures of ultramafic rocks [Huang, 1978] assumed to be ophiolitic in character and subduction zone mélange [Yin, 1980; Feng and Tang, 1984]. Although structural relationships reveal a record of multiple accretion including Recent deformation and uplift [Y. Zhao et al., 1985], it is an important Triassic structural, paleogeographic and metamorphic facies boundary which separates strongly deformed and regionally metamorphosed Triassic deep water marine clastics of the Bayanhar Terrane from the mildly deformed and unmetamorphosed coeval platform carbonates of the Yangzi Craton [Yin, 1980; Y. G. Wang et al., 1983].

The Yidun-Derong Terrane (Ic), confined by the east-vergent Moli (MO) and Kekexili-Jinsha (KJ) sutures, reveals a Mesozoic stratigraphy similar to that of the Bayanhar Terrane, intruded by Late Triassic-Jurassic granite and granodiorite. A satisfactory pre-Mesozoic stratigraphic sequence has not been established.

The South China Superterrane (II) consists of at least three Mesozoic Circum-Pacific terranes underlain by a basement complex of predominantly continental character, the Yangzi Craton (IIa), the Jiangnan (IIb) and the Cathaysian (IIc).

The crystalline nucleus of the Yangzi Craton (IIa), confined by the south-vergent South Qinling (SQ) suture to the north, the west-vergent Longmenshan-Kangdian (LKG) suture to the west and the sinuous, south- and east-vergent Jiangnan (JI) suture to the south and east comprises the oldest and broadest expanse of Proterozoic rocks in southern China [Sun, 1985]. It is overlain by platform carbonates and clastics that range from Proterozoic through Triassic and Early Jurassic age with a characteristic Devonian through Carboniferous hiatus in Sichuan and northern Guizhou, and an extensive Upper Silurian-Lower Devonian (post-accretionary?) molasse-facies sequence that appears to have been derived from the Jiangnan uplift. The Late Permian Omeishan basalt comprises an areally extensive accumulation of basalt and minor andesite flows of variable thickness which rests upon Early Permian platform carbonates and lays astride the Jiangnan (JI) suture which is interpreted to be the effusive locus. It appears to be confined to the western edge of the Yangzi Craton and the adjacent western section of the Jiangnan (IIb) Terrane, where it is regarded as Early Permian. Upper Permian paralic clastics with coal seams succeeded in marginal areas by platform carbonates and clastics rest atop the Yangzi Omeishan basalt. Triassic through Jurassic cover sequences are characterized by a peripheral décollement fold system of shallow-level detachments from underlying Paleozoic shales and Triassic evaporites, the trends of which parallel that of surrounding sutures. Jurassic and Cretaceous sequences of the Sichuan Basin are comprised primarily of terrestrial facies clastics and unconformably overlie the Triassic. The Jurassic of the lower Yangzi River region are represented by terrestrial facies clastics interbedded with coal seams and volcaniclastics and are unconformably overlain by Cretaceous redbeds and thick accumulations of andesite, rhyolite and associated volcaniclastics [Liu and Chai, 1984; Huo, 1985]. The Jurassic and Cretaceous unconformably overlie deformed Devonian through Lower Triassic carbonates.

The basement of the Jiangnan Terrane (IIb) consists of Late Precambrian gneisses unconformably overlain by Late Proterozoic (Sinian) tillites and shallow marine clastics and carbonates. These are in turn overlain by variably metamorphosed platform deposits of Cambrian through Early Ordovician age and paralic Silurian shales. Unconformably overlying Lower Devonian molasse facies clastics, Middle Devonian through Middle Triassic platform deposits and Upper Triassic flysch are in turn unconformably overlain by Lower Jurassic molasse, and are succeeded unconformably by Cretaceous-Tertiary terrestrial facies clastics that are equivalent to those of the Yangzi Craton. The Late Permian Omeishan basalt occupies an extensive area in the far west of the Jiangnan Terrane and adjacent Yangzi Craton. Consisting of basalt and andesite flows of variable thickness which overlie Early Permian platform carbonates and clastics, its relationship to the rifting of the Jiangnan Terrane from the Yangzi Craton and subsequent back-arc spreading within the locus of the future Jiangnan suture is well established [Fan, 1978]. The sinuous Jiangnan (JI) and the Cao Bang-Cathaysia (CBC) sutures define the northwestern and southeastern perimeters of the terrane, respectively, and consist of an imbricate assemblage of sheared granite and granodiorite, dismembered meta-ophiolite, radiolarian chert and strongly deformed deep water marine clastics of Triassic age.

The Cathaysian Terrane (IIc) consists of a pre-Devonian metamorphic complex unconformably overlain by Devonian through Carboniferous platform carbonates and Permian through Lower Triassic platform clastics and carbonates. These are unconformably overlain by Lower Jurassic paralic and terrestrial facies clastics which are similarly overlain by an unconformable sequence of Cretaceous rhyolites, andesites and terrestrial facies clastics. Unconformable Tertiary red beds and basalts cap the sequence.

The Tethyan Indosinian Terrane (III) is confined by the southwest-vergent Ailaoshan (AI) and Konvoi (KO) sutures to the northeast and the west-vergent Dien Bien Phu-Pak Lay-Luang Prabang (DPL) (a.k.a. Pha Som) and Medial Malaya (MM) (a.k.a. Bentong) sutures to the west. It appears to consist of two distinct terranes, the Kontum-Khorat (IIIa) to the north and the Cardamome (IIIb) to the south. The northwest-southeast trending Central Kampuchea (CK) suture, a problematic zone of strong Late Triassic-Early Jurassic deformation, silicic to intermediate magmatism, ophiolite (?), and subduction zone mélange, is located adjacent to the Early and Medial Carboniferous Dalat and Pailin foldbelts and buttressed against the southern periphery of the Kontum massif [Fontaine and Workman, 1978].

The basement of the Kontum-Khorat Terrane (IIIa) consists of Archean (?) to Proterozoic high-grade metamorphic rocks and are exposed in the Kontum massif and as marginal uplifts adjacent to the Konvoi (KO) and Dien Bien Phu-Pak Lay-Luang Prabang sutures. Unconformably overlying Upper Proterozoic and Lower Cambrian sequences consist of mildly metamorphosed phophorites and alternating marine and terrestrial facies clastics. These in turn are overlain by Middle Cambrian to Lower Ordovician terrestrial and marine clastics, Upper Ordovician and Lower Silurian marine clastics and interbedded volcaniclastics, Upper Silurian terrigenous and marine deposits and Devonian through Middle Permian platform carbonates. Unconformably overlying Upper Permian terrestrial deposits are overlain by Triassic platform carbonates and clastics with rhyolite and dacite interbeds. Near the northern and southern margins of the terrane, a thick accumulation of Triassic rhyolite unconformably overlies strongly deformed Carboniferous andesites and flysch. Unconformably overlying Upper Triassic through Cretaceous terrestrial facies clastics with minor intercalations of shallow water marine clastics and evaporites are exposed in the Khorat plateau of eastern Thailand. Jurassic marine deposits are exposed near the southern margin of the terrane and terrestrial deposits interbedded with rhyolites are known from the north and both are intruded by Early Cretaceous granite.

The Cardamome Terrane (IIIb) consists essentially of deformed Devonian through Middle Carboniferous platform deposits unconformably overlying a high-grade metamorphic basement complex. These in turn are unconformably overlain by strongly deformed Upper Carboniferous-Lower Permian flysch, unconformably succeeded by deformed Triassic platform deposits and Upper Triassic-Liassic flysch. This sequence is succeeded by an unconformable series of Jurassic terrestrial facies sediments and a thick accumulation of rhyolite. The entire stratigraphic sequence is intruded by Jurassic-Early Cretaceous granite.

The Shan-Thai-Malaya Block (IV) is confined by the west-vergent Kekexili-Jinsha (KJ), Ailaoshan (AI), Dien Bien Phu-Pak Lay-Luang Prabang (DPL) and Medial Malaya (MM) sutures to the east and the west-vergent Nujiang-Mogok-Mandalay (NMM) suture to the west. Whether or not it is an appendage of the Qantang Terrane (North Tibet Block) remains to be established. A basement complex comprised of regionally metamorphosed Precambrian clastic rocks are unconformably overlain by Lower and Middle Cambrian andesites, Upper Cambrian and Lower Ordovician platform clastics, Middle Ordovician platform carbonates and Upper Ordovician to Upper Silurian platform carbonates and clastics. These are unconformably overlain by Devonian through Permian carbonates in northern regions or by Lower and Middle Carboniferous flysch overlain by Upper Carboniferous through Middle Triassic platform carbonates in southern regions. Lower Triassic platform deposits, unconformably overlain by Upper Triassic evaporites and shales and in turn unconformably overlain by Middle and Upper Jurassic flysch collectively rest unconformably on Permian carbonates in the north, whereas Lower and Middle Jurassic turbidites and Upper Jurassic olistostromes unconformably overlain by Cretaceous molasse facies clastics collectively rest unconformably atop Middle Triassic carbonates in the south. The Nujiang ultramafic belt of western Yunnan and the north-south oriented Mogok-Mandalay uplift of northern and central Burma collectively comprise the Nujiang-Mogok-Mandalay (NMM) suture. It is a narrow zone of dismembered ophiolite, glaucophane schist, calc-alkaline igneous rocks and subduction zone mélange, associated with strong Medial to Late Jurassic deformation. The Dien Bien Phu-Pak Lay-Luang Prabang (DPL) suture appears to be a southwest-trending bifurcation and prolongation of the Ailaoshan (AI) suture. It extends from southwesternmost Yunnan through northwesternmost Vietnam, Laos and eastern Thailand and across the Gulf of Thailand to central Peninsular Malaysia. The Medial Malaya (MM) suture is presumed to be its southernmost prolongation. It is comprised of a linear belt of ophiolites, granites, granodiorites, subduction zone mélange, calc-alkaline volcanics, deep water marine clastics and cherts strongly deformed at the end of the Triassic and unconformably overlain by Rhaetian and younger terrestrial facies clastics [Fontaine and Workman, 1978; Bunopas, 1982; Bunopas and Vella, 1983; Panjasawathong and Yaowanoiyothin, 1983]. The Early Mesozoic serpentinized ultramafic (ophiolitic?) rocks [Khoo and Tan, 1983] and subduction zone mélange of the Medial Malaya (MM) suture, alternately referred to as the Bentong-Raub Line and Bentong suture, are exposed along the eastern margin of the Main Range Batholith which dissects Peninsular Malaysia. Widespread Mesozoic granite intrusion in western Malaya has been investigated radiometrically by Liew and Page [1985] and reveal Triassic through Cretaceous ages.

Medial Jurassic-Early Cretaceous

The Circum-Pacific Sino-Korea/Songliao-Bureya Terrane (I) is a Permo-Triassic superterrane comprised of two distinct, areally extensive, and remarkably dissimilar Paleozoic terranes, the Sino-Korea and Songliao-Bureya. It is circumferentially confined by the east-vergent Yinshan-Holanshan-Great Khingan (YHG) suture to the west, the south-vergent South Mongol-Okhotsk (SMO) suture to the north, the east-vergent Ussuri River (OR) suture to the east and the south-vergent North Qinling (NQ) suture to the south. The nature of its contact with the northwesternmost appendage of the Yangzi Craton remains unclear, however it is presumed to be a sinistral transcurrent fault associated with Late Triassic-Cretaceous displacements across the Tancheng-Lujiang Fault System [Wu et al., 1981; D. Wang, 1985; Xu, I985; Xu et al., 1985, Zhou, 1985]. Also known as the North China Block, the Sino-Korea terrane comprises approximately one-half of northeastern China, and consists of an Archean to Proterozoic crystalline core unconformably overlain by a succession of Paleozoic platform-type sediments from which Upper Ordovician through Lower Carboniferous are characteristically absent. Uppermost Paleozoic and almost the entire Mesozoic sequence appear to consist of terrestrial facies clastics and evaporites interbedded with coal seams. Upper Permian and Triassic sequences are also interbedded with volcaniclastics. Terrestrial facies Jurassic deposits contain thick flows of andesite and rhyolite. The entire succession is extensively intruded by Late Triassic through Cretaceous granite and granodiorite. The Songliao-Bureya Terrane, also referred to as the Northeast China Fold Zone [Klimetz, 1983b], occupies the remaining half of northeastern China. The full extent of Precambrian basement beneath the terrane is not known with certainty as the only confirmed exposures of Archean (?) to Proterozoic rocks are sporadic outcrops along the Yinshan-Tumen (YT) suture, at Lake Khanka and as slivers in the Ussuri River (UR) suture and in other marginal uplifts of the Bureya massif which extends from the Lesser Khingan Range into the lower Amur River valley. Most lithologies comprising the Songliao-Bureya Terrane are Late Paleozoic and younger and consist of Permo-Carboniferous granites, Mesozoic terrestrial and shallow marine facies sediments and Jurassic-Cretaceous intrusives and volcanics. Upper Devonian through Permian deposits are sparse and consist of deformed platform-type clastics and carbonates. These in turn are intruded by Late Paleozoic silicic plutons and, along the Yinshan-Tumen (YT) suture, are associated with serpentinized ophiolitic peridotites. There is a marked absence of Triassic deposits. Unconformably overlying Jurassic and Cretaceous sequences consist of terrestrial facies clastics, rhyolite, andesite, evaporite, and coals interbedded with shallow marine clastics which collectively reach considerable thickness in the Medial Jurassic-Early Cretaceous Songliao Basin. Along the eastern margin of the Bureya massif near Lake Khanka rest an unconformable sequence of

Proterozoic through Middle Cambrian platform clastics, intruded by granite, which in turn are unconformably overlain by Medial Devonian calc-alkaline volcanics and volcaniclastics, intruded by Late Devonian-Carboniferous granite and granodiorite. An unconformably overlying sequence comprised of Carboniferous molasse facies clastics is in turn unconformably overlain and intruded by Permian rhyolite and granite respectively. An unconformable sequence of Middle Triassic platform clastics and carbonates and Upper Triassic turbidites are in turn unconformably overlain by an imbricate, east-vergent thrust assemblage of Jurassic through Lower Cretaceous flysch, calc-alkaline volcanics, glaucophane schists and ophiolitic olistostromes and are intruded by Late Cretaceous silicic to intermediate plutons, collectively comprising the Ussuri River (UR) suture. Middle to Upper Cretaceous molasse-facies clastics unconformably cap the sequence. The Yinshan-Holanshan-Great Khingan (YHG) suture occupies a sinuous belt of Late Jurassic to Early Cretaceous deformation that extends from Ningxia Hui through Nei Monggol, China and follows the eastern margin of the Great Khingan Range to the Amur River and a presumed juncture with the South Mongol-Okhotsk (SMO) suture. It is associated with scattered ultramafics [Huang, 1984] and glaucophane schists of uncertain age, widespread Jurassic-Cretaceous granites and silicic to intermediate volcanics [Takahashi, 1983], minor Jurassic marine clastics [C. Wang and X. Zhang, 1984] and Late Jurassic-Early Cretaceous thrust faulting [Liu and Kong, 1984; C. Yang et al., 1984].

The Mongol-Okhotsk Terrane (II) comprises a sinuous corridor of strongly deformed and metamorphosed Triassic to Medial Jurassic oceanic and island arc assemblages that extends from the Hangay and Hentiy Ranges of central and eastern Mongolia through the Borshchovochnyi, Gazimurski, Tukuringra and Dzhagdy Ranges of eastern Transbaikalia to the Uda River delta at the Sea of Okhotsk. It constitutes the southwesternmost appendage of the Mongol-Chukotka Orogenic Complex [Parfenov et al., 1983; Parfenov and Natal'in, 1985], a remarkably continuous feature within the extensive, anastomosing network of Mesozoic orogenesis that transects eastern Asia from the Indonesian Archipelago to Siberia. The northern edge of the terrane is buttressed against the Proterozoic crystalline plinth of the Yablonovy and Stanovoi uplifts at the southern periphery of the Siberian Craton which constitutes the North Mongol-Okhotsk (NMO) suture. Its southern margin is confined by the Great Khingan Range and the Bureya massif at the South Mongol-Okhotsk (SMO) suture. South- and southeast-vergent imbricate assemblages of crystalline basement, dismembered ophiolite, glaucophane schist, subduction zone mélange, radiolarian chert and strongly deformed Triassic through Middle Jurassic deep water marine clastics, intruded by Triassic-Jurassic granodiorite and unconformably overlain by Upper Jurassic-Lower Cretaceous terrestrial deposits are observed in the Galimsk and Tukuringra-Dzagdy segments in Transbaikalia [Parfenov, et al., 1983; Cherepovskii, 1984; Parfenov and Natal'in, 1985], particularly along the Shilka River near Sretensk where they comprise a continuous, 100-kilometer-long exposure [Misnik and Shevchuk, 1977]. Similarly, a broad swath of Jurassic-Early Cretaceous calc-alkaline volcanic and plutonic rocks, greenschist facies metamorphism and northwest-vergent deformation occupy the northern perimeter of the terrane and involve adjacent Proterozoic basement. The puzzling Mongolian segment, associated with widespread Triassic-Early Cretaceous silicic to intermediate magmatism, appears to truncate continuous Paleozoic tectonic trends yet vanishes into the Permo-Carboniferous Ih Bogd-Undershilin (IBU) suture.

The Circum-Pacific Jiangfudong Terrane (III) of southeastern China is confined by the southeast-vergent Heyuan (HE) and Lishui-Haifeng (LH) sutures. The basement rocks of the terrane are comprised of Precambrian gneisses and schists [T. Yang et al., 1985] unconformably overlain by Devonian through Lower Carboniferous platform-type clastics which in turn are unconformably overlain by Upper Devonian through Lower Permian platform carbonates and Upper Permian through Lower Triassic alternating shallow marine carbonates and paralic clastics with interbeds of coal. An unconformably overlying sequence of Lower Jurassic alternating marine and paralic clastics is in turn unconformably overlain by a sequence of Cretaceous molasse facies red beds, volcaniclastics and thick flows of rhyolite. Unconformable Tertiary red beds and basalts cap the sequence. The Heyuan (HE) suture consists of a strongly deformed and metamorphosed assemblage of calc-alkaline volcanics, basalts, radiolarian cherts, and serpentinized peridotites of presumed ophiolitic affinity.

The Tethyan West Burma Terrane (IV) is confined by the west-vergent Nujiang-Mogok-Mandalay (NMM) suture to the east and the Naga-Arakan Yoma (Tertiary) suture to the west. A crystalline basement sequence consisting of granite-gneisses and schists of uncertain age is unconformably overlain by Lower Cretaceous platform carbonates and thick, Upper Cretaceous flysch which in turn are unconformably overlain by Tertiary molasse facies clastics and Quaternary fluvial clastics.

Cretaceous

The oldest known rocks in the Sikhote-Alin Terrane (I) consist of strongly deformed and metamorphosed Devonian (?) granitic gneisses unconformably overlain by Carboniferous through Permian platform carbonates which are in turn unconformably overlain by Middle to Upper Triassic carbonate turbidites, Middle Jurassic to Lower Cretaceous flysch and calc-alkaline volcanics. East of the Central Sikhote-Alin Fault, a Late Cretaceous-Tertiary sinistral transcurrent shear zone, Early Cretaceous imbricate thrust packages are west-vergent and involve Late Paleozoic and Early Mesozoic platform sequences, Jurassic through Lower Cretaceous flysch and are collectively intruded by Medial to Late Cretaceous granite and granodiorite stitching plutons.

The Ogcheon Terrane (II), situate at the southern end of the Korean Peninsula, is confined to the north by the southeast-vergent Ogcheon (OG) suture where it is buttressed against the Sino-Korea/Songliao-Bureya Superterrane. Nearly the entire pre-Lower Jurassic strata of the Ogcheon Terrane have been strongly deformed and metamorphosed. Archean to Middle Proterozoic polyphase deformed gneisses and schists underlie the entire region and are exposed in a broad northeast-southwest trending swath adjacent to the Ogcheon (OG) suture, the Ryeongam massif. Unconformably overlain by Upper Proterozoic through Middle Ordovician platform carbonates and clastics with intercalated tillites near the base, these are disconformably overlain by Upper Carboniferous through middle Permian platform deposits succeeded by a thick Upper Permian through Lower Triassic terrestrial facies clastic sequence with coal seams, intruded by Medial Triassic granite. A sequence of Upper Triassic-Lower Jurassic molasse facies clastics rests unconformably atop the Lower Triassic. The entire pre-Lower Jurassic sequence is strongly deformed and metamorphosed and intruded by widespread Medial and Late Jurassic granite and granodiorite stitching plutons of presumed oceanic island-arc affinity [Park and So, 1972]. These are succeeded by an unconformable sequence of Cretaceous terrestrial facies clastics, tuffs, rhyolites and andesites and are intruded by Late Cretaceous granite. Despite the tectonostratigraphic evidence which suggests a Late Triassic-Early Jurassic docking of the Ogcheon Terrane to Eurasia, recent paleomagnetic investigations indicate significant Cretaceous rotation of the Ogcheon Terrane with respect to Sino-Korea [Lee et al., 1983] and subsequently demand a Cretaceous (or later) accretion. The southeast-vergent Ogcheon (OG) suture, situate at the southern margin Of the Gyonggi massif, is comprised of strongly deformed and metamorphosed Jurassic-Cretaceous calc-alkaline volcanics, pelagic and calcareous metasediments, flysch, sheared metagabbros and epidote-amphibolites [Lee, 1980], and appears to define a paired metamorphic belt. The southeasternmost edge of the Ogcheon Terrane is also incised by the Late Cretaceous-Paleogene sinistral transcurrent Gyongsang Fault, interpreted to be the southern prolongation of the Central Sikhote-Alin Fault, both of which appear related to oblique subduction of Pacific oceanic lithosphere beneath eastern Asia.

The Circum-Pacific Fujian Terrane (III), accreted to the Jiangfudong Terrane at the southwest-vergent Lishui-Haifeng (LH) suture, consists of a Precambrian (?) metamorphic basement complex of gneisses and schist unconformably overlain by Upper Carboniferous through Lower Permian platform carbonates and Upper Permian alternating paralic and marine clastics. An unconformable series of Lower Jurassic marine and paralic clastics intruded by granite are in turn unconformably overlain by thick Lower Cretaceous rhyolites and Upper Cretaceous marine and paralic shales. The sequence is unconformably capped by Tertiary redbeds. The Lishui-Haifeng (LH) suture is comprised of a linear belt of sheared granite, strongly deformed Lower Jurassic marine clastics and serpentinized ultramafics of presumed ophiolitic origin and may be the southwestern prolongation of the Ogcheon (OG) suture of southern Korea. The southeastern coastal region of the Fujian Terrane is incised by the Cretaceous-Tertiary sinistral transcurrent Amoy Fault (the Zhenhai-Wenzhou suture of Zhou and Zhang [1983]), which appears to be closely associated with a linear group of small, lensoid masses of sheared peridotite. Whether or not these ultramafic rocks represent dismembered ophiolite emplaced by the accretion of an island arc remains to be established.

Discussion and Conclusions

The extant continental configuration of southeastern Mainland Asia, a complex collage of Tethyan and Circum-Pacific suture-bounded tectonostratigraphic terranes, reveals a predominantly Mesozoic accretionary heritage and clearly refutes the Wegenerian concept of a bipartite regionalization of one supercontinent (Pangaea) and one superocean (Panthalassa) sensu stricto. This rapidly expansive period of growth of the southeast Asian margin commenced with the Permo-Triassic accretion of the Tarim, Qaidam-Alashan Blocks and the Inner Mongolian-Great Khingan Arc followed by the Late Triassic-Early Jurassic accretion of terranes of the Songban-Ganzi Accretionary Complex and South China Superterrane including the Indosinia and Shan-Thai-Malaya Blocks. The Medial Jurassic through Early Cretaceous accretion of the Sino-Korea/Songliao-Bureya Superterrane and Jiangfudong Terrane of eastern China, the Mongol-Okhotsk Terrane of Mongolia and Transbaikalia and the West Burma Terrane was subsequently followed by the Cretaceous accretion of the Sikhote-Alin Terrane of the southern Soviet Far East, the Ogcheon Terrane of southern Korea and the Fujian Terrane of southeastern coastal China. The accretion of each terrane or groups of terranes was accompanied by intense silicic to intermediate igneous activity, regional metamorphism, widespread penetrative deformation and regional unconformity. A satisfactory evaluation of the magnitude of post-accretionary deformation and obscurement of original accretionary geometries and tectonic relationships by a combination of Mesozoic-Cenozoic marginal-Pacific transcurrent faulting and lithospheric extension and regional Paleogene to Recent intracontinental deformation has yet to be established, but is unquestionably responsible for creating the complex architectonic infrastructure as it is presently understood. Correlations between these and other Circum-Pacific terranes are tempting but clearly require further investigation and more detailed tectonostratigraphic analyses.

Acknowledgments. The author's participation in the Third Circum-Pacific Conference was made possible by an invitation and a generous travel stipend arranged by David G. Howell. Critical appraisal of the manuscript by Evan Leitch, Erwin Scheibner, Edmund Zhang (Zhang Zhi-meng) and an anonymous reviewer are gratefully acknowledged. The immeasureable contributions of Sally Stevenson, I.L.L., S.U.N.Y.A., Patrick J. McGrath and Barbara A. Liskin, C.-P.M.C., New York, both to the author and this work are sincerely appreciated. Michael Waters patiently and efficiently prepared the camera-ready typescript. This paper is respectfully dedicated to Nicholas Klimetz, Professor Akiho Miyashiro and Dr. Nikolai Dahl.

References

Achache, J., V. Courtillot and J. Besse, Paleomagnetic constraints on the Late Cretaceous and Cenozoic tectonics of Asia, Ear. Plan. Sci. Lett., 123-136, 1983.

Achache, J., A preliminary Upper Triassic paleomagnetic pole for the Khorat plateau (Thailand): consequences for the accretion of Indochina against Eurasia, Ear. Plan. Sci. Lett., 73, 147-157, 1985.

Argand, E., La tectonique de l'Asie, Comptes Rendues, Congres Geologique International, XIII, Bruxelles v. 1, 171-372, 1924.

Bai, Y. et al., Late Paleozoic polar wander path for the Tarim block and tectonic significance (in Chinese), Seismol. Geol., 7 (1), 71-80, 1985.

Bally, A. W. et al., Report to the American plate tectonics delegation to the People's Republic of China, U. S. Geol. Surv. Open File Rep., 80-501, 1980.

Bogolepov, K. V. et al., On relations and evolution of Sikhote-Alin geosyncline and east part of central Asia orogenic belt (in Russian), in Tectonics of the Mesozoic Orogenic Belt of Central Asia, K. V. Bogolepov (editor), Nauka Publishing, Novosibirsk, 257-269, 1973.

Bunopas, S., Paleogeographic history of western Thailand and adjacent parts of southeast Asia - a plate tectonics interpretation, Geol. Surv. Thailand Spec. Pap., no. 5, 810pp., 1982.

Bunopas, S. and P. Vella, Tectonic and geologic evolution of Thailand, in Proceedings, Workshop on Stratigraphic Correlation of Thailand and Malaysia, Haad Yai, Thailand, v. 1, 307-322, 1983.

Burrett, C. F., Plate tectonics and the fusion of Asia, Ear. Plan. Sci. Lett., 21, 181-189, 1974.

Chan, L. S. et al., Paleomagnetic results from some Permian-Triassic rocks from southwest China, Geophys. Res. Lett., 11 (11), 1157-1160, 1984.

Cherepovskii, V. F., Coal basins of the Baikal-Amur region (in Russian), Nedra Publishing, Moscow, 157pp., 1984.

Cui, K., On origin and development of Chinese continent (in Chinese), Research on Geology (I), Inst. Geol. Acad. Sin., 22-33, 1982.

Cui, K., On the origin and development of China's continent, Developments in Geoscience, Science Press, Beijing, China, 495-508, 1984.

Duan, Z. and H. Zhao, The Ailaoshan-Tengtaiohe fracture - the subduction zone of an ancient plate (in Chinese), Acta Geol. Sin., 55 (4), 258-266, 1981.

England, P. and G. Houseman, Finite strain calculations of continental deformation 2. Comparison with India-Asia collision zone, J. Geophys. Res., 91 (B3), 3664-3676, 1986.

Ermikov, V. D., Tectonics of Mesozoic orogenic complex of the Hangai-Hentiy-Yablon-Stanovoi mountain region (in Russian), in Tectonics of the Mesozoic Orogenic Belt of Central Asia, K. V. Bogolepov (editor), Nauka Publishing, Novosibirsk, 185-256, 1973.

Fan, P. F., Outline of the tectonic evolution of southwestern China, Tectonophysics, 45, 261-267, 1978.

Feng, Y. and Y. Tang, Chaotic geological bodies in China (in Chinese), Bull. Chin. Acad. Geol. Sci., no. 10, 77-92, 1984.

Feng, Y. and B. Zhu, Mélanges and tectonic development of the west Qinling mountains (in Chinese), Acta Geol. Sin., 54 (1), 34-43, 1980.

Fontaine, H. and D. R. Workman, Review of geology and mineral resources of Kampuchea, Laos and Vietnam, in Geology and Mineral Resources of Southeast Asia, Proceedings, Third Reg. Conf. Geol. Min. Res. SE Asia, Bangkok, P. Nutalaya (editor), Asian Inst.Tech., Bangkok, 541-603, 1978.

Haile, N.S., Fragmentation of Asia in the Permian: Pangaea at risk?, Nature, 293, 210, 1981.

Helmcke, D., The Permo-Triassic "Paleotethys" in mainland southeast Asia and adjacent parts of China, Geol. Rund., 74 (2), 215-228, 1985.

Howell, D. G., D. L. Jones and E. R. Schermer, Tectonostratigraphic terranes of the Circum-Pacific region, in Tectonostratigraphic Terranes of the Circum-Pacific region, D. G. Howell (editor), Circum-Pacific Council for Energy and Mineral Resources, Earth Science Series, 1, 3-30, 1985.

Huan, W., Z. Shi and J. Yan, Meso-Cenozoic tectonic evolution of eastern China and adjacent areas and movement of the Pacific plate (in Chinese), Sci. Geol. Sin., no. 2, 179-190, 1982.

Huang, T. K.. An outline of the tectonic characteristics of China, Eclogae Geol. Helv., 71 (3), 611-635, 1978.

Huang, Y., The aureoles of the Songshugou alpine-type massif (in Chinese), Geochimica, no. 3, 206-216, 1984.

Huo, Y., Tectonic environment of formation of Mesozoic volcanic rocks in the coastal areas of southeast China (in Chinese), Geochimica, no. 3, 255-263, 1985.

Ji, X., Suspect terranes of China, in Proceedings, Second Circum-Pacific Terrane Conference, D. G. Howell, D. L. Jones, A. Cox and A. Nur (editors), Stanford Univ. Pub. Ear. Sci., 17, 212-215, 1984.

Ji, X. and P. J. Coney, Accreted terranes of China, in Tectonostratigraphic Terranes of the Circum-Pacific Region, D. G. Howell (editor), Circum-Pacific Council for Energy and Mineral Resources, Earth Science Series, 1, 349-361, 1985.

Jiang, C. F. and C. C. Chu, On the Indosinian movement in eastern Tsinling miogeosyncline, Sci. Sin., 13 (8), 1346-1347, 1964.

Kent, D. V. et al., Paleomagnetic results from the Upper Cretaceous of South China, EOS Trans. AGU, 65 (45) 862, 1984.

Khoo, T. T. and B. K. Tan, Geologic evolution of Peninsular Malaysia, in Proceedings, Workshop on Stratigraphic Correlation of Thailand and Malaysia, Haad Yai, Thailand, v. 1, 253-290, 1983.

Kimura, G. and K. Tamaki, Collision, rotation, and back-arc spreading in the region of the Okhotsk and Japan Seas, Tectonics, 5 (3), 389-401, 1986.

Klimetz, M. P., Mesozoic accretionary events in south-eastern Asia, EOS, Trans. AGU, 64 (18), 320, 1983a.

Klimetz, M. P., Speculations on the Mesozoic plate tectonic evolution of eastern China, Tectonics, 2 (2), 139-166, 1983b.

Klimetz, M. P., The pre-Tertiary geology and Mesozoic tectonic evolution of eastern China, southeast Asia and adjacent regions, M. S. Thesis, State University of New York at Albany, 216pp., 1983c.

Klimetz, M. P., An outline of the plate tectonics of China: Discussion and reply: Discussion, Geol. Soc. Am. Bull., 96 (3), 407, 1985.

Klimetz, M. P., Remnants of a Triassic-Jurassic ocean basin and convergent plate margin processes in central east Asia: the enigmatic Mongol-Okhotsk foldbelt, EOS, Trans. AGU, 67 (16), 363, 1986.

Kovalenko, V. I., M. I. Kuz'min and V. S. Antipin, Mesozoic magmatism of the Mongol-Okhotsk belt and its possible geodynamic interpretation, Int. Geol. Rev., 26 (9), 1021-1035, 1984.

Lee, G., J. Besse and V. Courtillot, Geodynamic evolution of the Korean peninsula in the last 120 million years, EOS, Trans. AGU, 64 (45), 871, 1983.

Lee, D. S., An ophiolitic suite in the Ogcheon geosynclinal zone, Korea, Abstracts, 26th Int Geol. Cong. (Paris), v. 1, 61, 1980.

Li, C. Y. et al., Preliminary study of the plate tectonics of China (in Chinese), Bull. Chin. Acad. Geol. Sci., 1, (2), 11-22, 1980.

Li, C. Y. et al., Explanatory notes to the tectonic map of Asia, Res. Inst. Geol., Chin. Acad. Geol. Sci., Cartographic Publishing House, Beijing, China, 45pp., 1982.

Li, C. Y. and Y. Tang, Some problems on subdividing palaeo-plates in Asia (in Chinese), Acta Geol. Sin., 57 (1), 1-10, 1983.

Li, J. and S. Zhang, Discovery of Muli mélange in western Sichuan province (in Chinese), Sci. Geol. Sin., no. 3, 209-210, 1985.

Li, S. J. and X. M. Xiao, Ashgillian limestone with trilobites discovered from the Triassic strata in northwestern Sichuan and their relation to plate tectonic movements (in Chinese), Acta Paleont. Sin., 23 (6), 742-754, 1984.

Liew, T. C. and R. W. Page, U-Pb zircon dating of granitoid plutons from the west coast of Peninsular Malaysia, J. Geol. Soc. London, 142, 515-526, 1985.

Lin, J. L., APW paths for the North and South China Blocks (in Chinese), Seismol. Geol., 7 (1), 81-83, 1985.

Lin, J. L., M. Fuller and W. Y. Zhang, Preliminary Phanerozoic polar wander paths for the North and South China Blocks, Nature, 313, 444-449, 1985.

Liu, C. and J. Chai, Numerical classification of Mesozoic volcanic rocks of eastern part of China and its geological significance, Developments in Geoscience, Science Press, Beijing, China, 353-362, 1985.

Liu, Y. and Z. Kong, A prospect of oil and gas in the overthrust belt on western border of Orduosi basin (in Chinese), Pet. Expl. Dev., 11 (1), 33-40, 1984.

Mazarovich, A. O., Tectonic development of south Primor'ye in Paleozoic and early Mesozoic (in Russian), Trans. Akad. Nauk. SSSR, 392, Nauka Publishing, Moscow, 101pp., 1985.

McElhinny, M. W. et al., 1981, Fragmentation of Asia in the Permian, Nature, 293, 212-216, 1981.

Meyerhoff, A. A. and J.-O. Willums, Petroleum geology and industry in the People's Republic of China, United Nations E. S. C. A. P., C.C.O.P. Tech. Bull., 10, 212pp., 1976.

Misnik, Yu. F. and V. V. Shevchuk, Late Mesozoic serpentinite mélange in the Shilka zone, eastern Transbaikal, Dokl. Akad. Nauk. SSSR Ear. Sci. Sect. Engl. Trans., 232 (6), 131-136, 1977.

Mitchell, A. H. G., Phanerozoic plate boundaries in mainland southeast Asia, the Himalayas and Tibet, J. Geol. Soc. London, 138, 109-122, 1981.

Molnar, P. and P. Tapponnier, Cenozoic tectonics of Asia: effects of a continental collision, Nature, 189, 419-425, 1975.

Molnar, P. and W.-P. Chen, Evidence for large Cainozoic crustal shortening in Asia, Nature, 273, 218-220, 1978.

Molnar, P. and P. Tapponnier, Active tectonics of Tibet, J. Geophys. Res., 83 (B11), 5361-5375, 1978.

Nagibina, M. S., Mesozoic structures of east Mongolia (in Russian), in Tectonics of the Mesozoic Orogenic Belt of Central Asia, K. V. Bogolepov (editor), Nauka Publishing, Novosibirsk, 169-174, 1973.

Nagibina, M. S. et al., The main features of stratigraphy and history of development of Mongolian Mesozoic structures (in Russian), Trans. Jt. Sov.-Mong. Sci. Res. Geol. Exp., 22, 76-92, 1977.

Nagibina, M. S., Evolution of orogenesis of

Mongolia in the Mesozoic (in Russian), in A. V. Peive and Yu. M. Pushcharovskiy (editors), Problems of Earth Crust Tectonics, Nauka Publishing, Moscow, 194-206, 1981.

Opdyke, N. D. et al., Paleomagnetism of the Lower Triassic Yelang Formation, Guizhou, People's Republic of China, EOS Trans. AGU, 65 (45), 862, 1984.

Opdyke, N. D. et al., Paleomagnetic results from the Triassic of the Yangtze platform, J. Geophys. Res., 91, (B9), 9553-9568 1986.

Panjasawathong, Y. and W. Yaowanoiyothin, Igneous rocks of Thailand: a review of plutonic rocks with intermediate-ultrabasic compositions and volcanic rocks, in Proceedings, Workshop on Stratigraphic Correlation of Thailand and Malaysia, Haad Yai, Thailand, v. 1, 233-243, 1983.

Parfenov, L. M. et al., Tectonics of the eastern Mongol-Okhotsk fold system and its relationship to the Mesozoic tectono-magmatic activation zone of the Aldan-Stanovoi belt (in Russian), Tikh. Geol., no. 2, 26-33, 1983.

Parfenov, L. M. and B. A. Natal'in, Mesozoic accretion and collision tectonics of north-eastern Asia, in Tectonostratigraphic Terranes of the Circum-Pacific Region, D. G. Howell (editor), Circum-Pacific Council for Energy and Mineral Resources, Earth Science Series, 1, 363-373, 1985.

Park, B.-Y. and C.-S. So, The Ogcheon system in the central part of the southern Korean peninsula as an ancient island arc, J. Geol. Soc. Korea, 8 (3), 156-161, 1972.

Reedman, A. J. and S. H. Um, The Geology of Korea, Geological and Mineral Institute of Korea, Seoul, Korea, 131pp., 1975.

Ridd, M. F., Possible Palaeozoic drift of SE Asia and Triassic collision with China, J. Geol. Soc. London, 132, 635-640, 1980.

Roy, S. S., Eastern Tethys and microplates framed in Himalayan, central and southeast Asian geology, in Geology and Mineral Resources of Southeast Asia, Proceedings, Third Reg. Conf. Geol. Min. Res. SE Asia, Bangkok, P. Nutalaya (editor), Asian Inst. Tech., Bangkok, 165-172, 1978.

Schermer, E. R., Tectonostratigraphic terranes of China and Mongolia, in Proceedings, Second Circum-Pacific Terrane Conference, D. G. Howell, D. L. Jones, A. Cox and A. Nur (editors), Stanford Univ. Pub. Ear. Sci. 17, 175-179, 1984.

Sengor, A. M. C., The Cimmeride orogenic system and tectonics of Eurasia, Geol. Soc. Am. Spec. Pap., 195, 82pp., 1984.

Stauffer, P. H., Unravelling the mosaic of Paleozoic crustal blocks sin SE Asia, Geol. Rund., 72 (3), 1061-1080, 1983.

Sun, H., The evolution of the basement of the Yangzi faulting block (in Chinese), Sci. Geol. Sin., no. 4, 334-341, 1985.

Takahashi, M., Space-time distribution of Late Mesozoic to Early Cenozoic magmatism in east Asia and its tectonic implications, in Accretion Tectonics in the Circum-Pacific Regions, M. Hashimoto and S. Uyeda (editors), Terra Publishing, Tokyo, 69-90, 1983.

Tang, Y. and Y. Feng, Some advanves in plate tectonics of China (in Chinese), Bull. Chin. Acad. Geol. Sci., no. 10, 49-60, 1984.

Tapponnier, P. and P. Molnar, Slip-line field theory and large scale continental tectonics, Nature, 264, 319-324, 1976.

Tapponnier, P. and P. Molnar, Active faulting and tectonics of China, J. Geophys. Res., 82 (20), 2905-2930, 1977.

Tapponnier, P. and P. Molnar, Active faulting and Cenozoic tectonics of the Tien Shan, Mongolia and Baykal regions, J. Geophys. Res., 84 (B7), 3425-3459, 1979.

Tapponnier, P. et al., Propagating extrusion tectonics in Asia: new insights from simple experiments with plasticine, Geology, 10, 611-616, 1982.

Terman, M. J., The last 200 million years in eastern Asia: Yanshanian subduction and post-Yanshanian extension, in Proceedings, Second Circum-Pacific Terrane Conference, D. G. Howell, D. L. Jones, A. Cox and A. Nur (editors), Stanford Univ. Pub. Ear. Sci., 17, 192, 1984.

Wang, C. and X. Zhang, On the stratigraphical age of the Linxi formation (in Chinese), Oil Gas Geol., 5 (4), 403-407, 1984.

Wang. D., Meso-Cenozoic rift volcanic zone in eastern China (in Chinese), J. Changchun Coll. Geol., 41 (3), 49-56, 1985.

Wang, Q., Recognition of the suture between the Sino-Korean and Siberian paleoplates in the middle part of Inner Mongolia (in Chinese), Acta Geol. Sin., 60 (1), 31-43, 1986.

Wang, Y. G.. et al., An outline of the marine Triassic in China, Pub. Int. Un. Geol. Sci., no. 7, E. D. Tozer (editor), 21pp., 1981.

Wu, F. T. et al., On the activity of the Tancheng-Lujiang fault zone in China (in Chinese), Seismol. Geol., 3 (4), 15-26, 1981.

Wu, H., The glaucophane schists of eastern Qinling and northern Qilian mountains in China (in Chinese), Acta Geol. Sin., 54 (3), 195-207, 1980.

Xiang, D. et al., The origin of intermediate-acid intrusive rocks in west Qinling with special reference to their petrochemical characteristics (in Chinese), Geochimica, no. 1, 66-75, 1984.

Xu, J., Some advances in the study of the great horizontal displacement along the northern Tancheng-Lujiang fracture zone (in Chinese), Geol. Rev., 31 (1), 83-86, 1985.

Xu, J. et al., Mesozoic sinistral transcurrent faulting along the continental margin in east Asia (in Chinese), Mar. Geol. Quat. Geol., no. 2, 51-64, 1985.

Yang, C. et al., The Hingshan-Balihan fault zone, Inner Mongolia - its geologic features and seismicity (in Chinese), J. Seis. Res., 7 (4), 391-398, 1984.

Yang, T. et al., New evidence for the presence of old basement in the coastal regions of Fujian (in Chinese), Geotect. et Metall., 9 (4), 372, 1985.

Yang, Z. and Y. Chen, Discussion on the relation between the development of paleotectonics of the peninsula of Korea, southwestern Japan and eastern China and contiguous seas (in Chinese), Mar. Geol. Quat. Geol., 3 (1), 85-97, 1983.

Yin, H., Triassic paleobiogeography and the continental drift (in Chinese), Sci. Geol. Sin., no. 3, 265-278, 1980.

Zhang, Q., The sedimentary features of the flysch formation of the Xikang group in the Indosinian Songban-Ganze geosyncline and its geotectonic setting (in Chinese), Geol. Rev., 27 (5), 405-412, 1981.

Zhang, Q. and Y. Zhu, The Paleozoic ophiolite in the northern east-Qinling (in Chinese), J. Changchun Coll. Geol., 37 (3), 1-13, 1984.

Zhang, W. et al., Evaluation of new paleomagnetic data from the Phanerozoic of the Yangtze platform, EOS, Trans. AGU, 65 (45), 863, 1984.

Zhang, Z. M. et al., An outline of the plate tectonics of China, Geol. Soc. Am. Bull., 95 (3), 295-312, 1984.

Zhang, Z. M. et al., An outline of the plate tectonics of China: Discussion and reply: Reply, Geol. Soc. Am. Bull., 96 (3), 408, 1985.

Zhang, Z. Z. et al., The Tongbaishan-Dabieshan paired metamorphic belt in Henan and Hubei provinces (in Chinese), Seismol. Geol., 3 (4), 27-36, 1981.

Zhao, M. and P. Zhang, Plate tectonics of Fujian province (in Chinese), Acta Geol. Sin., 57 (4), 369-378, 1983.

Zhao, X. and R. S. Coe, Preliminary results of a paleomagnetic study of Paleozoic rock units from eastern China, EOS, Trans. AGU, 65 (45), 863, 1984.

Zhao, Y. et al., A note on the Longmenshan nappe (in Chinese), Oil Gas Geol., 6 (4), 356-368, 1985.

Zhou, F., The relation of the Tancheng-Lujiang fault with the main faults in the southern part of northeast China (in Chinese), Seismol. Geol., 7 (2), 1-9, 1985.

Zonenshain, L. P., Ophiolites of Mongolia (in Russian), Trans. Jt. Sov.-Mong. Sci. Res. Geol. Exp., 22, 26-37, 1977.

THE HIGH P/T METAMORPHIC ROCKS OF CHINA

Zh. M. Zhang[1] and J. G. Liou

Department of Geology, Stanford University, Stanford, CA 94305

Abstract. Blueschists and/or C-type eclogites have been found in almost all the accretionary fold belts and major sutures of China, including the Circum-Pacific and Tethys-Himalayan belts and those between the Precambrian cratons. The ages of these high P/T metamorphic rocks range from Early Paleozoic to late Cenozoic. The studies of the high P/T metamorphic rocks have concentrated on the easily identified exotic blocks within either ophiolitic (or serpentinite) or olistostromic melanges. Little work has examined the possibility of more widespread high P/T metamorphism of clastic rocks. A majority of the blueschists and C-type eclogites have undergone multiple-stage metamorphism, commonly with increasing temperature and/or decreasing pressure at the later events. The mineral assemblages and chemical compositions of sodic and sodic-calcic amphiboles, garnet and clinopyroxene in some blueschists and C-type eclogites are discussed. The sodic amphiboles are commonly replaced by other amphiboles, but the reverse alteration occurs in some samples. The P/T conditions of blueschist and eclogite recrystallization are estimated for a number of areas based on the clinopyroxene-garnet geothermometer, phengite geobarometer, and mineral phase equilibria. The protoliths of the high P/T metamorphic rocks are mainly mafic volcanics and sedimentary rocks, with subordinate siliceous and carbonate rocks. However, studies of high P/T metamorphic rocks so far are very preliminary in nature. The metamorphic maps and information about metamorphic facies series are available in a few areas where more studies have been done. It is hoped that this synthesis will stimulate further investigation, which in turn will provide better interpretations of the tectonics of China.

Introduction

Blueschists have been described in regional geological reports of Tienshan, the Qilian Mountains and other areas of China during the late 1950's and the early 1960's. They were thought to be products of normal regional metamorphism; yet their tectonic significance has not been fully realized. The studies of high P/T metamorphic rocks in China was not initiated until the early 1970's when the concept of plate tectonics was accepted. Since then, many high P/T metamorphic rocks have been discovered and described [Xiao et al., 1974; Liou et al., 1975; Zhang and Jin, 1979; Ye et al., 1980; Zhang et al., 1980; Wu, 1980, 1982; Duan and Chao, 1981; Wang and Liu, 1981; Wang and Shi, 1981; Chi et al,, 1981; Ying et, 1981; Zhou and Lin, 1982; Peng and Lue, 1982; Xiao and Gao, 1982; Xiao, 1983; Hu, 1983; Yan and Tang, 1984]. However, the shortcomings of these studies are similar to those early studies of blueschists in California where initial attention was focused on only the high-grade coarse-grained blueschist (or eclogite) blocks and the matrix assemblage jadeite-lawsonite-aragonite was not discovered until years later.

In spite of the reconnaissance nature of these studies on the high P/T metamorphic rocks in China, they show that: (1) the high P/T metamorphic rocks are distributed along major fold belts in a similar pattern to other convergent zones around the world; and (2) many of the blueschists are Paleozoic in age and are confined well within the regions that separate the Precambrian cratons; hence their genesis may not be related to those in Circum-Pacific and Tethys-Himalayan zones.

The purposes of this paper are to synthesize and review the available publications on high P/T metamorphic rocks in China (most of them are in Chinese) and to present observations and new analytical data on some Chinese blueschists. We hope this synthesis will provide a better understanding of the complexity of Chinese tectonics and stimulate further investigations.

Plate Tectonic Evolution of China

The tectonic framework of China was recently described in detail by Zhang et al. [1984]. Figure 1 shows the distribution of the high P/T metamorphic rocks of China; it is a new version of the map in the 1984 paper. As a guide to Figure 1, the names of the major tectonic units of China are listed in Table 1.

China is principally a part of the Eurasian plate, but the margins of the Indian and Philippine Sea plates are involved respectively in the areas from the Himalayas to the Yarlung Zangpo-Indus suture and in the Coast Range of Taiwan. These two sutures are still active and regarded as sections of the actual plate boundaries. Within the Chinese part of the Eurasian plate, the Cathaysian paleoplate is separated from the Angaraian paleoplate by the Junggar-Hegen suture, which came into being at the end of Paleozoic as demonstrated by the latest paleomagnetic data [Lin et al., 1985]. The Angaraian paleoplate has developed with the Siberian craton as its nucleus since the Late Proterozoic. The nuclei of the Cathaysian paleoplate consist of the Tarim craton and the Sino-Korean craton. The amalgamation of these cratons and accretionary fold belts was completed by the end of Paleozoic. The Yangtze craton was the nucleus of the Southeast China microplate. The Qinling fold belt welded the Tarim-Sino-Korean microplate and the Southeast China microplate together at the end of Late Paleozoic or earlier. A combined supercontinent, the so-called paleo-Eurasian which embraces the Tarim-Sino-Korean microplate, the Southeast China microplate and the Angaraian paleoplate, had come into being at least since the end of Paleozoic.

In the southwest China, a series of younger accretionary fold belts attached themselves to the existing paleo-Eurasian supercon-

[1]Present address: British Petroleum Development Ltd., P. O. Box 509, Guangzhou, China)

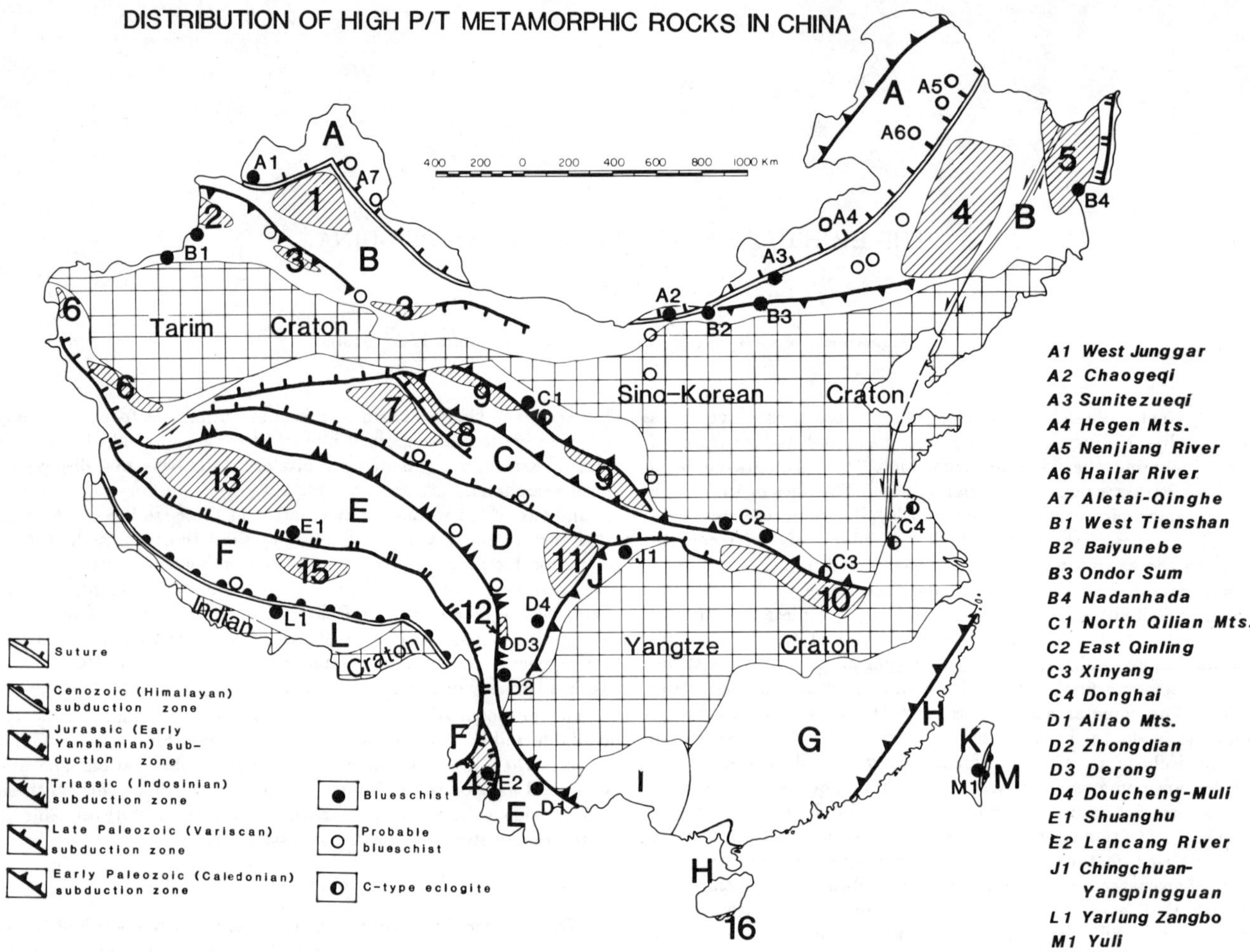

Fig. 1. Distribution of high P/T metamorphic rocks in China. Four cratons and localities, where high P/T metamorphic rocks occur, are shown on the map. For names of the major tectonic units of China see Table 1.

tinent during the Mesozoic. Many of the earlier microcontinents dispersed from the Yangtze craton or the Gondwana supercontinent during the Mesozoic were incorporated into the aforementioned accretionary fold belts. Later collision of the Indian plate and the Eurasian continent occurred during the Late Cretaceous-Paleogene.

From the distribution of blueschists and eclogites as shown in Figure 1, it is apparent that all of them are spatially related to the above-mentioned Phanerozoic accretionary fold belts or sutures.

Junggar-Xingan Accretionary Fold Belt

The West Junggar

The West Junggar is the westernmost section of the Junggar-Xingan accretionary fold belt and shows a northeast trend contrary to the other sections of this fold belt (Al in Figure 1). It consists mainly of Late Paleozoic volcano-sedimentary rocks with a few patches of Early Paleozoic complexes in the southern part. The blueschists occur as blocks in a melange belt to the north of Aibi Lake in the southern part of West Junggar. The melange belt trends E-W and locally WNW-ESE for about 80 km and is 8 km wide (Figure 2). The blueschist blocks range in size from 1-2 cm to tens of meters across; the biggest one is about 180 m long and 60 m wide. The weakly metamorphosed rocks in the melange belt consist of large and small blocks as well as finer grained clasts of spilite, variolitic pillow basalt, mafic pyroclastic, radiolarian chert, and minor turbidite and silty limestone. The radiolarians in the chert are reported to be Middle Ordovician [Feng et al., 1983; Wu, 1984] or Early Paleozoic in age [Wu, 1984]. Blocks of a dismembered ophiolite suite consisting of harzburgite, lherzolite, dunite, cumulate gabbro and diabase also occur in the ophiolitic melange. The ultramafics are partially to completely serpentinized. The blueschist blocks and especially the matrix rocks of the ophiolitic melange are severely deformed, twisted, broken, mixed and pervasively foliated.

The blueschist mineral assemblage includes combinations of sodic amphibole, sodic-calcic amphibole, epidote, albite, actinolite-tremolite, chlorite, and sphene with lesser amounts of phengite, calcite, clinozoisite and magnetite. Sodic amphibole compositions lie close to the glaucophane, crossite and ferro-glaucophane adjacent fields (A1 in Figure 3A) and the Na-Ca amphiboles are ferrowinchites (Figure 3C). Some of the blueschist blocks appear to be a

TABLE 1. Summary of Major Tectonic Units of China

Name of plate, paleoplate, and fold belt	Designation in Figures	
EURASIAN PLATE		
Angaraian paleoplate		
Junggar-Xingan accretionary fold belt	A	
Cathaysian paleoplate		
(1) Tarim-Sino-Korean microplate		
Tarim craton		
Sino-Korean craton		
Tienshan-Tumen accretionary fold belt	B	
Junggar microcontinent		1
Yili microcontinent		2
Central Tienshan microcontinent		3
Songliao microcontinent		4
Jiamusi microcontinent		5
Kunlun-Qinling accretionary fold belt	C	
Central Kunlun microcontinent		6
Chaidam microcontinent		7
Oulungbruk microcontinent		8
Central Qilian microcontinent		9
Huaiyang microcontinent		10
(2) Southwest China microplate		
Kekexili-Jinsha-Ailao accretionary fold belt	D	
Songpan microcontinent		11
Zhongza microcontinent		12
North Tibet-West Yunnan accretionary fold belt	E	
Qiangtang microcontinent		13
Gangdesi-Nienching Tangla accretionary fold belt	F	
Baoshan microcontinent		14
Namu Lake microcontinent		15
(3) Southeast China microplate		
Yangtze craton		
South China accretionary fold belt		
Maritime accretionary fold belt	H	
South China Sea microcontinent		16
Youjiang back-arc basin	I	
Motienling accretionary fold belt	J	
West Taiwan inland basin	K	
INDIAN PLATE		
Indian craton		
Yarlung Zangbo Indus accretionary fold belt	L	
PHILIPPINE SEA PLATE		
East Taiwan accretionary fold belt	M	

breccia that consists of fragments and matrix with different mineral assemblages and grain sizes. The amphibole compositions of the matrix vary more widely than those of the fragments (Figure 4). Thus, the fragments appear to be more stable during the overprinted greenschist facies metamorphism than the matrix. The protoliths of the blueschists are mainly mafic and intermediate volcanics.

The genesis of the blueschist was probably related to a Paleozoic collision between the West Junggar island arc on one side and the Junggar microcontinent and the Tienshan accretionary fold belt on the other side, as Paleozoic volcano-sedimentary formations and granitic intrusions are common in the West Junggar. As glaucophane has been found in the Lower Silurians basal conglomerate, the high P/T metamorphism is thought to be Early Paleozoic in age [Guo, 1983]. However, it is essential that more blueschists are carefully mapped and age dated before final conclusions can be made.

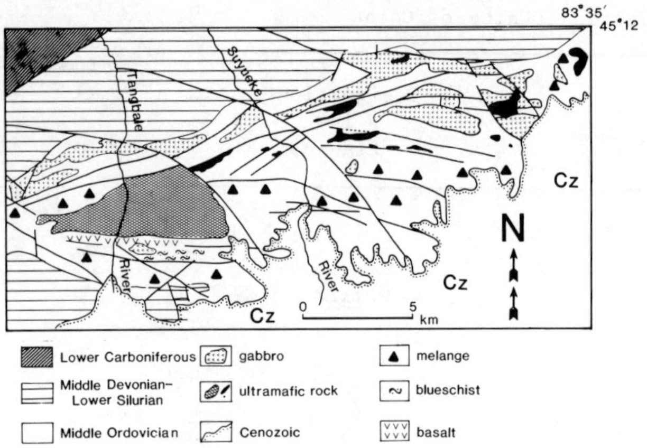

Fig. 2. Schematic geologic map of the southernmost part of West Junggar, showing the distribution of melange, blueschist and some of the larger fault-bounded slices of ultramafic and mafic rocks.

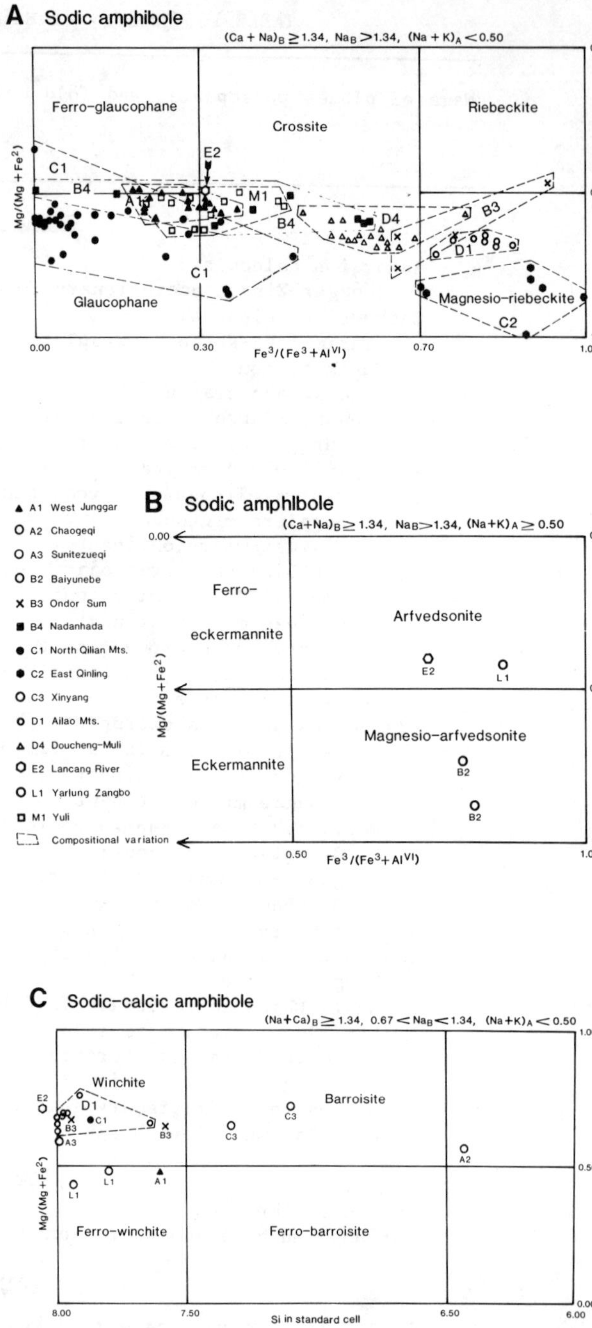

Fig. 3. The Na- and Na-Ca amphiboles in the blueschists and C-type eclogites of China. (A) sodic amphibole with Na + K in A-site less than 0.50, (B) sodic amphibole with Na + K in A-site equal or more than 0.50, (C) Sodic-calcic amphibole with Na + K in A-site less than 0.50. The plot is based on microprobe data. All the microprobe data are recalculated for ferric iron by MINFORM program [Baldwin and White, unpublished data] based on 23 oxygens and constrained by (1) the tetrahedral site $= 8 \pm 0.02$, (2) the octahedral site > 4.98, (3) the extra octahedral cations + Ca < 2.02, (4) M4 site $= 2 \pm 0.02$, (5) A-site < 1.02, and (6) the residual < 0.02.

The Suolun Mountains and Hegen Mountains

In the Inner Mongolia, blueschists have been found in the areas to the north of the Chaogeqi close to the China-Mongolia border and to the south of the Sunitezueqi area (A2 and A3 in Figure 1,5) along the Suolun-Hegen fault. In these two areas, blueschists occur as blocks or fault slices together with dismembered ophiolite suite and fault-bounded strata of Devonian, Carboniferous and Permian age. The country rocks have been metamorphosed under greenschist facies and amphibolite facies; the latter is characterized by kyanite, sillimanite, staurolite and garnet, with metamorphic ages of 298-355 m.y. in Chaogeqi area and 463-473 m.y. in Sunitezueqi area [Hu, 1983]. All the blocks and the matrix are strongly deformed and mixed, and clear cataclastic and mylonitic structures are commonly visible. The whole area is a tectonic melange zone consisting of both ophiolitic (or serpentinite) and olistostromic melanges, where blueschist occurs.

The major mineral assemblages are blue amphibole-albite-quartz-sericite-chlorite and blue amphibole-epidote-chlorite-albite-sericite. Although amphiboles from these areas are closer to sodic-calcic amphibole (winchite and barroisite) [Figure 3C, Hu, 1983], the existence of high P/T sodic amphibole is confirmed by X-ray powder diffraction data and optical properties [Chi et al., 1981; Hu, 1983]. Paleozoic granitic intrusions and associated low P/T metamorphic rocks occur to the north of the blueschist belt. This paired metamorphic belt reveals the last and northward subduction of the Late Paleozoic ocean crust between the Siberian craton in the north and the Tarim-Sino-Korean craton in the south; the Junggar-Hegen fault marks the suture between two cratons at the end of Late Paleozoic.

Further to the ENE in the Paleozoic strata of the Hegen Mountains (A4 in Figure 1,5) and along the eastern slope of the Great Xingan Range in the upper course of the Nenjiang River (A5 in Figure 1), blueschists have also been reported [Sobolev, 1982], but no data are available. To the north of the Hailar River (A6 in Figure 1), glaucophane, barroisite, jadeitic clinopyroxene and lawsonite have been found [Ma, 1983], but no further data are presented. In the western section of the Junggar-Xingan accretionary fold belt along the Altai Mountains from Aletai County to Qinghe County (A7 in Figure 1), blueschist facies metamorphism was shown on the metamorphic map of Asia [Zwart et al., 1978], but no description is available.

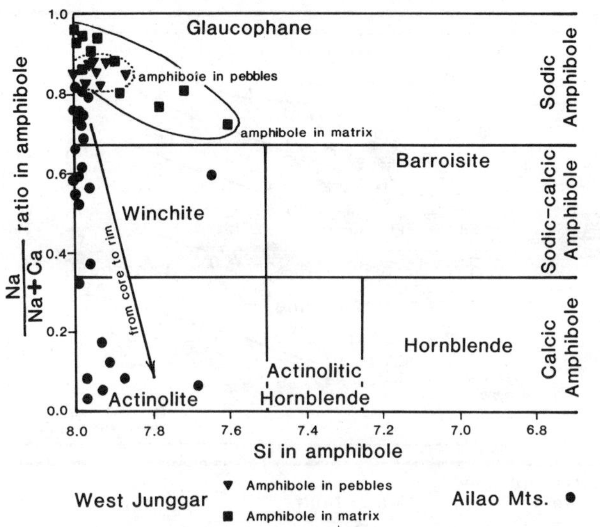

Fig. 4. Variation of amphibole composition in West Junggar and Ailao Mountains. Amphiboles in matrix in West Junggar show larger compositional variation than that of the pebbles. The amphibole composition changes from magnesio-riebeckite through winchite to actinolite from core to rim in blueschists of the Ailao Mountains.

Tienshan-Tumen Accretionary Fold Belt

The West Tienshan

Blueschists occur sporadically in the western section of the South Tienshan (B1 in Figure 1) along the Hantenggeli fault and Nalati fault for more than 200 km [Wu, 1982]. They are medium- to coarse-grained and contain mineral assemblages of garnet-glaucophane- phengite- quartz- rutile, epidote- garnet-glaucophane-quartz- sphene- tourmaline, chlorite- epidote- glaucophane- quartz-albite, and barroisite-phengite-sphene-quartz-albite. Their protoliths are mainly mafic volcanics in addition to some graywacke, carbonate and argillaceous rocks of Late Silurian age. Except for the blueschists, regional metamorphism ranges from the greenschist to amphibolite facies. A northward Paleozoic subduction is thought to occur along the southern margin of the Yili microcontinent (no. 2 in Figure 1).

The Baiyunebe and Ondor Sum Areas

In the Baiyunebe and the Ondor Sum (Wendurmiao) areas of Inner Mongolia (B2 and B3 in Figure. 1 and 5), blueschists occur as either thin layers or veins in the upper part of the Lower Paleozoic or as exotic blocks in the associated melange [Chi et al., 1981; Hu, 1983; Tang, 1983; Yan and Tang, 1984]. These blueschists occur in a zone trending ENE for more than 50 km. The country rocks were subjected to low to medium grade metamorphism but cataclastic metamorphism was also pervasive. The greenschist facies matrix of the melange is mixed with dismembered ophiolite suite, blocks of greywacke, limestone and cataclastic granite as well as blueschist lenses and blocks. However, in some areas, the surrounding matrix around the blueschist blocks is serpentinite. The assemblages of high P/T rocks depend on the protolith: crossite-epidote-actinolite-chlorite-albite with or without clinozoisite, calcite and quartz for metabasites, crossite-calcite-dolomite-quartz for metacarbonates, and crossite-quartz-magnetite-hematite-stilpnomelane-deerite for minor meta-ironstones [Hu, 1983]. The amphiboles in the Ondor Sum area include crossite, magnesio-riebeckite, riebeckite and winchite (Figure 3A, 3C) and the Baiyunebe area is characterized by high K-bearing magnesio-arfvedsonite (Figure 3B). The high potassium content might be related to the anomalous regional geochemistry which is also responsible for the genesis of the unusal iron deposits rich in rare earth elements [Qiu et al., 1981]. A recent study by Yan and Tang [1984] demonstrated the existence of lawsonite, aragonite and deerite in the Ondor Sum area. The glaucophane from the Ondor Sum yields an $^{39}Ar/^{40}Ar$ age of 426 ±15 m.y. which is comparable with the metamorphic ages of the country rocks at 489 and 435 m.y. [Hu, 1983]. The hornblende in the amphibolite is determined to be 458 and 434 m.y. [Gao, 1984]. The Early Paleozoic fossils and the metamorphic ages of the Ondor Sum Group reveal that the blueschist metamorphism was related to Early Paleozoic subduction.

The Heilungjing Group

Some blueschists (B4 in Figure 1) occur in the westernmost margin of the Nadanhada fold belt which extends northward to USSR as the Sikhote Alin accretionary fold belt. The Heilungjiang Group consists mainly of Precambrian pelitic schists and magnesian marble. These rocks have been subjected to regional greenschist- to amphibolite-facies metamorphism but the earlier blueschist minerals partially survive. The blueschist shows schistosity wrapping around euhedral to subhedral epidote and albite porphyroblasts. Glaucophane, crossite and chlorite predominate in the matrix together with minor sericite, sphene and hematite. Blue amphiboles are also enclosed in the albite and epidote porphyroblasts. Crossite and albite intergrowths are too fine-grained to be analyzed. The mineral assemblage of the blueschist reflects a mafic volcanic protolith. From these facts, the eastern part of the Heilungjing Group might not be Precambrian but a Mesozoic blueschist-bearing melange.

Kunlun-Qinling Accretionary Fold Belt

The North Qilian

In the North Qilian Mountains, tens of kilometers of volcanic, carbonate, and clastic rocks that contain abundant Early Paleozoic

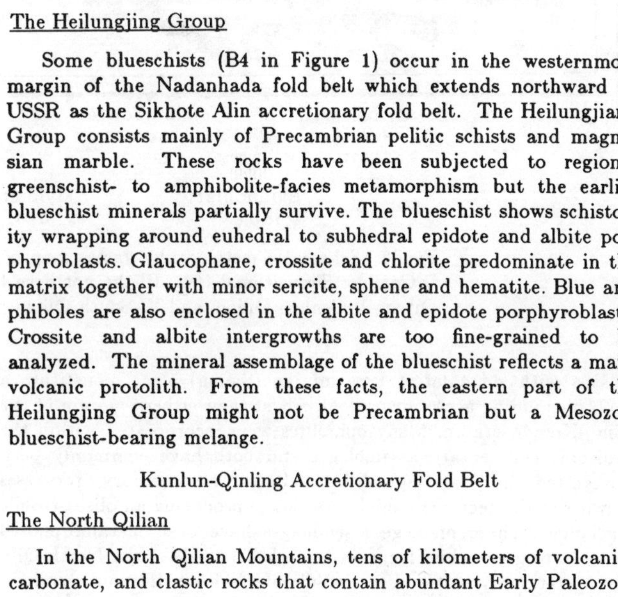

Fig. 5. Schematic geologic map of the middle part of Inner Mongolia, showing distribution of melange and blueschist.

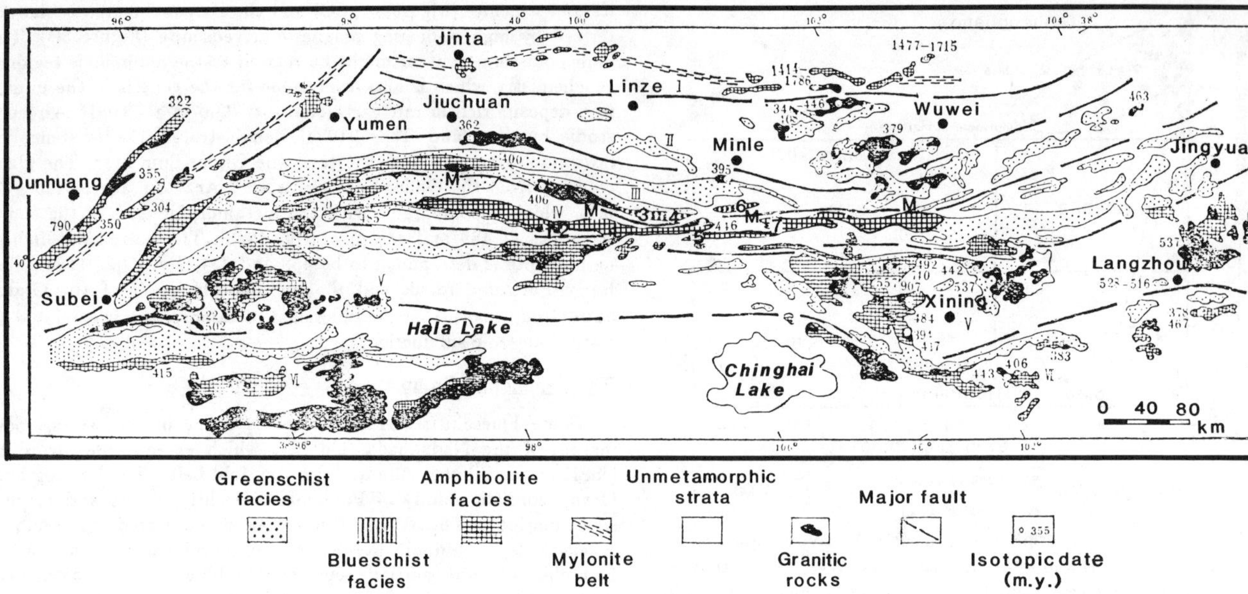

Fig. 6. Schematic metamorphic facies map of North Qilian Mountains. I-Alashan mylonite belt, II-Corridor Transitional Belt, III-the northern belt of North Qilian, IV-the southern belt of North Qilian, V-Central Qilian, and VI-South Qilian [revised from Wang and Liu, 1981].

fossils with some latest Precambrian (Sinian) algae, constitute a WNW-trending accretionary fold belt at the southern margin of the Sino-Korean craton. Many ophiolites were incorporated within the volcanic-sedimentary assemblages and both have commonly been converted into an ophiolitic melange. Sedimentary processes reworked the tectonic ophiolitic melange producing an olistostromic melange. These melange assemblages have been metamorphosed mainly to greenschist facies in the northern belt of the North Qilian accretionary belt and to amphibolite facies in the southern belt. The two belts are separated by a major fault which is parallel to the regional structural lines. Devonian red beds and younger marine and continental sedimentary formations unconformably overlie these metamorphic rocks. In the northern part of the belt (C1 in Figure 1, 6), blueschists and C-type eclogites occur as blocks and fault slices in the olistostromic melange and have survived from the overprinted regional greenschist metamorphism. They are generally less than a hundred meters across and at most several hundreds meters thick. Blueschist blocks have not yet been found in the ophiolitic melange.

The mineral assemblages of blueschists vary. The metachert consists of glaucophane (or crossite)-quartz-white mica-garnet, with or without tourmaline. The metabasites contain glaucophane-chlorite-phengite-albite, with or without epidote (or zoisite), stilpnomelane, garnet and quartz. The mineral assemblage of dolomite-crossite or dolomite-crossite-quartz-epidote-green amphibole (with over 50-90 % of dolomite) reflects a magnesio-carbonate origin. Glaucophane is the major sodic amphibole but ferro-glaucophane and crossite are also identified (Figure 3A). Sodic-calcic amphibole is mainly winchite (Figure 3C). The eclogite consists mainly of omphacite with 35-46 % jadeite and almandine-rich garnet (see Figure 7,8) with minor glaucophane, stilpnomelane, epidote, phengite, quartz and albite. Overprinting greenschist facies metamorphism of blueschist and C-type eclogite is commonly seen. The protolith of the blueschist is mafic volcanics, clastic rocks, chert and dolomitic limestone. The $^{39}Ar/^{40}Ar$ dating of phengite from the blueschist gives age of 448.6 ±1.1 m.y. [Dallmeyer, personal communication, 1986]. It reveals a Late Ordovician metamorphism which is consistent with the age relationship of the regional geology. The regional structural vergence and the abundance of granitic intrusions to the north imply that Early Paleozoic subduction was north-directed under the Sino-Korean craton.

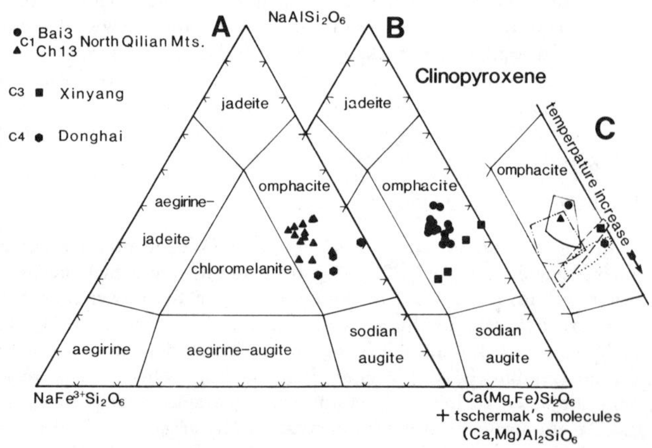

Fig. 7. Composition of clinopyroxene in eclogites of North Qilian Mountains (C1) (Bai3 = Baijingsi from the eastern section; Ch13 = Chingshuigou from the western section), Xinyang (C3) and Donghai (C4). (A) and (B) show the microprobe data. (C) The variation fields of omphacite from Baijingsi (Bai3) and Chingshuigou (Ch13) of North Qilian Mountains (C1), Xinyang (C3) and Donghai (C4). It reveals that the omphacite with lower jadeite content should form at a higher temperature.

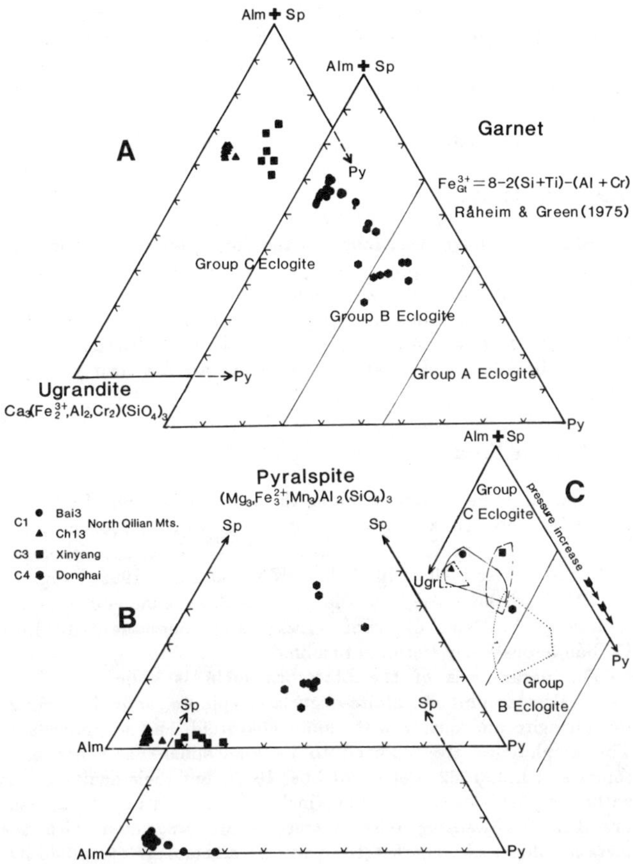

Fig. 8. Composition of garnet in eclogites of North Qilian Mountains (C1) (Bai3 = Baijingsi from the eastern section; Ch13 = Chingshuigou from the western section), Xinyang (C3) and Donghai (C4). (A) and (B) show the microprobe data. (C) The variation fields of garnet from Baijingsi (Bai3) and Chingshuigou (Ch13) of North Qilian Mountains (C1), Xinyang (C3) and Donghai (C4). It reveals that the garnet with higher pyrope content should form at a higher pressure.

The East Qinling

The blueschists in East Qinling (C2 in Figure 1) outcrop along two parallel WNW-ESE trending belts which mark the western section of the suture between the Sino-Korean craton in the north and the Yangtze craton in the south. The northern belt and the southern belt of blueschist are about 250 km and 100 km long respectively. Blueschists appear as exotic lenses, elongated blocks or boudins among the medium- to low-grade metamorphic blocks. The ages of the country rocks are thought to be Sinian (latest Precambrian), Silurian and Devonian. Cataclastic metamorphic structures are very common in these metamorphic complexes. The fault slices or blocks of metamorphosed ophiolitic suite can also be found in the metamorphic complexes.

In East Qinling, the metabasite consists of crossite-albite-chlorite-epidote-phengite-actinolite-quartz and magnesio-riebeckite (or crossite)-albite-chlorite-biotite-calcite. The crossite (or magnesio-riebeckite)-albite-quartz-aegirine-tourmaline assemblages of blueschist might reflect a volcanic origin [Wu, 1980]. The metacarbonate consists of crossite (or riebeckite)-calcite-albite-quartz-chloritoid. From the chemical data [Wu, 1980], only magnesio-riebeckite is confirmed to exist in East Qinling (Figure 3A). Whether glaucophane, crossite or sodic clinopyroxene are among the assemblages to prove that the blueschist facies metamorphism survived is open to question, although many authors [Wu, 1980, 1984, Gao, 1984] seem to believe their occurrence.

The Tongbue-Dabe Mountains

Along the northern slopes of the Tongbue Mountains and the Dabe Mountains, evidence of the high P/T metamorphism is not uncommon. The high P/T rocks form a east-west belt. Which is the eastern extension of the East Qinling belt. Many C-type eclogites (Xinyang area, C3 in Figure 1) and minor blueschists occur as blocks or lenses, which are several tens of meters to hundreds of meters across, along the belt presently considered to be Precambrian by most of the Chinese geologists. The eclogite consists mainly of omphacite and almandine-rich garnet (Figure 7,8) and minor phengite, barroisite, epidote, albite and quartz [Ye et al., 1980]. No sodic amphibole has been found in the C-type eclogite, but stilpnomelane, piedmontite, chlorite and pumpellyite are seen in small amounts in some blocks.

East of the Tancheng-Lujiang Fault

The Tancheng-Lujiang fault offsets the easternmost section of the suture between the Sino-Korean craton and the Yangtze craton in China. To the east of this fault, some eclogite blocks have been found in the Precambrian metamorphic complexes in Donghai County [Ying et al., 1981] (C4 in Figure 1). The eclogite contains garnet, omphacite (Figure 7), quartz, kyanite, hornblende, actinolite-tremolite, epidote, zoisite, chlorite, calcite, and white mica, and minor rutile, sphene, apatite, magnetite and ilmenite. The garnet composition lies close to the boundary between C-type and B-type eclogite field (Figure 8A), but mainly in the latter, and some garnets have higher spessartine content (Figure 8B). The mineral association of eclogite reveals that it might be originally C-type eclogite but overprinted by higher temperature metamorphism.

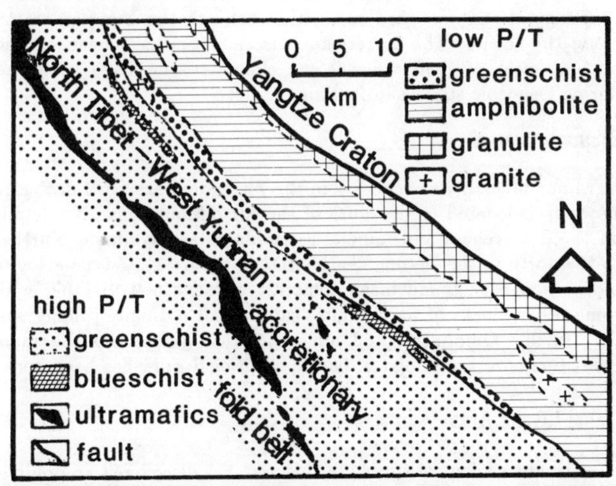

Fig. 9. Schematic metamorphic facies map of Ailao Mountains area, showing distribution of high P/T and low P/T metamorphic facies rocks [from Duan and Zhao, 1981].

Kekexili-Jinsha-Ailao Accretionary Fold Belt

The Ailao Mountains

Three closely spaced major faults running in NNW-SSE direction are located between the Yangtze Craton in the east and the North Tibet-West Yunnan accretionary fold belt in the west in the Ailao Mountains area of the southwestern China [Duan and Zhao, 1981] (D1 in Figure 1,9). The two elongated areas that are bounded by the three faults exhibit different tectonic features. The eastern strip seems to be part of the Yangtze craton because of its similarity in the Precambrian basement and the Ordovician to Triassic cover strata. The strip to the west consists of Paleozoic to Anisian slightly- metamorphosed marine deposits and calc-alkaline volcanics, which are unconformably overlain by Upper Triassic and younger red beds. The western strip is the relic of a broken-up Permo-Triassic volcanic arc along the southwestern border of the Yangtze craton. Its ophiolitic melange occurs as fault slices among the Paleozoic to Middle Triassic metamorphic rocks. The serpentinized dismembered harzburgite and dunite are mixed together with radiolarian chert, metabasite, sandstone, diabase, gabbro, marble and chromite bodies which have serpentine-chlorite slickensided rinds.

The eastern strip is characterized by low P/T metamorphism with parageneses of biotite-garnet, andalusite-staurolite-kyanite, sillimanite-alkali feldspar-cordierite, hypersthene-hornblende, clinopyroxene-garnet and wollastonite-chondrodite increasing in grade from the greenschist in the west through amphibolite in the middle to granulite facies in the east [Duan and Zhao, 1981]. The amphibole in the gneisses and the biotite in the schists give metamorphic ages of 171 and 135 m.y. [Wang et al., in preparation]. In the western strip, blueschists occur as blocks and fault slices within greenschists, and consist of magnesio-riebeckite, winchite (Figure 3A and 3C), stilpnomelane, sericite, epidote, chlorite, albite, potassium feldspar and phengite, with obvious mylonitic or cataclastic structure. From the microprobe data (Figure 4), it is apparent that magnesio-riebeckite changes through winchite to actinolite, probably due to a later thermal event. The protolith of the blueschist includes basaltic lava and pyroclastics.

The arrangement of the paired metamorphic belts, ophiolitic melange, volcano-sedimentary formations and regional tectonic setting suggests an eastward subduction beneath the Yangtze craton along the suture. The accretionary assemblages have been broken up and most of their volcanic arc characteristics have been lost during Cenozoic strike-slip faulting.

Zhongdian area

Minor blueschists occur within the Zhongdian area (D2 in Figure 1) which is located to the north of the Ailao Mountains. No details on the occurrences and mineral assemblages are available. Further to the north in the Derong County (D3 in Figure 1), C-type eclogite blocks had been found in an ophiolitic melange belt together with some other blocks of harzburgite, gabbro and rodingite [Zhang and Jin, 1979]. Omphacite and almandine-rich garnet are the main constituents of the eclogites, but no additional work has been done.

Ganzi-Litang Area

A recent geological survey in this area has discovered an ophiolitic melange that consists of ultramafic and mafic members of an ophiolite suite, and chert with Early Triassic radiolarians. Blueschist blocks in the melange are also reported but no research has yet been done. A sample from the southern extension of the belt in the Doucheng-Muli area (D4 in Figure 4) was studied by the authors. The mineral assemblage is sodic amphibole, epidote, phengite, albite, sphene and minor opaques. Sodic amphiboles are mainly crossite and minor magnesio-riebeckite (Figure 3A, 3C).

North Tibet-West Yunnan Accretionary Fold Belt

The North Tibet

Blueschists have been found in the Shuanghu area (E1 in Figure 1) of northern Tibet. From the data of the Tibet Geological Bureau, the blueschists in this region consist of glaucophane, barroisite, chlorite and quartz, with minor rutile and sphene. The blueschists are reported to be associated with Early Paleozoic rocks in fault slices but their modes of occurrence are not clear yet [Wu, 1984].

The Lancang River

A melange belt outcrops along the southern section of the Lancang River fault zone in a NW-SE direction for over 220 km (E2 in Figure 1, 10). The melange contains blocks of various lithologies and ages. These include blueschists [Zhou and Lin, 1982; Peng and Lue, 1982], dismembered ophiolite suite such as mafic volcanics, cumulate gabbro and ultramafic rocks, and limestone with Late Carboniferous-Early Permian fusulinids.

The paragenesis of the blueschist includes sodic amphibole, sodic-calcic amphibole, albite-oligoclase, epidote, actinolite, chlorite, phengite and quartz with minor chloritoid, and stilpnomelane. The amphiboles are reported to be glaucophane and barroisite [Zhou and Lin, 1982; Peng and Lue, 1982], but their analyses plot within the fields of crossite and winchite [Zhou and Lin, 1982] (Figure 3B,3C). Mesozoic felsic intrusives are associated with the metamorphic rocks characterized by the assemblage of andalusite, staurolite, cordierite and sillimanite. Whether these low P/T and high P/T metamorphic belts constitute a paired metamorphic belts and the subduction direction was toward the west is still open to question. Late Paleozoic-Triassic intrusives are widespread in this area and many sutures are attenuated by the later squeeze of the Assam "spur" of the Indian plate. It is difficult to determine which high P/T metamorphic belt matches with which low P/T metamorphic belt by the available data. A westward subduction beneath the Baoshan microcontinent is still possible as suggested by Zhang et al. [1984].

Motienling Accretionary Fold Belt

In the Chingchuan-Baishuijie-Yangpingguan area of the northeastern Motienling accretionary fold belt (J1 in Figure 1), blueschists have been found in the Bikou Gruop [Wu, 1984]. They occur in an assemblage of metamorphosed volcanic, siliceous and sedimentary rocks, unconformably overlain by unmetamorphosed Devonian strata. The geologic setting of blueschists has not been reported. The mineral assemblages consist of sodic amphibole-epidote- chlorite- albite- barroisite- stilpnomelane, garnet- sodic amphibole-quartz-deerite (?) and sodic amphibole-epidote-barroisite-albite-pumpellyite (?), with minor calcite and apatite. The sodic amphibole is reported to be mainly crossite and glaucophane with subordinate magnesio-riebeckite, but no chemical data are available. The blueschist commonly shows nematoblastic texture and cataclastic structure. This northeast-trending blueschist belt which extends more than 150 km marks the Early Paleozoic subduction along the northwestern margin of the Yangtze craton.

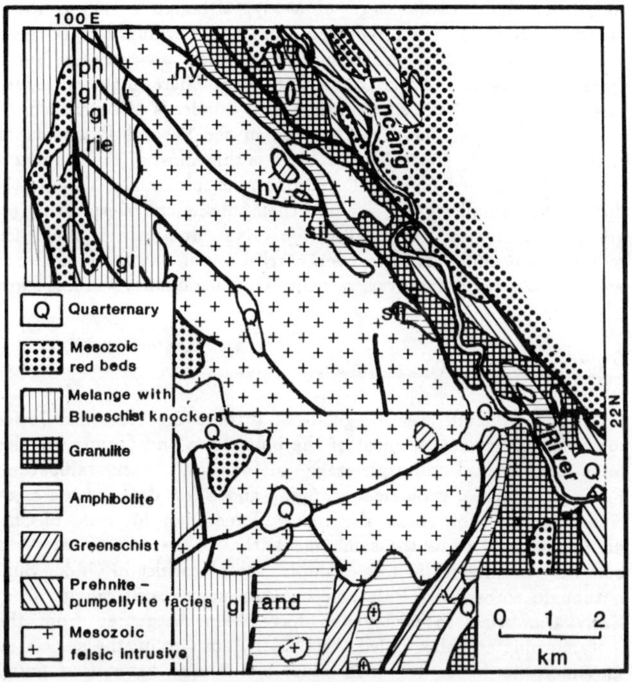

Fig. 10. Schematic metamorphic facies map of Lancang River area, showing distribution of blueschist. gl = glaucophane, rie = riebeckite, ph = phengite, and = andalusite, sil = sillimanite, hy = hypersthene [from Peng and Lue, 1983].

Yarlung Zangbo-Indus Accretionary Fold Belt

High P/T metamorphic rocks exist everywhere along the Yarlung Zangbo-Indus suture (L1 in Figure 1) [Jan and Kempe, 1973; Singh, 1977; Frank et al., 1977; Zhang et al., 1980; Xiao et al., 1983]. The blueschist belt extends for more than one thousand km from the Jhelum Syntaxis in the west to the Assam Syntaxis in the east and beyond, into Pakistan and Burma [Zhang, 1982]. North of the blueschist belt is an ophiolite belt, and south of the blueschist belt is a sedimentary melange (Figure 11). The blueschists occur commonly as blocks among ophiolitic melange but coherent beds and lenses of blueschist are incoporated within the neritic, fine-grained clastics of the Upper Jurassic to Lower Cretaceous and the turbidites of the Upper Triassic [Xiao et al., 1983]. The maximum width of a continuous blueschist outcrop is about 400 m near the section from Angren to Sangsang (Figure 11). In the vicinity of Jiajia and some other areas, blueschists occur as discontinuous blocks.

The high P/T minerals includes sodic amphibole and sodic-calcic amphibole (ferro-winchite), phengite, epidote, chlorite, actinolite, garnet, sericite, quartz, stilpnomelane, chloritoid and occasionally lawsonite, pumpellyite, prehnite, aragonite, calcite, and magnetite. The protolith of the high P/T metamorphics are mafic volcanics with subordinate siliceous and argillaceous rocks. In the middle section of the Yarlung Zangpo-Indus suture, the regional metamorphism to the south grades from the prehnite-pumpellyite through blueschist to greenschist facies. The corresponding low P/T metamorphic belt to the north of the suture is characterized by andalusite and sillimanite in association with the Gangdis magmatic arc of 50-100 m.y. The paleogeothermal gradient responsible for the high P/T metamorphism is estimated to be 10-14°C/km [Xiao and Gao, 1982].

East Taiwan Accretionary Fold Belt

The Dananao Complex of the Central Range represents the pre-Tertiary basement of the Taiwan Arc [Yen, 1963]. In this metamorphic complex, Permian fusulinids and corals have been found [Yen, 1953]. The western Tailuko belt of the low P/T type is charaterized by Permian and Lower Mesozoic pelitic schists, quartzofeldspathic gneisses, platform carbonate units, intercalated mafic

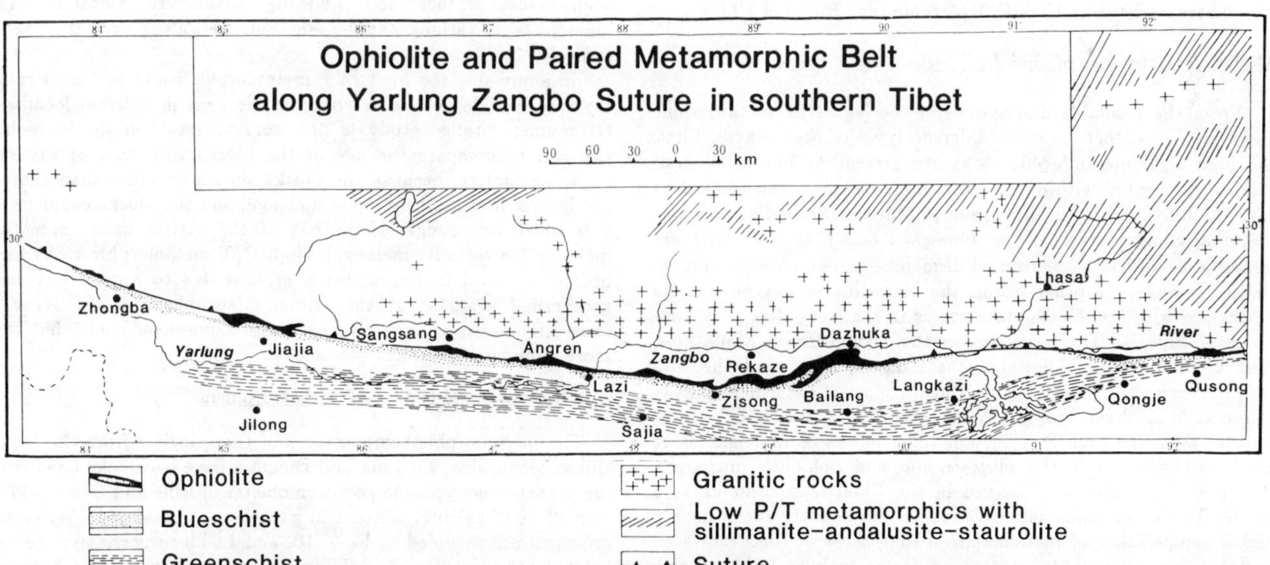

Fig. 11. Schematic geologic map of the Yarlung Zangbo suture, showing distribution of blueschist, granulite, amphibolite, and greenschist facies rocks.

and tuffaceous metasediments, greenschist and amphibolite, migmatite and S-type granitic rocks. East of the Tailuko belt lies a monotonous melange of pre-Cenozoic pelitic and mafic-ultramafic schists of oceanic affinity [Liou, 1981], collectively termed the Yuli belt. Rocks of the Yuli belt exhibit dominantly upper greenschist facies metamorphism, chiefly ascribable to the Plio-Pleistocene overprinting. However, the melange matrix and the enclosed ophiolitic tectonic blocks (associated with serpentinite lenses) attest to the late Cretaceous crystallization of relatively high-pressure barroisitic amphibolites and the late Miocene production of blueschists [Yen, 1966; Liou et al., 1975; Liou, 1981; Liou and Ernst, 1984]. It has been suggested that the two belts are separated by a fault [Yen, 1963], and the whole metamorphic complex of the Dananao Group is bounded on the east by the Longitudinal Valley Fault which marks the suture between the Philippine Sea plate and the Eurasian plate.

The effects of high P/T metamorphism are best exhibited in metabasaltic epidote amphibolite and mafic Mn-rich garnet amphibolite that are associated with serpentinite. In the Juisui area, these lithologies occur as large tectonic blocks. Two distinct metamorphic episodes have been delineated [Liou, 1981; Liou and Ernst, 1985]. The first stage is characterized by the development of barroisite + hornblende + rutile + garnet and is believed to have occurred about 80 m.y. ago at about 500°C with pressures in the order of 7-8 kb. The second-stage subduction zone metamorphism, approximately 8-14 m.y. ago, caused significant recrystallization in the ophiolite-bearing melange. The most significant features include: the formation of glaucophane at the expense of barroisite, and sphene after rutile for the Mn-rich tuff; the crystallization of Al-rich actinolite after barroisite in the mafic epidote amphibolite; and the formation of chlorite after garnet in the argillite matrix. The Tertiary subduction zone may have generated conditions of approximately 350-450°C and pressures of 4-5 kb or more during the closure of the South China Sea. The Plio-Pleistocene collision of the Eurasian basement and the Luzon volcanic arc since 4-5 m.y. has resulted in the later greenschist facies recrystallization and emplacement of high P/T metamorphic rocks along the Longitudinal Valley fault.

Characteristics of High P/T Metamorphic Rocks of China

Modes of Occurrence of High P/T Metamorphic Rocks

From the available descriptions of the high P/T metamorphic rocks of China, there are two different types of occurrences. First, the high P/T metamorphic rocks are present as blocks or fault slices incorporated within an olistostromic melange which could be converted by gravity sliding from wildflysch. Secondly, the surrounding country rocks of the blueschist and C-type eclogite are serpentinite and/or serpentinized dismembered ophiolite (or serpentinitic melange). In other words, the blueschist or C-type eclogite blocks are within ophiolitic melange as in the New Idria serpentinite of California. The situations in the Suolun-Hegen Mountains, East Qinling, Ailao Mountains, the Lancang River and the Yuli area belong to the first category. But for most of the other areas of China such as West Junggar, Baiyunebe, Ondor Sum and Yarlung Zangbo area, the high P/T metamorphic blocks or fault slices are associated with both the olistostromic and ophiolitic melanges, similar to the variation observed in the Franciscan complex as a whole. The main difference in China is that country rocks have a higher temperature of metamorphism than those of the Franciscan complex, reaching up to greenschist facies such as in the North Qilian Mountains.

Are there any regional terrains of blueschist except for blocks of high P/T metamorphic rocks in China? The answer seems negative from the present stage of study, although a 400-m thick blueschist bed has been reported from the Yarlung Zangbo area [Xiao and Gao, 1982] and a similar situation was seen in North Qilian Mountains and the Ondor Sum area. There might be bigger blocks or thicker fault slices rather than hundreds of square km of coherent blueschist such as the case in the Seward Peninsula of Alaska [Thurston, 1983]. The problem is that most of the geologists have not examined high P/T metamorphic minerals in the matrix due to their smaller grain size and their subsequent overprinting by greenschist facies recrystallization. No detailed studies for metamorphic minerals have been carried out in the metasediments and metabasites. Moreover, serpentinized ultramafics will not show high pressure metamorphic assemblages.

Age of High P/T Metamorphism

It is a difficult task for geologist to date the blueschist facies metamorphism because most of the blueschists and C-type eclogite have experienced multiple-stage metamorphic recrystallization. From the regional geological mapping and a few dates of the associated metamorphic rocks, the high P/T metamorphic rocks of China are assumed to be the same age as that of the accretionary fold belt they occur within. For example, the blueschist in Ondor Sum is thought to be Early Paleozoic because many fossils ranging from Early Cambrian to Ordovician have been discovered from the metamorphic complex called the Ondor Sum Group; the associated metamorphic rocks in which blueschist blocks have been found were dated radiometrically as 626, 536, 489 and 435 m.y. in different strata by different workers [Hu, 1983]. In addition, the Upper Silurian is deposited unconformably on the metamorphic complex. In the North Qilian Mountains, the country rocks of the blueschist and C-type eclogite yield many fossils from latest Proterozoic to Silurian and red beds in which Middle Devonian plant fossils have been found overlap the metamorphic complex. Therefore, the high P/T metamorphism is thought to be Early Paleozoic. The blueschists in Chaogeqi, Sunitezuegi, East Qilian, Xinyang and Donghai are thought to be Late Paleozoic. Those blueschists in the Southwest China microplate such as Ailao Mountains, Doucheng-Muli, North Tibet and Lancang River are Mesozoic. The blueschists in Yarlung Zangbo and Yuli are Paleogene and Neogene respectively.

In summary, the high P/T metamorphic rocks of China range from Early Paleozoic to as young as Neogene in different localities. Of course, detailed study is necessary to establish the blueschist age and to compare the age of the blocks with that of the surrounding matrix, because the blocks should be older than that of the matrix in an olistostromic melange, and the blocks could be either older or younger than that of the matrix in an ophiolitic melange (or tectonic melange). High P/T metamorphic rocks have not been found in Precambrian cratons due to a relatively high geothermal gradient at that time. Glaucophane might occur in some high-stress compressive or compressive-shear zones, but it is more abundant in typical blueschists.

P-T Conditions of High P/T Metamorphism

The metamorphic temperatures of the eclogites from the North Qilian Mountains, Xinyang and Donghia have been calculated from the garnet-clinopyroxene geothermometer of Ellis and Green [1979]. The effect of calcium content in garnet was considered. When the pressures are assumed to be 6, 10, and 14 kb respectively, the calculated temperatures of individual eclogites show a limited variaition of 8-9°C.

Three methods were used to estimate pressures: (1) the silicon content per unit formula of phengite [Massonne, 1981]; (2) the cros-

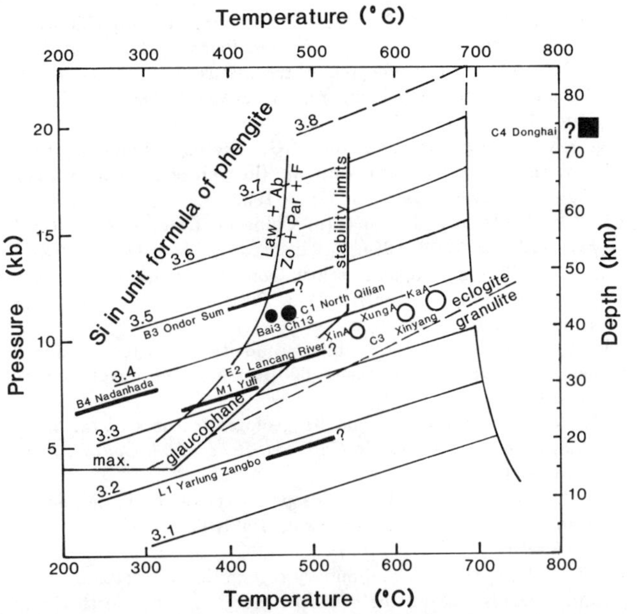

Fig. 12. The P-T conditions for formation of some high P/T metamorphic rocks. The pressure constraints are from the phengite geobarometer. The temperature constraints of eclogite from North Qilian Mountains (C1) (Bai3-Baijingsi, Ch13-Chingshuigou), Xinyang (C3) (XinA, XungA and KaA are from three different mineral pairs) and Donghai (C4) are from the Clinopyroxene-garnet geothermometer. The P-T condition of blueschist from Nadanhada (B4) is after the method of Maruyama et al., [1986]. The pressure constraint determined from the phengite geobarometer (6-7 kb) is higher than that deduced from the phases equilibrium (4-5 kb) in the blueschists of Yuli area (M1). The possible temperature ranges for Ordor Sum (B3), Lancang (E2) and Yarlung Zangbo (L1) are mainly based on the mineral assemblages in blueschists.

site component (Na in M4 site) of Ca-amphibole [Brown, 1977]; and (3) the jadeite content in clinopyroxene [Essene and Fyfe, 1967; Holland, 1980; Wood et al., 1980]. The estimated pressures for C-type eclogite of North Qilian based on these three methods are 10.7-11.7, 6.8-7.1 and 9.8-10.2 kb respectively, and for those from Xinyang are 10.5-2.0, 7.0-7.1 and 11.4-13.2 kb. The pressures estimated from phengite and jadeite contents are close to each other; whereas those from the crossite contents seem too low. The blueschists from Ondor Sum also yield a lower pressure using the crossite content compared to the phengite (Figure 12). The estimated pressures for blueschists from Nadanhada and Yuli are close to each other. Therefore, the phengite geobarometer appears to be reasonable for blueschists formed at pressures higher than 7 kb. No phengite and amphibole contents are available for the eclogite of Donghai area, but the metamorphic pressure based on the jadeite content is about 19-20 kb, which is consistent with the previous results of 17.25-24.75 kb [Ying et al., 1981]. In general, the estimated pressures for the C-type eclogite from North Qilian and Xinyang are much lower than that of the Donghai eclogite. On the other hand, the estimated temperatures for eclogites from these three areas increase from west to east. This relationship is consistent with the fact that the coexisting amphiboles are glaucophane-crossite, barroisite, and hornblende respectively for North Qilian, Xinyang, and Donghai.

The possible P-T ranges of recrystallization of blueschists in a number of areas are also shown in Figure 12. The occurrence of arfvedsonite and ferro-winchite in Yarlung Zangbo blueschist implies that the metamorphic temperature might be relatively high. The pressure estimates by phengite and crossite contents of Ca-amphibole in the Nadanhada area are consistent with each other (Figure 12) and are in agreement with the pressure of blueschist-greenschist facies transition.

It is important to note that the P-T conditions discussed above are simply the relationship between temperature and depth (or pressure) within the earth at any one time. In fact, the P-T arrays bear no relation to any one geotherm that existed during metamorphism, because they consist of a set of P-T conditions inferred from assemblages that were preserved at different times and different places during metamorphism [Thompson and England, 1984]. Therefore, more work has to be done in order to determine the P - T - t path for each of the areas discussed above.

Blue Amphiboles of High P-T Metamorphic Rocks

The compositions of sodic amphiboles in high P/T metamorphic rocks significantly affect pressure estimates for their occurrences [Maruyama et al., 1986]. Some blue amphiboles may be sodic-calcic amphibole like winchite and barroisite, such as the cases in the Ailao Mountains, Xinyang, and Yarlung Zangbo areas. The sodic-calcic amphiboles may be partly replaced by sodic or calcic amphibole as described in the Yuli and Ailao Mountains amphiboles. The common occurrence of sodic-calcic amphibole in blueschist suggests a higher temperature but similar pressure to that of glaucophane- and crossite-bearing blueschists. However, it should be emphasized that if the blue sodic amphiboles are mainly riebeckite, magnesio-riebeckite or arfvedsonite in composition, such as those in East Qinling and Baiyunebe, then their occurrences do not necessarily suggest high P/T metamorphism.

Protoliths of High P/T Metamorphic Rocks

The protoliths of the high P/T metamorphic rocks of China are mainly mafic rocks, especially for C-type eclogite, pelagic sediments and clastic rocks. But siliceous and carbonate rocks could also contain blueschist facies assemblages such as those in North Qilian Mountains and East Qinling which have similar litholigies to the Franciscan complex. The mineral assemblages of blueschists are significantly dependent on the bulk composition of rocks; hence the mineral assemblages provide information about blueschist protoliths.

Concluding Remarks

Nearly all of the presently reported occurrences of blueschists and C-type eclogites within China are found within sutures or accretionary fold belts. It is also significant that these high P/T metamorphic rocks are associated with other exotic blocks, some of which are metamorphosed and some not. Commonly, the associated rocks are mafic-ultramafics of an ophiolite suite; this indicates that oceanic crust was involved in the tectonism that produced the blueschists. The current consensus requires very high pressures and less than normal geothermal gradients to produce high P/T metamorphic rocks in the blueschist facies; these conditions are only found in subduction zones or collision zones. Therefore, the distribution of these blueschists in China described here follows a pattern that has been observed in other convergent zones around the world [Coleman, 1972; Ernst, 1973; Miyashiro, 1973].

An interesting aspect of our present knowledge of the distribution of blueschist in China is that nearly all the reported occurrences so far are from easily recognizable tectonic blocks. It is possible that high P/T index minerals such as lawsonite, jadeite, and aragonite may be present in regional metamorphic terranes where detailed mineralogical studies have not yet been attempted. The history of the discovery of jadeite-lawsonite-aragonite in the regionally metamorphosed Franciscan graywackes shows that blueschist blocks were well known in the early 1900's, but it was not until 1956 that jadeite was known to exist widely in the Franciscan graywacke [Bloxam, 1956], and a few years later aragonite + lawsonite was also found to be widespread [Coleman and Lee, 1962, McKee, 1962a, b]. These jadeite-lawsonite-aragonite graywackes do not have the characteristic blue color found in blueschists derived from metabasites, and these graywackes require thin section or X-ray diffraction studies before these minerals can be identified. The fact that many of the blueschists seem to be Paleozoic in age may preclude preservation of jadeite and lawsonite because subsequent high T/P metamorphism may have converted these high-pressure minerals into their higher temperature equivalents such as plagioclase, epidote, or calcite.

Future studies should concentrate on the structural evolution of the Chinese blueschists with regard to the major plate tectonic evolution of the region. Establishing the nature of the suture zones vis-a-vis subduction, collision, or obduction may be possible by carefully establishing poly-metamorphic and poly-structural histories within these blueschist belts. The apparent crustal P-T trajectories necessary to form blueschists and their return to the earth's surface require stress-strain parameters that are presently unknown.

Acknowledgement. Many Chinese geologists have provided the results of their field studies for summary and some samples for further analytical study. This manuscript has had critical reviews from R. G. Coleman, Heather Ponader, Moonsup Cho, and E. L. Miller. Mrs. C. Oh helped with typing the manuscript. This project is supported by the Department of Geology, Stanford University and NSF EAR-8214567 and EAR- 85-08247. We greatly appreciated the above named individuals and institutes for their assistance and financial support.

References

Bloxam, T. W., Jadeite-bearing metagraywackes in California, *Am. Mineral.*, *41* 488-496, 1956.

Brown, E. H., The crossite content of Ca-amphibole as a guide to pressure of metamorphism, *J. Petrol.*, *18*, 53-72, 1977.

Chai, B.H.T., Structural and tectonic evolution of Taiwan, *Am. J. Sci.*, *272*, 389-422, 1972.

Chi, S.J., Hu, X., Ding, F.C., Xu, J. and Shi, J.Z., The evidence for ancient plate movement in Inner Mongolia and its relationship to the evolution of geosyncline. in Selected papers from the second all-China congress on tectonics, (in Chineses), *Publishing House of Geology, Beijing, China*, *1*, 54-66, 1981.

Coleman, R. G., Blueschist metamorphism and plate tectonics, *24th Internat. Geol. Congr.* *2*, 19-26, 1972.

Coleman, R. G. and Lee, D. E., Metamorphic aragonite in the glaucophane schists of Cazadero area, California, *Am. J. Sci.*, *260*, 577-595, 1962.

Duan, X.H. and Zhao, H., The Ailaoshan-Tengtiaohe fracture: the subduction zone of an ancient plate, (in Chinese), *Acta Geol. Sinica*, *4*, 258-266, 1981.

Ellis, D.J. and Green, D.H., An experimental study of the effect of Ca upon garnet-clinopyroxene Fe-Mg exchange equilibrium, *Contrib. Mineral. Petrol.*, *71*, 13-22, 1979.

Ernst, W. G. Blueschist metamorphism and P-T regions in active subduction zones, *Tectonophysics*, *17* 255-272, 1973.

Essene, E. J., and Fyfe, W. G., Omphacite in California metamorphic rocks, *Contrib. Mineral. Petrol.*, *15*, 1-23, 1967.

Frank, W. and others, Geological observations in the Ladakh area (Himalayas), a preliminary report, *Schweiz. Mineral. Petrol. Mitt.*, *57*, 89-114, 1977.

Feng, Y.M., Zhang, Y.H. and Hue, Y.G., Tectonic evolution of the western part of northern Xinjiang, (in Chinese), *Contrib. Proj. Plate Tect. North China*, *1*, 17-33, 1983.

Gao, Y.L., Blueschist of China, (in Chinese), *Chinese Acad. Geol. Sci. Bull.*, *10*, 61-76, 1984.

Guo, Y. H., The glaucophane schist from Tangbalei district, Xinjiang, (in Chinese), *Contrib. Proj. Plate Tect. Northern China*, *1*, 89-104, 1983.

Holland, T.J.B., The reaction albite = jadeite + quartz determined experimentally in the range 600-1200°C, *Am. Mineral.*, *65*, 129-134, 1980.

Hu, X., Glaucophane schist and its geologic features of the Inner Mongolian geosyncline, (in Chinese), *Hebei Geol. College, Bull.*, *1*, 16-29, 1983.

Jan, M.Q. and Kempe, D.R.C., The petrology of the basic and intermediate rocks of upper Swat, Pakistan, *Geol. Mag.*, *11*, 285-300, 1973.

Li, D.Z. and Ye, D.N., A preliminary recognition of metamorphism and volcanism along the southern margin of the North China Fault Block Region. (in Chinese), in *Formation and Development of the North China Fault Block Region, Publishing House of Academia*, 133-142, 1980.

Lin, J.L., Fuller, M. and Zhang, W.Y., Preliminary Phanerozoic polar wander paths for the North and South China blocks, *Nature*, *313*, 444-449, 1985.

Liou, J.G., Ho, C.O. and Yen, T.P., Petrology of some glaucophane schists and related rocks from Taiwan, *J. Petrol.*, *16*, 80-109, 1975.

Liou, J.G., Petrology of metamorphosed oceanic rocks in the Central Range of Taiwan, *Geol. Soc. China Mem.*, *4*, 291-342, 1981.

Liou, J.G. and Ernst, W.G., Summary of Phanerozoic metamorphism in Taiwan, *Geol. Soc. China Mem.*, *6*, 133-152, 1984.

Ma, J.J., On the Paleozoic tectonics in Heilungjiang River Basin, (in Chinese), *Earth Sci., J. Wuhan College*, *2*, 49-62, 1983.

Maruyama, S., Cho, M., and Liou, J. G., Experimental investigations of blueschist-greenschist transition equilibria: Pressure dependence of Al_2O_3 contents in sodic amphiboles - a new geobarometer, *Geol. Soc. Am. Mem.*, *164*, 1-16, 1986.

Massonne, H.J., Phengite: Eine experimentelle Untersuchung ihres Druck-Temperatur-Verhaltens in System K_2O-MgO-Al_2O_3-SiO_2-HO_2, *Ph. D. dissertation, Ruhr Univ., Bochum, West Germany*, 1981.

Mckee, B., Widespread occurrence of jadeite, lawsonite, and glaucophane in Central California, *Am. J. Sci.*, *260*, 596-610, 1962a

McKee, B., Aragonite in the Franciscan rocks of the Pacheco Pass area, California, *Am. Mineral.*, *47*, 379-387, 1962b.

Miyashiro, A., Metamorphism and related magmatism in plate tectonics, *Am. J. Sci.*, *272*, 629-656, 1972.

Peng, X.J. and Lue, Y.L., Discovery and tectonic significance of glaucophane schist from the southern section of the Lancang River in the western Yunnan, (in Chinese), *Reg. Geol. China*, *2*, 1982.

Qiu, Y.Z., Wang, Z.G. and Zhao, Z.H., A preliminary study of REE iron formation, (in Chinese), *Geochemica*, *3*, 231-240, 1981.

Singh, L.C., Metamorphism and tectonism in the eastern Himalayas, (In Russian), *Geotectonics*, *2*, 66-74, 1977.

Sobolev, V.S. (Chief Editor), *Metamorphic Complexes of Asia, Pergamon Press, Oxford, Nauka Publishers, Moscow*, 1982.

Tang, K.D. and others, On Wundurmiao Group and its tectonic significance, (in Chinese), *Contrib. Proj. Plate Tect. Northern China, 1,* 186-208, 1983.

Thompson, A.B. and England, P.C., Pressure-temperature-time paths of regional metamorphism II. Their influence and interpretation using mineral assemblages in metamorphic rocks, *J. Petrol., 25,* 929-955, 1984.

Thurston, S. P., Structure, petrology and metamorphic history of the Nome Group blueschist terrane, Salmon Lake area, Seward Peninsula, Alaska, *Geol. Soc. Am. Bull., 96,* 600-617, 1985.

Wang, Q. and Liu, X.Y., On Caledonian polycyclic paired metamorphic belts of Qilian Mountains, northwest China. in Contribution to the tectonics of China and adjacent regions, (in Chinese), *Publishing House of Geology, Beijing, China,* 92-101, 1981.

Wang, K.Y. and Shi, G.X., The main characteristics of the structural metamorphic zones of western Yunnan, (in Chinese), in *Selected Papers from the Second all-China Congress on tectonics, Publishing House of Geology, Beijing, China, 1,* 79-89, 1981.

Wood, B. J., Holland, T. J. B., and Kleppa, O. J., Thermochemistry of jadeite-diopside pyroxenes, *Geochim. Cosmochim. Acta, 44,* 1363-1371, 1980.

Wu, H.Q., The glaucophane-schist of eastern Qiling and northern Qilian Mountains in China, (in Chinese), *Acta Geol. Sinica, 3,* 195-207, 1980.

Wu, H.Q., Petrology and mineralogy of the high pressure metamorphic zones in northern Qilian, (in Chinese), *Xian Inst. Geol. Min. Res. Bull., 4,* 5-21, 1982.

Wu, H.Q., Brief description of the high-pressure low-temperature metamorphic belts in China and discussions on some problems about glaucophane and barroisite, (in Chinese), *Xian Inst. Geol. Min. Res. Bull., 7,* 63-81, 1984.

Xiao, X.C., Chen, G.M. and Zhu, Z.Z., Some knowledges about the paleo-plate tectonics of the North Qilian Mountains, (in Chinese), *Geol. Sci. Tect., 3,* 73-78, 1974.

Xiao, X.C. and Gao, Y.L., New finds of the high-pressure and low-temperature greenschist facies from Yarlung Zangbo (Tsangpo) suture zone, Xizang (Tibet), (in Chinese), *Chinese Acad. Geol. Sci. Bull., 5,* 77-79, 1982.

Xiao, X.C., Wan, Z.Y., Li, G.C, Cao, Y.G. and Zhou, X., On the tectonic evolution of the Yarlung Zangbo (Tsangpo) suture zone and its adjacent areas, (in Chinese), *Acta Geologica Sinica, 2,* 205-212, 1983.

Yan, Z.Y. and Tang, K.D., The natures of several typomorphic minerals from high pressure metamorphic zone of Ondor Sum Group, Nei Mongol, (in Chinese), *Chinese Acad. Geol. Sci. Bull., 10,* 179-194, 1984.

Ye, D.N., Li, D.Z., Dong, G.F. and Chu, X.W., On the 3T-phengite and C-type eclogite in the Xinyang metamorphic belt and their tectonic significance. in Formation and development of the North China Fault Block Region, (in Chinese), *Publishing House of Academia, Beijing, China,* 122-132, 1980.

Yen, T.P., On the occurrence of the late Paleozoic fossils in the metamorphic complex of Taiwan, *Geol. Surv. Taiwan Bull., 4,* 23-29, 1953.

Yen, T.P., The metamorphic belts within the Tananao Schist terrain of Taiwan, *Geol. Soc. China Proc., 6,* 73-74, 1963.

Yen, T.P., Glaucophane schist of Taiwan, *Geol. Soc. China Proc., 9,* 70-73, 1966.

Ying, S.H., Yu, L.B. and Yang Z.E., The metamorphism and genetic problems of eclogites from Mts. Hongtushan and Qinglongshan of the Donghai region, Jiangsu Province, (in Chinese), *Seis. Geol., 3,* 17-30, 1981.

Zhang, Q., Zhang, Z.Y. and Li, S. H., Muscovite of metamorphic rocks in east and south Xizang and its petrologic significance, (in Chinese), *Sci. Geol. Sinica, 4,* 340-347, 1980.

Zhang, Zh.M. and Jin, M., Two kinds of melanges and their tectonic significance in Xiangcheng-Derong area, southwestern Sichuan, (in Chinese), *Sci. Geol. Sinica, 3,* 205-214, 1979.

Zhang, Zh.M., Contribution to tectonics of Himalayas. in contribution to the geology of the Qinghai-Xizang (Tibet) Plateau, (in Chinese), *Geol. Publishing House, Beijing, China, 1,* 296-317, 1982.

Zhang, Zh.M., Liou, J.G. and Coleman, R.G., An outline of the plate tectonics of China, *Geol. Soc. Am. Bull., 95,* 295-312, 1984.

Zhou, W.Q. and Lin, W.X., Characteristics of the glaucophane schist from the metamorphic belt in the southern section of the Lancang River, (in Chinese), *Reg. Geol. China, 2,* 1982.

Zwart, H.J., Sobolev, V.S. and Lepezin, G.G., *Metamorphic Map of Asia (1/5,000,000),* Inst. Geol. Geoph. Siberian Branch, USSR Acad. Sci, 1978.

ACCRETED TERRANES AND MINERAL RESOURCES OF EASTERN CHINA

Pow-Foong Fan

Hawaii Institute of Geophysics, University of Hawaii, Honolulu, Hawaii 96822

Abstract. The terranes of eastern China were formed mostly during late Paleozoic and early Mesozoic time. The Huabei terrane coalesced with the Yishan during the middle to late Carboniferous. The Inner Mongolia terrane accreted to the Siberia terrane during the late Paleozoic and collided with the Huabei-Yishan terrane during the Triassic. The Songliao terrane is probably a continental fragment that was caught between the Inner Mongolia and Dongbei terranes and coalesced with the Huabei terrane during the Triassic. The Yangzi terrane coalesced with the Huanan terrane during Silurian time. During the late Triassic, the Yangzi-Huanan terrane collided with the Qinba terrane to the north, and coalesced with the Huabei-Yishan terrane. The Southeast Maritime terrane was accreted against the Yangzi-Huanan terrane during the late Triassic.

In the Inner Mongolia terrane Variscan and Mesozoic mineralized belts contain pegmatitic rare earth, magmatic chromite, porphyry, skarn and vein copper and vein lead-zinc deposits. The Dongbei terrane contains mineralized belts of Variscan age. These comprise porphyry copper and molybdenum deposits, gold quartz-vein deposits, and skarn tungsten and vein lead-zinc deposits. There are three gold belts and two episodes of porphyry copper mineralization periods in the Huabei terrane (Fig. 1). The Mesozoic mineralized belt of the Qinba terrane contains carbonate-hosted lead-zinc, mercury, and antimony deposits. The lower Paleozoic mineralized belts in the Yangzi terrane are carbonate-hosted mercury and antimony, vein wolframite and cassiterite, and disseminated copper deposits. The mineralized belts in the Huanan terrane are Yanshanian skarn and vein wolframite and cassiterite, vein and disseminated copper, and skarn copper in the lower Yangzi Valley. The mineralized belts in the Southeast Maritime terrane resulted from the Yangshanian orogeny and are vein cassiterite, tin sulfide, porphyry, skarn and vein molybdenum, and vein and lead-zinc deposits.

Introduction

The terranes of eastern China were formed mostly during late Paleozoic and early Mesozoic time. They are fragments of continental, oceanic, and island arc crusts. There are ten accreted terranes in eastern China (Figure 1):--Inner Mongolia, Songliao, Dongbei, Jiamusi, Yishan, Huabei, Qinba, Yangzi, Huanan and Southeast Maritime [Xiong, 1983; Xiong and Coney, 1985; Schermer, 1983; Fan, 1985]. The accreted terranes and their tectonic evolution can provide information on the origin of the mineral deposits [Albers 1981, 1983].

Major Deposits in the Terranes of Eastern China

Inner Mongolia Terrane

The Inner Mongolia terrane accreted to the Siberia terrane during the late Paleozoic and later collided with the Huabei-Yishan terrane during the Triassic. The Inner Mongolia terrane has a late Paleozoic (Variscan) island arc assemblage built on a Precambrian to lower Paleozoic metamorphic basement. Variscan and Mesozoic mineralized belts in the Inner Mongolia terrane include pegmatitic rare earth, magmatic chromite, porphyry, skarn and vein copper and vein lead-zinc deposits.

Bayan Obo, Inner Mongolia, is an extraordinary rare earth deposit. The commercial reserves are estimated to be four times the combined reserves of the rest of the world (Wang, 1980). Bayan Obo is located in the 1500-m.y.-old continental margin at the northern edge of the Inner Mongolia terrane (Figure 2, no. 1). The stratified deposit occurs in a 3- to 4-km-thick sequence of slates and quartzites. The ore contains ooliths and other sedimentary textures. Ore minerals include magnetite, hematite, bastnaesite, aeschynite, and monazite. Some minerals have in excess of 70% rare earths; the Bayan Obo ores typically run about 10% total rare earths. Riebeckite and

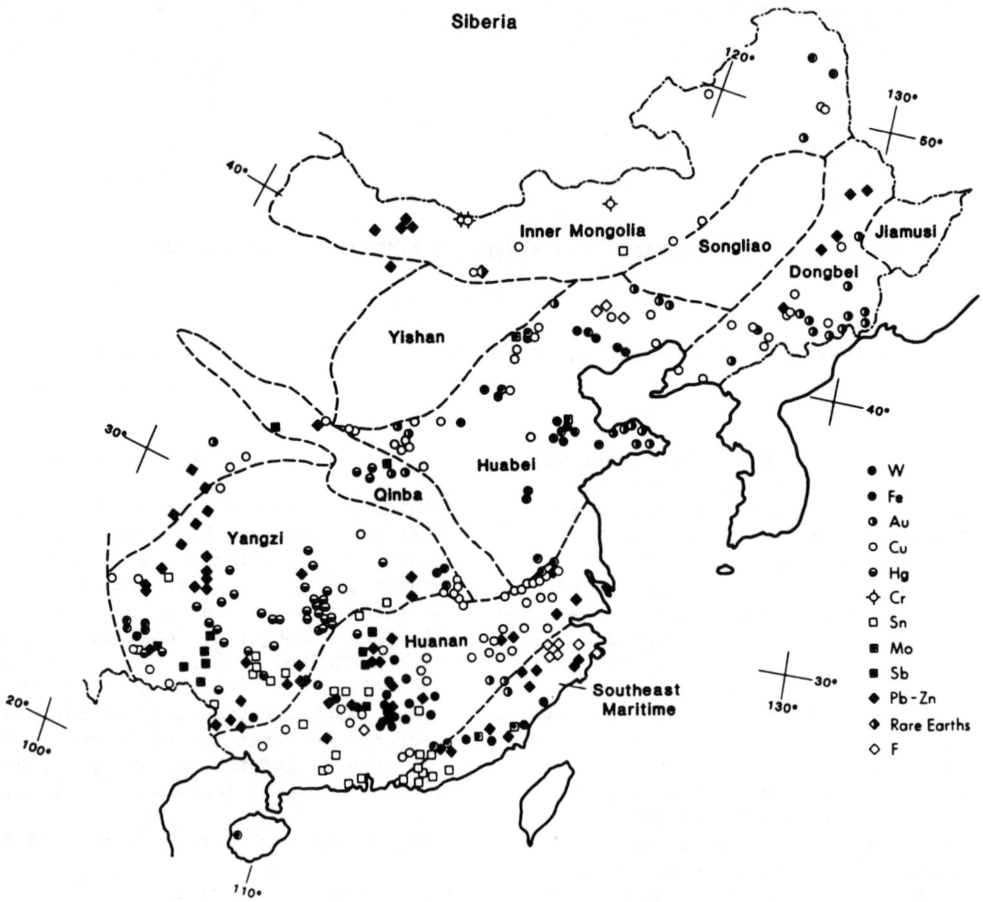

Fig. 1. Distribution of mineral deposits of eastern China. (Sources: Mining Industries of China [1980], Guo et al. [1982a], Li et al. [1983], Chen [1984], Fan [1986], and Xiong and Coney [1985]).

albite are associated with the ore and appear to be syngenetic with it although partly recrystallized. Aegirine-augite may have formed during Hercynian intrusion of granites [Lee, 1970, Wang et al., 1973]. The clastics indicate a shallow open sea environment of deposition, however the ore may have been deposited in a closed evaporite basin. The sedimentary sequence is not dissimilar to that of some other Precambrian iron formations [Tu, 1981]. The ore-forming processes may have been weathering of earlier rare earth mineralization, concentration and redeposition; (2) volcanic-sedimentary or (3) metasomatic. The probability that placer origin is enhanced by the presence of detrital uranothorite [Yang and Li, 1980].

There are two reported chromite deposits in Inner Mongolia (Figure 1)[Fan, 1985]. The northwestern Inner Mongolia chromite deposits [Wei, 1978], located along an east-west trending rift zone, are part of the eastern segment of the Hercynian Tienshan fold belt. The country rocks consist of hornblende-chlorite schist and quartz- mica schist formed through metamorphism of Ordovician-Carboniferous marine volcanics. Ultramafic rocks were emplaced during the middle Hercynian Orogeny. Gabbro intrusion occurred before the ultramafic intrusion which in turn was followed by Permian biotite granite intrusion. The ores occur as magmatic segregations in the form of lenses, veins, pods, stringers, and pillows.

Cao [1976] reported a chrome deposit in the western part of Inner Mongolia (Figure 1). This deposit is in an ultramafic body intruding Precambrian migmatite which is located in the continental margin and western Hercynian fold belt. Chromite and Cr-spinel are the major ore minerals.

There are six Paleozoic porphyry copper deposits in the Inner Mongolia terrane [Chen, 1984]. Four of the deposits are located in the northeast part of the Inner Mongolia terrane (Figure 1) and one is located in the southwest. All are located near major fault zones. The

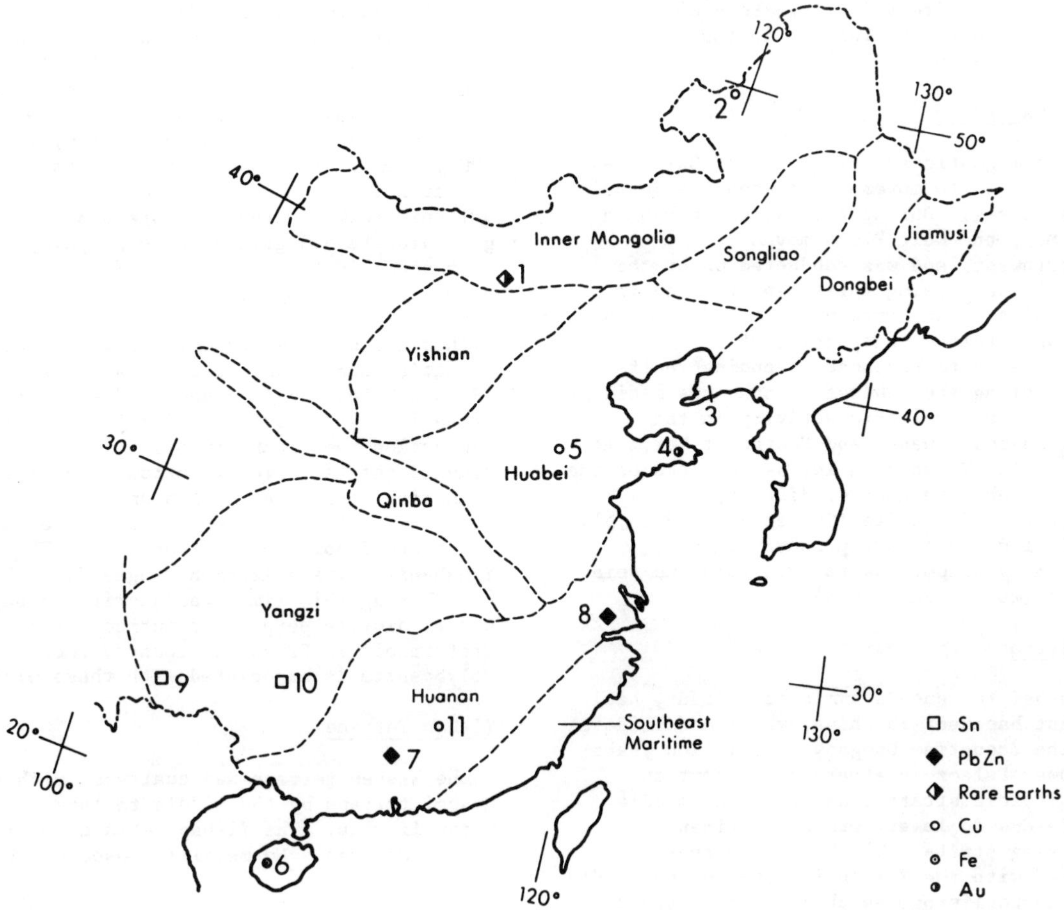

Fig. 2. Locations of selected mineral deposits: 1. Bayan Obo, 2. Dubaoshan, 3. Anshan-Benzi Basin, Liaoling Peninsula, 4. Zhaoyang County, Shandong Peninsula, 5. Zhongtiao-shan, Laoyan, 6. Shilu, 7, Fankou, 8. Qixiashan, 9. Gujin, 10. Dachang, 11. Xihushan, Dayu County.

Duobaoshan deposit (Figure 2, no. 2) is a typical Paleozoic porphyry copper deposit and is located southeast of a late Paleozoic subduction zone. The country rocks consist of Middle Ordovician Duobaoshan Formation (andesite and andesitic tuff interbedded with sandstone, sandy conglomerate and limestone) intruded by Carboniferous granodiorite [Du, 1980]. The deposit consists of a quartz core, an inner zone of quartz-perthite- biotite, a middle zone of sericite-chlorite-quartz and an outer zone of propylitic alteration. Ore mineralization occurs in the sericite-chlorite-quartz zone and the major ore minerals are bornite, chalcopyrite and molybdenite [Fan, 1986b].

Songliao Terrane

The Songliao terrane may be a continental block, and contains a thick sequence of Cretaceous lacustrine deposits. It may have been trapped during accretion of the Inner Mongolia terrane (to the west) and the Dongbei terrane (to the east). The Cretaceous lacustrine deposits contain large accumulations of hydrocarbons. No metallic deposits have been reported.

Dongbei Terrane

The Dongbei terrane consists of Late Jurassic continental sediments on Precambrian basement. It accreted westward and coalesced with the Songliao and Inner Mongolia terranes. During Triassic time it docked with the Huabei terrane.

In the Liaoling Penisula, Precambrian iron deposits are located along the margin of the Anshan-Benxi greenstone basin (Figure 2, no. 3), an area of 900 km^2. The country rocks consist of metamorphosed metavolcanics and sedimentary rocks. The iron formations dip toward the center of the basin [Xu and Chen, 1984]. Metamorphism has also altered Precambrian magnesium carbonates and surrounding country rocks into magnesite and talc. Several magnesite deposits are located in the Anshan-Benxi basin [Tao, 1985].

Mesozoic Magmatism

Mesozoic magmatic activity in East China was closely related to movement between the Eurasian and Kula Plates. During the Early Yanshanian (80-145 Ma), the Kula Plate moved north-northwest, and was subducted under the Eurasian Plate causing uplift and folding in eastern China. The north-northwest direction of subduction resulted in a series of north-northeast to northeast trending faults. Magmatism along the subduction zone was manifest by intrusive and volcanic activity in the Dongbei, Huabei, Huanan and Southeast Maritime terranes. The Yanshanian intrusive rocks of the Dongbei terrane are gabbro, diorite, granodiorite and granite [Guo et al., 1982a,b). The mineral deposits are porphyry copper, porphyry molybdenum, quartz vein gold and vein lead-zinc [Wu and Shen, 1984].

Huabei Terrane

The Huabei terrane is a microcontinent, having the oldest basement in China, which consolidated during the Zhongtiao Orogeny 1.7 billion years ago. Lower Paleozoic strata are absent and upper Paleozoic strata consist of the middle Carboniferous to lower Permian continental coal-bearing strata. The Huabei terrane was juxtaposed with the Yishan terrane by the middle to late Carboniferous as shown by the presence of similar late Carboniferous continental sediments in both terranes. McElhinny et al. [1981] showed that a relative rotation of 120° occurred between the Huabei and Yangzi-Huanan terranes.

There are three gold belts in the Huabei terrane: a northern, a western, and an eastern belt (Figure 1). The northern gold belt is associated with Mesozoic volcanics, which developed on Precambrian metamorphic rocks. The quartz vein gold and gold-bearing sulfide deposits are of Yenshanian age. The western gold belt contains quartz vein gold deposits in Precambrian metamorphic rocks. The eastern gold belt is characterized by the Linglong gold mine in Zhaoyang County on Shandong Peninsula [Yu, 1982] (Figure 2, no. 4). In the Linglong gold mine the gold-bearing quartz veins formed in the Precambrian basement during middle Yanshanian time (126-145 Ma) [Yen et al., 1979]. Three ore-forming stages are recognized: 1) An early quartz-pyrite stage; 2) a quartz-gold-metallic sulfide stage during which quartz, pyrite, chalcopyrite, galena, sphalerite, bornite, cosalite, native gold, electrum, and sericite were formed; and 3) a carbonate stage during which calcite, sericite and quartz were formed.

Porphyry copper deposits in the Huabei terrane were formed in two episodes: during the Middle Proterozoic Changcheng Period and during the Yanshanian [Chen, 1984; Rui et al., 1984; Ministry of Metallurgy, 1984]. The Proterozoic porphyry copper are represented by the Zhongtiaoshan deposit [Wang and Wen, 1957; Fan, 1984, 1985]. The country rocks consist of thick Proterozoic metamorphosed sedimentary and volcanic rocks intruded by pre-Sinian granodiorite and granodiorite porphyry. The ores occur in the upper part of the granodiorite porphyry and along the contact of the country rocks. The ores consist mainly of chalcopyrite, with minor bornite and molybdenite. The Zhongtiaoshan deposit located west of Laoyan (Figure 2, no. 5) is associated with diorite and probably with a north dipping Precambrian subduction zone [Fan, 1984]. Li et al. [1979] studied the tectonic evolution of a Precambrian magnetite deposit at Laoyan and came to the same conclusion, that the deposit developed above a northward dipping subduction zone. The Yanshanian mineralization ranges from 110-173 Ma. During this time granodiorite porphyry and quartz diorite porphyrite intruded the Proterozoic to Paleozoic country rocks. Molybdenite is associated with these plutons.

Yishan Terrane

The Yishan terrane had coalesced with the Huabei terrane by the middle to late Carboniferous. The Yishan terrane is rich in coal resources. No metallic resources have been reported.

Qinba Terrane

The Qinba terrane, a west-northwest to east-southeast oriented, elongate terrane, consists of Precambrian basement covered by Paleozoic and early Mesozoic marine limestone and cherty limestone. In late Triassic time, the Yangzi-Hunan terrane collided with the Qinba terrane, which also collided with the Huabei terrane, and by Jurassic time had been stitched by Cretaceous plutons. North-dipping subduction in the Qinba terrane is indicated by a paired metamorphic belt, ophiolites, flysch, and a line of Late Triassic and Early Cretaceous granites and granodiorites [Klimetz, 1983]. Mercury deposits (Figure 1) occur in the Indosinian Qinling fold belt. The mercury deposits are related to regional fracture zones and are associated with carbonate hosted antimony, lead, zinc, and polymetallic ore zones [Zen, 1983].

Yangzi Terrane

The Yangzi terrane consists of Late Proterozoic crystalline basement, (700-800 Ma) covered by Sinian to Middle Triassic marine deposits and a thick pile of Permian basalt. It coalesced with the Huanan terrane during the

Silurian. McElhinny et al. [1981] suggested that it was located at about 2°N latitude during Permian time.

Mercury deposits concentrated in the south of the Yangzi terrane (Figure 1) account for over 70% of China's total output of mercury and over 91% of total reserves. The ores consist mainly of cinnabar, with minor metacinnabar, native mercury, realgar, stibnite and sphalerite. They are carbonate-hosted strata-bound deposits [Zhou et al., 1982; Zen, 1983] of Cambrian age. Hua [1981] studied fluid inclusions and the geological setting of the cinnabar deposits and concluded that they might be of sedimentary origin and that the mineralization has undergone a late, low-temperature hydrothermal solution stage probably resulting from heated confined groundwater. Xiang [1984] suggested that the cinnabar deposits were related to Caledonian magma (318-439 Ma) and later migrated into the Cambrian carbonate host rocks.

There are three Paleozoic lead-zinc strata-bound ore belts in Yangzi terrane: eastern, southern, and western. They all occur in carbonates along paleocontinental margins [Wang, Y.Q., 1985, Zhu et al., 1985]. The eastern lead-zinc belt occurs in Lower Cambrian to Upper Cambrian carbonates. The southern mineralized belt occurs in Devonian carbonates [Feng, 1984]. The most favorable strata are dolomites, algal limestones and silicified carbonates. The ore minerals are galena, sphalerite and pyrite; and the gangue minerals are quartz, calcite, dolomite, barite and fluorite. The western lead-zinc belt occurs in Upper Sinian to Lower Cambrian carbonates, and has a north-south trend with a width of 100 km and a length of 300 km. The lead content of the country rocks of all three mineralized belts is 17-18 times higher than the world average for these types of rocks [Wang, 1985].

Huanan Terrane

The Huanan Terrane consists of Precambrian basement and Silurian metamorphic rocks intercalated with volcanic rocks which were tightly folded during the Caledonian orogeny. The Huanan terrane amalgamated with the Yangzi terrane during Silurian time and formed a composite terrane which continued to travel northward. Most of the mineral deposits were formed after the Silurian.

The Shilu iron deposit is one of the major rim deposits in the Huanan terrane (Figure 2, no. 6). The hematite- and magnetite-bearing Shilu Group of unknown age has a thickness of 2,000 m and consists of sericite-quartz schist with quartzite and dolomite. The hematite and magnetite occur in dolomitic layers. They are sedimentary in origin. Their association with anhydrite suggests that they formed in high salinity basins which were intermittently isolated in a hot, dry climate. Later the sedimentary sequences were subjected to regional metamorphism and granitic intrusion. In general, the Shilu deposit is very similar to the Superior-type iron deposits, except that the Shilu deposit was formed in closed or semi-closed basins [Tu, 1981].

Yangzi-Huanan

During the late Paleozoic, both the Yangzi and Huanan terranes had the same geological setting and shallow marine sediments were deposited, followed by Triassic continental deposits. Triassic plutons then stitched the two terranes. In Late Triassic time the Yangzi-Huanan terrane collided with the Qinba terrane, which also collided with the Huanbei terrane at the same time. Triassic to Cretaceous plutons of the Yanshinian Orogeny stitched the Yangzi-Huanan and Qinba terranes. North dipping subduction occurred in the Qinba terrane with development of a linear belt of granodiorites and granites [Klimetz, 1983].

The lead-zinc deposits in the Huanan terrane are late Paleozoic in age and strata-bound in form [Fu, 1984]. They are represented by the Fankou deposit, in the south of the terrane (Figure 2, no. 7) and the Qixiashan deposit in the east (Figure 2, no. 8). The Fankou deposit occurs in Devonian and Carboniferous carbonates. There are two types of ores, primary sedimentary and late stage replacement. The ores consist of galena, sphalerite, siderite and pyrite, stibnite and argentite. A regional geochemical survey indicates that the lead-zinc content of the Fankou region is 1.5 to 3 times higher than the world average [Lu, 1984]. The Qixiashan deposit occurs in Carboniferous limestones. The ores consist of galena, sphalerite, pyrite and marcasite. The ores were emplaced after lithification of the host rocks. Dolomitization occurred before and after the mineralization. This deposit is similar to the Mississippi Valley type lead-zinc deposits in the U.S.A. [Nakajima and Tao, 1985].

Strata-bound antimony is present in the Huanan terrane (Figure 1), especially in Huanan province. Sixty-seven percent of these deposits are in Devonian strata. Others occur in Presinian, Sinian and Cambrian strata [Xiao and Li, 1984]. The ores consist of stibnite with minor pyrite. The gangue minerals are quartz, calcite, fluorite, barite, and gypsum.

The porphyry and skarn copper deposits in eastern Huanan and eastern Yangzi terranes (Figure 1) are related to Mesozoic Yanshanian granites [Wang, 1982; Dong, 1984], which are the product of northwest-dipping subduction zones located along the Chinese continental margin.

The tin deposits of the Yangzi-Huanan Terrane can be divided into two belts: the southwestern and the southeastern. The southwestern tin belt is located in Yunnan province at the southwestern edge of the Yangzi terrane (Figure

2, no.9). Tin deposits are associated with granitic intrusions resulting from Proterozoic Jinning, Caledonian, Variscan, Indosinian and Yanshanian movements [Liu, H., 1983]. The highest tin concentrations are associated with the Yanshanian movement. Multiple magmatic activity during the Jinning to Indosinian movements and melting of the older crust helped to elevate tin content for later magmatic and hydrothermal mineralization during the Yanshanian movement. Wu et al. [1984] studied the whole rock Rb-Sr isochron dating of Sn-bearing granites and alkaline rocks at Gujiu (Figure 2, no. 9). From their emplacement ages and initial $^{87}Sr/^{86}Sr$ ratios they concluded that the rocks are the product of multiple magmatic cycles.

The Yanshanian granites are probably part of the back-arc magmatic belts resulting from north- and northeastern dipping subduction of the Indian plate during Cretaceous time [Fan, 1977, 1984; Li, 1980]. The Gujiu tin deposit is one of the largest examples of the Yanshanian tin deposits of the southwestern Yangzi terrane. It is a placer cassiterite occurrence derived from calcic skarn deposits that are the product of metamorphism of Middle Triassic carbonates by Upper Yanshanian biotite granites (81-147 Ma) [Wu, Q., et al., 1984]. The tin-bearing skarns are concentrated along north-northeasterly trending anticlines [Yen et al., 1979].

The southeast tin belt is situated in the Yangzi terrane (Figure 1). The Yanshanian deposit at Dachang (Figure 2, no. 10) is one of the largest tin deposits in China and is perhaps the largest hard-rock tin resource in the world. The southeast China deposits are located in the Yangzi, Huanan and Southeast Maritime terranes. Yanshanian granites constitute about 60% of the granites of southeast China. They are associated with the Mesozoic northwest-dipping subduction zone located along the Chinese continental margin. Tin and tungsten deposits are located farther landward from the subduction zone. In the Huanan terrane, cassiterite-quartz vein, and greisen-bearing and skarn associated Hercynian granites are found in a tungsten belt in southern Jiangxi, southern Hunan and northern Guangdong provinces. In the same region, Yanshanian tin skarn deposits occur at the contact between granite and upper Paleozoic carbonates. In the Southeast Maritime terrane, Upper Yanshanian (70-100 Ma) intrusions are present but no evidence exists of multicycle granites. Tin deposits are few and scattered, and most of them are of no economic importance.

The geology of the Dachang tin deposit is well described in the Excursion Guide of the Dachang Tinfield prepared by the Organizing Committee of the 1984 International Symposium on Geology of Tin Deposits at Nanning, China (ESCAP-RMRDC, 1984). The deposit is located in a northwestern-trending, fault-controlled depression containing Middle and Upper Devonian carbonates which have been intruded by Yanshanian (73-115 Ma) granite. The granite is leucocratic and belongs to the alumina-oversaturated series with tourmaline and fluorite.

The paragenesis of the tin and associated ore minerals of the Dachang tin deposit can be outlined in the following sequence:

1) During the Caledonian orogenic event the initial deposits were formed. These were enriched in sulfur and certain metallic elements.

2) Tectonic activity during the Indosinian-Yanshanian mobilized the initially enriched metallic elements, forming Hg-As and Sb-Pb-Zn deposits.

3) Late Yanshanian biotite granite intrusion gave rise to stanniferous skarn-type deposits containing zinc and copper.

4) Cassiterite-sulfide deposits resulted from metasomatism by mineralizing fluids during late Yanshanian granite porphyry intrusion and related hydrothermal activity which also resulted in tungsten-antimony enrichment.

5) Placer deposits have developed since the Yanshanian event.

In Guangxi province, pre-Caledonian cassiterite deposits show north-northeast and northeast structurally controlled trends and post-Caledonian deposits show northeast and northwest trends [Xian, 1984; Chen et al., 1985].

The granites of southern China cover approximately 20% of the total area. Their ages range from Proterozoic to Cretaceous [Tu et al., 1980]. The Proterozoic to Caledonian granites probably are "metasomatic"; the Hercynian granites are anatectic. The Yanshanian granites are mostly biotite granites, derived from remelting of rare-earth bearing, low grade metamorphic rocks. They are salic, and are enriched in alkalis, volatiles and Sn, W, Nb and Ta [Tu et al., 1980].

The tin and tungsten deposits in southern China are located on the innermost side of inner arcs. The Yanshanian ilmenite-bearing S-type biotite granites were generated above shallow subduction zones resulting in formation of tin deposits, whereas the tin in magnetite-bearing I-type granites could substitute for iron in sphene and magnetite, resulting in low concentration in the residual liquid [Ishihara, 1977].

Liu et al. [1982] studied the abundance of tungsten and other elements of 472 granitoids from southern China of Sipu-Zuefengian, Caledonian, Hercynian, Indosinian, and Yanshanian age. They noted that the tungsten content increases from 2.1 to 4.1 ppm (Table 1). The tin content also increases from 5 to 42

TABLE 1. Contents of Tin, Tungsten, and Other Elements in Granites Formed During Different Orogenic Events From South China (after Liu et al., 1982)

Orogenic Events	%				ppm			
	SiO_2	K_2O	Na_2O	Fe_2O_3	Sn	W	Nb	Ta
Yanshanian	70.02	4.71	3.29	0.90	42	4.1	35	8
Indosinian	71.23	4.36	2.23	1.20	13	2.5	25	3
Hercynian					25		21	4
Caledonia	70.53	3.86	2.78	0.86	15	2.7	21	6
Sipu-Xuefengian (Proterozoic)	63.31	3.92	2.85	0.82	5	2.1	15	3

ppm. The igneous rocks of the region show increases of SiO_2, K_2O, and Na_2O from Precambrian to Yanshanian. The igneous rocks evolve from poorly differentiated basic magmatic rocks in the Precambrian to well-differentiated Na-Si-rich magmatic rocks in the Cretaceous. Therefore, polycyclic tectonic movements can give rise to ore deposits enriched in tin or tungsten.

The extensive belt of Mesozoic granites in the Huanan terrane, which contains the world's major tungsten deposits, has been suggested as belonging to continental margin magmatic arcs where Mesozoic granites were emplaced above northwestward-subducting ocean floor [Jahn, 1974]. The 700-km width of the belt and the presence to the west of an eastward-dipping thrust suggest that the plutons in the western part of the belt where the tungsten and antimony deposits are located may be related to back-arc thrusting and folding. The belt may have developed inland of an easterly rather than westerly facing continental margin (Mitchell and Garson, 1981).

The Xihuashan mine in the Huanan terrane located in Dayu County is representative of the tungsten deposits (Figure 2, no.11). Early Yanshanian (140-148 Ma) granites intruded weakly metamorphosed Upper Cambrian, sandstone and slate country rocks. The granite body is a cupola projecting from a long, concealed batholith ridge trending northeast under a cover of metamorphosed flyschoid beds [Wu and Mei, 1982]. Multistage tectonic activity gives rise to faults striking northeast-southwest, northwest-southeast, and north-south. A large northeast-southwest fault separates the southeast part of an uplifted block from a basin. Six tungsten deposits are clustered in two northeast-trending parallel zones along the edge of a granite stock. The Xihuashan mine is located in the southern part of the stock.

Wolframite occurs in quartz veins associated with cassiterite, scheelite, beryl, topaz, fluorite, sulfides and carbonates [Wang and Zhou, 1982; Mu et al., 1982]. Mu et al. [1982] suggested that the Xihuashan granites are S-type. Oxygen isotopic studies show that the granites in Xihuashan are neither the product of autochthonous magmatism nor the result of the partial melting of lower crust (I-type) but are of anatectic origin derived from Precambrian sedimentary strata, the Proterozoic Banxi Group of submarine volcanic-sedimentary origin. Certain horizons with anomalous abundances of tungsten might be the source beds of the tungsten mineralization [Xu et al., 1982].

Other types of tungsten deposits are scheelite skarn deposits, strata-bound scheelite-stibnite-gold deposits, porphyry tungsten deposits, and tungsten contained in hematite beds of submarine exhalative origin [Liu and Shen, 1982; Li and Xu, 1982].

Southeast Maritime Terrane

The Southeast Maritime terrane is the Zhemingyanhai terrane of Xiong and Coney [1985]. It accreted against the Huanan terrane during Late Triassic time along a terrane boundary fault marked by metamorphosed intermediate-basic volcanic rocks, and layered mafic and ultramafic rocks with jasper deposits [Li, 1980]. The Southeast Maritime terrane developed as an active continental margin and island arc-type orogenic belt. A Mesozoic northwest-dipping subduction zone was probably located along the Chinese continental margin. The Yanshanian granites of the Southeast Maritime terrane were generated from this subduction zone.

The presence of Mesozoic vein fluorite deposits in the northern Southeast Maritime terrane (Figure 1) indicates that fluorite may have concentrated in alkalic magma generated along the subduction zone, carried through the crust as SiF_4, and later deposited as CaF_2 as observed by Lamarre and Hodder [1978] and Cumming et al. [1979] in the western United States and in northern Mexico. The Mesozoic lead-zinc deposits in northern and central parts of the Southeast Maritime terrane (Figure 1) are mostly hydrothermal vein lead-zinc deposits and some skarn-type deposits.

Molybdenum deposits in southern Southeast Maritime terrane (Figure 1) are of Mesozoic porphyry type. Tin deposits in the southeast of the terrane (Figure 1) are Mesozoic cassiterite-quartz vein and stannite deposits. Iron deposits in central Southeast Maritime terrane (Figure 1) are usually situated between limestones and intruding Mesozoic granodiorites and are of skarn type. The ore minerals consist of hematite and magnetite. Ge et al., [1981] suggested that some of this iron skarn type could be of marine volcanic-sedimentary origin.

The gold deposit located at the western border of the Southeast Maritime terrane is a placer deposit [Conway, 1986].

Conclusion

The distribution of the mineral deposits in ten terranes in eastern China indicates a distinct pattern.

Chromite deposits are associated with ultramafic bodies and ophiolites, such as the chromite deposits occurring in the northern accreted island arc of the Inner Mongolia terrane.

Gold-quartz veins in the Huanbei and Dongbei terranes were formed in thick continental cratons.

Carbonate-hosted mineral deposits are commonly found in cratons and platforms. Carbonate-hosted lead-zinc deposits are found in the Yangzi, Huanan, and Qinba terranes and carbonate-hosted mercury is concentrated in the southern Yangzi terrane.

Vein-type lead-zinc deposits are generally confined to island arc terranes and are present in accreted island arcs in the Inner Mongolia and Southeast Maritime terranes.

Porphyry copper and molybdenum deposits are related to subduction processes, such as occurred in the Precambrian belt in Huanan terrane, the Paleozoic belt in Inner Mongolia terrane, and the Mesozoic belts in Huanbei, Huanan, Yangzi, and Southeast Maritime terranes.

Vein and skarn tungsten and tin deposits formed in cratons and along cratonic margins. The Huanan deposits are located on the innermost side of inner arcs of subduction zones. They were derived from the mantle and added to the crust and recycled by magmatic and hydrothermal activity.

Mineralization episodes in eastern China occurred in Precambrian, lower Paleozoic, upper Paleozoic, Mesozoic (Indosinian and Yanshanian) times. Of these the Yanshanian episode is the most widespread.

Acknowledgements. This is Hawaii Institute of Geophysics Contribution no. 1781.

References

Albers, J.P., A lithologic-tectonic framework for the metallogenic provinces of California, Econ. Geol. 76, 765-790, 1981.

Albers, J.P., Distribution of mineral deposits inaccreted terranes and cratonal rocks of western United States. Canadian Jour. of Earth Sciences 20, 1019-1029, 1983.

Chen, Wenming, Ore sources and evaluation indicators of porphyry copper deposits in China, Chinese Acad. Geol. Sci. Inst. of Mineral Deposits Bull., 12, 78, 1984.

Chen, Y.C., M.Z. Huang, J. Xu, Y.D. Ai, X.M. Li, S.H. Tang and L.K. Meng, Geological features and metallogenetic series of the Dachang cassiterite-sulfide polymetallic belt, Acta Geologica Sinica, 229-240, 1985.

Cao, R., On the genesis of the alpine-type ultramafic complex and related chromite deposits in the western part of the Inner Mongolia orogenic belt, Geochimica 2, 113-125, 1976.

Conway, M.H., Gold mining and production, The China Business Rev., May-June, 8-12, 1986.

Cumming, G.L., S.E. Kesler, D. Krstic, Isotopic composition of lead in Mexico mineral deposit, Econ. Geol. 74, 1395-1407, 1979.

Dong, R.P., A tentative plate model for the rock-forming and ore-forming processes in the middle Yangtze Valley, Mineral Deposits, 3-2, 28-38, 1984.

Du, Qi, The alteration and mineralization features of the Duobaoshan porphyry copper deposits, Acta Geol. Sinica, 310-323, 1980.

ESCAP-RMRDC (Economic and Social Commission for Asia and the Pacific Regional Mineral Resources Development Center) 1984, Report of the International Symposium of Geology of Tin Deposits, Nanning, China, 26 Oct.-7 Nov., 1984, 28 p., 1984.

Fan, P.F., Outline of tectonic evolution of southwestern China, Tectonophysics, 45, 261-267, 1977.

Fan, P.F., Geologic setting of selected copper deposits of China, Econ. Geol., 79, 9-19, 1984.

Fan, P.F., Accreted terranes and mineral resources of eastern China, Third Circum. Pacific Terrane Conf. Extended Abstracts, Geol. Soc. Australia, 14, 62, 1985.

Fan, P.F., Geologic setting of selected chromium and nickel deposits of China, 179-193, In: Geology and Metallogeny of Copper Deposits, Edited by G. Friedrich et al., Springer-Verlag, Berlin-Heideberg, 1986a.

Fan, P.F., Porphyry copper deposits in East and

Southeast Asia, Geol. Soc. of China Mem. No. 7, (in press), 1986b.

Feng, Qunyao, A discussion on lead-zinc deposits in Guangxi--Their genetic types, the major ore controlling elements and ore distribution regularity, Mineral Deposits, 3-3, 47-56, 1984.

Fu, L.Z., A preliminary investigation into the ore-forming mechanism of mercury and lead-zinc stratabound deposits in Guizhou province: Mineral Deposits, 3-1, 83-92, 1984.

Ge, Chaohua, Fa Han, Tianren Zou and Deqian Chen, Geological characteristics of the Makeng iron deposit of marine volcano-sedimentary origin, Chinese Acad. Geol. Sci. Bull. 3-1, 47-69, 1981.

Guo, W.K., L.S. Liu, and Z.G. Yu, The fundamental features of metallogenic megaprovince and epochs of eastern China, Mineral Deposits, 1, 1-14, 1982a.

Guo, W.K., L.S. Liu and Z.G. Yu, The fundamental features and epochs of East China metallogenic domain, Bull. Chinese Acad. Geol. Sci., 5, 1-29, 1982b.

Hua, Y.F., Preliminary study on the origin of mercury deposits in Guizhou and its vicinity, Acta Geol. Sinica, 139-148, 1981.

Ishihara, S., The magnetite-series and ilmenite series granitic rocks, Min. Geol., 27, 293-305, 1977.

Jahn, B.M., Mesozoic thermal events in China, Nature 248, 480-483, 1974.

Klimetz, M.P., Speculations on the Mesozoic plate tectonic evolution of eastern China: Tectonics, 2, 139-166, 1983.

Lamarre, A.L. and R.N. Hodder, Distribution and genesis of fluorite deposits in western United States and their significance to metallogeny, Geology 6-4, 236-238, 1978.

Lee, K.Y., Some rare-element deposits in mainland China, U.S. Geol. Survey Bull. 1312-N, 34, 1970.

Li, C., A preliminary study of plate tectonics of China, Chinese Acad. Geol. Soc. Bull., 1-2, 11-22, 1980.

Li, Chongyan and Jing Xu, Genetic classification of tungsten deposits of Jiangxi and neighboring provinces, p. 287-299, In Tungsten Geology Symposium, Jiangxi, China, ESCAP/RMRDC, Bandung, Indonesia and Geol. Pub. House, Beijing, China, 583, 1982.

Li, H.Z., Y.Q. Wang, G. Xu, J.Y. Zhang, and L.S. Liu, On the metallogenic domains and distributive characteristics of nonferrous metal deposits in east Asia: Geol. Rev., 29, 341-350, 1983.

Li, Jiliang, Bolin Cong and Wenhua Zhang, A preliminary study on the tectonic evolution of the late Archean iron formation in the SW of North China fault block, Scientia Sinica 22, 1430-1442, 1979.

Liu, H., The distribution and genesis of metallic ore deposits in Yunnan province, Mineral Deposit, 2-4, 19-25, 1983.

Liu, J. and J. Shen, Metallogenetic magmatic systems of tungsten in Jiangxi province, p. 313-325, In: Tungsten Geology Jiangxi, China, ESCAP/RMRDC, Bandung, Indonesia and Geol. Pub. House, Beijing, China, 583, 1982.

Liu, Y., Q. Yang, Y. Zhu, and Q. Jiang, Tungsten abundance and its evolution in granitoid rocks of South China, In: Tungsten Geology Symposium, 349-357, Jiangxi, China. ESCAP/RMRDC, Bandung, Indonesia and Geological Pub. House, Beijing, China, 583 pp., 1982.

Lu, H., On the genesis of Fankon Pb-Zn ore deposit in Guangdong province, Geochimica, 357-365, 1984.

McElhinny, M.W., B.J.J. Embleton, X.H. Ma and Z.K. Zhang, Fragmentation of Asia in the Permian: Nature, 293, 212-216, 1981.

Mining Industries of China, Chinese Economic Research Department of Fuji Journal, 128-156, 1980.

Ministry of Metallurgy, Porphyry copper deposits in China, Science Press, Beijing, 240, 1984.

Mitchell, A.H.G. and M.S. Garson, Mineral Deposits and Global Tectonic Settings, Academic Press, 405 pp., 1981.

Mu, Z.G., F.S. Huang, C.Y. Chen, S.H. Zheng, S.L. Fen, D.R. Liu and Y.W. Mei, Oxygen, hydrogen and carbon isotope studies of Piaotang and Xihuashan quartz vein-type tungsten deposits, Jiangxi province, p. 385-401, In: Tungsten Geology, Jiangxi, China, Edited by J.V. Hepworth and Y.H. Zhang, ESCAP/RMRDC, Bandung, Indonesia and Geol. Pub. House, Beijiang, China, 583, 1982.

Nakajima, T. and Kuiyuan Tao, Geology of Qixianshan, A carbonate-hosted lead-zinc deposits, Jiangsu province, southeast China, pp. 49-84, Research on mineral resources associated with volcanic and plutonic rocks, Report of International Res. and Dev. Cooperation ITIT Project 8113. Geol. Survey of Japan and Nanjing Inst. Geol. and Mineral Resources, China, 191 pp., 1985.

Rui, Zongyao, Chongke Huang, Guoming Qi, Jue Xu, Hongtao Zhang, Porphyry copper (molybdenum) deposits of China. Geological Pub. House, Beijing, 350, 1984.

Schermer, E.R., Tectonostratigraphic terranes of China and Mongolia, in Howell, D.G., Cox, A. and Nur, A., eds., Proc. of the Circum-Pacific Terrane Conference, Stanford Univ. Pub. Geol. Ser., XVII, 175-179, 1983.

Tao, W.P., The distribution regularity of some nonmetallic deposits along the circum-Pacific belt of east China, Mineral Deposits, 4-3, 61-70, 1985.

Tu, Guangzhi, On the mineral deposits of Bayan Obo and Shilu, Scientia Sinica, 24, 228-236, 1981.

Tu, K.C., Z.G. Wang and X.Y. Yu, Genesis of granitic rocks in South China and related mineralization, p 189-196, In: Granitic Magmatism and Related Mineralization, Edited by Ishihara, S. and S. Takenouchi, Mining Geology Spec. Iss., no. 8, The Society of Mining Geologists of Japan, 247, 1980.

Wang, C., S.P. Li and S.C. Su, Genetic characters of a metasediment-pyrometasomatic rare-earth deposits, Geochimica, 1, 5-11, 1973.

Wang, J. and K. Wen, Porphyry copper deposit of Zhongtiaoshan, Acta Geol. Sinica 37, 401-415, 1957.

Wang, K.P., Rare-earth, Far East-China, Mining Annual Review, 451, 1980.Wang, Y.Q., Lead-zinc deposits in China and their relationship with geotectonics, Sci. Papers on Geology for International Exchange--Prepared for the 27th Intl. Geol. Congress. 2, Geol. Pub. House, Beijing, 11-18, 1985.

Wang, Zehua and Yuzhen Zhou, Major features of two-layered type mineralization in Xihuashan mine and its genetic model, p. 427-436, In: Tungsten Geology Symposium, Jiangxi, China, ESCAP/RMRDC, Bandung, Indonesia and Geol. Pub. House, Beijing, China, 583, 1982.

Wang, Zhitian, Complex model of copper metallogeny in Middle-lower Yangtze Valley and its significance for ore exploration, Sci. Sinica (Ser. B) 25, 765-776, 1982.

Wei, W., On the genetic types of a chromite deposit in Kansu, Acta Geol. Sinica, 269-280, 1978.

Wu, Liren and Yuanchao Shen, Mesozoic magmatites and related ore deposits in East China, 343-351, Academia Sinica Developments in Geosciences, Contribution to 27th Intl. Geol. Congress, Moscow, Sci. Press, Beijing, 1984.

Wu, Q., J. Xu, and Z. Yang, Sr isotopic characteristics of Geiju Sn-bearing granites and a study of ore search indicators, Geochimica, 293-302, 1984.

Wu, Yongle and Yongwen Mei, Multi-phase intrusion and multi-phase mineralization in Xihuashan tungsten ore field, p. 437-449, In: Tungsten Geology Symposium, Jiangxi, China, ESCAP/RMRDC, Bandung, Indonesia and Geol. Pub. House, Beijing, China, 583, 1982.

Xian, B., A discussion on the ore-forming conditions and distributional regularity of the tin ore deposits of Guangxi, Acta Geol. Sinica, 49-62, 1984.

Xiang, M.M., Doubts concerning the paper "A preliminary study of the origin of mercury deposits in Guizhou and its neighboring areas", Geol. Rev., 30, 373-376, 1984.

Xiao, Qiming, and Diankui Li, An investigation in to the genesis of the antimony deposits in Hunan, Mineral Deposits, 3-3, 12-26, 1984.

Xiong, Ji, Suspect terranes of China, in Howell, D.G., Cox, A. and Nur, A., eds., Proc. of the Circum-Pacific Terrane Conference, Stanford Univ. Pub. Geol. Ser., XVIII, 212-215, 1983.

Xiong, Ji and Peter Coney, Accreted terranes of China, Tectonostratigraphic Terranes of the Circum-Pacific Region, Edited by D.G. Howell, Circum-Pacific Council for Energy and Min. Res. Earth Sci. Ser. 1., 349-362, 1985.

Xu, G.G. and H.J. Chen, Metallogenetic characteristic of iron deposits in Archean Greenstone basin of Anshan-Benxi area, northeast China, Mineral Deposits, 3-2, 1-10, 1984.

Xu, Kegin, Shouci Hu, Minzhi Sun, Jingrong Zhang, Jun Ye and Huiping Li, Regional factors controlling the formation of tungsten deposits in south China, p. 473-488, In: Tungsten Geology Symposium, Jiangxi, China, ESCAP/RMRDC, Bandung, Indonesia and Geol. Pub. House, Beijing, China, 583, 1982.

Yang, Minzhi and Yongchun Li, Uranothorite discovered in rare-earth, niobium, iron deposits at Bayan Obo, Inner Mongolia, China, Kexue Tongbao, 25, 594-596, 1980.

Yen, J.C., S.Q. Chu and Y.S. Ho, eds., Ore Deposits, Beijing Geological Press, 443, 1979.

Yu, C., The nature of fluid inclusions in quartz crystals from the gold-bearing quartz veins and ore-forming temperature at Linglong gold mine, Zhaoyuan county, Shandong province, Sci. Geol. Sinica, 309-314, 1982.

Zen, R.L., A preliminary observation on the regional distribution regularity and the main characteristics of mercury mineralizations in China, Mineral Deposits, 2-1, 30-37, 1983.

Zhou, D.H., J.Q. Mao and G.Z. Yang, The classification of structural domains and structural zones of mercury ore deposits in Guizhou province, Mineral Deposits, 1-1, 43-50, 1982.

Zhu, S.Q., H.S. Mao, S.C. Chi and Q.H. Shi, On the temporal and spatial evolution of strata-bound copper, lead and zinc deposits of China, Sci. Papers on Geology for Intl. Exchange---Prepared for the 27th Intl. Geol. Congress. 4. Geol. Pub. House, Beijing, 75-84, 1985.

TECTONOSTRATIGRAPHIC TERRANES OF THE COASTAL REGION OF SOUTHEAST CHINA

Yangshen Shi, Lingzhi Guo, Ruishi Ma, Huafu Lu, and Huogen Dong

Department of Geology, Nanjing University, Nanjing, China

Abstract. The coastal region of southeast China is a complex orogenic region, a collage of terranes consisting of folded basement rocks of Proterozoic, Paleozoic and Mesozoic ages. Terranes of different ages have remarkable differences in paleontology, stratigraphy, sedimentation, igneous activity, metamorphism and crustal thickness. Their boundaries are subduction zones, sutures or deep fault zones.

The Jiangnan Proterozoic terranes accreted to the pre-Sinian Yangzi continental margin and gave rise to the Proterozoic Jiangnan arc during the Xuefengian orogeny. During the Caledonian orogeny the south China Caledonian terranes accreted to the southeast margin of the arc. Then the coastal Hercynian-Indosinian terranes in Zhejiang and Fujian Provinces accreted to the Caledonian orogenic belts. Consequently continental accretion displays a spatial polarity from northwest to southeast, comparable with that in southeastern Australia.

Taiwan Island of China consists of an eastern and a western terrane separated by a longitudinal valley fault zone. The eastern terrane, originally located north of the Philippines, was displaced northward and accreted to the continental margin of southeast China during the early Miocene.

Terranes of Hainan Island and the Xisha and Dongsha Islands were dispersed from the coast of Guangdong Province. The folded basement of Hainan Island is closely comparable with the Caledonian rocks of western Guangdong (Yunkai terrane).

Introduction

Plate tectonics provides a conceptual model for moving lithospheric plates across the surface of the earth. As a consequence of plate movements disparate lithospheric fragments are brought into contact along plate boundaries. These fragments, which are fault-bounded geologic entities of regional extent, comprise tectonostratigraphic terranes [cf. Howell et al., 1985]. At convergent plate margins, where orogenic belts are formed, the terranes accrete and a collage of lithospheric fragments results.

Circum-Pacific continental margins consist of terranes of different ages. Paleomagnetic studies have shown that a part of the Asian continent (including China) is a collage of continental fragments, some of which were originally part of Gondwanaland and originated in the southern hemisphere or in equatorial regions in the Late Paleozoic - Mesozoic period [McElhinny et al., 1981]. The application of the terrane concept of the study of Asia, especially its continental margins, is a major advance.

Because of the complex tectonics of south China, the range in structural styles, and the timing of orogenic movements, we have tried to find regularities in its tectonic development in terms of the terrane concept. This paper deals with the tectonostratigraphic terranes of the coastal region of southeast China, some of which must be regarded as suspect terranes.

At least 10 tectonostratigraphic terranes of different ages can be recognised in southeast China. Their recognition is based on differences in sedimentation, fauna, metamorphism, and igneous and structural features (Figure 1)[Guo et al., 1984].

Late Proterozoic Terranes

Kuaiji Terrane

The northeast trending Kuaiji terrane in Zhejiang Province is composed of ophiolite rocks of oceanic affinity and a calc-alkaline dacite tuff unit named the Shuangxiwu Group, part of the Dong'anian cycle of age greater than 1400 Ma, as well as calc-alkaline volcanoclastic flysch of the Loujiamen Formation, and andesite, basalt, and rhyolite of the Shanshu Formation both belonging to the Xuefengian cycle (1400 Ma-800 Ma). The ophiolite suite found in the Kuaiji Mountains mainly comprises pyritic copper deposit-bearing sodium-rich spilite-keratophyre, jasper and turbidites, and spheroidal pyroxene hornblendite, titanomagnetite-bearing pyroxenite, gabbro and diorite. It is thrust over mixed gneiss along the Shauxing-Jiangshan fault. The Shuangxiwu Group occurs in the Longmen Mountains. The Loujiamen Formation and the Shanshu Formation

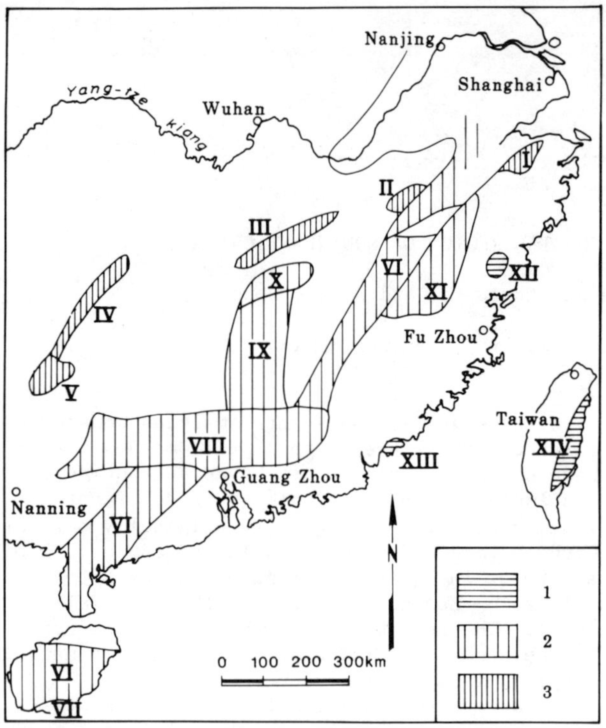

Fig. 1. Sketch map showing distribution of the tectonostratigraphic terranes of southeast China [from Guo Lingzhi et al., 1984]. 1. Terranes of coast of southeast China, 2. Terranes of Caledonian of south China, 3. Terranes of Jiangnan Proterozoic. I. Kuaiji terrane, II. Huaiyu terrane, III. Jiuling terrane, IV. Xuefen terrane, V. Jinwandashan terrane, VI. Yunkai terrane, VII. Nanhai terrane, VIII. Dayaoshan terrane, IX. South Jiangxi terrane, X. Wugongshan terrane, XI. North Fujian terrane, XII. Nanxi terrane, XIII. Dongshan terrane, XIV. Dananao terrane.

overlie the volcanic rocks of Dong'anian cycle unconformably. The intervening orogenic event is named the Shengong movement.

Huaiyu Terrane

The northeast trending Huaiyu terrane of Jiangxi Province is characterized by oceanic ophiolite melange of the Xuefengian cycle associated with a flysch sequence equivalent to the Banxi Group. The melange was obducted northwestward, on the fault of northeastern Jiangxi, over greenschist facies metamorphic rocks of the flyschoid Jiuling Group of the Dong'anian cycle. The southwest and northeast terminations of the terrane are transform faults. The ophiolite melange is younger than the Dong'anian ophiolite suite of the Kuaiji terrane.

Jiuling Terrane

The Jiuling terrane trends northeast in the Jiuling Mountains of western Jiangxi Province. It is composed of ophiolite of the Longjiaxi Group that belongs to the Dong'anian cycle. The ophiolite consists of metabasalt spilite-keratophyre and mafic-ultramafic rocks. It was subducted to the north along the Yifen-Wanzai fault zone. Flysch of the Banxi Group and sandstone and conglomerate interbedded with volcanic rocks of the continental facies Laokedong Formation, both of which belong to the Xuefengian cycle, overlie the Longjiaxi Group unconformably. An unconformity resulting from the Laokedong movement can be found locally.

Xuefen Terrane

Xuefen terrane is characterized by ophiolite suite rocks and flysch of the Xuefengian cycle Banxi Group. The main components of the ophiolite are harzburgite, peridotite, olivine pyroxenite, gabbro, gabbro-diabase, spilite and keratophyre with pillow structure, and bedded or laminated siliceous sedimentary rocks. The ophiolite is controlled by northnortheast faults and forms a thrust nappe directed towards the eastsoutheast. Fragments of ophiolite, brecciated ophiolite, and mylonitized or metamorphosed country rocks are found along the fault zone, and drilling indicates that the ultrabasic rock bodies are rootless wedges. The Xuefen terrane is characterized by the northeast structural grain that characterizes the Xuefengian cycle, in contrast to the east-west trend of structures of the Dong'anian cycle. The geochemistry of the ultrabasic rocks of the Xuefengian cycle differs from those of the Xuefengian cycle in the Huaiyu terrane, the latter being magnesium-rich and the former containing more iron.

Jinwandashan Terrane

Ophiolite suite rocks in the Jinwandashan terrane occur mainly in association with the upper part of the turbidite Sibao Group. Rocks along the contact zone between the mafic-ultramafic rocks and their country rocks possess a schistose structure and greenschist and tremolite schist occur commonly. Accompanying the mafic-ultramafic rock complex are well developed pillowed spilite and keratophyre, diorite and siliceous sedimentary rocks. The ophiolite suite together with the turbidites have an east-west trend and occur in a south dipping intensely compressed imbricate structural zone. The flysch of the Banxi Group unconformably overlies rocks of the Sibao Group.

To summarise, the composition, age of emplacement and mineralization of ophiolite suites, and the structural configuration and age of orogenic movements in the terranes that make up the ancient Jianan arc zone, appear to differ one from another. Thus the Jianan arc consists of exotic terranes of different ages which are mainly of oceanic affinity. At the end of the Xuefengian orogenic movement, about 800 Ma ago, these late Proterozoic terranes were accreted to the Yangzi continental margin. Evidence of this accretion is provided by the occurrence of Sinian System rocks which are relatively undeformed and not only overlie elements of the ancient arc system unconformably but also rest unconformably on the Yangzi continental margin.

Terranes of the Caledonian Folded Zone

The Caledonian folded zone in southeast China has a complex mosaic configuration.

Yunkai Metamorphic Terrane

The main part of the Yunkai terrane comprises Upper Sinian-Cambrian metamorphosed graywackes overlain by early Ordovician strata. The northwestern and southeastern flanks of the terrane are bounded by northeast trending faults. Amphibolite facies metamorphic rocks, migmatites, and intrusive granite bodies are developed in this terrane. The southwestern extension of Yunkai terrane is located on Hainan Island which was dispersed from the North Gulf and continental margin of China during the Cenozoic. The folded basement of Hainan Island is comparable with the Caledonian metamorphic rocks of the Yunkai metamorphic terrane on the mainland.

Nanhai Terrane

Nanhai terrane comprises siliceous carbonate rock, graptolitic shale and quartz sandstone of middle Cambrian and middle Ordovician age unconformably overlain by the late Ordovician Shanpo Group. This terrane has continental affinities. It was dispersed from mainland China during the Cenozoic.

Dayaoshan Terrane

The Dayaoshan terrane is an east-west elongated block composed of incipiently metamorphosed Cambrian graywacke flysch. The northern flank of the terrane comprises intermediate-basic volcanic rocks and siliceous rocks. The terrane is separated from the adjacent Yunkai metamorphic terrane by an east-west trending fault.

South Jiangxi Terrane

The South Jiangxi Terrane is north-south elongate and composed of Upper Sinian and Cambrian graywacke and basalt metamorphosed to greenschist facies and unconformably overlain by flysch of Upper Ordovician age. The terrane is characterized by a longitudinally oriented gravity gradient.

Wugongshan Terrane

The Wugongshan terrane, elongated near east-west, is composed of magnetite-bearing Late Proterozoic sandy shale and metamorphosed basic volcanic rocks and flyschoid rocks of the Sinian System. The central part of the terrane appears to be a dome structure of migmatite and granite.

North Fujian Metamorphic Terrane

North Fujian terrane is characterized by strongly metamorphosed schists named the Jian'ou Group and the Chencai Group. The terrane is bounded by the Shangyu-Zhenghe-Dapu fault in the southeast. There is an ophiolite suite along the fault, the age of which is still argued, but the main part of which may comprise an early Caledonian plate suture.

In summary, clastic rocks either unmetamorphosed or of greenschist grade and of Cambrian, Ordovician and Silurian age, developed in depressions among early Caledonian tectonostratigraphic terranes. The history of individual depressions differs, but the Cambrian fauna is one of mainly non-agnostid trilobites and brachiopods. Ordovician and Silurian faunas are planktonic. These lithologic and faunal features are different from those of the Yangzi and Jiangnan regions.

The Caledonian terranes, of differing structural configuration and nature and age of deformation, accreted to the Yangzi continent at different times but had accreted to the Proterozoic arc by the end of the early Paleozoic. Clastic rocks of Devonian age rest unconformably on the Caledonian orogenic collage and provide an overlap sequence.

Hercynian-Indosinian Terranes

Nanxi Terrane

The Nanxi terrane, located in a vast region largely covered by calc-alkaline volcanic rocks of a Mesozoic arc, is composed of Carboniferous flysch. This flysch differs from the platform carbonate facies rocks of the Carboniferous elsewhere in southern China. Thus the coastal area east of the Shangyu-Zhenghe-Dapu fault zone comprises a miogeosynclinal orogenic belt [Guo et al., 1983].

Dongshan Island Terrane

The Dongshan Island terrane, located within the Hercynian-Indosinian orogenic zone, is composed of gneiss about (?) 3000 Ma-4000 Ma.

Dananao Terrane

The Dananao terrane, located in the Cenozoic orogenic zone of Taiwan, is a Hercynian terrane.

In summary, Hercynian-Indosinian terranes were accreted to the Caledonian orogenic margin as a consequence of Hercynian-Indosinian orogenic movements, and as a result the continental margin of southeast China grew gradually southeastward.

Cenozoic Dispersion

Since the beginning of the Cenozoic, the Japanese Islands, Sakhalin Island, the Ryukyu Islands, Taiwan and the Philippines Islands, with trenches to the east and marginal seas to the west, have evolved into the well-known Cenozoic trench/arc/back-arc basin system by rifting close to the Asian continental margin. We agree that Taiwan, the Philippines Islands and Kalimantan are continental fragments (i.e. discrete terranes) dispersed from the continental margin of East Asia during the Mesozoic-Cenozoic. This conclusion is supported both by regional tectonic data and crustal thicknesses (average about 30 km).

The average crustal thickness in continental shelf areas of the East China Sea and the Strait of Taiwan, some 27 km, and Bouger gravity anomalies in this area, 40-50 mgal, indicate that the crust here is of continental character. Therefore we suggest that Taiwan and the Ryukyu Islands are part of an active continental margin linked to the east of China.

The Nanhai and Yunkai terranes of continental affinity on Hainan Island were dispersed from Guandong Province during the Cenozoic.

In summary, from the beginning of the Cenozoic, the continental margin of southeast China appears to have been the site of terrane dispersion with fragments now found in Taiwan and Hainan.

Conclusions

The terranes of the coastal region of southeast China can be divided into three groups according to their age of accretion: Jiangnan Proterozoic terranes, Caledonian terranes, and Hercynian-Indosinian terranes. The Proterozoic terranes were accreted to the Yangzi continental margin during the Xuefengian orogeny; the early Caledonian terranes were accreted to the Jiangnan arc, the new margin of the Yangzi continent, during the Caledonian orogeny; and the Hercynian-Indosinian terranes were accreted to the Caledonian folded zone during the Hercynian-Indosinian orogeny.

Subsequently, in the late Mesozoic and the Cenozoic, the terranes of Taiwan Island, Hainan Island and Xisha Island were dispersed from the continental margin of southeast China.

Accretion and dispersion at the continental margin of the coastal region of southeast China occurred at different times, and controlled the configuration of the margin. This accords with the point of view expressed by Howell et al. [1985] who stated that "The growth and shaping of continents may be viewed as a result of both terrane accretion and terrane dispersion".

References

Guo, Lingzhi, Yangshen Shi and Ruishi Ma, On the formation and evolution of the Mesozoic-Cenozoic active continental margin and island arc tectonics of the western Pacific ocean (in Chinese with English abstract), Acta Geologica Sinica, 1, 12-21, 1983.

Guo, Lingzhi, Yangshen Shi, Ruishi Ma, Shangfu Ye, and Huafu Lu, Tectonostratigraphic terranes of southeast China (in Chinese with English abstract), Acta Sci. Nat. Univ. Nanjing, 20, 732-739, 1984.

Howell, D.G., D.L. Jones, and E.R. Schermer, Tectonostratigraphic terranes of the Circum-Pacific region, in Tectonostratigraphic terranes of the Circum-Pacific region, edited by D.G. Howell, pp.3-30, Circum Pacific Council for Energy and Mineral Resources, Earth Science Series, No.1, 1985.

McElhinny, M.W., B.J.J. Embleton, Xinghua Ma, and ZhengKun Zhang, Late Paleozoic plate tectonics of Asia, Nature, 293, 212-216, 1981.

MESOZOIC TERRANES IN THE JAPANESE ISLANDS AND NEIGHBOURING EAST ASIA

Shinjiro Mizutani

Department of Earth Sciences, Nagoya University, Chikusa, Nagoya, 464, Japan

Abstract. Studies of radiolarian biostratigraphy, especially over the last decade, have led to a considerable revision of ideas concerning the Mesozoic geology of the Japanese Islands. The studies have revealed that most of the sedimentary complexes which were previously believed to be of Permian or Permian to Triassic age are of Jurassic age. The Tamba-Mino-Ashio terrane, for example, consists chiefly of an olistostromal complex made up of blocks of upper Paleozoic limestone locally associated with greenstone, Permian and Triassic bedded chert, and Jurassic siliceous shale, embedded within Jurassic clastics. Biostratigraphic data and structural trends indicate that the Tamba-Mino-Ashio terrane extends north to Sikhote Alin and northeast China, where a Jurassic sedimentary complex very similar to that of the Japanese Islands is found. Lithologic and paleontologic studies and paleogeographic reconstructions, lead to the conclusion that the Tamba-Mino-Ashio terrane and the sedimentary complex in the southern part of Sikhote Alin and in the Nadanhada Range in China formed a single large terrane along the continental margin of East Asia prior to opening of the Sea of Japan.

Introduction

Since the first report on radiolarian fossils was published in Japan by Yehara [1926], more than 300 papers have been written by Japanese paleontologists and biostratigraphers. Of these, three fourths have been published in the last five years [Mizutani, 1986]. Dealing with radiolarian skeletons extracted from shale, tuffaceous shale, tuff, siliceous shale and chert, most of the papers discuss the age, biostratigraphy and geologic structure of Late Paleozoic and Mesozoic units. Twenty radiolarian assemblages, ranging in age from Permian to Cretaceous, have been identified. Published data are summarized by Nakaseko et al. [1983], Ichikawa et al. [1985], Matsuoka and Yao [1986], and have been correlated with isotopic ages by Shibata and Mizutani, [1980, 1982].

These data have necessitated a reappraisal of ideas about the older sedimentary complexes of the Japanese Islands. They indicate that almost all of the sedimentary complexes which were once believed to be Permian, or Permian and Triassic, are in fact Jurassic and are composed of olistostromes made up of blocks of Permian limestone, sometimes associated with greenstone, Permian and Triassic bedded chert, and Jurassic siliceous shale, enclosed within Jurassic shale and sandstone [e.g., Mizutani et al., 1981; Ishiga, 1983; Yamamoto, 1985]. An olistostrome complex, entirely composed of Permian limestone, tuff and chert blocks embedded in Upper Permian shale, has also been recognised in southwest Japan [Miyake, 1985].

Based on this new information, terrane analyses have been carried out [Saito and Hashimoto, 1982; Mizutani and Hattori, 1983; Saito, 1985], together with the re-investigation of intraterrane features in Paleozoic - Mesozoic sedimentary units of the Japanese Islands.

The increased understanding of the Mesozoic geology of Japan, particularly of Jurassic radiolarian biostratigraphy, allows investigation of interterrane correlation between the Japanese Islands and the continental margin of East Asia. In this paper the geological similarities of some of the Mesozoic terranes in these two regions are outlined, and their paleogeographic relations prior to the opening of the Sea of Japan are discussed.

Mesozoic Geology of the Japanese Islands

The tectonic framework of the Japanese Islands is well illustrated in Honshu and Shikoku in southwest Japan where a number of tectonostratigraphic terranes have been recognized (Fig. 1) as outlined from north to south below.

The Hida terrane consists of metamorphosed Paleozoic and possibly Precambrian rocks, and Triassic granite, unconformably overlain by Jurassic and Cretaceous clastics. The Circum-Hida belt, which occupies a narrow suture zone to the south of the Hida terrane, comprises a complicated structural unit including blocks of lower to upper Paleozoic rocks, serpentinite, and metamorphic rocks yielding dates of between 310 and 360 Ma [Nozawa and Shibata, 1982]. These are

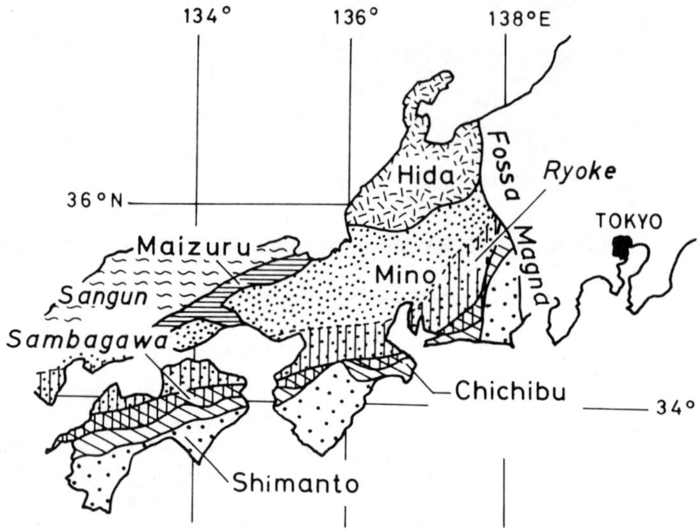

Fig. 1. Tectonic divisions of central Japan.

unconformably overlain by Jurassic and Cretaceous clastics. The Sangun terrane is made up of an upper Paleozoic accretionary complex and high P/T metamorphics derived therefrom. The Maizuru terrane is an ophiolite complex overlain by Triassic shallow marine clastics. Gabbroic and metamorphic rocks from this terrane have isotopic ages ranging from 240 to 330 Ma [Nozawa and Shibata, 1982; Shibata and Nozawa, 1982]. The Mino terrane comprises blocks of upper Paleozoic limestone and greenstone, Permian and Triassic bedded chert, and Jurassic siliceous shale, embedded within Jurassic clastics. The Ryoke metamorphic complex consists of low P/T metamorphics associated with Cretaceous granitic rocks regarded as the metamorphic equivalent of the sedimentary formations of the Mino terrane. The Chichibu terrane is characterised by blocks of upper Paleozoic limestone, Triassic bedded chert, and Jurassic siliceous shale set in Jurassic clastics, a lithofacies very similar to that of the Mino terrane. The Sambagawa metamorphic rocks are believed to be the metamorphic equivalent of the Chichibu terrane. The Kurosegawa belt, which is interleaved within the Chichibu terrane, contains blocks of lower to upper Paleozoic and Triassic stratified rocks, sheared granite that yields isotopic dates of 400-450 Ma, and high P/T metamorphics, embedded in serpentinite. The Shimanto terrane of Cretaceous greenstone and chert intercalated in thick late Cretaceous clastics extends for much of the length of the Japanese Islands.

The Tamba-Mino-Ashio Terrane

The Mino terrane extends to the west where the tectonostratigraphically equivalent unit, the Tamba terrane, is found. It is separated from its eastern equivalent, the Ashio terrane, by the Tertiary basin of the Fossa Magna (Fig. 1). The geology of the Neo area described by Yamamoto [1985], is an example of the complicated lithology, biostratigraphy and geologic structure found in the Mino terrane of central Japan (Fig. 2 and Fig. 3). Exotic blocks or sheets of a greenstone-limestone complex, bedded chert, mudstone and siliceous shale, are included in late Jurassic, or possibly even early Cretaceous, clastic rocks. Permian fusulinaceans (*Acervoschwagerina endoi*, *Neoschwagerina simplex*, *Pseudofusulina kraffti*, *P. vulgaris*, and others) are found in the limestone; Permian radiolarians (*Albaillella levis*, *A. triangularis*, *Follicucullus monacanthus*, *F. scholasticus*, *F. ventircosus*, *Pseudoalbaillella rhombothoracata* and others) and Triassic radiolarians (*Canoptum triassicum*, *Pseudoheliodiscus gracilis*, *Triassocampe deweveri*, *T. japonica* and others) occur in the bedded chert, and Jurassic radiolarians (*Cyrtocapsa mastoidea*, *Dictyomitrella* (?) *kamoensis*, *Eucyrtidiellum ptyctum*, *Mirifusus baileyi*, *M. guadalupensis*, *Pantanellium foveatum*, *Parvicingula mashitaensis*, *Protunuma fusiformis*, *Pseudodictyomitra minoensis*, *Stichocapsa japonica*, *Tricolocapsa plicarum*, *T. tetragona*, *Unuma echinatus*, *Xitus gifuensis* and others) are found in the mudstone and siliceous shale. The disorganized succession and olistostromal nature of these fossiliferous rocks are well displayed in a route map of Yamamoto [1985] as shown in Fig. 4. No entirely coherent sequence from Permian to Jurassic age has been observed in the Neo area. The fossil evidence and the nature of the rocks suggest that two lithostratigraphic assemblages can be distinguished; assemblage "A" consists mainly of an olistostrome complex in which exotic blocks of Permian greenstone-limestone and

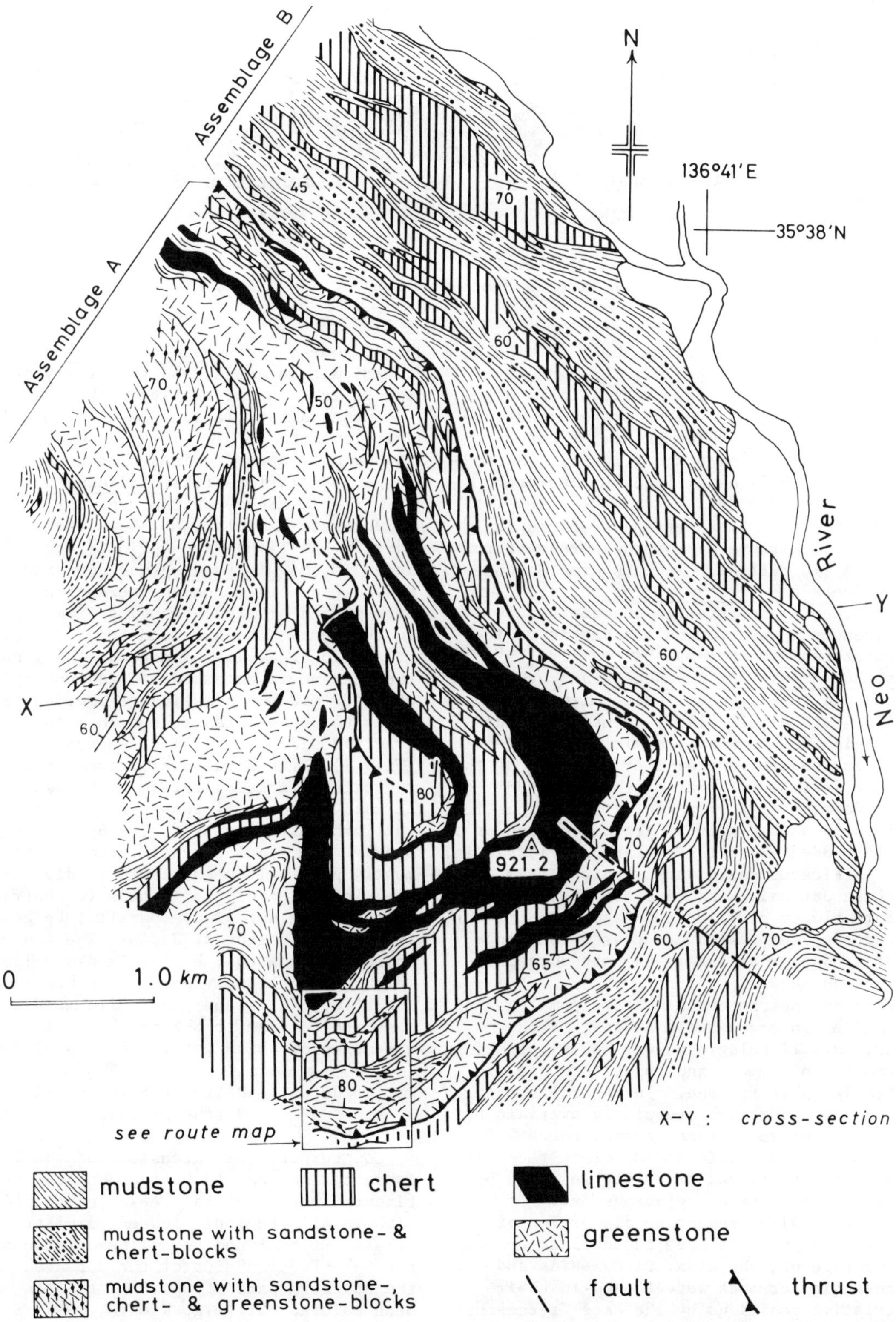

Fig. 2. Geologic map of the Neo area in the Mino terrane (redrawn from Yamamoto [1985]).

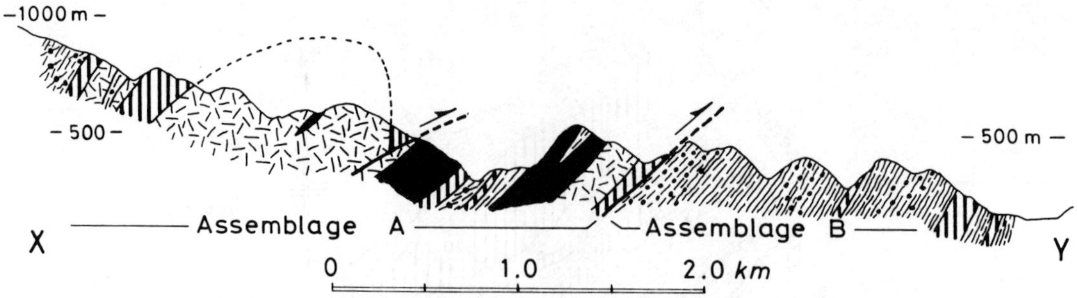

Fig. 3. Cross-section on X - Y in the geologic map of the Neo area (Fig. 2) (redrawn from Yamamoto [1985]).

Triassic bedded chert are set in shale of Middle Jurassic age, and assemblage "B", also an olistostrome complex, which carries exotic blocks of Triassic bedded chert and Upper Jurassic mudstone and shale embedded in a muddy matrix of Late Jurassic - Early Cretaceous age (Fig. 5). According to Yamamoto [1985] assemblage "A" has been thrust over assemblage "B".

Another example of intraterrane features is shown in Fig. 6, a route map of the northeastern part of the Ashio terrane. In this area Carboniferous conodonts and Neoschwagerina sp. are found in limestone, and late Triassic conodonts in chert. Recently however, middle Jurassic radiolarians (Cyrtocapsa mastoidea, Dictyomitrella (?) kamoensis, Eucyrtidiellum unumaensis and others) were discovered in siliceous shale [Mizutani et al., 1984]. Sato et al. [1985) pointed out that the lithologic and biostratigraphic assemblages in the Ashio terrane are very similar to those of the Mino terrane in that (i) most of the limestone blocks are Permian or Permian to Carboniferous in age and locally accompanied by greenstone and/or chert, (ii) bedded chert is Permian or Triassic in age, (iii) siliceous shale ranges in age from Early to Late Jurassic, and (iv) all these lithologies form blocks or separated sheets that occur in a disrupted fashion within a clastic sequence made up of sandstone and shale of Late Jurassic - Early Cretaceous age.

Paleozoic and Mesozoic sedimentary complexes of the Mino and Ashio areas and also part of the Tamba area, have lithologic as well as biostratigraphic traits in common, and constitute one tectonostratigraphic terrane. In central Japan, the Mino and Hida terranes are widely overlain by late Cretaceous effusive rocks termed the Nohi rhyolite, and the Paleozoic and Mesozoic rocks in the Ashio area are covered or partly intruded by acidic rocks also of Late Cretaceous age. Although lateral displacement or dispersion of these terranes might well have taken place before the Late Cretaceous, the Hida, Circum-Hida and Tamba-Mino-Ashio terranes were lodged in their present relative positions by the Late Cretaceous.

Radiolarian biostratigraphic zonation of the Jurassic in the Japanese Islands, as summarized by Yao [1986], involves seven or eight assemblage zones; the Unuma echinatus zone represents the middle Jurassic, and the overlying Guexella nudata zone is correlative with a deep-sea core including a radiolarian association of Callovian age [Yamamoto et al., 1985]. The upper Jurassic is commonly characterized by the presence of large radiolarians like Mirifusus guadalupensis or M. baileyi. Study of the regional distribution of Jurassic radiolarians shows that the northern part of the Tamba-Mino-Ashio terrane is impoverished in the Mirifusus group compared to the southern part. Although the entire sequence is extremely disrupted, younging from north to south in the Tamba-Mino area and from west to east in the Ashio area, is indicated. The inner part of the terrane is composed of sedimentary successions of up to middle Jurassic age whereas successions of up to upper Jurassic age form the outer part (Pacific side) of the Tamba-Mino-Ashio terrane (Fig. 7). The structural trend, as denoted by the axial traces of folds in the sedimentary formations, overall runs parallel to the trend of the Tamba-Mino terrane. The distribution of sedimentary units and of radiolarian fossils are also mostly parallel to the tectonic divisions and to the elongation of the islands in central Japan, but they are not always parallel to geographic features in northeast Japan. The Ashio terrane is bounded to the east by the Tanakura Tectonic Line which runs obliquely across northeast Honshu. The structural trend twists from E-W to SW-NE in the Mino terrane to N-S in the Ashio terrane, the extension of which enters the Sea of Japan region.

Jurassic Units in Sikhote Alin and Northeast China

Counterparts or extensions of the Tamba-Mino-Ashio terrane should be found on the present continental margin of East Asia, particularly in the Sikhote Alin region. Indeed, similarities of Triassic floral and faunal associations between the Sea of Japan side of the Japanese Islands and the southern part of the Sikhote Alin region have been pointed out [e.g. Kimura, 1984; Silberling, 1985]. Telling evidence is found in the comprehensive description by Zhamoida [1972] of the distribution of Mesozoic radiolarians and the

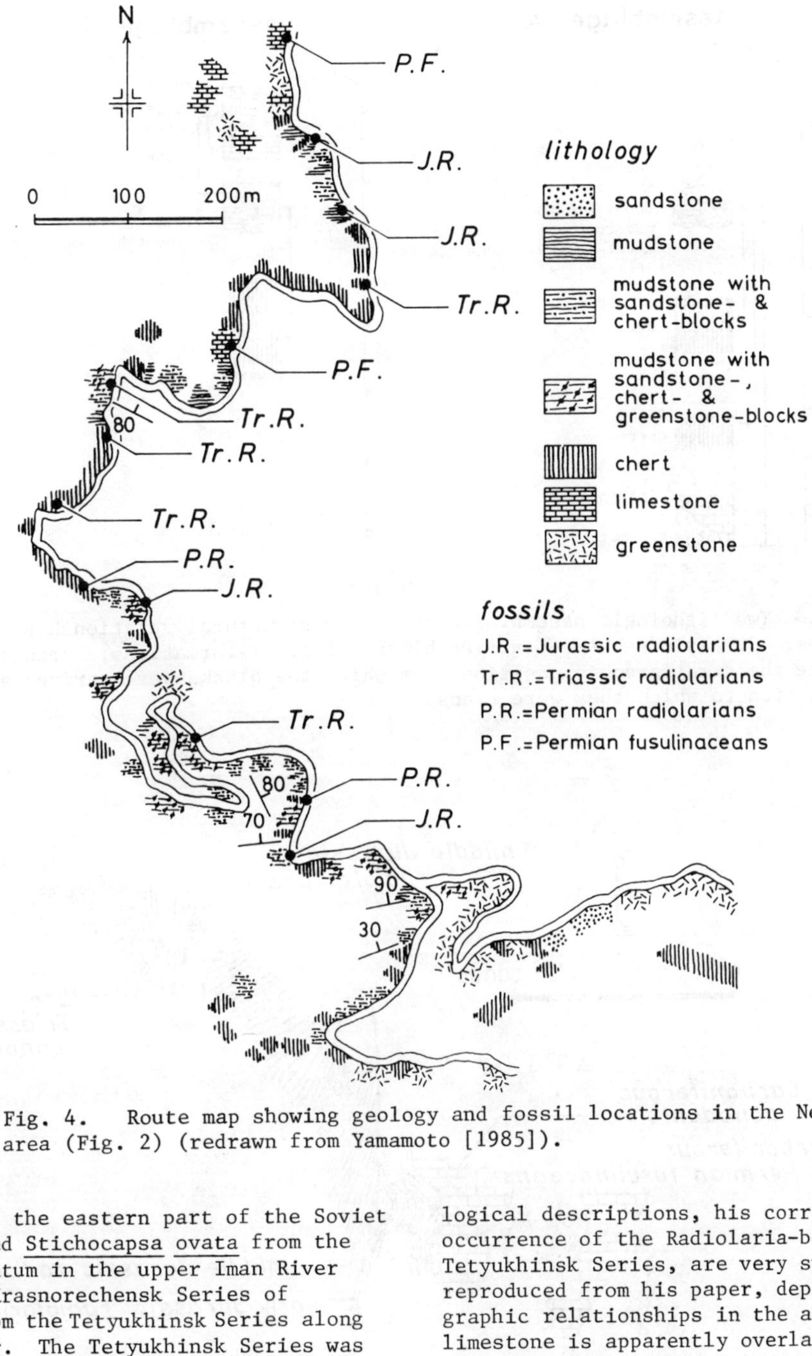

Fig. 4. Route map showing geology and fossil locations in the Neo area (Fig. 2) (redrawn from Yamamoto [1985]).

biostratigraphy of the eastern part of the Soviet Union. He reported Stichocapsa ovata from the Nikolaevskaya Stratum in the upper Iman River region, from the Krasnorechensk Series of Khekhtsir, and from the Tetyukhinsk Series along the Tetyukhe River. The Tetyukhinsk Series was referred to the Late Triassic because it appears to have intercalated limestone masses which yield Late Triassic bivalves. On the other hand, Zhamoida [1972] described Stichocapsa plicatoderma and S. ovatoidea from the Erdagonsk Series, which he assigned to the Jurassic. All these species of Stichocapsa have similarities in size, external shape and skeletal structure to the Mirifusus group of Pessagno [1977], one of the characteristic Jurassic genera commonly found in the circum-Pacific region. Zhamoida's paleontological descriptions, his correlations, and the occurrence of the Radiolaria-bearing rocks in the Tetyukhinsk Series, are very suggestive. Fig. 8, reproduced from his paper, depicts the stratigraphic relationships in the area: upper Triassic limestone is apparently overlain by a thick clastic sequence containing many lenticular bodies of limestone, limestone-breccia, siliceous shale, mudstone and tuff. Radiolarian fossils were extracted from the siliceous shale and the mudstone. These rock types, their ages, structural characteristics and biostratigraphy are similar to those of the Mesozoic sedimentary complexes of the Tamba-Mino-Ashio terrane in the Japanese Islands.

From the Nadanhada Range, located in the eastern part of Heilongjiang Province of northeast China, Li et al. [1979] have reported a limestone

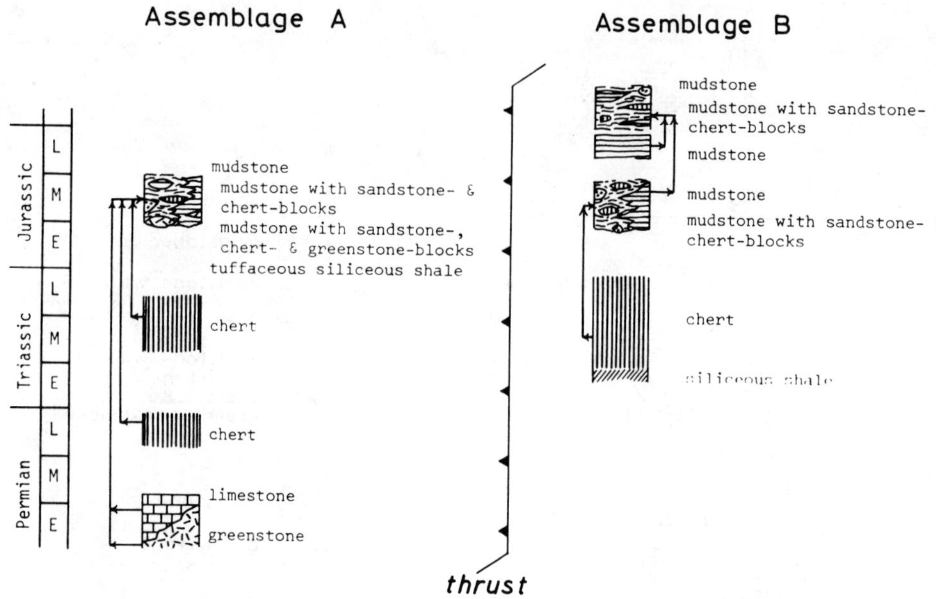

Fig. 5. Two lithologic assemblages and their structural relationship in the Neo area. Jurassic sequence carries blocks of older formations; arrows indicate the stratigraphic position from which the blocks were derived and the position to which they were transported.

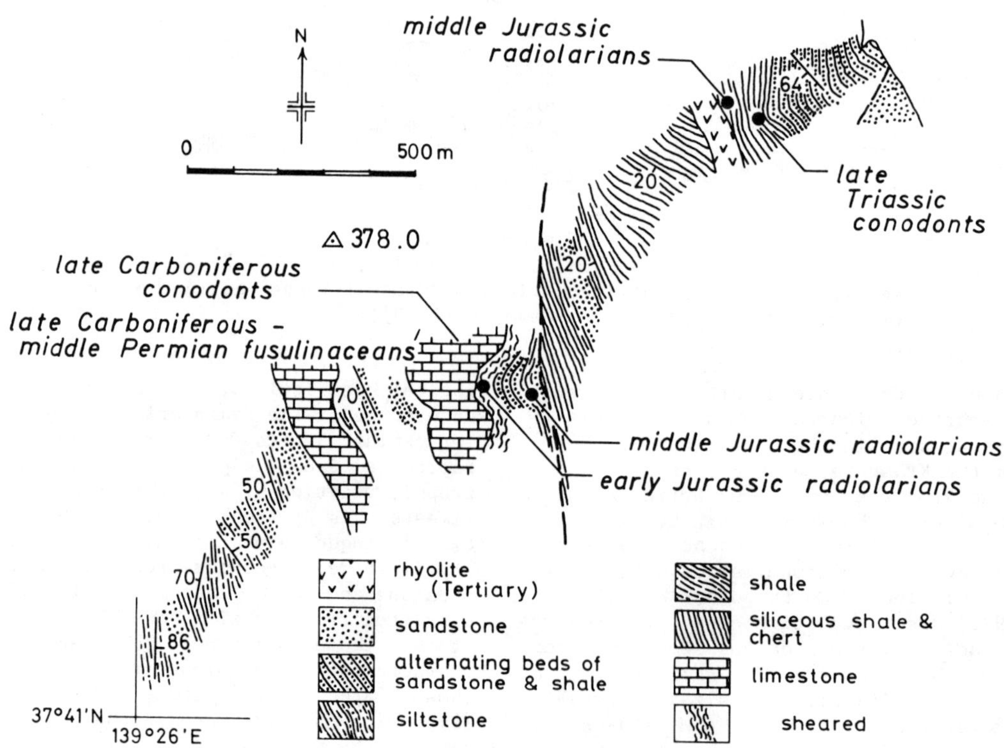

Fig. 6. Route map showing geology and fossil locations in the Tsugawa area in the Ashio terrane (redrawn from Mizutani et al. [1984]).

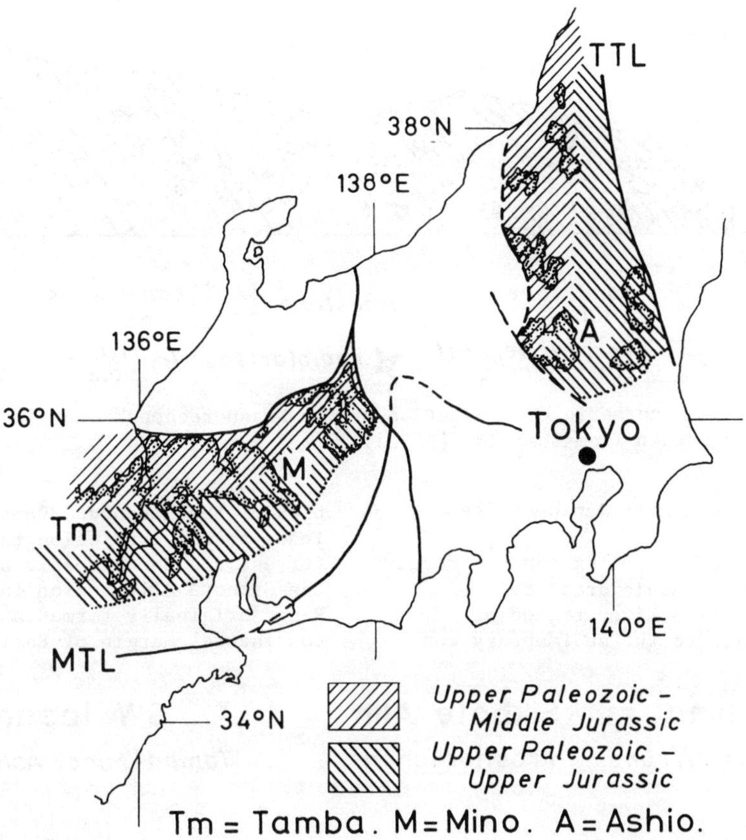

Fig. 7. Tamba (Tm) - Mino (M) - Ashio (A) Terrane. Dotted areas show exposed sedimentary complex of Upper Paleozoic to Jurassic age. MTL - Median Tectonic Line; TTL = Tanakura Tectonic Line.

that contains Carboniferous and Permian fusulinacean faunas. However, their report also described the occurrence of radiolarian fossils of Jurassic type [Li et al., 1979, Plate I, Figs. 1-5]. Furthermore, their geologic sections strongly suggest that the main part of the sedimentary complexes which constitute the Nadanhada Range and the Tamba-Mino-Ashio terrane have much in common both in their lithologies and their biostratigraphic assemblages (Fig. 9). Very recently upper Triassic red chert, almost identical to that of central Japan, has been found in the Nadanhada Range (Dr. J. A. Shao, Beijing University, personal communication, 1986). Undoubtedly the Nadanhada area is composed of an olistostromal complex of Jurassic age joined to an equivalent part of the Sikhote Alin region. Feng and Yang [1984] pointed out the chaotic nature of geologic formations in the Nadanhada Range, and stated that, from a tectonic viewpoint, it represents a mobile belt not correlative with those of the mainland of China but related to the circum-Pacific region. As illustrated in Fig. 9, the stratigraphic sequence of the Nadanhada Range preserves a sedimentary history quite different from that of other areas in northeast China, and it constitutes a separate tectonostratigraphic terrane, called the Nadanhada terrane. In Late Jurassic to Early Cretaceous time, the Nadanhada terrane was located to the east of, or formed part of, the Jamusi terrane described by Xiong and Coney [1985].

Discussion and Conclusion

The structural trend and spatial distribution of the Jurassic olistostrome complex of the Tamba-Mino-Ashio terrane suggest that its northern extension is located in the Sikhote Alin region.

Although a number of small blocks of continental affinity have been recognized in the Sea of Japan, most of this region is believed to be underlain by oceanic crust [e.g. Tamaki, 1985]. Paleomagnetic studies by Otofuji and Matsuda [1983] and Otofuji et al. [1985] on the Late Cretaceous and Tertiary rocks of Japan show that southwest Japan was rotated clockwise 58 degrees between 28 Ma and 12 Ma, while northeast Japan was rotated counterclockwise 47 degrees in the past 14 Ma. On the basis of these studies paleogeographic maps showing the relationship of the Japanese Islands and Korea to Sikhote Alin

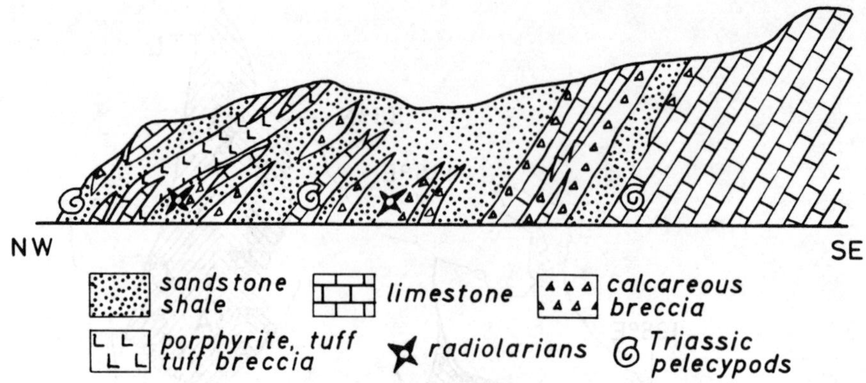

Fig. 8. Schematic cross-section of the Krasnorechensk area in Sikhote Alin (redrawn from Zhamoida [1972]).

prior to opening of the Sea of Japan can be reconstructed (Fig. 10).

A Jurassic olistostrome complex can be traced from the Tamba, Mino and Ashio areas to Krasnorechensk in the Iman River region of Sikhote Alin, and north to the sedimentary complex of the Nadanhada Range (Fig. 10), which leads to the conclusion that the Tamba-Mino-Ashio terrane and the Mesozoic sedimentary complex in the Sikhote Alin region including the Nadanhada Range originally formed a single terrane on the continental margin of East Asia. The occurrence

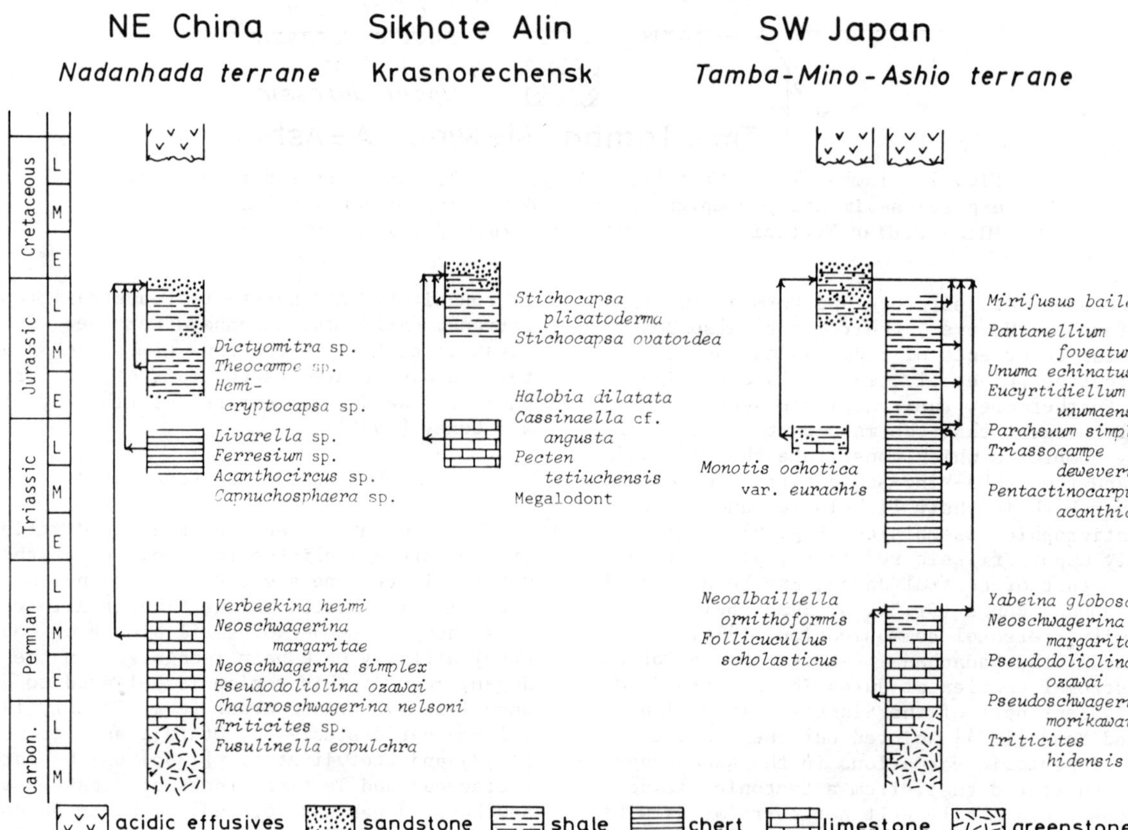

Fig. 9. Litho- and biostratigraphic interterrane correlation in East Asia. Jurassic sequence in each region carries blocks of older formations; arrow indicates the stratigraphic position from which the blocks were derived and the position to which they were transported.

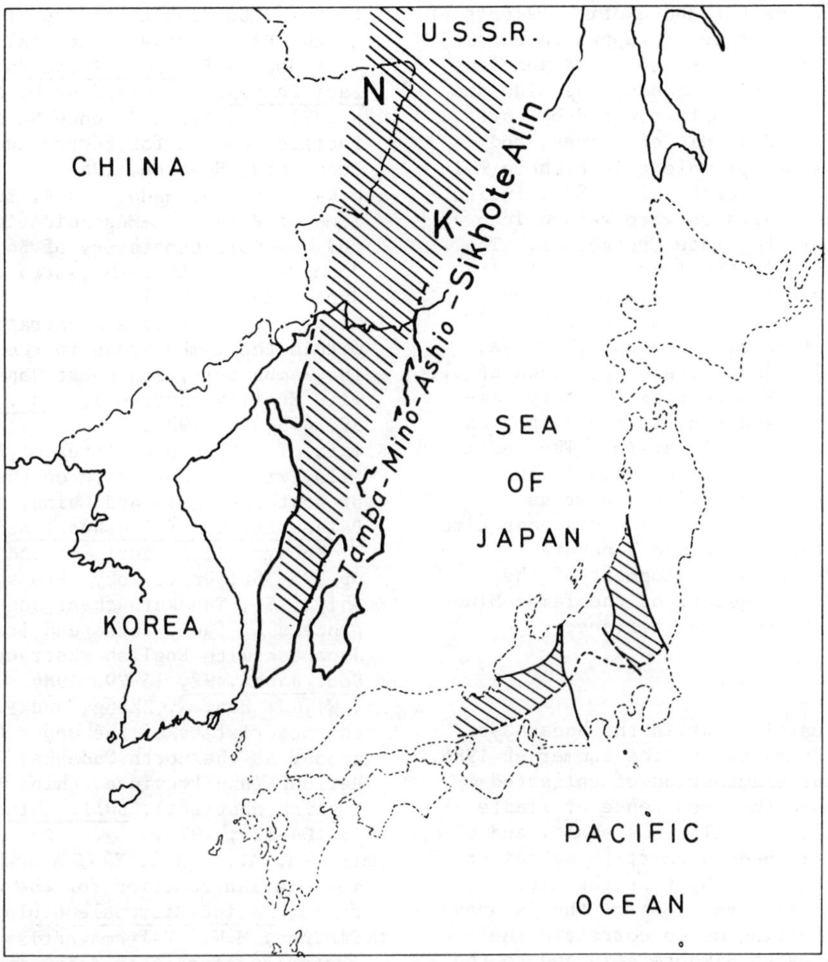

Fig. 10. Reconstructed paleogeographic relation of Tamba-Mino-Ashio-Sikhote Alin terrane based on paleomagnetic data by Otofuji et al. [1983]. N = Nadanhada range in Heilongjiang Province of China; K = Krasnorechensk of Sikhote Alin.

of large-sized upper Jurassic radiolarians in the southern part of Sikhote Alin suggests a probable correlation between this area and the outer part of the Tamba-Mino-Ashio terrane.

In the Nadanhada Range, a Jurassic ophiolite belt consisting of metamorphic peridotite, cumulate ultramafic and mafic rocks, and pillow lavas extends for more than 200 km [Zhang et al., 1984]. Judging from the spatial relationship of the Tamba-Mino-Ashio terrane to the Maizuru terrane and the Circum-Hida belt, all the areas including mafic, ultramafic and greenstone complexes may have constituted a large suture zone along the northwestern (continental) side of the Tamba-Mino-Ashio-Sikhote Alin terrane.

Paleomagnetic studies by Hattori [1982] and Hirooka et al. [1985] show that Triassic bedded chert and Jurassic siliceous shale in the Mino terrane, which have shallow initial inclinations, were originally magnetized in an equatorial region. A sedimentary complex including these equatorial elements, was built at an ancient continental margin where oceanic sediments were tectonically intermingled with coarse clastics of continental provenance. It is highly probable that many isolated or fragmented blocks composed of pre-Jurassic rocks impinged onto this margin and olistostromal complexes including exotic blocks of Permian or Triassic age were produced. According to the paleomagnetic study of McElhinny [1973], some geologic elements now found in the eastern part of the Soviet Union were once situated in equatorial regions and were subsequently accreted to a continental margin. Such accretion occurred during the Late Jurassic or Early Cretaceous, for upper Jurassic radiolarian shale is an element of the Mesozoic sedimentary complex. Along with or after that event, dispersion of the Tamba-Mino-Ashio-Sikhote Alin terrane took place along the Circum-Hida tectonic belt, where many fragmental and exotic elements of oceanic affinity and of lower to upper Paleozoic, and partly of Middle Jurassic age, were jumbled together. In the Japanese Islands, the Hida and the Mino terranes are commonly covered by late Cretaceous rhyolite and acidic effusives, and granitic intrusions are prevalent in Sikhote Alin and in northeast China [Takahashi, 1983), indicating that these older terranes were welded into a single terrane before the late Cretaceous. The Median Tectonic Line and the Tanakura Tectonic Line, major structures in southwest Japan and northwest Japan respectively, show evidence of late Mesozoic sinistral displacement [Koshiya, 1986; Takagi, 1986]. As discussed by Ozawa et al. [1985], the Paleozoic-Mesozoic sedimentary complexes were disrupted and sinistrally dispersed along an ancient continental margin. The sedimentary manifestation of that process is the olistostromal Jurassic formations, whereas ensuing tectonism through to late Cretaceous time was intense along the Median and Tanakura Tectonic Lines and led to development of the presently observed arrangement of the Tamba-Mino-Ashio terrane and the Chichubu terrane.

Postscript

Field work was carried out in the Nadanhada area of northeast China during the summer of 1986. Micropaleontological examination of collected samples has disclosed the occurrence of middle Jurassic radiolarians in siliceous shale, and of late Triassic ones in bedded chert [Mizutani et al., 1986]. It is noteworthy that the microfossils are very similar to those of the Japanese Islands, thereby enabling us to correlate the Tamba-Mino-Ashio area to Sikhote Alin and to the Nadanhada area as discussed in this paper.

Acknowledgements. I would like to express my sincere appreciation to Drs. Y. Saito, A. Taira, I. Hattori, M. Adachi, and S. Kojima for fruitful discussions and invaluable suggestions given to me during all phases of this work. My studies on the tectonics of the circum-Pacific region have been financially supported by grants from the Ministry of Education, Science and Culture of the Japanese Government (Grant Nos. 5041130, 57340047 and 61042007) and by the Japan Society for the Promotion of Sciences, for which I record my thanks.

References

Feng, Y., and Y. Yang, Chaotic geological bodies in China (in Chinese with English abstract), Bull. Chin. Acad. Geol. Sci., 10, 77-92, 1984.

Hattori, L., The Mesozoic evolution of the Mino terrane, central Japan: A geologic and paleomagnetic synthesis, Tectonophysics, 85, 313-340, 1982.

Hirooka, K., S. Uchiyama, T. Date, H. Kanai, and T. Nakajima, Paleomagnetic evidence for accretion and bending tectonics of the Hida and the circum-Hida Terranes, central Japan, in Tectonostratigraphic Terranes of the Circum-Pacific Region, edited by D. G. Howell, pp.391-399, Earth Science Series, no.1, Circum Pacific Council for Energy and Mineral Resources, Houston, 1985.

Ichikawa, K., S. Hada, and A. Yao, Recent problems of Paleozoic-Mesozoic microbiostratigraphy and Mesozoic geohistory of Southwest Japan (in Japanese with English abstract), Mem. Geol. Soc. Japan, 25, 1-18, 1985.

Ishiga, H., Two suites of stratigraphic succession within the Tamba Group in the western part of the Tamba belt, Southwest Japan (in Japanese with English abstract), Jour. Geol. Soc. Japan, 89, 443-454, 1983.

Kimura, T., Mesozoic floras of East and Southeast Asia, with a short note on the Cenozoic floras of southeast Asia and China, in Geology and Palaeontology of Southeast Asia, edited by T. Kobayashi, R. Toriyama and W. Hashimoto, pp.325-350, Univ. Tokyo Press, 1984.

Koshiya, S., Tanakura shear zone: the deformation process of fault rocks and its kinematics (in Japanese with English abstract), Jour. Geol. Soc. Japan, 92, 15-29, 1986.

Li, W., J. Han, S. Zhang, and F. Meng, The main characteristics of the upper Palaeozoic stratigraphy at the north Nadanhada Range, Heilongjiang Province, China (in Chinese with English abstract), Bull. Chin. Acad. Geol. Sci., 1, 104-120, 1979.

Matsuoka, A., and A. Yao, A newly proposed radiolarian zonation for the Jurassic of Japan, Marine Micropaleontology, in press, 1986.

McElhinny, M.W., Paleomagnetism and plate tectonics of eastern Asia, in The Western Pacific: Island Arcs, Marginal Seas, Geochemistry, p.407-414, Univ. Western Australia Press, 1973.

Miyake, K., Permian olistostrome complex in the Katsuyama area, Okayama Prefecture, Southwest Japan (in Japanese with English abstract), Jour. Geol. Soc. Japan, 91, 463-475, 1985.

Mizutani, S., List of papers on radiolarian fossils and radiolarian biostratigraphy by Japanese paleontologists and stratigraphers, News of Osaka Micropaleontologists, Spec. Vol., 7, 327-361, 1986.

Mizutani, S., and I. Hattori, Hida and Mino: tectonostratigraphic terranes in central Japan, in Accretion Tectonics in the circum-Pacific Regions, edited by M. Hashimoto and S. Uyeda, pp.169-178, Terra Sci. Publ. Co., Tokyo, 1983.

Mizutani, S., I. Hattori, M. Adachi, K. Wakita, Y. Okamura, S. Kido, I. Kawaguchi, and S. Kojima, Jurassic formations in the Mino area, central Japan, Proc. Japan Acad., 57(B), 194-199, 1981.

Mizutani, S., S. Kojima, J.A. Shao, and Q.L. Zhang, Mesozoic radiolarians from the Nadanhada area area, northeast China, Proc. Japan Acad., 62(B), in press, 1986.

Mizutani, S., T. Uemura, and H. Yamamoto, Jurassic radiolarians from the Tsugawa area, Niigata Prefecture, Japan, Earth Science Jour. Assoc. Geol. Collab. Japan, 38, 352-358, 1984.

Nakaseko, K., S. Mizutani, and A. Yao, Radiolarian fossils and Mesozoic geology of the Japanese Islands (in Japanese), Kagaku (Science), 53, 177-183, 1983.

Nozawa, T., and K. Shibata, Radiometric Age Map 3. Metamorphic rocks, Scale 1:4,000,000. Geological Atlas of Japan, 74-75, Geological Survey of Japan, 1982.

Otofuji, Y., and T. Matsuda, Paleomagnetic evidence for the clock-wise rotation of Southwest Japan, Earth Planet. Sci. Let., 62, 349-359, 1983.

Otofuji, T., T. Matsuda, and S. Nohda, Paleomagnetic evidence for the Miocene counterclockwise rotation of Northeast Japan - rifting process of the Japan Arc, Earth Planet. Sci. Let., 75, 265-277, 1985.

Ozawa, T., A. Taira, and F. Kobayashi, Zonal structures of Southwest Japan - How formed? (in Japanese), Kagaku (Science), 55, 4-13, 1985

Pessagno, E.A., Jr., Upper Jurassic Radiolaria and radiolarian biostratigraphy of the California Coast Ranges, Micropaleontology, 23, 56-113, 1977.

Saito, Y., Jurassic geologic framework in the Japanese Islands, in Tectonostratigraphic Terranes of the Circum-Pacific Region edited by D. G. Howell, pp.401-407, Earth Science Series, No.1, Circum-Pacific Council for Energy and Mineral Resources, Houston, 1985.

Saito, Y., and M. Hashimoto, South Kitakami region: an allochthonous terrane in Japan, Jour. Geophys. Res., 87, 3691-3696, 1982.

Sato, T., S. Takizawa, K. Sashida, Y. Katsura, and H. Aono, Guidebook for Excursion: Jurassic in Japan, IGCP 171, Circum-Pacific Jurassic, Field Conference III, 1-59, Univ. Tsukuba, 1985.

Shibata, K., and S. Mizutani, Isotopic ages of siliceous shale from Hida-Kanayama, central Japan, Geoch. Jour., 14, 235-241, 1980.

Shibata, K., and S. Mizutani, Isotopic ages of Jurassic siliceous shale and Triassic bedded chert in Unama, central Japan, Geoch. Jour. 16, 213-223, 1982.

Shibata, K., and T. Nozawa, Radiometric age map 1. Granitic rocks, Scale 1:4,000,000. Geological Atlas of Japan, 66-67, Geological Survey of Japan, 1982.

Silberling, N.J., Biogeographic significance of the upper Triassic bivalve Monotis in circum-Pacific accreted terranes, in Tectonostratigraphic Terranes of the circum-Pacific Region, edited by D. G. Howell, pp.63-70, Earth Science Series, No.1, Circum-Pacific Council for Energy and Mineral Resources, Houston, 1985.

Takagi, H., Implications of mylonitic microstructures for the geotectonic evolution of the Median Tectonic Line, Central Japan, Jour. Struct. Geol., 8, 3-14, 1986.

Takahashi, M., Space-time distribution of late Mesozoic to early Cenozoic magmatism in East Asia and its tectonic implications, in Accretion Tectonics in the Circum-Pacific Regions, edited by M. Hashimoto and S. Uyeda, pp.69-88, Terra Sci. Publ. Co., Tokyo, 1983.

Tamaki, K., Two models of back-arc spreading, Geology 13, 475-478, 1985.

Xiong, J., and P.J. Coney, Accreted terranes of China, in Tectonostratigraphic Terranes of the Circum Pacific Region, edited by D. G. Howell, pp.349-361, Earth Science Series. No.1, Circum-Pacific Council for Energy and Mineral Resources, Houston, 1985.

Yamamoto, H., Geology of the late Paleozoic-Mesozoic sedimentary complex of the Mino Terrane in the southern Neo area, Gifu Prefecture and the Mt. Ibuki area, Shiga Prefecture, central Japan (in Japanese with English abstract), Jour. Geol. Soc. Japan, 91, 353-369, 1985.

Yamamoto, H., S. Mizutani, and H. Kagami, Middle Jurassic radiolarians from Blake Bahama Basin, West Atlantic Ocean, Bull. Nagoya Univ. Museum 1, 25-49, 1985.

Yao, A., Geological age of Jurassic radiolarian zones in Japan and their international correlations (in Japanese with English abstract), News of Osaka Micropaleontologists, Spec. Vol., 7, 63-74, 1986.

Yehara, S., On the Monobegawa and Shimantogawa-Series in southern Shikoku, Jour. Geogr. Japan, 38, 1-10, 1926.

Zhamoida, A. I., Biostratigraphy of the Mesozoic siliceous strata of the east of the USSR (as based on the study of radiolarians), Min. Geol., USSR, Trans. Geol. Sci. Res. Inst., New Ser., 183, 1-99, 1972.

Zhang, Z. M., J.G. Liou., and R.G. Coleman. An outline of the plate tectonics of China, Geol. Soc. Am. Bull., 95, 295-312, 1984.

COMPLEX FRAGMENTATION OF PERMO-TRIASSIC AND JURASSIC ACCRETED TERRANES IN THE CHUGOKU REGION, SOUTHWEST JAPAN AND THE FORMATION OF THE SANGUN METAMORPHIC ROCKS

Teruo Watanabe, Takao Tokuoka, and Takahito Naka

Department of Geology, Shimane University, Matsue, 690, Japan

Abstract. In the Chugoku region both Permo-Triassic and Jurassic terranes are recognised. The former includes the Akiyoshi-Nishiki and Maizuru terranes and the latter, a complex olistostromal terrane incorporating rocks of the Tsunotani, Nichihara and Kuga-Mikawa belts which may be an extension of the Tamba terrane. The high pressure Sangun metamorphic rocks were formed during accretion of both the Permo-Triassic and Jurassic terranes and associated subduction processes. Rocks of an older group (250-310 Ma) occur adjacent to serpentinite masses and older metagabbros and/or gneisses (370-470 Ma) in the Nagato, Wakasa, and Oosayama areas and comprise the newly defined Nagato-Wakasa terrane. This terrane is probably the western extension of the serpentinite-melange zone called the Hida marginal belt. The rocks evolved in the subduction zone of a Permian arc-trench system. In contrast younger metamorphic rocks (160-225 Ma), forming the Sangun terrane, occur widely in the Chugoku region, are rarely associated with serpentinite, and are associated with Jurassic thrust sheets. Part of the Sangun terrane, for example schists of the Gotsu area, is of polymetamorphic origin. Terranes of the Chugoku region have a fragmented and scattered distribution as compared with those of the region south of the Median Tectonic Line and were formed during middle-late Mesozoic time in a suture zone with significant strike-slip movement in the northwestern Pacific rim, adjacent to the Jurassic granitic terrane in Korea, and Jurassic oceanic terranes including those of the Sanbagawa and Chichibu belts of southwest Japan.

Introduction

The Median Tectonic Line divides southwest Japan into two regions of contrasting geology. The region south of the Line, on the Pacific Ocean side, comprises the Outer Zone of southwest Japan and makes up the Shikoku region, whereas to the north, on the Japan Sea side, is the Inner Zone which makes up the Chugoku region. Some attempts at terrane recognition in southwest Japan have been based on recent biostratigraphic studies [Kanmera and Nishi, 1983; Saito, 1983; Taira, 1983; Ichikawa, 1984; Ozawa et al., 1985]. The discovery of radiolarian fossils in the Chugoku region has enabled the identification of Jurassic and Permian olistostromes and demonstrates that terrane distribution is complicated and disharmonic when compared with that in the Shikoku region which shows the juxtaposition of subparallel terranes, i.e., the Sanbagawa, Chichibu and Shimanto terranes. This poses the question: how has the complicated terrane distribution in the Chugoku region arisen?

In this paper we re-examine terrane recognition and distribution, and discuss the formation of terranes of both Permo-Triassic and Jurassic age, and the significance of the Sangun metamorphic rocks. The nature of terranes is discussed, as is their relationship to the Sangun rocks. The Sangun metamorphic belt was once considered to be the high pressure partner of a Permian-Jurassic paired metamorphic belt [Miyashiro, 1973]. However we distinguish older Sangun rocks (250-310 Ma) and younger Sangun rocks (160-225 Ma). The older rocks are associated with serpentinite masses and in places with olistostrome units the matrix of which contains middle Permian radiolarians. The younger rocks are associated with thrust sheets of Permo-Triassic rocks and Jurassic olistostromes.

The main purpose of this paper is to outline the complicated distribution of both the Permo-Triassic and Jurassic rocks and the related metamorphic belts. The processes involved in the fragmentation and mixing of the Permo-Triassic and Jurassic terranes and the Sangun rocks are difficult to understand, but some preliminary ideas are discussed.

Tectonostratigraphic Units

Prior to the introduction of the terrane concept, the major geotectonic units recognised in Japan comprised fault-bounded belts the distribution of which approximately paralleled the trend of the Japanese Islands [Tanaka and Nozawa, 1977]. Recent attempts to convert these into terranes have met with varying degrees of success. In the

TABLE 1. Correlation Between Belts and Zones and Terranes in the Chugoku Region (The original relative positions of units are indicated)

<table>
<tr><td>Hida Terrane</td><td colspan="6">H i d a B e l t</td></tr>
<tr><td>Nagato-Wakasa Disrupted Terrane</td><td>Nagato Tectonic Zone</td><td>Oosayama Area</td><td>Wakasa Area</td><td colspan="2"></td><td>Hida Marginal Belt</td></tr>
<tr><td>Permian-Triassic Composite Terrane</td><td>Akiyoshi-Nishiki Belt
(A.-N. Terrane)</td><td colspan="2">Maizuru Belt
(M. Terrane)</td><td colspan="2">Ultra Tamba Belt
(UT. Terrane)</td><td></td></tr>
<tr><td>Sangun Metamorphic Terrane</td><td colspan="4">Sangun Metamorphic Belt</td><td></td><td></td></tr>
<tr><td rowspan="2">Jurassic Olistostromal Terrane</td><td>Nichihara Belt</td><td></td><td colspan="2">Tsunotani Belt</td><td rowspan="2">Tamba Belt
(Tamba Terrane)</td><td rowspan="2"></td></tr>
<tr><td colspan="4">K u g a - M i k a w a B e l t</td></tr>
</table>

Chugoku region the relationship we recognise between belts (or zones) and terranes is indicated in Table 1. To the south of the Hida terrane Cretaceous-Paleogene granites and associated volcanic and volcaniclastic rocks are widespread and the recognition of older terranes is made difficult by the scattered distribution of pre-Cretaceous rocks. However both Permo-Triassic and Jurassic stratigraphic terranes can be distinguished, each composed of a few fragmented belts. Accretion of the Permo-Triassic and Jurassic terranes to the Hida Precambrian continent, and their fragmentation, resulted from complex processes occurring in pre-Cretaceous times. The distribution of terranes is shown in Figure 1, drawn on the basis of the latest data. The distribution of the Tsunotani and Nichihara belts of the Jurassic olistostromal terrane differs from that presented by Saito [1983], Taira [1983], Ozawa et al. [1985], Ichikawa et al. [1985] and Hara et al. [1985]. Except for the Hida terrane all were included in the Akiyoshi terrane by Zhang [1985]. Saito [1985] grouped the Nichihara, Tsunotani and Akiyoshi-Nishiki belts of this paper in his Chugoku terrane, possibly because of the difficulty of discriminating the boundaries of these units. The term Sangun terrane is used here for the terrane of younger Sangun rocks. The older Sangun rocks are included in the Nagato-Wakasa terrane.

Aspects of individual terranes are discussed below. Terranes have been discriminated on the basis of similarities of lithology and age of sedimentary rocks for stratigraphic terranes, and similarities of metamorphic age and metamorphic facies series for metamorphic terranes.

Hida Terrane

The Hida terrane occurs in the extreme north of the Chugoku region and is represented by gneisses and amphibolites on Oki Island, 60 km north of the Chugoku mainland. Two metamorphic episodes have been recognised in these rocks, an early one, probably of granulite facies, and a later retrogressive amphibolite facies metamorphism [Hoshino, 1979]. K-Ar and Rb-Sr ages of from 165 Ma to 180 Ma [Nozawa, 1977] and Sm-Nd hornblende ages of 166 Ma [Tanaka and Hoshino, 1986] have been recorded from the Oki metamorphic rocks and correspond with the age of intrusion of the Funatsu granites in the Hida Mountains located further east. However Tanaka and Hoshino [1986] have also reported Sm-Nd whole rock ages of about 2000 Ma. In the Hida Mountains at least two stages of metamorphism are distinguished [Suzuki, 1977] and igneous rocks record multiple episodes of plutonism, the earliest of which is assigned to the Pre-Cambrian [Shibata and Nozawa, 1980].

Nagato-Wakasa Terrane

This terrane forms a very narrow and discontinuous belt that lies to the south of the Hida terrane. It includes the Nagato tectonic zone [Kawano et al., 1966; Murakami and Nishimura, 1979] and older Sangun metamorphic rocks and associated metagabbro in the Wakasa and Oosayama districts. The Nagato zone comprises a serpentinite melange which, in the Nagato district, includes Paleozoic (Upper Carboniferous-Permian) sedimentary units the ages of which were determined from fusilinids and radiolarians [Murakami and Mikami, 1975; Isozaki, 1983; Igi et al., 1986], and older metamorphic rocks such as orthogneiss and amphibolite (370-420 Ma) and schist (260-300 Ma) [Nishimura et al., 1983; Igi, 1985].

In the Wakasa district schists of the Shitani Formation [Uemura et al., 1979] and metagabbro surrounded by serpentinite belong to this terrane. K-Ar ages of the schists range from 279 Ma to 292 Ma [Shibata and Nishimura, 1983] and those of the metagabbro from 430 to 470 Ma [Nishimura and Shibata, 1986].

Blocks of glaucophane-bearing mica schists, including one of K-Ar age 251 Ma, occur in ultramafic rocks in the Oosayama district [Watanabe,

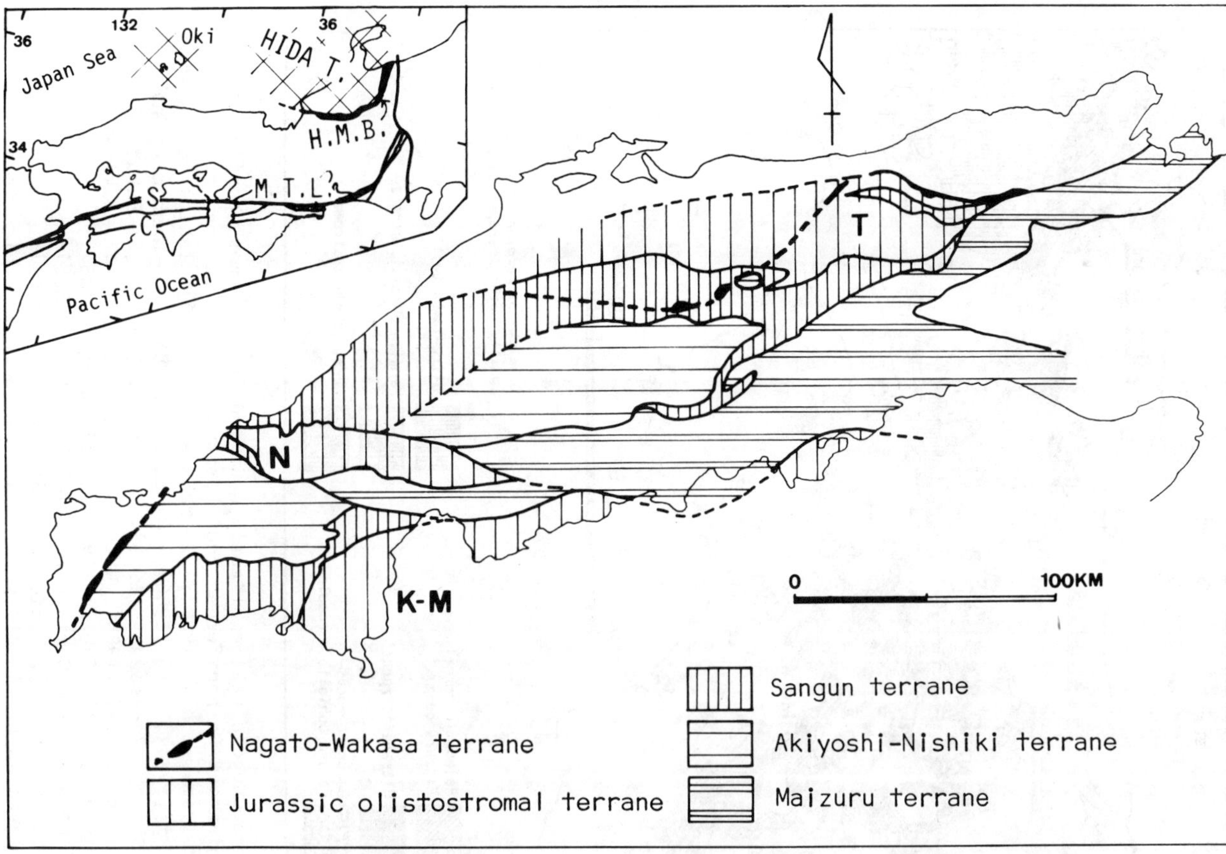

Fig. 1. Distribution of pre-Cretaceous tectonic elements in the Chugoku region. Hida T. = Hida terrane; H.M.B. = Hida marginal belt; M.T.L. = Median Tectonic Line; T = Tsunotani belt; N = Nichihara belt; K-M = Kuga-Mikawa belt; S = Sanbagawa belt (or terrane); C = Chichibu belt (or terrane).

1984; unpublished data). In the Ashidachi ultramafics, located to the west of the Oosayama ultramafics, brown hornblende dated at 350 Ma occurs in a gabbroic mass [Shibata et al., 1977]. These two areas are parts of the zone characterized by the occurrence of older metamorphic and igneous rocks.

The Nagato-Wakasa tectonic zone is considered to be the western extension of the Hida marginal belt. The zone, although narrow, is regarded as an example of a disrupted terrane with a serpentinite matrix [Jones et al., 1983], formed during accretion of Permo-Triassic terranes.

Sangun Terrane

In this paper the Sangun terrane is regarded as composed of metamorphic rocks of ages less than 225 Ma. Their complex distribution is shown in Figures 1 and 2. We do not have any fossil age determinations for the protoliths of these metamorphic rocks except for a report of *Follicucullus* and *Pseudoalbeillella* of Middle Permian age from weakly metamorphosed rocks in the Tanoharagawa Formation of the Gotsu area [Takeshita et al., 1986]. It is noteworthy that the Sangun terrane contains metamorphosed acidic tuffs [Mitsuno, 1984], for acidic tuff horizons occur in the Akiyoshi-Nishiki and Maizuru terranes. It is also possible that some of the Sangun siliceous rocks were derived from correlatives of the Jurassic olistostromal terrane, as first pointed out by Hayasaka [1985], although this would apply only to the youngest rocks indicated in Table 2.

Metamorphic grades range from pumpellyite-actinolite zone through epidote-crossite zone to barroisite zone, indicating high-pressure intermediate type metamorphism [Miyashiro, 1961; 1973]. Basic studies of the metamorphic zonation were reported by Hashimoto [1968] and Nishimura [1971] in the Katsuyama and Nishiki-cho districts respectively. The lowest grade areas occur adjacent to prehnite-pumpellyite rocks of the Akiyoshi-Nishiki belt, and a gradual change in metamorphic grade between these rocks has been described in the Nishiki-cho district [Nishimura, 1971; Nishimura et al., 1985], from where K-Ar ages of 210-225 Ma have been reported from micas in pelites of the Paleozoic Nishiki Group

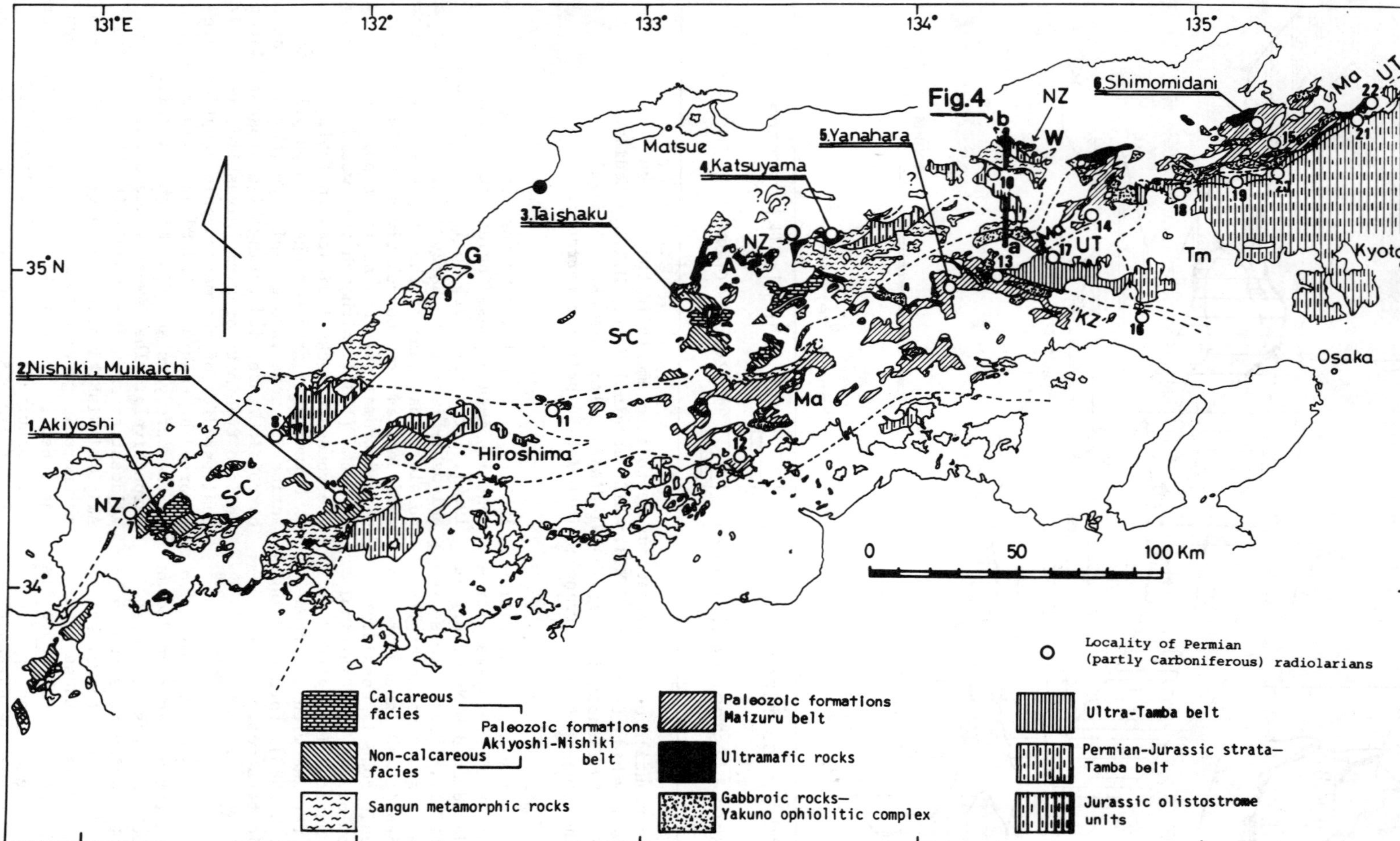

Fig. 2. Geologic map of the pre-Cretaceous rocks of the Chugoku region slightly modified from the map compiled by Yamada et al. [1982]. Numbers show localities of Permian and Carboniferous radiolarians. Data sources of radiolarian localities: 1 - Kanmera and Sano [1985], Uchiyama et al. [1986]; 2 - Naka et al. [1986], Ishiga et al. [1986], Naka [1986], Nishimura et al. [1985]; 3 - Isozaki [1984], Goto [1986]; 4 - Miyake [1985]; 5 - Nishimura and Tokuoka [1986]; 6 - Ishiga and Suzuki [1984], Ishiga [1986c]; 7 - Isozaki [1983], Igi et al. [1986]; 8 - Fukutomi and Isozaki [1986]; 9 - Takeshita et al. [1985]; 10 - Yogoro et al. [1986], Watanabe et al. [1987]; 11 - Wakita [1985]; 12 - Ishiga and Sato [1985]; 13 - Goto and Hori [1985]; 14 - Ishiga [1986a]; 15 - Ishiga [1984]; 16 - Ishiga [1986a]; 17 - Caridroit and De Wever [1984], De Wever and Caridroit [1984], Ishiga [1986a]; 18 - Ishiga [1986a]; 19 - Ishiga [1986a]; 20 - Caridroit et al. [1985], Kurimoto [1986], Ishiga [1986a]; 21 - Ishiga [1985]; 22 - Ishiga [1985]. The solid circle shows the locality of Triassocampe from pebbles in a Neogene formation.

TABLE 2. Comparison of K-Ar and Rb-Sr Ages for Rocks From the Same Area of the Sangun Belt

Location	K-Ar Age (Ma)	Rb-Sr Age (Ma)
Katsuyama	191	183
Tsukita	165	183
Tari	173 - 179	165 - 183
Gotsu	191 - 195	186
Asahi-cho	120	185
Asahi-cho	181	178
Hatto	154	174
Hatto	129	158
Nishiki-cho	175	224

Data from Shibata and Nishimura (1985).

[Nishimura and Isozaki, 1984]. These ages may indicate an early stage of uplift of the lower grade rocks. K-Ar ages in the Sangun belt mostly fall between 160 Ma and 195 Ma but some are younger, reflecting the thermal effect of Cretaceous granites. Rb-Sr ages often coincide with K-Ar ages as shown by Shibata and Nishimura [1985], although in some cases there are differences (Table 2), which possibly reflect a stepwise uplift of the metamorphic rocks. The significance of some Rb-Sr ages is open to question, judging from the closure temperature of minerals (400-450°C for muscovite) compared with the maximum temperature attained during metamorphism (400°C or less).

Jurassic Olistostromal Terrane

The Tsunotani, Nichihara, and Kuga-Mikawa belts are all characterised by the presence of Jurassic olistostromes. These rocks now have a very scattered distribution (Fig. 3) and hence, although they probably constitute a single now-disrupted terrane, we do not feel justified in proposing a formal name.

The Tsunotani and Nichihara belts occur on the northern side of the Maizuru belt and its western extension. The Jurassic age of the muddy matrix of the olistostromes here has been determined from radiolaria including *Hsuum* sp. [Hayasaka et al., 1983]. Chert olistoliths in the Nichihara area of western Chugoku are Triassic [Tanaka, 1980], and range from Permian to Triassic in the Tsunotani area of eastern Chugoku [Yogoro et al., 1986; Watanabe et al., 1987]. Lenticular bodies of quartz sandstone and greenstone are also embedded in the muddy matrix. Hayasaka [1986] has reported Early Jurassic radiolarians from pelitic rocks of the Hatto Formation, and Yogoro et al. [1986] and Watanabe et al. [1987] have found *Triassocampe* sp. of Middle Triassic age in chert masses and Jurassic radiolarians in pelitic rocks of the Chizu Formation which is found further south (Fig. 3). These two formations are also Jurassic olistostrome units. Greenschists containing sodic amphibole, winchite and lawsonite in the Hatto Formation [Hayasaka, 1986] are considered to have been "injected" as metamorphic septas or walls in thrust sheet zones (Fig. 4B). Olistostrome units are widely distributed in the Chugoku region for similar lithologic assemblages have been reported from several areas by Yamada [1972]. *Triassocampe* found in pebbles in Tertiary rocks [Tamaki, Tukuoka and Naka, unpublished data] suggests that the olistostrome units may have extended to the west.

The Kuga-Mikawa belt situated on the southernmost part of the Chugoku region is composed mainly of chert, sandstone, greenstone and pelitic rocks often showing diamictic facies. Permian and Triassic conodonts [Toyohara, 1974; 1977] and Early and Late Jurassic radiolarians [Hayasaka et al., 1983; Takata and Isozaki, 1986] have been reported from chert blocks and mudstones, in addition to Triassic and Jurassic megafossils (*Monotis* and ammonites) from mudstone [Hase, 1961; Higashimoto et al., 1986]. The greenstones have mostly suffered pumpellyite-chlorite grade metamorphism.

Akiyoshi-Nishiki Terrane

This terrane is composed of weakly metamorphosed Paleozoic rocks including pelitic, psammitic, and siliceous sedimentary rocks, greenstone, acidic tuff and limestone. From a metamorphic view-point, Nishimura et al. [1977] considered the belt to have suffered the Sangun metamorphism and proposed the term "Sangun-Chugoku belt". The Nishiki-cho district is a type locality of gradually changing metamorphic grade as mentioned above, but as a tectonic slide is recognized between crystalline schists and Paleozoic Nishiki group [Nishimura, 1971] they are considered to belong to different terranes.

Two contrasting lithofacies, a calcareous (limestone) facies and a non-calcareous facies, are discriminated in the Akiyoshi-Nishiki terrane. Precise biostratigraphic studies in the limestone areas have been carried out by Ota [1977] and Hase et al. [1974] amongst others. Greenstones, interpreted as fragments of ancient sea-mounts, form the basal part of typical calcareous facies sequences in the Akiyoshi and Taishaku districts and elsewhere. Carboniferous to Early Permian limestones depositionally overlie the greenstones, and are succeeded by terrigenous sediments. In non-calcareous sequences Early-Middle Permian cherts are developed, overlain by terrigenous rocks. Middle Permian acidic tuff horizons are often intercalated in the rocks, especially in the Nishiki Group. The Permian sequences and their radiolarian biostratigraphy in several representative non-calcareous areas are summarised in Figure 5. Permian olistostrome units containing limestone, greenstone, and chert blocks are recognised in some areas [Kanmera and Nishi, 1983; Uchiyama et al., 1986; Miyake, 1985). The presence of

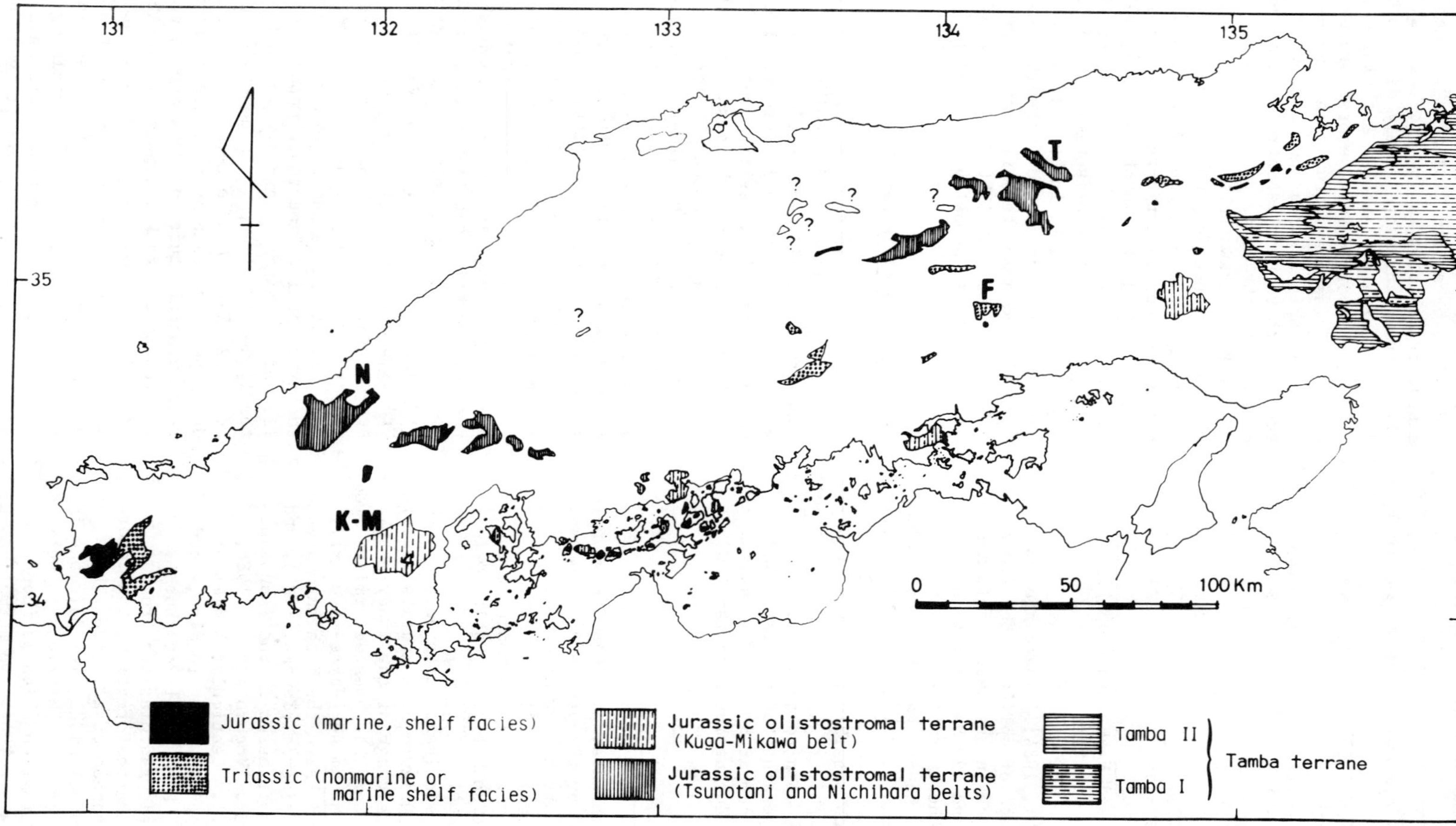

Fig. 3. Distribution of Triassic and Jurassic rocks in the Chugoku region based mainly on Yamada et al. [1982], Ishiga [1983], Imoto [1984], and Watanabe et al. [1987]. T = Tsunotani; F = Fukumoto; N = Nichihara; K-M = Kuga-Mikawa. Question marks show suspected Jurassic olistostromal units.

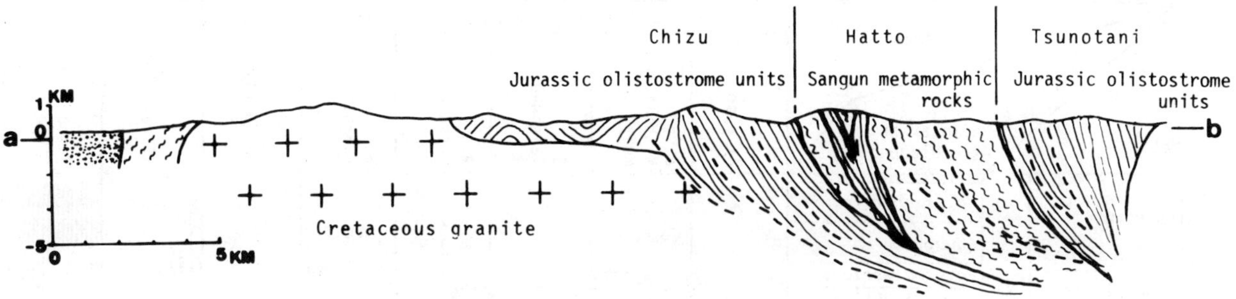

Fig. 4. Structural section across the Tsunotani belt in the Wakasa area (see Fig. 2 for location).

olistostrome units and acidic tuffs leads us to propose the presence of an arc-trench system during Permian time. The Mugitani Formation of Middle-Upper Permian age found in the western Chugoku region contains a pebble of gneiss dated at 359 Ma [Nishimura et al., 1981], and the Tsunemori Formation (Middle-Upper Permian) in the Akiyoshi area contains many pebbles of granodiorite-quartz diorite, granite porphyry, and dacite-rhyolite [Murakami, 1978]. The Paleozoic rocks are unconformably overlain by shallow marine and non-marine Upper Triassic rocks. Early Jurassic shelf sediments also are sporadically distributed within the belt (Fig. 3).

Maizuru Terrane

The Maizuru terrane comprises rocks of the Maizuru belt and likely correlatives found further west. It is composed mainly of Middle Permian pelitic rocks often intercalated with acidic tuffs and mafic lithologies and the terrane includes the Yakuno ophiolitic rocks [Shimizu, 1962].

The Maizuru belt forms a fault-bounded belt in the eastern part of the Chugoku region and further east. The western extension of the belt is not clear, but similar lithologies and successions are traceable further west [Hase, 1964; 1984]. Typical successions in the terrane are summarized in Figure 5. In some districts, including Kamigori and Shimomidani, radiolarians indicating uppermost Carboniferous to lowest Permian ages [Goto and Hori, 1985; Ishiga, 1986] as well as Middle Permian age [Ishiga, 1986c) have been found. Carboniferous fossils occur in the Kamigori area. A regular change from chert facies to sandstone facies through pelitic and subsequent acidic tuff facies, as recognised in the Akiyoshi-Nishiki terrane, also occurs in the Maizuru terrane. However the pelitic facies became predominant from the lower middle Permian in the Maizuru terrane, whereas the chert facies was still developed at this time in the Akiyoshi-Nishiki terrane. Comparatively extensive acidic tuffs occur in horizons of upper Middle Permian age in both terranes. The chemical composition of greenstones in the terrane matches that of ocean floor basalt with slightly low TiO_2 (marginal sea type [Koide, 1986]). Pumpellyite-actinolite facies metamorphic rocks are reported in some areas [Igi, 1967; Kinzoku-Tanko-Jigyodan, 1977]. The Paleozoic rocks are considered to be part of a subduction complex. Part of the Yakuno ophiolite is believed to have been derived from unusually thick oceanic crust like that which underlies the Black Sea today [Ishiwatari, 1985], whereas another part may have been derived from the crust of an island arc [Ishiwatari, 1986].

The Lower Triassic Fukumoto Formation (shallow marine facies) unconformably covers the Paleozoic rocks [Nakazawa et al., 1954]. Granitic clasts containing considerable amounts of potassium feldspar have been reported from conglomerate in the Maizuru belt [Kano et al., 1961].

Ultra-Tamba and Tamba Terranes

The Ultra-Tamba belt was defined by Caridroit et al. [1985] for rocks mapped as belonging to the Tamba belt but which contain only Permian faunas. Although olistostromes are present, both blocks and matrix are Permian. These rocks are the basis for the Ultra-Tamba terrane which has been described in detail by Ishiga [1986a].

Two olistostrome units, Tamba I and II, have been identified in the Tamba terrane [Ishiga, 1983, 1986b; Imoto, 1984]. The Tamba II has been thrust over the Tamba I and both were then folded together. The Tamba I unit contains Upper Jurassic shale or mudstone and Jurassic chert but no Permian chert, whereas the Tamba II unit does not contain Upper Jurassic mudstones or Jurassic chert but does contain Carboniferous to Permian chert. Ishiga [1983] pointed out that clearly contrasting radiolarian assemblages occur in clastic rocks of Tamba I and Tamba II and eventually the two Tamba units may be identified as discrete terranes.

Formation of a Permo-Triassic Composite Terrane

Differences exist between the lithological sequences in the Akiyoshi-Nishiki and Maizuru terranes. Limestones are very common in the

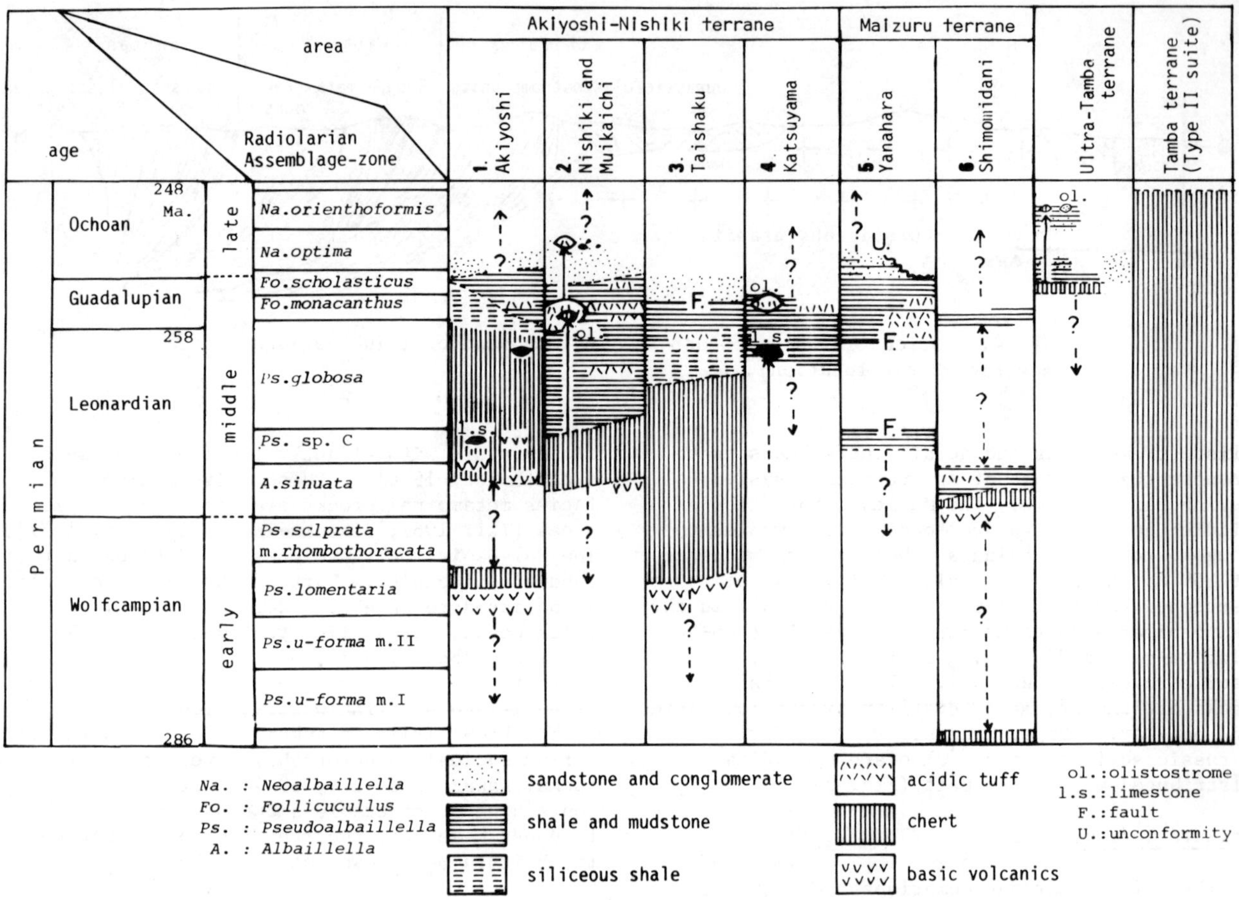

Fig. 5. Permian stratigraphic columns for the non-calcareous facies of the Permo-Triassic terranes based on radiolarian biostratigraphy [from Naka 1986 and Naka 1987]. Locations shown in Fig. 2. Data for the Ultra-Tamba belt are from Caridroit et al. [1985] and Ishiga [1986a]. Data for the Tamba belt are from Ishiga [1983]. Numerical ages are based on the data of Harland et al. [1982].

Carboniferous to Early Permian of the Akiyoshi-Nishiki terrane whereas chert is not rare in Maizuru rocks of this age. In the Middle Permian, a chert facies became predominant in the Akiyoshi-Nishiki whereas a pelitic facies characterised the Maizuru belt. However, in the late Middle to Late Permian, pelitic rocks and acidic tuffs accumulated in both terranes and were succeeded by sandstones.

Such overall coarsening-upward sequences have been assigned to trench-fill sedimentation by Lash [1985] and Matsuoka [1984; 1985]. A typical sequence in the Nishiki Group comprises Middle Permian chert (120 m), overlain by Middle to Upper Permian mudstone with acidic tuff (200 m), and then by massive sandstone (500-1000 m) of probable Upper Permian age [Naka et al., 1986]. This succession strongly suggests progressive accretion from an oceanic basin to a trench. Such accretion has been suggested in the Akiyoshi and surrounding areas by Kanmera [1983] and Kanmera and Nishi [1983] on the basis of lithological studies. A typical sequence in the Maizuru Group of the Yanahara area [Nishimura and Tokuoka, 1986] has thick acidic tuffs, commonly intercalated with monotonous pelitic rocks of Middle Permian age, overlain by Upper Permian thick-bedded sandstone with some conglomerate.

Rocks of the two terranes become similar in the upper Middle to Upper Permian and similar Upper Permian fusulinid assemblages occur in both [Ishii et al., 1985]. In both terranes Permian strata are unconformably overlain by Triassic non-marine and/or shallow marine sediments and they must have formed a composite terrane [Jones et al., 1983; Howell et al., 1985] prior to the Late Triassic. The presence of thick-bedded acidic tuffs and coarser terrigenous detritus, apparently of acidic volcano-plutonic provenance, strongly suggests that sedimentation in the Middle to Upper Permian occurred close to a volcanic arc.

Most of the ages of the older Sangun metamorphic rocks correspond with the age of the accretion and amalgamation of these terranes, and rocks of the Nagato-Wakasa terrane may have originally formed at the margin of a volcanic arc, and have been displaced during formation of the composite terrane.

Characteristics of Jurassic Terranes

Rocks of the Jurassic olistostromal terrane in the Tsunotani, Nichihara and Kuga-Mikawa belts can be correlated with those of the Tamba terrane Hara et al., 1985; Hayasaka et al., 1983]. However both the Tsunotani and Kuga Formations show elements of Tamba I and Tamba II units as the Tsunotani Formation contains Permian and Jurassic cherts, and the Kuga formation, as pointed out by Takata and Isozaki [1986], contains upper Jurassic radiolarian fossils and Canoptum sp. in siliceous mudstones, characteristic of Tamba II. Accordingly precise matching of the Jurassic rocks of the Chugoku region with the Tamba terrane is not possible at present. All Jurassic olistostrome units of the Chugoku district require further detailed investigation. Mixing of elements of Tamba I and Tamba II may be a characteristic of the Jurassic olistostromal terrane. Radiometric ages of the younger Sangun metamorphic rocks would correspond well with this postulated Jurassic disturbance.

Fragmentation and Mixing of Permo-Triassic and Jurassic Terranes

In the Chugoku region all pre-Cretaceous terranes have a scattered distribution (Fig. 2). The Nagato-Wakasa terrane mostly occurs as disrupted fragments at the northern edge of the Permo-Triassic composite terrane. The Akiyoshi-Nishiki terrane tends to be distributed on both the north and south sides of the Maizuru terrane and forms several blocks ranging in area from a few square kilometres to several tens of square kilometres. Fragmented Jurassic terranes are recognised north and south of the Permo-Triassic composite terrane, in blocks ranging up to several square kilometres in area.

A characteristic of terrane distribution in the Chugoku region is that the Sangun metamorphic terrane often separates stratigraphic terranes, occurring as fault-bounded blocks (walls or septas) (Figs. 2 and 4). As the Sangun rocks have suffered high-pressure metamorphism their occurrence is considered to be evidence of "injection" events into the stratigraphic terranes from the deeper part of a subduction zone perhaps during the Jurassic (or Early Cretaceous). The entire imbricated thrust sheet structure formed around Jurassic olistostromal units and is possibly similar to that of the duplex accretion complex in Alaska described by Sample and Fisher [1986]. This structure may be indicative of underplating by stacked thrust slices [Cowan, 1985] although the metamorphic grade of the Sangun rocks is higher than that of the first-described underplated deposits which were of zeolite and prehnite-pumpellyite grade [Moore et al., 1982].

We postulated a Permian arc-trench system related to the formation of the older Sangun metamorphic rocks and to amalgamation of the Permo-Triassic terranes. How was the Permian arc-trench system related to the intense deformation in the Jurassic just mentioned? It is difficult to give a full explanation but it is certain that the Permian system had been destroyed, and the terranes fragmented into several blocks before the Jurassic. There are two opposing ideas on the relationship between the Permo-Triassic and Jurassic terranes, either (i) the Jurassic basins in which the olistostromal units accumulated were "in situ" basins developed on both the northern and southern sides of Triassic land [Kimura, 1980] or (ii) the Jurassic olistostromal units have been thrust as nappes over the Permo-Triassic composite terrane from the north [Hara et al., 1983]. The latter idea is supported by Ozawa et al. [1985], Ichikawa et al. [1985] and Maruyama and Seno [1985]. Such a nappe structure would be comparable with the structure in the Tamba region, where the Ultra-Tamba nappe overthrusts the two-storied Tamba I and II nappe [Caridroit et al., 1985; Ishiga, 1986a]. Thus nappe tectonics may have played an essential role in producing the complicated distribution of Permo-Triassic and Jurassic terranes. However there are differences from the situation in the Tamba region that are worthy of note. These include the complicated distribution of the Sangun metamorphic rocks and the mixed Tamba I and II character of some of the Jurassic olistostromal units. These differences suggest that pre-Cretaceous terranes in the Chugoku region suffered comparatively severe deformation, probably related to subduction zone metamorphism.

A fuller understanding of fragmentation and mixing in the subduction zone during the final stages of accretion are considered to be the key to interpreting the pre-Cretaceous structure of the Chugoku region. This intensely deformed zone was situated on the southeastern side of the Jurassic orogenic belt in Korea (Daebo orogeny), and to the northwest of the Sanbagawa belt and the probably continuously accreted Chichibu belt. Jurassic plutonism in the Hida terrane corresponds with that in Korea. According to Takahashi [1983], the Jurassic magmatic belts of Korea and Hida are continental margin type, the rocks having moderate to high K_2O, $K_2O + Na_2O$ and K_2O/Na_2O. If this is correct then it is difficult to envisage an arc-trench system in the Chugoku district and simultaneous Jurassic accretion of the Sanbagawa

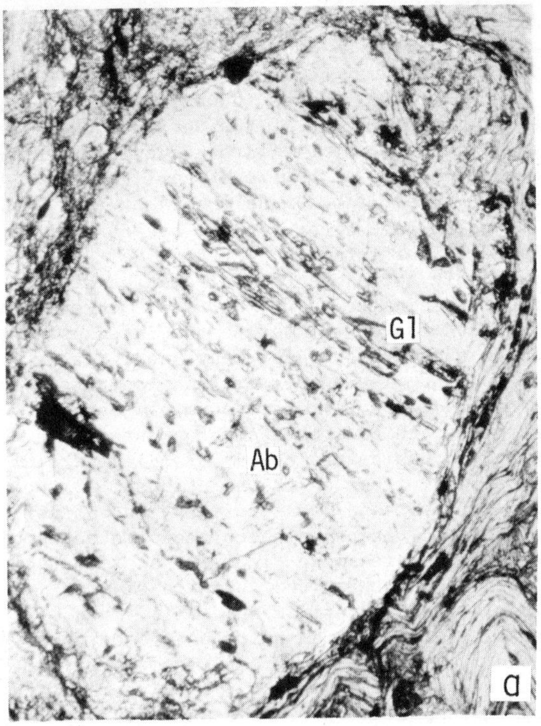

Fig. 6a. Porphyritic albite relics containing glaucophane. Width of field of view 1 mm.

belt without invoking strike-slip disruption, as speculated by Takahashi [1983], to explain the present day distribution of these rocks. Strong east-west trending stretching lineations in the Sangun terrane in the Nishiki-cho and Gotsu areas support this hypothesis. As we consider the younger Sangun metamorphism to have occurred in the Jurassic, the strike-slip movements are considered to have been associated with an oblique subduction zone of moderate dip angle, judging from the high pressure intermediate metamorphic facies series. Pull-apart basins related to the strike-slip movement have not been identified.

High Pressure Poly-Metamorphic Rocks of the Sangun Terrane

As described above, uplift of the Sangun metamorphic rocks has played an important role in forming the complicated terrane distribution of the Chugoku region and has disrupted an earlier Permo-Triassic arc-trench system. The younger Sangun metamorphic rocks may have been in part derived from older high-pressure metamorphic rocks, as for example some of the schists in the Gotsu area [Sengan, 1985]. These rocks comprise pelitic schists, green/blueschists, minor quartzose schists and strongly foliated serpentinite. Metamorphic grade is mostly pumpellyite-actinolite schist facies to blueschist facies (epidote-crossite zone), though an exceptional occurrence of higher grade metamorphic rocks including barroisite schist around a serpentinite mass has been recognised [Sengan, 1985]. Lawsonite is stable in the pelitic schists [Watanabe et al., 1983]. K-Ar ages for one of the barroisite schists, a pelitic schist, and a blueschist are 165 Ma, 180 Ma, 185 Ma in turn [Watanabe, Nishido and Nagao, unpublished data]. Around the serpentinite mass pelitic schists containing porphyritic albite relics are also found and these, and some of the barroisite schists, show evidence of poly-metamorphic episodes. Some barroisite crystals have replaced glaucophane (sensu stricto) suggesting decreased pressure conditions. Porphyritic albite relics also contain glaucophane (Fig. 6a). The age of the glaucophane has not yet been determined, but two phases of metamorphism are confirmed. Repeated zoning from glaucophane to crossite/magnesio-riebeckite in sodic amphibole crystals (Fig. 6b) is found in blueschist around the serpentinite mass but not elsewhere. Assuming deep-seated flow in a broad forearc ridge as described by Pavlis and Bruhn [1983], it is possible that the younger Sangun rocks picked up older glaucophane schist masses during the Jurassic Sangun metamorphism. Thus we consider the Sangun rocks have suffered a complicated mixing process that included flow in the subduction zone in the

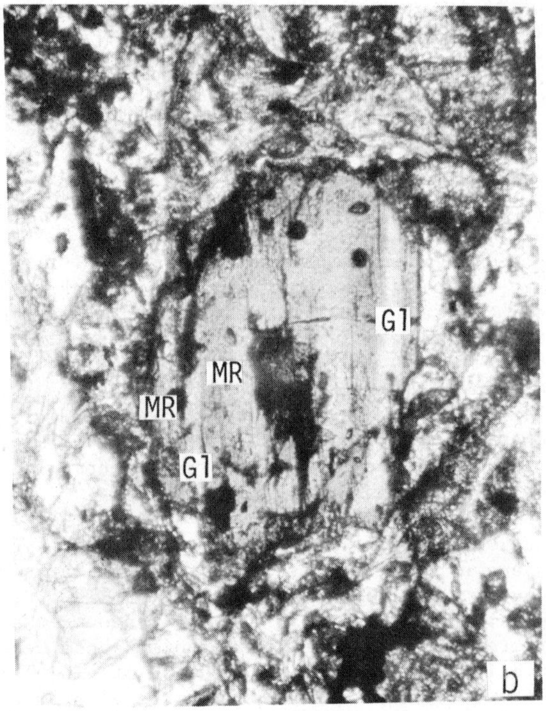

Fig. 6b. Dual zoning pattern in a sodic amphibole crystal (Mr=magnesio-riebeckite; Gl=glaucophane). Width of field of view 1 mm.

Jurassic following their formation during the accretion, amalgamation and mixing of the Permo-Triassic and Jurassic terranes.

Concluding Remarks

In this paper we have discriminated several stratigraphic terranes which can be placed into two major groups, Permo-Triassic terranes including the Akiyoshi-Nishiki and Maizuru, and Jurassic terranes including rocks of the Tsunotani, Nichihara and Kuga-Mikawa belt. We recognise metamorphic terranes associated with both groups, the Nagato-Wakasa terrane with the Permo-Triassic terranes, and the Sangun terrane with the Jurassic terranes. On the basis of recent radiolarian biostratigraphy in the Akiyoshi-Nishiki and Maizuru terranes, the process of amalgamation of the Permo-Triassic terranes is described. The complicated distribution of the fragmented terranes is tentatively explained by deformation in a Jurassic oblique subduction zone. One of the important characteristics of the pre-Cretaceous terranes is the presence of "injected" Sangun metamorphic rocks within and between stratigraphic terranes. In addition to terrane classification, examination of the process of terrane fragmentation is important in understanding the pre-Cretaceous geologic framework of the Chugoku region.

Acknowledgments. We would like to express our thanks to Dr. E. C. Leitch, the University of Sydney, for his critical reading of the manuscript and discussion. We would like to express our thanks to Ms Dhana Pillai, Shimane University, for her improvement of our English and her comments. Our thanks are extended to Professor Y. Nishimura, Yamaguchi University, for discussions concerning the Sangun metamorphic rocks, although the authors accept all responsibility for the interpretation given in this paper.

References

Caridroit, M., and P. De Wever, Description de quelques nouvelles especes de Follicucullidae et d'Entactiniidae (radiolaires polycystines) du Permien du Japon, Geobios, 17, 639-644, 1984.

Caridroit, M., K. Ichikawa, and J. Charvet, The Ultra-Tamba Zone, a new unit in the Inner Zone of Southwest Japan, Earth Science (Chikyu Kagaku), 39, 210-219, 1985.

Cowan, D. S., The origin of some common types of melange in the Western Cordillera of North America, in Formation of active ocean margins 257-272, edited by N. Nasu, K. Kobayashi, S. Uyeda, I. Kushiro and H. Kagami, Terra, Tokyo, 1985.

De Wever, P., and M. Caridroit, Description de quelques nouveaux Latentifistulidae (radiolaires polycystines) Paleozoiques du Japon. Revue de Micropal., 27, 98-106, 1984.

Fukutomi, T., and Y. Isozaki, Relationship among Mesozoic rocks, Paleozoic rocks and crystalline schists in the "Kanoashi Group", Shimane Prefecture, (in Japanese), Abst. 93rd Ann. Meeting Geol. Soc. Japan, 227, 1986.

Goto, H., and R. Hori, Geology of northern part of the Kozuki-Tatsuno Belt (Kamigori Belt), Southwest Japan, (in Japanese), Abst. 92nd Ann. Meeting Geol. Soc. Japan, 189, 1985.

Goto, N., Stratigraphy of Permian-Carboniferous siliceous rock and its relation to coarse grained clastic rock, Taishaku district, Hiroshima Prefecture, (in Japanese), Abst. 93rd Ann. Meeting Geol. Soc. Japan, 245, 1986.

Hara, I., Y. Hayasaka, M. Maeda and T. Miyamoto, Some problems of Paleozoic-Mesozoic tectonics in Southwest Japan - Tectonics of metamorphic belts of high-pressure type, (in Japanese with English abstract), Mem. Geol. Soc. Japan, 25, 109-126, 1985.

Hara, I., K. Hide, and Y. Nishimura, Problems of Paleozoic Mesozoic tectonics in Southwest Japan, (in Japanese with English abstract), Jour. Research Group of Structural Geology, 25, 1-12, 1983.

Harland, W.B., A.V. Cox, P.G. Llewellyn, C.A.G. Pickton, A.G. Smith and R. Walters, A geologic time scale, Cambridge University Press, 131p, 1982.

Hase, A., A find of Monotis (Entomonotis) from eastern Yamaguchi Prefecture, Japan, Trans. Proc. Palaeont. Soc. Japan, N.S., 42, 79-87, 1961.

Hase, A., Geologic map of the Hiroshima Prefecture and its explanatory text, (in Japanese), 1964.

Hase, A., Regional geology of Japan, "Chugoku District", (in Japanese), 8-15, Asakura Shoten, Tokyo, 1984.

Hase, A., Y. Okimura, and T. Yokoyama, The Upper Paleozoic formations in and around Taishadu-dai, Chugoku Massif, Southwest Japan, with special reference to the sedimentary facies of limestone, (in Japanese with English abstract), Geol. Rept. Hiroshima Univ., 19, 1-39, 1974.

Hashimoto, M., Glaucophanitic metamorphism of the Katsuyama district, Okayama Prefecture, Japan, Jour. Fac. Sci. Univ. Tokyo, II 17, 99-162, 1968.

Hayasaka, Y., Re-examination of the geologic structure of the Sangun Belt, (in Japanese), Memorial volume of Professor H. Yoshida, 339-354, 1985.

Hayasaka, Y., Discovery of Triassic conodont and Jurassic type radiolarians from the Hato formation of "Sangun Metamorphic Rocks" (in Japanese), Abst. 93rd Ann. Meeting Geol. Soc. Japan, 465, 1986.

Hayasaka, Y., Y. Isozaki, and I. Hara, Discovery of Jurassic radiolarians from the Kuga and Kanoashi groups in the western Chugoku district, Southwest Japan, (in Japanese), Jour. Geol. Soc. Japan, 89, 527-530.

Higashimoto, S., Y. Takashi, H. Makimoto, K. Wakita, and E. Tsukuda, Geology of the Otake district, with geological sheet map at

1:50,000, (in Japanese), Geol. Surv. Japan, 70p, 1986.

Hoshino, M., Two-pyroxene amphibolites in Oki-Dogo Island, Shimane-ken, Japan. Jour. Japan. Assoc. Mineral. Petrol. Econ. Geol., 74, 87-99, 1979.

Howell, D.G., D.L. Jones, and E. R. Schermer, Tectonostratigraphic terranes of the Circum-Pacific region, in Tectonostratigraphic terranes of the Circum Pacific region, edited by D.G. Howell, 3-30, Earth Science Series No. 1, Circum-Pacific Council for Energy and Mineral Resources, Houston, 1985.

Ichikawa, K., Mesozoic accretion of eastern continental margin of Asia (Tectonic development of basement structure of Eastern Asia I), (in Japanese), in Tectonic Belts of Asia, edited by K. Huzita, 223-238, Kaibundo Shuppan, Tokyo, 1984.

Ichikawa, K., S. Hada, and A. Yao, Recent problems of Paleozoic-Mesozoic microbiostratigraphy and Mesozoic geohistory of Southwest Japan, (in Japanese with English abstract), Mem. Geol. Soc. Japan, 25, 1-18, 1985.

Igi, Y., Influence of the "Sangun-metamorphism" on the Maizuru Structural Belt, Japan, Earth Science (Chikyu Kagaku), 21, 34-38, 1967.

Igi, Y., Review on the problems of the high-pressure metamorphic belt in the Inner Side of Southwest Japan, (in Japanese), Abst. 92nd Ann. Meeting Geol. Soc. Japan, 11-12, 1985.

Igi, Y., K. Kiminami, G. Shibuya, Y. Nishimura, and Y. Isozaki, Serpentine melange zone and geologic structure of related Paleozoic-Mesozoic Groups in the western part of Chugoku District, (in Japanese), Rept. of Grant-in-Aid for Scientific Research (No. 9460045), 16p., 1986.

Imoto, N., Late Paleozoic and Mesozoic cherts in the Tamba belt, Southwest Japan, Bull. Kyoto Univ. Education, B 65, 1-71, 1984.

Ishiga, H., Two suites of stratigraphic successions within the Tamba group in the western Tamba Belt, Southwest Japan, (in Japanese with English abstract), Jour. Geol. Soc. Japan, 89, 443-454, 1983.

Ishiga, H., Follicucullus (Permian radiolaria) from Maizuru group in Maizuru Belt, Southwest Japan, Earth Science (Chikyu Kagaku), 38, 427-434, 1984.

Ishiga, H., Discovery of Permian radiolarians from Katsumi and Li formations along south of Maizuru Belt, Southwest Japan, Earth Science (Chikyu Kagaku), 39, 175-185, 1985.

Ishiga, H., Ultra-Tamba Zone of Southwest Japan, Jour. Geosci. Osaka City Univ., 29, 45-88, 1986a.

Ishiga, H., Late Carboniferous and Permian radiolarian biostratigraphy of Southwest Japan, Jour. Geosci. Osaka City Univ., 29, 89-100, 1986b.

Ishiga, H., Re-examination of the Shimomidani formation, western part of Maizuru district, Kyoto Prefecture, (in Japanese), Rept. Nishinihon Branch Geol. Soc. Japan, 101, 1986c.

Ishiga, H., and M. Sato, Paleozoic clastic formation in Numakura Peninsula, southeastern part of Hiroshima Prefecture, (in Japanese), Proc. Kansai Branch Geol. Soc. Japan, 98, 2-3, 1985.

Ishiga, H., and S. Suzuki, Discovery of Permian radiolarians and conodonts from the Shimomidani Formation in the "Maizuru Belt", Southwest Japan and its significance, Earth Science (Chikyu Kagaku), 38, 197-206, 1984.

Ishiga, H., H. Watase, and T. Naka, Permian radiolarians from Nishiki Group in Sangun Chugoku Belt, Southwest Japan, Earth Science (Chikyu Kagaku), 40, 124-136, 1986.

Ishii, K., Y. Okimura and K. Ichikawa, Notes on Tethys biogeography with reference to Middle Permian Fusulinaceans, in The Tethys, edited by K. Nakazawa and J.M. Dickins, 139-155, Tokai Univ. Press, Tokyo, 1985.

Ishiwatari, A., Granulite-facies metacumulates of the Yakuno Ophiolite, Japan: evidence for unusually thick oceanic crust, Jour. Petrol., 26, 1-30, 1985.

Ishiwatari, A., Are the Yakuno Ophiolites of Ohara in Okayama Pref. and Kamigori in Hyogo Pref. the product at the front of volcanic arcs? (in Japanese), Abst. 93rd Ann. Meeting Geol. Soc. Japan, 431, 1986.

Isozaki, Y., Occurrence of Paleozoic radiolarians from Nagato Tectonic Zone in Yamaguchi Pref. and Taishaku-dai area in Hiroshima Pref., Southwest Japan, (in Japanese), Proc. Nishinohon Branch Geol. Soc. Japan, 77, 7-8, 1983.

Isozaki, Y., Occurrence of Permian radiolarians from the clastic formation in Taishaku-dai area Hiroshima Prefecture and re-examination of its stratigraphy, (in Japanese), Report of Co-operative Research on DESK, 28-34, 1984.

Jones, D.L., D.G. Howell, P.J. Coney and J.W. Moner, Recognition, character and analysis of tectonostratigraphic terranes in western North America, in Accretion Tectonics in the Circum-Pacific Regions, edited by M. Hashimoto and S. Uyeda, Terra, Tokyo, 21-35, 1983.

Kanmera, K., A problem on the sedimentation and tectonics of the Upper Paleozoic rocks in Southwest Japan - Accretion of allochthonous sedimentary bodies, (in Japanese), in Geol. Develop. of the Kyushu Island with special reference to tectonic relation of island arcs and continent, Special Publication Committee of Nishinihon Branch Geol. Soc. Japan, 67-76, 1983.

Kanmera, K., and H. Nishi, Accreted oceanic reef complex in the Southwest Japan, in Accretion Tectonics in the Circum-Pacific Regions, edited by M. Hashimoto and S. Uyeda, Terra, Tokyo, 195-206, 1983.

Kanmera, K., and H. Sano, Sedimentation and tectonics in the inner zone of Southwest Japan, (in Japanese), Abst. 92nd Ann. Meeting Geol. Soc. Japan, 28-29, 1985.

Kano, H., K. Nakazawa, and T. Shiki, Considerations on the Permian backgrounds of the Maizuru districts judging from the conglomerates (in Japanese with English abstract), Jour. Geol. Soc. Japan, 67, 463-475, 1961.

Kawano, Y., Y. Uyeda, and N. Murakami, K-Ar dating on the granitic rock from Mine City, Yamaguchi Prefecture, Jour. Jap. Assoc. Mineral. Petrol. Econ. Geol., 56, 183-186, 1966.

Kimura, T., Japanese Islands - Their developments, (in Japanese), Vol. 2b, 586-614, 1980.

Kinzoku-Tanko-Jigyodan, Reports of the detail geologic survey (Bantan district), Showa, 50, 54p, 1977.

Koide, Y., Origin of the Ibara metabasalts from the Maizuru Tectonic Belt, Southwest Japan, Jour. Geol. Soc. Japan, 92, 329-348, 1986.

Kurimoto, C., Ultra-Tamba zone in the Fukuchiyama area, Kyoto Prefecture - its constituent rocks and distribution - (short report), (in Japanese), Earth Science (Chikyu Kagaku), 40, 64-67, 1986.

Lash, G.G., Recognition of trench fill in orogenic flysch sequences, Geology, 13, 867-870, 1985.

Maruyama, S., and T. Seno, Relative motion of plates and orogeny around Japanese Islands, (in Japanese), Kagaku, 55, 32-41, 1985.

Matsuoka, A., Togano group of the southern Chichibu Terrane in the western part of the Kochi Prefecture, Southwest Japan, (in Japanese with English abstract), Jour. Geol. Soc. Japan, 90, 455-477, 1984.

Matsuoka, A., Middle Jurassic Keta formation of southern part of the middle Chichibu Terrane in the Sakawa area, Kochi Prefecture, Southwest Japan, (in Japanese with English abstract), Jour. Geol. Soc. Japan, 91, 411-420, 1985.

Mitsuno, T., Sangun Metamorphic Rocks in the eastern part of the Hiroshima Prefecture and Okayama Prefecture, (in Japanese), Regional Geology of Japan, "Chugoku District", Asakura Shoten, Tokyo, 105-111, 1984.

Miyake, K., Permian olistostrome complex in the Katsuyama area, Okayama Prefecture, Southwest Japan, (in Japanese with English abstract), Jour. Geol. Soc. Japan, 91, 463-475, 1985.

Miyashiro, A., Evolution of metamorphic belts, Jour. Petrol., 2, 277-311, 1961.

Miyashiro, A., Metamorphism and metamorphic belts, George Allen & Unwin, London, 492p, 1973.

Moore, J.C., J.S. Watkins, and T.H. Shipley, Summary of accretionary processes, Deep Sea Drilling Project Leg 66, offscraping, underplating, and deformation of the slope apron. Initial Repts. Deep Sea Drilling Project, 66, 825-836, 1982.

Murakami, M., Unconformity at the bottom of Upper Triassic Mine group and its geologic significance suggested by the igneous rock pebbles, (in Japanese with English abstract), in Symposium Unconformity, edited by M. Hoshino, 117-126, Tokai University, 1978.

Murakami, M., and T. Mikami, Finding of Fusulinids from Nagato Tectonic Zone, Jour. Geol. Soc. Japan, 81, 331-332, 1975.

Murakami, M., and Y. Nishimura, Nagato tectonic zone, (in Japanese), Memorial Vol. Prof. Kano, H., The Basement of the Japanese Islands, 153-181, 1979.

Naka, T., Permian Nishiki group in the Kidani river area, Yamaguchi Prefecture, (in Japanese with Englist abstract), Abst. 93rd Ann. Meeting Geol. Soc. Japan, 246, 1986.

Naka, T., Permian Nishiki group in the Nishiki-cho, Southwest Japan, M.Sc. thesis, Shimane University, 1987.

Naka, T., H. Watase, and T. Tokuoka, Permian Nishiki group in the Muikaichi-cho, western part of the Sangun-Chugoku Belt, Shimane Prefecture, Southwest Japan, Earth Science (Chikyu Kagaku), 40, 160-170, 1986.

Nakazawa, K., T. Shiki, and D. Shimizu, Paleozoic and Mesozoic formations in the vicinity of Fukumoto, Okayama Prefecture, Japan, (in Japanese with English abstract), Jour. Geol. Soc. Japan, 63, 455-464, 1954.

Nishimura, K., and T. Tokuoka, Permian rocks and Triassic Fukumoto group in the Yanahara area, Okayama Prefecture, (in Japanese), Abst. 93rd Ann. Meeting Geol. Soc. Japan, 244, 1986.

Nishimura, Y., Regional metamorphism of the Nishiki-cho district, Southwest Japan, Jour. Hiroshima Univ., C 6, 203-268, 1971.

Nishimura, Y., T. Inoue, and H. Yamamoto, Sangun belt, with special reference to the stratigraphy and metamorphism, (in Japanese with English abstract), in The Sanbagawa Belt, edited by K. Hide, 257-282, 1977.

Nishimura, Y., and Y. Isozaki, Radiolarians and radiometric age of Nishiki group in eastern part of Yamaguchi Prefecture, Southwest Japan, (in Japanese), in Report of co-operative research on high pressure metamorphism in Inner Zone of Southwest Japan, 2, 37-40.

Nishimura, Y., Y. Isozaki, and T. Nureki, Sangun-Chugoku Belt and Ryoke Belt in the eastern part of Yamaguchi Prefecture, (in Japanese), Abst. 92nd Ann. Meeting Geol. Soc. Japan, Field Guide Report, 17-49, 1985.

Nishimura, Y., T. Mikami, and M. Suzuki, Metamorphic pebbles in the Permian Mugitani formation, South West Japan, (in Japanese), Report of Co-operative study on Lower Paleozoic-Upper Proterozoic in Japan, 33-36, 1981.

Nishimura, Y., E. Nakamura, and I. Hara, K-Ar ages of Sangun metamorphic rocks in Yamaguchi Prefecture and their geologic significance, Jour. Japan Assoc. Mineral. Petrol. Econ. Geol., 78, 11-20, 1983.

Nishimura, Y., and K. Shibata, Occurrence of metagabbro and amphibolite in the Wakasa area, Tottori Prefecture, (in Japanese), Abst. 93rd Ann. Meeting Geol. Soc. Japan, 466, 1986.

Nozawa, T. (compiler), Radiometric age map of Japan, Metamorphic rocks, Geological Survey of Japan, 1977.

Ota, M., Geological Studies of Akiyoshi, Part 1,

General geology of the Akiyoshi limestone group, Bull. Akiyoshi-dai Mus. Nat. Hist., 12, 1-33, 1977.

Ozawa, T., A. Taira, and H. Kobayashi, How have the zonal structures of Southwest Japan been formed? (in Japanese), Kagaku, 55, 4-13, 1985.

Pavlis, T.L., and R.L. Bruhn, Deep-seated flow as a mechanism for the uplift of broad forearc ridges and its role in the exposure of high P/T metamorphic terranes, Tectonics, 2, 473-497, 1983.

Saito, Y., Pre-Tertiary geotectonic units in Japan, Abst. Circum-Pacific Terrane Conf., Stanford Univ., 3p, 1983.

Saito, Y., Jurassic geologic framework in the Japanese Islands, in Tectonostratigraphic terranes of the Circum-Pacific region, edited by D.G. Howell, 401-407, Earth Science Series No. 1, Circum-Pacific Council for Energy and Mineral Resources, Houston, 1985.

Sample, J.C., and D.M. Fisher, Duplex accretion and underplating in an ancient accretionary complex, Kodiak Islands, Alaska, Geology, 14, 160-163, 1986.

Sengan, H., Petrology of the Sangun metamorphic rocks in the Hazumi area, Gotsu City, Shimane Prefecture, (in Japanese), Geol. Rept. Shimane Univ., 4, 41-59, 1985.

Shibata, K., and Y. Nishimura, Radiometric ages of the Sangun Metamorphic rocks, (in Japanese), Abst. 90th Ann. Meeting Geol. Soc. Japan, 385, 1983.

Shibata, K., and Y. Nishimura, Radiometric ages of the Sangun-Chugoku belt, (in Japanese), Abst. 92nd Ann. Meeting Geol. Soc. Japan, 13-14, 1985.

Shibata, K., Igi, S., and S. Uchiumi, K-Ar ages of hornblendes from gabbroic rocks in Southwest Japan, Geochem. Jour., 11, 57-65, 1977.

Shibata, K., and T. Nozawa, Rb-Sr ages of the Hida Metamorphic rocks at Kagasawa, Toyama Prefecture, (in Japanese), Jour. Japan. Assoc. Mineral. Petrol. Econ. Geol., 75, 130, 1980.

Shimizu, D., The Permian Maizuru Group, its stratigraphy and syntectonic faunal succession through the Latest Paleozoic Orogeny, Mem. Coll. Sci., Kyoto Univ., B 28, 571-609, 1962.

Suzuki, M., Polymetamorphism in the Hida Metamorphic Belt, Central Japan, Jour. Sci., Hiroshima Univ., C 7, 217-296, 1977.

Taira, A., Plate tectonic evolution of Japan, Abst. Circum-Pacific Terrane Conf., Stanford Univ., 2p, 1983.

Takahashi, M., Space-Time distribution of late Mesozoic to early Cenozoic magmatism in east Asia and its tectonic implication, in Accretion Tectonics in the Circum-Pacific Regions, edited by M. Hashimoto and S. Uyeda, Terra, Tokyo, 69-88, 1983.

Takata, M., and Y. Isozaki, Occurrence of Jurassic radiolarians from the "Mikawa Group" in eastern Yamaguchi Prefecture, (in Japanese), Jour. Geol. Soc. Japan, 92, 447-450, 1986.

Takeshita, H., T. Watanabe, and H. Ishiga, Discovery of Permian radiolarians from the Tanoharagawa formation (the Sangun metamorphic rocks), Gotsu, Shimane Prefecture, Japan, (in Japanese), Jour. Geol. Soc. Japan (in Press) 1987.

Tanaka, K., Kanoashi Group, an olistostrome, in the Nichihara area, Shimane Prefecture, (in Japanese with English abstract), Jour. Geol. Soc. Japan, 86, 613-628, 1980.

Tanaka, K., and T. Nozawa (eds.), Geology and mineral resources of Japan, 3rd ed., 1 Geology, 430p, Geological Survey of Japan, 1977.

Tanaka T., and M. Hoshino, Sm-Nd whole rock ages of the Oki metamorphic rocks, (in Japanese), Abs. Ann. Meeting Geochem. Soc. Japan, 169, 1986.

Toyohara, F., On the age of the Kuga group and Ryoke metamorphic rocks in eastern Yamaguchi Prefecture, (in Japanese), Jour. Geol. Soc. Japan, 80, 51-53, 1974.

Toyohara, F., Early Mesozoic development of the northwestern Chichibu geosyncline in west Chugoku, Japan, Jour. Fac. Sci., Univ. Tokyo, II 19, 253-336, 1977.

Uchiyama, T., H. Sano, and K. Kanmera, Depositional and tectonic settings of cherts around the Akiyoshi Limestone Group, Southwest Japan, Mem. Fac. Sci., Kyushu Univ., D 26, 51-68, 1986.

Uemura, F., T. Sakamoto, and N. Yamada, Geology of the Wakasa district, (in Japanese with English abstract), Quadrangle series, scale 1:50,000, Geol. Surv. Japan, 91p, 1979.

Wakita, K., Discovery of Middle-Late Permian radiolarians in the Karita and Hachimandai formations in the northeast environs of Hiroshima, Japan, (in Japanese), Earth Science (Chikyu Kagaku), 39, 237-240, 1985.

Watanabe, T., Glaucophane and ferroglaucophane in the Sangun Metamorphic Terrane of the Oosa district, Okayama Prefecture, Japan, Mem. Fac. Sci. Shimane Univ., 18, 69-74, 1984.

Watanabe, T., A note on the metamorphic ages and rocks in the Sangun terrane, (in Japanese with English abstract), Geol. Rept. Shimane Univ., 4, 33-39, 1985.

Watanabe, T., Y. Harada, T. Kogoro, Y. Kosuge, S. Fujie, M. Takamatsu, and A. Yao, Jurassic olistostrome distributed in the southeastern part of the Tottoki Prefecture - the Chizu formation, (in Japanese with English abstract), Jour. Geol. Soc. Japan, (in press), 1987.

Watanabe, T., H. Kobayashi, and H. Sengen, Lawsonite from quartzofeldspathic schist in the Sangun Metamorphic Belt, Shikuma, Shimane Prefecture, Mem. Fac. Sci. Shimane Univ., 17, 81-86, 1983.

Yamada, N., Sangun metamorphic rocks of the Tottori-Ohara district, Southwest Japan, (in Japanese with English abstract), Bull. Geol. Surv. Japan, 23, 525-537, 1972.

Yamada, N., Y. Teraoka, and M. Hata (chief eds.), Geological map of Japan, scale 1:1,000,000, (in Japanese with English abstract), Geological Atlas of Japan, 3-19, 22-25, Geol. Surv. Japan, 1982.

Yogoro, T., T. Watanabe, Y. Harada, S. Fujie, M. Takamatsu, Y. Kosuge and Research group of the Chizu Formation, Mesozoic Nassellaria (radiolarians) from the Chizu formation (the Sangun Metamorphic Belt), Wakasa, (in Japanese), Abst. 101st Meeting Kansai Branch, Geol. Soc. Japan, 1986.

Zhang, Z.M., Tectonostratigraphic terranes of Japan that bear on the tectonics of Mainland Asia, in Tectonostratigraphic terranes of the Circum-Pacific Region, edited by D.G. Howell, 409-420, Earth Science Series No. 1, Circum-Pacific Council for Energy and Mineral Resources, Houston, 1985.

PALEOMAGNETIC CONSTRAINTS ON THE EVOLUTION OF THE UPPER CRETACEOUS IZUMI BASIN IN SHIKOKU, SOUTHWEST JAPAN

Kazuto Kodama

Department of Geology, Faculty of Science, Kochi University, Kochi 780, Japan

Abstract. The Izumi belt forms a narrow and extremely elongated basin extending continuously 300 km from northwestern Shikoku to the western Kii Peninsula. Generally the succession in the basin youngs eastwards, showing a synclinal structure with an axis plunging 30-50 degrees east. The composite stratigraphic thickness has been suggested to exceed 100 km but despite this value no significant difference in geologic age is indicated by paleontological studies of a few scattered faunas which are dated as Campanian to Lower Maastrichtian. Paleomagnetic investigations were carried out in two regions 100 km apart in northwest and northeast Shikoku. A total of 28 reliable mean directions were obtained from 33 sites. The 16 directions from the eastern region show normal polarity and, with one exception, those from the western region show reversed polarity. Apart from anomalous declinations from southern sites in the eastern region, which may have been rotated by sinistral drag on the Median Tectonic Line which forms the faulted southern margin of the basin, northeast-southwest declination concentrations are consistent with the clockwise tectonic rotation of Southwest Japan. Inclinations are compatible with the present-day latitude of the paleomagnetic sites. The fossil age constraints suggest the normal polarity in the east may be correlated with anomaly number 33 in the Upper Campanian, and the reversed polarity in the west with the preceding reversed polarity chron after the Cretaceous long normal polarity epoch. The age of the polarity boundary is estimated to be 80 Ma, from the latest geomagnetic polarity time-scale.

Introduction

Southwest Japan consists of a variety of geologic units distributed almost parallel to the trend of the arc. On Shikoku Island they are, from south to north, the Shimanto belt, the Chichibu belt, the Sanbagawa metamorphic belt, the Izumi belt, and the Ryoke metamorphic belt (Fig. 1). The outermost Shimanto belt consists of highly deformed sedimentary rocks of Cretaceous and Tertiary age, recently interpreted as a subduction complex [Taira et al., 1982]. The Chichibu belt is possibly a Paleozoic to Early Mesozoic subduction complex, although less geologic evidence has been reported than for the Shimanto [Taira et al., 1983]. By contrast the Izumi belt comprises a coherent sequence of Late Cretaceous sedimentary rocks distributed in a very narrow (5 to 10 km wide) basin which extends for 300 km from northwestern Shikoku to the western Kii Peninsula. Immediately south of the belt is the Median Tectonic Line (MTL) which separates it from the Sanbagawa metamorphic belt. The dominant sense of strike-slip displacement on the MTL in the geological past is inferred to have been left-lateral, and the cumulative displacement since the Cretaceous amounts to several tens of kilometers [Ichikawa, 1980]. To the north the Izumi belt succession, most of which comprises thick alternating sandstone and shale, lies unconformably on metamorphic rocks and granite of the Ryoke metamorphic belt. Overall the structure of the basin is synclinal with an axis plunging 30° to 50° east [Kanmera and Hashimoto, 1980]; the succession dips southeast in the north and northeast in the south. This structure suggests that horizons are structurally higher toward the east, and the apparent total thickness of the entire succession amounts to over 100 km [Nishimura, 1984]. Although systematic age-lithology relationships appear never to have been determined, the Izumi Group is generally assigned to the Campanian to Lower Maastrichtian on the basis of a few megafossil occurrences [e.g. Matsumoto et al., 1980] and recent radiolarian age determinations [Okamura et al., 1984; Suyari and Hashimoto, 1985].

The aim of the present study was to try to combine magnetostratigraphic data with previous fossil age control to help date the rocks and, using paleomagnetic constraints, to clarify possible movement on the Median Tectonic Line which it is suspected exerted tectonic control on the evolution of the Izumi basin. This article briefly reports preliminary results of a combined paleomagnetic and geologic study carried out in two regions approximately 100 km apart in northeastern and northwestern Shikoku.

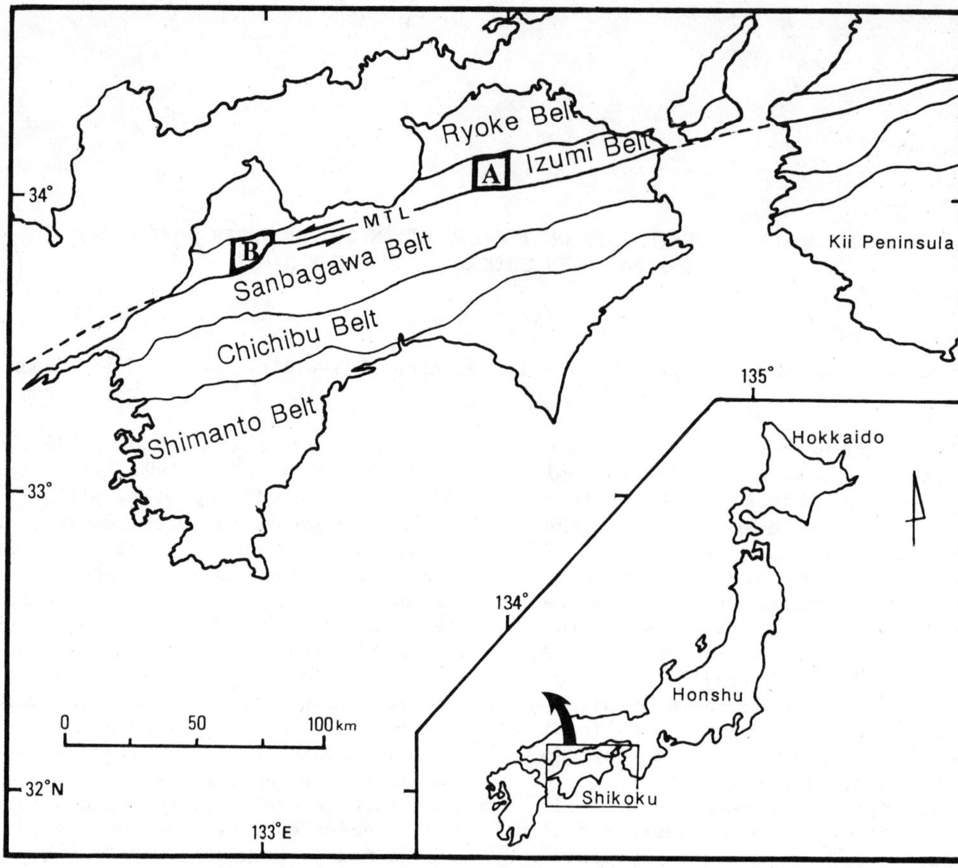

Fig. 1. Map showing major geologic belts in Shikoku, Southwest Japan and the location of the two regions (A, B) for paleomagnetic studies of the Izumi Group.

Geology and Paleomagnetic Sampling

The first region, of about 10 km^2 areal extent, lies in the western part of the Asan Mountains in the midst of the Izumi belt (region A in Fig. 1), where representative facies of the Izumi Group outcrop well (Fig. 2). Bedding orientation was measured in as many places as possible (76 sites) to help determine the orientation of the fold axis. An optimum plunge of 30° east was determined, which was used for tilt corrections applied to the in situ remanent magnetization vectors. A preliminary microfossil examination [Nozoe, 1985] found a few species of radiolaria at five localities close to the northern margin which can be assigned to the Upper Campanian. More accurate age constraints cannot be given because of the limited distribution of the species. A summary of the geology of this region and sample localities for the paleomagnetic study, are shown in Figure 2. A total of 16 sites were visited from the northern and southern margins, and one to four blocks oriented individually by magnetic compass were collected from each site. The rocks are shaly to sandy mudstone (9 sites), fine-grained sandstone (5 sites), and acidic tuff (2 sites). The last lithology, although not dominant, is very commonly intercalated in thin layers in the Izumi group.

The second region (B in Fig. 1) is located close to the western end of the Izumi belt where the succession consists of alternating sandstone and shale (Fig. 3). Despite lithological similarity to other areas of Izumi group rocks, the structure here seems to be different from the general structure of the Izumi belt. The beds to the north of the axial zone show a synclinal fold with an ENE-WSW directed horizontal axis, while toward the southern margin the beds are folded and commonly locally faulted [Nishimura, 1984; Takahashi, 1986]. It is notable that in this area the Median Tectonic Line appears to be bent (Figs. 1, 3). The apparent bending has been investigated by previous authors [e.g. Takahashi, 1986] in terms of basin analysis. The age of the sediments here is not accurately known, but megafossils in the lowest horizons suggest Lower Campanian [Noda and Tashiro, 1973], and radiolarians from the lower to middle horizons also suggest a Campanian age [Okamura et al., 1984]. Samples for paleo-

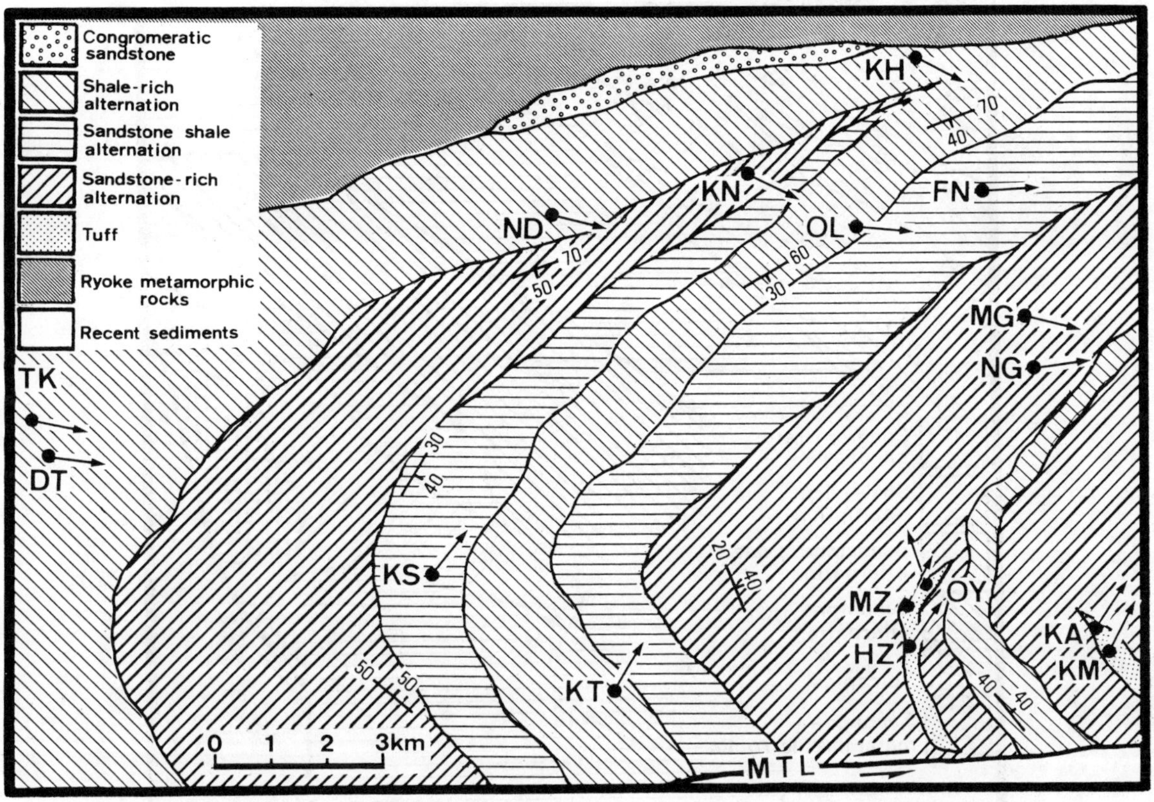

Fig. 2. Localities of 16 sites giving stable remanent magnetizations, and geologic outline of region A in Figure 1. Arrows indicate declinations of magnetization vector after tilt correction.

magnetic measurements were collected from a total of 17 sites (Fig. 3) the rock types of which are fine-grained sandstone (9 sites) and shale (8 sites).

Paleomagnetic Results

Recent magnetizations of 309 specimens prepared from 33 sites in the two regions were measured with a SCT cryogenic magnetometer (Kyoto University) because of their weak intensities. The magnitude of the natural remanent magnetization (NRM) is somewhat different for different rock types; $1x10^{-4}$ to $4x10^{-3}$ A/m for shale, mudstone and tuffaceous siltstone, and $3x10^{-3}$ to $4/10^{-3}$ A/m for sandstone. After measurements of the NRM of all specimens, one or two were chosen from each site to examine their stability using progressive alternating field (a.f.) demagnetization with peak intensities of from 5 to 100 mT. Typical examples of changes in magnetization vectors are illustrated in Figure 4. These results suggest that the rocks carry a single stable magnetization component, with median destructive field intensity of 15 to 25 mT. Therefore, all remaining specimens were subjected to a.f. demagnetization of 15, 20, 25 and, if necessary, 30 to 40 mT peak intensity. Then an optimum demagnetization step to calculate a site mean direction was defined site by site as that producing the minimum dispersion in direction. Rock-magnetic studies including Curie temperature analysis and saturation magnetization acquisition tests are in progress and will be reported elsewhere. The magnetic mineral responsible for the stable remanent magnetization is very likely to be magnetite not hematite because of the lower magnetic coercivity shown by a.f. demagnetization, microscopic observation, and chemical analysis of heavy minerals [Nishimura, 1984]. Mean directions of five sites in region B (SG-2, 10, 11, 16, 17) were rejected because of within-site dispersion equivalent to a precision parameter k less than 10.

Table 1 summarizes the site mean directions before and after tilt correction, and associated statistical parameters. Tilt correction for the sites in region A was performed carefully taking the observed structure into consideration. Since folding took place about an axis plunging 30° east, the conventional way of restoring strata to the horizontal, by rotation about the line of strike, is incorrect. Generally a simple unfolding procedure produces declination error or apparent tectonic rotation about a vertical axis

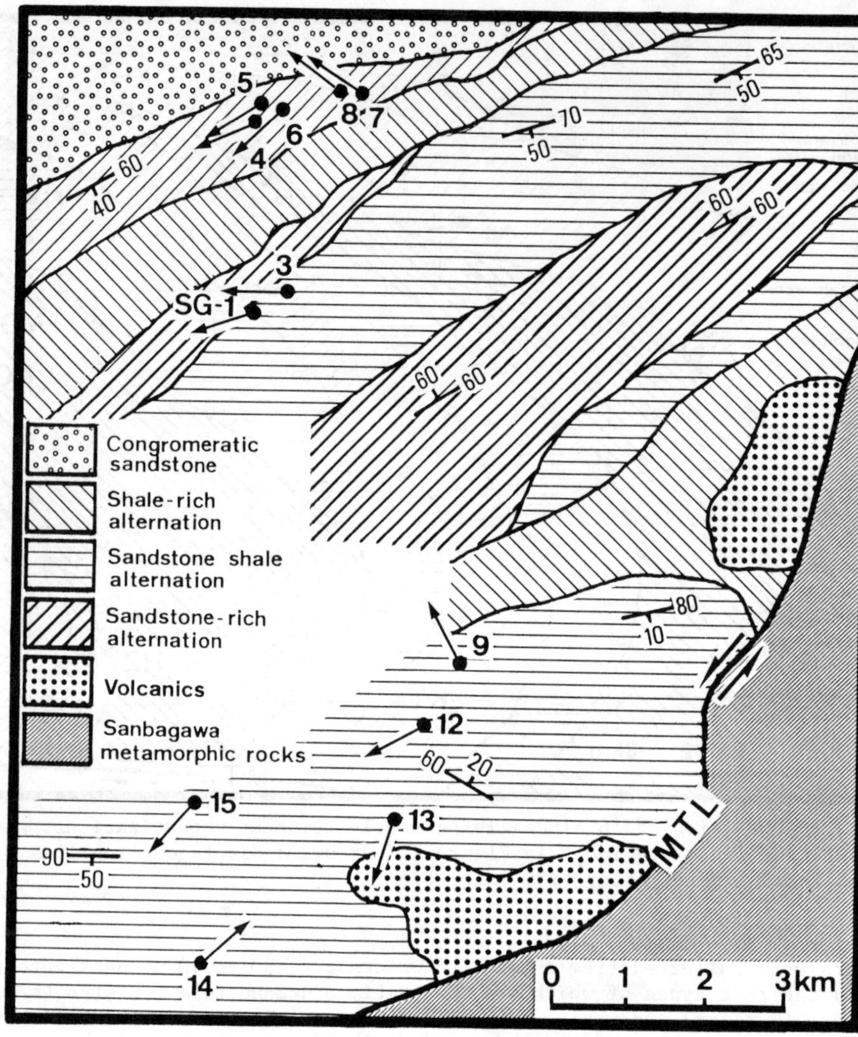

Fig. 3. Localities of 12 sites giving stable remanent magnetizations, and geologic outline of region B in Figure 1, modified in part after Okamura et al. [1984]. Arrows indicate declinations of magnetization vector after tilt correction.

[Macdonald, 1980]. Therefore tilt correction for this region was carried out by two successive procedures; first, rotating the initial paleomagnetic directions and bedding pole directions through the plunge angle of the fold axis (30°) about the north-south horizontal axis, and secondly, rotating the respective paleomagnetic direction through the dip angle of the reoriented bed about the new strike direction. The composite unfolding procedure is illustrated schematically in Figure 5. Site mean directions corrected in this way are listed in Table 1 and illustrated in Figure 6a.

Mean directions for 12 sites in region B were subjected to conventional tilt correction because the succession here is unlikely to be folded about an inclined axis, and unresolvable minor folding and faulting with diverse orientations are frequently observed between the axial zone and the southern margin.

The corrected site mean directions fall in two groups in terms of magnetic polarity. An apparent normal polarity is recognizable in all the sites in region A, and a reverse polarity in the sites in region B except for SG-14, samples from which are normally polarised. It is important to note that site SG-14 seems to be located in the stratigraphically highest rocks of the region. This suggests that a geomagnetic polarity change may have occurred during the latest stage of sedimentation in this region. If this is indeed the case then the age of the polarity boundary could be determined by correlating the observed polarities with a standard geomagnetic reversal

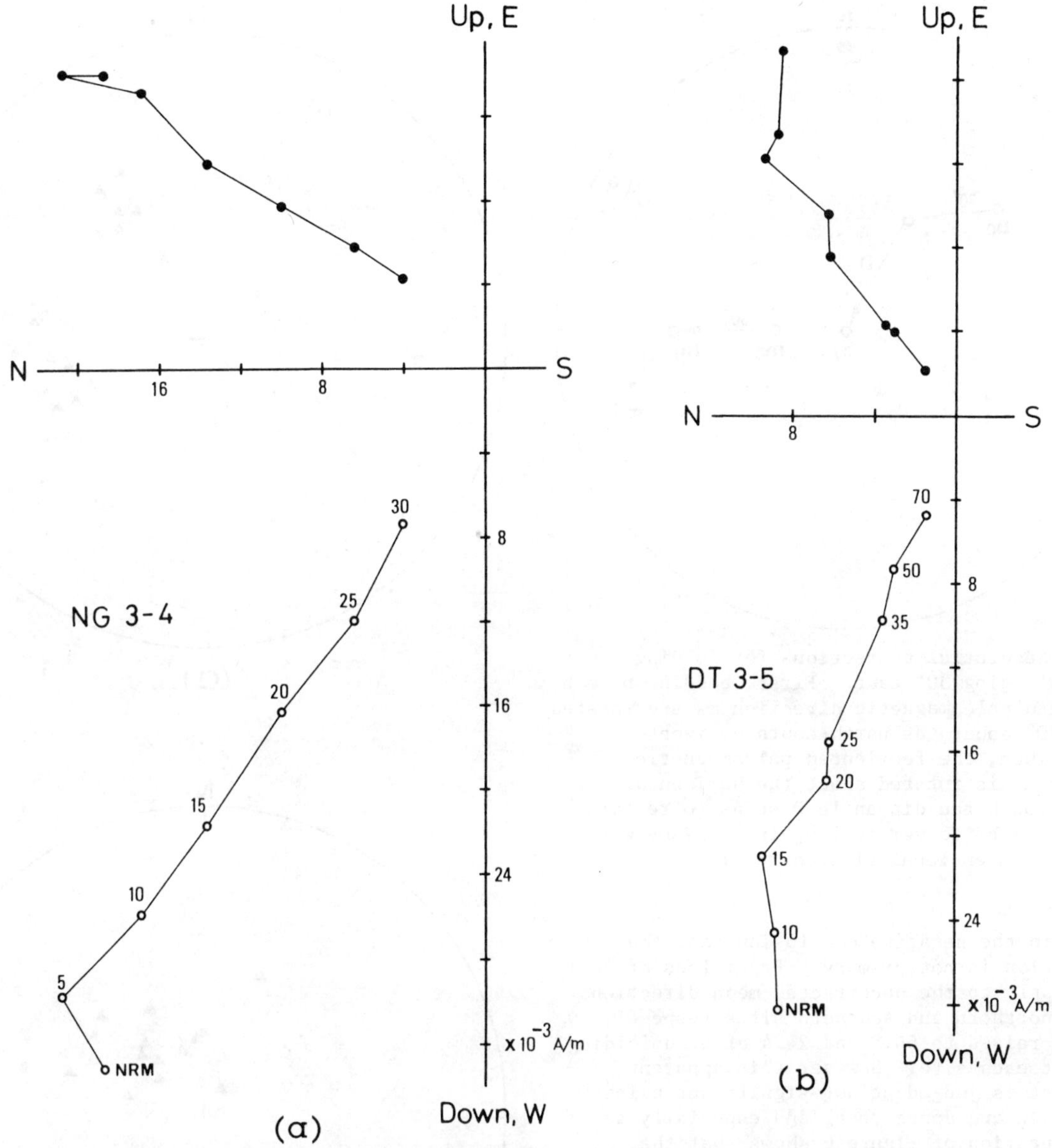

Fig. 4. Typical examples of orthogonal plots of successive endpoints of magnetization vector during progressive alternating field demagnetization for specimens from Site NG (a) and Site DT (b). Numbers are intensities of peak fields in mT. Solid (open) circles are projections onto horizontal (vertical) plane.

time-scale. Although no paleomagnetic data is yet available from anywhere between regions A and B, it is possible that the one normal polarity site in the west, and all those in the east, correspond with the same Upper Cretaceous polarity zone as biostratigraphical control indicates no difference in age between the two regions.

Discussion

The corrected directions from region A shown in Figure 6a seem to be divisible into two distinct groups depending on declination. One group, with declinations over 90° to the east, corresponds to the nine sites on the northern limb of the fold and the other group to the seven sites on the southern limb. The divergence of declination after the application of tilt corrections suggests that the magnetization was acquired after folding. Applying the statistical analysis of McFadden and Jones[1981], it is concluded that directions before correction belong to one population, but after tilt correction consist of two different populations, meaning that the fold test

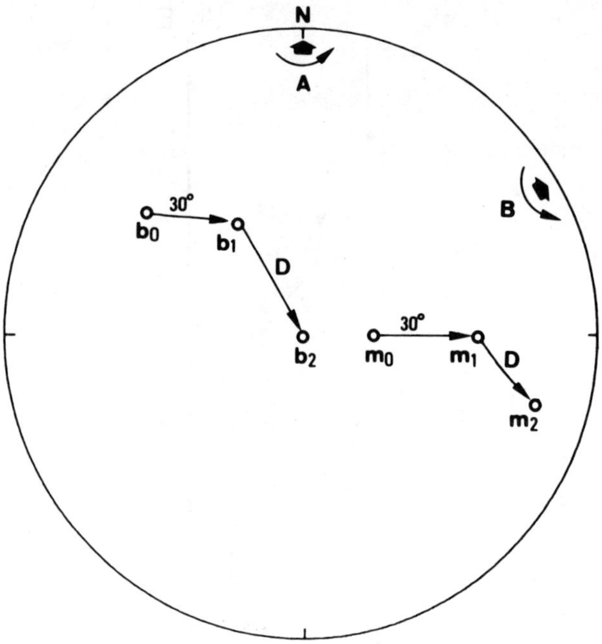

Fig. 5. Sequential corrections for folding about an axis plunging 30° east. First, bedding pole b_0 and in situ paleomagnetic direction m_0 are rotated through 30° about the north-south horizontal axis A. Then, the reoriented paleomagnetic direction m_1 is rotated about the horizontal axis B through the dip angle D so as to restore bedding pole b_1 to vertical b_2 in the same way as in the conventional tilt correction.

resulted in the negative conclusion that the magnetization is not primary. Precisions of 40.1 and 3.7 apply to the uncorrected mean directions from the northern and southern sites respectively, which are raised to 66.5 and 24.4 after unfolding bedding at each site. However this apparent improvement is judged as not significant using the McFadden and Jones test, and especially as close inspection of Figure 6 shows that the increase of precision of the southern sites is due to a great change in direction of only one site (OY). The precision of the other six sites actually decreases from 68.9 to 62.4, judged statistically as approximately constant.

Although the statistical invalidity of the fold test applied to the two subareas may not support interpretation of the magnetization as of depositional origin, the observation that reverse polarity is dominant in the stratigraphically lower horizons (region B), while normal polarity dominates stratigraphically higher horizons (region A), suggests it is unlikely that the sediments were remagnetized contemporaneously after folding. Considering that most of the rocks in the Izumi basin, including those sampled in this study, are of turbidite origin [Kanmera and Hashimoto, 1980],

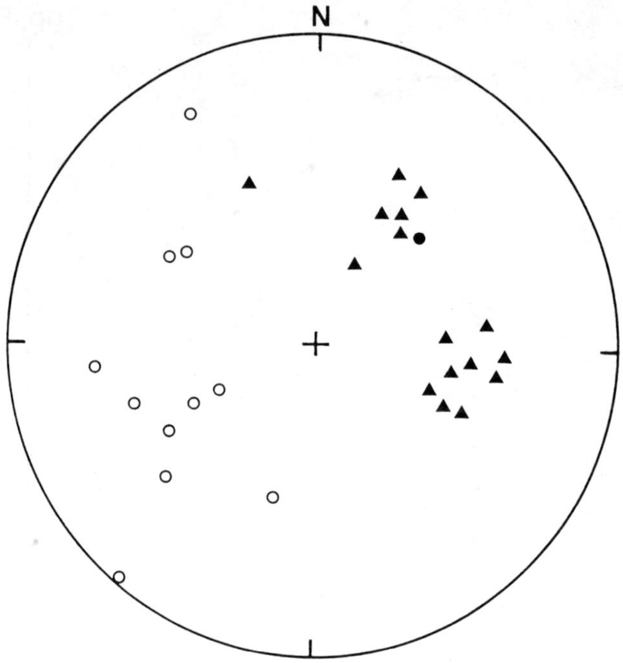

(a)

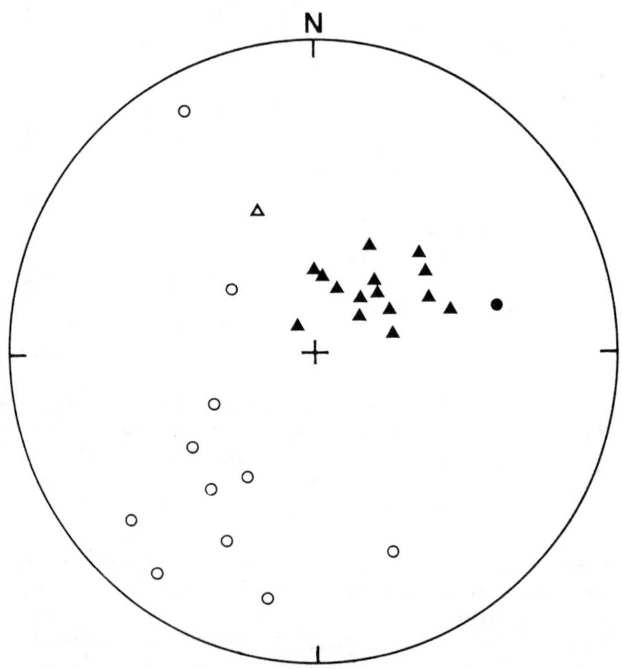

(b)

Fig. 6. Equal-area projection of all the site mean directions from the two regions after (a) and before (b) tilt correction, on the lower (solid) and upper (open) hemisphere. Triangles are from region A, circles from region B.

Table 1. Paleomagnetic Data From the Izumi Group in Shikoku, Southwest Japan

Site	N_s	N_b	I	D	I_c	D_c	k	α_{95}
KH	12	2	67.3	60.9	47.9	113.8	105.0	4.3
OL	11	2	57.0	65.9	38.5	93.0	379.7	2.3
KN	5	1	68.6	75.9	51.7	115.5	490.1	3.5
ND	18	4	72.1	46.2	52.6	101.1	49.5	5.0
DT	16	3	53.8	54.8	48.0	96.7	181.4	2.7
TK	12	4	52.2	73.5	40.0	101.3	113.2	4.1
MG	11	3	71.1	2.7	58.0	111.2	311.8	2.6
NG	12	2	52.8	48.1	43.1	84.9	104.5	4.3
FN	18	3	59.5	27.9	55.9	88.3	211.7	2.4
KT	16	2	82.2	325.1	67.2	27.4	344.1	2.0
KS	16	2	73.9	49.1	41.1	35.0	255.2	2.3
MZ	7	2	70.7	8.9	52.1	27.0	98.0	6.1
OY	12	3	-50.5	337.7	44.5	337.5	78.4	4.9
KA	10	3	66.4	46.4	49.5	33.8	404.3	2.4
HZ	5	4	71.7	17.1	53.7	38.0	23.3	16.2
KM	12	3	65.6	42.3	40.6	26.2	281.3	2.6
SG-1	5	4	-10.4	215.8	-39.3	251.2	15.8	19.8
SG-3	6	3	-18.0	227.7	-30.1	263.8	22.5	14.4
SG-4	7	2	-42.3	217.3	-54.9	242.6	49.0	8.7
SG-5	8	4	-29.6	159.2	-45.2	238.4	83.7	6.1
SG-6	5	2	-17.3	193.9	-36.7	227.5	42.9	11.8
SG-7	7	2	-59.4	242.9	-45.8	300.6	46.2	9.0
SG-8	9	3	-51.6	228.2	-48.3	304.7	138.8	4.4
SG-9	9	3	-10.0	332.0	-17.4	331.3	23.9	10.8
SG-12	8	3	-62.4	306.3	-63.0	243.6	22.3	12.0
SG-13	5	3	-31.4	205.4	-48.1	194.6	21.4	16.9
SG-14	5	3	36.6	76.4	50.2	44.2	13.8	21.4
SG-15	7	3	-47.4	232.3	-1.9	218.9	29.5	11.3

N_s, number of specimens. N_b, number of blocks collected and individually oriented. I, D, in situ inclination and declination. I_c, D_c, inclination and declination after correction for bedding attitude. k, α_{95}, Fisher's precision parameter and semi-angle of 95% confidence cone.

the invalid results of the fold test may be due to synsedimentary slump folding, with remanent magnetization locked in sometime after slump deformation. If the magnetizations acquired after slumping were not reset during subsequent tectonic folding, the dispersion of declinations peculiar to the northern and southern subareas after structural correction may arise as a consequence of tectonic rotation about a vertical axis of one subarea relative to the other. One possible interpretation is to attribute the declination anomaly to a tectonic rotation arising from transcurrent movement on the MTL [Kodama, 1986]. Since left-lateral displacement is inferred to be dominant for the MTL, it is speculated that the southern margin of the Izumi belt may have been affected more strongly by horizontal drag, resulting in a counterclockwise tectonic rotation as observed in the declination differences in region A. Reversed magnetization directions in region B do not show a subgrouping in declination but tend to concentrate in the southwest, although a few divergent directions remain after tilt correction. Simple tilt correction was applied to all the directions, resulting in a somewhat better grouping (precision parameter k increases from 4.4 to 5.9) and an eastwards shift in mean-site direction from D=228°, I=43° to D=252°, I=-47°. The declination anomaly observed in the southern margin of the eastern region is not recognizable in the western region.

It is important to point out that the declinations in general suggest clockwise rotation, consistent with numerous paleomagnetic data reported for Southwest Japan and currently interpreted as a consequence of the opening of the Japan Sea in the Early Miocene [e.g. Otofuji et al., 1986]. The inclinations after bedding correction are compatible with the present latitude of the sites, suggesting that the sediments were deposited in place, in remarkable contrast to several contemporaneous rock bodies in the Outer Zone which were emplaced in their present positions after northward migration from distant origins [Kodama et al., 1983].

Using the paleontologically determined age of the studied areas (approximately Campanian) it is

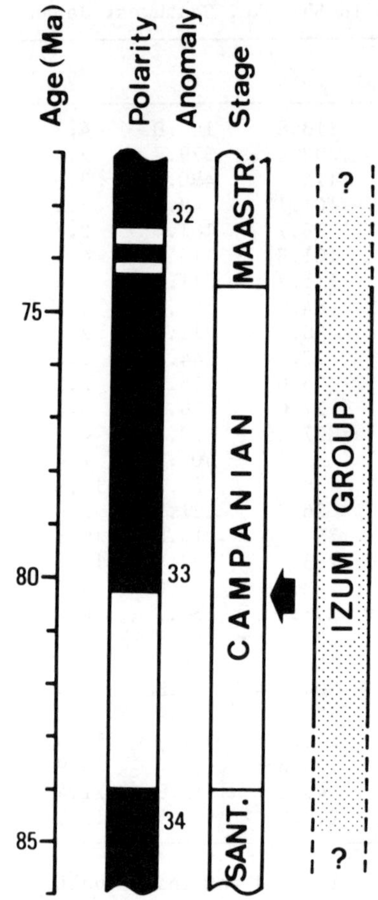

Fig. 7. Correlation of the Upper Cretaceous geomagnetic polarity sequence (black=normal; white=reversed) to the Izumi Group. Polarity time-scale is partly from Kent and Gradstein [1985]. A possible polarity-zone boundary found in this study is marked by an arrow.

possible to correlate the normal and reversed polarities with a standard geomagnetic polarity time-scale. According to the Cretaceous polarity time-scale, the Campanian stage consists of one reversed polarity chron [e.g. Alvarez and Lowrie, 1984; Kent and Gradstein, 1985] commencing at the Campanian-Santonian boundary, and one normal chron extending into the lowermost Maastrichtian numbered as Anomaly 33 (Fig. 7). It is possible to assign the reversed polarity found in the stratigraphically lower successions to the reversed polarity chron of the Lower Campanian, and the normal polarity from the higher successions to the following normal polarity chron. From the time-scale of Kent and Gradstein [1985] an absolute age of 80 Ma is estimated for the polarity-zone boundary, although the precise location of the boundary in the basin succession was not determined in this study.

The paleomagnetic stratigraphy discussed above is preliminary and tentative, requiring further studies of a number of sections under accurate fossil age control, but it indicates at least the usefulness of magnetostratigraphic studies in understanding the development of the extremely thick sequence of the Izumi belt, in which deposition was probably under the dominant tectonic control of the MTL. Combined paleomagnetic, sedimentological, structural, and chronological studies would make it possible not only to determine the depositional history of the sediments in more detail, but also to establish the Upper Cretaceous magnetic polarity sequence for the first time from the Japanese arc.

Acknowledgments. I thank Dr. M. Torii (Kyoto University) for allowing me to use the cryogenic magnetometer, and Mr. M. Nozoe (Kochi University) for help with the field work and laboratory measurements. Discussion with Professor M. Tashiro (Kochi University) on paleontological data was very helpful. This work was supported in part by grants from the Itoh Science Foundation, and from Dynamics and Evolution of the Lithosphere Project (DELP) 1985.

References

Alvarez, W., and W. Lowrie, Magnetic stratigraphy applied to synsedimentary slumps, turbidites, and basin analysis: the Scaglia limestone at Furlo (Italy), Geol. Soc. Am. Bull., 95, 324-336, 1984.

Ichikawa, K., Geohistory of the Median Tectonic Line of Southwest Japan, Mem. Geol. Soc. Japan, no. 18, 187-212, 1980.

Kanmera, K., and M. Hashimoto, The Izumi belt and its lateral extension (in Japanese), in The Geology of Japan, edited by K. Kanmera, M. Hashimoto, and T. Matsuda, pp.37-39, 15 Iwanami Shoten, Tokyo, 1980.

Kent, D.V., and F.M. Gradstein, A Cretaceous and Jurassic geochronology, Geol. Soc. Am. Bull., 96, 1419-1427, 1985.

Kodama, K., Two different paleomagnetic directions from the Izumi Group in Shikoku, Southwest Japan, J. Geomag. Geoelectr., 38, 279-284, 1986.

Kodama, K., A. Taira, M. Okamura, and Y. Saito, Paleomagnetism of the Shimanto belt in Shikoku, Southwest Japan, in Accretion tectonics in the Circum-Pacific regions, edited by M. Hashimoto and S. Uyeda, pp.231-241, Terra Sci. Publ. Tokyo., 1983.

MacDonald, W.D., Net tectonic rotation, apparent tectonic rotation, and the structural tilt correlation in paleomagnetic studies, J. Geophys. Res., 85, 3659-3669, 1980.

Matsumoto, T., H. Hashimoto, and M. Furuichi, An interesting species of Baculites (Ammonoidea) from the Cretaceous Izumi Group of Shikoku, Proc. Japan Acad., ser.B., 56, 408-413, 1980.

McElhinny, M.W., Statistical significance of the fold test in palaeomagnetism, Geophys. J. Roy. Astron. Soc., 8, 338-340, 1964.

McFadden, P.L., and D.L. Jones, The fold test in palaeomagnetism, Geophys. J. Roy. Astron. Soc., 67, 53-58, 1981.

Nishimura, T., Basin analysis of the Upper Cretaceous Izumi Group in Western Shikoku, Japan (in Japanese with English abstract), J. Geol. Soc. Japan, 90, 157-174, 1984.

Noda, M., and M. Tashiro, The fauna from Dogo-Himezuka, Matsuyama City and its stratigraphic significance (in Japanese), J. Geol. Soc. Japan, 79, 493-495, 1973.

Nozoe, M., Sedimentary structure of the Izumi Group in Western Asan Mountains (in Japanese), M.S. Thesis, Kochi Univ., 1985.

Okamura, M., Y. Kagawa, and M. Tashiro, Geology and radiolarians of the Izumi Group in the eastern part of Matsuyama city, Ehime prefecture (in Japanese with English abstract), Res. Rep. Kochi Univ., 32, 339-347, 1984.

Otofuji, Y., T. Matsuda, and S. Nohda, Brief review of Miocene opening of the Japan Sea: paleomagnetic evidence from the Japan arc, J. Geomag. Geoelectr., 38, 287-294, 1986.

Suyari, K., and H. Hashimoto, A radiolarian assemblage from the Izumi Group of Eastern Shikoku (in Japanese with English abstract), J. Sci. Univ. Tokushima, 18, 103-127, 1985.

Taira, A., H. Okada, J.H.M. Whitaker, and A.J. Smith, The Shimanto Belt of Japan: Cretaceous-lower Miocene active-margin sedimentation, in Trench-Forearc Geology: Sedimentation and tectonics on modern and ancient active plate margins, edited by J.K. Leggett, pp.5-26, Geol. Soc. London, Spec. Publ., no.10, 1982.

Taira, A., Y. Saito, and M. Hashimoto, The role of oblique subduction and strike-slip tectonics in the evolution of Japan, in Geodynamics of the Western Pacific-Indonesian Regions, edited by S. Uyeda and T.W.C. Hilde, pp.303-316, Geophys. Monogr. 11, Am. Geophys. Union, 1983.

Takahashi, J., On "The Median Tectonic Line" in Matsuyama city and the neighbouring areas, Ehime prefecture, Japan, Memoirs of the Faculty of Education, ser.III, Nat. Sci., 6, 1-44, 1986.

PRELIMINARY GEOPHYSICAL INTERPRETATION OF THE OCEANIC TERRANES OF INTERIOR AND WESTERN ALASKA: EVIDENCE FOR THICK CRUST OF INTERMEDIATE DENSITY

John W. Cady

U. S. Geological Survey, Denver, Colorado 80225

Abstract. Interior and western Alaska is an area characterized by oceanic terranes, low, irregular topography, and moderate Bouguer gravity highs. Digital gravity and aeromagnetic data are being compiled, and quantitative analysis and modeling of these data are underway to determine geologic structure in the Yukon-Koyukuk province (YKP) and Tozitna terrane. The YKP contains varied tectonic elements that include thrust sheets of ophiolite, pillow lavas, a magmatic arc, and graywacke basins. The elements have distinctive topographic relief and geophysical signatures. The preferred gravity model for the YKP involves oceanic crust that has been thickened by island arc magmatism and sedimentation and probable imbricate faulting; only at the province margins has oceanic crust been thrust over continental crust. The Tozitna terrane contains sills of layered gabro and cumulus wehrlite intruding and faulted against diabase, rare basalt, and marine sedimentary rocks. Unlike the YKP, the Tozitna terrane contains no exposed ophiolites. It is interpreted to be oceanic crust thickened by the intrusion of laccoliths in an oceanic plateau or primitive magmatic arc.

Regional Setting

Interior and western Alaska is a vast region of low, irregular topography drained primarily by the Yukon, Koyukuk, and Kuskokwim Rivers. Paleozoic and Mesozoic rocks, both igneous and sedimentary, are generally oceanic in character. The region is characterized by moderate (>-20 mGal complete Bouguer anomaly) gravity highs indicating denser crust or thinner crust than is found in the Brooks Range, Alaska Range, and Yukon-Tanana terrane. Digital gravity and aeromagnetic data are being compiled and quantitative analysis and modeling of these data are underway to determine geologic structure in the Yukon-Koyukuk province and Tozitna terrane (Figure 1). Seismic refraction and deep reflection studies will be made in the next few years as part of the Trans Alaska Crustal Transect (TACT) program and the gravity and magnetic data and their interpretation will help to plan optimal locations for TACT seismic profiles. This paper presents a preliminary interpretation of the subsurface geology and crustal structure of the Yukon-Koyukuk province and Tozitna terrane. Data compilation is still incomplete; therefore, an index map is used in place of gravity and aeromagnetic maps in this preliminary report. Previously published geophysical maps of the area include the Alaska Bouguer anomaly map of Barnes [1977] and the Alaska aeromagnetic map of Godson [1984].

Rocks of the oceanic Angayucham, Koyukuk, and Tozitna terranes [Jones and others, 1984; Howell and others, 1985] and an area herein referred to as the Kuskokwim-Tanana province produce Bouguer gravity anomaly highs. Bouguer anomaly lows occur over old continental crust of the Brooks Range and most of the Yukon-Tanana terrane and tectonically thickened rocks of the Alaska Range. Gravity gradients are steep and coincide with terrane boundaries along the south flank of the Brooks Range and along the northwest-trending part of the Tintina fault. Such boundaries are interpreted to be relatively steep and to separate regions of differing crustal structure. In contrast, along the northwest boundary of the Yukon-Tanana terrane west of Fairbanks and along the Denali fault, gravity gradients are not present over terrane boundaries, suggesting that here terranes have subhorizontal boundaries or do not contrast in density structure. A relatively minor gravity low over the Ruby geanticline indicates that it is only a continental sliver.

Gravity and magnetic highs form a V-shaped pattern open to the southwest, coincident with the northern and southeastern margins of the Yukon-Koyukuk province (YKP). These anomalies are asymmetrical, with steeper gradients to the outside of the province, and can be modeled by assuming dense, magnetic sources that dip 30-70 degrees inward beneath the YKP. Oceanic rocks of both the northern and the southeastern

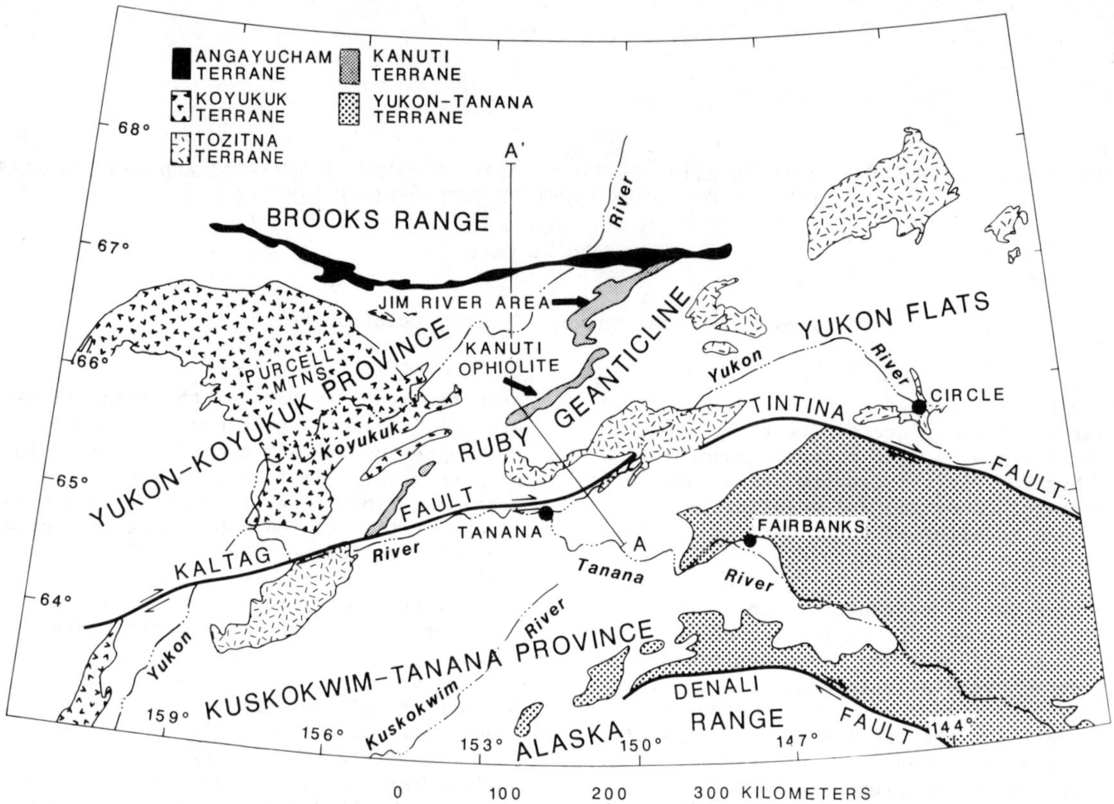

Fig. 1. Geologic index map of interior Alaska [after Beikman, 1980; Jones and others, 1984] - Line A-A' is the gravity model section line shown in Figure 2.

margins of the V are assigned to the Angayucham terrane by Jones and others [1984]. Geologically and geophysically, however, the arms of the V are different, suggesting that they should be assigned to different lithotectonic terranes. In this discussion, the term Angayuchamn terrane refers to the northern arm of the V, while the term Kanuti terrane refers to the southeast margin. The boundary between the Kanuti and Angayucham terranes is not yet determined, but it is tentatively shown north of the Jim River area.

The predominant rock type of the Angayucham terrane is pillow basalt formed in an oceanic plateau and (or) oceanic island setting [Pallister, 1985]. The amplitude of the magnetic anomaly over the Angayucham terrane is generally less than 500 nT. In contrast, the Kanuti terrane contains a complete dismembered ophiolite sequence. Both the cumulus mafic-ultramafic rocks and ultramafic tectonite of the informally designated Kanuti ophiolite cause a magnetic anomaly of more than 2000 nT. Magnetic modeling shows that rocks lithologically equivalent to the Kanuti ophiolite cannot be present at depth beneath the Angayucham terrane.

In contrast to the nearly continuous gravity high over the Angayucham terrane, the gravity high associated with the Kanuti terrane is interrupted by lows over granitoids that have intruded the Kanuti ophiolite and the mafic-ultramafic rocks exposed en echelon to the north along the Jim River. The magnetic high, however, is continuous over the Kanuti ophiolite and the plutons that intrude it. These plutons of the Ruby geanticline are normally non-magnetic S-type granites [Miller, 1985] that occur in magnetic lows with local relief of less than 50 nT. Where they have intruded mafic and ultramafic rocks, however, the plutons are magnetic and cause magnetic highs with relief of more than 1000 nT. Perhaps the plutons are magnetic because they were contaminated by the mafic crust that they assimilated or from which they were derived.

Within the YKP are U-shaped gravity and magnetic highs open to the southwest produced by an Early Cretaceous intraoceanic island arc [Box and Patton, 1985] labeled the Koyukuk terrane (Figure 1). This island arc divides mid-Cretaceous clastic rocks of the YKP into two subbasins marked by gravity and magnetic lows: an outer, remnant-forearc basin and an inner, remnant-backarc basin [Box and Patton, 1985]. Gravity lows (data are terrain corrected) also

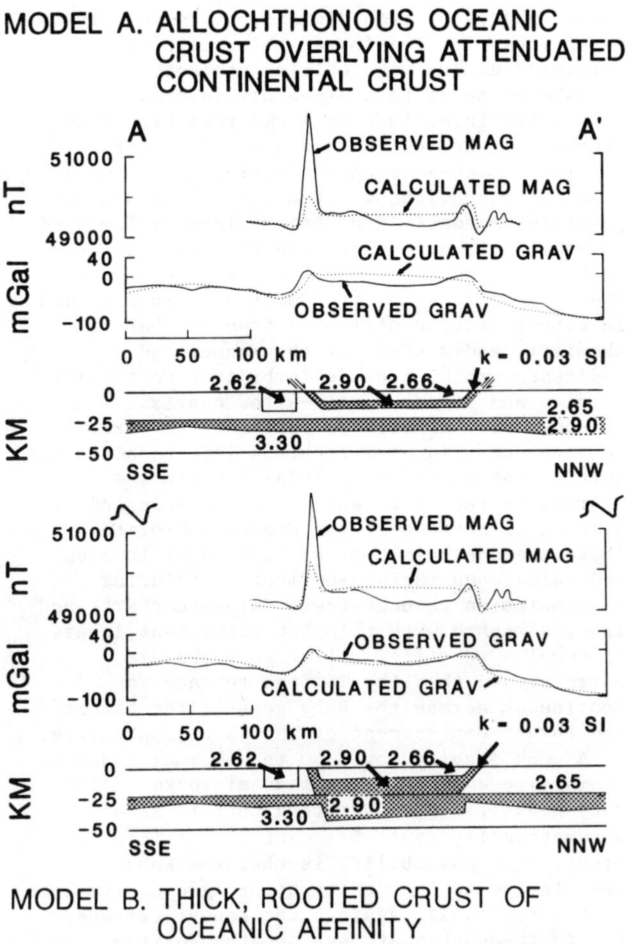

Fig. 2. Gravity and partial magnetic models across the eastern Yukon-Koyukuk province along line A-A' (Figure 1) Densities in g/cm^3. Densities of upper crustal rocks are averages of wet bulk densities of rock samples collected at several hundred gravity stations. Magnetic susceptibility (k) is in the SI system. Patterned area with density of 2.90 g/cm^3 includes the informally designated Kanuti ophiolite (in south), Angayucham terrane (in north), and mafic lower crust. Area with density 2.66 g/cm^3 is graywacke and volcanic rocks of Yukon-Koyukuk province. Area with density 2.62 g/cm^3 is granite. Upper crust and mantle have densities of 2.65 and 3.30 g/cm^3 respectively.

occur over I-type [Miller, 1985] granitoids of the Purcell Mountains which have summit elevations of about 1000 m, or about 800 m above valley floors. Thus, the YKP contains varied tectonic elements including thrust sheets of ophiolite, abundant pillow basalts, a magmatic arc containing both volcanic and granitic rocks, and two graywacke basins. The elements have distinctive topographic relief and geophysical signatures. Although the entire assembly has often been termed the Yukon-Koyukuk "basin", the term Yukon-Koyukuk province implies more tectonic variability and is preferred.

Modeling for deep crustal structure is ambiguous in the absence of seismic refraction or deep seismic reflection data. An unreversed seismic refraction profile [Hanson and others, 1968] indicates continental crust thinning from 48 km in the Alaska Range to 32 km near Fairbanks. Marine seismic studies off Kuskokwim Bay southwest of figure 1 show oceanic crustal velocities of 6.7-6.9 km/s at a depth of 10 km, but mantle velocities are not observed [Cooper and Marlow, 1984]. Oceanic rocks, including oceanic crust and upper mantle of the Kanuti ophiolite, are exposed at the surface in the YKP. Since the area is above sea level, isostacy requires that the crust of the YKP must be thicker than oceanic crust and have a lower mean density. How has the crust been thickened?

Figure 2 shows two regional crustal models across the eastern YKP (Figure 1) that satisfy the gravity data and illustrate contrasting hypotheses about how the crust was thickened. Model A shows allochthonous oceanic crust thrust over attenuated continental crust. Although it satisfies the gravity data, isotopic data [Arth, 1985] argue against the presence of old continental crust beneath the YKP. Preferred model B shows thick, rooted crust of oceanic affinity. Because the crust has relatively high density, it must have a root protruding into the mantle to maintain isostatic equilibrium. The magnetic data were used only to constrain the near-surface configuration of dense, magnetic rocks. Note that if a similar source is used to explain anomalies over both the Kanuti ophiolite and Angayucham terrane, the calculated anomaly is too small over the Kanuti ophiolite. Short-wavelength magnetic anomalies at the right side of the profile are caused by low-density magnetic schists of the Brooks Range not represented in the magnetic model.

Geophysical data are compatible with the interpretation [Box, 1985] that the YKP contains oceanic crust that remained unsubducted when an island arc collided with an irregular continental margin in Early Cretaceous time and was oroclinally bent. Further tightening of the orocline accompanied by probable imbricate thrusting occurred when terranes were accreted from the southeast in Late Cretaceous time. Thus, the YKP is underlain by oceanic crust that has been thickened by island arc magmatism and sedimentation and by probable imbricate faulting; only at the province margins has oceanic crust been thrust over continental crust.

Tozitna Terrane

East of the Ruby geanticline is an area of moderately high gravity coincident with Yukon

Flats and surrounding outcrops of the Rampart Group [the Rampart Group was included in the Tozitna terrane by Howell and others, 1985]. Short wavelength magnetic highs indicate that Rampart Group basement occurs at shallow depths beneath Quaternary sediments of Yukon Flats. Hence I will refer to both the exposed Rampart Group and the basement of Yukon Flats as the Tozitna terrane.

Unlike the Yukon-Koyukuk province, the Tozitna terrane contains no exposed ophiolites. Ultramafic tectonites like those exposed in the Kanuti ophiolite have not been reported. No magnetic highs with amplitudes approaching that over the Kanuti ophiolite have been observed in the Tozitna terrane. The Tozitna terrane has been less studied than the YKP, however, and many intriguing gravity and magnetic anomalies have not been investigated in the field.

I have studied rocks of the Tozitna terrane along a nearly continuous belt of gravity and magnetic highs that trends east-northeast from west of Tanana to east of Circle, north of the strike-slip Kaltag and Tintina faults. The gravity high, which is poorly defined by the available data, is broad and coincides generally with exposed rocks of the Tozitna terrane. Within the gravity high is a narrow (typically 5 km wide) nearly continuous belt of magnetic highs caused by layered gabbro and cumulus wehrlite that occur in sills up to 500 m thick. Intruded by and faulted against the gabbro and wehrlite are diabase, minor basalt, and marine sedimentary rocks. Associated with the narrow belt of magnetic layered gabbro and wehrlite is a wider zone of less magnetic diabase, diorite, and rare pillow basalt. Ovoid or irregular magnetic highs within the wider zone are thought to mark intrusive centers that were the source of the mafic to ultramafic sills. Alternatively, they may indicate buried remnants of submarine volcanoes.

The Tozitna terrane is best exposed in synclinal remnants of nappes thrust over older Paleozoic metamorphic rocks. Within the Tozitna terrane, the magnetic layered gabbro and wehrlite appear to belong to a higher nappe, whereas the surrounding, less magnetic zone belongs to a lower nappe. In the west, the Tozitna nappes overly older Paleozoic rocks on the south flank of the Ruby geanticline. In the east, between the Tintina fault and Yukon Flats, the Tozitna nappes occur above the southern thrust contact of an extensive body of similar rock thought to underlie Yukon Flats.

I propose a laccolith model for the emplacement of the intrusive rocks of the Tozitna terrane. The creation of oceanic plateaus by laccoliths is an attractive concept, for laccoliths can thicken the oceanic crust without causing abundant volcanism. Differentiation within thick sills formed cumulus wehrlite and layered gabbro that in places grades up into diorite. Less voluminous injections of magma formed thin sills of unlayered gabbro with chilled contacts of diabase adjacent to sedimentary layers. Successive injections of magma rose higher and higher, thickening the oceanic crust, but only rarely did magma reach the surface to form basalt. Laccoliths might also form in a primitive magmatic arc, before large volumes of intermediate magma penetrate the crust to form batholiths and volcanoes. Clues about whether the igneous rocks of the Tozitna terrane formed in either setting will come from the basalt chemistry and a study of the associated sediments, which include turbidites containing calcite and abundant silt-sized quartz.

Rocks very similar to parts of the Tozitna terrane occur in an allochthon [the basalt-chert unit of Patton and Box, 1985] beneath the ultramafic tectonite and cumulate mafic and ultramafic section of the Kanuti ophiolite. These include basalt or diabase, argillaceous and calcareous marine sediments, including Mississippian through Lower Jurassic chert, and layered gabbro with thin but persistant layers of cumulus wehrlite. If these rocks are correlative, then the Tozitna terrane was continuous across the Ruby geanticline before the thrust emplacement of the Kanuti ophiolite.

A weak gravity low (-40 to -60 mGal Bouguer anomaly) occurs over continental rocks of the Ruby geanticline, suggesting that it is a volumetrically small fragment of continental crust. One possibility is that the Ruby geanticline is bounded by high-angle faults both to the northwest, against the Kanuti terrane, and to the southeast, against the Tozitna terrane. In this case, the Tozitna terrane could be thick crust of an oceanic plateau, rooted under Yukon Flats, that has locally been thrust over the Ruby geanticline. An alternate possibility is that the continental rocks of the Ruby geanticline are thin because they are attenuated continental crust of a west-facing Paleozoic passive margin. In this model, the Tozitna terrane is the crust of an oceanic plateau, delaminated from the mantle, that was thrust eastward out of the Yukon-Koyukuk province over the leading edge of this continental margin, now exposed as the basement of the Ruby geanticline. According to this hypothesis, attenuated continental crust underlies the Tozitna terrane in Yukon Flats and may be continuous at depth with cratonal North America.

Conclusions and Suggestions for Future Research

Modeling of regional gravity and magnetic anomalies cannot resolve between allochthonous and rooted models of oceanic terranes in the Yukon-Koyukuk province and Yukon Flats. From interpretation of isotopic, geologic, and geophysical data, I favor a model for the Yukon-

Koyukuk province in which an island arc was built upon rooted oceanic crust. The Tozitna terrane and Yukon Flats contain a probable oceanic plateau. The geophysical data permit thick, rooted crust of an oceanic plateau, or allochthonous crust of a plateau delaminated from the mantle and thrust over attenuated continental crust of a passive margin.

These interpretations will be be tested by seismic and magnetotelluric studies of the Trans Alaska Crustal Transect (TACT). Digital compilation of aeromagnetic and gravity data is nearing completion in the Yukon-Koyukuk province and Tozitna terrane, although some of the aeromagnetic data are of poor quality. Several hundred new gravity stations, with density samples, have been obtained in the YKP and combined with previous data sets. Additional gravity data are needed, especially in the Tozitna terrane. These data will allow quantitative analysis and modeling of the potential field data for geologic structure, especially in the top 10 to 15 km. Using surface geology and potential field data to constrain seismic and magnetotelluric interpretations should yield a less ambiguous picture of crustal structure.

Acknowledgements. The manuscript was reviewed by D. L. Campbell, Arthur Grantz, Warren Hamilton, J. W. H. Monger, and S. S. Webster.

References

Arth, J. G., Neodymium and strontium isotopic composition of Cretaceous calc-alkaline plutons of the Yukon-Koyukuk basin, Ruby geanticline, and Seward Peninsula, Alaska, EOS, 66, 1102, 1985.

Barnes, D. F., Bouguer gravity anomaly map of Alaska, U. S. Geol. Surv. Geophys. Invest. Map GP-913, 1977.

Beikman, H. M., Compiler, Geologic map of Alaska, U.S. Geol. Surv., Scale 1:2,500,000, 1980.

Box, S. E., Early Cretaceous orogenic belt in northwestern Alaska: internal organization, lateral extent, and tectonic interpretation, in Tectonostratigraphic Terranes of the Circum-Pacific Region, Earth Sci. Ser., 1, edited by D. G. Howell, pp. 137-145, Circum-Pacific Council for Energy and Mineral Resources, Houston, 1985.

Box, S. E., and Patton, W. W., Jr., Collided intraoceanic volcanic arc in the Yukon-Koyukuk basin, western Alaska, EOS, 66, 1102, 1985.

Cooper, A. K., and Marlow, M. S., Geophysical and geological studies of the Bering Sea shelf, U. S. Geol. Surv. Circ. 939, 123-127, 1984.

Godson, R. H., compiler, Composite magnetic anomaly map of the United States, Part B: Alaska and Hawaii, U. S. Geol. Surv. Geophys. Invest. Map GP-954-B, 1984.

Hanson, Kenneth, Berg, Eduard, and Gedney, Larry, A seismic refraction profile and crustal structure in central interior Alaska, Bull. Seismol. Soc. Am., 58, 1657-1665, 1968.

Howell, D. G., Schermer, E. R., Jones, D. L., Ben-Avraham, Zvi, and Scheibner, Erwin, Preliminary tectonostratigraphic terrane map of the Circum-Pacific region, in Tectonostratigraphic Terranes of the Circum-Pacific Region, Earth Sci. Ser., 1, edited by D. G. Howell, (in pocket), Circum-Pacific Council for Energy and Mineral Resources, Houston, 1985.

Jones, D. L., Silberling, N. J., Coney, P. J., and Plafker, George, Lithotectonic terrane maps of the North American Cordillera, Part A - Lithotectonic terrane map of Alaska, in Lithotectonic Terrane Maps of the North American Cordillera, edited by N. J. Silberling and D. L. Jones, U.S. Geol. Surv. Open-File Rep. 84-523, 1984.

Miller, T. P., Petrologic character of the plutonic rocks of the Yukon-Koyukuk basin and its borderlands, EOS, 66, 1102, 1985.

Pallister, J. S., Pillow basalts from the Angayucham Range, Alaska: chemistry and tectonic implications, EOS, 66, 1102, 1985.

THERMAL MATURITY OF TECTONOSTRATIGRAPHIC TERRANES WITHIN THE FRANCISCAN COMPLEX, CALIFORNIA

Michael B. Underwood

Department of Geology, University of Missouri, Columbia, MO 65211

M. C. Blake, Jr., and David G. Howell

U.S. Geological Survey, 345 Middlefield Road, Menlo Park, CA 94025

Abstract. Indicators of organic metamorphism provide valuable tools for analyzing the thermal history of tectonostratigraphic terranes. Paleotemperature estimates derived from vitrinite reflectance, for example, are more precise than values based upon inorganic mineral assemblages in low-grade rocks. Isothermal geometries must be interpreted within the context of structural and stratigraphic data, but, by doing so, direct comparisons of thermal history can be made between adjacent terranes. Data from the Franciscan Complex of California show that three basic types of thermal anomalies are related to terrane amalgamation: (1) maturity inversion across terrane boundaries, (2) shear heating along terrane boundaries, and (3) resetting of maturity gradients via thrust loading. In addition, thermal overprints result if there is a substantial rise in local or regional geothermal gradient following terrane amalgamation; under these circumstances, homogenization of thermal maturity may be evident across terrane boundaries. If associated magmatic rocks or hydrothermal veins can be dated, an upper limit is established on the suturing event. Thermal data are also useful for resolving structural ambiguities on a local scale; clear distinctions can be made, for example, between faulted contacts, diapirs, and mildly tectonized depositional contacts or unconformities.

Introduction

Tectonostratigraphic terranes are fault-bounded geologic units of regional extent; terranes are discriminated from one another by their differing geologic histories [Schermer et al., 1984]. Under the simplest circumstances, the contrasting geologic histories are expressed stratigraphically; in other cases, a superimposed structural fabric or metamorphic grade may characterize a terrane. The key question in terrane analysis is whether or not the inferred geologic histories of adjacent rock units are compatible with the present spatial relations.

By strict definition, metamorphic terranes display "a regional, terrane-wide penetrative metamorphic fabric and development of metamorphic minerals to such a degree that original stratigraphic features and relations are obscured" [Jones et al., 1983]. Sharp contrasts in metamorphic grade also signify fundamental differences in geologic history. Moreover, discordant relations across structural boundaries (or between the boundaries and metamorphic isograds) require some type of relative tectonic displacement following metamorphism. Spatial patterns in metamorphism, therefore, can be used to confirm suspicions regarding possible allochthonous relationships between adjacent rock units.

Most metamorphic data cited in terrane research have been derived from inorganic mineral assemblages. In particular, regional variations in metamorphic grade have been used both to reconstruct tectonic events and to model processes associated with terrane emplacement [Blake et al., 1969; Ernst, 1983; Pavlis and Bruhn, 1983; Cloos, 1985]. Many experimental studies have been completed to pinpoint phase equilibria for common metamorphic minerals such as laumontite, prehnite, pumpellyite, and lawsonite [e.g., Boles, 1977; Ernst, 1977; Liou et al., 1985]. Unfortunately, the stability fields for all of these minerals are broad, and the metamorphic reactions are sensitive to variables other than temperature and lithostatic pressure, including primary mineralogy, H_2O and CO_2 pressure, and time. Consequently, even if the P-T boundaries for a metamorphic facies are known, it is seldom possible to arrive at precise estimates of burial depth or burial temperature for individual samples using inorganic phases alone, especially where strata have been affected by retrograde reactions and/or metastable mineral assemblages.

Additional techniques, such as thermal alteration index, conodont coloration, vitrinite reflectance, illite crystallinity, fission-track annealing, fluid inclusions, and organic geochemistry can be used to help define burial conditions during diagenesis and metamorphism [Scholle and Schluger, eds., 1979]. Although there are problems associated with each technique, the precision of corresponding paleotemperature estimates generally is superior to estimates based upon inorganic mineral assemblages.

Vitrinite reflectance is a particularly powerful method because it is reliable over a wide range of temperature conditions from early diagenesis to low-temperature metamorphism (25°C to at least 400°C)[Price and Barker, 1985]. Vitrinite is one type of organic matter derived from terrestrial plants, and transported particles are a nearly ubiquitous component of terrigenous sediment. Vitrinite is also found in some limestones and cherts. The method is based upon the empirical observation that reflectance values (Ro) increase exponentially with increasing burial depth and temperature [e.g., Dow, 1977; Bostick, 1979]. Reflectance values are a function of the maximum temperature applied to enclosing rocks, and there are no problems with retrograde reactions. The technique is employed routinely on strata ranging in age from middle Paleozoic to Quaternary, although older strata containing algae remains can be studied in a similar manner.

Regional maps of paleo-isotherms can be constructed for exposed surface strata based upon measurements of vitrinite reflectance [e.g., Bostick, 1979]. Moreover, borehole profiles of vitrinite reflectance versus depth clearly show the anomalous effects of major erosional unconformities, contact metamorphism, and faulting [e.g., Dow, 1977]. It should be possible, therefore, to document similar anomalies using a plan-view perspective.

The intent of this paper is to illustrate various types of thermal anomaly patterns in accreted terranes using vitrinite reflectance. The examples discussed herein have been chosen from the Franciscan Complex of California. Similar analyses have been completed in the Ouachita Mountains of Arkansas and Oklahoma [Houseknecht and Matthews, 1985; Underwood et al., in press]. In addition to relatively straightforward interpretations of peak paleotemperature, we demonstrate how thermal data can be used to constrain models of structural evolution and terrane amalgamation. Of particular importance is the timing of a peak heating event relative to terrane-wide structural deformation or terrane amalgamation. In addition, we show how thermal data help resolve ambiguities in structural relations both within terranes and along terrane boundaries.

Methods

The long-held view in studies of organic metamorphism is that vitrinite reflectance is dependent upon both maximum temperature and duration of heating [e.g., Karweil, 1955; Connan, 1974; Hood et al., 1975; Waples, 1980]. More recently, however, several workers have challenged the validity of this contention, suggesting instead that vitrinite reflectance equilibrates with peak temperatures within less than 1 Ma [Suggate, 1982; Barker, 1983; Price, 1983]. The subject remains controversial [e.g., Bostick, 1984; Kohsmann, 1985], but we are more concerned in this study with the relative differences in paleotemperature rather than absolute values. Consequently, we use an empirical correlation between mean vitrinite reflectance (Rm) and borehole temperature established by Price [1983].

Values of vitrinite reflectance can be influenced by a number of internal and external variables other than time and temperature. For example, vitrinite is anisotropic in its reflectance above about 1.0% Rm, especially in strongly deformed strata and at higher levels of thermal maturation [Stach et al., 1975; Stone and Cook, 1979]. Values can be suppressed by high concentrations of hydrogen-rich organic macerals (Type I and Type II kerogens) and/or paleoenvironmental factors such as anaerobic conditions during deposition [Hutton and Cook, 1980; Newman and Newman, 1982; Price and Barker, 1985]. Weathering and oxidation of rocks at or near the surface can have deliterious effects [Marchioni, 1983], and the inclusion of unrecognized second-cycle particles (if more mature than the indigenous population) may result in inaccurate calculations of mean reflectance.

The potential problems outlined above, as well as laboratory and statistical techniques, are discussed in greater detail in Underwood and Strong [1986a]. In brief, laboratory processing consisted of mechanical crushing of shale samples, maceration with HCl and HF, random mounting of kerogen residue in epoxy, and polishing. Values of mean vitrinite reflectance (Rm) are based upon measurements in oil (Ro) of approximately 50 randomly oriented particles per sample. Data were plotted in histogram form to help identify the indigenous vitrinite population, but "recycled" grains were eliminated from calculations only if separated from the main population by at least 0.1%Ro (one sampling interval on histograms). Direct comparisons also were made between our results and those of two commercial geochemical laboratories; reproducibility is typically 0.1%Rm or less [Underwood and Strong, 1986a] As an additional check on reliability we employed the technique of illite crystallinity [Kubler, 1968] on all samples yielding unusually high values of mean reflectance (see Underwood and Strong [1986b] for X-ray techniques). Almost without exception, the illite peaks on X-ray diffractograms confirmed the elevated grade of metamorphism suggested by vitrinite reflectance, and the correlation between Rm and IC is particularly good above 1.0% Rm [Strong, 1986]. We believe changes in Rm of more than 0.15% definitely are geologically significant.

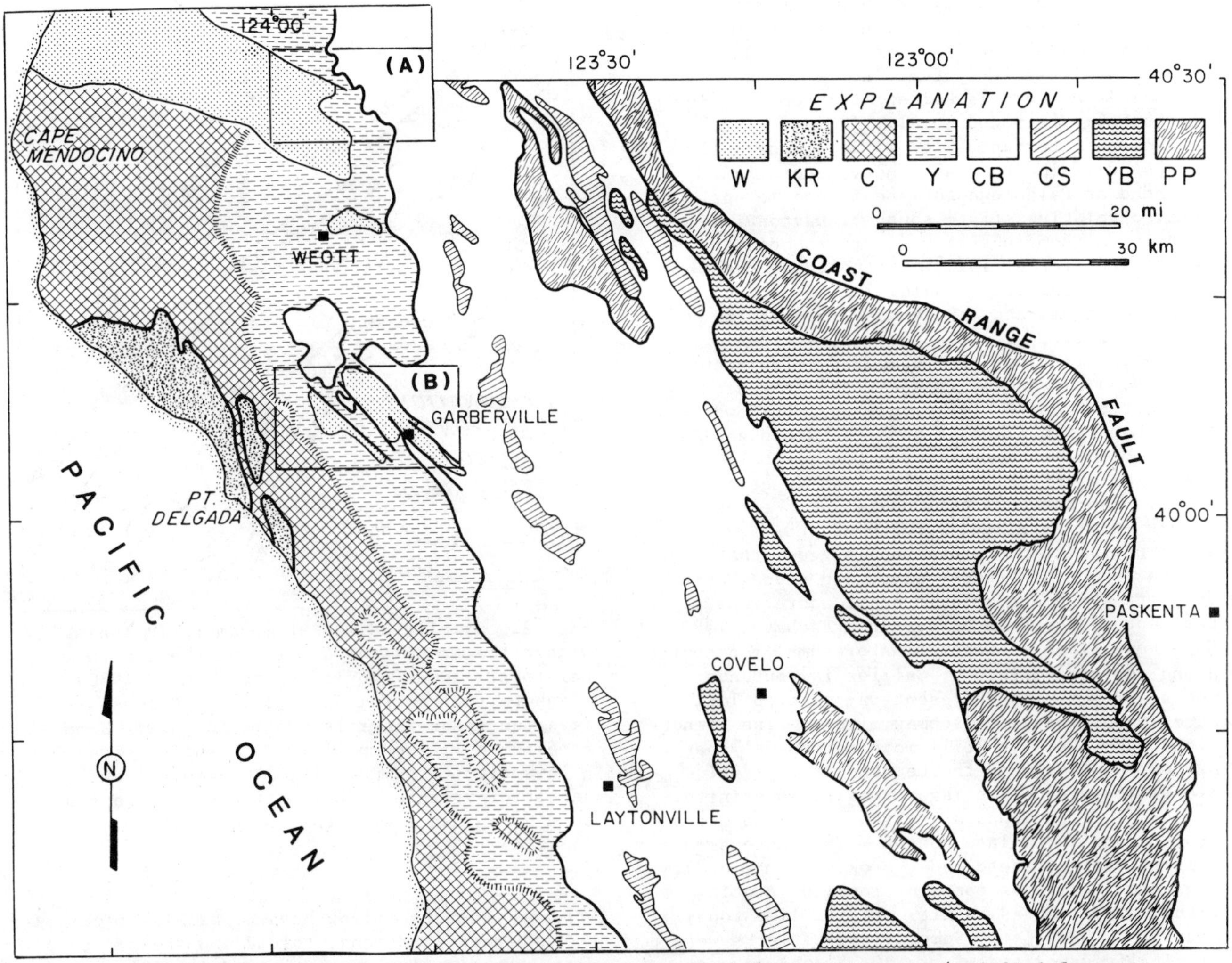

Fig. 1. Simplified map of Franciscan terranes in northern California (modified from Blake et al. [1985]). PP = Picket Peak, YB = Yolla Bolly, CS = coherent strata of Central belt, CB = melange of Central belt, Y = Yager, C = Coastal, KR = King Range, W = Wildcat Group.

Geologic Background

The stratigraphy, age, metamorphic grade, and structural style of Franciscan terranes in northern California have been outlined in a number of previous papers [Jones et al., 1978; Bachman, 1982; McLaughlin et al., 1982; Blake et al., 1985]. Strata of the Sur-Obispo terrane in central California have been described by Page [1982], Vedder and others [1982], and Champion and others [1984].

Northern California

Blake and others [1985] recognized six discrete terranes within the Franciscan Complex of northern California. Most of the thermal maturity data we have collected come from the four westernmost terranes: the Central, Yager, Coastal, and King Range terranes (Fig. 1).

The Franciscan Central belt is a highly dismembered, regional-scale melange. Detailed mapping allows the discrimination of numerous smaller tectonostratigraphic units of diverse age, lithology, and metamorphic grade. Rocks range in age from Middle Jurassic to Late Cretaceous, and the metamorphic grade varies between zeolite and blueschist facies. Kilometer-scale blocks of previously-accreted Franciscan lithologies are common, as are blocks of pillow lavas with associated pelagic limestone or chert. Smaller fragments of high-grade blueschist, amphibolite, and eclogite also are immersed in the matrix of pervasively sheared argillite and graywacke. Polyphase stratal disruption is typical, and the most recent fault activity and folding is related to the San Andreas strike-slip system. Long-distance tectonic transport is evident for some exotic blocks [e.g., Tarduno et al., 1985].

The Yager terrane (Fig. 1) is a moderately disrupted sedimentary sequence, generally regarded as Paleocene to late Eocene in age. Some correlative strata south of Laytonville contain palynomorphs of Late Cretaceous age [Evitt and Pierce, 1975]. Lithologies comprise sandstone turbidites, shale, limy mudstone, and pebbly conglomerate. Polyphase folding is characteristic, but the degree of stratal disruption is minor when compared to adjacent Franciscan terranes. In addition, tectonically-incorporated exotic lithologies are unknown, and the metamorphic grade is no higher than zeolite facies. The contact with the Central belt is the east-dipping Eel River fault [Bachman et al., 1984], also known as the Coastal belt thrust [Jones et al., 1978]. Structural and sedimentologic interpretations suggest that Yager strata represent trench- slope and slope-basin deposits which accumulated above accretionary basement of the Coastal terrane [Bachman, 1982; Underwood, 1985]. Deposition presumably occurred along the west margin of California, but not necessarily near the present-day outcrop latitudes [Underwood and Bachman, 1986]. The trench-slope hypothesis is still more conceptual than proven, however, and the tectonostratigraphic classification of the Yager remains indefinite.

Strata of the Coastal terrane range in age from Campanian to late Eocene. Turbidite sequences are strongly deformed ("broken formation"), and some localities contain blocks of exotic lithologies such as basalt, pelagic limestone, and rare blueschist. Some of these blocks were originally deposited within 20° of the equator [Sliter et al., 1986]. In most cases, however, the lithologies of the Coastal terrane are virtually identical to adjacent Yager strata, and the distinction between the two broadly coeval geologic units is based almost exclusively on structural style [Bachman, 1982]. Consequently, the Yager-Coastal contact (Fig. 1) is regarded as a structural transition rather than a discrete fault [Underwood, 1985].

The King Range terrane (Fig. 1) consists largely of highly folded turbidites and mudrocks of Paleogene(?) to middle Miocene age [McLaughlin et al., 1982]. There are also rare blocks of pelagic limestone, basalt, and chert. Polyphase folding is evident, and aeromagnetic data suggest that King Range strata were thrust over the Coastal terrane along a southwest-dipping Neogene fault. The boundary is difficult to map, however, because of the lithologic similarities between the two terranes. Rocks at Point Delgada (Fig. 1) are classified as a separate sub-terrane [McLaughlin et al., 1982]. The Point Delgada rocks comprise a fault-bounded melange containing

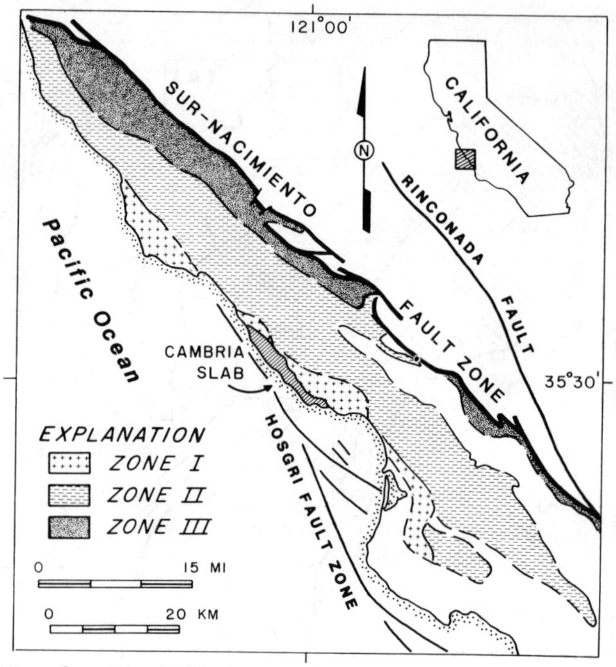

Fig. 2. Simplified map of metamorphic zones within the Franciscan Complex of the central California coast (modified from Ernst, 1980). Offshore faults from Hall [1981]. Recrystallization index minerals are as follows: Zone I = none; Zone II = pumpellyite ± prehnite; Zone III = lawsonite (± pumpellyite), aragonite, and jadeitic pyroxene. Strata of the Cambria slab show no evidence of metamorphic recrystallization.

Upper Cretaceous pillow basalt, basalt breccia, thick-bedded sandstone, folded and sheared turbidites, debris-flow deposits, and rare blocks of chert and blueschist. The composite King Range terrane is bounded on the west by the San Andreas fault zone.

Central California

Thermal-maturity data from central California are restricted to coastal exposures in the vicinity of San Simeon (Fig. 2). Of primary interest are the well-bedded and variably deformed Upper Cretaceous strata of the Cambria slab, which are in contact with more highly deformed melange of the Franciscan Complex. Structural analyses completed by Cowan [1978, 1982] demonstrate that multiple episodes of deformation have affected the melange, including an early phase of axially symmetric extension. The metamorphic grade of sandstones within the melange increases from west to east (Fig. 2), and pumpellyite is a common constituent of metagraywackes near San Simeon [Ernst, 1980]. Strata of the Cambria slab are essentially unmetamorphosed; turbidite lithofacies and structural style collectively support the contention that deposition took place within a restricted basin perched on the Cretaceous trench slope [Howell et al., 1977; Smith et al., 1979]. Regional geologic relations and paleomagnetic data indicate these strata were translated approximately 2500 km to the north as part of the Sur-Obispo terrane [Page, 1982; Champion et al., 1984].

Types of Thermal Anomalies

Several types of anomalies in thermal maturity are possible within any type of fault-bounded stratal unit (Fig. 3). Considerable attention has been focused on perturbations in geothermal gradient caused by the emplacement of thrust sheets [Angevine and Turcotte, 1983; Edman and Surdam, 1984]. In essence, the hanging wall of a thrust couplet serves as an insulation blanket, and heat is radiated downward from the hanging wall into the footwall. Moreover, additional heat may be generated by friction along fault surfaces [Scholz, 1980; Brewer, 1981]. The direct thermal effects of thrusting will dissipate with time, but thrust-induced anomalies will be preserved in the strata unless a subsequent rise in geothermal gradient causes an overprint of the earlier temperature regime.

Maturity inversion refers to cases in which hanging-wall strata display higher levels of maturation than underlying footwall sequences (Fig. 3a). This relationship requires hanging-wall sediments to reach peak burial temperatures prior to the initiation of thrusting. A second type of anomaly (thermal aureole) is restricted to the immediate vicinity of a fault surface, where abnormally high temperatures can be caused by shear heating (Fig. 3b). Typically, thrust emplacement will also change the temperature gradient in accordance with the depth of structural burial rather than stratigraphic position. As a consequence, variations in thermal maturity over a given stratigraphic interval are commonly abnormal (Fig. 3c). Given the polyphase history of deformation common to many accreted terranes, the ultimate reflectance gradients may be a consequence of the final phase of deformation; for example, a broad regional-scale warping of a terrane following thrust emplacement may produce reflectance anomalies that mimic the patterns expected with simple synclines and anticlines.

Patterns of thermal maturity can become chaotic if a terrane is slivered excessively by faulting after peak maturation. The possibility also exists for high-temperature magmatic or hydrothermal events to overprint completely all previous structural and/or stratigraphic gradients (Fig. 3d). These overprint events are particularly important, however, for placing upper limits on the timing of terrane amalgamation, in much the same way as "stitching" plutons or overlap sequences in successor basins [Jones, et al., 1983]. If peak temperatures are attained following terrane amalgamation, then isoreflectance contours should cut across terrane boundaries, and concordance will be evident between two adjacent terranes.

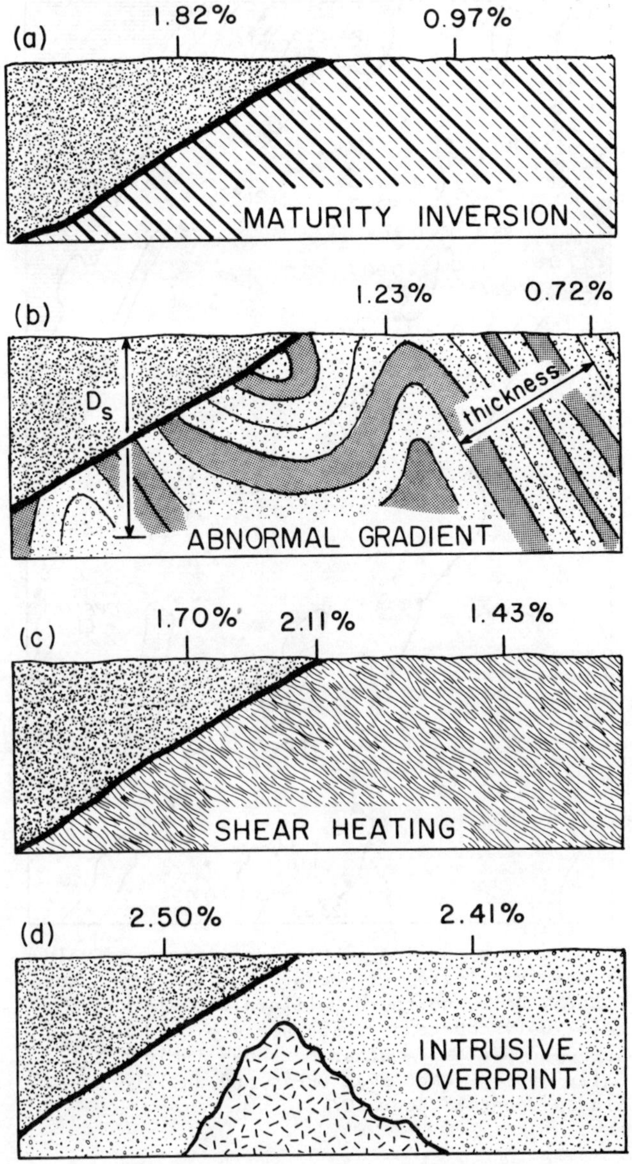

Fig. 3. Schematic diagrams illustrating four principal types of thermal maturity anomalies within thrust-faulted terranes. Numbers refer to hypothetical values of mean vitrinite reflectance. See text for descriptions of each type of anomaly.

One additional benefit of thermal-maturity data lies in their potential to resolve structural and/or stratigraphic ambiguities on a local scale. For example, mesoscopic structural features may not provide the evidence needed for the proper interpretation of contact relationships between two geologic units. Sharp

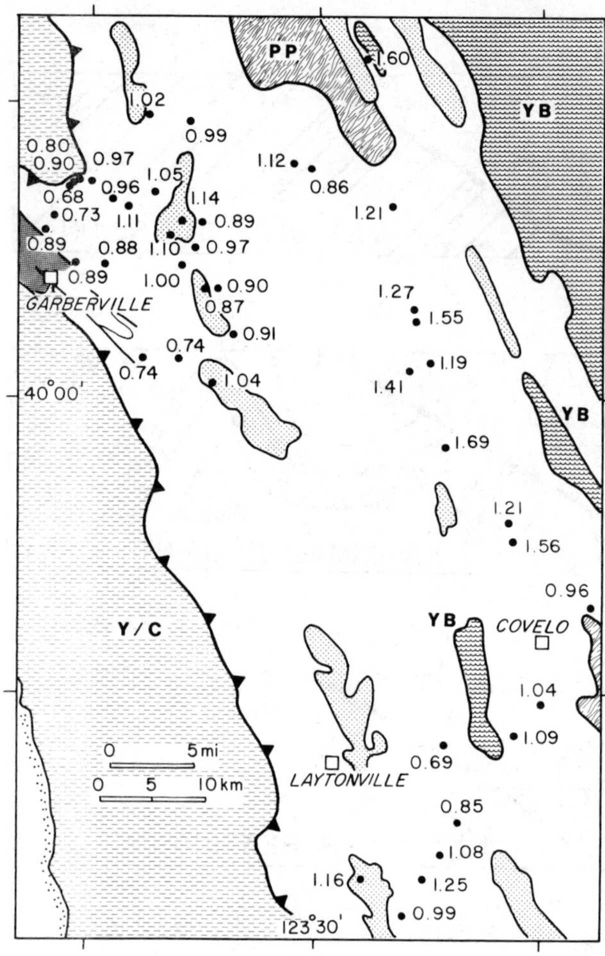

Fig. 4. Map showing sample localities and values of mean vitrinite reflectance for the Franciscan Central belt, northern California. Blocks of relatively coherent strata are depicted with stippled pattern. Other terranes include Yolla Bolly (YB), Picket Peak (PP), and undifferentiated Coastal/Yager (C/Y).

contrasts in thermal maturity across a contact require juxtaposition by faulting and, hence, possible subdivision of two units into discrete terranes. Conversely, overlap in levels of thermal maturity would support the interpretation of a depositional contact; stratal subdivision into two terranes would not be warranted under these circumstances.

Maturity Inversion

There are many convincing example of maturity inversion within the Franciscan Complex; the most dramatic examples occur where slices of blueschist-grade strata of the Yolla Bolly terrane are thrust over Central belt melange [see Blake et al., 1985]. The paleotemperature contrasts associated with these thrust slices are poorly documented, however, because few samples from the Yolla Bolly terrane have been analyzed for vitrinite reflectance.

Pelitic rocks extracted from the matrix of the Central belt (as well as shales from relatively coherent blocks within the melange) display higher levels of thermal maturity than adjacent Yager beds (Figs. 4 and 5). The immediate contrast across the Eel River fault (or Coastal belt thrust) is not large, however, because mean vitrinite reflectance decreases progressively in a westward direction within the Central terrane (Fig. 4). From this trend, we infer that the magnitude of uplift and erosion within the melange increases from west to east.

Substantial uplift of the Central belt melange must have occurred prior to peak maturation of the Yager terrane. Some inorganic mineral phases from the melange matrix are suggestive of low geothermal gradients (less than 15°C/km) and high-pressure conditions (3-6 kbar); maximum burial depths may have reached 15-20 km [Cloos, 1982]. In contrast, metamorphism within the Yager terrane is no higher than zeolite grade; calcite is a common cement and vein mineral, and laumontite occurs locally. Consequently, Yager sediments attained peak burial temperatures at relatively shallow levels of the crust, certainly no deeper than 11 km (3 kbar). Paleotemperature estimates range from 136-256°C (Central) to 68-175°C (Yager), and the geothermal gradient probably increased to about 25°C/km by the end of Eocene time [Underwood, 1985]. The expression of the structural inversion is reduced when viewed only in terms of organic metamorphism and paleotemperature, and some overlap in Rm is evident locally along the contact. This example illustrates some of the ambiguities in assessing burial history when adjacent terranes have experienced different burial trajectories under conditions of changing geothermal gradient.

Shear Heating Aureoles

Many theoretical and empirical studies have demonstrated that substantial temperature anomalies can be caused by shear heating along thrust faults [Oxburgh and Turcotte, 1974; Cardwell et al., 1978; Hoffman and Hower, 1979; Scholz, 1980; Brewer, 1981; Bustin, 1983; Johnston and White, 1983; Johnson et al., 1985]. Anomalous values of vitrinite reflectance (0.82-1.08%) are evident at two localities along the Eel River fault (Fig. 6), and the Rm anomalies are confirmed by abnormally narrow illite peaks on X-ray diffractograms ($\Delta°2\theta$ = 0.50-0.65). These anomalies may be the result of shear heating during the emplacement of Yager beds beneath the Central belt melange. An additional anomaly (Rm = 1.50%) is evident near Leggett (Fig. 5); that sample locality is well removed from the fault contact, however, and the cause of the anomaly remains unknown [Underwood and O'Leary, 1985].

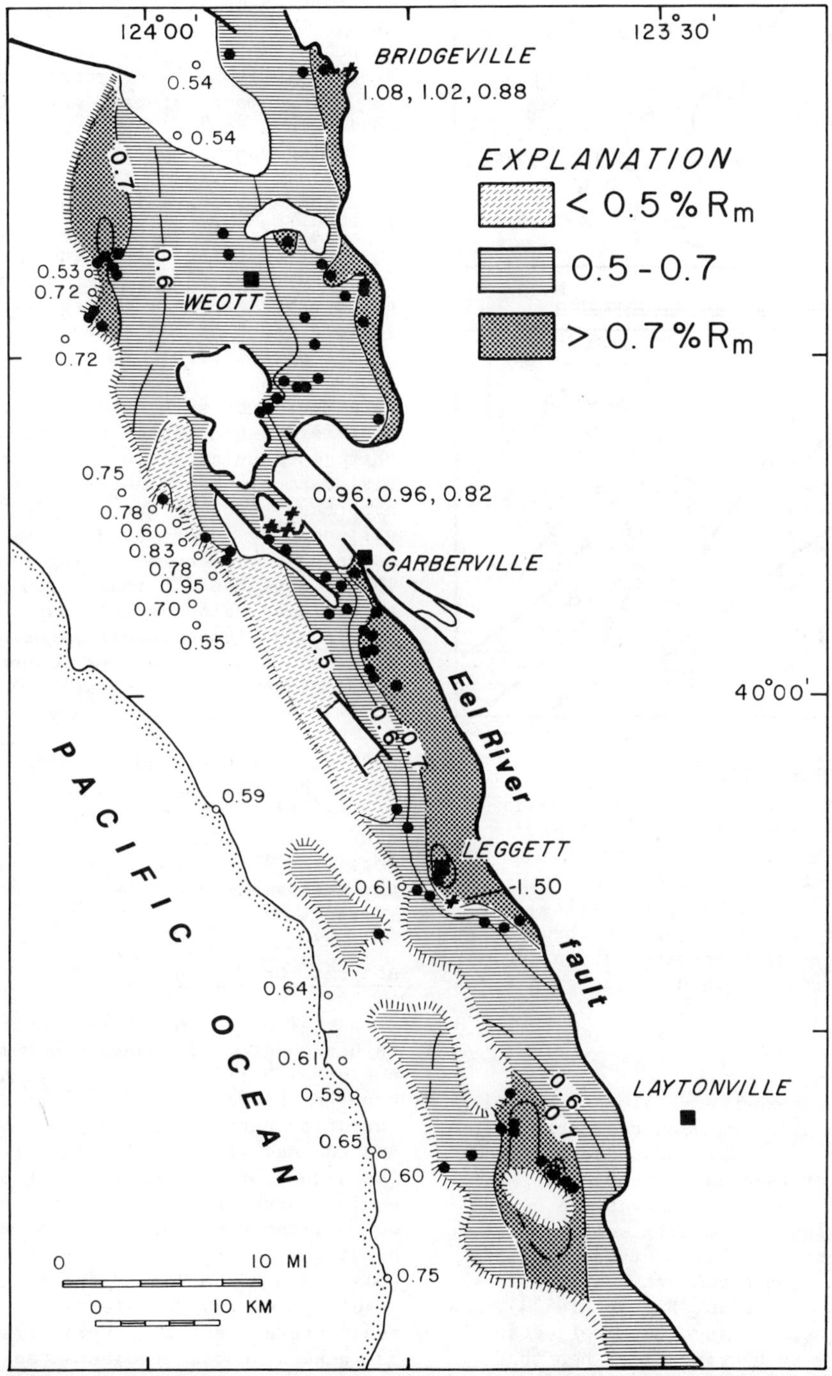

Fig. 5. Isoreflectance contour map for the Yager terrane, northern California. Solid dots indicate sample localities. Numbers refer to mean Ro values for adjacent strata of the Coastal terrane. Anomalous values near Bridgeville and Garberville are illustrated in greater detail in Fig. 6. Modified from Underwood and O'Leary [1985] and Underwood and Strong [1986a].

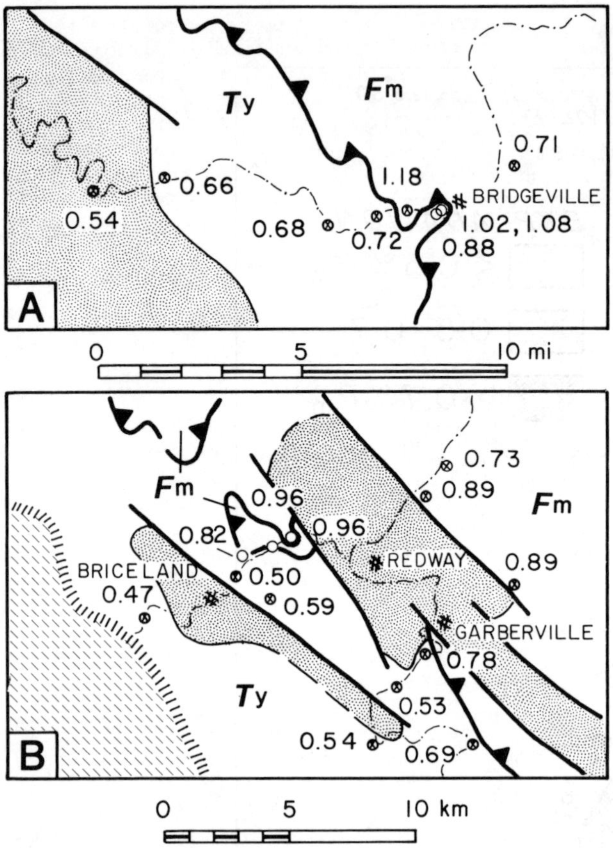

Fig. 6. Detailed maps of the Bridgeville and Garberville regions (Fig. 1) showing sample localities and values of mean vitrinite reflectance. Abnormally high values adjacent to the Eel River fault (shown by open circles) are attributed to shear heating along the thrust. Fm = Central belt, Ty = Yager terrane, stippled pattern = Wildcat Group (Neogene), dashed pattern = Coastal terrane.

Estimates of paleotemperature along the fault zone are 182 to 205°C (based upon the correlation established by Price [1983]), and these values are 50 to 85°C higher than estimates for background samples. If shear heating was the cause, the absolute temperatures probably were even higher; the correlation of Price [1983] applies only to equilibrium conditions and heating times of 10^4 to 10^6 years (see also, Barker [1983]). Bostick [1979] has demonstrated that reflectance values are sensitive to heating time when the time frame is brief. A value of 1.0%Rm, for example, can be generated by heating a lignitic shale to 300°C for one month. We view this value as a reasonable upper limit for shear-heating temperatures along the Eel River fault.

It is important to note that Rm anomalies are not apparent across all transects of the Eel River fault. This may be because the effects of shear heating are restricted to narrow zones (no more than a few meters), and other sampling traverses simply did not encounter outcrops close enough to the fault trace. Moreover, the terrane boundary south of Garberville has been modified substantially by late Cenozoic strike-slip faulting. The amount of heat generated during fault motion is also influenced by a host of variables, including confining (lithostatic) pressure, pore-fluid pressure, fluid temperature, rate of displacement, thermal conductivity, coefficient of friction, distance of movement, thickness of the overthrusting sheet, and the geometry of the fault surface. Shear heating is most likely to occur along geometric irregularities in the fault surface, such as ramps and lateral ramps.

The possibility also remains that the Rm anomalies associated with the Eel River fault were caused by some other process, such as transfer of heat from the hanging wall, or hydrothermal circulation during or following fault dislocation. We do not favor the heat-transfer hypothesis because many samples from the hanging wall display lower values of vitrinite reflectance than the anomalous Yager samples (Fig. 6). The role of hydrothermal circulation is difficult to evaluate using organic data only. Additional studies should be completed using fluid inclusions and stable isotopes to determine how vein temperatures relate to data from organic metamorphism. An unequivocal example of hydrothermal overprinting is discussed in the following section, however, and the effects of that event are much more widespread than the anomalies along the Eel River fault. Consequently, we believe the available data are best explained by the localized generation of frictional heat during the amalgamation of the Yager and Central terranes.

Abnormal Gradients

Burial of a sedimentary section beneath a thrust sheet will change the temperature distribution with respect to the primary stratigraphic gradient [Angevine and Turcotte, 1983]. The variation between the original burial gradient and the new structural gradient will depend upon the amount of deformation that precedes or accompanies thrust emplacement, the depth of thrust burial, the temperature at the base of the hanging wall, and the elapsed time prior to uplift and unroofing of the thrust belt.

Clear regional gradients in vitrinite reflectance are evident in several Franciscan terranes, but none of these gradients can be related to stratigraphic position. A well documented example is the Yager terrane, which has been subjected to at least three phases of folding [Underwood, 1985]. Values of mean reflectance decrease systematically to the west, away from the Eel River fault (Fig. 5). Preliminary interpretations attributed this westward decrease in Rm to a westward tapering of the hanging wall of the Eel River fault, thereby reducing the depth of tectonic burial [Underwood, 1985]. Additional data show that this relatively straightforward pattern breaks down west of Weott, where Rm values increase to over 0.80% (Fig. 5). Isoreflectance contours near Weott mimic the pattern expected with a simple syncline. We believe that maturation of the Yager strata is best explained by thrust burial and rise in geothermal gradient during post-Eocene time; however, the surface anomaly pattern is also a product of Quaternary tectonic events which generated broad, regional folds. Late-stage tectonism folded both the Eel River fault and Miocene to Pliocene beds of the overlying Wildcat Group of Ogle [1953]. This example provides a vivid illustration of why interpretations of maturation trends must be made within the context of the entire tectonic evolution of a terrane.

Magmatic and Hydrothermal Overprints

A substantial rise in regional or local geothermal gradient following terrane amalgamation can obscure or completely overprint previous structural or stratigraphic gradients in thermal maturity. Strata of the King Range terrane display the highest level of thermal maturity of any Franciscan section examined to date (Fig. 7), yet these rocks are the youngest and most recently accreted part of the Franciscan Complex [McLaughlin et al., 1982]. The King Range thermal anomaly is centered at Point Delgada, where mean reflectance values are as high as 2.50%. According to Price [1983], this degree of maturation corresponds to a peak paleotemperature of 308°C. Isoreflectance contours define a radiating pattern of progressive reduction in Rm away from Point Delgada (Fig. 7). Significantly, the contours cut obliquely across the terrane boundary mapped by McLaughlin et al. [1982], such that strata assigned to the Coastal terrane also display elevated levels of thermal maturity.

McLaughlin et al. [1985] presented independent evidence of hydrothermal activity at Point Delgada in their study of vein mineralization. Fluid-inclusion data indicate vein homogenization temperatures ranging between 194°C and 287°C for three separate stages of vein emplacement; most of the temperatures for stage-one veins are above 231°C. A decrease in temperature is one of the main causes for deposition of mineral phases in hydrothermal systems, so fluid inclusions seldom indicate the maximum temperature to which a system has been exposed. In essence, the fluid-inclusion data compliment our vitrinite data. McLaughlin et al. [1985] also dated the age of mineralization at 13.8 ± 0.4 Ma based upon K-Ar dating of adularia.

Collectively, the organic metamorphism and vein mineralization prove that the local geothermal gradient rose substantially following the amalgamation of the King Range and Coastal terranes. The K-Ar data place an upper limit on the terrane amalgamation event, in much the same fashion as radiometric dates obtained from a stitching pluton. The cumulative thickness of the overlying Wildcat Group north of the King Range is approximately 2500 m. The lowest Rm value within the King Range (0.77%) corresponds to a paleotemperature of 153°C, so the geothermal gradient must have been at least 60°C/km. The source of the thermal anomaly, however, remains open to question. McLaughlin et al. [1985] attributed the heat to intrusion of asthenospheric material into a slab window associated with the Mendocino triple junction (see also Zandt and Furlong, [1982]; Furlong [1984]). If this interpretation is correct, the thermal overprint must have occurred much farther south, because the triple junction was located at approximately 36°N during middle Miocene time [Dickinson and Snyder, 1979]. Obviously, this scenario also requires a sizeable component of right-slip motion following the hydrothermal event. A less mobilistic explanation places the heat source in the subducting Farallon plate. Recent plate-tectonic reconstructions show the Farallon-Pacific spreading ridge in very close proximity to the Franciscan trench during middle Miocene time [Engebretson et al., 1986]. Subduction of a hot slab must have increased the regional geothermal gradient, and off-axis volcanism or instability in the Blanco fracture zone ("leaky transform") could have generated localized magmatic pulses within the developing accretionary prism at the present latitude of Point Delgada (40°N).

Structural Ambiguities

Thermal-maturity data are also useful for resolving structural ambiguities on a local scale. As an example, past studies have drawn an analogy between the depositional facies and structural style of the Cambria slab (Fig. 8) and the restricted intraslope basins common to modern trench slopes [Smith et al., 1979]. Cambria turbidites clearly overlie melange of the Franciscan Complex, but it is not clear whether the contact is entirely structural or a tectonized unconformity. If the slope-basin model is correct, Cambria strata should display lower levels of thermal maturity compared to the inferred basement. Local field relations also indicate that parts of the Franciscan melange basement were remobilized by diapirism [Becker and Cloos, 1985]; one manifestation of the diapirism is the incorporation of sandstone blocks from the Cambria slab into the sheared matrix of the melange. According to Becker and Cloos [1985], diapirs rose from considerable depth (10 km or more) from an actively deforming flow-melange channel near the base of the Franciscan accretionary wedge (see also Cloos [1982]). An alternative interpretation simply places the melange within the core of an anticlinal structure that

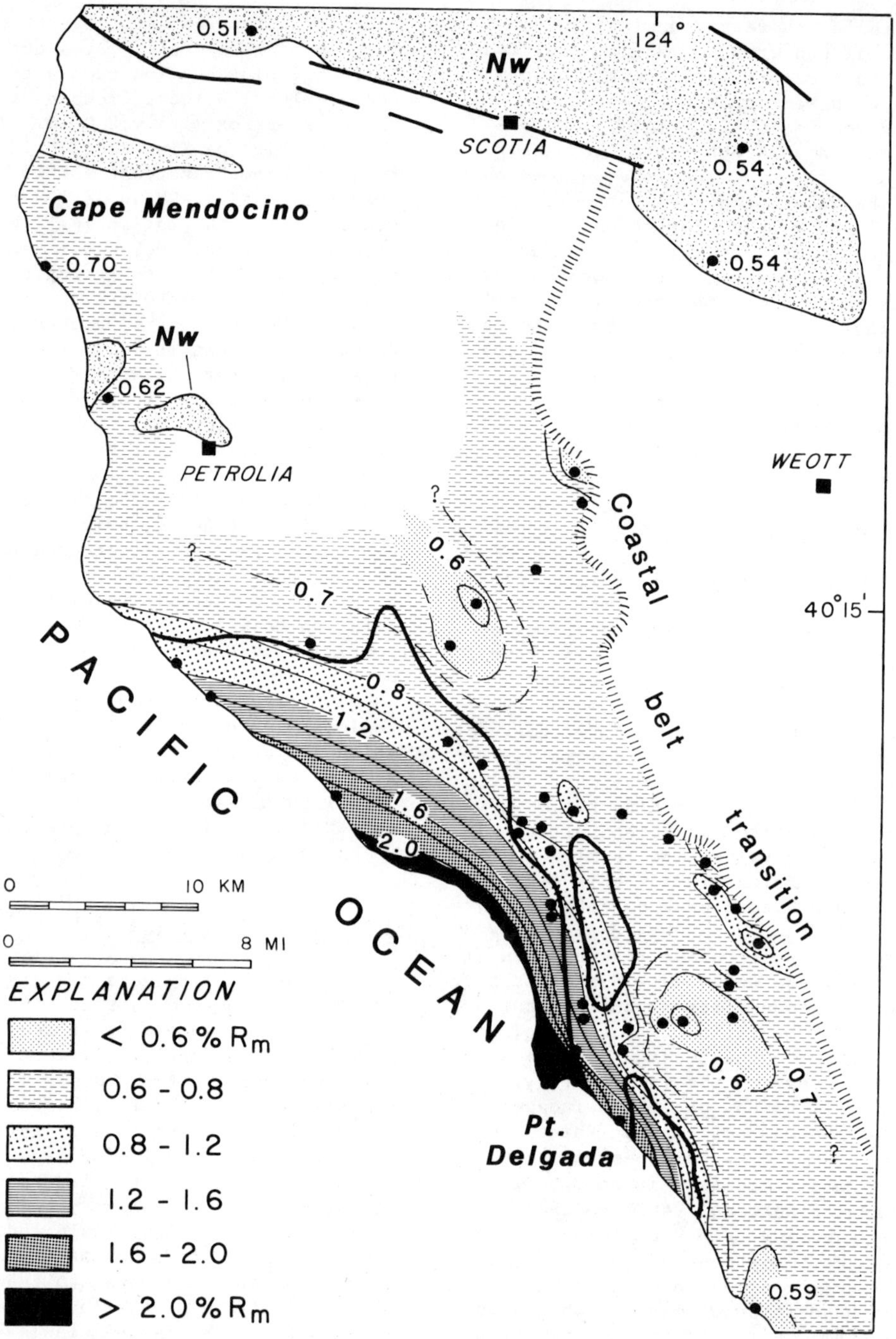

Fig. 7. Isoreflectance contour map for the Coastal terrane and King Range terrane. Solid dots indicate sample localities. Nw = Wildcat Group (Neogene). Boundary of the King Range terrane is from McLaughlin et al. [1982]. Modified from Underwood and Strong [1986a].

has been truncated by a high-angle fault [Howell et al., 1977].

The thermal evolution of the Cambria region is discussed in greater detail by Underwood and Strong [1986b] and Underwood and Howell [in press]. Vitrinite-reflectance data support the trench-slope depositional model. There is considerable overlap in Rm values between Cambria shales and melange matrix, but all of the lowest values (below 0.70%Rm) occur within the Cambria

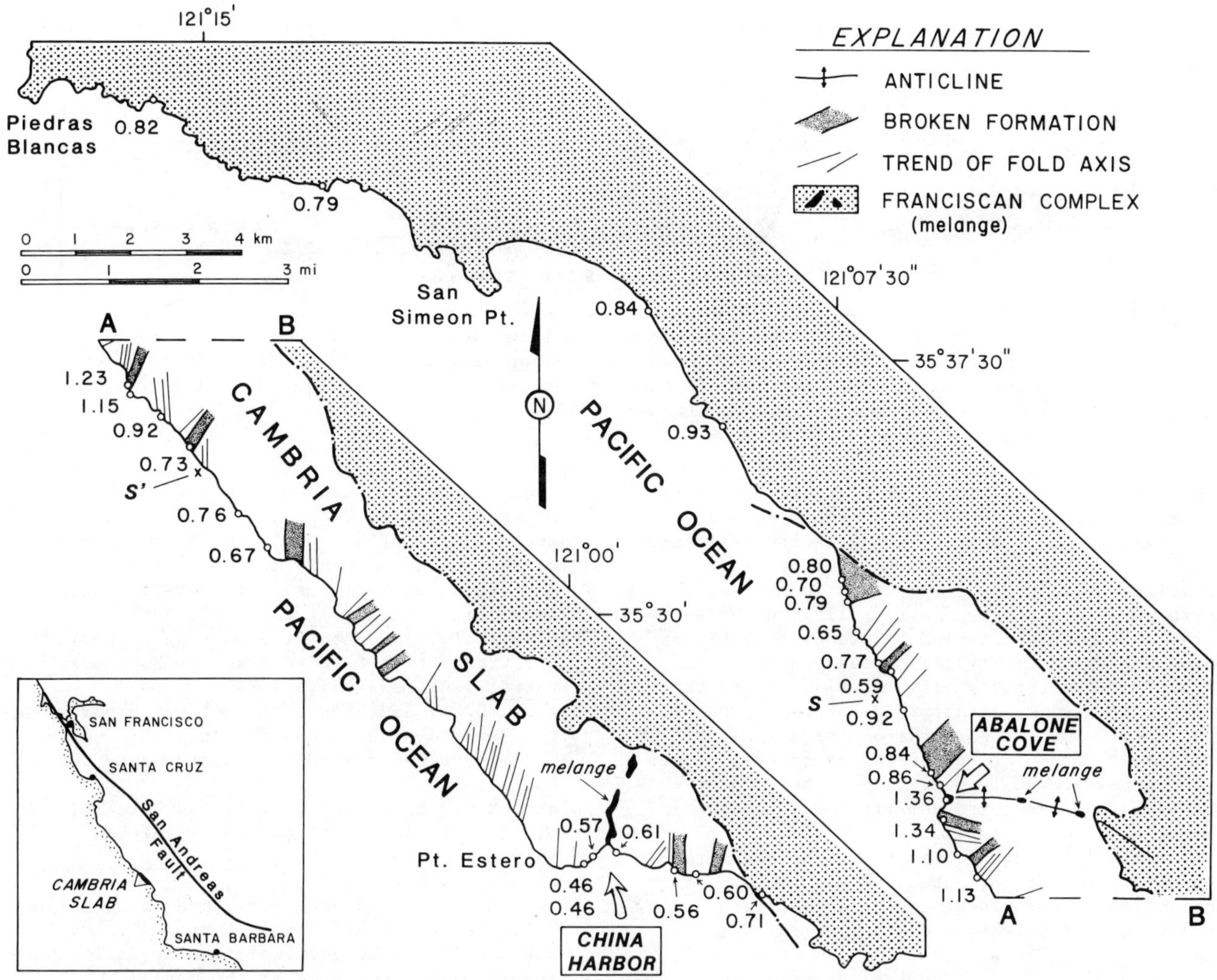

Fig. 8. Simplified structural map of the Cambria slab together with sample localities and corresponding values of mean vitrinite reflectance. Zones of broken formation and trends of fold axes are from Smith [1978]. See Fig. 2 for regional geologic context. From Underwood and Strong [1986b].

slab (Fig. 8). Moreover, at both the northern and southern contacts of the slab, Cambria shales display lower maturity levels than immediately adjacent melange (Fig. 8). On the other hand, Rm data fail to support the melange-diapir model proposed by Becker and Cloos [1985]. Maximum values for the inferred diapir at Abalone Cove are virtually identical to values for surrounding Cambria strata (Fig. 9). Some heat transfer should be expected with a diapiric intrusion, but the aureole should be limited in width because of the low thermal conductivity of the wall rocks. Instead of this limited aureole, the thermal-maturity gradient decreases very gradually away from the southeastern flank of the melange outcrop, and there is an abrupt decrease in Rm across the northwestern flank (Fig. 9). This gradient is consistent with the development of an anticlinal structure. The anticline is cored by melange and breached along its northern limb by a high-angle fault with substantial vertical displacement (Fig. 9).

Summary

The value of thermal-maturity data in terrane analysis lies in the ability to quantify physical conditions associated with peak burial events. It is possible, furthermore, to map and compare paleo-isotherms within neighboring tectonostratigraphic terranes. Peak burial depths can be calculated, but the reliability of such calculations will depend upon the accuracy of inferred geothermal gradients. Several types of Rm

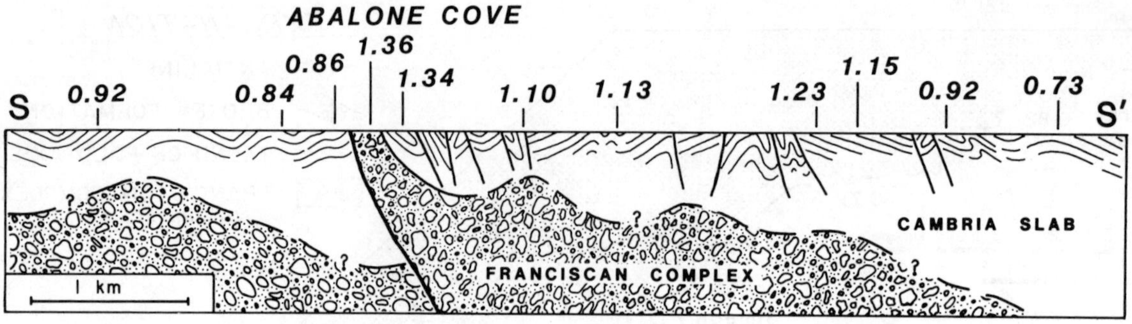

Fig. 9. Cross-section through Abalone Cove melange locality (see Fig. 8 for section points S and S'). Numbers refer to values of mean vitrinite reflectance. Rm gradients away from Abalone Cove support the faulted-anticline hypothesis [Howell et al., 1977] rather than the melange-diapir hypothesis [Becker and Cloos, 1985]. From Underwood and Strong [1986b].

anomalies are possible along terrane boundaries, including maturity inversion across terrane boundaries, shear heating along fault surfaces, and the generation of abnormal (structural) gradients via thrust burial. If geothermal gradients rise substantially following terrane amalgamation (accompanied by either magmatic or hydrothermal activity), a complete thermal overprint may result. If overprint events can be dated, limits are established on the timing of suturing. Spatial patterns of thermal maturity display the cumulative effects of several phases of tectonism, and in some cases, the Rm gradients preserved at the surface mimic the final-phase structural geometries. Thermal data must be carefully integrated with structural and stratigraphic information, but, by doing so, many types of ambiguous structural and/or depositional relations can be resolved.

Acknowledgements. Funding was provided through a USGS/DOE interagency agreement (#DE-A121-83-MC20422). We thank D.W. Houseknecht for advise concerning organic petrography, P. Quinn and C. Gillett for laboratory assistance, J. O'Leary and R. Strong for petrographic analyses, and R. McLaughlin and S. Ellen for collecting some of the samples. Manuscript preparation was payed for by the University of Missouri Geology Development Fund.

References

Angevine, C.L., and Turcotte, D.L., Oil generation in overthrust belts: Am. Assoc. Petrol. Geol. Bull., 67, 235-241, 1983.

Bachman, S.B., The Coastal Belt of the Franciscan: youngest phase of northern California subduction, in Trench-forearc geology, edited by J. K. Leggett: Geol. Soc. London Spec. Pub., 10, 401-417, 1982.

Bachman, S.B., Underwood, M.B., and Menack, J.S., Cenozoic evolution of coastal northern California, in Tectonics and sedimentation along the California margin, edited by J.K. Crough and S.B. Bachman: Soc. Econ. Paleo. Mineral, Pacific Section, 38, 55-66, 1984.

Barker, E.C., Influence of time on metamorphism of sedimentary organic matter in liquid dominated geothermal systems, western North America: Geology, 11, 348-388, 1983.

Baskin, D.K., A method of preparing phytoclasts for vitrinite reflectance analysis: Jour. Sed. Petrol., 49, 633-635, 1979.

Becker, D.G., and Cloos, M., Melange diapirs into the Cambria slab: A Franciscan trench slope basin near Cambria, California: Jour. Geol., 93, 101-110, 1985.

Blake, M.C., Jr., Irwin, W.P., and Coleman, R.G., Blueschists-facies metamorphism related to regional thrust faulting: Tectonophysics, 8, 237-246, 1969.

Blake, M.C., Jr., Jayko, A.S., and McLaughlin, T.J., Tectonostratigraphic terranes of the northern Coast Ranges, California, in Tectonostratigraphic terranes of the Circum-Pacific region, edited by D.G. Howell: Circum-Pacific Council for Energy and Mineral Resources, Earth Science Series, 1, 159-171, 1985.

Boles, J.R., Zeolites in low-grade metamorphic grades, in Mineralogy and geology of natural zeolites, edited by F.A. Mumpton: Mineral. Soc. Amer., Short Course Notes, 4, 103-135, 1977.

Bostick, N.H., Microscopic measurement of the level of catagenesis of solid organic matter in sedimentary rocks to aid exploration for petroleum and to determine former burial temperatures - a review, in Aspects of diagenesis, edited by P.A. Scholle and P.R. Schluger: Soc. Econ. Paleo. Mineral. Spec. Pub. 26, 17-43, 1979.

Bostick, N.H., Comment on "Influence of time on metamorphism of sedimentary organic matter in liquid-dominated geothermal systems, western North America": Geology, 12, 689-690, 1984.

Brewer, J., Thermal effects of thrust faulting: Earth Planet. Sci. Let., 56, 233-244, 1981.

Bustin, R.M., Heating during thrust faulting in the Rocky Mountains: friction of fiction: Tectonophysics, 95, 309-328, 1983.

Cardwell, R.K., Chinn, D.S., Moore, G.F., and Turcotte, D.L., Frictional heating on a fault zone with finite thickness: Geophys. Jour. Royal Astron. Soc., 52, 525-530, 1978.

Champion, D.E., Howell, D.G., and Gromme, C.S., Paleomagnetic and geologic data indicating 2500 km of northward displacement for the Salinian and related terranes, California: Jour. Geophys. Res., 89, 7736-7752, 1984.

Cloos, M., Flow melanges: numerical modeling and geologic constraints on their origin in the Franciscan subduction complex, California: Geol. Soc. Am. Bull., 93, 330-345, 1982.

Cloos, M., Thermal evolution of convergent plate margins: thermal modeling and reevaluation of isotopic Ar-ages for blueschists in the Franciscan Complex of California: Tectonics, 4, 421-434, 1985.

Connan, J., Time-temperature relation in oil genesis: Am. Assoc. Petrol. Geol. Bull., 58, 2516-2521, 1974.

Cowan, D.S., Origin of blueschist-bearing chaotic rocks in the Franciscan Complex, San Simeon, California: Geol. Soc. Am. Bull., 89, 1415-1423, 1978.

Cowan, D.S., Deformation of partly dewatered and consolidated Franciscan sediments near Piedras Blancas Point, California, in Trench-forearc geology, edited by J.K. Leggett: Geol. Soc. London Spec. Pub., 10, 439-457, 1982.

Dickinson, W.R., and Snyder, W.S., Geometry of triple junctions related to San Andreas transform: Jour. Geophys. Res., 84, 561-572, 1979.

Dow, W.G, Kerogen studies and geological interpretations: Jour. Geochem. Expl., 7, 79-99, 1977.

Edman, J.D., and Surdam, R.C., Influence of overthrusting on maturation of hydrocarbons in Phosphoria Formation, Wyoming-Idaho-Utah Overthrust belt: Am. Assoc. Petrol. Geol. Bull., 68, 1803-1817, 1984.

Engebretson, D.C., Cox, A., and Gordon, R.G., Relative motions between oceanic and continental plates in the Pacific Basin: Geol. Soc. Am. Spec. Paper, 290, 1986.

Ernst, W.G., Mineral paragenesis and plate tectonic settings of relatively high-pressure metamorphic belts: Fortschr. Miner., 54, 192-222, 1977.

Ernst, W.G., Mineral paragenesis in Franciscan metagraywackes of the macimiento block, a subduction complex of the southern California Coast Ranges: Jour. Geophys. Res., 85, 7045-705, 1980.

Ernst, W.G., Phanerozoic continental accretion and the metamorphic evolution of northern and central California: Tectonophysics, 100, 287-320, 1983.

Evitt, W.R., and Pierce, S.T., Early Tertiary ages from the coastal belt of the Franciscan complex, northern California: Geology, 3, 433-436, 1975.

Furlong, K.P., Lithospheric behavior with triple junction migration: an example based on the Mendocino triple junction: Physics Earth Planet. Interiors, 36, 213-223, 1984.

Hall, C.A., Jr., San Luis Obispo transform fault and Middle Miocene rotation of the western Transverse Ranges, California: Jour. Geophys. Res., 86, 1015-1031, 1981.

Hoffman, J., and Hower, J., Clay mineral assemblages as low-grade metamorphic geothermometers: application to the thrust faulted disturbed belt of Montana, U.S.A., in Aspects of diagenesis, edited by P.A. Scholle and P.R. Schluger: Soc. Econ. Paleo. Mineral Spec. Pub., 26, 55-79, 1979.

Hood, A., Gutjar, C.C.M., and Heacock, R.L., Organic metamorphism and the generation of petroleum: Am. Assoc. Petrol. Geol. Bull., 59, 986-996, 1975.

Houseknecht, D.W., and Matthews, S.M., Thermal maturity of Carboniferous strata, Ouachita Mountains: Am. Assoc. Petrol. Geol. Bull., 69, 335-345, 1985.

Howell, D.G., Vedder, J.G., McLean, H., Joyce, J.M., Clarke, S.H., Jr., and Smith, G., Review of Cretaceous geology, Salinian and Nacimiento blocks, Coast Ranges of central California, in Cretaceous geology of the California Coast Ranges west of the San Andreas fault, edited by D.G. Howell, J.G. Vedder, and K.A. McDougall, Soc. Econ. Paleo. Mineral., Pacific Section, Pacific Coast Paleogeography Field Guide 2, 1-42, 1977.

Hutton, A.C., and Cook, A.C., Influence of alginite on the reflectance of vitrinite from Joadja, New South Wales, and some other coals and oil shales containing alginite: Fuel, 59, 711-714, 1980.

Johnson, M.R.W., Kelley, S.P., Oliver, G.J.H., and Winter, D.A., Thermal effects and timing of thrusting in the Moine thrust zone: Jour. Geol. Soc. London, 142, 863-873, 1985.

Johnston, D.C. and White, S.H., Shear heating associated with movement along the Alpine Fault, New Zealand: Tectonophysics, 92, 241-252, 1983.

Jones, D.L., Blake, M.C., Jr., Bailey, E.H., and McLaughlin, R.J., Distribution and character of Upper Mesozoic subduction complexes along the west coast of North America: Tectonophysics, 47, 207-222, 1978.

Jones, D.L., Howell, D.G., Coney, P.T., and Monger, J.W.H., Recognition, character, and analysis of tectonostratigraphic terranes in western North America, in Accretion Tectonics in the Circum-Pacific Regions, edited by M. Hashimoto and S. Uyeda, Terra Scientific, Tokyo, 21-35, 1983.

Karweil, J., The metamorphasis of coals from the standpoint of physical chemistry: Z. Deut. Geol., 107, 132-139, 1955. (Translated by Associated Technical Services Inc., East Orange, N.J.).

Kohsmann, H.J., Geologic time as a parameter in organic metamorphism and vitrinite reflectance as an absolute paleogeothermometer: discussion: Jour. Petrol. Geol., 8, 229-232, 1985.

Kubler, G., Evaluation quantitative du metamorphisme par la cristallinite de l'illite: Societe Nationale des petroles d'Aquitaine Centre de Recherches Bulletin, 2, 385-37, 1968.

Liou, J.G., Maruyama, S., and CHo, M., Phase equilibria and mineral parageneses of Metabasites in low-grade metamorphism: Mineral. Mag., 49, 321-333, 1985.

Marchioni, D.L., The detection of weathering in coal by petrographic, rheologic, and chemical methods: International Jour. Coal Geol., 2, 231-259, 1983.

McLaughlin, R.J., Kling, S.A., Poore, R.Z., McDougall, K., and Beutner, E.C., Post-middle Miocene accretion of Franciscan rocks, northwestern California: Geol. Soc. Am. Bull., 93, 595-605, 1982.

McLaughlin, R.J., Sorg, D.H., Morton, J.L., Theodore, R.G., and Meyer, C.E., Paragenesis and tectonic significance of base and precious metal occurrences along the San Andreas Fault at Point Delgada, California: Econ. Geol., 80, 344-359, 1985.

Newman, J., and Newman, N.A., Reflectance anomalies in Pike River coals: evidence of variability in vitrinite type, with implications for maturation studies and "Suggate rank": New Zealand Jour. Geol. Geophys., 25, 233-243, 1982.

Ogle, B.A., Geology of Eel River Valley area, Humboldt County, California: Calif. Div. Mines and Geol. Bull. 164, 128, 1953.

Oxburgh, E.R., and Turcotte, D.L. Thermal gradients and regional metamorphism in overthrust terrains with special reference to the Eastern Alps: Schweiz. Mineral. Petrogr. Mitt., 54, 641-661, 1974.

Page, B.M., Migration of Salinian composite block, California, and disappearance of fragments: Am. Jour. Sci., 282, 1694-1734, 1982.

Pavlis, T.L., and Bruhn, R.L., Deep-seated flow as a mechanism for the uplift of broad forearc ridges and its role in the exposure of high P/T metamorphic terranes: Tectonics, 2, 473-497, 1983.

Price, L.C., Geologic time as a parameter in organic metamorphism and vitrinite reflectance as an absolute paleogeothermometer: Jour. Petrol. Geol., 6, 5-38, 1983.

Price, L.C., and Barker, C.E., Suppression of vitrinite reflectance in amorphous rich kerogen - a major unrecognized problem: Jour. Petrol. Geol., 8., 59-84, 1985.

Schermer, E.R., Howell, D.G., and Jones, D.O., The origin of allochthonous terranes: perspectives on the growth and shaping of continents: Ann. Rev. Earth Planet. Sci., 12, 107-131, 1984.

Scholle, P.A., and Schluger, P.R. (editors), Aspects of diagenesis: Soc. Econ. Paleo. Mineral., Spec. Publ. 26, 443 pp, 1979.

Scholz, C.H., Shear heating and the state of stress on faults: Jour. Geophys. Res., 85, 6174-6184, 1980.

Sliter, W.V., McLaughlin, R.J., Keller, G., and Evitt, W.R., Paleogene accretion of Upper Cretaceous oceanic limestone in northern California: Geology, 14, 350-353, 1986.

Smith, G.W., Stratigraphy, sedimentology, and petrology of the Cambria slab, San Luis Obispo County, California (unpublished M.S. thesis): University of New Mexico, Albuquerque, 123 pp., 1978.

Smith, G.W., Howell, D.G., and Ingersoll, R.V., Late Cretaceous trench slope basins of central California: Geology, 7, 303-306, 1979.

Stach, E., and others, Stach's textbook of coal petrology: Gebruder, Borntraeger, 428 pp., 1975.

Stone, I.J., and Cook, A.C., The influences of some tectonic structures upon vitrinite reflectance: Jour. Geol., 87, 497-508, 1979.

Strong, R.H., Thermal maturity of Franciscan and related strata, southern Humboldt County, northern California (unpublished M.S. thesis): University of Missouri, Columbia, 138 pp., 1986.

Suggate, R.P., Low-rank sequences and scales of organic metamorphism: Jour. Petrol. Geol., 4, 377-392, 1982.

Tarduno, J.A., McWilliams, M., Debiche, M.B., Sliter, W.V., and Blake, M.C., Jr., Franciscan Complex Calera limestones: Accreted remnants of Farallon plate oceanic plateaus: Nature, 317, 345-347, 1985.

Underwood, M.B., Sedimentology and hydrocarbon potential of Yager structural complex, possible Paleogene source rocks in Eel River basin, northern California: Am. Assoc. Petrol. Geol. Bull. 69, 1088-1100, 1985.

Underwood, M.B., and O'Leary, J.D., Vitrinite reflectance and paleotemperature within Franciscan terranes of coastal northern California: 38°45'N to 40°00'N: U. S. Geological Survey Open-File Report 85-663, 32 pp., 1985.

Underwood, M.B., and Strong, R.H., Vitrinite reflectance and paleotemperature within the Franciscan terranes of coastal northern California: 40°00'N to 40°35'N: U. S. Geological Survey Open-File Report 86-258, 41 pp., 1986a.

Underwood, M.B., and Strong, R.H., Vitrinite reflectance and illite crystallinity, Cambria slab and Franciscan melange, central California: U. S. Geological Survey Open-File Report 86-295, 19 pp., 1986b.

Underwood, M.B., and Bachman, S.B., Sandstone petrofacies of the Yager complex and the Franciscan Coatal belt, Paleogene of northern California: Geol. Soc. Am. Bull., 97, 809-817, 1986.

Underwood, M.B., and Howell, D.G., Thermal maturity of the Cambria slab, an inferred trench-slope basin in central California: Geology, in press, 1986.

Underwood, M.B., Fulton, D.H., and McDonald, K.W., Thrust control on thermal maturity of the frontal Ouachita Mountains, central Arkansas: Jour. Petrol. Geol., in press, 1986.

Vedder, J.G., Howell, D.G., and McLean, H., Stratigraphy, sedimentation, and tectonic accretion of exotic terranes, southern Coast Ranges, California, in Studies in continental margin geology, edited by J.S. Watkins and C.L. Drake: Am. Assoc. Petrol. Geol. Memoir 34, 471-498, 1982.

Waples, D.W., Time and temperature in petroleum formation: application of Lopatin's method to petroleum exploration: Am. Assoc. Petrol. Geol. Bull., 64, 916-926, 1980.

Zandt, G., and Furlong, K.P., Evolution and thickness of the lithosphere beneath coastal California: Geology, 10, 376-381, 1982.

BASEMENT FRAGMENTS, ACCRETED TECTONOSTRATIGRAPHIC TERRANES, AND OVERLAP SEQUENCES: ELEMENTS IN THE TECTONIC EVOLUTION OF THE ARABIAN SHIELD

Peter R. Johnson[1], Erwin Scheibner[2], and E. Alan Smith[3]

[1]JV Associates, Ltd, 1242 Tenth St. N.W., Washington, DC 20001, USA
[2]NSW Dept. Mineral Resources, Sydney, NSW 2001, Australia
[3]Riofinex Limited, London W1R 5PA, England

Abstract. The layered rocks of the Arabian Shield comprise stratotectonic units that are diagnostic of particular tectonic events contributory to Proterozoic crustal growth in Arabia, and constitute stages in the tectonic evolution of the Shield. Middle Proterozoic or older (2000 to 1630 Ma) paraschist, paragneiss, and correlative metagranitic rocks, which underlie parts of the central Shield, represent an initial "basement" stage. The rocks form basement fragments; they probably originated by extension and dispersion of a pre-existing craton, and were incorporated as allochthonous microplates into younger tectonostratigraphic units. Volcanic complexes, ophiolitic complexes, and marginal-basin and fore-arc stratotectonic units deposited between 950 to 670 Ma form a "precratonic" stage in Shield evolution. The rocks accumulated in the intraoceanic to continental-marginal environments which resulted from crustal dispersion subsequent to the basement stage. The precratonic rocks constitute as many as ten tectonostratigraphic terranes, which accreted between 770 and 620 Ma to form a neocraton on which younger rocks were deposited. Volcanosedimentary overlap sequences deposited (760-570 Ma) in downwarp basins and in ensialic successor basins constructed above the joins between older terranes represent the succeeding "transitional-tectonic" stage. Volcanosedimentary rocks deposited in pull-apart basins which resulted from intraplate movements associated with the formation of transcurrent faults of the Najd system, constitute a "reactivation-tectonic" stage (670-570 Ma). By about 550 Ma, these events had resulted in the formation of a stable block of continental lithosphere. Phanerozoic sedimentary and Cenozoic volcanic rocks, deposited during an "epicratonic" stage of tectonic evolution, partly cover the continental block. It is currently being dispersed by ocean-floor spreading in the Red Sea, an event that may mark the onset of a new cycle of crustal growth.

Introduction

The Arabian Shield comprises an assemblage of middle to late Proterozoic rocks that are exposed in the western third of the Arabian Peninsula and underlie the peninsula elsewhere (see Figure 1). They are overlapped to the north, east, and south by a 'cover rock' succession of flat-lying Phanerozoic (Cambrian to Tertiary) sedimentary formations, and are partly concealed beneath fields of Cenozoic lava flows that were extruded concurrent with the opening of the Red Sea. Prior to this event, the Shield rocks were joined to Proterozoic formations in Egypt, Sudan, and Ethiopia as part of the Arabian-Nubian Shield.

Middle Proterozoic or older rocks crop out in limited parts of the central Shield. Much more conspicuous are assemblages of late Proterozoic volcanosedimentary and associated mafic to intermediate intrusive rocks. In the literature, these assemblages have been interpreted as the result of volcanism and magmatism either in ensialic basins [Kazmin et al., 1978], or above late Proterozoic subduction zones [Schmidt et al., 1979]. However, more recent work has indicated that many of the assemblages belong to late Proterozoic volcanic-arc systems that comprise distinct tectonic units, each having a distinctive geologic history and tectonic environment, and each separated by major structural discontinuities. Following definitions established in the North American cordillera [Coney et al., 1980], it has been recognized that these units represent distinct tectonostratigraphic terranes, and, as a consequence, the plate-tectonic models utilized in earlier accounts of Shield evolution have given way to ones that draw on the concept of terrane accretion [Camp, 1984; Johnson and Vranas, 1984a; Stoeser and Camp, 1985; Vail, 1985].

These recent studies also indicate the presence of stratotectonic units in the Shield additional to volcanosedimentary assemblages in

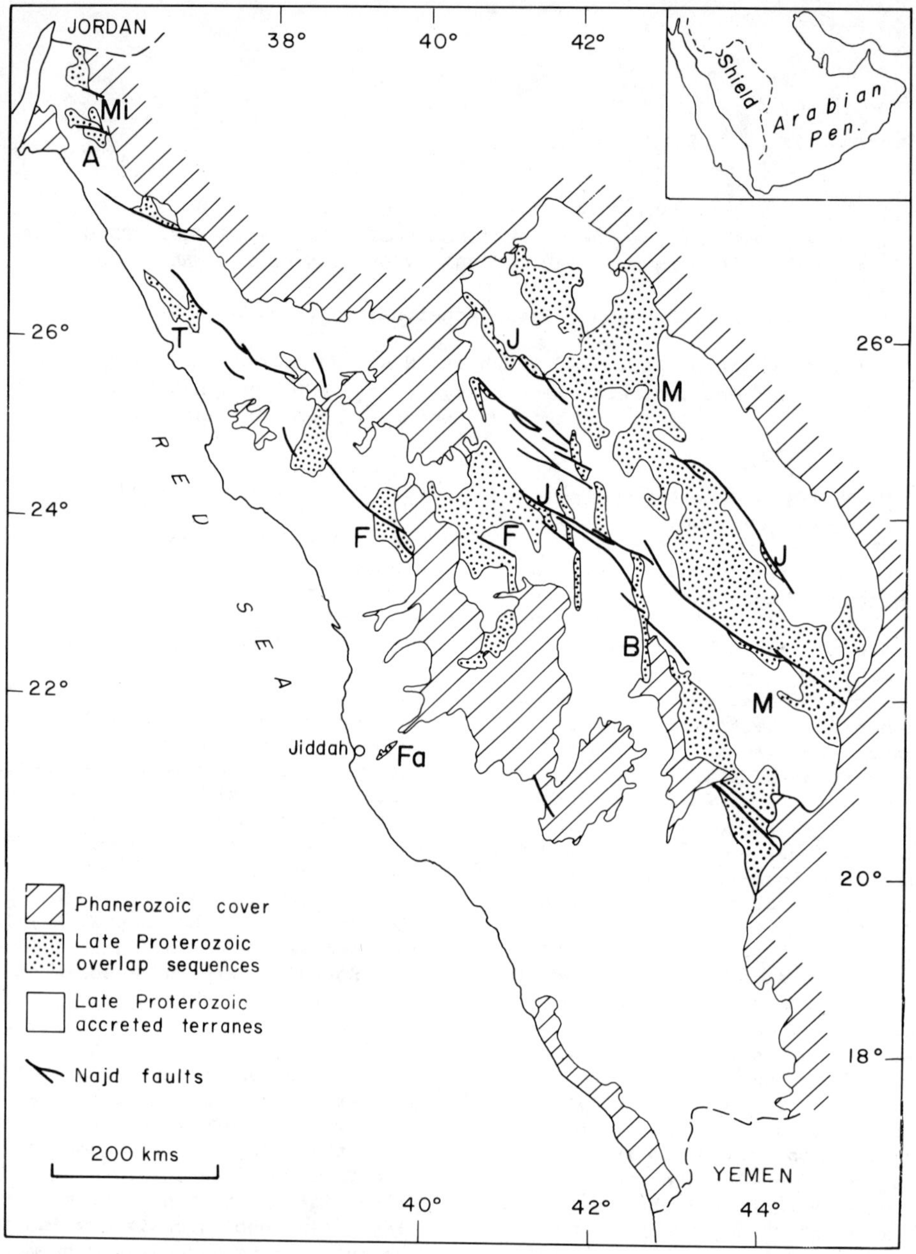

Fig. 1. Simplified geologic map of the Arabian Shield showing the older late Proterozoic accreted tectonostratigraphic terranes, younger late Proterozoic overlap sequences, Phanerozoic cover, and faults of the late Proterozoic Najd fault system. Overlap sequences named in the text include: A - Amlas formation, B - Bani Ghayy group, F - Furayh group, Fa - Fatimah group, J - Jibalah group, M - Murdama group and correlatives, Mi - Minawah formation, T - Thalbah group.

accreted terranes, including basement fragments, overlap sequences, and post-accretionary granitic rocks. As discussed by Johnson and Vranas [1984a], the stratotectonic units reflect a variety of formational environments and distinct tectonic settings diagnostic of the tectonic processes contributing to crustal growth in Arabia. This paper focuses on the layered stratotectonic units of the Shield, discusses their tectonic settings and, based on the

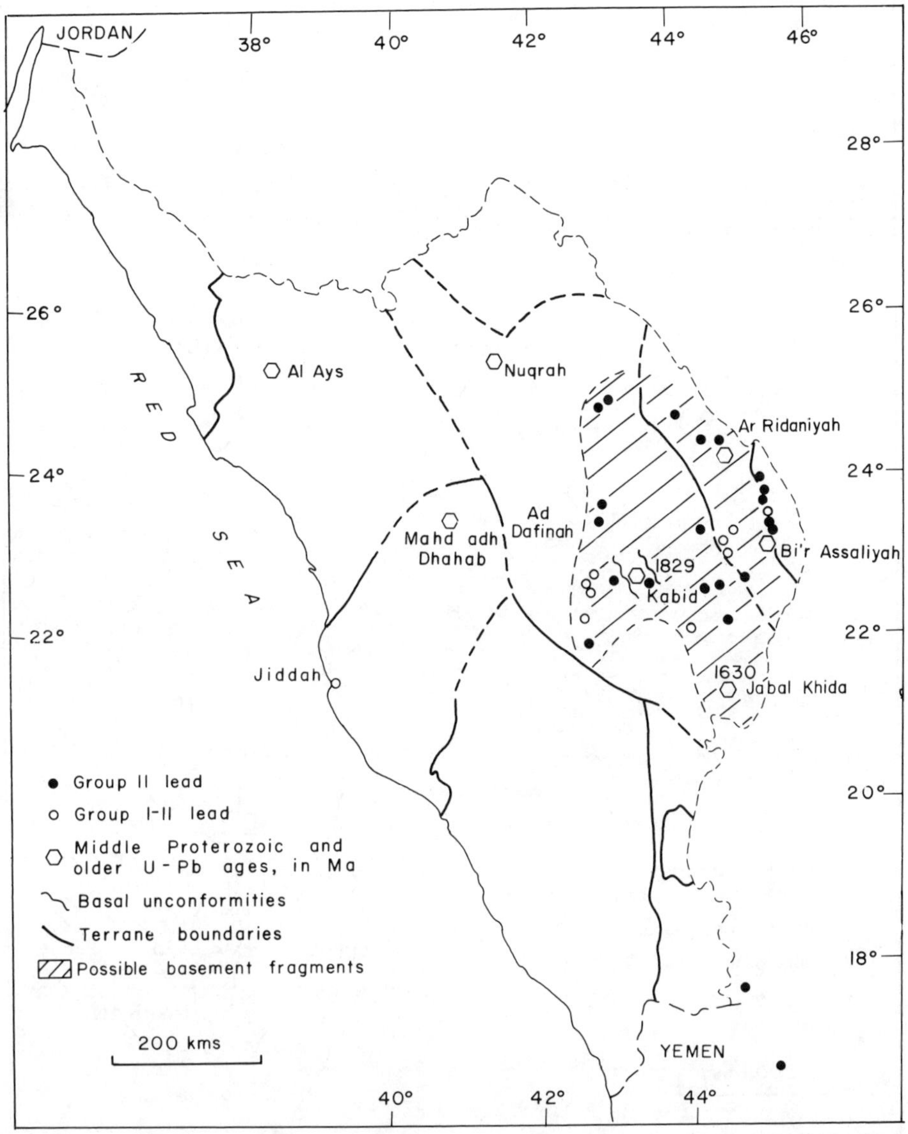

Fig. 2. Map of the Arabian Shield showing the distribution of data indicating middle Proterozoic or older rock units, superimposed on an outline of late Proterozoic accreted tectonostratigraphic terranes (lead data from Stacey and Stoeser, 1983; age data from Calvez et al., 1985, Stacey and Agar, 1985, Stacey and Hedge, 1984; unconformity data from Agar, 1985).

perceived temporal relationship of the tectonic processes so identified, presents an evolutionary tectonic model for the Shield.

Basement Fragments

The results of recent geologic mapping and geochronologic investigations establish that basement rocks occur in parts of the central Shield (see Figure 2). The oldest rocks mapped include the Kabid paragneiss and its possible correlative, the Khida formation. They consist of pelitic, felsic, arkosic, and amphibolitic gneiss, and are separated by unconformities from younger volcanosedimentary formations [Agar, 1985; Thieme, in press]. Structural data show that the Kabid paragneiss and the Khida formation had been deformed and metamorphosed prior to deposition of the volcanosedimentary units, and together with high-grade paragneiss in the Ad Dafinah area, are inferred to represent older basement. This inference is supported by U-Pb,

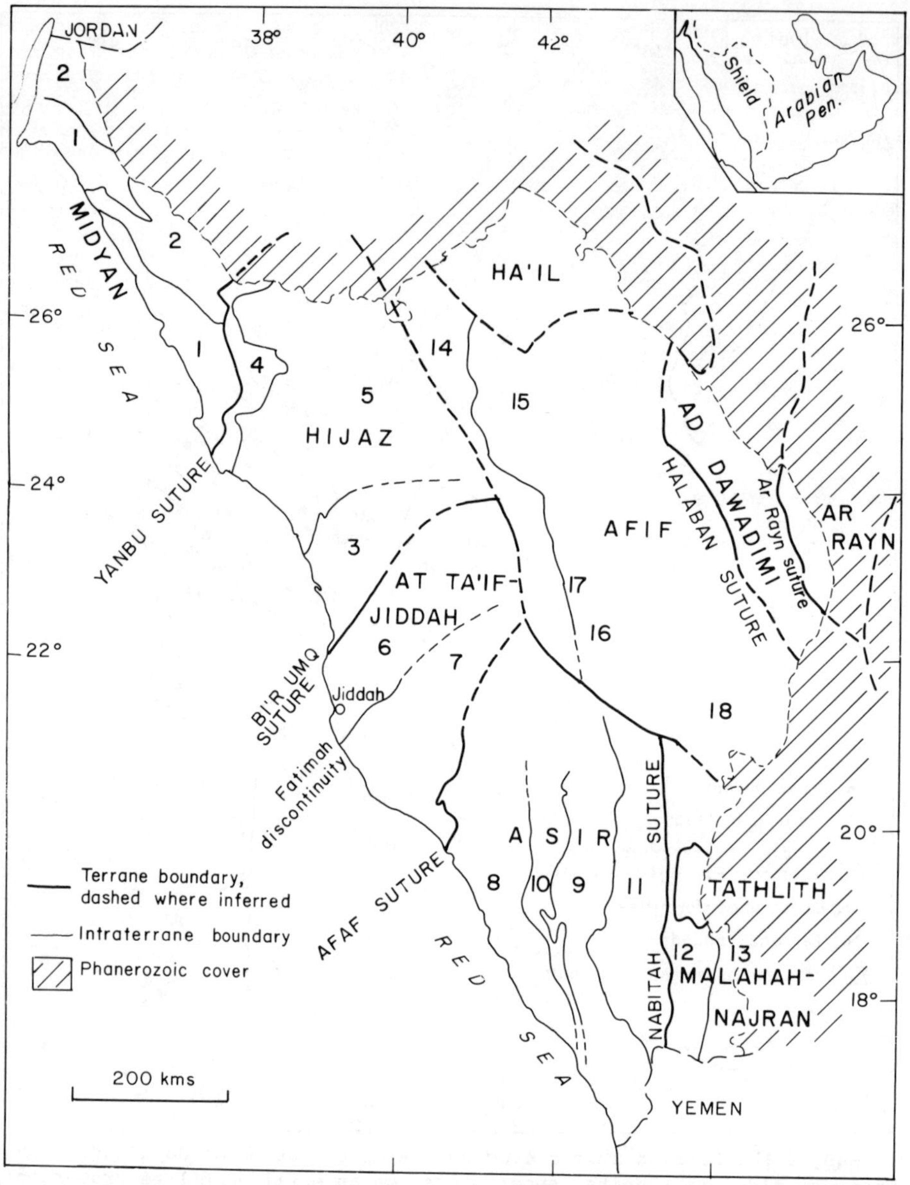

Fig. 3. Late Proterozoic accreted tectonostratigraphic terranes in the Arabian Shield. Stratotectonic units within composite terranes include: 1- Zaam basin; 2- Bayda-Hegaf volcanic complex; 3- Birak volcanic complex; 4- Farri marginal basin; 5- Al Ays volcanic complex; 6- Jiddah volcanic complex; 7- Al Lith complex; 8- Bidah-Hali volcanic complex; 9- Abha-Bishah volcanic complex; 10- Ablah volcanic rift; 11- Khadra-Ishmas volcanic complex; 12- Malahah-Yafikh volcanic complex; 13- Wassat volcanic complex; 14- Hulayfah volcanic complex; 15- Nuqrah volcanic complex; 16- Siham arc and basin; 17- Ad Dafinah ophiolitic complex; 18- Farshah volcanic complex.

Sm-Nd, and Rb-Sr ages of approximately 1830 Ma and 1630 Ma [Stacey and Agar, 1985; Stacey and Hedge, 1984] obtained from the area. It is also supported by independently derived evidence of moderately to strongly elevated $^{207}Pb/^{204}Pb$ and $^{208}Pb/^{204}Pb$ ratios in the eastern Shield (Group I-II and Group II respectively, shown in Figure 2) [Stacey and Stoeser, 1983]. The lead ratios were obtained from galena and feldspar samples, and require a lead contribution from sialic crust of early Proterozoic or even Archean crust. Elsewhere in the Shield, middle to early Proterozoic inherited zircon ages have been obtained from rock samples in the Bi'r Assaliyah, Ar Ridaniyah, Nuqrah, Mahd adh Dhahab, and Al Ays areas [Calvez et al., 1985], and Group II lead ratios are known from the Saudi Arabia-Yemen border region.

Utilizing this body of data, Stacey and Stoeser [1983] and Stoeser and Camp [1985] concluded that much of the eastern Shield was underlain by continental basement. At this time, the precise extent and continuity of the basement is unknown because of the dispersive effects of subsequent accretionary and other tectonic processes in the evolution of the Shield. Nevertheless, regions of distinctive subdued magnetic relief shown on aeromagnetic maps of the Shield suggest that basement rocks of probable granitoid composition are likely to be present beneath parts of the Afif terrane, and continuous beneath most of the Ad Dawadimi terrane [Georgel et al, 1985]. Elsewhere, a concealed block of continental crust is inferred from recently obtained aeromagnetic data in central Arabia, east of the exposed margin of the Shield [Phoenix Corp., 1985].

Volcanic-arc and Terrane Formation

The late Proterozoic volcanosedimentary formations of the Shield are divisible into two overlapping groups: 1.-assemblages formed between approximately 950 and 700 Ma and, 2.-younger assemblages formed between 760 and 570 Ma. The older assemblages constitute separate systems of volcanic arcs and other stratotectonic units [in the sense of Rast, 1969; see also Scheibner, 1976] which, based on differences in age, chemistry, depositional environment, and structural history, and on evident structural and magnetic boundaries, are further divisible into tectonostratigraphic terranes. In contrast, the younger assemblages represent overlap sequences that evolved subsequent to terrane accretion.

Following Johnson and Vranas [1984a], this paper describes ten tectonostratigraphic terranes; five larger and five smaller (see Figure 3 and Table 1). In an independent study, Stoeser and Camp [1985] recognized only five Proterozoic terranes in Arabia; three ensimatic island-arc terranes in the western part of the Shield and two terranes of continental affinity farther east. However, we believe that the available geologic and geophysical data support the more complex interpretation presented here. The smaller terranes consist of single depositional units, whereas the larger ones are composite, comprising two or more amalgamated stratotectonic rock units (see Figure 4). The lithostratigraphic units that form these terranes are described in the following paragraphs; terrane boundaries and associated accretionary structures are described in a later section.

Midyan Composite Terrane

The Midyan composite terrane is exposed in the northwestern Shield, extends into the Eastern Desert of Egypt, and continues beneath the Phanerozoic cover for as much as 100 km to the northeast [Phoenix Corp., 1985]. It includes two strongly deformed amalgamated units, the Zaam basin and the Bayda-Hegaf volcanic complex; their contact is largely obscured by younger rocks, but in part is an unconformity, and elsewhere a thrust, volcanic arc on the Zaam basin. A third, ophiolitic, unit, not described in this paper, may be represented by slices of mafic-ultramafic rock widely exposed in the Egyptian Eastern Desert.

Zaam basin. The Zaam basin comprises a distinctive succession of volcanosedimentary and chemical (chert and iron formation) deposits [Davies, 1985] that, based on age data obtained from associated mafic intrusions [Hedge, 1984], probably formed between about 800 and 725 Ma. The rocks are inferred to represent a basin environment at the edge of a continental block. Oxide-facies, iron-rich chemical deposits within the succession probably formed in local basins of shallow to moderate depth, whereas pyritic, graphitic shale, black chert, and fetid limestone indicate basins of greater depth.

Bayda-Hegaf volcanic complex. The Bayda-Hegaf volcanic complex is an assemblage of volcanic, volcaniclastic and sedimentary rocks [Clark, 1985]. The succession includes low-potassium basalt and andesite, intercalated with rhyolite, typical of island-arc volcanic complexes. However, the presence of epiclastic sedimentary rock indicates a facies change to a continental-marginal environment in the far north of the exposed complex. An assemblage of arc-related calc-alkaline diorite and tonalite, and I-type granodiorite and granite intrude the complex, and yield dates which suggest that the volcanic complex is older than 710 Ma [Hedge, 1984].

Hijaz Composite Terrane

The Hijaz composite terrane occupies the west-central part of the Shield. Towards the east it is concealed by younger Proterozoic and Phanerozoic sedimentary basins and by Cenozoic lava flows and, to the north, is overlapped by Phanerozoic cover, although the terrane can be traced as a high-frequency aeromagnetic signature 115 km north beneath the cover [Phoenix Corp., 1985]. Stoeser and Camp [1985] extend the Hijaz terrane into the northeastern part of the Shield. However, we limit it to west of a structural and magnetic boundary in the central Shield. To the southwest, the terrane extends into Sudan [Vail, 1985].

Birak volcanic complex. Where exposed in Arabia, the terrane includes three amalgamated stratotectonic units. The oldest, cut by intrusive rocks dated between 870 and 800 Ma [C. E. Hedge, written commun., 1983; Al-Shanti et al., 1984b], is discontinuously exposed along the southern margin of the terrane as the Birak

TABLE 1. Summary of Precratonic Tectonostratigraphic Terranes in the Arabian Shield (see text for sources of data)

Terrane	Component Stratotectonic Rock Units and Related Age Dates	Lithology
MIDYAN	Zaam basin, 800-725 Ma	Mafic, felsic, and abundant volcaniclastic rocks, interbedded with turbidity-current deposited sandstone, siltstone, pyritic shale, and banded iron formation (Zaam group, Silasia formation, Alqunnah and As Safra and part of Farri groups); associated diorite, tonalite and trondhjemite.
	Bayda and Hegaf volcanic complex, pre-710 Ma	Mafic to felsic volcanic and coarse- to fine-grained volcaniclastic rocks (Bayd group, Hegaf formation); arc-related calc-alkalic diorite, tonalite, granodiorite and minor I-type granite.
HIJAZ	Farri marginal basin 796-740 Ma	Low-K tholeiitic basalt, pillow basalt and breccia, subordinate felsic volcanic rocks, interbedded sandstone, siltstone and conglomerate (Farri group); serpentinized gabbro, dunite and peridotite (Jabal al Wask and Jask Ess ophiolites).
	Birak volcanic complex 870-800 Ma	Low-K tholeiitic volcanic and volcaniclastic sedimentary rocks (Birak group); arc-related calc-alkalic diorite, tonalite, granodiorite and serpentinized gabbro and peridotite (Bi'r Umq and Jabal Thurwah ophiolites).
	Al Ays volcanic complex 735-705 Ma	Low- to high-K alkalic volcanic rocks, predominantly felsic in composition, including abundant ignimbrite, interbedded with volcaniclastic sedimentary rocks (Al Ays group); arc-related calc-alkalic granodiorite and tonalite.
AT-TA'IF-JIDDAH	Jiddah volcanic complex pre-816 to 760 Ma	Low-K tholeiitic and alkalic basalt and rhyolite, interbedded with volcaniclastic and minor epiclastic rocks (Samran, Arj and Mahd group); arc-related tonalite and granodiorite.
	Al Lith volcanic complex ?890-?820 Ma	Abundant subalkalic gabbro, diorite, tonalite and quartz diorite; subordinate bimodal assemblage of tholeiitic basalt and soda dacite/rhyolite, interbedded with volcaniclastic and locally epiclastic continental sedimentary rocks (Arafat and Baish groups, Sadiyah formation).
ASIR	Bidah-Hali volcanic complex, pre-900 to ?830 Ma	Bimodal, predominantly tholeiitic basalt and subordinate soda dacite/rhyolite interbedded with volcaniclastic and rare epiclastic rocks (Baish, Bahah groups); calc-alkalic arc-related diorite, tonalite and granodiorite.
	Abha-Bishah volcanic complex, ?912-797 Ma	Tholeiitic basalt, pillow basalt, andesite and dacite, interbedded with abundant volcaniclastic and subordinate epiclastic rocks (Baish, Bahah and Jiddah groups, Sabya formation); arc-related diorite, tonalite, and trondhjemite.
	Ablah volcanic rift ?795-720 Ma	Phyllite, slate, graywacke, marble, quartzite, and conglomerate passing NE-wards into assemblages of low-K tholeiitic and calc-alkalic basalt to rhyolite interbedded with volcaniclastic rocks (Ablah group, Qirshah and Khutnah formations); associated diorite and tonalite.

TABLE 1. (Continued)

Terrane	Component Stratotectonic Rock Units and related Age Dates	Lithology
	Khadra-Ishmas volcanic complex, 785-725 Ma	Tholeiitic basalt, andesite, dacite and abundant volcaniclastic sedimentary rocks (Halaban group); abundant arc-related diorite, quartz diorite and tonalite.
MALAHAH-NAJRAN	Malahah-Yafikh volcanic complex, pre-730 Ma	Basalt, pillow lava, andesite and subordinate dacite/rhyolite, interbedded with abundant volcaniclastic sedimentary rocks (Malahah formation, Halaban group); arc-related sills, dikes and plutons of diorite and tonalite
	Wassat volcanic complex (approx. 730 Ma)	Low- to high-K calc-alkalic basalt, andesite, pillow lava and breccia, interbedded with abundant fine- to coarse-grained volcaniclastic and subordinate epiclastic continental sedimentary rocks (Wassat formation); arc-related mafic sills, dikes and plutons.
TATHLITH	Tathlith melange and basin, pre-664 Ma	High-grade metamorphosed quartzo-feldspathic and calcareous sedimentary rocks interbedded with subordinate mafic volcanic rocks, underlain (or ?intruded) by serpentinized peridotite and minor pyroxenite.
HA'IL	Ha'il volcanic complex pre-735 Ma	Moderately to strongly metamorphosed basaltic to andesitic volcanic and subvolcanic dioritic rocks (Nuf formation); abundant arc-related gabbro, diorite tonalite, granodiorite and granite.
AFIF	Hulayfah volcanic complex, pre-720 Ma	Tholeiitic and calc-alkalic basalt to dacite volcanic, abundant volcaniclastic sedimentary rocks (Hulayfah and Mahd groups); arc-related diorite, tonalite, and granodiorite.
	Nuqrah volcanic complex pre-820 Ma	Low-K tholeiitic and calc-alkalic mafic and abundant felsic volcanic, and fine-grained volcaniclastic rocks (Nuqrah formation); arc-related mafic to intermediate plutonic rocks.
	Siham arc and basin pre 720-690 Ma	Magmatic-arc facies comprising andesite, basalt, shale, and rhyolite passing eastwards into marginal-basin facies comprising marble, quartzite, conglomerate and rhyolite unconformably overlying basement remnants.
	Ad Dafinah ophiolite various ages: 830 Ma, 770-750 Ma	Assemblages of strongly deformed and metamorphosed serpentinite, gabbro, diorite, paraschist, orthoschist and gneiss (?disrupted ophiolite complex).
	Farshah volcanic complex (ages unknown)	Complex region, including remnants of middle Proterozoic continental microplate, metamorphosed and deformed mafic and ultramafic rocks, and assemblage of basaltic to dioritic volcanic and volcaniclastic rocks.
AD DAWADIMI	Abt marginal basin pre-695 Ma	Metasedimentary rocks including schist, phyllite, slate, meta-sandstone, conglomerate and dolomitic limestone; underlain by ophiolite assemblage of serpentinite, gabbro, diorite and pillow basalt.
AR RAYN	Ar Rayn volcanic complex pre-670 Ma	Transitional tholeiitic and calc-alkalic basalt, andesite, dacite and soda rhyolite, interbedded with volcaniclastic and subordinate epiclastic rocks (Al Amar group); abundant arc-related tonalite, trondhjemite and granodiorite.

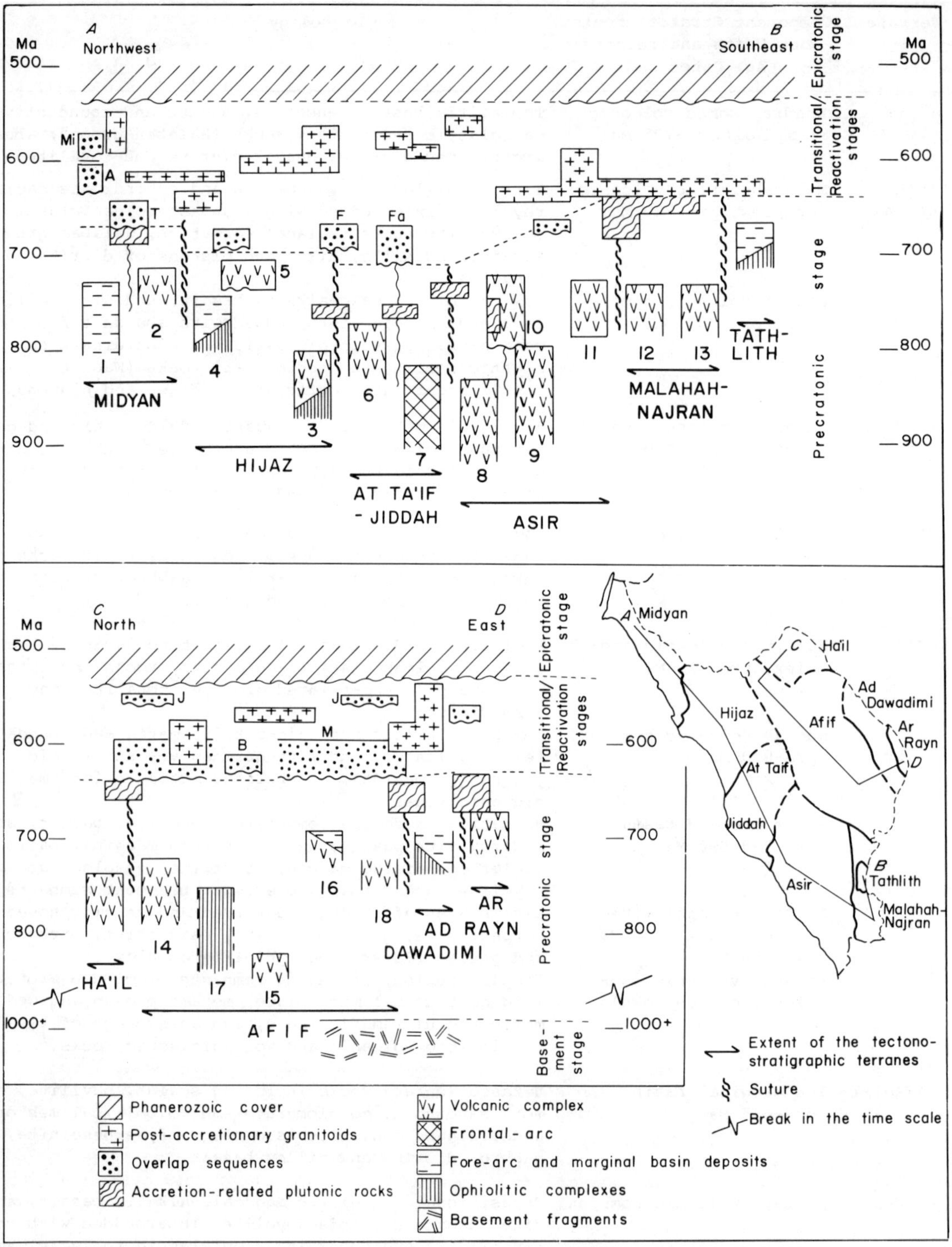

Fig. 4. Flow diagram illustrating accretion history of the Arabian Shield. For lettered overlap sequences see Figure 1, and for numbered stratotectonic units see Figure 3.

volcanic complex. The abundance of pillow basalt in the assemblage, its low-potassium to tholeiitic composition [Ramsay, 1983], and the presence of tectonically emplaced bodies of ultramafic rock, attest to an intraoceanic origin. The ultramafic rocks probably represent a dismembered ophiolite complex that has been emplaced along north- and northwest-dipping thrusts [Naseef et al., 1984].

Farri marginal basin. A younger stratotectonic unit, the Farri marginal basin, crops out on the northwestern margin of the Hijaz composite terrane as a strongly deformed accretionary prism [Camp, 1984]. It is intruded by a suite of preaccretionary tonalite, dated at 796 and 743 Ma old [Kemp et al., 1980], and contains tectonic slices of ophiolitic ultramafic rock. The marginal basin includes rhyolite and low-potassium tholeiitic basalt which probably reflect a transition from felsic volcanism associated with the break-up of an earlier continental crust, to mafic volcanism associated with later sea-floor spreading. The ophiolitic rocks yield U-Pb and Sm-Nd ages of 780 to 740 Ma [Claesson et al., 1984], and their trace-element geochemistry suggests that they formed as juvenile oceanic lithosphere during back-arc spreading [Pallister et al., 1985].

Al Ays volcanic complex. Most of the Hijaz terrane is occupied by the Al Ays volcanic complex, approximately 735 to 705 Ma old [Kemp et al., 1980]. In the west, the complex unconformably overlies the Farri marginal basin, and probably forms a fore-arc succession [Camp, 1984]; a volcanic-arc facies predominates in the east, differentiated from adjacent stratotectonic units by its low- to high-potassium contents and alkalic geochemistry. These features, and the abundance of rhyolite and welded tuffs in the succession, have been interpreted as indicating formation of the volcanic facies in rifts above continental crust or older arc complexes [Calvez et al., 1983]. The recently described indication of 1400 Ma and older zircons in the Al Ays group [Calvez et al., 1985] (see Figure 2) is consistent with, though not confirmatory of, such an environment.

At Ta'if-Jiddah Composite Terrane

The At Ta'if-Jiddah composite terrane described in this paper corresponds to the northern part of the Asir terrane of Stoeser and Camp [1985], but is treated by us as a distinct crustal unit, separated from the Asir terrane proper by the Afaf suture [Johnson and Vranas, 1984a]. To the northeast the terrane is covered by a Proterozoic overlap succession and by Cenozoic lava flows, but is inferred to extend as far as an aeromagnetic boundary approximately at long. 41° 30' E. In Sudan, the terrane extends almost 200 km inland from the Red Sea [Vail, 1985].

Jiddah volcanic complex. The Jiddah volcanic complex is the more northerly of two amalgamated stratotectonic units exposed within the terrane in Arabia. It is associated with plutons of mafic to intermediate intrusive rock, dated variously between 816 and 760 Ma [Calvez and Kemp, 1982]. The Samran group crops out in the southwestern part of the complex as a bimodal assemblage of low-potassium, tholeiitic composition comparable to lavas of present-day oceanic arcs [Ramsay, 1983]. To the northeast is an assemblage of sub-alkalic and alkalic volcanic rocks, the Arj and Mahd adh Dhahab groups, that probably formed in volcanic rifts [Kemp et al., 1982].

Al Lith complex. The Al Lith complex, in the southern portion of the terrane, chiefly consists of preaccretionary mafic to intermediate plutonic rocks, and is similar to frontal arcs found in modern arc systems [Camp, 1984]. Age data from these rocks are chiefly unreliable but suggest that the complex is older than the Jiddah volcanic complex and possibly as old as approximately 890 Ma [Fleck et al., 1980; Fleck, 1981]. Volcanosedimentary layered rocks are preserved as strongly deformed, relatively narrow synclinoria between the plutons. Some of the rocks are bimodal in composition and comparable to immature island-arc successions; others resemble mid-ocean floor or ocean plateau assemblages, whereas some parts of the complex contain quartzite and kyanite-rich schists that probably reflect a periodic influx of sediment from a continental margin [Pallister, in press; Reischmann et al., 1984].

Asir Terrane

The Asir terrane is one of the larger accreted crustal units in the Arabian Shield, and includes four amalgamated stratotectonic units [Johnson and Vranas, 1984a, b]. It comprises the southern portion of the Asir terrane of Stoeser and Camp [1985], essentially conterminous with the arc terrane shown by them as greater than 900 to 800 Ma old; it also corresponds to the south and west Asir terranes of Vail [1985]. Cenozoic lava flows and extensive Quaternary cover obscure the northern part of the terrane, but north-trending aeromagnetic anomalies delineate its probable continuation beneath this cover as far north as approximately lat 22°30' N. To the south, exposures of late Proterozoic rock of the Asir terrane extend a short distance into the Yemen Arab Republic, beyond which the terrane is overlapped by Phanerozoic sedimentary rocks and the Cenozoic volcanic Trap Series.

Bidah-Hali volcanic complex. The Bidah-Hali volcanic complex, the oldest of the four amalgamated stratotectonic units in the terrane, contains plutons dated between 901 and 838 Ma, and is characterized by very low initial-strontium ratios and calc-alkalic fractionation trends, indicating derivation from a parent magma of mantle origin [Fleck et al., 1980; Marzouki et al., 1982]. Much of the complex was deposited in a distal volcanic environment and consists of volcaniclastic rocks, although proximal volcanic assemblages are locally present that are comparable to bimodal assemblages found in present-day intraoceanic-arc complexes [Roobol et al., 1983]. Ocean-floor ultramafic rocks occur in parts of the complex, interbedded with pillow basalt, chert and carbonaceous schist [Greenwood et al., 1982], and a continental provenance has been inferred for quartz-rich metasedimentary rocks in the northern part of the complex [Ramsay et al., 1981].

Abha-Bishah volcanic complex. The Abha-Bishah volcanic complex, in the central part of the Asir terrane, is lithologically similar to the rocks of Bidah-Hali volcanic complex. It consists of mafic to felsic volcanic rocks, volcaniclastic and epiclastic rocks of local continental provenance, and arc-related intrusive rock [Greenwood et al., 1985]. An Rb-Sr date of approximately 912 Ma, derived from metavolcanic rock, and Rb-Sr and U-Pb dates of between 837 and 797 Ma, obtained from associated intrusive rocks [Cooper et al., 1979; Fleck et al., 1980], indicate that the Abhah-Bishah complex is probably coeval with, to slightly younger than, the Bidah-Hali volcanic complex. Associated intrusive rocks are typified by the diorite to trondhjemite complex of the An Nimas batholith, and are marked by an immature geochemistry and low initial-strontium ratio indicative of the oceanic origin of the complex.

Ablah volcanic rift. The Ablah volcanic rift [Greenwood et al., 1982] is a distinctive north-trending stratotectonic unit in the west-central part of the Asir terrane, between 800 and 780 Ma old in the south, and as young as 721 Ma in the north [Bokhari and Kramers, 1981; Cooper et al., 1979; Fleck et al., 1980; Marzouki et al., 1982]. Shallow-marine metasedimentary rocks predominate in the southern part of the stratotectonic unit. They pass north into a volcano-sedimentary succession deposited in a transitional, continental-deltaic and subaerial-volcanic environment, and pass to the northeast into a succession that resembles deposits found in present-day island-arc complexes [Roobol et al., 1983]. Much of the stratotectonic unit is fault bounded. However, metasedimentary rocks of the unit locally rest unconformably on the older Bidah-Hali and Abha-Bishah volcanic complexes, and the unit possibly originated as a volcanic rift by the splitting of these older complexes. Subsequent closure of the southern portion of the rift may be indicated by the presence of synkinematic gneiss in the Baqarah dome, dated by the U-Pb method at 778 and 746 Ma [Cooper et al., 1979].

Khadra-Ishmas volcanic complex. Preaccretionary rock units of the Khadra-Ishmas volcanic complex, in the eastern part of the Asir terrane, are represented by belts of multiply deformed and metamorphosed volcanic, volcaniclastic, and mafic to intermediate intrusive rocks preserved between batholiths of accretion-related granitic rock. The assemblage is dated between 785 and 725 Ma [Fleck et al., 1980; 1980; Stoeser et al., 1984]. A number of volcanic centers are defined by vertical and lateral facies [Greenwood et al., 1982], and the low-alkali content and tholeiitic composition of the rocks indicate that they likely originated as an island-arc complex. The associated intrusive rocks, typified by the Tarib batholith, are geochemically unevolved and contain low initial-strontium ratios, features which indicate crystallization from a magma of immature composition, and are consistent with the inferred arc-related origin of the intrusive rocks.

Malahah-Najran Composite Terrane

The Malahah-Najran composite terrane comprises two assemblages of metavolcanic, metasedimentary, and preaccretionary intrusive rocks about 732 Ma or older [Stoeser et al., 1984] They are exposed east of the Nabitah fault zone, and constitute the Malahah-Yafikh and the Najran volcanic complexes [Johnson and Vranas, 1984a]. The layered formations in the terrane are approximately contemporary with, and petrologically similar to, the layered rocks of the Khadra-Ishmas volcanic complex of the adjoining Asir terrane. This suggests that the Malahah-Yafikh and the Najran complexes may, in fact, constitute part of the Asir terrane. However, they are assigned by us to a separate terrane because of gravity and seismic evidence that the Nabitah fault zone is the surface expression of a major crustal boundary [Gettings et al., 1986], and because of the abundance and size of sulfide mineral deposits contained in both complexes, which imply that the stratotectonic units were affected by a metallogenic event not evidenced in the contemporary volcanic rocks of the Asir terrane.

In outcrop, the Malahah-Najran terrane is elongate north-south. It is covered in the south, 50 km south of the Yemen border, by Phanerozoic sandstone, and truncated in the north by strike-slip faults, part of the late Proterozoic Najd system of transcurrent faults that traverse the Shield (see Figure 1). Phanerozoic rocks also delimit exposure of the terrane in the east; however, by means of its distinctive aeromagnetic signature, the terrane can be traced a 125 km farther east, beneath the Phanerozoic cover.

The Malahah-Yafikh volcanic complex includes volcanic rocks of probable calc-alkalic composition, and abundant volcaniclastic rocks. They formed at a number of volcanic centers, and were intruded by mafic dikes and a suite of arc-related plutonic rocks.

The Wassat volcanic complex, separated from the Malahah-Yafikh complex by the Ashara fault zone, is an assemblage of high-alumina, calc-alkalic basalt and andesite, immature sedimentary rocks, sub-volcanic mafic dikes, and bodies of mafic to intermediate intrusive rock. It formed in the axial part of the complex, in a rapidly sinking trough; to the east, a stable-shelf environment was present, marked by quartzite, and quartzo-feldspathic schist and gneiss derived from protoliths of continental provenance [Greenwood, 1985b; Sable, 1985].

Tathlith Terrane

The Tathlith terrane is a strongly deformed and metamorphosed assemblage of layered and intrusive rocks that, because of its lithology, inferred provenance, and metamorphic grade, and because of indications of thrust faulting at its base, is distinguished from adjacent units of Shield rocks as a separate, allochthonous terrane [Johnson and Vranas, 1984a]. On its southern and western margins, the terrane is separated by a thrust from the underlying Malahah-Najran terrane. The rocks include serpentinized peridotite and pyroxenite, intruded by gabbro dated at 664 Ma [Cooper et al., 1979], and are overlain by a mélange, comprising disrupted ultramafic, metasedimentary and metavolcanic rocks [Greenwood, 1985a]. The interior parts of the terrane are occupied by amphibolite-grade metamorphic rocks that were derived from an epiclastic succession of continental provenance. They are associated with subordinate volcanic and volcaniclastic rocks, and may have been deposited in a shelving basin [Riddler, et al., 1983].

Ha'il Terrane

Discontinuously exposed accretionary rocks in the northeastern part of the Shield are tentatively assigned to the Ha'il terrane [Johnson and Vranas, 1984a]. Stoeser and Camp [1985] treat the rocks as part of the Hulayfah magmatic arc extended from its type area along the western flank of the Afif terrane. This interpretation is supported by an age of 735 Ma or older reported for the preaccretionary rocks in the Ha'il region [C. E. Hedge, written commun. 1985], and by the similar inferred volcanic-arc environments of the two areas. However, the accretionary rocks in the Ha'il area are separated from adjacent parts of the Arabian Shield by major magnetic boundaries and are characterized by a distinctive northeast-trending magnetic grain [Blank and Andreasen, in press.; Phoenix, 1985], on which basis the region is delineated as a separate crustal unit. The preaccretionary assemblage of the Ha'il terrane includes strongly deformed and metamorphosed volcanic and sedimentary rocks intruded by broadly coeval mafic to intermediate plutonic rocks.

Afif Composite Terrane

The Afif terrane, a name introduced by Stoeser and Camp [1985], and adopted here in place of our earlier term, the Zalm terrane [Johnson and Vranas, 1984a], underlies the north-central part of the Arabian Shield. Its western and southern margins are a complex accretionary zone, modified by younger faults; its eastern margin is bounded by the Halaban suture.

The terrane comprises remnants of middle Proterozoic or older basement microplates, and synclinorial and fault-bounded belts of metavolcanic and metasedimentary rocks that we tentatively assign to five stratotectonic units. These are extensively intruded by synkinematic and late- to postkinematic granitic rocks, and are overlain by the Murdama and equivalent groups that fill the largest of the overlap basins in the Shield.

Hulayfah volcanic complex. The Hulayfah volcanic complex is a belt of deformed volcano-sedimentary rocks that are provisionally interpreted as a sliver of volcanic-arc amalgamated to the leading, western, edge of the Afif terrane. The complex includes the Hulayfah group and part of the Mahd group, and is approximately 770 Ma to a minimum of 720 Ma old [Calvez et al., 1983; Fleck and Hadley, 1982]. Tholeiitic and calc-alkalic mafic to felsic volcanic rocks are common in the northern part of the complex, and resemble lavas associated with present-day island arcs or continental margins. They are cut by heterogeneous bodies of mafic to intermediate intrusive rock, and contain serpentinized and silicified peridotite. In the central part of the complex, the rocks are interpreted as having formed in tensional basins and shallow rifts [Kemp et al., 1982].

Nuqrah volcanic complex. The Nuqrah volcanic complex consists of north-trending belts of layered and intrusive rocks that discontinuously crop out between plutons of synkinematic and younger granite. The layered formations include low-potassium, tholeiitic to calc-alkalic, mafic and felsic volcanic rocks that have been dated at 839 and 821 Ma [Calvez et al., 1983]. They are associated with a heterogeneous suite of preaccretionary mafic to intermediate and granitic rocks, and geochemically resemble deposits in present-day mature island arcs or continental-marginal arcs.

Siham arc and basin. In their original tectonic analysis, Johnson and Vranas [1984a] extended the Nuqrah volcanic complex south as far as the Zalm area. However, it is clear from work by Agar [1985] that the preaccretionary rocks in the Zalm area constitute a separate stratotectonic unit, the Siham magmatic arc and marginal basin, welded to the western edge of the Afif terrane. The Siham group, and its associated intrusive rocks, comprise a magmatic-arc facies in the west, and a marginal-basin and continental-shelf facies underlain by middle Proterozoic or older basement rocks farther east; the marginal basin was probably generated by back-arc crustal extension and was probably ensialic. They date from before 720 to 690 Ma, and are therefore about 100 million years younger than the Nuqrah volcanic complex.

Ad Dafinah ophiolite complex. Linear to discontinuous belts of mafic to ultramafic rocks, schist and gneiss crop out between the Hulayfah, and the Nuqrah and Siham, stratotectonic units. The belts are cut by syn- to late-kinematic intrusions, and are extensively dislocated by strike-slip faults of the Najd system.

Since the mid-1970s, these rocks have been interpreted as allochthonous, ophiolitic remnants of ocean crust [Frisch and Al-Shanti, 1977]. They have also been treated as part of a once-continuous late Proterozoic suture that extends as a linear zone north-south through the Shield, and forms the approximate axis of the Nabitah orogenic belt. However, the strong deformation and metamorphism of these rocks obscures their precise stratigraphic and structural relations to other stratotectonic units in the Shield, and makes their Shield-wide correlation difficult to sustain. In this context, it is important to note, additionally, that current research interprets bodies of mafic-ultramafic rock in the northern part of the Nabitah orogenic belt as dismembered ophiolite, approximately 830 Ma old [Pallister et al., 1985]; ultramafic rocks in the Zalm area are treated as intrusions emplaced during a period of back-arc spreading between approximately 750 and 770 Ma [Agar, 1985]; and farther south, along the Nabitah fault zone, the ultramafic bodies form Alpine-type intrusions located in relatively high-level fault structures that post-date active accretion in the region. It would appear, therefore, that the mafic-ultramafic rocks along the Nabitah orogenic belt, including the Ad Dafinah ophiolite complex discussed here, are neither coeval nor similar in origin and, as a consequence, the conventional rock correlations and interpretation requires reconsideration.

Farshah volcanic complex. Preaccretionary rocks in the southeastern part of the Afif terrane are assigned to the Farshah volcanic complex [Johnson and Vranas, 1984a]. The region includes remnants of continental basement [Thieme, in press], and a succession of younger volcanosedimentary rocks of volcanic-arc facies that has been complexly deformed. These rocks have been intruded by large plutons of synkinematic and postaccretionary granitic rock, and are extensively covered by younger sedimentary basins.

Ad Dawadimi Terrane

The Ad Dawadimi terrane constitutes one of the most homogeneous crustal units in the Arabian Shield. It is exposed in the east-central Shield, between the Afif and Ar Rayn terranes, and can be traced by its distinctive, subdued, aeromagnetic signature, some 300 km north beneath the Phanerozoic Cover rocks [Phoenix Corp., 1985]. The terrane is approximately 600 km long and between 100 and 200 km wide and, on the basis of available seismic and structural data, is believed to form an allochthonous tectonostratigraphic unit composed of slices of rock bound by listric thrusts [Schmidt et al., 1979; Gettings et al., 1986]. At surface, the preaccretionary rocks of the terrane principally consist of low-grade metasedimentary rocks of the Abt group; they represent an assemblage of distal turbidite deposits that were likely to have accumulated in a marginal basin.

Mafic-ultramafic rocks structurally underlie the Abt group, and crop out as linear belts along the margins of the terrane. However, although the mafic-ultramafic rocks are of possible ophiolite affinity, it has yet to be established whether they formed in a back-arc or fore-arc setting [Al-Shanti and Gass, 1983; Camp et al., 1984]. Likewise, it is unresolved whether the Abt group was deposited on the margin of the Afif terrane, on the margin of the Ar Rayn terrane, or whether it represents a stratotectonic unit unrelated to either terrane. The ophiolitic rocks are intruded by gabbro dated at 695 Ma and trondhjemite at 620 Ma, and the Abt group contains zircons dated at 710 Ma [Al-Shanti et al., 1984b; Stacey et al., 1984].

Ar Rayn Terrane

The Ar Rayn terrane crops out at the eastern margin of the Shield, and extends beneath the Phanerozoic cover rocks, to form a crustal unit over 400 km long and as much as 175 km wide [Phoenix Corp., 1985]. It is bounded on the west by the Al Amar-Idsas fault, and on the east by an unnamed magnetically defined fault zone. The terrane includes a bimodal volcanic assemblage, and is cut by large plutons of mafic to intermediate intrusive rock. In composition the volcanic rocks are transitional between the tholeiitic and calc-alkalic series, and probably formed in a mature island arc and/or active continental margin environment [Coulomb et al., 1980-1981]. The volcanic rocks were intruded by synkinematic plutons dated between 667 and 640 Ma [Stacey et al., 1984], and formed prior to about 670 Ma. This suggests that the terrane is one of the younger accretionary units in the Shield. However, the formation age of the layered rocks is poorly constrained, in addition to which, inherited, middle Proterozoic, zircon ages and the character of lead isotopes in the region indicate that basement rocks may be locally present [Stacey and Stoeser, 1983; Calvez et al., 1985].

Terrane Boundaries and Accretion

In western North America, where the concept was first developed, the boundaries of terranes are drawn along major faults and complex fault zones, or along linear belts of melange, blueschist, and ophiolite [Howell et al., 1985]; these zones constitute crustal discontinuities and are inferred to represent accretionary sutures between adjacent terranes. Crustal discontinuities also bound the late Proterozoic terranes described in this paper which, because of their associated structural features, are likewise interpreted to represent accretionary sutures.

We have drawn a number of terrane boundaries along linear zones of mafic-ultramafic rock. Additionally interterrane boundaries have been drawn along contacts between domains of contrasted structural grain, along zones of major thrusts and faults, and along zones of intense penetrative deformation and high-grade metamorphism. In many instances, the terrane boundaries are also marked by elongate bodies of accretion-related synkinematic granitic rock, or gneiss domes [Stoeser et al., 1984]. In some cases, terrane boundaries determined by surface geologic features coincide with major crustal boundaries inferred from magnetic, seismic and gravity data [Blank and Andreasen, in press; Gettings et al., 1986, Phoenix Corp., 1985]. However, local divergences between the geophysically and the geologically derived boundaries suggest places where the present tectonic analysis of the Shield is inadequate, and provide a stimulus to further investigations.

In the North American cordillera and other areas that contain late Phanerozoic accreted terranes, the amount of lateral movement that resulted from accretion can be estimated by the difference between the present-day and the original geographic positions of the accreted rock units inferred from paleontological and paleomagnetic data. However, such estimates are not possible in the Arabian Shield because reliable paleomagnetic and diagnostic paleontologic evidence of original terrane positions are not available.

Yanbu Suture

The Yanbu suture [Camp, 1984] joins the Midyan and Hijaz terranes. It is a zone of intensely deformed volcanosedimentary and ultramafic rocks offset along left-lateral strike-slip faults of the later Najd system. The structural orientation changes across the suture, from a northwest trend in the Midyan terrane, to a north-to-northeast trend in the Hijaz terrane. In Egypt and Sudan, a 200-km strike extension of the suture is present between Sol Hamed and Wadi Onib [Nassief et al., 1984].

Accretion along the Yanbu suture has not been directly dated. However, a maximum age is provided by dates of 676 and 672 Ma [Hedge, 1984] obtained from synkinematic granite emplaced along the join between the Zaam basin and Bayda-Hegaf volcanic complex in the Midyan terrane (see Figure 4), the amalgamation of which probably occurred prior to accretion of the Midyan and Hijaz terranes. A minimum age is provided by a suite of calc-alkalic stitching granites, 640 to 630 Ma old [Calvez et al., 1982; Kemp et al., 1980].

Bi'r Umq Suture

The Bi'r Umq suture, between the Hijaz and At Ta'if-Jiddah terranes, is marked by north- to northwest-dipping thrusts at the base of a deformed ophiolite succession [Camp, 1984; Nassief et al., 1984]. In the Nubian Shield, the suture extends 200 km inland from the vicinity of Port Sudan, southwest to Wadi Amur [Nassief et al., 1984; Vail, 1985]. The age of accretion along the suture is not closely constrained, but suturing is likely to have occurred between approximately 775-750 Ma, the age of synkinematic gneiss bodies in the region [Fleck, 1981; Duyverman et al., 1982], and 700-675 Ma, the probable age of the Furayh group which unconformably overlies both the Midyan and At Ta'if-Jiddah terranes.

Afaf Suture

The north-trending Afaf suture [Johnson and Vranas, 1984a], between the At Ta'if-Jiddah and Asir terranes, is drawn along the western margin of a zone of strongly deformed and metamorphosed layered rocks and elongate batholiths of synkinematic granite exposed in the Jabal Afaf region of Saudi Arabia. Similar rocks are also exposed as the 'Lower Complex' along the coastal part of northern Ethiopia [Kazmin, 1973], and possibly mark a continuation of the suture. The suture separates regions of northeast-trending structures and magnetic anomalies in the At Ta'if-Jiddah terrane from north-south trends in the Asir terrane [Georgel et al., 1985]. A provisional accretion age of 730 Ma has been obtained from strongly foliated granite along the Afaf suture [Kröner et al, 1984]. This age is compatible with an episode of amalgamation within the At Ta'if-Jiddah terranes, indicated by dates of 775 and 755 Ma [Fleck, 1981; Duyverman et al., 1982] obtained from synkinematic granite plutons along the Fatimah discontinuity joining the Jiddah and Al Lith stratotectonic units.

Nabitah Suture

The boundary between the Asir and Malahah-Najran terranes is drawn along the Nabitah fault. The fault itself is a relatively young structure which cuts post-accretionary granite, but occurs within a zone of older synkinematic granitic intrusions and multiply deformed pre-accretionary rocks that identify the Nabitah suture and are part of the Nabitah orogenic belt, a north-trending accretionary zone that extends the length of the exposed Shield [Stoeser and Camp, 1985].

In its central part, the Nabitah suture coincides with crustal discontinuities marked by a seismically determined lateral velocity contrast at depths of between 20 and 40 km [Gettings et al., 1986], and by steep east-dipping thrust faults extending from surface to a depth of approximately 30 km, that are modeled from gravity data [Gettings, 1984]. Accretionary processes along the southern part of the suture are dated by synkinematic plutons emplaced between 680 and 640 Ma old [Stoeser et al., 1984].

Farther north, the Nabitah orogenic belt extends into the western margin of the Afif terrane. Terrane boundaries here are largely concealed by Proterozoic overlap sequences, Phanerozoic sedimentary basins, and Cenozoic lava flows. They are drawn on the basis of limited surface evidence, augmented by aeromagnetic data, particularly a north-northwest-trending aeromagnetic boundary at approximately 41° E and 24°30' N, separating regions of northeast- and north-northwest-trending magnetic grain [Georgel et al., 1985]. These regions of contrasted magnetic signature correspond to the Hijaz and At Ta'if-Jiddah, and Afif terranes respectively. The western portion of the Afif terrane itself is a complex zone of intraterrane amalgamation that has been drastically modified by later Najd strike-slip faulting (see Figure 1). The age of accretionary events along the Nabitah orogenic belt in this part of the Shield has not been clearly established, but is likely to range from early events, prior to about 700 Ma, dated by the onset of Furayh group overlap sequence deposition, to final events, prior to about 630 Ma, the age of local postkinematic granite emplacement.

Halaban and Ar Rayn Sutures

The Halaban and Ar Rayn sutures form the western boundaries of the Ad Dawadimi and Ar Rayn terranes respectively. They are defined by thrusts or high-angle faults and are associated with linear belts of mafic-ultramafic rock. The boundaries coincide with prominent magneto-tectonic province boundaries [Blank and Andreasen, in prep.], and have been modified by Najd faults. Geologic mapping in the central part of the terrane [Delfour et al., 1982] suggests that the Al Amar-Idsas fault bounding the Ar Rayn terrane is vertical to steeply dipping, although aeromagnetic data indicate that, at depth, the fault dips gently west between 12° and 45° [Griscom, 1982; Phoenix Corp., 1985]. This variation in fault-plane attitude is consistent with proposals that the region west of the Al Amar-Idsas fault may contain unidentified thrust faults and may represent a zone of imbricate structures [Schmidt et al., 1979]. It is also consistent with a suggestion, based on seismic data [Gettings et al., 1986], that the Al Amar-Idsas fault itself is a high-level listric structure, rather than the paleo-subduction zone envisaged by Al-Shanti and Mitchell [1976]. The seismic data implies that the Ad Dawadimi and Ar Rayn terranes are relatively thin, allochthonous units thrust over deeper crustal units. Emplacement of the Ad Dawadimi and Ar Rayn terranes occurred during the Al Amar orogeny, and has been dated by Stacey et al. [1984] between 670 and 640 Ma, contemporary with accretion along the southern part of the Nabitah orogenic belt.

Ha'il Terrane Boundaries

The inferred boundaries of the Ha'il terrane are concealed by postaccretionary rock units, or have been modified by postaccretionary Najd faulting. In the southwest, the boundary is drawn along a northwest-trending magnetic contact that, at surface, coincides with Najd faults. To the east, the boundary is composite, in part a magnetic boundary that coincides with the northwestern edge of the Murdama overlap basin, and in part an east-west-trending, post-Murdama group fault zone. The fault zone contains carbonate-altered gabbro and is inferred to have resulted from reactivation of the sub-Murdama Ha'il terrane boundary [Johnson and Vranas, 1984a]. The concealed terrane boundaries, shown in Figure 3 north of the exposed Shield, are magnetic boundaries which, in places, have been modeled as west-dipping faults [Phoenix Corp., 1985]. Collisional events in the region are represented by bodies of synkinematic mafic to calc-alkalic intrusive rock, dated at approximately 650 Ma [Cole and Hedge, 1985].

Overlap Sequences

Separate basins filled by successions of relatively young, late Proterozoic rock occur in several parts of the Shield. The successions include locally abundant formations of volcanic rock, and a range of volcaniclastic to epiclastic sedimentary rocks; in places they are moderately to strongly deformed and are locally metamorphosed. The successions unconformably overlie older units of accreted rock, and represent "overlap sequences": autochthonous covering strata that obscure previously accreted terranes and terrane boundaries [Howell et al., 1985].

The overlap sequences range in age from approximately 760 to 570 Ma, and, as a whole, were deposited diachronously across the Shield.

Because of their age range, the older of the sequences overlapped in time with the formation of the younger of the volcanic arcs contained within the accreted terranes. However, the overlap sequences are recognized as distinct stratotectonic components in the tectonic evolution of the Shield because of their unconformable relation to adjacent accreted rocks; they represent the first assemblages that were deposited on newly accreted continental crust.

In the Shield, the overlap sequences accumulated in a variety of structural environments. A number of sequences were deposited in "successor basins" constructed above interterrane sutures, or above the join between component tectonostratigraphic units within composite terranes; others filled basins formed by large-scale downwarping in newly created continental lithosphere, whereas some sequences accumulated in pull-apart basins associated with zones of strike-slip faulting.

Successor Basins

A typical successor basin occurs unconformably above the join between the Zaam basin and Bayda-Hegaf volcanic complex in the Midyan composite terrane (see Figures 1, 3, and 4). It was filled between approximately 670 and 640 Ma by polymict conglomerate, sandstone, and shale of the Thalbah group [Davies, 1985]. Farther north, fine-grained sedimentary and subordinate andesite and felsic tuffs of the Amlas formation occupy a successor basin constructed above the join between the same two stratotectonic units. The formation is as much as 2000m thick, and is cut by granites between 625 and 600 Ma old [Clark, 1985]. In the west-central Shield, the Furayh group (700 to 675 Ma) was deposited in a successor basin above the sutures joining the Hijaz, At Ta'if-Jiddah, and the Afif terranes. It comprises conglomerate and sandstone, interbedded with andesite and basalt [Delfour, 1981; Pellaton, 1981]. Elsewhere, similar volcano-sedimentary rocks of the Fatimah group (704 to 675 Ma) fill a successor basin along the join between the Jiddah and Al Lith volcanic complexes within the At Ta'if-Jiddah composite terrane.

The overlap sequences deposited in successor basins were deformed by movements that probably reflect final reactivation, or "tightening", of the underlying sutures. The Fatimah group, for example, was overthrust by the Al Lith preaccretionary complex, and contains subhorizontal folds that are upright or overturned to the northwest. In other parts of the Shield, the movements were periodic, and were favorable for the formation and preservation of additional, younger, deposits. The join between the Zaam basin and Bayda-Hegaf volcanic complex, the site of the Thalbah successor basin, was reactivated during the Mesozoic and Cenozoic and formed the site of the Azlam graben; the northeastern part of the Furayh basin was the subsequent site of deposition of the Jibalah group, one of the youngest late Proterozoic formations in the Shield, of Cambro-Ordovician epiclastic rocks, and of Cenozoic lava flows.

Crustal Downwarp Basins

The Murdama basin contains the most extensive overlap sequence in the Arabian Shield. The exposed portion, which covers parts of the Afif, Ad Dawadimi, and Ha'il terranes, as an elongate body of rock, 600 km long and 150 km wide, is filled by the Murdama, Jurdhawiyah and Hibshi groups. Beneath the Phanerozoic cover to the southeast, where it has been traced by its aeromagnetic signature, the basin is probably filled by a succession similar to the Murdama group [Phoenix Corp., 1985]. The basin-fill includes a thick succession of epiclastic and volcaniclastic rocks deposited in fluviatile and shallow-marine environments. Volcanic rocks along the southwestern margin of the basin, dated at 631 Ma, have a calc-alkaline affinity and are comparable to Cenozoic lavas erupted in mature island arcs [Jackson et al., 1983]. On the northwestern margin of the basin a separate volcanic succession of between 646 and 616 Ma old was deposited in a rift [Johnson and Williams, 1984].

Geophysically, the basin is characterized by a signature of subdued magnetic relief, the extent and continuity of which suggest that the causative rocks are relatively homogeneous. These are inferred to be mainly granitic in composition [Blank and Andreasen, in prep.], and clearly form a sub-Murdama basement that is distinct from the heterogeneous crust formed by accreted rock units in adjacent parts of the Afif terrane. Part of this distinctive basement may consist of middle Proterozoic granitic rock, representing some of the basement fragments in the region (see Figure 2), but Smith and Johnson [1986] also proposed that much of the sub-Murdama crust resulted from large-scale emplacement of granite, during or soon after accretion of the Afif terrane. Thermal expansion during this magmatic event, and subsequent contraction and downwarp of the basement rocks, would have provided a likely mechanism for the creation of the Murdama basin.

Pull-apart Basins

Other types of overlap sequences in the Shield were deposited in sub-aerial and shallow-water environments in fault-controlled, ensialic pull-apart basins. They include the Jibalah and Bani Ghayy groups [Delfour, 1970, Agar, 1986], and the Minawah formation [Clark, 1985]. The basins are typically associated with post-accretionary faults that constitute the Najd system of strike-slip faults, and are present in several parts of the northern Shield.

TECTONIC STAGE	GEOLOGIC EVENTS
EPICRATONIC	Separation of Arabia from Africa; formation of the Red Sea and Gulf of Aden spreading centers; extrusion of Cenozoic lava flows (24 Ma to the present)
	Deposition of the Phanerozoic "cover rock" succession on the stable Arabian craton
REACTIVATION	Final cratonization of the Arabian Shield (about 550 Ma)
	Intrusion of alkali granitoids (625-570 Ma) in conjunction with intraplate transcurrent movement of the Najd fault system (670-550 Ma) and deposition of volcanosedimentary rocks of the Jibalah and Bani Ghayy groups, and Minawah formation in pull-apart basins
TRANSITIONAL	Intrusion of late- to postkinematic calc-alkalic and alkali granitoids (680-610 Ma)
	Deposition of volcanosedimentary rocks in successor and downwarp basins (Thalbah, Furayh, Fatima, Murdama and correlative groups (760-570 Ma)
PRECRATONIC	Successive phases of accretion of individual terranes, associated with strong deformation and metamorphism and the emplacement of synkinematic granitoids (775-640 Ma)
	Evolution of precratonic terranes, mainly composed of volcanic arcs formed in intraoceanic to continental-marginal environments (950-670 Ma), and incorporating remnant microplates of older continental material
BASEMENT	? Rifting of a Proterozoic supercontinent and creation of an oceanic environment for subsequent arc evolution

Fig. 5. A tectonic model for the Arabian Shield

The Bani Ghayy group (630 to 610 Ma) unconformably overlies accreted rocks close to the western margin of the Afif terrane. It comprises a 6- to 7-km thick deposit of epiclastic rocks interbedded with or underlain by lava flows and volcaniclastic rocks. The group was deposited in three separate graben that contain marginal fanglomerate and fan delta deposits and axial graywacke. The volcanic rocks form a bimodal suite that geochemically is typical of lavas in continental rifting environments.

The Jibalah group crops out in fault-controlled basins in several parts of the northern and northwestern Shield. Individual basins contain as much as 2500 m of interbedded sedimentary and volcanic rocks. These unconformably overlie older accreted rock units and include basal conglomerate, sandstone and siltstone deposited as coalescing alluvial fans, littoral to shallow-marine limestone and submarine to subaerial deposits of andesitic basalt and rhyolite.

The Minawah formation, in the northwestern Shield, overlies older accreted rock units with a profound unconformity, and is present in small fault graben. It comprises as much as 500 m of clastic rocks and felsic volcanic and pyroclastic rocks, which are probably the extrusive equivalent of alkali granites dated at about 575 Ma.

Discussion

This paper describes layered Proterozoic stratotectonic units that comprise roughly 50 percent of the area of the exposed Shield. The remaining 50 percent of the exposed Shield consists chiefly of postaccretionary granitic rocks. However, the granitic rocks are not a specific topic of this paper; they are not discussed by us in any detail, and the interested reader is referred to Drysdall et al. [1986].

The layered stratotectonic units themselves display a great variety of ages, lithologies, formational environments, and structural histories diagnostic of the tectonic events which contributed to crustal growth in the Shield. The temporal relationships of these events are shown in Figure 5, and delineate a model for the tectonic evolution of the region.

The model utilizes concepts derived from work by Doutch [1972] and Scheibner [1976] in eastern Australia. It recognizes that, although terrane accretion is a powerful concept by which to account for the diversity of stratotectonic units in the Shield, terrane accretion was only one of several processes which operated during crustal growth in Arabia. The other processes included dispersion of pre-existing continental crust, the formation of overlap sequences during transitional-tectonic and reactivation-tectonic episodes, and major episodes of magmatism. Fundamental to the model is the concept that the boundaries between the five stages recognized are defined in terms of evolving tectonic characteristics, not necessarily geologic time; the boundaries reflect episodes of orogeny, or coincide with changes in the prevalent style of deformation and dominant crustal character. As a consequence, the various stages, particularly the transitional and reactivation stages, overlapped in time, and the boundaries are partially diachronous across the Shield (see Figure 5).

The earliest, basement, stage is represented by the middle Proterozoic or older rock units preserved within the accreted terranes of the central Shield. It is inferred that they constitute segments of continental lithosphere that had been formed during a pre-late Proterozoic cycle of cratonization and were fragmented as a result of reactivation of the earlier craton. The pieces of former crust were incorporated into newly accreted crust in Arabia as allochthonous blocks of "older basement" after re-establishment of the cratonizing process during the late Proterozoic. The tectonic events that generated these fragments of continental crust are poorly known. It is likely, as suggested by Kazmin et al. [1978], that they principally resulted from rifting of the Mozambique mobile belt and the break-up of the middle Proterozoic supercontinent which is believed to have existed between approximately 2600 and 1000 Ma [Piper, 1983].

Dispersion of this supercontinent, as envisaged by Stoeser and Camp [1985], would have resulted in the formation of ocean floor between rifted portions of the supercontinent, thereby creating the oceanic to continental-margin environments in which the accretionary arc systems of the Shield were subsequently deposited. These arc systems and associated rock units are termed "precratonic", because we infer that they formed at the active margins of the basement fragments or in mobile zones at or near the boundary of interacting oceanic lithospheric plates, prior to the evolution of any large-scale late Proterozoic craton in the region. Deposited between 950 and 670 Ma, they are characterized by an immature to calc-alkaline composition [Roobol et al., 1983], low initial-strontium ratios [Fleck et al., 1980], and unevolved lead contents [Stacey and Stoeser, 1983], and are generally inferred to have been derived from oceanic lithosphere or primitive upper mantle material. In chemistry and lithology, many of the assemblages resemble modern intraoceanic and island-arc volcanic complexes; others represent dismembered ophiolite complexes, or marginal basins and fore-arc basins.

The precratonic assemblages combined to form tectonostratigraphic terranes, and the terranes subsequently accreted between 770 and 620 Ma to form a neocraton on which the younger Proterozoic rocks of the Shield were deposited. Prior to, and during, their accretion the tectonostratigraphic terranes were affected by multiple deformation, metamorphism and both arc-related and synkinematic plutonic activity.

The "transitional-tectonic" stage succeeded the precratonic stage, and yielded bodies of rock deposited on newly accreted crust. Their lithologic and structural features indicate that the rocks accumulated in stable to unstable environments; these environments represent a transition between the mobile environments of the precratonic stage and the stable environments of the epicratonic stage. The layered rocks of the transitional-tectonic stage comprise the successor-basin and downwarp-basin sequences described in this paper. They were deposited between 760 and 570 Ma. As a group, the sequences include shallow-marine to terrestrial, compositionally immature clastic rocks and marble, and high-potassium alkalic mafic-to-felsic volcanic rocks that are transitional in composition between lavas of island-arc and continental-margin complexes [Roobol et al., 1983]. They were typically affected by late-stage deformation, and were intruded by large amounts of postaccretionary I- and S-type granitic rock.

The pull-apart overlap sequences of the Shield are assigned to the "reactivation-tectonic" stage of Shield evolution. As studies elsewhere in the world indicate [Tapponnier and Molnar, 1976; Howell et al., 1985], initial stages of crustal accretion are commonly followed by, and are often concurrent with, a long and still poorly known history of tectonic reactivation. Reactivation episodes in crustal evolution include complex strike-slip faulting and thrust faulting and may result, in extreme cases, in the break-up of previously accreted tectonostratigraphic terranes. Reactivation in Arabia overlapped in time with the transitional-tectonic stage of crustal evolution. Its effect was the formation of the Najd system of transcurrent faults, deposition of volcanic and sedimentary rocks in pull-apart basins or graben, intraplate S- and A-type magmatism, and a "tightening" of the component rock units of the Shield.

The Najd fault system is a zone of northwest-trending faults, as much as 400 km wide and over 1100 km long, in which the dominant sense of movement is left-lateral. The system reflects a complex evolution that included both extensional and compressive phases over a period of about 90 million years, between 670 and 550 Ma [Cole and Hedge, 1985; Stacey and Agar, 1985]. Reactivation rock units that clearly record episodes of crustal extension concurrent with the Najd system include the pull-apart basins described in this paper, volcanosedimentary formations in northern Egypt [Stern, 1985], and plutons of alkalic granite and coeval caldera deposits present in many parts of the northern Arabian Shield. The cumulative effect of reactivation was the formation of a stable lithospheric plate welded to Africa.

At the close of the Proterozoic, the newly cratonized Arabian Shield was eroded and overlapped by a succession of Phanerozoic rocks. The Phanerozoic rocks form a platform cover and comprise the "epicratonic" stage in the tectonic model shown in Figure 5. The cover as such is not discussed in this paper, but consists of sedimentary rocks east and north of the craton, and sedimentary and igneous rocks along the Red Sea and its margins. The latter include fields of alkali basalt lava flows extruded on the craton and along the Red Sea as a result of intraplate magmatism associated with the Cenozoic break-up of the Afro-Arabian plate.

Overall, the Proterozoic rock units and geologic events discussed by us represent episodes of volcanism, sedimentation, metamorphism and deformation during a relatively short-lived but successful cycle of crustal growth. The cycle began approximately 950 Ma ago with the formation of precratonic volcanosedimentary successions in intraoceanic to continental-marginal and other environments, and culminated 400 million years later with the accretion to Africa of stable continental crust between 35 and 45 km thick. Prior to the onset of precratonic deposition, the region had undergone an episode of crustal extension and dispersion of pre-existing continental lithosphere. Subsequent to the culmination of cratonization, the region underwent epeirogenic movements that lasted until the Tertiary onset of Arabian-Nubian crustal extension. Cenozoic crustal extension continues to the present, and has resulted in the formation of new ocean floor in the Red Sea. This represents a current episode of continental dispersion in the area, and may well mark the onset of a renewed phase in crustal evolution.

Acknowledgments. The research and compilation behind the model described in this paper was mainly carried out under the auspices of the Saudi Arabian Deputy Ministry for Mineral Resources, and we thank HE Ghazi Sultan, the Deputy Minister, for his support of our work. Much of the geologic information utilized by us was first available in open-file reports of the Deputy Ministry; we gratefully acknowledge access to these reports, and we wish to thank our many colleagues in Saudi Arabia whose work stimulated our investigations. We also thank G. J. Vranas for his careful editing of the manuscript, and D.G. Howell, P. J. Coney, and C. McA. Powell for their review of an earlier version of the paper. Erwin Scheibner publishes with the permission of the Secretary, New South Wales Department of Mineral Resources, Sydney, N.S.W.

References

Agar, R. A., The stratigraphy and paleogeography of the Siham group; direct evidence for a late Proterozoic continental microplate and active continental margin in the Arabian Shield, J. Geol. Soc. London, 142, 1205-1220, 1985.

Agar, R. A., The Bani Ghayy group; sedimentation and volcanism in pull-apart grabens of the Najd strike-slip orogen, Saudi Arabian Shield, Precambrian Res., 31, 259-274, 1986.

Al-Shanti, A. M. S., A. A. Abdel, and A. A. Radain, Rb-Sr dating and petrochemistry of Um Gerad granitic rocks (Rabigh area), western Saudi Arabia, Pan-African crustal evolution in the Arabian-Nubian Shield, I.G.C.P. Project 164, King Abdulaziz University, Jiddah, Faculty of Earth Sciences Bull. 6, 221-232, 1984a.

Al-Shanti, A. M. S., A. A. Abdel-Monem, and F. H. Marzouki, Geochemistry, petrology and Rb-Sr dating of trondhjemite and granophyre associated with Jabal Tays ophiolite, Idsas area, Saudi Arabia, Precambrian Res., 24, 321-334, 1984b.

Al-Shanti, A. M., and I. G. Gass, The Upper Proterozoic ophiolite mélange zones of the easternmost Arabian shield, J. geol. Soc. London, 140, 867-876. 1983.

Al-Shanti, A. M. S., and A. H. G. Mitchell, Late Precambrian subduction and collision in the Al Amar-Idsas region, Arabian Shield, Kingdom of Saudi Arabia, Tectonophysics, 30, T41-T47, 1976.

Blank, H. R., and G. E. Andreasen, Explanatory text for the total-intensity residual aeromagnetic map of the Arabian Precambrian Shield, Kingdom of Saudi Arabia, Saudi Arabian Deputy Ministry for Mineral Resources Open-File Report, in press.

Bokhari, F. Y., and J. D. Kramers, Island arc character and late Precambrian age of volcanics at Wadi Shwas, Hijaz, Saudi Arabia: geochemical and Sr and Nd isotopic evidence, Earth and Planetary Science Letters, 54, 409-422, 1981.

Calvez, J. Y., C. Alsac, J. Delfour, J. Kemp, and C. Pellaton, Geologic evolution of western, central and eastern parts of the northern Precambrian Shield, Kingdom of Saudi Arabia, Saudi Arabian Deputy Ministry for Mineral Resources Open-File Report BRGM-OF- 03-17, 57 pp., 1983.

Calvez, J. Y., J. Delfour, J. Kemp, and P. Elsass, Pre-Pan-African inherited zircons from the northern Arabian Shield, Saudi Arabian Deputy Ministry for Mineral Resources Open-File Report BRGM-OF-05-13, 22 pp., 1985.

Calvez, J. Y., and J. Kemp, Geochronologial investigations in the Mahd adh Dhahab quadrangle, Central Arabian Shield, Saudi Arabian Deputy Ministry for Mineral Resources Technical Record BRGM-TR-02-5, 41 pp., 1982.

Calvez, J. Y., C. Pellaton, C. Alsac, and M. Tegyey, Geochronology and geochemistry of plutonic rocks in the Umm Lajj and Jabal al Buwanah areas, Saudi Arabian Deputy Ministry for Mineral Resources Open-File Report BRGM-OF-02-36, 36 pp., 1982.

Camp, V. E., Island arcs and their role in the evolution of the western Arabian Shield, Geol. Soc. Amer. Bull., 95, 8, 913-921, 1984.

Camp, V. E., N. J. Jackson, C. R. Ramsay, M. J. Roobol, D. B. Stoeser, and D. L. White, Discussion on the Upper Proterozoic ophiolite mélange zones of the easternmost Arabian Shield, J. geol. Soc. London, 141, 1083-1087, 1984.

Claesson, S., J. S. Pallister, and M. Tatsumoto, Samarium-neodymium data on two late Proterozoic ophiolites of Saudi Arabia and implications for crustal and mantle evolution, Contrib. to Mineralogy and Petrology, 85, 244-252, 1984.

Clark, M. D., Late Proterozoic crustal evolution of the Midyan region, northwestern Saudi Arabia, Geology, 13, 611-615, 1985.

Cole, J. C., and C. E. Hedge, Geochronologic investigation of Late Proterozoic rocks in the northeast Arabian Shield, Saudi Arabian Deputy Ministry for Mineral Resources Technical Record USGS-TR-05-5, 66 pp., 1985.

Coney, P. J., D. L. Jones, and J. W. H. Monger, Cordilleran suspect terranes, Nature, 288, 329-333, 1980.

Cooper, J. A., J. S. Stacey, D. B. Stoeser, and R. J. Fleck, An evaluation of the zircon method of isotopic dating in the southern Arabian craton, Contrib. Mineral. Petrol., 68, 429-439, 1979.

Coulomb, J. J., J. Felenc, and J. Testard, Volcanisme et minéralisations à Zn-Cu de la ceinture d'Al Amar (Royaume d'Arabie Saoudite), Bull. du BRGM (2me série), Section II, 1-2, 41-71, 1980-81.

Davies, F. B., Geology of the Al Wajh quadrangle, sheet 26B, Kingdom of Saudi Arabia: Saudi Arabian Deputy Ministry for Mineral Resources Geologic Map GM-83, scale 1:250,000, 1985.

Delfour, J., Le groupe de J'Balah: une nouvelle unité du Bouclier Arabe, Bull. du BRGM (2me série), 4, 19-32, 1970.

Delfour, J., Geologic map of the Al Hissu quadrangle, Sheet 24E, Kingdom of Saudi Arabia: Saudi Arabian Deputy Ministry for Mineral Resources Geologic Map GM-58, scale 1:250,000, 1981.

Delfour, J., R. Dhellemmes, P. Elsass, D. Vaslet, J-M. Brosse, Y-M. Le Nindre, and O. Dottin, Geologic map of the Ad Dawadimi quadrangle, sheet 24 G, Kingdom of Saudi Arabia, Saudi Arabian Deputy Ministry for Mineral Resources Geologic Map GM-60, scale 1:250,000, 1982.

Doutch, H. F., The concept of transitional tectonism in Australia, (Abs.) Joint Specialists Group Meetings, Geol. soc. Aust., Canberra, E10-E12, 1972.

Drysdall, A. R., C. R. Ramsay, and D. B. Stoeser, (eds.), Felsic plutonic rocks and associated mineralization of the Kingdom of Saudi Arabia, J. of African Earth Sciences, Special Vol., 4, 291 pp., 1986.

Duyverman, H. J., N. B. W. Harris, and C. J. Hawksworth, Crustal accretion in the Pan African: Nd and Sr isotope evidence from the Arabian Shield, Earth and Planetary Science Letters, 59, 315-326, 1982.

Fleck, R. J., Age of diorite-granodiorite gneisses of the Jiddah-Makkah region, Kingdom of Saudi Arabia, Saudi Arabian Deputy Ministry for Mineral Resources Open-File Report USGS-OF-01-1, 18 pp., 1981.

Fleck, R. J., W. R. Greenwood, D. G. Hadley, R. E. Anderson, and D. L. Schmidt, Rubidium-strontium geochronology and plate-tectonic evolution of the southern part of the Arabian Shield, U. S. Geological Survey Professional Paper 1131, 38 pp., 1980.

Fleck, R. J., and D. G. Hadley, Ages and strontium initial ratios of plutonic rocks in the transect of the Arabian Shield, Saudi Arabian Deputy Ministry for Mineral Resources Open-File Report USGS-OF-03-38, 43 pp., 1982.

Frisch, W., and A. M. Al-Shanti, A. M., Ophiolite belts and the collision of island arcs in the Arabian Shield, Tectonophysics, 43, 293-306, 1977.

Georgel, J-M., J. Bobillier, J. Delom, and M. Bourlier, Total-intensity aeromagnetic maps of the Arabian Shield, Saudi Arabian Deputy Ministry for Mineral Resources Technical Records BRGM-TR-05-1 to 53, scale 1:250,000, 1985.

Gettings, M. E., The isostatic gravity anomaly field of southwestern Saudi Arabia and its interpretation, Saudi Arabian Deputy Ministry for Mineral Resources Technical Record USGS-TR-04-22, 107 pp., 1984.

Gettings, M. E., H. R. Blank, Jr., W. D. Mooney, and J. H. Healey, Crustal structure of southwestern Saudi Arabia, J. of Geophysical Res., 91, B6, 6491-6512, 1986.

Greenwood, W. R., Geological map of the Jibal al Qahr quadrangle, sheet 19G, Kingdom of Saudi Arabia, Saudi Arabian Deputy Ministry for Mineral Resources Geologic Map GM-76, scale 1:250,000, 1985a.

Greenwood, W. R., Geologic Map of the Bi'r Idamah quadrangle, sheet 18G, Kingdom of Saudi Arabia, Saudi Arabian Deputy Ministry for Mineral Resources Geologic Map GM-79, scale 1:250,000, 1985b.

Greenwood, W. R., R. O. Jackson, and P. R. Johnson, Geologic map of the Jabal al Hasir quadrangle, sheet 19F, Kingdom of Saudi Arabia, Saudi Arabian Deputy Ministry for Mineral Resources Geologic Map GM-94, scale 1:250,000, 1985.

Greenwood, W. R., D. B. Stoeser, R. J. Fleck and J. S. Stacey, Late Proterozoic island-arc complexes and tectonic belts in the southern part of the Arabian Shield, Saudi Arabian Deputy Ministry for Mineral Resources Open-File Report USGS-OF-02-8, 46 pp., 1982.

Griscom, A., An aeromagnetic interpretation of eleven map sheets, scale 1:250,000, in the southern Najd and part of the southern Tuwayq quadrangles, Kingdom of Saudi Arabia, Saudi Arabian Deputy Ministry for Mineral Resources Open-File Report USGS-OF-02-70, 21 pp., 1982.

Hedge, C. E., Precambrian geochronology of part of northwestern Saudi Arabia, Saudi Arabian Deputy Ministry for Mineral Resources Open-File Report USGS-OF-04-31, 12 pp., 1984.

Howell, D. G., Jones, D. L., and E. R. Schermer, Tectonostratigraphic terranes of the Circum-Pacific region, in D. G. Howell (ed.), Tectonostratigraphic terranes of the Circum-Pacific region, CPCEMR Earth Science Series, 1, 3-30, 1985.

Jackson, N. J., A. N. Basahel, P. G. Cooray, and D. P. F. Darbyshire, Late Precambrian (Sequence A) rocks of the east-central Shield of Saudi Arabia, King Abdulaziz University, Jiddah, Faculty of Earth Sciences Bull. 5, 119-134, 1983.

Johnson, P. R., and G. J. Vranas, The origin and development of late Proterozoic rocks of the Arabian Shield: an analysis of terranes and mineral environments, Saudi Arabian Deputy Ministry for Mineral Resources Open-File Report RF-OF-04-32, 96 pp., 1984a.

Johnson, P. R., and G. J. Vranas, The geotectonic environments of Late Proterozoic mineralisation in the southern Arabian Shield: Precambrian Res., 25, 329-384, 1984b.

Johnson, P. R., and P. L. Williams, Precambrian geology of the Jabal Habashi quadrangle, sheet 26F, Kingdom of Saudi Arabia, Saudi Arabian Deputy Ministry for Mineral Resources Open-File Report USGS-OF-04-10, scale 1:250,000, 1984.

Kazmin, V., Geological map of Ethiopia, Ministry of Mines, Energy and Water Resources, Geological Survey of Ethiopia, scale 1:2,000,000, 1973.

Kazmin, V., Schifferaw, A., and Balcha, T., The Ethiopian basement: stratigraphy and possible manner of evolution, Geol. Rundschau, 67, 2, 531-546, 1978.

Kemp, J., Y. Gros, and J-P. Prian, Geologic map of the Mahd adh Dhahab quadrangle, Sheet 23E, Kingdom of Saudi Arabia, Saudi Arabian Deputy Ministry for Mineral Resources Geologic Map GM-64, scale 1:250,000, 1982.

Kemp. J., C. Pellaton, and J. Y. Calvez, Geochronological investigations and geologic history in the Precambrian of northwestern Saudi Arabia, Bureau de Recherches Géologiques et Minières (Jiddah) Open-File Report BRGM-OF-01-1, 120 pp., 1980.

Kroner, A., M. Halpern, and A. Basahel, Age and significance of metavolcanic sequences and granitoid gneisses from the Al Lith area, southwestern Arabian Shield, Pan-African crustal evolution in the Arabian-Nubian Shield I. G. C. P. Project 164, King Abdulaziz University, Jiddah, Faculty of Earth Sciences Bull. 6, 379-388, 1984.

Marzouki, F. M. H., N. J. Jackson, C. R. Ramsay, and D. P. F. Darbyshire, Composition, age and origin of two Proterozoic diorite-tonalite complexes in the Arabian Shield, Precambrian Res., 19, 31-50, 1982.

Nassief, M. O., R. MacDonald, and I. G. Gass, The Jebel Thurwah Upper Proterozoic ophiolite complex, western Saudi Arabia, J. geol. Soc. London, 141, 537-546. 1984.

Pallister, J. S., Geologic map of the Lith quadrangle, sheet 20D, Kingdom of Saudi Arabia, Saudi Arabian Deputy Ministry for Mineral Resources Geologic Map GM-95, scale 1:250,000, in press.

Pallister, J. S., J. S. Stacey, and L. B. Fischer, Precambrian ophiolites of Saudi Arabia: geochronology and implications for continental accretion, (Abs.) EOS, 66, 18, 420, 1985.

Pellaton, C., Geologic map of the Al Madinah quadrangle, sheet 24D, Kingdom of Saudi Arabia, Saudi Arabian Deputy Ministry for Mineral Resources Geologic Map GM-52, scale 1:250,000, 1981.

Phoenix Corp., The interpretation of an airborne geophysical survey of the Cover Rocks region,

Kingdom of Saudi Arabia, Saudi Arabian Deputy Ministry for Mineral Resources Report, 76 pp, 1985.

Piper, J. D. A., Dynamics of the continental crust in Proterozoic times, in G. L. Medaris, Jr., (ed.), Proterozoic Geology: Selected Papers from an International Proterozoic Symposium, Geol. Soc. America Memoir 161, 11-34, 1983.

Ramsay, C. R., Geology of the Rabigh quadrangle, sheet 22D, Kingdom of Saudi Arabia, Saudi Arabian Deputy Ministry for Mineral Resources Open-File Report DGMR-OF-03-15, 90 pp., scale 1:250,000, 1983.

Ramsay, C.R., A. M. Basahel, and N. J. Jackson, Petrography, geochemistry, and origin of the volcanosedimentary succession between Jabal Ibrahim and Al Aqiq, King Abdulaziz University, Jiddah, Institute of Applied Geology Bull. 4, 1-24, 1981.

Rast, N., Orogenic belts and their parts, in, P. E. Kent et al., eds., Time and Place in Orogeny, Geol. Soc. London Special Publication, 3, 197-213, 1969.

Reischmann, Th., A. Kröner, and A. Basahel, Petrography, geochemistry and tectonic setting of metavolcanic sequences from the Al Lith area, southwestern Arabian Shield, Pan-African crustal evolution in the Arabian-Nubian Shield, I. G. C. P. Project 164, King Abdulaziz University, Jiddah, Faculty of Earth Sciences Bull. 6, 365-378, 1984.

Riddler, G. P., M. R. Worsley, T. V. Harvey, N. R. Sheppy, J. R. Bavin, and Y. T. Saleh, Mineral exploration in the Ash Sha'ib-Tathlith area, southern Najd, Saudi Arabian Deputy Ministry for Mineral Resources Open-File Report RF-OF-03-13, 144 pp., 1983.

Roobol, M. J., C. R. Ramsay, N. J. Jackson, and D. P. F. Darbyshire, Late Proterozoic lavas of the central Arabian Shield -- evolution of an ancient volcanic arc system, J. geol. Soc. London, 140, 185-202, 1983.

Sable, E. G., Geologic map of the Najran quadrangle, sheet 17G, Kingdom of Saudi Arabia, Saudi Arabian Deputy Ministry for Mineral Resources Geologic Map GM-78, scale 1:250,000, 1985.

Scheibner, E., Explanatory notes on the tectonic map of New South Wales, Department of Mines, Geological Survey of New South Wales, 283 pp., 1976.

Schmidt, D. L., D. G. Hadley, and D. B. Stoeser, Late Proterozoic crustal history of the Arabian Shield, southern Najd Province, Kingdom of Saudi Arabia, in, A. M. S. Al-Shanti (Convenor), Evolution and mineralization of the Arabian-Nubian Shield, 2, Oxford-New York, Pergamon, 41-58, 1979.

Smith, E. A., and P. R. Johnson, Selected mineral occurrences of the Arabian Shield, showing their relationship to major Precambrian tectonostratigraphic entities, Saudi Arabian Deputy Ministry for Mineral Resources Technical Record RF-TR-06-1, 143 pp., scale 1:1,000,000, 1986.

Stacey, J. S., and R. A. Agar, U-Pb isotopic evidence for accretion of a continental microplate in the Zalm region of the Saudi Arabian Shield, Saudi Arabian Deputy Ministry for Mineral Resources Open-File Report USGS-OF-05-24, 24 pp., 1985.

Stacey, J. S., and C. E. Hedge, Geochronologic and isotopic evidence for early Proterozoic crust in the eastern Arabian Shield, Geology, 12, 310-313, 1984.

Stacey, J. S., and D. B. Stoeser, Distribution of oceanic and continental leads in the Arabian-Nubian Shield, Saudi Arabian Deputy Ministry for Mineral Resources Open-File Report USGS-OF-03-55, 36 pp., 1983.

Stacey, J. S., D. B. Stoeser, W. R. Greenwood, and L. B. Fischer, U-Pb zircon geochronology and geological evolution of the Halaban-Al Amar region of the eastern Arabian Shield, Kingdom of Saudi Arabia, J. geol. Soc. London, 141, 6, 1043-1057, 1984.

Stern, R. J., The Najd fault system, Saudi Arabia and Egypt: a late Precambrian rift-related transform system?, Tectonics, 4, 497-511, 1985.

Stoeser, D. B., and V. E. Camp, Pan-African microplate accretion of the Arabian Shield, Geol. Soc. America Bull., 96, 7, 817-826, 1985.

Stoeser, D. B., J. S. Stacey, W. R. Greenwood, and L. B. Fischer, U/Pb zircon geochronology of the southern portion of the Nabitah mobile belt and Pan-African continental collision in the Saudi Arabian Shield, Saudi Arabian Deputy Ministry for Mineral Resources Technical Record USGS-TR-04-5, 88 pp., 1984.

Tapponnier, P. and P. Molnar, Slip-line field theory and large scale continental tectonics, Nature, 264, 319-324, 1976.

Thieme, J., Geologic map of the Jabal Khida quadrangle, sheet 21G, Kingdom of Saudi Arabia, Saudi Arabian Deputy Ministry for Mineral Resources Geologic Map GM-90, scale 1:250,000, in press.

Vail, J. R., Pan-African (late Precambrian) tectonic terranes and the reconstruction of the Arabian-Nubian Shield, Geology, 13, 839-842, 1985.